P9-AQK-504

ANNUAL REVIEW OF NEUROSCIENCE

ANNUAL REVIEW OF NEUROSCIENCE

VOLUME 10, 1987

W. MAXWELL COWAN, *Editor*
Washington University

ERIC M. SHOOTER, *Associate Editor*
Stanford University School of Medicine

CHARLES F. STEVENS, *Associate Editor*
Yale University School of Medicine

RICHARD F. THOMPSON, *Associate Editor*
Stanford University

ANNUAL REVIEWS INC. 4139 EL CAMINO WAY P.O. BOX 10139 PALO ALTO, CALIFORNIA 94303-0897

ANNUAL REVIEWS INC.
Palo Alto, California, USA

International Standard Serial Number: 0147-006X
International Standard Book Number: 0-8243-2410-2

TYPESET BY AUP TYPESETTERS (GLASGOW) LTD., SCOTLAND
PRINTED AND BOUND IN THE UNITED STATES OF AMERICA

Annual Review of Neuroscience
Volume 10, 1987

CONTENTS

(Note: Titles of chapters in Volumes 6–10 are arranged by category on pages 712–16.)

(*Continued*)

SOME RELATED ARTICLES IN OTHER *ANNUAL REVIEWS*

From the *Annual Review of Biochemistry*, Volume 56 (1987)

Inositol Trisphosphate and Diacylglycerol: Two Interacting Second Messengers, M. J. Berridge

Molecular Genetics of Myosin, C. P. Emerson and S. I. Bernstein

Receptors for Epidermal Growth Factor and Other Polypeptide Mitogens, G. Carpenter

From the *Annual Review of Cell Biology*, Volume 2 (1986)

Cell-Matrix Interactions and Cell Adhesion During Development, P. Ekblom, D. Vestweber, and R. Kemler

Cell Adhesion Molecules in the Regulation of Animal Form and Tissue Pattern, G. M. Edelman

G Proteins: A Family of Signal Transducers, L. Stryer and H. R. Bourne

Microtubule-Associated Proteins, J. B. Olmsted

From the *Annual Review of Entomology*, Volume 32 (1987)

Insects as Models in Neuroendocrine Research, B. Scharrer

From the *Annual Review of Medicine*, Volume 38 (1987)

Endogenous Digitalis-like Natriuretic Factors, S. W. Graves and G. H. Williams

Prions Causing Degenerative Neurological Diseases, S. B. Prusiner

Anorexia Nervosa and Bulimia, K. A. Halmi

Affective Disorders in the Elderly, R. E. Neshkes and L. F. Jarvik

Clinical Pharmacology of Benzodiazapines, M. Lader

Clinical Perspectives on Neuropeptides, D. A. Lewis and F. E. Bloom

From the *Annual Review of Physiology*, Volume 49 (1987)

Lateral Diffusion of Proteins in Membranes, K. Jacobson, A. Ishihara, and R. Inman

Thyroid Hormones and Brain Development, J. H. Dussault and J. Ruel

Sex Steroids and Afferent Input: Their Roles in Brain Sexual Differentiation, C. Beyer and H. H. Feder

Adrenocortical Steroids and the Brain, J. W. Funder and K. Sheppard

Functions of Angiotensin in the Central Nervous System, M. I. Phillips

Insulin in the Brain, D. G. Baskin, D. P. Figlewicz, S. C. Woods, D. Porte, Jr., and D. M. Dorsa

Gastroenteropancreatic Peptides and the Central Nervous System, D. P. Figlewicz, F. Lacour, A. Sipols, D. Porte, Jr., and S. C. Woods

Organization of Central Control of Airways, M. P. Kalia

The Nature and Identity of the Internal Excitational Transmitter of Vertebrate Phototransduction, E. N. Pugh, Jr.

Ionic Conductances in Rod Photoreceptors, W. G. Owen

ANNUAL REVIEWS INC. is a nonprofit scientific publisher established to promote the advancement of the sciences. Beginning in 1932 with the *Annual Review of Biochemistry*, the Company has pursued as its principal function the publication of high quality, reasonably priced *Annual Review* volumes. The volumes are organized by Editors and Editorial Committees who invite qualified authors to contribute critical articles reviewing significant developments within each major discipline. The Editor-in-Chief invites those interested in serving as future Editorial Committee members to communicate directly with him. Annual Reviews Inc. is administered by a Board of Directors, whose members serve without compensation.

Berta Scharrer

Ann. Rev. Neurosci. 1987. 10: 1–17

NEUROSECRETION: Beginnings and New Directions in Neuropeptide Research

Berta Scharrer

Department of Anatomy and Structural Biology and Department of Neuroscience, Albert Einstein College of Medicine, Bronx, New York 10461

The Early Years

During the past decade, spectacular advances in our knowledge of the far-reaching significance of neuropeptides have ushered in a new era in neurobiology. It may be of interest, therefore, in this prefatory chapter to present some personal recollections of the gradual evolvement of these new insights, which, at the outset, were entirely unforeseen and difficult to accept.[1] In 1928, when Ernst Scharrer reported his discovery of gland-like nerve cells in the hypothalamus of a teleost fish, *Phoxinus laevis*, neurobiologists considered neuronal signaling largely as an electrophysiological process. What made this small group of unusual hypothalamic neurons so striking, therefore, was their content of impressive amounts of a secretory material comparable to that found in typical protein-secreting gland cells, such as those of the pancreas. For this reason, they were called "neurosecretory neurons." E. Scharrer made the bold proposal that their role may be endocrine, and at the outset suggested a functional relationship with the pituitary gland. He was the first to point out that the neurosecretory neuron, because of its dual (neural and glandular) capacities, seems to be ideally suited for the very special task of

[1] This account is by necessity selective. Much of the information not documented in detail will be found in the following comprehensive publications: E. Scharrer & B. Scharrer 1963, B. Scharrer 1970, Meites et al 1975, 1978, Gainer 1977, Barker & Smith 1980, Bloom 1980, Burgen et al 1980, Costa & Trabucchi 1980, Krieger et al 1983, Bloom et al 1985, Kobayashi et al 1985, White et al 1985.

0147–006X/87/0301–0001$02.00

conveying neural directives to the endocrine system in its own "language" (E. Scharrer 1952). These pioneering ideas foreshadowed the growing realization of the close interdependence of the two systems of integration, and the central role played by the neuroendocrine axis. In the early 1960s, when neuroendocrinology, i.e. the study of the interactions between the nervous and the endocrine systems, became established as a discripline in its own right (see E. Scharrer & B. Scharrer 1963), the examination of mutually "understandable" signals received due attention.

However, initially the idea that neurons may be capable of dispatching neurohormonal, i.e. blood-borne, signals, an activity heretofore attributed only to endocrine cells proper, met with powerful resistance. It did not conform with the tenets of the neuron doctrine, and the fact that it was proposed on the basis of cytological evidence was considered preposterous.

The theory of the chemical transmission of the nervous impulse proposed by Loewi in 1921 provided no tangible support for the concept of neurosecretion. The type of neurohormonal messenger substances envisioned to function in these special hypothalamic cells appeared to differ significantly from "neurohumors" (classical neurotransmitters) with respect to their proteinaceous character, the amount of material produced, and the extracellular pathway from site of release to site of action.

When I joined forces with Ernst Scharrer in the search for the functional significance of the phenomenon of neurosecretion, which became a lifetime endeavor, we had the territory practically to ourselves for quite a long time. Eventually, disbelief gave way to mild interest, as demonstrated by a memorable meeting on the hypothalamus of the Association for Research in Nervous and Mental Disease in 1939, at which we were invited to present our views (E. Scharrer & B. Scharrer 1940). Nevertheless, for a number of reasons, among them a lack of facilities and appropriate techniques, progress was slow.

A search of the literature yielded very few reports that appeared to be relevant, and it was not until much later that such reports found their proper place in the emerging saga (see below). One was the description of "gland-cells of internal secretion" in the caudal spinal cord of skates (Speidel 1919). Another called attention to the presence of a protein material, called "Substance P," extractable from mammalian gut as well as nervous tissue (von Euler & Gaddum 1931). A third proposed, on the basis of extirpation experiments in a lepidopteran insect, *Lymantria*, that pupation is controlled by a brain hormone (Kopeć 1917, 1922).

What is so fascinating, in retrospect, is that each milestone reached in the elucidation of neurosecretory activities is marked by the breakdown of once powerful conceptual barriers and their replacement by major new and seminal insights. Within the space of 50 years, there occurred not only the emergence of the new discipline of neuroendocrinology but, as an outgrowth of these endeavors, the discovery of new modes of interneuronal communication that reach far beyond the confines of neuroendocrine phenomena.

To return to the early phase of our search, as a first step a broadly based comparative approach was undertaken in which E. Scharrer focused on vertebrates and I on invertebrates (see E. Scharrer & B. Scharrer 1937, 1945, 1954). It revealed the virtually universal occurrence of distinctive peptide-producing neurosecretory cell groups throughout the animal kingdom, up to mammals, including man. This laid to rest recurrent criticisms suggesting their pathological nature. It also showed that neurosecretory neurons are not restricted to a small magnocellular component of the hypothalamus of vertebrates.

The first invertebrate ganglia in which impressive signs of secretory activity were detected were those of the opisthobranch snail *Aplysia* (B. Scharrer 1935), an animal that has since proved to be an excellent model for experimentation in this area of research (e.g. Scheller 1984, Strumwasser 1985). What turned out to be of even greater heuristic value was the study of the brain-corpus cardiacum-corpus allatum system of insects, because of its remarkable analogy with the hypothalamic-hypophysial system of vertebrates (B. Scharrer & E. Scharrer 1944). A number of new insights gained from the comparison of the structure–function relationships observed in these two neuroendocrine organ complexes could not have been obtained from mammalian material alone.

Extensive cytophysiological studies, undertaken by E. Scharrer and his collaborators in vertebrates and by myself and others in invertebrates, were focused on structural differences in the neuroendocrine systems of animals examined under different physiological or experimentally altered conditions. These efforts provided indirect, though persuasive, evidence in support of a functional role of neurosecretory products in the control of developmental, reproductive, and metabolic functions. The search for structural parameters of known functional states remained an absolute requirement even after increasingly sophisticated methods permitted a more direct approach to the elucidation of neuroglandular function. In fact, it is difficult to estimate where this field of research would stand today without the continued essential contributions made by morphologists.

The advent of electron microscopy made possible a precise characterization of the sites of synthesis and release of peptidergic neurosecretory products. As shown in early contributions by Palay, Bargmann, Bodian and others, as well as our own, this material consists of strikingly electron-dense, membrane-bounded granules that are easily identified and localized. The demonstration of their formation in packaged form by the Golgi apparatus (E. Scharrer & S. Brown 1961), as well as their axonal release by exocytosis, provided further evidence in support of a close similarity with the products of other protein-secreting gland cells. In combination with sensitive and selective immunocytochemical procedures, electron microscopy now permits us to determine not only the distribution but also the chemical nature of neurosecretory messenger substances in various parts of the organism. In this program, the pioneering contributions by Sternberger (1986), especially his peroxidase-antiperoxidase method, have proved to be most productive in the hands of many investigators, including those searching for analogous phenomena in invertebrates (B. Scharrer 1981).

The Hypothalamic-Neurohypophysial System

The first documentation of a neurohormonal activity attributable to peptides released by neurosecretory neurons did not pertain to the postulated role of the hypothalamus in the control of adenohypophysial function. Instead it solved an enigma of long standing by demonstrating the hypothalamic origin of the so-called posterio-lobe hormones, vasopressin and oxytocin. Here again, a morphological observation led the way. A new histological staining procedure, developed by Gomori for the demonstration of pancreatic beta-cells, the chrome-alum hematoxylin phloxin technique, turned out to be selective for neurosecretory products, and for the first time it made possible the visualization of their presence throughout the entire neuron. This convenient marker enabled Bargmann (1949) to trace the neurosecretory fiber tract connecting the hypothalamic cell bodies, where the material is synthesized, with the posterior pituitary, where it is released into the circulation. These cells constitute the magnocellular component of the nuclei supraopticus and paraventricularis of the mammalian hypothalamus, which in fishes and amphibians are represented by the singular nucleus praeopticus. The correct interpretation of this topographic relationship (Bargmann & E. Scharrer 1951) was subsequently substantiated by appropriate experimental procedures. Evidence supporting the concept of an intraneuronal transport of such neuron-derived hormonal messengers to their release sites in analogous neurohemal organs located outside of the brain (posterior lobe and corpus cardiacum, respectively) was obtained by surgical interruption of the neurosecretory pathway. In mammals (Hild 1951) as well as insects (B. Scharrer 1952) the active material accumulated proximal and became depleted distal to the level of severance. Hild & Zetler (1953) further showed a clear correlation between the functional potency of tissue extracts, determined by pharmacological tests, and the amount of selectively stainable neurosecretory material present in the tissues under investigation.

This breakthrough, i.e. the establishment of valid criteria for the existence of a new class of neurochemical mediators, for which the term "neurohormone" is appropriate, posed a conceptual problem, namely the rationale for such one-step neurohormonal activities. The question of why neurons should deviate so profoundly from the norm to function as gland cells of internal secretion in organisms endowed with endocrine glands proper was eventually resolved by the examination of the evolutionary history of bilogically active neuropeptides (see below).

The Hypothalamic-Adenohypophysial System

The search for a solution of the central tenet, i.e. the operation of neural mechanisms controlling adenohypophysial function, was encouraged once a case was established for the existence of neurohormonal regulators. Not only were neurohormones now considered to be the most likely candidates for neuroendocrine mediation at the level of the hypothalamic-adenohypophysial axis, but the first active principles suspected to serve in this capacity were the "posterior-lobe hormones" themselves. Such a role has, in fact, been demonstrated quite recently for vasopressin, which participates in the control of corticotropin release.

However, the majority of stimulatory and inhibitory neuropeptides directed to the anterior lobe of the pituitary gland was found to originate in hypothalamic centers other than those known to provide vasopressin and oxytocin. The operation of a "semiprivate" vascular pathway for the conveyance of these hypophysiotropic messenger substances, i.e. the hypophysial portal system, was established by the pioneering work of Harris, in collaboration with Green and Jacobsohn (Harris 1955). They demonstrated that reproductive function can be interrupted by section of the hypophysial stalk and restored after the regeneration of the portal vessels. The median eminence was recognized as the neurohemal release organ for hypophysiotropic neurohormones, and an explanation had thus been found for the virtual absence of a nerve supply to the adenohypophysial parenchyma. The rich contributions by Everett, Gorski, Martini, McCann, Reichlin, Saffran, Sawyer, and others brought further advances in the clarification of the neuroendocrine control of reproductive phenomena.

These investigations made use of various techniques, among them organ culture, electrical stimulation, or lesions in discrete hypothalamic centers (e.g. Szentágothai et al 1962), and transplantation experiments (e.g. Nikitovitch-Winer & Everett 1958), and showed a comparable dependency on hypothalamic directives for other adenohypophysial, including thyrotropic, adrenocorticotropic and metabolic functions. Further contributions demonstrating neuroendocrine control systems in nonmammalian vertebrates were made by, for example, Benoit and Assenmacher (1959) and Oksche and Farner (Oksche et al 1964) in birds and by Etkin (1963) and his collaborators in amphibians.

Parallel studies established that two-step neuroendocrine control systems are not restricted to vertebrates. In insects, cerebral neuropeptides, such as the pupation hormone discovered by Kopeć (1917, 1922), have been shown to act via the regulation of non-neural glands of internal secretion, i.e. the corpus allatum and the prothoracic gland. Evidence for an inhibitory effect of the brain on the corpus allatum, an analogue of the adenohyphysis (B. Scharrer 1952), parallels that for the inhibitory neural control of the pars intermedia of tadpoles reported by Etkin (1962).

The isolation, chemical characterization, and de novo synthesis of several of these hypothalamic regulatory principles, referred to as "releasing" or "regulating" factors (RFs) or hormones (RHs), had to await the development of an appropriate biochemical technology. Achieved largely by Guillemin (1978) and Schally (1978) and their collaborators, this momentous accomplishment placed neurosecretion and neuroendocrinology on firm ground. An intensive multidisciplinary effort, aimed at the investigation of the functional capacities and modes of operation of these active principles, became a major concern of many laboratories.

At this juncture, the widely held view was that, in addition to a vast majority of conventional, synaptically transmitting neurons, a small and quite separate class of peptide-producing neurosecretory cells, which dispatch their signals exclusively via circulatory channels, had to be accommodated in the spectrum of neuronal elements. Their deviation from the norm was then generally understood as being related to their unique role in the transduction of cues to the endocrine system.

Nonendocrine Functions of Neural Peptides

Another quite unexpected turn of events necessitated a shift in our interpretation of neurosecretory phenomena. A variant by which neurosecretory signals may address endocrine cells at close range was discovered by means of electron microscopy. The observation of synapse-like ("synaptoid") terminals on cells of the mammalian pars intermedia led Bargmann et al (1967) to introduce the term "peptidergic neuron," which today extends far beyond its original conception. The same localization of neurosecretory release sites in close spatial relationship to effector cells was found in an analogous endocrine organ of insects, the corpus allatum (B. Scharrer 1972), and subsequently in several nonendocrine tissues of vertebrates as well as invertebrates.

What had become increasingly clear was that peptidergic neurons have available several alternate, i.e. non-neurohormonal, possibilities for neurochemical communication. Some of them come very close to the pattern of classical synaptic intervention, others are neither strictly "private" (in loco) nor neurohormonal. The latter case, in which a relatively narrow zone of extracellular stroma intervenes between nerve terminal and effector cell, represents an intermediate mode of action, both in space and time. The recognition of this interstitium has given rise to the concept of "synapse à distance."

The most intriguing outcome of these studies was the realization that not only somatic cells but even neurons can be addressed by neuropeptides in a nonconventional manner. One of the possible pathways is the cerebrospinal fluid. The dispatch of a neuroregulatory peptide into an extracellular compartment wider than the synaptic gap may operate in a special, nonsynaptic form of neurocommunication known as "peptide-mediated neuromodulation" (see below). Other variants are synapse-like contacts in which either the presynaptic partner of a pair of neurons or else two (or more) neurons in close contact are of peptidergic nature. Some junctional complexes of the latter type (B. Scharrer 1974) give the impression of providing for a reciprocal exchange of signals. It has been suggested that peptidergic synaptic transmission closely resembles that involving nonpeptidergic transmitters, the difference being a slower time course, which is necessitated by the different mechanism of inactivation of a peptidic neuroregulator.

The Spectrum of Neuropeptides

The new direction in neurobiological research discussed in the preceding section was greatly stimulated by yet another spectacular, almost explosive, step forward, i.e. the discovery, in rapid succession, of a multitude of "new neuropeptides." A current estimate of their number is about 100, and the search still continues. By the same token, the number of known peptide-producing cell types has steeply risen.

The family of peptidergic neurons now encompasses, in addition to the classical or archetypal neurosecretory cells, a long list of others known to synthesize and make use of biologically active peptides in one form or another. Sites of neuropeptide production are widely distributed in the central and peripheral nervous systems. Maps prepared by means of immunocytochemical visualization, at both the light and electron microscopic levels, show a distinctive pattern in the distribution of cell bodies and their axonal projections containing specific neuropeptides. Noteworthy examples are the precise laminar organization of several different neuropeptides in the optic tectum (Kuljis & Karten 1982) and their localization in distinct retinal cell populations, primarily amacrine cells (Brecha & Karten 1983).

Moreover, the same or very closely related chemical principles are now known to be manufactured by various non-neuronal cells, e.g. those of the "diffuse endocrine cells" of the digestive apparatus (Pearse & Takor Takor 1979). An example of the commonality of such peptides is somatostatin, first known as a hypophysiotropic neurohormone, now identified, and functionally characterized, also in the endocrine pancreas and in other non-neural tissues (Patel & Reichlin 1978). Conversely, substances heretofore better known as digestive hormones (e.g. gastrin and cholecystokinin) or as adenohypophysial products (e.g. ACTH, α-MSH, and β-lipotropin) have now been identified also in the brain.

Much has been learned about the chemical identities of numerous neuropeptides, as well as the biosynthetic process that gives rise to them, by the use of strategies developed by molecular genetics. Virtually all of these bioactive peptides (and other proteins manufactured by the body) are known or postulated to be derived from high-molecular-weight precursor substances. These inactive pro-proteins (or pre-pro-proteins) are synthesized under the direction of mRNA templates on ribosomes bound to the endoplasmic reticulum (rough surfaced ER) of the perikaryon. Examples of such precursor molecules are pro-opiomelanocortin (containing the amino acid sequences of ACTH, α-MSH, β-endorphin, and β-lipotropin), pro-enkephalin (containing multiple copies of met-enkaphalin plus leu-enkephalin), and pro-pressophysins (precursors of vasopressin and oxytocin plus their specific neurophysins, respectively).

Posttranslational processing of pro-neuropeptides includes sequential proteolytic cleavage of the precursor as well as additional steps of conversion resulting in active substances, among them glycosylation, phosphorylation, acetylation, and amidation. Synthesis and processing of pro-opiomelanocortin in the arcuate nucleus of the hypothalamus are similar, though not identical, chemical steps as compared with those occurring in the adenohypophysis.

All of these processes occur within membrane-bounded compartments (organelles) of the cytoplasm, presumably during the axoplasmic transport of these products to their sites of release, primarily at the axon terminal.

The degradation of the active principles is considered to be brought about by the action of specific peptidases.

Evolutionary History of Neuropeptides

Immunobiological tests, i.e. both radioimmunoassays and immunocytochemical probes, have shown that neuropeptides are widely distributed in the animal kingdom, including unicellular organisms (Roth &

Le Roith 1984). The use of antisera raised against a number of mammalian-type neuropeptides has yielded distinctly localized reaction products in the nervous systems of lower phyla, including a variety of invertebrates (see, for example, Rémy & Dubois 1981, Hansen et al 1982). Conversely, such positive responses have been elicited in vertebrates with antisera raised against a few physiologically and chemically identified invertebrate neuropeptides, among them a molluscan cardioexcitatory factor, the tetrapeptide FMRFamide, which resembles gastrin/cholecystokinin. Perhaps the most striking parallelism demonstrated is the occurrence of an undecapeptide, identical with the growth-promoting "head activator" of the coelenterate *Hydra*, in the human hypothalamus (Bodenmüller & Schaller 1981).

This similarity, or even identity, in molecular configuration indicates that biologically active neuropeptides are stable elements with a long evolutionary history. Their origins seem to reach back to when the first primitive nervous systems began to form and even farther. Apparently, when no endocrine cells proper were as yet in existence, the nervous system was in charge of all required integrative functions. In fact, in primitive nerve cells endocrine functions must have taken precedence over synaptic activities, which seem to have taken over at a later date.

It is tempting to speculate that the bioactive neuropeptides of today are derived from ancestral protein precursors (see B. Scharrer 1978) which, step by step, have developed the capacity of splitting off active principles in a manner illustrated by the present mechanism of peptide biosynthesis from pro-proteins. As the demand for increasingly complex signaling devices arose, the assignment of specific physiological functions to selected amino-acid sequences had to evolve in concert with that of corresponding receptor molecules. Some of the neuropeptides identified in various tissues may be functionless, being mere signposts of a process of molecular evolution.

It makes sense to propose that, after the appearance in more advanced organisms of an endocrine system proper, the pluripotential neurosecretory neurons have taken over a new and highly significant role, that of presiding over the endocrine system. However, their capacity for direct communication with terminal effector cells has not been lost. One-step neurohormonal activities, exemplified by the antidiuretic effect of vasopressin, make sense when viewed as carryovers from a time in the distant past when neurons had to perform all integrative functions directly. In this evolutionary framework, the neurosecretory neuron, far from being a newcomer and rare exception, can be considered as ancestral. It seems to have more in common with the nerve cell precursor than has the "conventional" neuron with its specialization for synaptic transmission.

The Neurosecretory Neuron in a New Light

Quite obviously, the original definition of the neurosecretory neuron, based on the hypothalamic prototype dispatching its signal via the general circulation, no longer suffices. The fact that this classical type, which was the first discovered, represents the most unorthodox known to date accounts for the difficulty with which the scientific community accepted it. This difficulty has been largely overcome by the subsequent demonstration of a spectrum of transitional peptidergic neuronal types bridging the gap between the classical neurosecretory and the conventional nonpeptidergic neurons. Now the multiplicity of neurons known or presumed to be engaged in peptidergic activity poses a quite different problem, that of where to draw a sharp line of demarcation between them and neurons synaptically transmitting by means of nonpeptidergic conventional neurohumors.

It is a problem that has become accentuated by one of the most recent and quite challenging developments in neuropeptide research. There is rising evidence for the coexistence in one and the same neuron of two or more potentially neuroactive mediators (Hökfelt et al 1984, Chan-Palay et al 1978, Chan-Palay & Palay 1984). This principle applies to the colocalization, not only of different neuropeptides, but also of peptides and classical neurotransmitters. The presence of, for example, ACTH, α-MSH, and β-endorphin in neurons (even within the same secretory vesicles) of the arcuate nucleus can be explained in biosynthetic terms, i.e. by their origin from the common precursor, pro-opiomelanocortin. That the same or a very similar combination of neuropeptides also occurs in endocrine cells of the adenohypophysis is an example of the commonality of these chemical mediators. What is still unclear, however, is how these multiple peptides operate after being released.

Even less can be said about the function of neuropeptides found to be colocalized with conventional neurotransmitters. The number of neurons shown to contain combinations such as catecholamine and enkephalin, serotonin and Substance P, acetylcholine and vasoactive intestinal peptide (VIP) has risen sharply within recent years. There is some speculation that in this situation conventional neuroregulators may function as primary neurotransmitters, and peptides present in the same neuron may function in an auxiliary capacity as "cotransmitters." However, as long as this role remains unclear, the question of whether or not a "peptidergic activity" should be attributed to these neurons remains in abeyance. Be this as it may, these cases of colocalization contribute much to the general conclusion that neuropeptides are now a solid majority in the armamentarium of neurochemical messenger substances.

By the same token, neurons engaged in internal secretion can no longer be considered to be the only nerve cells releasing functionally important glandular products. With the exception of those engaged in electrotonic transmission, not involving neurochemical mediators, the vast majority of neurons must be accorded at least a small measure of secretory capacity. How large a proportion of these may operate entirely without the participation of neuropeptides remains yet to be determined. Therefore, the terms "neurosecretory" and "peptidergic" should be understood to refer to the degree to which they apply rather than to the characterization of strictly separate cell types.

Even neurons as far removed from the conventional end of the spectrum as classical hormone-producing neurosecretory cells show a number of structural and functional features characteristic of neurons in general. They are capable of generating action potentials. They are known to respond to afferent stimulatory as well as inhibitory neuronal signals. In the case of neurons dispatching hypophysiotropic signals, afferent directives are provided by various intero- and exteroceptive factors, which, in integrated form, result in the "final common pathway" to the endocrine apparatus (E. Scharrer 1966). Major examples of exteroceptive influences on hypophysiotropic activity are olfactory and photoperiodic stimuli. Afferent interoceptive directives of importance are contributed by circulating hormones that become part of the three-step feedback mechanism by which the dispatch from their sites of production is controlled.

The release of neuropeptides at axon terminals, like that of "regular" neurotransmitters such as acetylcholine, occurs by the process of exocytosis. This process is triggered by an influx of Ca^{2+} through voltage sensitive channels following depolarization of the axon terminal, and presumably involves the interaction of Ca^{2+} with calmodulin and specific protein kinases (Thorn et al 1985).

In both situations, the operation of stereospecific receptors at the respective sites of action of the neuroregulators has been demonstrated. This concept, based on the recognition between conformations of two reacting molecules, is instrumental in the determination of putative sites of action of neuropeptides reaching them via circulatory pathways. The available information refers primarily to receptors for opioid peptides, e.g. enkephalin, and for somatostatin, which have been identified in mammals and several nonmammalian vertebrates (Way 1980). Among invertebrates, high-affinity binding sites that suggest the operation of such receptors in a mollusc and an insect have been demonstrated with the use of a radiolabeled enkephalin analogue, selected as an exogenous ligand because of its stability (see Stefano & Scharrer 1981). These binding sites, demonstrated in the insect brain, may be involved in a modulatory role played by "enkephalinergic" neurons identified by Rémy & Dubois (1981) immunocytochemically in this organ.

The wide variety of available neuropeptides, outnumbering all other, previously recognized neuroregulators combined, goes hand in hand with the greater versatility in their mode of operation. Depending on the extracellular pathway used (ranging from general circulation to intersynaptic gap), the amount of neuropeptide released varies, and so does the type of signal. It may closely resemble "standard" synaptic transmission, as is the case with Substance P or gastrin.

A different form of close-range activity appears to be that of neuromodulation. In this new and intriguing mode of neurochemical intervention, neuropeptides may either augment or depress synaptic signals passed on between two synaptically joined "conventional" neurons. This role, as well as the still enigmatic function of neuropeptides colocalized with other neuroregulators, has been referred to above.

As to the neural directives aimed at the adenohypophysis, the type of operation is flexible. A given signal to one of the specific "target cells" may be accomplished by the collaboration of several hypophysiotropins. Conversely, one hypophysiotropic factor may contribute to the regulation of more than one type of pituitary cell.

The principle that one and the same neuropeptide can function in different capacities, depending on where it is put to use, is best illustrated by vasopressin. This neuroregulator, primarily known for its antidiuretic neurohormonal role, also participates in hypophysiotropic signaling, specifically in collaboration with the corticotropin-releasing factor, CRF (e.g. Blume et al 1978). A third role of vasopressin is its presumed effect on extrahypothalamic neurons that are contacted by terminals containing this nonapeptide. The relationship between this ultrastructural and immunocytochemical evidence (Buijs et al 1978, Dogterom et al 1978, Sofroniew & Weindl 1978) and the demonstration by De Wied and his collabrators (De Wied 1978) of vasopressin effects on certain memory and learning functions remains to be clarified.

Reference should be made here also to interactions between peptidergic neural centers and two regulatory systems that have recently attracted much attention.

One of these links is with the immune system (see Goetzl 1985), which shares with neuroendocrine structures a set of regulatory substances, including ACTH, TRH, and endorphins, as well as the corresponding receptors (Blalock et al 1985a,b). Neuroimmunologic communication is bidirectional.

Numerous studies indicate that immune functions are subject to neuroendocrine regulation, as shown by their susceptibility to stress. Such external regulatory signals received by the immune system are now considered to operate in addition to its well-documented autoregulatory mechanism (Besedovsky et al 1985). Peptide signals dispatched by neuronal or endocrine cells can reach lymphoid tissues in two ways: (*a*) These centers have access to blood-borne messengers, such as vasopressin and oxytocin, that are known to act as "helper signals" in the regulation of lymphokine production (Johnson & Torres 1985); (*b*) some lymphoid structures, e.g. thymus and spleen, have been shown to receive peptidergic innervation, in particular by fibers showing vasoactive intestinal peptide (VIP)-like, neuropeptide-Y-like, met-enkephalin-like, cholecystokinin (CCK)-like, and neurotensin-like immunoreactivity (Felten et al 1985).

Conversely, there is increasing evidence that signals dispatched by immunomodulator agents reach the nervous system and thus provide part of the information recorded and processed by the neuroendocrine regulatory apparatus. For example, microiontophoretic application of α-interferon to various parts of the rat brain, combined with simultaneous single neuron recordings, revealed alterations in cellular activity in all brain structures examined. Moreover, systemic administration of α-interferon resulted in certain behavioral responses (Dafny et al 1985).

A second relationship involves the control over fluid and electrolyte balance (and blood pressure) by vasopressin and by the atrial natriuretic factors (ANFs, atriopeptins) produced by the heart. These recently discovered polypeptide hormones are synthesized by atrial cardiocytes that exhibit ultrastructural features of protein-secreting cells (De Bold 1985). Their potent diuretic and hypotensive effects counteract those of vasopressin. The existence of a feedback mechanism between these antagonistic regulatory substances is demonstrated by the fact that the administration of arginine vasopressin stimulates the release of atriopeptins from the cardiac endocrine system (Manning et al 1985). Moreover, the presence of atriopeptin-immunoreactive neurons in the hypothalamus has been revealed by tests with antisera to atriopeptin III and to the precursor molecule, atriopeptigen (Saper et al 1985).

In light of our present insights, the three early, seemingly disparate, contributions mentioned at the beginning of this chapter fall into line. The study of the gland-like neurons in the spinal cord of fishes discovered by Speidel (1919) was followed up by several groups of investigators (Bern et al 1985, Lederis et al 1985). These cells are now recognized to be part of a neurosecretory system whose neurohemal release site is the urophysis. The chemical and physiological properties of the neurohormones produced by this system, neurotensin I and neurotensin II, have much in common with those of the products of the hypothalamic-hypophysial system of higher vertebrates.

Substance P, discovered by von Euler & Gaddum (1931), has been chemically identified by Leeman (1980) and found to be a multifunctional neuropeptide. The initially puzzling fact that it is extractable from the gut as well as the brain now places it in line with an array of other neuropeptides.

The report on the "pupation hormone" produced by the insect brain (Kopeć 1917, 1922), initially viewed with much skepticism, is now fully supported by multiple studies on the endocrine role as well as the structural properties of the neurosecretory cells of this group of invertebrates.

Concluding Remarks

The elucidation of the phenomenon of neurosecretion has become an enormously productive area of research. New insights gained with accelerating speed from multidisciplinary studies focused on peptide-producing neurons have wrought major changes in neurobiological concepts long considered to be firmly established. Once interpreted as a small, possibly aberrant minority among a well established group of neuroregulators, neuropeptides have now moved to center stage. Their importance in the control over a variety of fundamental biological processes can hardly be overestimated. Their versatility seems to account for much of the complexity and subtlety in neurochemical signaling that could not be accomplished by classical synaptic transmission alone.

The wide distribution of neuropeptides within and outside of the central nervous system, including their coexistence with nonpeptidergic neurotransmitters in many nerve cells, leaves much less room for "conventional neurons" than that assigned to them in the past. The borderline between neurons and other cellular species engaged in chemical communication, especially endocrine cells, has become somewhat indistinct. The focus of attention on features shared by neurons and non-neurons producing bioactive peptides has resulted in two proposals for classification: "APUD cells" (Pearse & Takor Takor 1979) and "paraneurons" (Fujita 1985).

The time has come to make use of this new knowledge in clinical neurology and neuroendocrinology, especially in areas dealing with stress, pain, biological rhythms, memory loss, and depression. The future pace of advances in this field promises to be lively.

Literature Cited

Bargmann, W. 1949. Über die neurosekretorische Verknüpfung von Hypothalamus und Neurohypophyse. *Z. Zellforsch.* 34: 610–34

Bargmann, W., Lindner, E., Andres, K. H. 1967. Über Synapsen an endokrinen Epithelzellen und die Definition sekretorischer Neurone. Untersuchungen am Zwischenlappen der Katzenhypophyse. *Z. Zellforsch.* 77: 282–98

Bargmann, W., Scharrer, E. 1951. The site of origin of the hormones of the posterior pituitary. *Am. Scientist* 39: 255–59

Barker, J. L., Smith, T. G. Jr., eds. 1980. *The Role of Peptides in Neuronal Function*. New York: Dekker

Benoit, J., Assenmacher, I. 1959. The control by visible radiations of the gonadotropic activity of the duck hypophysis. *Recent Progr. Hormone Res.* 15: 143–64

Bern, H. A., Pearson, D., Larson, B. A., Nishioka, R. S. 1985. Neurohormones from fish tails: The caudal neurosecretory system. I. "Urophysiology" and the caudal neurosecretory system of fishes. *Recent Progr. Hormone Res.* 41: 533–52

Besedovsky, H. O., Del Rey, A. E., Sorkin, E. 1985. Immune-neuroendocrine interactions. *J. Immunol. Suppl.* 135: 750s–54s

Blalock, J. E., Bost, K. L., Smith, E. M. 1985a. Neuroendocrine peptide hormones and their receptors in the immune system. Production, processing and action. *J. Neuroimmunol.* 10: 31–40

Blalock, J. E., Harbour-McMenamin, D., Smith, E. M. 1985b. Peptide hormones shared by the neuroendocrine and immunologic systems. *J. Immunol. Suppl.* 135: 858s–61s

Bloom, F. E., ed. 1980. *Peptides: Integrators of Cell and Tissue Function. Soc. General Physiol. Ser.* Vol. 35. New York: Raven

Bloom, F. E., Battenberg, E., Ferron, A., Mancillas, J. R., Milner, R. J., Siggins, G., Sutcliffe, J. G. 1985. Neuropeptides: Interactions and diversities. *Recent Progr. Hormone Res.* 41: 339–67

Blume, H. W., Pittman, Q. J., Renaud, L. P. 1978. Electrophysiological indications of a 'vasopressinergic' innervation of the median eminence. *Brain Res.* 155: 153–58

Bodenmüller, H., Schaller, H. C. 1981. Conserved amino acid sequence of a neuropeptide, the head activator, from coelenterates to humans. *Nature* 293: 579–80

Brecha, N. C., Karten, H. J. 1983. Identification and localization of neuropeptides in the vertebrate retina. In *Brain Peptides*, ed. D. T. Krieger, M. J. Brownstein, J. B. Martin, pp. 437–62. New York: Wiley

Buijs, R. M., Swaab, D. F., Dogterom, J., van Leeuwen, F. W. 1978. Intra- and extrahypothalamic vasopressin and oxytocin pathways in the rat. *Cell Tiss. Res.* 186: 423–33

Burgen, A., Kosterlitz, H. W., Iversen, L. L., eds. 1980. Neuroactive Peptides. *Proc. R. Soc. London Ser. B* 210: 1–195

Chan-Palay, V., Jonsson, G., Palay, S. L. 1978. Serotonin and substance P coexist in neurons of the rat's central nervous system. *Proc. Natl. Acad. Sci.* 75: 1582–86

Chan-Palay, V., Palay, S. L. 1984. Cerebellar Purkinje cells have GAD, motilin and CSAD immunoreactivity. Existence and coexistence of GABA, motilin and taurine. In *Coexistence of Neuroactive Substances*, ed. V. Chan-Palay, S. L. Palay, pp. 1–22. New York: Wiley

Costa, E., Trabucchi, M., eds. 1980. *Neural Peptides and Neuronal Communication*. New York: Raven

Dafny, N., Prieto-Gomez, B., Reyes-Vazquez, C. 1985. Does the immune system communicate with the central nervous system? Interferon modifies central nervous activity. *J. Neuroimmunol.* 9: 1–12

De Bold, A. J. 1985. Atrial natriuretic factor: A hormone produced by the heart. *Science* 230: 767–70

De Wied, D. 1978. Pituitary peptides and adaptive behavior. See Meites et al 1978, pp. 383–400

Dogterom, J., Snijdewint, F. G. M., Buijs, R. M. 1978. The distribution of vasopressin and oxytocin in the rat brain. *Neurosci. Lett.* 9: 341–46

Etkin, W. 1962. Hypothalamic inhibition of pars intermedia activity in the frog. *Gen. Comp. Endocrinol. Suppl.* 1: 148–59

Etkin, W. 1963. Metamorphosis. In *Physiology of the Amphibia*, ed. J. Moore, Chapt. 8. New York: Academic

Felten, D. L., Felten, S. Y., Carlson, S. L., Olschowka, J. A., Livnat, S. 1985. Noradrenergic and peptidergic innervation of lymphoid tissue. *J. Immunol. Suppl.* 135: 755s–65s

Fujita, T. 1985. Neurosecretion and new aspects of neuroendocronology. See Kobayashi et al, eds., 1985, pp. 521–28

Gainer, H., ed. 1977. *Peptides in Neurobiology*. New York/London: Plenum

Goetzl, E. J., ed. 1985. Neuromodulation of immunity and hypersensitivity. *J. Immnol. Suppl.* 135: 739s–863s

Guillemin, R. 1978. Peptides in the brain: The new endocrinology of the neuron. *Science* 202: 390–402

Hansen, B. L., Hansen, G. N., Scharrer, B. 1982. Immunoreactive material resembling vertebrate neuropeptides in the corpus cardiacum and corpus allatum of the insect *Leucophaea maderae*. *Cell Tissue Res.* 255: 319–29

Harris, G. W. 1955. *Neural Control of the Pituitary Gland*. London: Arnold

Hild, W. 1951. Experimentell-morphologische Untersuchungen über das Verhalten der "Neurosekretorischen Bahn" nach Hypophysenstieldurchtrennungen, Eingriffen in den Wasserhaushalt und Belastung der Osmoregulation. *Virchows Arch.* 319: 526–46

Hild, W., Zetler, G. 1953. Experimenteller Beweis für die Entstehung der sog. Hypophysenhinterlappenwirkstoffe im Hypothalamus. *Pflügers Arch.* 257: 169–201
Hökfelt, T., Johansson, O., Goldstein, M. 1984. Chemical anatomy of the brain. *Science* 225: 1326–34
Johnson, H. M., Torres, B. A. 1985. Regulation of lymphokine production by arginine vasopressin and oxytocin: Modulation of lymphocyte function by neurohypophyseal hormones. *J. Immunol. Suppl.* 135: 773s–75s
Kobayashi, H., Bern, H. A., Urano, A., eds. 1985. *Neurosecretion and the Biology of Neuropeptides. Proc. 9th Int. Symp. on Neurosecretion.* Tokyo: Japan Sci. Soc. Press, and Berlin/Heidelberg/New York/Tokyo: Springer-Verlag
Kopeć, S. 1917. *Bull. Acad. Sci. Cracovie, Classe Sci. Math. Nat. Sér. B*, pp. 57–60
Kopeć, S. 1922. Studies on the necessity of the brain for the inception of insect metamorphosis. *Biol. Bull.* 42: 323–42
Krieger, D. T., Brownstein, M. J., Martin, J. B., eds. 1983. *Brain Peptides.* New York: Wiley
Kuljis, R. O., Karten, H. J. 1982. Laminar organization of peptide-like immunoreactivity in the anuran optic tectum. *J. Comp. Neurol.* 212: 188–201
Lederis, K., Fryer, J., Rivier, J., MacCannell, K. L., Kobayashi, Y., Woo, N., Wong, K. L. 1985. Neurohormones from fish tails. II: Actions of urotensin I in mammals and fishes. *Recent Progr. Hormone Res.* 41: 553–76
Leeman, S. E. 1980. Substance P and neurotensin: Discovery, isolation, chemical characterization and physiological studies. *J. Exp. Biol.* 89: 193–200
Manning, P. T., Schwartz, D., Katsube, N. C., Holmberg, S. W., Needleman, P. 1985. Vasopressin-stimulated release of atriopeptin: Endocrine antagonists in fluid homeostasis. *Science* 229: 395–97
Meites, J., Donovan, B. T., McCann, S. M., eds. 1975. *Pioneers in Neuroendocrinology*, Vol. 1. New York/London: Plenum
Meites, J., Donovan, B. T., McCann, S. M., eds. 1978. *Pioneers in Neuroendocrinology*, Vol. 2. New York/London: Plenum
Nikitovitch-Winer, M., Everett, J. W. 1958. Functional restitution of pituitary grafts retransplanted from kidney to median eminence. *Endocrinology* 63: 916–30
Oksche, A., Wilson, W. O., Farner, D. S. 1964. The hypothalamic neurosecretory system of *Coturnix coturnix japonica*. *Z. Zellforsch.* 61: 688–709
Patel, Y. C., Reichlin, S. 1978. Somatostatin in hypothalamus, extrahypothalamic brain, and peripheral tissues of the rat. *Endocrinology* 102: 523–30
Pearse, A. G. E., Takor Takor, T. 1979. Embryology of the diffuse neuroendocrine system and its relationship to the common peptides. *Federation Proc.* 38: 2288–94
Rémy, C., Dubois, M. P. 1981. Immunohistological evidence of methionine enkephalin-like material in the brain of the migratory locust. *Cell Tissue Res.* 218: 271–78
Roth, J., Le Roith, D. 1984. Intercellular communication: The evolution of scientific concepts and of messenger molecules. In *Medicine, Science, and Society*. Symposia celebrating the Harvard Medical School Bicentennial, ed. K. J. Isselbacher, pp. 425–47. New York: Wiley
Saper, C. B., Standaert, D. G., Currie, M. G., Schwartz, D., Geller, D. M., Needleman, P. 1985. Atriopeptin-immunoreactive neurons in the brain: Presence in cardiovascular regulatory areas. *Science* 227: 1047–49
Schally, A. V. 1978. Aspects of hypothalamic regulation of the pituitary gland. Its implications for the control of reproductive processes. *Science* 202: 18–28
Scharrer, B. 1935. Über das Hanströmsche Organ X bei Opisthobranchiern. *Pubbl. Staz. Zool. Napoli* 15: 132–42
Scharrer, B. 1952. Neurosecretion. XI. The effects of nerve section on the intercerebralis-cardiacum-allatum system of the insect *Leucophaea maderae*. *Biol. Bull.* 102: 261–72
Scharrer, B. 1970. General principles of neuroendocrine communication. In *The Neurosciences: Second Study Program*, ed. F. O. Schmitt, pp. 519–29. New York: Rockefeller Univ. Press
Scharrer, B. 1972. Neuroendocrine communication (neurohormonal, neurohumoral, and intermediate). *Progr. Brain Res.* 38: 7–18
Scharrer, B. 1974. New trends in invertebrate neurosecretion. In *The Final Neuroendocrine Pathway. Proc. 6th Int. Symp. on Neurosecretion*, pp. 285–87. Berlin/Heidelberg/New York: Springer-Verlag
Scharrer, B. 1978. *An evolutionary interpretation of the phenomenon of neurosecretion*, pp. 1–17. New York: Am. Museum Natural History
Scharrer, B. 1981. Neuroendocrinology and histochemistry. In *Histochemistry: The Widening Horizons*, ed. P. J. Stoward, J. M. Polak, pp. 11–20. Chichester, UK: Wiley
Scharrer, B., Scharrer, E. 1944. Neurosecretion. VI. A comparison between the intercerebralis-cardiacum-allatum system of the insects and the hypothalamo-hypo-

physeal system of the vertebrates. *Biol. Bull.* 87 : 243–51

Scharrer, E. 1928. Die Lichtempfindlichkeit blinder Elritzen (Untersuchungen über das Zwischenhirn der Fische). *Z. Vergl. Physiol.* 7 : 1–38

Scharrer, E. 1952. The general significance of the neurosecretory cell. *Scientia* 46 : 177–83

Scharrer, E. 1966. Principles of neuroendocrine integration. In *Endocrines and the Central Nervous System. Res. Public. Assoc. Nerv. Ment. Dis.* 43 : 1–35

Scharrer, E., Brown, S. 1961. Neurosecretion. XII. The formation of neurosecretory granules in the earthworm, *Lumbricus terrestris* L. *Z. Zellforsch.* 54 : 530–40

Scharrer, E., Scharrer, B. 1937. Über Drüsen-Nervenzellen und neurosekretorische Organe bei Wirbellosen und Wirbeltieren. *Biol. Rev.* 12 : 185–216

Scharrer, E., Scharrer, B. 1940. Secretory cells within the hypothalamus. *Res. Public. Assoc. Res. Nerv. Ment. Dis.* 20 : 170–94

Scharrer, E., Scharrer, B. 1945. Neurosecretion. *Physiol. Rev.* 25 : 171–81

Scharrer, E., Scharrer, B. 1954. Hormones produced by neurosecretory cells. *Recent Progr. Hormone Res.* 10 : 183–240

Scharrer, E., Scharrer, B. 1963. *Neuroendocrinology.* New York : Columbia Univ. Press

Scheller, R. H., Kaldany, R.-R., Kreiner, T., Mahon, A. C., Nambu, J. R., Schaefer, M., Taussig, R. 1984. Neuropeptides: Mediators of behavior in *Aplysia. Science* 225 : 1300–8

Sofroniew, M. V., Weindl, A. 1978. Extrahypothalamic neurophysin-containing perikarya, fiber pathways and fiber clusters in the rat brain. *Endocrinology* 102 : 334–37

Speidel, C. C. 1919. Gland-cells of internal secretion in the spinal cord of the skates. *Carnegie Inst. Washington Publ. No.* 13 : 1–31

Stefano, G. B., Scharrer, B. 1981. High affinity binding of an enkephalin analog in the cerebral ganglion of the insect *Leucophaea maderae* (Blattaria). *Brain Res.* 225 : 107–14

Sternberger, L. A. 1986. *Immunocytochemistry.* New York : Wiley. 3rd ed.

Strumwasser, F. 1985. The structure of the neuroendocrine commands for egg-laying behavior in *Aplysia.* In *Comparative Neurobiology*, ed. M. J. Cohen, F. Strumwasser, pp. 169–79. New York : Wiley

Szentágothai, J., Flerkó, B., Mess, B., Halász, B. 1962. *Hypothalamic Control of the Anterior Pituitary. An Experimental Morphological Study.* Budapest: Akadémiai Kiadó

Thorn, N. A., Chenoufi, H.-L., Tiefenthal, M. 1985. The calcium-calmodulin-proteinkinase system and the mechanism of release of neurohypophysial hormones. *Acta Physiol. Scand. Suppl. 542* 124 : 300

Von Euler, U. S., Gaddum, J. H. 1931. An unidentified depressor substance in certain tissue extracts. *J. Physiol.* 72 : 74–87

Way, E. L., ed. 1980. *Endogenous and Exogenous Opiate Agonists and Antagonists.* New York/Oxford/Toronto/Sydney/Frankfurt/Paris : Pergamon

White, J. D., Stewart, K. D., Krause, J. E., McKelvy, J. F. 1985. Biochemistry of peptide-secreting neurons. *Physiol. Rev.* 65 : 553–97

Ann. Rev. Neurosci. 1987. 10 : 19–40

PERSPECTIVES ON THE DISCOVERY OF CENTRAL MONOAMINERGIC NEUROTRANSMISSION

Arvid Carlsson

Department of Pharmacology, University of Göteborg,
S-400 33 Göteborg, Sweden

The concept of chemical neurotransmission goes back to the beginning of this century but was confined to the peripheral nervous system for a long time. A number of parallel events in the 1950s triggered its penetration through the blood-brain barrier: these included the discovery of the catecholamines and 5-hydroxytryptamine (5-HT, serotonin) in the central nervous system, the introduction of the modern psychotropic drugs, and the development of sensitive and specific biochemical and histochemical methods for the detection of the monoamines and their precursors and metabolites in tissues and body fluids.

In the 1950s very little was known about information transfer between nerve cells in the central nervous system. Electrical transmission seemed to be a likely mechanism in many instances, given the generally close synaptic contacts and sometimes short synaptic delays in the central nervous system. In his monograph, *Synaptic Transmission*, McLennan (1963) concludes: "In the vertebrate central nervous system there is only one synapse identified whose operation can, with assurance, be ascribed to acetylcholine." He was referring to the synapses where the Renshaw cells in the spinal cord receive innervation from axon collaterals of the motoneurons. Since these neurons are undoubtedly cholinergic, the conclusion was reasonable. However, no generalizations from this special case could of course be made. As to the monoamines and other putative neurotransmitters, McLennan did not find a single case in which any of them

0147–006X/87/0301–0019$02.00

could be ascribed a role as a neurotransmitter in the vertebrate central nervous system.

"Sympathin," i.e. a mixture of noradrenaline and adrenaline, was shown to be a normal constituent of brain tissue by Marthe Vogt in 1954. Simultaneously, 5-HT was discovered in the brain (Twarog & Page 1953, Amin et al 1954).

Impact of Psychopharmacology and New Analytical Techniques

The ability of the hallucinogenic agent LSD to block peripheral 5-HT receptors led to speculations about a role for 5-HT in maintaining sanity (Gaddum 1954, Woolley & Shaw 1954). These speculations prompted Brodie and Shore to study the interactions among 5-HT, LSD, and the newly discovered antipsychotic agent, reserpine. From observations on sleeping times in mice they concluded that LSD could antagonize both 5-HT and reserpine (Shore et al 1955).

At this time an important methodological innovation was being developed in Dr. Brodie's laboratory at the National Heart Institute in Bethesda, Maryland: Dr. Bowman, in collaboration with Drs. Brodie and Udenfriend, was constructing the first prototype of a spectrophotofluorometer. This instrument proved to be extremely useful especially in biochemical research, e.g. for the analysis of monoamines and their precursors and metabolites in tissues and body fluids. It was soon to replace the previous bioassay techniques. Using the new instrument, Brodie and Shore were able to demonstrate the virtually complete disappearance of 5-HT from tissues, including the brain, and the simultaneous increase in the urinary excretion of the 5-HT metabolite, 5-hydroxyindoleacetic acid, following treatment with reserpine (Shore et al 1955, Pletscher et al 1955).

The first demonstration of the effect of a psychotropic agent on an endogenous agonist in the central nervous system had an enormous impact on the research field. Already the first attempts to explain the actions of reserpine on 5-HT were based on the tacit assumption that a neurotransmitter mechanism was involved. Brodie and his colleagues put forward the hypothesis that reserpine, by blocking the storage mechanism without causing any inhibition of synthesis, induced a continuous, uninhibited release of 5-HT onto receptor sites. This interpretation seemed logical in view of the interaction experiments carried out by these workers, mentioned above. However, their interpretation of these experiments was questionable in view of the poor penetration of systemically administered 5-HT into the brain.

I had the great privilege to spend six months in 1955–1956 in Dr. Brodie's laboratory, while on sabbatical leave from my assistant professorship at the University of Lund, Sweden. My contact with Drs. Brodie and Udenfriend was established through Dr. Sune Bergström, who was then Chairman of the Department of Physiological Chemistry at the University of Lund. I am indebted to Dr. Bergström for a lot of support and encouragement during this early part of my research career. Coming from an entirely different field (bone-mineral metabolism), I was introduced by Drs. Brodie and Shore to the fascinating field of biogenic amines and to the use of the spectrophotofluorometer. I was given the opportunity to demonstrate, in collaboration with Brodie and Shore, that reserpine, added in low concentration to platelets in vitro, was capable of releasing 5-HT from the platelets by a stereospecific mechanism (Carlsson et al 1957c).

I suggested to Drs. Brodie and Shore that it might be worthwhile to study the action of reserpine on the catecholamines as well, but they did not consider such an approach very promising, given the lack of sympathomimetic actions of this agent. While still in Bethesda, however, I had read with great interest a paper from my home University by Hillarp et al (1953) that showed the existence of specific storage organelles, called granules or vesicles, in the adrenal medulla, and a subsequent paper (Hillarp et al 1955) demonstrating the co-existence of catecholamines and ATP in these organelles. In preliminary experiments, I found large amounts of ATP also in platelets, as was shortly afterwards reported by Born (1956). Thus a link between the storage of 5-HT and catecholamines was suggested.

Immediately after my return to Lund, Hillarp and I started to collaborate. We very soon discovered the depletion of adrenal catecholamine stores by reserpine (Carlsson & Hillarp 1956). Shortly afterwards, together with my graduate students Bertler and Rosengren, I demonstrated the depletion of noradrenaline stores in the heart and brain by this agent (Bertler et al 1956, Carlsson et al 1957a). Moreover, we showed that the sympathetic nerves ceased to respond to stimulation following depletion of catecholamines by reserpine. Similar observations were made by Holzbauer & Vogt (1956) and by Muscholl & Vogt (1958).

Thus, my colleagues and I had to disagree with my highly esteemed mentors and friends, Drs. Brodie and Shore, on two essential points. We concluded (*a*) that not only 5-HT but also the catecholamines had to be considered in attempts to explain the mode of action of reserpine, and (*b*) that rather than the proposed continuous agonist release onto receptors, one should consider lack of neurotransmitter as the functionally crucial result of reserpine's action. In fact, the release of the adrenergic transmitter by reserpine did not seem to occur into the synaptic cleft, because it was not accompanied by any sympathomimetic actions. It was thus reasonable to assume that the release occurred from the storage organelles into the cytoplasm and was followed by intracellular deamination by monoamine oxidase (MAO). In support of this proposal, reserpine had already been shown to cause a sympathomimetic response as well as central stimulation following pretreatment with a MAO inhibitor, thus suggesting that amine release onto receptors did indeed take place when deamination was prevented.

In order to test the amine-deficiency hypothesis we tried to replenish the stores of reserpine-treated animals by the systemic administration of precursors, which in contrast to the amines themselves are capable of penetrating from the blood into the brain. We thus discovered the dramatic antireserpine action of the catecholamine precursor, DOPA, as well as its centrally stimulating action in nonpretreated animals; the 5-HT precursor, 5-hydroxytryptophan, was not an effective reserpine antagonist, however, thus suggesting that lack of catecholamines rather than 5-HT was responsible for the gross behavioral actions of reserpine (Carlsson et al 1957b).

Dopamine: A Central Agonist in Its Own Right

Since the action of DOPA was strongly potentiated by the MAO inhibitor, iproniazide, we were convinced that the catecholamine(s) formed from DOPA mediated the effect. We were hoping that noradrenaline, which was then recognized as the major central catecholamine, would prove to be responsible for the effect. Much to our regret, however, we found that despite the virtual elimination of the behavioral action of reserpine after DOPA treatment, the noradrenaline levels in the brain remained at or below the limit of detection by the methods then available. We then turned our attention to dopamine, which at that time was recognized as a poor sympathomimetic agonist and believed to be just a precursor of noradrenaline and adrenaline. We had first to develop a spectrophotofluorometric method for detection and quantitative measurement of dopamine (Carlsson & Waldeck 1958). This method was used after the purification of tissue extracts on Dowex 50 columns. The procedure could thus be adapted to our previous method for the determination of adrenaline and noradrenaline in tissues (Bertler et al 1958). These methods soon became widely used in catecholamine research. We were pleased to find that dopamine did accumulate in the brains of animals treated with DOPA following reserpine pretreatment and that this accumulation coincided with the antireserpine response (Carlsson et al 1958). Moreover, we found that dopamine occurs normally in the brain, as shortly before suggested by Montagu (1957). Since the amounts were comparable to those of noradrenaline and since dopamine, too, was made to disappear almost completely by reserpine, we suggested that dopamine besides being an intermediate in catecholamine synthesis, is an agonist in its own right in the central nervous system.

To investigate the regional distribution of dopamine seemed to be an appropriate part of the thesis work conducted by Bertler & Rosengren. They (1959) thus discovered the unique accumulation of dopamine in the basal ganglia. This region had long been recognized as an important component of the extrapyramidal system. Moreover, since the extrapyramidal, Parkinson-like side effects of reserpine were known, it did not seem farfetched to propose a role for dopamine in the control of extrapyramidal motor functions, a deficiency of which would lead to Parkinsonism, while excessive function would give rise to chorea. These ideas were first proposed at the First International Catecholamine Symposium in Bethesda in 1958 (Carlsson 1959).

My colleagues and I were quite excited by these observations. We felt that the opposite behavioral actions of the monoamine depletor, reserpine, and the catecholamine precursor, DOPA, made a strong case for the catecholamines as agonists in the central nervous system, controlling important functions such as motor activity and alertness. We were very disappointed by the scepticism with which our interpretations were received. Thus the Ciba Foundation Symposium in London in 1960 (Vane et al 1960) revealed considerable disagreement on various points, as evident from the recorded discussions. Marthe Vogt in particular was reluctant to accept the available evidence for a role of catecholamines (or 5-HT) in behavior. Sir Henry Dale was surprised to hear that an amino acid such as DOPA could be a "poison." Unpublished data were quoted, indicating that the catecholamines in the brain were located in glia cells. Sir John Gaddum, in his final remarks, concluded: "The meeting was in a critical mood, and no-one ventured to speculate on the relation between catechol amines and the function of the brain." This was indeed a puzzling comment, since I for one had speculated quite a lot on precisely this issue.

When rereading the proceedings of this symposium today I still find it difficult to understand the resistance to our interpretations of the reserpine-DOPA data. There was little or no disagreement about the actual observations. In fact, others had confirmed or independently demonstrated the most salient findings. As mentioned, Marthe Vogt and her colleagues had also observed the depletion of noradrenaline by reserpine and the resultant loss of adrenergic nerve function. At the meeting Drs. Blaschko and Chrusciel described the antireserpine action of DOPA. As it turned out, our disagreement with Brodie was minor in comparison with these other controversies, which raised doubts about any role for the monoamines as agonists in the brain.

Visualization of the Monoamines in the Fluorescence Microscope

The acceptance of the monoamines as neurotransmitters in the central nervous system was a slow process. Certainly the important discovery of low dopamine levels in the basal ganglia of Parkinson patients (Ehringer & Hornykiewicz 1960) and the demonstration of a therapeutic effect of L-dopa in Parkinson patients (Birkmayer & Hornykiewicz 1962, Barbeau et al 1962) contributed to bringing the central monoamines into focus. However, the real breakthrough came with the histochemical techniques visualizing the cellular localization of the monoamines by means of the fluorescence microscope. This fascinating story, which has recently been reviewed in some detail (Dahlström & Carlsson 1986), is briefly summarized below.

I had the privilege to collaborate for many years with the late Dr. Nils-Åke Hillarp, a highly talented histologist who died at the age of 48 in 1965. Hillarp had already in the 1940s and 1950s acquired a considerable research experience in the area of the autonomic nervous system and made several important discoveries. Our collaboration was not confined to the mechanism of storage of catecholamines; we also engaged in attempts to visualize the catecholamines under the microscope. At that time such visualization was possible only in the chromaffin cells of, for example, the adrenal medulla, by using the so-called chromaffin reaction or, as shown by Eränko (1955), by the green fluorescence occurring in sections of the adrenal medulla following formalin fixation. The high concentrations of dopamine in the basal ganglia seemed encouraging, however, and Falck & Hillarp (1959) had actually tried the chromaffin reaction to visualize the catecholamines in the central nervous system, though without success. Evidently, more sensitive methods had to be developed. It was logical to turn our attention to fluorescence, given the tremendous success of this principle in the biochemical methods recently developed for the analysis of the monoamines.

When I was appointed to the chair of pharmacology at the University of Göteborg in 1959, I was delighted to learn that Hillarp wanted to join me. This was made possible through a grant from the Swedish Medical Research Council, which enabled Hillarp to take leave from his associate professorship at the University of Lund. This was probably one of the most profitable investments of the Council to date. In 1960 we moved into the newly built department of pharmacology in Göteborg. The University had allowed fairly generous funding for the equipment of the new institution, and thus Hillarp and I could start our research in Göteborg without too much delay.

Our first attempt to visualize catecholamines in the fluorescence microscope utilized the principle of the so-called trihydroxyindole method for quantitative analysis of catecholamines, i.e. the principle used, for example, by Bertler et al (1958). Tissue sections were first exposed to iodine to oxidize the catecholamines to adrenochromes, and then exposed to ammonia to rearrange the adrenochromes to fluorescent adrenolutines. Our efforts were very successful insofar as the adrenal medulla was concerned: a highly intensive fluorescence developed; it was reduced, though still clearly visible, after removing more than 95% of the catecholamines by reserpine treatment (Carlsson et al 1961). However, for some unknown reason, perhaps diffusion of the amines out of the tissue, the adrenergic transmitter could not be visualized in nerve terminals or cell bodies. Nevertheless, the results were considered very encouraging, and Hillarp was convinced that the project would ultimately prove successful.

Hillarp decided to try another principle for the development of fluorescence from monoamines, based on the analytical method of Hess & Udenfriend (1959) for measurement of tryptamine. In this method tryptamine is condensed with formaldehyde to form a highly fluorescent product. Together with his skillful Research Engineer, Georg Thieme, Hillarp started to investigate systematically this reaction in histochemical model experiments. Various amines were dissolved in solutions of serum albumin, sucrose, gelatin, or gliadin, spotted on glass slides and air-dried. Upon treatment with formaldehyde vapor, generated from a 35% solution of formaldehyde, a very strong fluorescence developed in spots of noradrenaline or dopamine. Protein catalyzed the reaction. The fluorescence products were identified as tetrahydroisoquinolines, as verified by Corrodi & Hillarp (1964).

The report of these fundamental model experiments, all of which were performed by Hillarp and Thieme in the Department of Pharmacology at the University of Göteborg, was authored by Falck, Hillarp, Thieme & Torp (1962), and was one of the 100 most cited publications in 1961–1982. When Hillarp moved to Göteborg, Falck remained at the Department of Histology, University of Lund. After the model experiments were completed, a considerable part of the experiments on various tissue specimens were performed in collaboration with Falck in Lund, since a histology department was of course better equipped for this purpose. Various attempts were made, though without any clearcut success, until one day in late August 1961, when Hillarp visited Falck and his old department, Hillarp proposed that they try air-dried stretch preparations of thin tissue specimens, such as rat iris and mesentery. Hillarp was very familiar with these preparations from his previous work (1946, 1959), now considered classical treatises on the functional organization of the autonomic ground plexus. Such preparations were exposed to dry formaldehyde gas generated from paraformaldehyde powder. The outcome was dramatic: in the fluorescence microscope Hillarp and Falck saw the same nerve-plexus pattern as previously observed by Hillarp following staining with methylene blue. But this time it was the adrenergic transmitter, which showed up as green fluorescence as a consequence of treatment with formaldehyde. In addition, yellow fluorescence derived from mast-cell 5-HT could be seen in the mesenterium preparations. In principle, a great discovery had thus been made, but needless to say, a lot of work remained. For example, the technique had to be adapted to embedded tissue specimens. This work was first performed in Lund and later, after Hillarp's move to take over the chair in histology at the Karolinska Institute in Stockholm in 1962, also by a rapidly growing group of young, enthusiastic students in Hillarp's new working place. For a detailed account of the further development of the histofluorescence techniques, see Dahlström & Carlsson (1986) and Björklund & Hökfelt (1984). The new technique was a powerful tool that helped to solve a large number of important problems in the monoamine field.

Cellular Localization and Mapping of the Monoamines

A major, initial step in cellular localization and mapping of the monoamines was the demonstration of the monoamines in nerve cell bodies and nerve terminals of the central nervous system (Carlsson et al 1962). The neuronal localization of the central monoamines was confirmed by lesion experiments. In fact, the first lesion experiments, which demonstrated the virtually complete disappearance of monoamines from the spinal cord below a transection, utilized biochemical techniques only (Magnusson & Rosengren 1963, Carlsson et al 1963b), shortly before the demonstration of the descendent bulbospinal monoaminergic pathways by means of the histofluorescence technique (Carlsson et al 1964). The first lesion experiments to confirm the neuronal localization of a monoamine in the brain demonstrated the disappearance of dopamine, measured biochemically, and of green-fluorescent nerve terminals from the rat neostriatum following a lesion of the substantia nigra (Andén et al 1964a). Moreover, removal of the striatum was followed by an accumulation of fluorescent material in the cell bodies of the substantia nigra and of axons proximal to the lesion, thus demonstrating the existence of a nigrostriatal dopamine pathway. Independent work in Lund by Bertler et al (1964) reported similar findings.

But these were only beginnings. An enormous amount of work remained to map out all the monoaminergic pathways in the central nervous system. A large number of workers became engaged in this important mapping.

The first systematic studies were performed by Dahlström & Fuxe (1964) and Fuxe (1965), followed by Ungerstedt (1971) and many others (see Lindvall 1974, Björklund & Hökfelt 1984). Needless to say, the detailed knowledge of the central monoaminergic pathways has been of fundamental importance for the further development of research in this field. For example, it enabled Andén and his colleagues (see Andén et al 1969) to develop simple and useful functional models for the individual monoamines, and Aghajanian and his colleagues (see e.g. Aghajanian & Bunney 1974, 1977, Bloom 1984) to embark on their pioneer studies of the electrophysiological activity of monoaminergic neurons.

Fitting Together the Pieces of the Synapse Puzzle

The impact of the histochemical visualization of the monoamines of course extends far beyond the mapping of monoamine-carrying neuronal pathways. After the introduction of the new technique, previous doubts expressed about the transmitter function of the monoamines gradually changed into a debate on the complex issue of neurotransmitter versus neuromodulator function. Today nobody appears to question the neurotransmitter function of the monoamines in the central nervous system, at least in a broad sense of this term. In fact, the monoamines have become "spearheads" in neurotransmission research, especially in the central nervous system. In particular, the monoaminergic synapse has become a very useful model.

After the localization of the central monoamines to special neurons had been established, it remained to demonstrate their subcellular distribution. In the fluorescence microscope the accumulation of monoamines in the so-called varicosities of nerve terminals was obvious. This corresponded to the distribution of synaptic vesicles, as observed in the electron microscope. In fact, Hökfelt (1968) was able to demonstrate the localization of central as well as peripheral monoamines to synaptic vesicles in the electron microscope.

In 1960, Axelrod et al (see Axelrod 1964) discovered the uptake of circulating labeled catecholamines by adrenergic nerves. This uptake proved to be an important inactivation mechanism. According to Axelrod it could be blocked by a large number of drugs, for example, reserpine, chlorpromazine, cocaine, and imipramine, thus leading to supersensitivity to catecholamines and acceleration of their metabolism. This interpretation was not entirely in line with ours, especially insofar as reserpine was concerned; we considered the action of reserpine on the adrenergic transmitter to be a strictly intraneuronal event. The discrepancy could be resolved by combined biochemical (Carlsson et al 1963a, Kirshner 1962) and histochemical studies (Malmfors 1965). Two different amine-concentrating mechanisms were detected: uptake at the level of the cell membrane, sensitive for example to cocaine and imipramine; and uptake by the storage granules or synaptic vesicles, sensitive for example to reserpine. Blockade of the former but not the latter mechanism leads to catecholamine supersensitivity (although secondary receptor supersensitivity may develop after blockade of the latter). In the presence of reserpine, extracellular amine is still pumped into the cytoplasm with unabated efficiency. However, since it cannot be stored, it is deaminated by intraneuronal MAO (see also Carlsson 1966).

Another issue emerged after the discovery by Axelrod and his colleagues of catechol-O-methyl-transferase (COMT). Axelrod (1960) proposed that COMT was mainly responsible for the metabolism of catecholamines, whereas MAO was thought to be primarily involved in the metabolism of the O-methylated metabolites of the catecholamines. Axelrod referred to the intracellular geography: since COMT occurs in the cell sap, newly released catecholamines will be primarily exposed to this enzyme and only secondarily to the mitochondrially located MAO. However, our reserpine data (Carlsson et al 1957a), as well as our observations on the accumulation of catecholamines and their metabolites (Carlsson et al 1960), suggested to us that intraneuronally released catecholamines would be primarily metabolized by MAO; only after release into the extracellular space would they be exposed to COMT (Carlsson 1960, Carlsson & Hillarp 1962). This concept was later generally accepted (see e.g. Axelrod 1964, Jonason & Rutledge 1968, Westerink & Spaan 1982).

In the mid-1960s several controversies still existed in the area of monoaminergic synaptology. This is evident from the recorded discussions of the symposium, "Mechanisms of Release of Biogenic Amines," held in Stockholm in February 1965 (von Euler et al 1966). One of the major issues dealt with the release mechanism in relation to the subcellular distribution of the transmitter. According to Drs. Axelrod and von Euler, a considerable part of the transmitter was located outside the granules (vesicles), mainly in a bound form. This fraction was proposed to be more important than the granular fraction, since it was more readily available for release. Quoting a conversation with Udenfriend, von Euler underlined this by facetiously referring to the granules as "garbage cans." Our group had arrived at a different view, based on a variety of biochemical, histochemical, and pharmacological data. We felt that the granules were essential in transmission, in that the transmitter had to be taken up by them in order to become available for release by the nerve impulse. In favor of this contention was our finding, mentioned above, that reserpine's site of action is the amine uptake mechanism of the granules. The failure of adrenergic transmission as well as the behavioral actions of reserpine were closely correlated to the blockade of granular uptake induced by the drug (Lundborg 1963). Moreover, extragranular noradrenaline (accumulated in adrenergic nerves by pretreatment with reserpine, followed by an inhibitor of MAO and systematically administered noradrenaline) was found to be unavailable for release by the nerve impulse, as observed histochemically (Malmfors 1965). We proposed that under normal conditions the extragranular fraction of monoaminergic transmitters was very small, owing to the presence of MAO intracellularly. Subsequently work in numerous laboratories has lent support to these views. Already, at the Symposium, Douglas had presented evidence suggesting a Ca^{2+}-triggered fusion between the granule and cell membranes, preceding the release. The release is now generally assumed to take place as "exocytosis," even though the complete extrusion of the granule content may still be debatable.

Another, much debated area was the regulation of monoamine synthesis. Shortly after the discovery of DOPA decarboxylase (Holtz 1939), Blaschko (1939) formulated the main pathway for catecholamine synthesis: tyrosine, DOPA, dopamine, noradrenaline, adrenaline. The above-mentioned rapid conversion of administered DOPA to dopamine indicated that the first step in the pathway was rate limiting. Perhaps the first evidence for a regulation of catecholamine synthesis came from the observation that the accumulation of catecholamines and their O-methylated basic metabolites in mouse brain following inhibition of MAO leveled off within a few hours, suggesting that the synthesis was brought to a standstill when the stores had been filled (Carlsson et al 1960). It should be noted that the highly polar catecholamines and their basic O-methylated metabolites do not seem to escape easily through the blood-brain barrier. Subsequent, more sophisticated work in a large number of laboratories has revealed that catecholamine synthesis is regulated by several independent mechanisms. After the isolation of tyrosine hydroxylase, Udenfriend and his colleagues discovered an inhibitory action of catechols on this enzyme (Nagatsu et al 1964), whose affinity for the tetrahydropteridine co-enzyme was reduced. End-product inhibition was thus demonstrated, and evidence that this mechanism operates under physiological conditions was later presented (see Carlsson et al 1976). For several years, end-product inhibition was believed to be the only mechanism of short-term control of catecholamine synthesis. For long-term control, enzyme induction was shown to be responsible (Thoenen et al 1973). However, short-term activation of tyrosine hydroxylase can take place in dopamine neurons in vivo despite an increase in dopamine levels, thus indicating the existence of an additional control mechanism (Carlsson et al 1972). Phosphorylation of the enzyme seems to be involved in this reglation (Lovenberg et al 1975).

Several in-vitro and in-vivo approaches have been used to demonstrate and measure the release of amine transmitters and their dependence on the nerve impulses. Release in vivo has been shown by using push-pull cannulas (Cheramy et al 1981), semi-permeable tubes (Ungerstedt et al 1983), and voltammetry (Adams & Marsden 1982). Indirect but non-traumatic biochemical approaches have also proven useful, such as measurements of the rate of transmitter disappearance following inhibition of its synthesis (see Andén et al 1969) and the rate of accumulation of O-methylated basic metabolites following inhibition of MAO (Kehr 1976). The data thus obtained demonstrate that dopamine, noradrenaline, and serotonin are released from nerve terminals by nerve impulses. However, in the somatodendritic region of the nigral dopamine neurons, a rapid release and turnover of dopamine appears to occur more or less independently of nerve impulses, beyond the reach of receptor-mediated control (see Cheramy et al 1981, Nissbrandt et al 1985). The physiological significance of this observation is not known, but the possible implications are intriguing.

The False-Transmitter Concept

The "false-transmitter" concept is another spin-off of the early attempts to elucidate the mode of action of antipsychotic agents. Udenfriend and his colleagues (Hess et al 1961) had discovered that alpha-methyl-meta-tyrosine is capable of depleting central noradrenaline stores, while leaving the 5-HT stores intact. Costa et al (1962) pointed out that no sedation occurred after treatment with this agent despite the virtually complete depletion of noradrenaline stores; thus the central action of reserpine was unrelated to catecholamine deficiency (dopamine was ignored in this context) but rather was induced by 5-HT release. At the First International Catecholamine Symposium in Stockholm in August 1961, we challenged this interpretation (Carlsson 1962). We had found that alpha-methyl-meta-tyrosine and alpha-methyl-DOPA yield decarboxylation products that displace the catecholamines from their storage sites stoichiometrically. To explain the lack of sedation by alpha-methyl-meta-tyrosine and the fact that only very mild sedation was induced by alpha-methyl-DOPA, we pointed out that, unlike reserpine, these agents do not appear to block the storage mechanism. We proposed that the amines formed from these amino acids are able to take over the functions of the displaced endogenous amines (Carlsson & Lindqvist 1962). The "false-transmitter" concept thus formulated was then thoroughly investigated in numerous laboratories (for review, see Kopin 1968). Not only the aforementioned decarboxylation products but several other amines were found to be taken up in monoamine stores, thereby displacing the endogenous transmitters, and to be released by nerve impulses, thus causing postsynaptic effects.

The false-transmitter concept, as it is nowadays often understood, assumes that the false transmitter is less active than the endogenous transmitter. Thus, the displacement of the latter by the former amine should lead to a deficient transmission mechanism. To what extent this actually occurs, however, seems doubtful. To account for the lack of sedative action of alpha-methyl-metatyrosine, the possibility should be considered that despite the pronounced displacement of the endogenous catecholamines by the alpha-methylated decarboxylation products, newly synthesized endogenous transmitter may still be available for release in sufficient amounts to keep the function intact. The reduction of blood pressure by alpha-methyl-DOPA appears to be induced by a complex mechanism. It is central in origin: the effect persists after pretreatment with a peripheral decarboxylase inhibitor but is prevented by a centrally acting inhibitor (Henning 1969). The hypotensive action may be analogous to that of clonidine, i.e. it may be due to preferential activation of alpha-2 receptors by alpha-methyl-noradrenaline (see Andén 1979). Differences in pharmacological profile between the endogenous and the false transmitter should thus be taken into account.

Receptor Studies

For a long time the presynaptic events attracted most attention in this area of research. During the last 10 to 15 years, however, the monoaminergic receptors have attracted an ever increasing interest. These studies have utilized, for example, the biochemical changes induced by receptor manipulation in vivo, the electrophysiological changes caused by such manipulations in both pre- and postsynaptic neurons, and in-vitro binding studies to characterize and subclassify receptors. Much progress has also been made in elucidating the events occurring beyond the receptors (see Nestler & Greengard 1984).

Not unexpectedly, the discovery of the dopamine-receptor blocking action of the major neuroleptic (antipsychotic) agents has attracted a lot of interest. Our first study in this area (Carlsson & Lindqvist 1963) was undertaken in the hope that our recently improved fluorimetric methods for the determination of basic catecholamine metabolites would enable us to solve the riddle as to why the major antipsychotic agents, such as chlorpromazine and haloperidol, had a reserpine-like pharmacological and clinical profile and yet lacked the monoamine-depleting properties of the latter drug. Earlier, an inhibitory action of chlorpromazine on the turnover of monoamines in the brains of small rodents had been reported. However, when the chlorpromazine-induced hypothermia was prevented, the effect was no longer detectable. We found that chlorpromazine and haloperidol actually enhanced the turnover of dopamine and noradrenaline in the brain: they accelerated the formation of the dopamine metabolite, 3-methoxytyramine, and of the noradrenaline metabolite, normetanephrine, while leaving the neurotransmitter levels unchanged. In support of the specificity, promethazine, a sedative phenothiazine lacking antipsychotic and neuroleptic properties, did not change the turnover of the catecholamines. It did not seem farfetched, then, to propose that rather than reducing the availability of monoamines, as does reserpine, the major antipsychotic drugs block the receptors mediating dopamine and noradrenaline neurotransmission. This would explain their reserpine-like pharmacological profile. To account for the enhanced catecholamine turnover we proposed that neurons can increase their physiological activity in response to receptor blockade. This, I believe, was the first time that a receptor-mediated feedback control of neuronal activity was proposed. These findings and interpretations have been amply confirmed and extended by numerous workers, using a variety of techniques. In the following year, three of my students discovered the neuroleptic-induced increase in the concentrations of deaminated dopamine metabolites (Andén et al 1964b). Despite confirmatory work by others, our findings did not receive much attention until several years later. A possible explanation for this was that in the 1960s most workers in this field were focusing on other aspects of neurotransmission. Since the early 1970s, however, receptors have attracted an ever increasing interest. Moreover, the ability of catecholaminergic, especially dopaminergic agonists and antagonists, to induce and alleviate, respectively, psychotic symptoms, led to the much debated "dopamine hypothesis of schizophrenia." For a review of the historical background and recent developments, see Carlsson (1983).

The further analysis of receptor-mediated feedback control of neuronal activity, which, incidentally, soon proved to occur also in noradrenergic and serotonergic systems (see Andén et al 1969) and in other systems as well, revealed that this control was largely, if not entirely, mediated by a special population of receptors, apparently located on the monoaminergic neuron itself. These receptors have been called *presynaptic receptors* or, perhaps preferably, *autoreceptors*, since they are characterized by being sensitive to the neuron's own neurotransmitter. The first suggestion of the existence of such receptors came from studies on brain tissue slices (Farnebo & Hamberger 1971) that demonstrated inhibition and stimulation of nerve-impulse-induced dopamine release by dopamine agonists and antagonists, respectively. Subsequent in vivo studies demonstrated inhibition of striatal dopamine synthesis by the dopamine-receptor agonist, apomorphine, and blockade of this action by the neuroleptic agent, haloperidol; moreover, this effect persisted after cutting the dopaminergic axons, thus demonstrating that this feedback control was not loop-mediated but was restricted to the nerve-terminal area (Kehr et al 1972). Finally, Aghajanian & Bunney (1974, 1977) demonstrated a similar control in the somatodendritic part of dopamine neurons that causes decreased firing by dopamine-receptor agonists and a blockade of this action by dopamine-receptor antagonists. Further work along this line has led to the discovery of selective dopamine-autoreceptor agonists and antagonists with very interesting pharmacological properties and with potential clinical utility (see Clark et al 1985a,b, Svensson et al 1986a,b).

Speculations on the Molecular Requirements of a Neurotransmitter

In this prefatory chapter I may be permitted to indulge in some speculations on the molecular properties required of a neurotransmitter. A general requirement for any chemical to serve as a messenger must be that it can be readily identified with some degree of accuracy, for example, by binding to a specific receptor molecule. More than one recognition site would of course be an advantage. The monoamines contain a 2-carbon aliphatic chain with an amino group and an aromatic component—benzene or indole—with a phenolic hydroxyl group in a critical position, i.e. they apparently possess at least two binding sites. In noradrenaline and adrenaline, a hydroxyl group is added on the side chain, which confers chirality and an additional binding site to the molecules, and in adrenaline a further methyl group is added on the nitrogen, thus increasing the affinity especially to the beta-adrenoceptor. Why take the trouble to go through these cumbersome synthetic steps, starting out from sometimes scarcely available essential amino acids, when other, apparently satisfactory solutions of the problem are so readily within reach? Cells have to synthesize proteins anyway, for example, and proteins or peptides would obviously serve this purpose. Indeed, dozens of proteins and peptides are used as chemical messengers, especially as hormones.

Whereas proteins and peptides play a prominent role as hormones, the biogenic amines occupy a similar position among the neurotransmitters. A possible rationale for this apparent specialization could be that proteins and peptides represent a phylogenetically older group of chemical messengers that suffered from a serious drawback at the evolutionary stage, however, when some cells started to evolve toward becoming neurons: the messenger molecules could no longer be manufactured close to the site of release but had to be transported from the cell bodies through the axons to reach this site. Thus, the availability might have become inadequate. Smaller molecules would offer the advantage of being producible in the nerve terminals. Nonessential amino acids would fulfill these requirements.

It seems logical that not a single essenial amino acid appears to serve as a neurotransmitter.

To pursue the speculation one step further, we may assume that the amino acids, though undoubtedly useful, do not satisfy all needs for neurotransmitters. The amino acids may be just insufficient in number, or the biogenic amines may possess some useful property not shared by the amino acids. One conspicuous difference between the two groups of transmitters pertains to the storage mechanism: the amines are fairly strong bases that can be stored in high concentration, together with an acid such as ATP. This may provide the basis for an efficient storage-release mechanism. That the amines would lend themselves to a more rapid inactivation seems less likely.

Another entirely speculative possibility to consider in this context would be that the interaction between an agonist and its receptor does not always merely involve a conformational change of the receptor; in addition, the agonist and the receptor might interact like a substrate with an enzyme, hence leading to a chemical conversion of both molecules. Since such a mechanism would probably involve but a small number of transmitter molecules, it might prove fruitful to search for minor metabolic pathways of neurotransmitters. The phenolic hydroxyl groups of the monoamine neurotransmitters might be involved in such a mechanism. Our laboratory has actually engaged in investigating a newly discovered metabolic pathway of catecholamines, occurring in the mammalian brain and apparently involving autoxidation to quinones followed by coupling to glutathione (Rosengren et al 1985, Fornstedt et al 1986). Whether this pathway is at all related to the transmitter-receptor interaction is, however, entirely unknown at present.

The hypothesis that proteins and peptides represent a phylogenetically older type of chemical messengers than the "classical," small neurotransmitter molecules, receives some indirect support from the fact that certain neuropeptides occur abundantly in embryonic and neonatal brain regions but disappear at a later stage (for references, see Bloom 1984). As a corollary, some neuropeptides coexisting with classical neurotransmitters may be rudimentary phenomena of limited functional significance. Shortage of adequate research tools may thus not afford the only explanation of the many failures thus far to demonstrate beyond doubt a transmitter function of neuropeptides.

Functional and Clinical Implications

The functions of the central monoaminergic neurons and their role in neurological and psychiatric disorders are more poorly understood than the morphology and synaptology of these systems. Perhaps the functional and pathophysiological aspects of dopamine are least obscure. That the nigrostriatal dopamine system is involved in extrapyramidal functions and disorders, especially Parkinson's disease, is clearly established. The role of the tuberoinfundibular dopamine pathway for the control of prolactin secretion is also obvious, and even though the pathogenetic role of dopamine in clinical hyperprolactinemia is doubtful, the use of dopamine agonists in such cases is a significant advance. That dopamine plays a role in important mental functions is evident from the therapeutic actions of dopamine antagonists, as well as from the psychotomimetic effects of dopamine agonists. However, no disturbance in dopamine function has as yet been demonstrated beyond doubt in schizophrenia or any other psychotic condition (see Carlsson 1983).

Likewise, our knowledge of the physiology and pathology of central noradrenaline, not to speak of adrenaline, is very fragmentary. Like dopamine, noradrenaline seems to be somehow involved in arousal, and the locus coeruleus seems to respond very actively to incoming stimuli, especially when they signify novelty. Stimulation of noradrenaline neurons, for example by electrical stimulation of the locus coeruleus or by yohimbine-induced blockade of alpha-2-adrenergic autoreceptors, seems to cause strong arousal accompanied by severe anxiety. Conversely, stimulation of alpha-2-adrenergic receptors by clonidine causes, in addition to a decrease in blood pressure, sedation and an anxiolytic response. Bilateral destruction of the locus coeruleus has a similar effect. Both dopamine and noradrenaline appear to be somehow involved in the positive reinforcing action of dependence-producing drugs, and clonidine appears to be capable of alleviating certain withdrawal reactions (for review, see Elam 1985, Engel & Carlsson 1977).

The physiology and pathogenetic significance of 5-HT are also poorly comprehended. Animal data support the contention, however, that serotonergic systems exert an inhibitory function on aggressive behavior (Valzelli 1974). The serotonergic system appears to be more strongly developed in female than in male rats (Carlsson et al 1985), a finding that may at least partly account for the well-known sex difference in aggressive behavior among vertebrates. The serotonergic systems seem to control the mating behavior of both sexes, but here the situation may be more complex, with notable species differences. That 5-HT also plays a part in the control of motor functions is suggested by the "5-HT motor syndrome," but the physiological implications of this phenomenon are obscure. Similarly, the role of 5-HT and the other monoamines in various aspects of sleep seems to be established, although the available data are contradictory in certain respects.

Evidence for a role of 5-HT, noradrenaline, and dopamine in the control of mood and psychomotor activity and in affective disorders and anxiety has also been presented, and some impressive observations relating 5-HT to suicidal behavior have been published (Träskman et al 1981). Moreover, there seems to be no doubt that all the three major monoamines play a role in various neuroendocrinological and autonomic nervous functions.

Enormous efforts are obviously required to obtain a reasonably complete picture of the biological and clinical significance of the monoaminergic systems. More sophisticated animal models are needed, and human studies using, for example, modern imaging techniques will no doubt prove helpful. Although pharmacological tools have contributed much, better ones are needed. Especially, more selective monoaminergic agonists and antagonists, applicable also to humans, would prove useful.

Literature Cited

Adams, R. N., Marsden, C. A. 1982. Electrochemical detection methods for monoamine measurements *in vitro* and *in vivo*. In *Handb. Psychopharmacol.* 15: 1–74

Aghajanian, G. K., Bunney, B. S. 1974. Pre- and postsynaptic feedback mechanisms in central dopaminergic neurons. In *Frontiers of Neurology and Neuroscience Research*, ed. P. Seeman, G. M. Brown, pp. 4–11. Toronto: Univ. Toronto Press

Aghajanian, G. K., Bunney, B. S. 1977. Dopamine autoreceptors: Pharmacological characterization by microiontophoretic single cell recording studies. *Naunyn-Schmiedeberg's Arch. Pharmacol.* 297: 1–8

Amin, A. H., Crawford, T. B. B., Gaddum, J. H. 1954. The distribution of substance P and 5-hydroxytryptamine in the central nervous system of the dog. *J. Physiol.* 126: 596–618

Andén, N.-E. 1979. Selective stimulation of central alpha-autoreceptors following treatment with alpha-methyldopa and FLA 136. *Naunyn-Schmiedeberg's Arch. Pharmacol.* 306: 263–66

Andén, N.-E., Carlsson, A., Dahlström, A., Fuxe, K., Hillarp, N.-Å., Larsson, K. 1964a. Demonstration and mapping out of nigro-neostriatal dopamine neurons. *Life Sci.* 3: 523–30

Andén, N.-E., Roos, B.-E., Werdinius, B. 1964b. Effects of chlorpromazine, haloperidol and resperine on the levels of phenolic acids in rabbit corpus striatum. *Life Sci.* 3: 149–58

Andén, N.-E., Carlsson, A., Häggendal, J. 1969. Adrenergic mechanisms. *Ann. Rev. Pharmacol.* 9: 119–34

Axelrod, J. 1960. Discussion remarks. See Vane et al 1960, pp. 558–59

Axelrod, J. 1964. The uptake and release of catecholamines and the effect of drugs. In *Progress Brain Res.* 8: 81–89

Barbeau, A., Sourkes, T. L., Murphy, G. F. 1962. Les catecholamines de la maladie de Parkinson. In *Monoamines et Systeme Nerveux Central*, ed. J. de Ajuriaguerra, pp. 247–62. Geneve/Paris: Georg/Masson

Bertler, Å., Rosengren, E. 1959. Occurrence and distribution of dopamine in brain and other tissues. *Experientia* 15: 10

Bertler, Å., Carlsson, A., Rosengren, E. 1956. Release by reserpine of catecholamines from rabbits' hearts. *Naturwissenschaften* 22: 521

Bertler, Å., Carlsson, A., Rosengren, E. 1958. A method for the fluorimetric determination of adrenaline and noradrenaline in tissues. *Acta Physiol. Scand.* 44: 273–92

Bertler, Å., Falck, B., Gottfries, C. G., Ljunggren, L., Rosengren, E. 1964. Some observations on adrenergic connections between mesencephalon and cerebral hemispheres. *Acta Pharmacol. (Kbh.)* 21: 283–89

Birkmayer, W., Hornykiewicz, O. 1962. Der L-Dioxyphenylalanin (=L-DOPA)-Effekt beim Parkinson-Syndrom des Menschen: Zur Pathogenese und Behandlung der Parkinson-Akinese. *Arch. Psychiat. Nervenkr.* 203: 560–74

Björklund, A., Hökfelt, T., eds. 1984. *Handbook of Chemical Neuroanatomy*, Vol. 2, *Classical Transmitters in the CNS*, Pt. 1, pp. 1–463. Amsterdam/New York/Oxford: Elsevier

Blaschko, H. 1939. The specific action of L-

dopa decarboxylase. *J. Physiol.* 96: 50P–51P
Bloom, F. E. 1984. General features of chemically identifiable neurons. See Björklund & Hökfelt, pp. 1–22
Born, G. V. R. 1956. Adenosinetriphosphate (ATP) in blood platelets. *Biochem. J.* 62: 33P
Carlsson, A. 1959. The occurrence, distribution and physiological role of catecholamines in the nervous system. *Pharmacol. Rev.* 11: 490–93
Carlsson, A. 1960. Discussion remark. See Vane et al 1960, pp. 558–59
Carlsson, A. 1962. Discussion. See Costa et al 1962, pp. 71–74
Carlsson, A. 1966. Physiological and pharmacological release of monoamines in the central nervous system. See Euler et al 1966, pp. 331–46
Carlsson, A. 1983. Antipsychotic agents: Elucidation of their mode of action. In *Discoveries in Pharmacology*, Vol. 1: *Psycho- and Neuro-pharmacology*, ed. M. J. Parnham, J. Bruinvels, pp. 197–206. Amsterdam/New York/Oxford: Elsevier
Carlsson, A., Hillarp, N.-Å. 1956. Release of adrenaline from the adrenal medulla of rabbits produced by reserpine. *Kgl. Fysiogr. Sällsk. Förhandl.* 26(8)
Carlsson, A., Hillarp, N.-Å. 1962. Formation of phenolic acids in brain after administration of 3,4-dihydroxyphenylalanine. *Acta Physiol. Scand.* 55: 95–100
Carlsson, A., Lindqvist, M. 1962. In-vivo decarboxylation of alpha-methyldopa and alpha-methyl metatyrosine. *Acta Physiol. Scand.* 54: 87–94
Carlsson, A., Lindqvist, M. 1963. Effect of chlorpromazine and haloperidol on the formation of 3-methoxytyramine in mouse brain. *Acta Pharmacol.* 20: 140–44
Carlsson, A., Waldeck, B. 1958. A fluorimetric method for the determination of dopamine (3-hydroxytyramine). *Acta Physiol. Scand.* 44: 293–98
Carlsson, A., Rosengren, E., Bertler, Å., Nilsson, J. 1957a. Effect of reserpine on the metabolism of catecholamines. In *Psychotropic Drugs*, ed. S. Garattini, V. Ghetti, pp. 363–72. Amsterdam: Elsevier
Carlsson, A., Lindqvist, M., Magnusson, T. 1957b. 3,4-Dihydroxyphenylalanine and 5-hydroxytryptophan as reserpine antagonists. *Nature* 180: 1200
Carlsson, A., Shore, P. A., Brodie, B. B. 1957c. Release of serotonin from blood platelets by reserpine *in vitro*. *J. Pharmacol. Exp. Ther.* 120: 334–39
Carlsson, A., Lindqvist, M., Magnusson, T., Waldeck, B. 1958. On the presence of 3-hydroxytyramine in brain. *Science* 127: 471
Carlsson, A., Lindqvist, M., Magnusson, T. 1960. On the biochemistry and possible functions of dopamine and noradrenaline in brain. See Vane et al 1960, pp. 432–39
Carlsson, A., Falck, B., Hillarp, N.-Å., Thieme, G., Torp, A. 1961. A new histochemical method for visualization of tissue catechol amines. *Med. Exp.* 4: 123–25
Carlsson, A., Falck, B., Hillarp, N.-Å. 1962. Cellular localization of brain monoamines. *Acta Physiol. Scand.* 56(Suppl. 196): 1–27
Carlsson, A., Hillarp, N.-Å., Waldeck, B. 1963a. Analysis of the Mg^{++}-ATP dependent storage mechanism in the amine granules of the adrenal medulla. *Acta Physiol. Scand.* 59(Suppl. 215): 1–38
Carlsson, A., Magnusson, T., Rosengren, E. 1963b. 5-Hydroxytryptamine of the spinal cord normally and after transection. *Experientia* 19: 359
Carlsson, A., Falck, B., Fuxe, K., Hillarp, N.-Å. 1964. Cellular localization of monoamines in the spinal cord. *Acta Physiol. Scand.* 60: 112–19
Carlsson, A., Kehr, W., Lindqvist, M., Magnusson, T., Atack, C. V. 1972. Regulation of monoamine metabolism in the central nervous system. *Pharm. Rev.* 24: 371–84
Carlsson, A., Kehr, W., Lindqvist, M. 1976. The role of intraneuronal amine levels in the feedback control of dopamine, noradrenaline and 5-hydroxytryptamine in rat brain. *J. Neural Transm.* 39: 1–19
Carlsson, M., Svensson, K., Eriksson, Carlsson, A. 1985. Rat brain serotonin: Biochemical and functional evidence for a sex difference. *J. Neural Transm.* 63: 297–313
Cheramy, A., Leviel, V., Glowinski, J. 1981. Dendritic release of dopamine in the substantia nigra. *Nature* 289: 537–42
Clark, D., Carlsson, A., Hjorth, S. 1985a. Dopamine receptor agonists: Mechanisms underlying autoreceptor selectivity. I. Review of the evidence. *J. Neural Transm.* 62: 1–52
Clark, D., Hjorth, S., Carlsson, A. 1985b. Dopamine receptor agonists: Mechanisms underlying autoreceptor selectivity. II. Theoretical considerations. *J. Neural Transm.* 62: 171–207
Corrodi, H., Hillarp, N.-Å. 1964. Fluoreszenzmethoden zur histochemischen Sichtbarmachung von Monoaminen. 2. Identifizierung des fluoriszierendes Productes aus Dopamin und Formaldehyd. *Helv. Chim. Acta* 47: 911–18
Costa, E., Gessa, G. L., Kuntzman, R., Brodie, B. B. 1962. In *Pharmacological Analysis of Central Nervous Action*, ed. W. D. M. Paton, P. Lindgren, pp. 43–71. Oxford: Pergamon
Dahlström, A., Carlsson, A. 1986. Making

visible the invisible. Recollections of the first experiences with the histochemical fluorescence method for visualization of tissue monoamines. In *Discoveries in Pharmacology*, Vol. 3: *Chemical Pharmacology and Chemotherapy*, ed. M. J. Parnham, J. Bruinvels, pp. 97–125. Amsterdam/New York/Oxford: Elsevier

Dahlström, A., Fuxe, K. 1964. Existence of monoamine-containing neurons in the central nervous system. I. Demonstration of monoamines in the cell bodies of brain stem neurons. *Acta Physiol. Scand.* 62(Suppl. 232): 1–55

Ehringer, H., Hornykiewicz, O. 1960. Verteilung von Noradrenalin und Dopamin (3-Hydroxytyramin) im Gehirn des Menschen und ihr Verhalten bei Erkrankungen des extrapyramidalen Systems. *Klin. Wschr.* 38: 1236–39

Elam, M. 1985. In *On the physiological regulation of brain norepinephrine neurons in rat locus ceruleus*, pp. 1–43. Thesis, Univ. Göteborg, Sweden

Engel, J., Carlsson, A. 1977. Catecholamines and behavior. *Curr. Dev. Psychopharmacol.* 4: 3–32

Eränkö, O. 1955. Distribution of fluorescent islets, adrenaline and noradrenaline in the adrenal medulla of the cat. *Acta Endocrinol.* 18: 180–88

Falck, B., Hillarp, N.-Å. 1959. On the cellular localization of catecholamines in the brain. *Acta Anatomica* 38: 277–79

Falck, B., Hillarp, N.-Å., Thieme, G., Torp, A. 1962. Fluorescence of catecholamines and related compounds condensed with formaldehyde. *J. Histochem. Cytochem.* 10: 348–54

Farnebo, L.-O., Hamberger, B. 1971. Drug-induced changes in the release of ^{3}H-monoamines from field stimulated rat brain slices. *Acta Physiol. Scand. Suppl.* 371: 35–44

Fornstedt, B., Rosengren, E., Carlsson, A. 1986. Occurrence and distribution of 5-S-cysteinyl derivatives of dopamine, dopa and dopac in the brains of eight mammalian species. *Neuropharmacology* 25: 451–54

Fuxe, K. 1965. Evidence for the existence of monoamine neurons in the central nervous system. IV. Distribution of monoamine nerve terminals in the central nervous system. *Acta Physiol. Scand.* 64(Suppl. 247): 37–84

Gaddum, J. H. 1954. Drugs antagonistic to 5-hydroxytryptamine. In *Ciba Found. Symp. on Hypertension. Humoral and Neurogenic Factors*, ed. J. H. Gaddum, pp. 75–77. Boston: Little, Brown

Henning, M. 1969. Studies on the mode of action of alpha-methyldopa. *Acta Physiol. Scand.* 76(Suppl. 322): 1–37

Hess, S. M., Udenfriend, S. 1959. A fluorimetric procedure for the measurement of tryptamine in tissues. *J. Pharmacol. Exp. Ther.* 127: 175–77

Hess, S. M., Connamacher, R. H., Ozaki, M., Udenfriend, S. 1961. The effects of alpha-methyldopa and alpha-methyl-m-tyrosine on the metabolism of norepinephrine and serotonin *in vivo*. *J. Pharmacol. Exp. Ther.* 134: 129–38

Hillarp, N.-Å. 1946. Structure of the synapse and the peripheral innervation apparatus of the autonomic nervous system. *Acta Anatomica Suppl.* 4: 1–153

Hillarp, N.-Å. 1959. The construction and functional organization of the autonomic innervation apparatus. *Acta Physiol. Scand.* 46(Suppl. 157): 1–38

Hillarp, N.-Å., Lagerstedt, S., Nilsson, B. 1953. The isolation of a granular fraction from the suprarenal medulla containing the sympathomimetic catecholamines. *Acta Physiol. Scand.* 28: 251–63

Hillarp, N.-Å., Högberg, B., Nilsson, B. 1955. Adenosine triphosphate in the adrenal medulla of the cow. *Nature* 176: 1032–33

Holtz, P. 1939. Dopadecarboxylase. *Naturwissenschaften* 27: 724–25

Holzbauer, M., Vogt, M. 1956. Depression by reserpine of the noradrenaline concentration in the hypothalamus of the cat. *J. Neurochem.* 1: 8–11

Hökfelt, T. 1968. *Electron microscopic studies on peripheral and central monoamine neurons*, pp. 1–30. Thesis, Karolinska Inst., Stockholm

Jonason, J., Rutledge, C. O. 1968. Metabolism of dopamine and noradrenaline in rabbit caudate nucleus *in vitro*. *Acta Physiol. Scand.* 73: 411–17

Kehr, W. 1976. 3-Methoxytyramine as an indicator of impulse-induced release in rat brain in vivo. *Naunyn-Schmiedeberg's Arch. Pharmacol.* 293: 209–15

Kehr, W., Carlsson, A., Lindqvist, M., Magnusson, T., Atack, C. 1972. Evidence for a receptor-mediated feedback control of striatal tyrosine hydroxylase. *J. Pharm. Pharmacol.* 24: 744–47

Kirshner, N. 1962. Uptake of catecholamines by a particular fraction of the adrenal medulla. *J. Biol. Chem.* 237: 2311–17

Kopin, I. J. 1968. False adrenergic transmitters. *Ann. Rev. Pharmacol.* 8: 377–94

Lindvall, O. 1974. *The glyoxylic acid fluorescence histochemical method for monoamines. Chemistry, methodology and neuroanatomical application.* Thesis, Dept. Histology, Univ. Lund, Lund

Lovenberg, W., Bruckwick, E. A., Hanbauer, I. 1975. ATP, cyclic AMP, and magnesium increase the affinity of rat striatal tyrosine hydroxylase for its cofactor. *Proc. Natl. Acad. Sci. USA* 72: 2955–58

Lundborg, P. 1963. Storage function and amine levels of the adrenal medullary granules at various intervals after reserpine treatment. *Experientia* 19: 479

Magnusson, T., Rosengren, E. 1963. Catecholamines of the spinal cord normally and after transection. *Experientia* 19: 229

Malmfors, T. 1965. Studies on adrenergic nerves. The use of rat and mouse iris for direct observations on their physiology and pharmacology at cellular and subcellular levels. *Acta Physiol. Scand.* 64(Suppl. 248): 1–93

McLennan, H. 1963. In *Synaptic Transmission*, pp. 1–134. Philadelphia/London: Saunders

Montagu, K. A. 1957. Catechol compounds in rat tissues and in brains of different animals. *Nature* 180: 244–45

Muscholl, E., Vogt, M. 1958. The action of reserpine on the peripheral sympathetic system. *J. Physiol.* 141: 132–55

Nagatsu, T., Levitt, M., Udenfriend, S. 1964. Tyrosine hydroxylase: The initial step in norepinephrine biosynthesis. *J. Biol. Chem.* 238: 2910–17

Nestler, E. J., Greengard, P. 1984. Protein phosphorylation in nervous tissue. In *Catecholamines*, Pt. A: *Basic and Peripheral Mechanisms*, ed. E. Usdin, A. Carlsson, A. Dahlström, J. Engel, pp. 9–22. New York: Liss

Nissbrandt, H., Pileblad, E., Carlsson, A. 1985. Evidence for dopamine release and metabolism beyond the control of nerve impulses and dopamine receptors in rat substantia nigra. *J. Pharm. Pharmacol.* 37: 884–89

Pletscher, A., Shore, P. A., Brodie, B. B. 1955. Serotonin release as a possible mechanism of reserpine action. *Science* 122: 374–75

Rosengren, E., Linder-Eliasson, E., Carlsson, A. 1985. Detection of 5-S-cysteinyldopamine in human brain. *J. Neural Transm.* 63: 247–53

Shore, P. A., Silver, S. L., Brodie, B. B. 1955. Interaction of reserpine, serotonin, and lysergic acid diethylamide in brain. *Science* 122: 284–85

Svensson, K., Hjorth, S., Clark, D., Carlsson, A., Wikström, H., Andersson, B., Sanchez, D., Johansson, A. M., Arvidsson, L.-E., Hacksell, U., Nilsson, J. L. G. 1986a. (+)-UH 232 and (+)-UH 242: Novel stereoselective DA receptor antagonists with preferential action on autoreceptors. *J. Neural Transm.* 65: 1–27

Svensson, K., Carlsson, A., Johansson, A. M., Arvidsson, L.-E., Nilsson, J. L. G. 1986b. A homologous series of *N*-alkylated *cis*-(+)-(1S,2R)-5-methoxy-1-methyl-2-aminotetralins: Central DA receptor antagonists showing profiles ranging from classical antagonism to selectivity for autoreceptors. *J. Neural Transm.* 65: 29–38

Thoenen, H., Otten, U., Oesch, F. 1973. Trans-synaptic regulation of tyrosine hydroxylase. In *Frontiers in Catecholamine Research*, ed. E. Usdin, S. Snyder, pp. 179–85. Oxford: Pergamon

Träskman, L., Åsberg, M., Bertilsson, L., Sjöstrand, L. 1981. Monoamine metabolites in CSF and suicidal behavior. *Arch. Gen. Psychiat.* 38: 631–36

Twarog, B. M., Page, I. H. 1953. Serotonin content of some mammalian tissues and urine and a method for its determination. *Am. J. Physiol.* 175: 157–61

Ungerstedt, U. 1971. *On the anatomy, pharmacology and function of the nigrostriatal dopamine system.* Thesis, Karolinska Inst., Stockholm

Ungerstedt, U., Herrera-Marschitz, M., Ståhle, L., Tossman, U., Zetterström, T. 1983. Dopamine receptor mechanisms correlating transmitter release and behavior. In *Dopamine Receptor Agonists*, ed. A. Carlsson, J. L. G. Nilsson, 1: 165–81. Stockholm: Swedish Pharmaceutical Press

Valzelli, L. 1974. 5-Hydroxytryptamine in aggressiveness. *Adv. Biochem. Psychopharmacol.* 11: 255–64

Vane, J. R., Wolstenholme, G. E. W., O'Connor, M., eds. 1960. *Ciba Found. Symp. on Adrenergic Mechanisms*, pp. 1–632. London: Churchill

Vogt, M. 1954. The concentration of sympathin in different parts of the central nervous system under normal conditions and after the administration of drugs. *J. Physiol.* 123: 451–81

Von Euler, U. S., Rosell, S., Uvnäs, B., eds. 1966. *Mechanisms of Release of Biogenic Amines*, pp. 469–77. Oxford: Pergamon

Westerink, B. H. C., Spaan, S. J. 1982. On the significance of endogenous 3-methoxytyramine for the effects of centrally acting drugs on dopamine release in the rat brain. *J. Neurochem.* 38: 680–86

Woolley, D. W., Shaw, E. 1954. A biochemical and pharmacological suggestion about certain mental disorders. *Science* 119: 587–88

Ann. Rev. Neurosci. 1987. 10: 41–65

COMPUTATIONAL MAPS IN THE BRAIN

Eric I. Knudsen, Sascha du Lac, and Steven D. Esterly

Department of Neurobiology, Stanford University School of Medicine, Stanford, California 94305

INTRODUCTION

Computation is the essence of brain function. Through computation, the nervous system sorts and evaluates sensory information that is of biological importance. Based on the results, decisions are made and executive commands for behavioral responses are issued; these commands in turn require further computation to produce the spatial and temporal patterns of motor neuron activity that mediate behavior.

The "computational map" is a key building block in the infrastructure of information processing by the nervous system. For the purposes of this chapter, we define "computation" as any transformation in the representation of information. In a computational map, there is a systematic variation in the value of the computed parameter across at least one dimension of the neural structure. The neurons that make up such a map represent an array of preset processors or filters, each tuned slightly differently, that operate in parallel on the afferent signal. Consequently, they transform their input almost instantaneously into a place-coded probability distribution that represents values of the mapped parameter as locations of peaks in activity within the map. Sorted in this fashion, the derived information can be accessed readily by higher order processors (which themselves may be computational maps) using relatively simple schemes of connectivity.

Many computational processes in the nervous system are not mapped. For example, neurons selective for a particular parameter value or set of parameter values often are clustered to form functional modules in the brain, but there is no systematic variation in their tuning, i.e. there is no

0147–006X/87/0301–0041$02.00

map. Conversely, some neural maps are not computational, such as maps of sensory epithelia, which simply reproduce the peripheral representation of the sensory information. In this chapter, we restrict our discussion to maps that are derived through computation.

Most of the computational maps discovered so far are involved in processing sensory information. As is discussed in ensuing sections, they map parameters of the sensory world that are crucially important to the animal. Computational maps in the visual system include those of line orientation preference and ocular dominance in the primary visual cortex (Hubel & Wiesel 1963, 1977) and a map of movement direction in cortical middle temporal visual area MT[1] (Albright et al 1984). A map of color in cortical area V4 has been suggested, but its topography has not been demonstrated (Zeki 1974, 1980). Computational auditory maps include maps of interaural delay, interaural intensity difference, and sound source location in the brainstem (Knudsen & Konishi 1978, Knudsen 1982, King & Palmer 1983, Middlebrooks & Knudsen 1984, Wenstrup et al 1986, Sullivan & Konishi 1986, Konishi et al 1986) and maps of amplitude spectrum and time interval in the auditory cortex (Suga 1977, O'Neill & Suga 1982). In all cases, these maps are most precisely elaborated and occupy the greatest portions of the brain in species that have highly developed sensory capacities, such as monkeys (vision) and owls and bats (audition). However, similar visual and auditory maps have been found to exist in more generalized species such as the cat, ferret, and guinea pig.

Computational maps are also involved in motor programming. The best known is the map of motor space, or gaze direction, in the superior colliculus (optic tectum in nonmammals) of the midbrain (e.g. Roucoux et al 1980, Wurtz & Albano 1981). In computational motor maps, systematic variations of a movement are represented topographically across the neural structure. These maps are computational in that the topographically organized command must be transformed into spatio-temporal patterns of motor neuron activity, and it is the site of peak activity in the map that dictates the transformation that is to be made. Thus, computational motor maps demonstrate the principles of computational sensory maps operating in reverse: instead of representing the results of systematically varying computations (as do sensory maps), they represent the source code that commands systematically varying computations.

The realization that the nervous system makes use of computational

[1] Abbreviations used: FM, frequency modulation; ICc, central nucleus of the inferior colliculus; IID, interaural intensity difference; ICx, external nucleus of the inferior colliculus; LSO, lateral superior olive; MSO, medial superior olive; MT, middle temporal visual area; n. Lam, nucleus laminaris; V1, primary visual cortex; VLVp, posterior division of the ventral nucleus of the lateral lemniscus.

maps leads to such questions as, under what conditions does the nervous system map a computational process, and what advantages are derived from such maps? In this chapter, we discuss properties common to known computational maps and consider the implications of neural activity that is topographically organized. The need for brevity requires that many details of specific maps be omitted, and the reader is directed to original papers and reviews for such information. As our focus of discussion, we use the synthesis of the auditory map of space in the brainstem. The auditory space map is a high order map that is itself dependent upon lower order mapped and nonmapped computations. As such, its synthesis demonstrates fundamental properties of computational maps as well as interactions of parallel and serial maps in a hierarchy of information processing.

EXAMPLES OF COMPUTATIONAL MAPS

The most familiar computational maps were discovered in the primary visual cortex (V1) of cats and monkeys by Hubel & Wiesel (1963, 1968). They found that V1 contains two kinds of repeating computational maps: maps of preferred line orientation, which represent the angle of tilt of a line stimulus, and maps of ocular dominance, which represent the relative strengths of excitatory influence of each eye (Figure 1). (Ocular dominance only becomes a computational map with the creation of binocular cells. In primates, this occurs in layers outside of layer 4C.) Orientation and ocular dominance are mapped across the cortical surface along independent, though often not perpendicular, axes. Arrays of neurons which together represent all values of ocular dominance or line orientations have been termed *hypercolumns*. Each hypercolumn processes information from a small region of visual space; neighboring hypercolumns process information from overlapping but systematically shifted portions of visual space. Thus, hypercolumns are examples of computational maps that are repeated within a larger, noncomputational (retinotopic) map.

Cortical area MT, or V5, in the macaque monkey is a higher order visual area that contains repeating maps of movement direction, again organized within the context of retinotopy (Albright et al 1984). The tuning of individual neurons for direction of motion is broad, but changes systematically across the cortical surface such that 180° of movement directions are represented across a 0.5 mm slab of cortex. Adjacent slabs represent motion in opposite directions.

A greater number of computational maps have been discovered in the auditory system. Three of these, the maps of interaural delay, interaural intensity difference, and space, contribute to the spatial analysis of sound and are described in detail in the next section. Two others have been found in the auditory cortex of the mustache bat. In a portion of the bat's primary auditory cortex, neurons are tuned for the amplitude as well as for the frequency of sound (Suga 1977, Suga & Manabe 1982). The best amplitude for exciting neurons varies systematically across the cortical surface along a dimension that is perpendicular to the tonotopic (receptotopic) axis. Thus, the map represents the amplitude spectrum of an acoustic stimulus topographically.

A high order area of the bat's auditory cortex maps the time interval between acoustic events (Suga & O'Neill 1979, Suga et al 1983). Neurons in this area (the "frequency modulation processing area") are tuned to the interval between two frequency modulated (FM) sounds, and the best time interval for exciting neurons varies systematically across the cortical

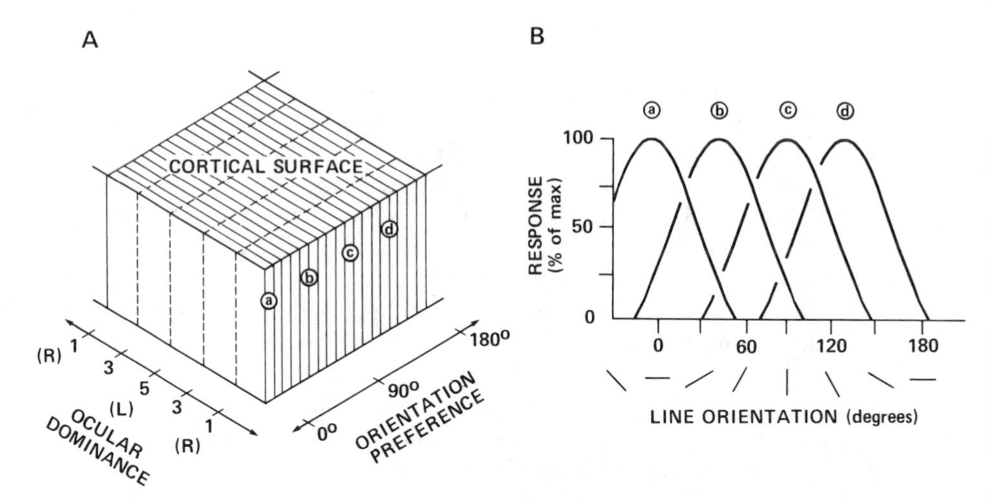

Figure 1 Computational maps in the primary visual cortex (V1) of monkeys. *A*. A schematic diagram of hypercolumns for ocular dominance and line orientation in V1. The axes of these maps are independent, but often are not perpendicular. The ocular dominance map represents systematically the relative strengths of the excitatory influence of each eye. The numerical scale refers to the ocular dominance groups defined by Hubel & Wiesel. Here, units excited exclusively by the right eye are assigned a value of 1, and units excited exclusively by the left eye are assigned a value of 5. In the map of orientation preference, neurons are tuned to the angle of tilt of a line stimulus; the tuning of neurons varies systematically across the map over a 180° range of orientations. Adapted from Hubel & Wiesel (1977).

B. Representative tuning curves of neurons in V1 for line orientation. Response is plotted as a function of stimulus orientation. The circled letters identify tuning curves with the positions indicated in the map in *A*. Neuronal tuning in the orientation map, as in other computational maps, can be quite broad: While each neuron responds maximally to a narrow range of orientations, a substantial response can be elicited by stimuli over a 50° range of orientations. Based on data from Schiller et al (1976).

surface. In this map, tonotopy is absent, and instead neurons are grouped in modules according to the ranges of frequencies contained in the FM component of the sonar pulses. This is a case of computational maps occurring within functional clusters. The bat uses time interval information to determine the delay of echoes relative to emitted sonar pulses and, therefore, the distance of objects (the same type of information could be used by other species to categorize and identify vocalizations; Liberman et al 1954, Margoliash 1983).

The superior colliculus (optic tectum) in all species tested contains a computational motor map that directs orienting movements of the eyes, head, and/or ears (Robinson 1972, Stryker & Schiller 1975, Roucoux et al 1980, Stein & Clamann 1981). For example, in primates, neurons in the intermediate and deep collicular layers discharge prior to saccadic eye movements, and the magnitude and direction of the eye movement associated with the discharge of neurons varies topographically with location in the colliculus (Sparks et al 1976). Moreover, focal electrical simulation of the colliculus elicits saccadic eye movements, the magnitude and direction of which also vary systematically with the site of stimulation (Robinson 1972, Stryker & Schiller 1975). The data indicate that the output of the superior colliculus (optic tectum) initiates specific motor programs that depend on the location of maximum activity in the motor map.

COMPUTATIONAL MAPS IN AUDITORY SPACE ANALYSIS

Cues for the Analysis of Auditory Space

The map of auditory space represents the apex of a hierarchy of sensory information processing (Figure 2). In the auditory system, "space" must be computed from a variety of cues that are only indirectly related to the location of a sound source. These include binaural differences in the timing and intensity of spectral components of sound, and monaural cues that result from direction-dependent spectral shaping by the head and external ears. The computational process involves two steps: (*a*) the determination of cue values, and (*b*) the association of particular sets of cue values with appropriate locations in space. Both steps are handled in computational maps.

The values of the cues and their relationships to sound source location depend upon the size and shape of the external ears, the position of the external ears on the head, and the mode of operation of the eardrum, i.e. whether it operates as a pressure or a pressure-gradient receiver (Searle et al 1976, Colburn & Durlach 1978, Coles et al 1980, Knudsen 1980). These physical properties vary across the animal kingdom, and consequently different species use different cues and ranges of cue values to generate maps of auditory space. However, the basic task of evaluating binaural and monaural cues is the same for all animals.

The cues for sound localization are frequency-specific. For example, an interaural intensity difference (IID) of 6 dB at 2 kHz signifies locations in space different from those signified by the same interaural intensity difference at 8 kHz (Shaw 1974, Knudsen 1980). Even interaural delay exhibits some frequency dependence (Kuhn 1977, Roth et al 1980). For this reason, localization cues must be evaluated in a frequency-specific manner (for a review see Knudsen 1983a). Conveniently, the cochlea breaks down the signal into its frequency components even before the signal is transduced.

Information about the timing and intensity of each frequency component is conveyed to the cochlear nuclei by frequency-tuned fibers in the auditory nerve. There, timing and intensity information is processed by neurons that project in parallel pathways to pontine nuclei specialized for comparing signals from the two ears (Figure 2; Sullivan & Konishi 1984, Cant & Morest 1984). Although major transformations in the representations of the afferent signal occur by the level of the cochlear nuclei, none of the computational processes appear to be mapped. Only a map of frequency tuning, which is a topographic representation of the sensory epithelium, is apparent.

The Computation of Interaural Delay

The first computational map appears at the initial site of binaural convergence. Neurons in the medial superior olive (MSO) in mammals and in the nucleus laminaris (n. Lam) in birds compute the interaural delay of frequency-specific signals from each ear. They receive information about the timing of individual frequency components at each ear from neurons in the cochlear nuclei that discharge in phase with frequency components of the stimulus. When sounds are presented binaurally, the response rate of MSO (n. Lam) neurons varies dramatically as a function of interaural delay (Goldberg & Brown 1969, Moushegian et al 1975, Moiseff & Konishi 1983). The most and least favorable interaural delays (i.e. those eliciting maximum and minimum responses, respectively) can be predicted from the difference in the latencies of a neuron's responses to stimulation of each ear alone. The optimal interaural delays are those that cause the phase-locked excitatory input from each ear to arrive simultaneously, while the least favorable delays are those that cause the excitatory inputs to arrive 180° out of phase with each other.

That these neurons compute interaural delay on the basis of the relative timing of phase-locked afferent signals has an important consequence for higher order neurons that must interpret their outputs: Different interaural

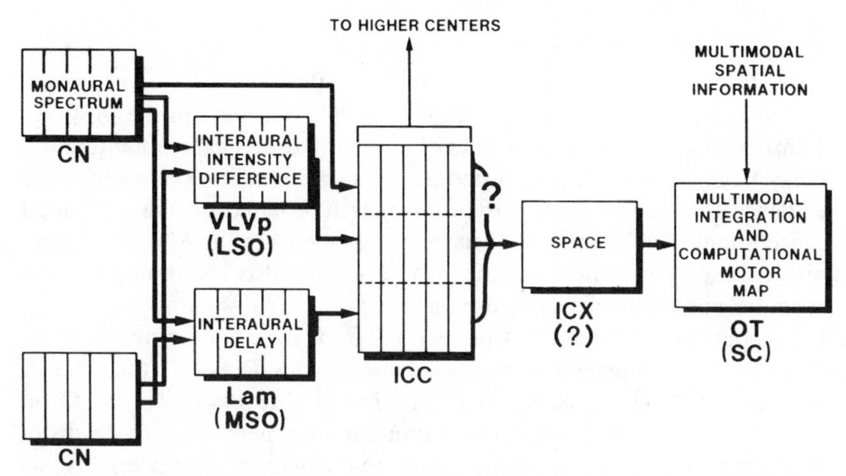

Figure 2 Steps in the synthesis of a map of auditory space in the barn owl; note that not all connections of the auditory pathway are shown. The names of analogous structures in mammals are shown in parentheses. Monaural intensity and phase spectra of acoustic stimuli are represented in the firing rate and time of firing of neurons in the tonotopically organized cochlear nuclei (CN) (tonotopic organization is indicated throughout the figure by *vertical stripes*). The CN project to both the nucleus laminaris (Lam) and the posterior division of the ventral nucleus of the lateral lemniscus (VLVp). Neurons in the VLVp in birds and in the lateral superior olive (LSO) in mammals compute frequency-specific interaural intensity differences by combining inhibitory input from one CN with frequency-matched excitatory input from the other CN. Neurons in the n. Lam in birds and in the medial superior olive (MSO) in mammals compute frequency-specific interaural delays by cross-correlating signals from each CN. The VLVp (LSO) and Lam (MSO) project in parallel to the central nucleus of the inferior colliculus (ICc). The ICc also receives direct input from the cochlear nucleus. Neurons in the external nucleus of the inferior colliculus (ICx) receive convergent projections from the ICc. Sharp frequency tuning and tonotopic organization disappears in the ICx. Instead, neurons are tuned for source location and, in the barn owl, they form a map of auditory space. This spatial information is sent to the optic tectum (OT) in birds and to the superior colliculus (SC) in mammals. The OT (SC) also receives topographically ordered somatosensory and visual input and contains a computational motor map as described in the text.

delays that give rise to the same interaural phase cannot be distinguished (Rose et al 1966, Kuwada et al 1984, Takahashi & Konishi 1986). Thus, the responses of these neurons are said to be phase-ambiguous. In the derivation of sound source location, phase ambiguity can equate with spatial ambiguity (Knudsen 1984a, Esterly 1984). As we discuss, the resolution of phase ambiguity to yield an unambiguous measure of interaural delay is a computational problem that is solved by higher order neurons.

Maps of Interaural Delay

MAPS OF INTERAURAL DELAY IN THE NUCLEUS LAMINARIS AND MEDIAL SUPERIOR OLIVE Although it is clear that different units in the MSO (n. Lam) are tuned for different values of interaural delay, a map of interaural delay has yet to be demonstrated at the single unit level. However, the evidence that such a map exists is strong; it is based on the structured morphology of afferent and postsynaptic elements in the MSO (n. Lam), and on the systematic variation in the interaural delay tuning of field potentials in this nucleus (Figure 3).

The pattern of the afferent projections from the cochlear nuclei to the MSO (n. Lam) suggests a systematically varying delay-line (Scheibel & Scheibel 1974, Young & Rubel 1983, Konishi et al 1986). The MSO (n. Lam) is a slab-shaped nucleus containing bipolar neurons with dendrites oriented oppositely and perpendicularly to the plane of the nucleus. Afferent axons from the cochlear nucleus bifurcate, with one branch of each axon projecting to the ipsilateral and the other to the contralateral MSO (n. Lam). Frequency-matched afferents from each cochlear nucleus converge systematically on opposing dendrites of MSO (n. Lam) neurons, establishing a tonotopic organization along one axis of the nucleus. The afferent axons enter at opposite edges of the nucleus and run in a straight line along an iso-frequency plane (Figure 3A). Presynaptic arbors peel off from the stem axons as they pass through the nucleus (in chickens, the contralateral afferents are as described, but the ipsilateral afferents form multiple arbors of equal length; Young & Rubel 1983). The progressive increase in axonal length to successive portions of an afferent's terminal

---

Figure 3 Evidence for a map of interaural delay in the n. laminaris of the barn owl. *A.* Camera lucida drawings of cochlear nucleus (CN) axons filled with horseradish peroxidase. The axons project from each CN to an isofrequency lamina in n. laminaris (Lam). Axons from each cochlear nucleus enter Lam from opposite sides of the nucleus and form presynaptic arbors as they penetrate the nucleus. This geometry suggests that the axons are acting as oppositely oriented delay-lines; this predicts that the positions of neurons receiving simultaneous input from the two ears vary as a function of interaural delay. Courtesy of Dr. Catherine E. Carr.

B. Systematic variation in the tuning of the field potential for interaural delay within an isofrequency (5 kHz) lamina of Lam. *Dashed lines* indicate recording locations exhibiting the same interaural delay tuning, indicated as microseconds of ipsilateral (*i*) or contralateral (*c*) delay. *Circled numbers* indicate locations where responses in *C* were recorded.

C. Tuning of field potentials in Lam for interaural delay. Each *curve* indicates the normalized response magnitude as a function of interaural delay for the recording sites indicated in *B*. The optimal delay shifts systematically as a function of position in the nucleus. *Double peaks* reflect delays that differ by one period of the stimulus frequency (5 kHz). These data are from Sullivan & Konishi (1986).

field is ideally suited to act as a delay-line. The axonal geometry predicts that afferent action potentials sweep systematically through the MSO from one edge to the other, the waves of excitation from opposite ears traversing the nucleus in opposite directions.

Electrophysiological evidence supports this prediction. An electrode placed in n. Lam (MSO) records a field potential that is a spectrally filtered analog of the acoustic stimulus (Wernick & Starr 1968, Sullivan & Konishi 1986, Konishi et al 1986). The field potential can be elicited by stimulation of either ear. In the owl, as the electrode penetrates the nucleus along an

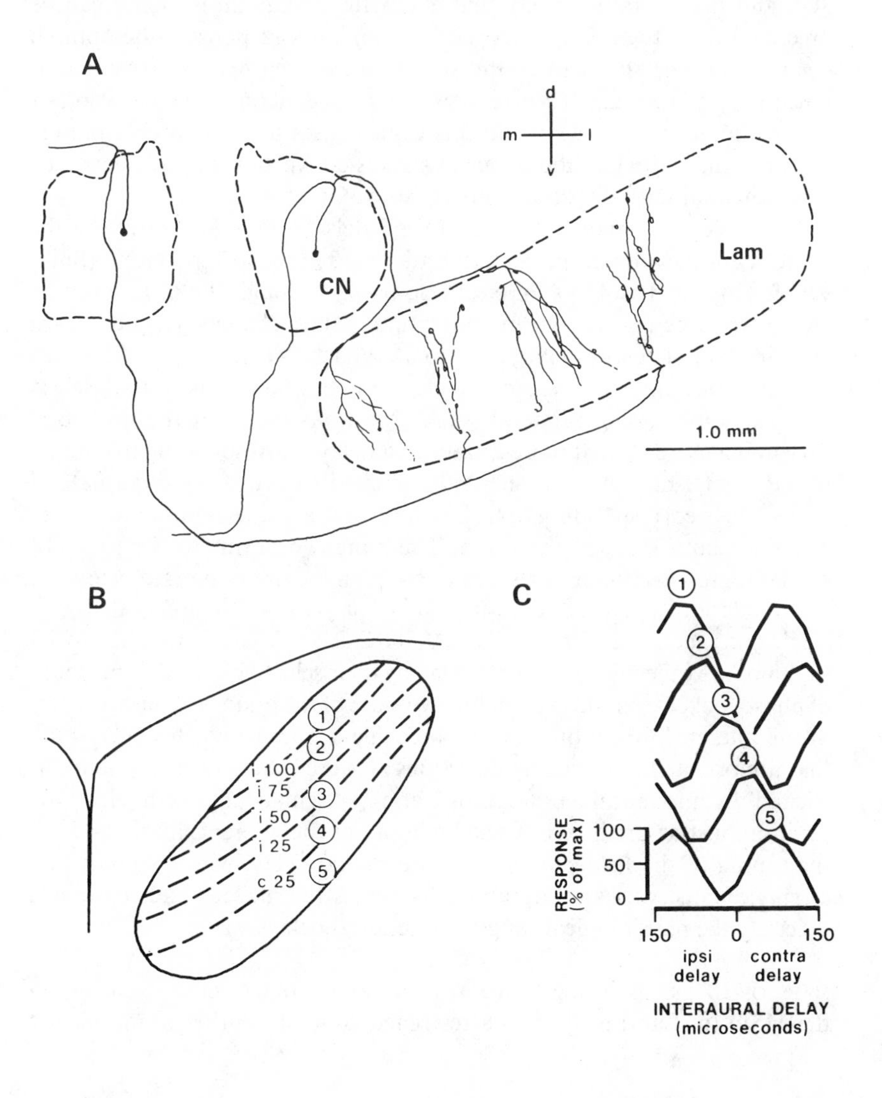

axis that is parallel to the afferent projection, the phase of the field potential (relative to the monaural stimulus) changes systematically, the sign of the change depending on which ear is stimulated (Sullivan & Konishi 1986). If the phase of the potential evoked by ipsilateral ear stimulation retards as the electrode progresses, then the phase of the potential evoked by contralateral ear stimulation advances—just as the opposing orientations of the axonal delay-lines predict.

As with single units in the n. Lam (MSO), the field potential at any given site is tuned for the interaural delay of a binaural stimulus (Figure 3C), and the interaural delay that elicits the maximum response can be predicted from the relative latencies of monaural responses: the optimal delay is the one that causes the signals from each ear to arrive simultaneously, or in phase (Wernick & Starr 1968, Konishi et al 1986). A change in interaural delay from this value causes the response amplitude to diminish, and when the binaural signals arrive 180° out of phase, the field potential virtually disappears (Figure 3C).

The systematic mapping of interaural delay that is suggested by the anatomy of the afferent projection is observed in the field potential (Sullivan & Konishi 1986). As an electrode passes through the n. Lam along an iso-frequency plane, optimal interaural delays change gradually from large ipsilateral delays to large contralateral delays (Figure 3). In the barn owl, each n. Lam contains a complete representation of interaural delays corresponding to contralateral source locations (delay at the ipsilateral ear), and a more limited representation of delays corresponding to sources in the frontal half of the ipsilateral hemifield (delay at the contralateral ear). Delays corresponding to frontal locations are thus represented in the nuclei on both sides of the brain. This range conforms to the range of spatial representation in the maps of space in the midbrain (see below).

To reiterate, a map of interaural delay has yet to be demonstrated with single unit recordings. The evidence that it exists is (*a*) neurons in the MSO (n. Lam) are tuned for interaural delay by their selectivity for a coincidence of phase-locked excitation from the two sides of the brain; (*b*) the geometry of the afferent projection from each cochlear nucleus creates delay-lines that are oriented in opposing directions; (*c*) the timing of field potentials elicited by monaural stimulation varies systematically with electrode position, just as predicted by the anatomy of the afferent fibers; and (*d*) the tuning of the field potential for interaural delay, which mimics that of single units, varies topographically across the nucleus and represents precisely the physiological range of interaural delays.

MAPS OF INTERAURAL DELAY IN THE INFERIOR COLLICULUS Neurons in the MSO (n. Lam) project to a restricted zone of another tonotopically organized nucleus, the central nucleus of the inferior colliculus (ICc) (Figure 2; Roth et al 1978, Semple & Aitkin 2979, Takahashi & Konishi 1985). Studies in the barn owl and in the cat demonstrate that neurons in this zone also are tuned for interaural delay, and that interaural delay tuning is mapped within iso-frequency planes (Yin et al 1983, Aitkin et al 1985, Wagner et al 1986), an organization that mimics that of the MSO (n. Lam). The data from barn owls indicate further that (*a*) neuronal selectivity for interaural delay has been sharpened considerably, probably by lateral inhibitory interactions within the delay map, and (*b*) the maps of interaural delay are aligned across iso-frequency planes (Wagner et al 1986). For example, a neuron in the ICc tuned to a delay of 100 μsec at 3 kHz is aligned with neurons tuned to the same delay at other frequencies. Although this organization reflects an organization that exists already in the MSO (n. Lam), that it exists in the ICc greatly simplifies the next step in the computational process: the integration of interaural delays across frequency. Before discussing this next step, we review the processing of the other binaural cue for sound localization: frequency-specific interaural intensity difference.

Maps of Interaural Intensity Difference

The frequency-by-frequency analysis of interaural intensity differences begins in the lateral superior olive (LSO) in mammals and in the posterior division of the ventral nucleus of the lateral lemniscus (VLVp) in birds (Figure 2; Boudreau & Tsuchitani 1970, Guinan et al 1972, Moiseff & Konishi 1983). In the LSO, excitatory input comes directly from the ipsilateral cochlear nucleus. Frequency-matched inhibitory input comes from the contralateral cochlear nucleus by way of the medial nucleus of the trapezoid body. The antagonistic influences of these binaural inputs vary in parallel over a wide range of sound levels, so that the response rate of an LSO (VLVp) neuron indicates the binaural difference in sound intensity, independent of the overall intensity. This simple subraction causes neurons to be sensitive to interaural intensity difference, but not tuned for interaural intensity difference: The neurons respond ever more strongly, up to a saturating level, as the interaural intensity difference increases beyond some threshold value. The important parameter that varies within the population is the interaural intensity difference threshold. Some neurons begin to respond even when the sound in the excitatory ear is weaker than sound in the inhibitory ear, while others require the sound in the excitatory ear to be much stronger than sound in the inhibitory ear before they respond. Preliminary data indicate that interaural intensity difference thresholds vary systematically across VLVp in the barn owl (J. A. Manley and M. Konishi, personal communication). Based on the highly regular cellular anatomy in the LSO (Scheibel & Scheibel 1974), it is possible that a similar map of thresholds exists there as well.

The earliest stage for which there is documentation of a map of interaural intensity difference is in the ICc, the next nucleus in the ascending pathway (Figure 2). Neurons in the LSO (VLVp) project to a zone in the ICc that adjoins the MSO (n. Lam) projection zone (Roth et al 1978, Semple & Aitkin 1979, Takahashi & Konishi 1985). Neurons in the LSO projection zone, like those in the LSO itself, are sharply tuned for frequency and are sensitive, but not tuned, for interaural intensity difference. (The interaural intensity difference sensitivity is opposite to that found in the LSO on the same side: ICc neurons are excited by stimulation of the contralateral ear and are inhibited by stimulation of the ipsilateral ear.) In the bat, neurons in this zone of the ICc are organized according to their sensitivity to interaural intensity difference, in addition to being tonotopically organized (Wenstrup et al 1986). At one edge of the zone, ipsilateral inhibition is weak and neurons respond even when the interaural intensity difference favors the ipsilateral ear. As an electrode moves across the zone, the efficacy of ipsilateral inhibition increases and the interaural intensity difference required to drive neurons shifts until, at the opposite edge of the zone, only large interaural intensity difference's favoring the contralateral ear are sufficient to drive the neurons. Thus, this computational map represents the interaural intensity difference of a given frequency topographically by the extent of maximally activated neurons across the LSO projection zone in the ICc.

Synthesis of a Map of Auditory Space

At the level of the ICc, the stage is set for the computation of sound source location. Frequency-specific interaural delays and intensity differences have been measured and mapped, and the peripherally filtered spectrum of the stimulus is represented along the tonotopic axis of the nucleus (Figure 2). Auditory space is computed by combining these cues. In the classical auditory pathway, this computation does not take place until sometime after the information passes through the primary auditory cortex (Middlebrooks & Pettigrew 1981, Jenkins & Merzenich 1984). However, in a secondary auditory pathway in the midbrain, the computation is performed immediately. In owls, this pathway leads from the ICc to the external nucleus of the inferior colliculus (ICx), then on to the optic tectum. (The analogous pathway has not been found in other species.)

To construct a map in which neurons are tuned to sounds from specific regions in space, auditory cue values must be associated with appropriate source locations. Although each of the cue values mapped in the ICc varies with source location, any single value can correspond with many possible locations (Knudsen 1980, Calford et al 1986). The spatial ambiguity associated with frequency-specific cue values can be eliminated by combining cues across frequencies. In barn owls, this integrative step occurs as the space map is synthesized in the ICx (Knudsen 1983a,b, 1984a). In cats, neurons broadly tuned for frequency and sharply tuned for space are concentrated in the rostral portion of the ICx (Aitkin et al 1984), but the first documented appearance of an auditory space map is in the superior colliculus (Middlebrooks & Knudsen 1984).

CONSTRUCTING A MAP FROM MAPS The maps of localization cues in the ICc simplify the circuitry for computing space. The systematic relationship between interaural delay and sound source azimuth enables a map of azimuth to be created in the barn owl by a topographic projection to the ICx from the delay sensitive zone of each iso-frequency plane. In addition, since these maps of interaural delay are aligned across iso-frequency planes, delay tuning that is independent of frequency would result if the ouputs of aligned neurons (from different iso-frequency planes) simply converged on a neuron in the space map. This systematic convergent projection occurs from the ICc to the ICx in the barn owl (Knudsen 1983b, Wagner et al 1986).

A map of space would also result from a roughly topographic projection from each frequency-specific interaural intensity difference map. This is because, in general, interaural intensity differences tend to increase with angular distance from the midline, i.e. as a function of source azimuth (in owls with asymmetrical ears, interaural intensity differences increase with azimuth for low frequencies and with elevation for high frequencies; Knudsen 1980). Due to this relationship, the iso-frequency maps of interaural intensity difference in the ICc each represent the azimuth (or elevation in barn owls) of the source in a roughly topographic fashion. In cats, the map of auditory space in the superior colliculus is based, in large part, on such a topographic variation in interaural intensity difference sensitivity (Wise & Irvine 1985).

Finally, a systematic projection from along the tonotopic axis of the ICc contributes to the construction of the space map. This projection reflects the systematic relationship between the spectral shaping of the signal by the external ears and source location: High frequency components of sound (wavelengths shorter than the dimensions of the external ears) are differentially enhanced when the source is in front of the external ear, and are increasingly attenuated as the source moves away from the axis of the external ear (Harrison & Downey 1970, Knudsen 1980, Phillips et al 1982). Accordingly, although frequency tuning is broad, neurons tuned to locations in front of the pinnae tend to respond to higher frequencies, while neurons tuned to locations away from the axis of the pinnae respond preferentially to lower frequencies (Knudsen 1984b, Middlebrooks 1986).

The Auditory Map of Space

The properties of the space map in the optic tectum (superior colliculus) are remarkably similar over a variety of species (Knudsen 1982, King & Palmer 1983, Middlebrooks & Knudsen 1984, Wong 1984). The map consists of neurons that respond strongly only when the source is in a particular region of space, called the neuron's "best area" (Figure 4B).

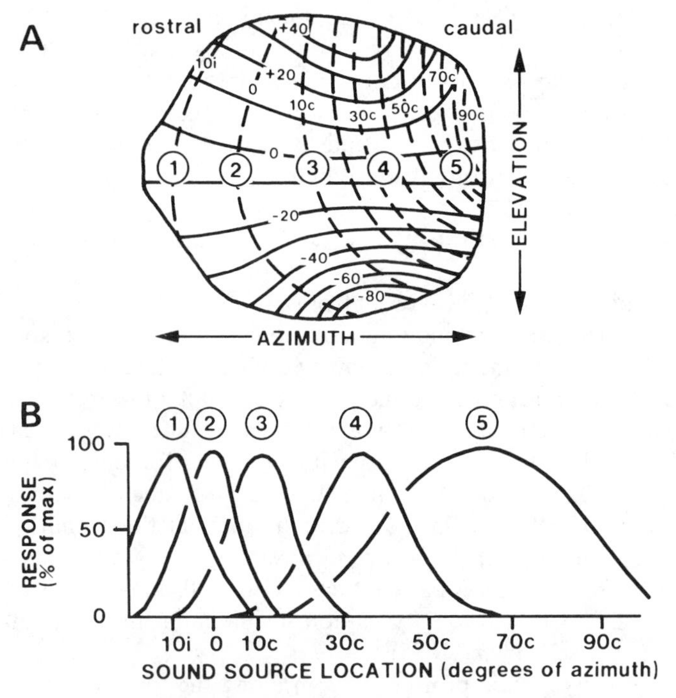

Figure 4 The representation of auditory space in the owl's optic tectum. *A*. The auditory map of space. In this flattened two-dimensional projection of the tectal surface, each *contour line* indicates the locations of units tuned to the same azimuth (*dashed lines*) or the same elevation (*solid lines*). The spatial tuning of units varies systematically across the tectum. Space is expressed in double pole coordinates, with 0° azimuth, 0° elevation directly in front of the animal; *i* signifies ipsilateral space, *c* signifies contralateral space. *Circled numerals* indicate the positions of units whose responses are shown in *B*.

B. Tuning of tectal units for sound source location. Each *curve* indicates the normalized response of a unit as a function of stimulus azimuth. Tuning is sharpest in the expanded representation of frontal space and broader in the representation of peripheral space. These data are from Knudsen (1982).

The location of a best area is independent of the spectral properties or the intensity of the sound source. As a stimulus moves away from a neuron's best area, the response of the neuron decreases and most neurons possess a receptive field beyond which stimuli are completely ineffective. The map of space is based on a systematic ordering of neurons according to the locations of their best areas. The map is primarily of contralateral space, and, at least in cats and owls, the map is two-dimensional, representing both the horizontal (azimuth) and vertical (elevation) location of the sound source (Knudsen 1982, Middlebrooks & Knudsen 1984).

Neurons in the auditory map of space respond best to broad-band stimuli (sounds containing multiple frequencies); their spatial tuning is a consequence of the convergence of multiple cue values across frequency. This convergence enables facilitatory and inhibitory interactions to sharpen spatial tuning and eliminate ambiguities (Wise & Irvine 1984, Hirsch et al 1985, Esterley 1984, Takahashi & Konishi 1986).

The sharpness of spatial tuning varies across species. In barn owls, the azimuthal dimension of best areas (defined as the area from which a stimulus elicits greater than 50% of the maximum response) ranges from 3 to 50° (Knudsen 1982). In cats, it ranges from 10 to more than 90° (Middlebrooks & Knudsen 1984), and in guinea pigs it is rarely less than 20° and is usually much larger (King & Palmer 1983). These species differences reflect differences in the spatial resolution of the cues available to each species and differences in the precision of the auditory system in processing the cues.

PROPERTIES OF COMPUTATIONAL MAPS

A number of fundamental properties shared by the auditory maps of interaural delay, interaural intensity difference, and sound source location are found in other computational maps as well. Regardless of the parameter being analyzed, neurons are typically quite broadly tuned for the mapped parameter, and the map's algorithm can operate independently of conscious guidance. Although details of topography vary, all maps represent ranges of parameter values that are biologically relevant, and they often magnify the representation of certain ranges of values that are of particular importance to the animal. In the following sections, we use examples from the auditory and visual systems to discuss the implications of these shared properties of computational maps.

Tuning for the Mapped Parameter Is Broad

In computational maps, neurons tend to respond to a broad range of parameter values. As a result, a stimulus will activate neurons across a large portion of a computational map. For example, in the maps of orientation preference found in the visual cortex, although any given neuron responds maximally to line stimuli over a small range of orientations, it may respond at least weakly to almost any orientation (Figure 1). In primates, the range of orientations to which the average neuron will respond at least half maximally is about 50–60° (Schiller et al 1976, Albright 1984). Since the entire range of orientations is 180°, neurons throughout almost one third of the map are activated strongly by any given stimulus orientation. Similarly, neurons in the visual map of movement direction in area MT of the macaque respond at least half maximally to moving stimuli over 90° of a 360° range of directions, and for any given stimulus, neurons throughout almost the entire map will be active at some level (Maunsell & Van Essen 1983a, Albright et al 1984).

Despite the fact that neurons in computational maps are broadly tuned, precise information about the values of parameters is contained in the output of these maps. Although the neurons respond to wide ranges of parameter values, their tuning curves are peaked (Figures 1, 3, and 4). Moreover, the tuning curves of neurons shift systematically across the map. These systematic differences in tuning curves give rise to systematic differences in response rates across the map for any given stimulus. Thus, high resolution information about the value of the mapped parameter is contained in the relative responses of neurons. To access this detailed information, a subsequent processor must be sensitive to relative levels of activity within a large population of neurons and must detect locations of peak activity within the map.

Computations Are Preset

The neurons that make up computational maps perform preset computations in parallel on the afferent signal. "Preset" does not mean that the computations cannot be modulated by descending influences, as indeed they must be in certain cases such as in the interpretation of interaural intensity difference cues by animals that can turn their ears (Middlebrooks & Knudsen 1986). Rather, it means that the neurons do not *need* input from higher centers in order to carry out their computations. The basic computations rely entirely on patterns of connectivity that are intrinsic to the processing circuit. Neither does "preset" imply genetically hardwired. Although the basic patterns of connectivity that underlie the maps are undoubtedly genetically determined, the details can be influenced extensively by experience, as in the auditory map of space and the visual map of line orientations (Knudsen 1985, Stryker et al 1978), or by patterns of correlated neuronal activity, as in the visual map of eye dominance (LeVay et al 1980, Stryker & Harris 1986). An important consequence of the computations being preset and executed in parallel is that information is processed rapidly in maps.

Factors Influencing Topography

A computational map is not necessarily an end-point, but usually an intermediate step in processing information. The generation of a map indicates that the parameter is being evaluated at that particular site in the nervous system, and that the value of the parameter is crucial for subsequent processing. The topography of a map may reflect the mechanism by which the computation is carried out (as in the map of interaural delay in n. Lam; Figure 3), economy of axonal and dendritic circuitry, developmental constraints, and/or the needs of subsequent processors. An example of map topography that seems to be regulated by a subsequent processor is the auditory space map in the owl's ICx (Knudsen & Konishi 1978). The magnification properties and the extent of spatial representation are not predicted by the properties of localization cues. Instead, they conform to the topography of the visual map of space in the optic tectum (Knudsen 1982). The topographical match of the auditory and visual maps in the optic tectum enables space-specific bimodal integration to occur throughout the tectum, and provides spatially consistent, modality-independent information to the computational motor map (Figure 2). The hypothesis that map topography conforms to the requirements of subsequent processors suggests that computational maps of the same parameter that feed into different processors may exhibit different topographies. For example, a cortical map of auditory space that subserves cognitive functions may exhibit different topographical properties from those of the tectal map.

The range of parameter values represented in a computational map corresponds with the needs of the animal. When all possible values of a parameter are important, as is the case for line orientation and movement direction, the entire range of values is represented in the map (e.g. Figure 1). However, when only a limited range of parameter values is biologically relevant, as is the case for parameters such as interaural delay and amplitude spectrum, the range of the map is restricted to those values (Figure 3; Suga 1977). For example, the maximum interaural delay that the barn owl experiences is approximately 170 μsec, which corresponds to the range of delays mapped in the owl's nucleus laminaris (Moiseff & Konishi 1981, Sullivan & Konishi 1986).

The relative magnification of the representation of parameter values can vary within a map. Where such anisotropies occur, they correlate with behavioral performance. Thus, the greatly expanded representation of frontal space in the owl's auditory space map (Figure 4) corresponds to a region of exceptionally high localization accuracy and precision (Knudsen et al 1979). Similarly, a disproportionately large representation of horizontal and vertical line orientations in the foveal region of V1 of the macaque monkey is reflected in an enhanced ability to discriminate stimuli at these angles (Mansfield 1974). These anisotropies do not arise as a result of variations in receptor densities (as do analogous anisotropies in maps of the sensory epithelium), but instead must be created through a differential scaling of the algorithms that generate the map.

Limits to the Number of Simultaneously Mapped Parameters

In theory, the number of parameters that can be mapped independently and continuously in one area is limited to the number of dimensions in the neural structure. In the cortex, where columnar organization uses one dimension for functional specialization, serial processing, and ordering the distribution of information (Maunsell & Van Essen 1983b, Bolz & Gilbert 1986), only two dimensions are left for mapping. The nervous system overcomes this limitation by organizing fine-grain maps within coarse-grain maps. For example, numbers and letters can be mapped simultaneously along a single dimension in the following way:

1A 1B 1C 2A 2B 2C 3A 3B 3C

The letters form a fine-grain map within a coarse-grain map of numbers. On the local scale, the relative activity of neurons is determined by their tuning for the parameter mapped in fine grain, while on a broader scale relative activity is determined by the neurons' tuning for the parameter mapped in coarse grain. This is, in essence, how the primary visual cortex simultaneously maps two dimensions of space, orientation preference, ocular dominance, and possibly other parameters across only two dimensions of the cortex (Hubel & Wiesel 1977).

Notice, however, that nesting one map within another erodes some of the advantages of mapping. Map-dependent neural interactions, described in the next section, operate optimally only for the parameter mapped in fine grain. Moreover, the circuitry necessary for processing and accessing specific information becomes more complicated. For example, separate regions of the map must be sampled to compare sequential values of the coarse-grain parameter having the same fine-grain parameter value (e.g. 1A, 2A, 3A). Only by merging, and thereby losing, the information contained in the fine-grain map, can the information in the coarse-grain map be accessed in a simple topographic manner. Thus, nesting maps within maps deteriorates to a complex, though systematic, representation requiring intricate connections to process and access the information. Perhaps this is one reason that different parameters are mapped in separate, functionally specialized areas of the brain.

ADVANTAGES OF MAPPING

Efficient Information Processing

The nervous system must continuously analyze complex events in a dynamic environment. This requires strategies of processing that are capable of handling large amounts of information rapidly. Computational maps, with their parallel arrays of preset processors, are ideally suited for such a task : They rapidly sort and process components of complex stimuli, and represent the results in a simple, systematic form.

Information that is represented in this manner is simple to access. When parameter values are presorted in a map, further processing of the information can be based on relatively straightforward schemes of connectivity. Recall, for example, that by mapping interaural delays and interaural intensity differences, the nervous system is able to derive sound source location by using simple patterns of convergence and gradients of projections. Similarly, the map of auditory space that results from these schemes of connectivity can itself be readily integrated with other topographic maps of sensory space from other modalities.

By encoding various kinds of stimulus parameters in a common, mapped code, the nervous system can employ a single strategy for reading the information. Whether the desired information is the orientation or the direction of motion of a visual stimulus, the time interval between two particular sounds, or the location of a sound source in space, the answer is always represented as the location of a peak of activity within a population of neurons.

Maps Enable Additional Interactions

When a parameter is represented in topographic form, a variety of neuronal mechanisms can operate to further sharpen tuning in ways not possible if the information is in a non-topographic code. One class of mechanisms is regional interactions such as local facilitation and lateral inhibition, which can only work on mapped information. A second class is nonspiking synaptic interactions which operate only over relatively short distances (e.g. Watanabe & Bullock 1960, Pearson 1979). Finally, there are nonsynaptic interactions that result from changes in the electrical and chemical environment caused by synchronously active neurons in the immediate vicinity (e.g. Haas & Jeffreys 1984).

DETECTING MAPS

Not all computations are mapped. However, concluding that a computational map does not exist in a given area of the nervous system or for a given parameter is difficult for several reasons. First, neurons are commonly tuned for a multitude of parameters (known and unknown), the vast majority of which will not be mapped. Discovering which of the parameters may be mapped becomes increasingly difficult as the response properties of neurons become more complex. Second, the size of the map may be small. A hypercolumn in the primate visual cortex occupies less than one square millimeter of cortical surface (Hubel & Wiesel 1977), and there is no reason that maps might not be even smaller. Theoretically, a single line of neurons could function as a map; such a map would be difficult to demonstrate. Third, complicated topography can make detection of the map difficult. The axes may be irregular, convoluted, or curved (as in the auditory amplitude map; Suga 1977, Suga & Manabe 1982). In addition, a map can be obscured by embedded fine-grain maps and functional clusters.

WHAT ABOUT HIGH ORDER COMPUTATIONAL MAPS?

The nervous system has been shown to use computational maps to derive and represent both basic sensory parameters, such as line orientation in vision and interaural delay in audition, and more abstracted parameters, such as sound source location, which depend on the interactions of multiple parameters. However, the computational maps discovered so far are of relatively low order: They map fundamental sensory or motor parameters. This does not mean that higher order maps are scarce. Rather, it reflects the experimental conditions under which most animals are studied. Low order maps can operate independently of attention, behavioral significance, or levels of consciousness, and they have been revealed in untrained, anesthetized animals. Only when cortical areas are explored under conditions that require animals to perform sophisticated analyses or behavioral responses (i.e. to use their brains) can high order computational maps, if present, be revealed. Another difficulty with uncovering high order maps is anticipating the organizational parameters of the maps. In general, however, high order maps might be be expected in areas of the brain that process aspects of perception or behavior for which the value of a parameter is the essence of the analysis, such as binocular disparity for visual depth (Poggio 1984), rise time or direction and rate of frequency modulation in vocalization analysis (Liberman et al 1967, Kuhl 1981), or the absolute positions of objects in personal and extrapersonal space (Anderson et al 1985).

SUMMARY

The nervous system performs computations to process information that is biologically important. Some of these computations occur in maps—arrays of neurons in which the tuning of neighboring neurons for a particular parameter value varies systematically. Computational maps transform the representation of information into a place-coded probability distribution that represents the computed values of parameters by sites of maximum relative activity. Numerous computational maps have been discovered, including visual maps of line orientation and direction of motion, auditory maps of amplitude spectrum and time interval, and motor maps of orienting movements. The construction of the auditory map of space is the most thoroughly understood: information about interaural delays and interaural intensity differences is processed in parallel by separate computational maps, and the outputs of these maps feed into a higher order processor that integrates sets of cues corresponding to sound source locations and creates a map of auditory space.

Computational maps represent ranges of parameter values that are relevant to the animal, and may differentially magnify the representation of values that are of particular importance. The tuning of individual neurons for values of a mapped parameter is broad relative to the range of the map. Consequently, neurons throughout a large portion of a computational map are activated by any given stimulus, and precise information about the mapped parameter is coded by the locations of peak activity.

There are a number of advantages of performing computations in maps. First, information is processed rapidly because the computations are preset and are executed in parallel. Second, maps simplify the schemes of connectivity required for processing and utilizing the information. Third, a common, mapped representation of the results of different kinds of computations allows the nervous system to employ a single strategy for reading the information. Finally, maps enable several classes of neuronal mechanisms to sharpen tuning in a manner not possible for information that is represented in a non-topographic code.

ACKNOWLEDGMENTS

We thank Drs. Masakazu Konishi, John Middlebrooks, Carla Shatz, David Van Essen, and Herman Wagner for reviewing the manuscript, and

Ms. Brenda Robertson for typing the manuscript. Preparation of the manuscript was supported by grants from the March of Dimes (1-863), the National Institutes of Health (R01 NS 16099-06 and 5T32 NS 07158-07), the National Institute of Mental Health (5T32 MH 17047-04), and a Neuroscience Development award from the McKnight Foundation.

Literature Cited

Aitkin, L. M., Gates, G. R., Phillips, S. C. 1984. Responses of neurons in inferior colliculus to variations in sound-source azimuth. *J. Neurophysiol.* 52: 1–17

Aitkin, L. M., Pettigrew, J. D., Calford, M. B., Phillips, S. C., Wise, L. Z. 1985. Representation of stimulus azimuth by low-frequency neurons in inferior colliculus of the cat. *J. Neurophysiol.* 53: 43–59

Albright, T. D. 1984. Direction and orientation selectivity of neurons in visual area MT of the macaque. *J. Neurophysiol.* 52: 1106–30

Albright, T. D., Desimone, R., Gross, C. G. 1984. Columnar organization of directionally selective cells in visual area MT of the macaque. *J. Neurophysiol.* 51: 16–31

Anderson, R. A., Essick, G. K., Siegel, R. M. 1985. Encoding of spatial location by posterior parietal neurons. *Science* 230: 456–58

Bolz, J., Gilbert, C. D. 1986. Generation of end-inhibition in the visual cortex via interlaminar connections. *Nature* 320: 362–65

Boudreau, J. C., Tsuchitani, C. 1970. Cat superior olive S-segment cell discharge to tonal stimulation. *Contrib. Sensory Physiol.* 4: 143–213

Calford, M. B., Moore, D. R., Hutchings, M. E. 1986. Central and peripheral contributions to coding of acoustic space by neurons in inferior colliculus of cat. *J. Neurophysiol.* 55: 587–603

Cant, N. B., Morest, D. K. 1984. The structural basis for stimulus coding in the cochlear nucleus. In *Hearing Sciences: Recent Advances*, ed. C. Berlin, pp. 371–421. San Diego: College-Hill Press

Colburn, H. S., Durlach, N. I. 1978. Models of binaural interaction. *Handb. Percept.* 4: 467–518

Coles, R. B., Lewis, D. B., Hill, K. G., Hutchings, M. E., Gower, D. M. 1980. Directional hearing in the Japanese quail (*Coturnix coturnix japonica*). II. Cochlear physiology. *J. Exp. Biol.* 86: 153–70

Esterly, S. D. 1984. Responses of space-specific neurons in the optic tectum of the owl to narrow-band sounds. *Soc. Neurosci.* 10: 1149 (Abstr.)

Goldberg, J. M., Brown, P. B. 1969. The response of binaural neurons of dog superior olivary complex to dichotic tonal stimuli: Some physiological mechanisms of sound localization. *J. Neurophysiol.* 32: 613–36

Guinan, J. J. Jr., Norris, B. E., Guinan, S. S. 1972. Single auditory units in the superior olivary complex. II: Locations of unit categories and tonotopic organization. *Intern. J. Neurosci.* 4: 147–66

Haas, H. L., Jeffreys, J. G. R. 1984. Low calcium field burst discharge of CA1 pyramidal neurons in rat hippocampal slices. *J. Physiol.* 354: 185–201

Harrison, J. M., Downey, P. 1970. Intensity changes at the ear as a function of the azimuth of a tone source: A comparative study. *J. Acoust. Soc. Am.* 47: 1509–18

Hirsch, J. A., Chan, J. C. K., Yin, T. C. T. 1985. Responses of neurons in the cat's superior colliculus to acoustic stimuli. I. Monaural and binaural response properties. *J. Neurophysiol.* 53: 726–45

Hubel, D. H., Wiesel, T. N. 1963. Shape and arrangements of columns in cat's striate cortex. *J. Physiol. London* 165: 559–68

Hubel, D. H., Wiesel, T. N. 1968. Receptive fields and functional architecture of monkey striate cortex. *J. Physiol.* 195: 215–43

Hubel, D. H., Wiesel, T. N. 1977. Functional architecture of macaque monkey visual cortex. *Proc. R. Soc. London Ser. B* 198: 1–59

Jenkins, W. M., Merzenich, M. M. 1984. Role of cat primary auditory cortex for sound-localization behavior. *J. Neurophysiol.* 52: 819–47

King, A. J., Palmer, A. R. 1983. Cells responsive to free-field auditory stimuli in guinea-pig superior colliculus: Distribution and response properties. *J. Physiol. London* 342: 361–81

Knudsen, E. I. 1980. Sound localization in birds. In *Comparative Studies of Hearing in Vertebrates*, ed. A. N. Popper, R. R. Fay, pp. 287–322. Berlin/Heidelberg/New York: Springer

Knudsen, E. I. 1982. Auditory and visual

maps of space in the optic tectum of the owl. *J. Neurosci.* 2: 1177–94

Knudsen, E. I. 1983a. Space coding in the vertebrate auditory system. In *Bioacoustics*, ed. B. Lewis, pp. 311–44. London: Academic

Knudsen, E. I. 1983b. Subdivisions of the inferior colliculus in the barn owl (*Tyto alba*). *J. Comp. Neurol.* 218: 174–86

Knudsen, E. I. 1984a. Synthesis of a neural map of auditory space in the owl. In *Dynamic Aspects of Neocortical Function*, ed. G. M. Edelman, W. M. Cowan, W. E. Gall, pp. 375–96. New York: Wiley

Knudsen, E. I. 1984b. Auditory properties of space-tuned units in owl's optic tectum. *J. Neurophysiol.* 52: 709–23

Knudsen, E. I. 1985. Experience alters the spatial tuning of auditory units in the optic tectum during a sensitive period in the barn owl. *J. Neurosci.* 5: 3094–3109

Knudsen, E. I., Konishi, M. 1978. A neural map of auditory space in the owl. *Science* 200: 795–97

Knudsen, E. I., Blasdel, G. G., Konishi, M. 1979. Sound localization by the barn owl measured with the search coil technique. *J. Comp. Physiol.* 133: 1–11

Konishi, M., Takahashi, T. T., Wagner, H., Sullivan, W. E., Carr, C. E. 1986. Neurophysiological and anatomical substrates of sound localization in the owl. In *Functions of the Auditory System*, ed. G. M. Edelman, W. E. Gall. New York: Wiley. In press

Kuhl, P. K. 1981. Discrimination of speech by nonhuman animals: Basic auditory sensitivities conducive to the perception of speech-sound categories. *J. Acoust. Soc. Am.* 70: 340–49

Kuhn, G. F. 1977. Model for the interaural time differences in the azimuthal plane. *J. Acoust. Soc. Am.* 62: 157–67

Kuwada, S., Yin, T. C. T., Syka, J., Buunen, T. J. F., Wickesberg, R. E. 1984. Binaural interaction in low-frequency neurons in inferior colliculus of the cat. IV. Comparison of monaural and binaural response properties. *J. Neurophysiol.* 51: 1306–25

LeVay, S., Wiesel, T. N., Hubel, D. H. 1980. The deevelopment of ocular dominance columns in normal and visually deprived monkeys. *J. Comp. Neurol.* 191: 1–51

Liberman, A. M., Delattre, P. C., Cooper, F. S. 1954. The role of selected stimulus-variables in the perception of the unvoiced stop consonants. *Am. J. Psychol.* 65: 497–516

Liberman, A. M., Cooper, F. S., Shankweiler, D. P., Studdart-Kennedy, M. 1967. Perception of the speech code. *Psychol. Rev.* 74: 431–61

Mansfield, R. J. W. 1974. Neural basis of orientation perception in primate vision. *Science* 186: 1133–35

Margoliash, D. 1983. Acoustic parameters underlying the responses of song-specific neurons in the white-crowned sparrow. *J. Neurosci.* 3: 1039–57

Maunsell, J. H. R., Van Essen, D. C. 1983a. Functional properties of neurons in middle temporal visual area (MT) of macaque monkey. I. Selectivity for stimulus direction, velocity and orientation. *J. Neurophysiol.* 49: 1127–47

Maunsell, J. H. R., Van Essen, D. C. 1983b. The connections of the middle temporal visual area (MT) and their relationship to a cortical hierarchy in the macaque monkey. *J. Neurosci.* 3: 2563–86

Middlebrooks, J. C. 1986. Binaural mechanisms of spatial tuning in the cat's superior colliculus distinguished using monoaural occlusion. *J. Neurophysiol.* In press

Middlebrooks, J. CC., Knudsen, E. I. 1984. A neural code for auditory space in the cat's superior colliculus. *J. Neurosci.* 4: 2621–34

Middlebrooks, J. C., Knudsen, E. I. 1986. Changes in external ear position modify the spatial tuning of auditory units in the cat's superior colliculus. *J. Neurophysiol.* In press

Middlebrooks, J. C., Pettigrew, J. D. 1981. Functional classes of neurons in primary auditory cortex of the cat distinguished by sensitivity to sound location. *J. Neurosci.* 1: 107–20

Moiseff, A., Konishi, M. 1981. Neuronal and behavioral sensitivity to binaural time differences in the owl. *J. Neurosci.* 1: 40–48

Moiseff, A., Konishi, M. 1983. Binaural characteristics of units in the owl's brainstem auditory pathway: Precursors of restricted spatial receptive fields. *J. Neurosci.* 3: 2553–62

Moushegian, G., Rupert, A. L., Gidda, J. 1975. Functional characteristics of superior olivary neurons to binaural stimuli. *J. Neurophysiol.* 38: 1037–48

O'Neill, W. E., Suga, N. 1982. Encoding of target-range information and its representation in the auditory cortex of the mustached bat. *J. Neurosci.* 2: 17–24

Pearson, K. G. 1979. Local neurons and local interactions in the nervous system of invertebrates. In *The Neurosciences Fourth Study Program*, ed. F. O. Schmitt, F. G. Worden. Cambridge: MIT Press

Phillips, D. P., Calford, M. B., Pettigrew, J. D., Aitkin, L. M., Semple, M. N. 1982. Directionality of sound pressure transformation at the cat's pinna. *Hearing Res.* 8: 13–28

Poggio, G. F. 1984. Processing of stereoscopic information in primate visual cortex. See Knudsen 1984a, pp. 613–35

Robinson, D. A. 1972. Eye movements evoked by collicular stimulation in the alert monkey. *Vision Res.* 12: 1795–1808

Rose, J. E., Gross, N. B., Geisler, C. D., Hind, J. E. 1966. Some neural mechanisms in the inferior colliculus of the cat which may be relevant to localization of a sound source. *J. Neurophysiol.* 29: 288–314

Roth, G. L., Aitkin, L. M., Anderson, R. A., Merzenich, M. M. 1978. Some features of the spatial organization of the central nucleus of the inferior colliculus of the cat. *J. Comp. Neurol.* 182: 661–80

Roth, G. L., Kochhar, R. K., Hind, J. E. 1980. Interaural time differences: Implications regarding the neurophysiology of sound localization. *J. Acoust. Soc. Am.* 68: 1643–51

Roucoux, A., Guitton, D., Crommelinck, M. 1980. Stimulation of the superior colliculus in the alert cat. II. Eye and head movements evoked when the head is unrestricted. *Exp. Brain Res.* 39: 75–85

Scheibel, M. E., Scheibel, A. B. 1974. Neuropil organization in the superior olive of the cat. *Exp. Neurol.* 43: 339–48

Schiller, P. H., Finlay, B. L., Volman, S. F. 1976. Quantitative studies of single-cell properties in monkey striate cortex. II. Orientation specificity and ocular dominance. *J. Neurophysiol.* 39: 1320–33

Searle, C. L., Braida, L. D., Davis, M. F., Colburn, H. S. 1976. Model for auditory localization. *J. Acoust. Soc. Am.* 60: 1164–75

Semple, M. N., Aitkin, L. M. 1979. Representation of sound frequency and laterality by units in central nucleus of cat inferior colliculus. *J. Neurophysiol.* 42: 1626–39

Shaw, E. A. G. 1974. Transformation of sound pressure level from the free field to the eardrum in the horizontal plane. *J. Acoust. Soc. Am.* 56: 1848–61

Sparks, D. L., Holland, R., Guthrie, B. L. 1976. Size and distribution of movement fields in the monkey superior colliculus. *Brain Res.* 113: 21–34

Stein, B. E., Clamann, H. P. 1981. Control of pinna movements and sensorimotor register in cat superior colliculus. *Brain Behav. Evol.* 19: 180–92

Stryker, M. P., Harris, W. A. 1986. Binocular impulse blockade prevents the formation of ocular dominance columns in cat visual cortex. *J. Neurosci.* 6: 2117–33

Stryker, M. P., Schiller, P. M. 1975. Eye and head movements evoked by electrical stimulation of monkey superior colliculus. *Exp. Brain Res.* 23: 103–12

Stryker, M. P., Sherk, H., Leventhal, A. G., Hirsch, H. V. B. 1978. Physiological consequences for the cat's visual cortex of effectively restricting early visual experience with oriented contours. *J. Neurophysiol.* 41: 896–909

Suga, N. 1977. Amplitude spectrum representation in the doppler-shifter-CF processing area of the auditory cortex of the mustache bat. *Science* 196: 64–67

Suga, N., Manabe, T. 1982. Neural basis of amplitude-spectrum representation in auditory cortex of the mustached bat. *J. Neurophysiol.* 47: 225–55

Suga, N., O'Neill, W. E. 1979. Neural axis representing target range in the auditory cortex of the mustache bat. *Science* 206: 351–53

Suga, N., O'Neill, W. E., Kujirai, K., Manabe, T. 1983. Specificity of combination-sensitive neurons for processing of complex biosonar signals in auditory cortex of the mustached bat. *J. Neurophysiol.* 49: 1573–1626

Sullivan, W. E., Konishi, M. 1984. Segregation of stimulus phase and intensity coding in the cochlear nucleus of the barn owl. *J. Neurosci.* 4: 1787–99

Sullivan, W. E., Konishi, M. 1986. A neural map of interaural phase difference in the owl's brainstem. *Proc. Natl. Acad. Sci.* In press

Takahashi, T. T., Konishi, M. 1985. Parallel pathways in the owl's brainstem auditory system. *Anat. Res.* 211: 191A

Takahashi, T. T., Konishi, M. 1986. Selectivity for interaural time difference in the owl's midbrain. *J. Neurosci.* In press

Wagner, H., Takahashi, T. T., Konishi, M. 1986. The central nucleus of the inferior colliculus as an input stage to the map of auditory space in the barn owl. *Abstr. Assoc. Res. Otolaryng.*, pp. 44–45

Watanabe, A., Bullock, T. H. 1960. Modulation of activity of one neuron by subthreshold slow potential in another in lobster cardiac ganglion. *J. Gen. Physiol.* 43: 1031–45

Wenstrup, J. J., Ross, L. S., Pollack, G. D. 1986. Binaural response organization within a frequency-band representation of the inferior colliculus: Implications for sound localization. *J. Neurosci.* 6: 962–73

Wernick, J. S., Starr, A. 1968. Binaural interaction in the superior olivary complex of the cat: An analysis of field potentials evoked by binaural-beat stimuli. *J. Neurophysiol.* 31: 428–41

Wise, L. Z., Irvine, D. R. F. 1984. Interaural intensity difference sensitivity based on facilitatory binaural interaction in cat superior colliculus. *Hearing Res.* 16: 181–87

Wise, L. Z., Irvine, D. R. F. 1985. Topographic organization of interaural intensity difference sensitivity in deep layers of cat superior colliculus: Implications for auditory spatial representation. *J. Neurophysiol.* 54: 185–211

Wong, D. 1984. Spatial tuning of auditory neurons in the superior colliculus of the echolocating bat, *Myotis lucifugus. Hearing Res.* 16: 261–70

Wurtz, R. H., Albano, J. E. 1981. Visual motor function of the primate superior colliculus. *Ann. Rev. Neurosci.* 3: 189–226

Yin, T. C. T., Chan, J. C. K., Kuwada, S. 1983. Characteristics, delays, and their topographical distribution in the inferior colliculus of the cat. In *Mechanisms of Hearing*, ed. W. R. Webster, L. M. Aitkin, pp. 94–99. Melbourne: Monash Univ. Press

Young, S. R., Rubel, E. W. 1983. Frequency-specific projections of individual neurons in chick brainstem auditory nuclei. *J. Neurosci.* 3: 1373–78

Zeki, S. M. 1974. Functional organization of a visual area in the posterior bank of the superior temporal sulcus of the rhesus monkey. *J. Physiol.* 236: 549–73

Zeki, S. M. 1980. The representation of colours in the cerebral cortex. *Nature* 284: 412–18

Ann. Rev. Neurosci. 1987. 10: 67–95

EXTRATHALAMIC MODULATION OF CORTICAL FUNCTION

Stephen L. Foote

Department of Psychiatry, School of Medicine, University of California, San Diego, California 92093, and Division of Preclinical Neuroscience, Research Institute of Scripps Clinic, La Jolla, California 92037

John H. Morrison

Division of Preclinical Neuroscience, Research Institute of Scripps Clinic, La Jolla, California 92037

INTRODUCTION

THE DEMONSTRATION OF EXTRATHALAMIC CORTICAL AFFERENTS As recently as 25 years ago, it was commonly assumed that essentially all afferents to the neocortex arose from the thalamus, With the development of methods for visualizing noradrenergic, serotonergic, and dopaminergic neuronal processes, it gradually became clear that axons arising from subthalamic cell groups that contain these putative transmitters travel rostrally through the brainstem and monosynaptically innervate the neocortex. With the more recent demonstration that acetylcholinergic (ACh) neurons in the basal forebrain project monosynaptically into the neocortex, it is now clear that there are at least four substantial extrathalamic projections to the neocortex. Recently, these afferents have been demonstrated to be much more dense and more highly organized than was initially thought.

The existence and nature of these cortical afferents raises major questions about cortical organization and function. Thalamocortical and corticocortical systems appear to be organized to implement both serial and parallel processing of information through modules that are interconnected with precise topography, thus permitting functional segregation.

0147–006X/87/0301–0067$02.00

This well-known columnar, radial organization of the neocortex is apparently violated by the tangentially organized extrathalamic afferents in which single axons may innervate not only different columns within a functional region but different functional areas. What are the functions of these systems? What information do they convey that could not be conveyed as readily and perhaps more precisely by thalamic afferents? How do these afferents alter cortical processing so that the neocortex can better accomplish its functions?

THE SCOPE OF THIS ARTICLE Answers to these questions are not presently available, but recent anatomic and physiologic observations offer tantalizing clues about how they might be experimentally addressed. The purpose of this article is to briefly review some of these recent findings and discuss their implications for future research. The literature review concentrates on primate neocortex, since it is clear that certain aspects of neocortical organization, including the characteristics of its extrathalamic innervation, are qualitatively different in primates and rodents. We review the current knowledge about the four extrathalamic cortical afferent systems that have been well documented: (*a*) the noradrenergic afferents arising from the pontine nucleus locus coeruleus (LC); (*b*) the serotonergic afferents arising from the mesencephalic raphe nuclei; (*c*) the dopaminergic afferents arising from the substantia nigra–ventral tegmental area complex (SN/VTA); and (*d*) the ACh afferents arising primarily from the nucleus basalis of the substantia innominata. (For descriptions of a less extensively studied system, see Saper 1985 and Saper et al 1986.) After presenting the anatomy and physiology for each system, we briefly discuss the functional implications of these data. In the concluding section, we discuss the possible general impact of these systems on cortical function.

The most recent data available suggest the following principles of organization for extrathalamic afferents to neocortex:

1. Each extrathalamic afferent system exhibits pronounced regional and laminar specialization.
2. These systems differ from each other in terms of which major cortical subdivisions, cytoarchitectonic areas, and cortical laminae constitute their principal termination sites.
3. The extensive phylogenetic enlargement, elaboration, and specialization of primate neocortex is paralleled by analogous development of extrathalamic afferent systems.
4. Regional innervation patterns of an extrathalamic system in neocortex may be matched by its preferential innervation of functionally related thalamic and other subcortical structures.

These principles of organization raise fundamental questions about the processes underlying the ontogenetic development of the systems. We have recently reviewed the available data bearing on the development of extrathalamic systems (Foote & Morrison 1987) and do not therefore deal with these issues in the present article. Also, due to space constraints, we do not review the currently available data about receptor localization for extrathalamic systems in neocortex.

NORADRENERGIC INNERVATION OF NEOCORTEX

Anatomy: Source Cells and Termination Patterns of Neocortical Afferents

The noradrenergic innervation of neocortex arises solely from the nucleus locus coeruleus, which is located in the pontine brainstem (reviewed in Foote et al 1983). This nucleus innervates every major region of the neuraxis, even though it is composed of a relatively small number of neurons: approximately 1600 per hemisphere in the rat, 5000 per hemisphere in the monkey, and 13,000 per hemisphere in the human. Individual LC neurons often innervate widely separated brain regions, thus indicating that these axons must be highly divergent. It has also been demonstrated that individual LC neurons innervate different cortical regions. Thus, after an LC axon has traveled through the brainstem to the frontal regions of cortex, it diverges as it sweeps in a rostrocaudal trajectory through the cortical hemisphere, sending off collaterals to many different cortical regions and to both superficial and deep cortical layers (Morrison et al 1978, 1979, 1981, Loughlin et al 1982). The noradrenergic fibers are primarily distributed in a tangential (i.e. parallel to the cortical surface) fashion, particularly in layer VI, where they are oriented predominantly in the anteroposterior plane, forming a continuous sheet of longitudinal fibers overlying the white matter. In rat, these very fine caliber fibers branch to innervate all six layers of the neocortex, and the pattern of noradrenergic axon distribution possesses a geometric orderliness and distinct laminar pattern that is consistent throughout the lateral neocortex (Morrison et al 1978, 1981), with significant regional specialization present only in cortical regions on the medial surface of the hemisphere (Morrison et al 1979). Layer I possesses a lattice of rostrocaudal and mediolateral tangential fibers. The layer VI axons branch profusely into upper layer V and layer IV; possibly they represent terminal axons. The fibers in layers II and III are predominantly radial in orientation. The layer I fibers do not necessarily all arise from local fibers in deeper layers; some fibers may run long distances within this layer.

In contrast, the noradrenergic innervation of primate cortex exhibits striking regional specialization in both density and laminar pattern of innervation (Morrison et al 1982a,b, 1984, Morrison & Foote 1986, Levitt et al 1984). For example, primary somatosensory and motor regions are densely innervated in all six laminae, whereas temporal cortical regions are very sparsely innervated. In primary visual cortex, the density of innervation is intermediate, but there is a striking absence of fibers in lamina IV. As in the rat, a strong tangential, intracortical trajectory is a dominant feature of the noradrenergic innervation of the much more convoluted primate cortex (Morrison et al 1982b).

The presence of regional specialization in noradrenergic, and other extrathalamic, innervation patterns raises the question of whether there are underlying organizing principles for these regional variations. The presently available rodent and primate data do not support a simple scheme for the distribution of noradrenergic fibers in the neocortex, such as preferential innervation of motor versus sensory structures, one sensory modality in preference to others, or primary sensory versus secondary sensory or association areas. Since a large portion of the primate neocortex is visual in nature, we have utilized immunohistochemical methods to characterize the noradrenergic innervation of several cortical and subcortical visual areas in monkey in order to search for possible organizing principles in these afferent systems (Morrison & Foote 1986). Cortical areas 17 and 18, as well as visual areas in the temporal and parietal lobe, were found to exhibit regional specialization of noradrenergic innervation. Precisely at the border between areas 17 and 18, the laminar innervation patterns characteristic of noradrenergic fibers in area 17 shift such that layer IV of area 18 contains more fibers than layer IV of area 17, and the overall density of fibers in area 18 is higher. The visual region of the inferotemporal cortex was found to be very lightly innervated by noradrenergic fibers, while area 7 of the parietal lobe was much more densely innervated. Visual thalamic nuclei exhibited pronounced regional differences in noradrenergic innervation density. The lateral geniculate was found to be virtually devoid of noradrenergic fibers, while the pulvinar-lateral posterior complex was densely innervated. In the mesencephalon, the superficial layers of the superior colliculus were found to be densely innervated by noradrenergic fibers.

The patterns of innervation indicate that, in these primate species, functionally related visual regions share common and distinguishable densities of noradrenergic innervation. Specifically, tectopulvinar-juxtastriate structures are more densely innervated than geniculostriate and inferotemporal structures. These relationships suggest that, within the visual system, noradrenergic fibers preferentially innervate regions involved in spatial analysis and visuomotor response rather than those involved in feature extraction and pattern analysis. This is especially interesting given the proposed involvement of the LC-noradrenergic system in attentional mechanisms.

Physiology

ACTIVITY OF SOURCE NEURONS LC neurons in unanesthetized, nonparalyzed rat, cat, and monkey have been shown to be most active during waking, less active during slow-wave sleep, and silent during rapid-eye-movement sleep (reviewed in Foote et al 1983). Within the waking state, the mean discharge rates of LC cells increase when enhanced levels of arousal or attentiveness are exhibited by the animal. In monkey, for example, discharge rates vary from second to second, and anticipate by several hundred milliseconds subsequent electroencephalographic changes indicating increased or decreased levels of alertness (Foote et al 1980).

CONDUCTION PROPERTIES OF AFFERENT AXONS The conduction properties of primate LC axons projecting to neocortex have recently been characterized (Aston-Jones et al 1985b). Such information is obviously crucial in delineating the mechanisms whereby neurons residing at such a great distance from the neocortex, but innervating it so substantially, could influence neocortical activity. Discharge activity was recorded extracellularly from individual, histologically verified LC-noradrenergic neurons in anesthetized squirrel monkeys. These neurons were found to exhibit several properties previously described for LC cells in rat, including slow (0.2–2 Hz) spontaneous discharge rates, distinctive impulse waveforms (notched, entirely positive in unfiltered recordings, and 2–3 msec in duration), antidromic activation from many target areas, a period of suppressed discharge activity following either antidromic or orthodromic driving, and responsiveness to noxious stimuli presented as subcutaneous electrical stimulation of a rear foot. Antidromically driven action potentials (verified by collision tests) were recorded from individual LC neurons in response to electrical stimulation of neocortical (frontal, somatosensory, and occipital sites) and thalamic areas. These cells reliably conducted impulses from various cortical sites, some of which were up to 100 mm from the LC. The very high reliability of collision testing that we observed indicates that orthodromic impulses were also conducted with high reliability and temporal fidelity. Many monkey LC neurons projecting to the thalamus and neocortex were found to exhibit more rapid conduction velocities than previously observed in rat (e.g. approximately 34% were greater than 1 M/sec), resulting in similar conduction latencies to distant target areas for the two species.

These observations indicating that action potentials are conducted from

LC to neocortex with high reliability are significant because they lend additional credence to the hypothesis that noradrenergic release is controlled by the rate of impulse production in LC-noradrenergic neurons, rather than by local factors within the neocortex, as has been postulated by others. The demonstration of faster conduction velocities for long LC axons in the monkey is also important since extrapolation from the slow conduction velocities found in rat to species with larger brains (such as primates) would predict that exceptionally long latencies (i.e. hundreds of milliseconds) would prevail for LC impulse activity to reach distant target areas, such as the neocortex. Such a property would have profound implications for proposed functions of the LC system in vigilance and sensory processing. The data indicate that the conduction times required for LC impulses to reach distant target areas may be conserved across different species and sizes of brains and suggest that these latencies play an important role in the general function of the LC system in brain and behavioral processes.

LC-NORADRENERGIC EFFECTS ON CORTICAL NEURONAL ACTIVITY There have been numerous studies of the possible effects of the LC-noradrenergic projection to the cortex on putative target cells in various neocortical regions. The initial study of noradrenergic effects in the neocortex of unanesthetized monkeys was performed several years ago (Foote et al 1975). Most studies of LC-noradrenergic effects had been performed in anesthetized animals, on target neurons that were not being activated (or inhibited) in a manner relevant to their function. Adequate characterization of the functional properties of the noradrenergic system, however, requires that its impact be considered with regard to other neural systems projecting to the same target neurons. That is, LC-noradrenergic effects on functional activity, rather than spontaneous activity, of the target neuron must be assessed. This requires simultaneous activation of other inputs to target neurons while activating the LC-noradrenergic system. In this experiment, the effects of iontophoretically applied noradrenaline were studied, using as test cells auditory cortex neurons that were activated acoustically by species-specific vocalizations in unanesthetized, unparalyzed squirrel monkeys. Poststimulus-time histograms and raster displays of neuronal responses to the vocalizations were computed before, during, and after iontophoresis. Noradrenaline caused dose-dependent inhibition of spontaneous and vocalization-evoked activity. A given dose of noradrenaline reduced spontaneous activity proportionately more than it reduced activity evoked by acoustic stimuli. During auditory responses, segments with lower discharge rates were reduced proportionately more than segments with higher discharge rates. These observations led the authors to propose that noradrenaline may act on such neurons to enhance elicited activity relative to spontaneous activity; in engineering terms, it enhances the "signal-to-noise" characteristics of the cells.

Many analogous experiments have been conducted in the past several years, aimed at determining whether LC-noradrenergic system differentially affects various aspects of target cell activity. Waterhouse, Woodward, and their colleagues have intensively investigated the effects of iontophoretic noradrenaline on neocortical neuronal activity in anesthetized rats (Waterhouse et al 1980, 1981, Woodward et al 1979). They have observed a similar reduction of background activity relative to stimulus-elicited activity in the somatosensory cortex. Additionally, in some cells, evoked excitatory responses were enhanced by noradrenaline application. Often, noradrenaline also augmented stimulus-bound inhibition and post-excitatory suppression of activity. Iontophoretic noradrenaline enhances responses to iontophoretic GABA and ACh, and enhanced excitatory sensory responses are produced and blocked by alpha- but not beta-adrenergic agents. The effects of LC electrical stimulation are generally a depression of spontaneous cortical neuronal activity (reviewed in Foote et al 1983; see also Jones & Olpe 1984).

HYPOTHESES OF LC-NORADRENERGIC EFFECTS ON TARGET NEURONAL ACTIVITY Various hypotheses have been advanced to describe the electrophysiological effects produced by the LC system. Some emphasize the enhancement of evoked activity relative to spontaneous activity (Foote et al 1975, Woodward et al 1979). Others describe this process as "enabling" (Bloom 1979), by which coactivity in LC terminals enables other systems converging on the same target neurons to transmit more effectively during the period of simultaneous activity. Yeh et al (1981) and Moises & Woodward (1980) present evidence that such enhancement by noradrenaline may occur in Purkinje cells only with certain transmitters and suggest more specific versions of these hypotheses (see also Moises et al 1983). The term "modulation" has often been used to describe these types of effects (see Dismukes 1979).

From the data, we draw the following conclusions: (*a*) the noradrenergic effects are too complex to describe as simple excitation or inhibition; (*b*) several studies find similar effects in different LC-noradrenergic terminal areas; (*c*) these effects constitute a substantial modification of the operation of the presumed target neurons.

Hypothesis

Taken together, the data concerning LC projections, the activity of source neurons, and the effects of the transmitter on target neurons indicate that the LC is activated during alerting or arousal and releases noradrenaline onto target neurons in many brain regions, including the neocortex. This transmitter then acts to enhance the selectivity and vigor of responses to subsequent sensory stimuli or other synaptic input to the target neurons. The LC may well also play a role in more tonic behavioral state changes, such as the sleep-wake cycle. It has been proposed that the function of LC could best be described as altering behavioral modes from internally oriented and generated states, such as sleep, grooming, and food consumption, to an externally oriented mode that involves active matching of appropriate behaviors with novel, stressful, or informative stimuli (e.g. Aston-Jones & Bloom 1981a, 1981b, Astron-Jones et al 1984c).

Recent evidence concerning the anatomy and physiology of the LC-noradrenergic cortical projection permits the development of more specific hypotheses dealing with the site, mode, and time-course of action of this system on target neurons (reviewed in Foote et al 1983). First, studies of the activity of LC neurons in behaving animals suggest environmental and behavioral conditions under which this system is active in releasing noradrenaline onto target neurons: Specifically, because LC neurons exhibit phasic responses to certain sensory stimuli and systematically alter their discharge in anticipation of phasic arousal, this should be reflected in response patterns of target cells. Second, light-microscopic studies demonstrate that in primate neocortex, unlike rodent neocortex, terminal arborization patterns of noradrenergic axons are distinctive for each cytoarchitectonic region in such a way as to suggest that within each area specific neuronal classes receive this innervation. Third, other studies (Olschowka et al 1981) have demonstrated that noradrenergic axon terminals form synaptic contacts onto their target neurons, thus suggesting that this system acts on target elements in a temporally and spatially restricted fashion. Fourth, the data just reviewed indicate that noradrenaline, either released from LC-noradrenergic fibers or applied by microiontophoresis, has specific effects on well-defined functional activity of target neurons. These four sets of data suggest a specific interaction of the LC-noradrenergic system with thalamocortical and corticocortical circuits to alter the latter's functioning via spatially localized and temporally discrete modification of neuronal information processing.

SEROTONERGIC INNERVATION OF NEOCORTEX

Anatomy: Source Cells and Termination Patterns of Neocortical Afferents

The serotonergic innervation of the rat neocortex arises primarily from the dorsal and median raphe nuclei (e.g. O'Hearn & Molliver 1984, Porrino

& Goldman-Rakic 1982). The serotonergic innervation of adult rat neocortex is more dense than is the noradrenergic innervation (Lidov et al 1980). The entire neocortical mantle is penetrated by these fine, varicose, and highly convoluted fibers, which appear in relatively uniform density across all cortical layers. The density and distribution of these fibers are such that they might innervate every neuron in the neocortex (Lidov et al 1980).

The serotonergic innervation of the monkey neocortex is also very dense, but in the monkey the innervation of different neocortical regions exhibits differences in density and laminar distribution of axons (Morrison et al 1982a, Takeuchi & Sano 1983, Kosofsky et al 1984, Morrison & Foote 1986). The serotonergic innervation of primary visual cortex (area 17) has been characterized in greater detail than that of any other region. The serotonin fibers here are very dense and are distributed in a strictly laminated fashion. Serotonin fibers show a strong preference for layer IV in area 17, a tendency that is also evident in other cortical areas. This tendency is especially interesting since this is the lamina that is the primary recipient of thalamocortical afferents. The morphology of serotonin fibers in the cortex is quite heterogeneous, with a mixture of thick, nonvaricose fibers, large, varicose fibers, and extremely fine, varicose fibers. Since the very thick fibers are most evident in white matter and the deep cortical laminae, and are more evident in young animals, it is highly likely that they are fibers of passage. The fibers are usually tangential in orientation. Differences between primate species are evident in the distribution and density of serotonin fibers in area 17.

In primary visual cortex (area 17) of the squirrel monkey, noradrenergic and serotonergic projections exhibit a high degree of laminar complementarity: Layers V and VI receive a dense noradrenergic projection and a very sparse serotonergic projection, whereas layer IV receives a very dense serotonergic projection and is largely devoid of noradrenergic fibers (Morrison et al 1982a, Kosofsky et al 1984, Foote & Morrison 1984). Also noradrenergic fibers manifest a geometric order that is not readily apparent in the orientation of serotonergic fibers. These patterns of innervation imply that the two transmitter systems affect different stages of cortical information processing. For example, the raphe-serotonergic projection may preferentially innervate the spiny stellate cells of layers IVa and IVc, whereas the LC-noradrenergic projection may innervate pyramidal cells.

We have examined the serotonergic innervation of several cortical regions subserving visual functions (Morrison & Foote 1986). Often a degree of complementarity exists between the density of noradrenergic and serotonergic innervation in these areas. For example, as mentioned above, the noradrenergic innervation of temporal lobe visual areas is very sparse. Serotonin innervates these same areas very densely.

Physiology

ACTIVITY OF SOURCE NEURONS The discharge activity of raphe neurons has been studied extensively in unanesthetized, behaving cats (reviewed in Jacobs et al 1984). Presumed serotonin neurons in the dorsal raphe nucleus have been found to exhibit slow, regular discharge patterns during quiet waking and to show increases in firing with phasic arousal. Presentation of a simple auditory or visual stimulus results in a single spike response, followed by a period of reduced activity lasting about 200 msec. Mean discharge rates decrease to about 1 Hz during slow-wave sleep, and the neurons become almost silent during REM sleep. With no external stimulation, these cells show very periodic activity, which various studies indicate is very likely caused by pacemaker activity intrinsic to the neurons themselves. Jacobs and colleagues (1984) have hypothesized that dorsal raphe discharge rates in behaving animals is closely related to activity in central motor systems, not in the sense that raphe neurons are motor or premotor neurons themselves but rather that their discharge rates appear to covary with general levels and intensity of movement.

RAPHE-SEROTONERGIC EFFECTS ON CORTICAL NEURONAL ACTIVITY The responses of cortical neurons to microiontophoretically applied serotonin have been much less extensively evaluated than for noradrenaline. Neuronal responses to iontophoretic application of serotonin have been found in virtually every brain region tested. Depending on the region tested, the responses of unidentified neurons can be either excitation or depression. However, in areas known to receive a dense and uniform serotonergic innervation, the most common response to iontophoretically applied serotonin (reviewed in Bloom et al 1972, Nelson et al 1973) or electrical stimulation of the median raphe (e.g. Wang & Aghajanian 1977) is marked inhibition of spontaneous activity. Particularly in cerebral cortex, the variability of serotonergic effects (excitation versus inhibition) has led to much controversy and questioning of methods. However, more recent studies suggest that such discrepancies may be due to the unconventional nature of serotonergic effects. It has been suggested that serotonin, like noradrenaline, may have primarily a "modulatory" role, rather than a classical inhibitory or excitatory action.

The most extensive evidence for a modulatory action of serotonin comes from the work of Aghajanian and co-workers in the facial motor nucleus (reviewed in Aghajanian 1981). Iontophoretically applied serotonin failed to excite motoneurons in the facial nucleus over a large dose range, but small amounts of serotonin were shown to markedly facilitate the excitatory effects of both threshold and subthreshold doses of iontophoretically applied glutamate and the excitation evoked by stimulation of trigeminal afferents (McCall & Aghajanian 1979, 1980).

Although substantial interlocking evidence exists to suggest a potent neuromodulatory role for noradrenaline in the neocortex, however, little evidence exists to substantiate or dismiss a similar role for serotonin.

Hypothesis

Strong analogies between the LC-noradrenergic system and the raphe-serotonergic system suggest that the system may perform similar but distinct functions in the neocortex. Single-cell recordings in freely moving animals indicate that the spontaneous firing of both LC-noradrenergic and raphe-serotonergic neurons fluctuates as a function of sleep-wake cycle, level of arousal, and phasic sensory stimulation. However, raphe neurons exhibit activity that is most strongly correlated with behavioral "state," whereas LC-noradrenergic neurons exhibit more phasic fluctuations, associated with "attentiveness." Both systems in turn innervate vast areas of neocortex via widely ramifying axonal projections. At the gross morphological level, both systems exhibit parallel trajectories and regions of innervation. However, more detailed analyses of their laminar distributions suggest that noradrenergic and serotonergic fibers may terminate on different cell types, or at least on distinctive portions of the same cell type. Taken together, these observations suggest that the raphe-serotonergic system may serve a similar, yet distinct (and perhaps complementary) role in cortical information processing from that proposed for the LC-noradrenergic system; namely, alteration of cortical neuronal responses to afferent input in response to changes in state. Since serotonergic innervation of the neocortex is even more dense than noradrenergic innervation, serotonin may have even more profound, or more widespread, effects on neocortical neuronal activity than does noradrenaline. A major question yet to be addressed is whether serotonin also plays a modulatory role in the neocortex, since it does play a modulatory role on target neurons in other brain areas.

DOPAMINERGIC INNERVATION OF NEOCORTEX

Anatomy: Source Cells and Termination Patterns of Neocortical Afferents

The first evidence for the existence of a dopaminergic projection to the neocortex was obtained through biochemical studies of dopamine synthesis in the neocortex of rats (Thierry et al 1973). This study was followed by histochemical confirmation of a dopaminergic projection to the cortex (Hökfelt et al 1974, 1977, Lindvall et al 1974) and the demonstration that the projection originated in the mesencephalon (Lindvall et al 1974, 1978, Fuxe et al 1974). The anatomic organization of the dopaminergic projection to the neocortex has been analyzed in great detail in the rat (see Lindvall & Bjorklund 1984 for review). This innervation originates exclusively from the substantia nigra–ventral tegmental (SN/VTA) area cell groups and, until recently (see below), was thought to be restricted to four discrete terminal fields: a prefrontal dopaminergic projection that is subdivided into (*a*) anteromedial and (*b*) suprarhinal terminal fields, (*c*) a supragenual terminal field that is coincident with the anterior cingulate cortex, and (*d*) a perirhinal terminal field. Specific laminar patterns of termination and topographically restricted cell bodies of origin exist for each terminal field. For example, the terminal fibers of the anterormedial and suprarhinal systems are primarily directed at layers V and VI and originate from the medial and dorsolateral ventral tegmental area (VTA), respectively.

The supragenual dopaminergic system is restricted to the anterior cingulate cortex, where fine terminal fibers are present in layers I–III. Unlike the prefrontal systems described above, the supragenual system of fibers originates largely from substantia nigra neurons, with a possible minor contribution from the lateral VTA (Swanson 1982), which is more likely to be related to the caudal extension of the anteromedial terminal field.

Although it is clear that the SN/VTA complex projects topographically to the four major neocortical terminal fields in rat, certain aspects of the relationship between cell bodies of origin and terminal fields in the ascending dopaminergic systems remain controversial. For example, the degree of collateralization of the dopamine fibers remains a controversial issue. Swanson (1982) maintains that a given dopamine cell has one major target, even though the cell bodies of origin of two discrete projections may be in the same portion of the cell group. However, Fallon & Loughlin (1982) maintain that single neurons in the central portion of the substantia nigra and VTA project to multiple targets, whereas the more peripherally situated neurons tend to project to a single target area. The issues of topographic organization (dopaminergic vs. nondopaminergic cortically projecting cells) and the degree of collateralization will be critically important to a characterization of this system in the primate, where we suspect that the terminal fields in the cortex are far more diverse and widespread than in the four target regions described above. In the monkey, Porrino & Goldman (1982) demonstrated a topographically organized projection from VTA to various regions of the frontal cortex (Porrino & Goldman

1982); however, the cells bodies of origin of the terminal fields outside of the frontal lobe have not been analyzed, and the Porrino & Goldman study did not differentiate dopaminergic from nondopaminergic projection neurons in VTA. In addition, no data are available on the collateralization of the dopamine neurons in primate VTA.

In addition, recent data suggest that the dopaminergic innervation of rat cortex may be more widespread than the four terminal regions described above. Recent anti-tyrosine-hydroxylase immunohistochemical studies (Berger et al 1985) suggest that additional, less densely innervated, regions exist. Berger and her colleagues demonstrated convincingly that well-defined portions of motor and visual association cortices receive a dopaminergic projection. In addition, evidence exists for a dopaminergic projection to the visual cortex in cat (Tork & Turner 1981).

It is difficult to accurately determine the regions in primate cortex that might be directly homologous to the neocortical regions in the rat that receive a dopaminergic projection. Biochemical studies (Bjorklund et al 1978, Brown & Goldman 1977, Brown et al 1979) demonstrated that dopamine levels as well as dopamine/noradrenaline ratios were elevated in various frontal and temporal regions of the monkey cortex. These careful biochemical studies offered the first evidence that the dopaminergic projection in the primate was likely to involve numerous discrete regions of association cortex within the frontal and temporal lobes, as well as the primary motor cortex, and that the density of innervation was likely to be highly region-specific. Based on endogenous dopamine levels in carefully dissected regions, Brown & Goldman (1977) proposed that within the frontal lobe, the densest dopaminergic innervation was in the prefrontal cortex and a rostrocaudal gradient of decreasing density existed. In fact, the rostrocaudal gradient could be extended into the parietal and occipital lobes, with low levels in parietal lobe relative to frontal, and only trace amounts of dopamine present in occipital cortex. Bjorklund et al (1978) relied more heavily on dopamine/noradrenaline ratios in their interpretation and supported the notion of a rostrocaudal gradient within the frontal lobe. Anatomic analysis of the dopaminergic system in primate neocortex has been hampered by the fact that the fluorescence histochemical method does not distinguish between noradrenaline and dopamine fibers. Immunohistochemical studies using antibodies to tyrosine-hydroxylase (TH) potentially suffer from the same problem. However, there are reports that TH levels are below the levels necessary for immunohistochemical detection in noradrenaline fibers. For example, this appears to be the case for the anti-TH that we have used (Lewis et al 1987) to study primate neocortex.

Levitt et al (1984) have completed an extensive fluorescence histochemical analysis of primate neocortex, in which they differentiated dopamine from noradrenaline fibers on the basis of morphological characteristics. Fibers with "dopamine-like characteristics" were present only in the frontal and temporal lobes, and were most numerous in the prefrontal cortex, precisely the region exhibiting the highest endogenous levels of dopamine. In addition, dopamine-like fibers were present in the motor cortex, thus supporting the earlier prediction of a dopaminergic innervation of motor cortex that was based on biochemical data. In frontal and cingulate cortex, most of the dopamine-like fibers were present in layers II and III; however, no details on the laminar or regional distribution of the dopaminergic innervation of temporal lobe were given, and no dopamine-like fibers were seen in the parietal or occipital lobes. The most important finding of Levitt and co-workers was the anatomic confirmation that in the primate, the dopamine system extended well beyond the medial, prefrontal, and perirhinal areas characterized in the rat, and was likely to innervate vast areas of frontal and temporal association cortices. Our preliminary data using anti-TH support this contention and, in fact, suggest that the dopaminergic innervation extends throughout all four lobes of the neocortex, with dramatic regional variations in density that coincide with cytoarchitectonic and functional boundaries (Lewis et al 1987). We have carefully compared our anti-DBH (dopamine-beta-hydroxylase) staining patterns with anti-TH staining patterns, and have analyzed the anti-TH staining pattern in monkeys in which the ascending noradrenergic projection to cortex has been ablated. Our immunohistochemical findings suggest that the differentiation of dopamine from noradrenaline fibers based on morphologic grounds in the histochemical studies of Levitt et al (1984) must have been misleading in several cases. Our data suggest that extensive regional heterogeneity of the dopaminergic innervation exists within frontal cortical areas, such that the variations in density are more complex than a simple rostrocaudal gradient, and that the primary motor cortex has the densest dopaminergic innervation of any frontal area. Also, there is an extensive dopaminergic innervation of the inferior parietal lobe (area 7), such that the innervation of this area is denser than several prefrontal areas (see Figures 1 and 2). Primary visual, auditory, and somatosensory cortices all exhibit a very sparse dopaminergic innervation. For all three modalities, density is significantly higher in related association areas. The laminar pattern of fibers in a given region is correlated with fiber density. In very sparsely innervated regions, dopamine fibers are limited to layer I, and areas of intermediate density have dopamine fibers in layers I, superficial III, and deep V–VI. The primary motor cortex displays fibers in all laminae. Throughout the neocortex, layer IV has the lowest density of innervation. In summary distribution patterns reveal a

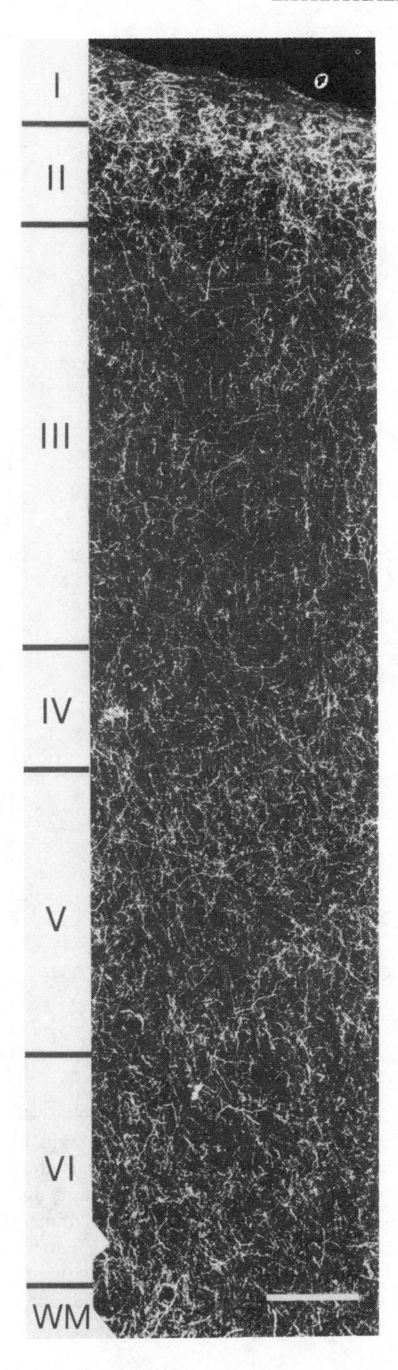

Figure 1 Darkfield photomicrograph of dopamine fibers visualized using TH immunohistochemistry. This image is from a coronal 40 μm section through area 7, the inferior parietal lobule, of a cynomolgus monkey. The numerals at *left* indicate the cortical laminae. Calibration bar = 200 μm. WM = white matter. From Lewis et al (1987).

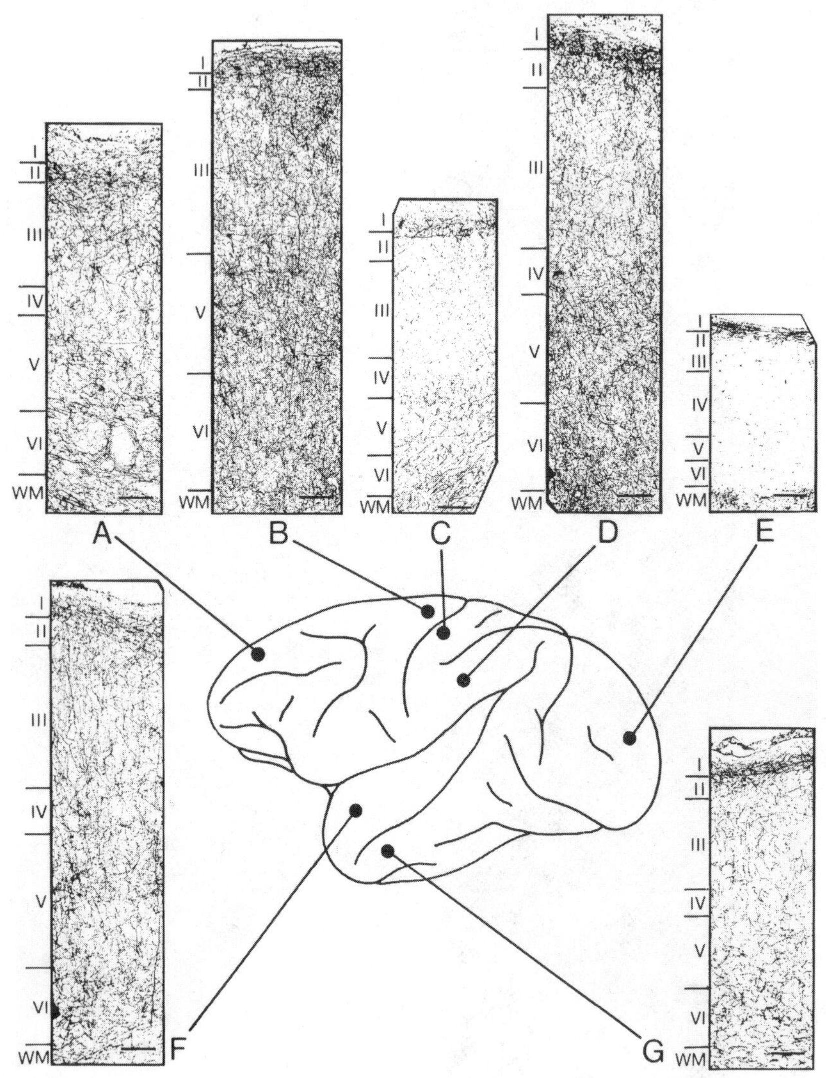

Figure 2 Reverse-image photographic reproductions of darkfield photomicrographs of dopamine fibers revealed by TH immunohistochemistry in various cytoarchitectonic areas of cynomolgus monkey cortex. Note the extensive regional heterogeneity in the density and laminar distribution of labeled fibers. *A*, dorsomedial prefrontal cortex (area 9); *B*, primary motor cortex (area 4); *C*, primary somatosensory cortex (area 3); *D*, posterior parietal cortex (area 7); *E*, primary visual cortex (area 17); *F*, rostral superior temporal gyrus; *G*, rostral inferior temporal gyrus. Calibration bars equal 200 μm. From Lewis et al (1987).

functional specialization of the dopaminergic innervation of primate cortex such that fibers preferentially innervate motor relative to sensory regions, sensory association relative to primary sensory areas, and auditory association relative to visual association areas.

Physiology

ACTIVITY OF SOURCE NEURONS There have been a limited number of studies of the discharge activity of SN/VTA neurons in unanesthetized, behaving animals. In cat, these neurons discharge more rapidly during active than quiet waking, but do not further decrease their activity during sleep (Trulson et al 1981, Steinfels et al 1983, see also Miller et al 1983). Brief excitatory and inhibitory responses to phasic auditory or visual stimuli are observed during quiet waking. Various stressful and arousing stimuli do not alter the discharge rates of substantia nigra neurons, although phasic responses to neutral auditory or visual stimuli are blocked by such manipulations (Strecker & Jacobs 1985). The most striking change in activity is a prolonged suppression of activity that accompanies orientation toward and fixation of a novel or meaningful stimulus in the environment (Steinfels et al 1983). In monkeys, substantia nigra dopamine neurons show little relation to the phasic movements or other aspects of an operant paradigm (DeLong et al 1983). It is not known whether the subset of dopamine neurons projecting to neocortex exhibits any of these properties.

EFFECTS OF DOPAMINERGIC SN/VTA ON CORTICAL NEURONAL ACTIVITY In iontophoretic tests, dopamine has been found to be inhibitory on neurons in the frontal and cingulate cortices. Bunney & Aghajanian (1976) observed in rat prefrontal cortex that cells in layers II and III are more sensitive to the inhibitory effects of noradrenaline than those of dopamine on spontaneous activity. The converse was true for cells in layers V and VI. This corresponds to the preferential innervation of superficial layers by noradrenaline fibers and of deep layers by dopamine. Ferron et al (1984) observed that stimulation in the region of dopamine cells projecting to neocortex in the anesthetized rat blocked the excitatory effect on cortical neurons of thalamic stimulation. In the orbitofrontal cortex of behaving monkeys, Aou et al (1983) observed that those cells most sensitive to microiontophoretically applied noradrenaline decreased their activity during a food-acquisition behavior whereas those most sensitive to dopamine increased their activity during this behavior.

Hypothesis

In both rodent and primate, the dopaminergic innervation of neocortex exhibits a far greater degree of regional heterogeneity than either the noradrenergic or serotonergic systems, with the highest densities occurring in limbic and association cortices. Laminar and regional patterns of innervation suggest that the dopaminergic system is in a position to influence the activity of corticocortical rather than thalamocortical circuits and higher-order integrative processes rather than the more analytic aspects of sensory processing. In addition, this system is likely to be involved in some aspect of cortical regulation of motor control and associated functions in the frontal lobe. It will be of interest to determine whether cortically projecting SN/VTA neurons respond to environmental and behavioral manipulations differently from dopamine neurons projecting to the basal ganglia. It may be that the mesocortical system exhibits quite different properties that have yet to be described.

CHOLINERGIC INNERVATION OF THE NEOCORTEX

Anatomy: Source and Termination Patterns of Neocortical Afferents

Putative acetylcholinergic (ACh) axons in the neocortex have been visualized by acetylcholinesterase (AChE) histochemistry, AChE immunohistochemistry, and choline acetyltransferase (ChAT) immunohistochemistry (see Wainer et al 1984a for review). The ACh extrathalamic innervation of rat neocortex appears to be widespread and to arise primarily from the nucleus basalis and portions of the diagonal band, although ACh neurons in other sites contribute a minor portion of this cortical projection (Lehmann et al 1980, Bigl et al 1982, Mesulam et al 1983). ChAT fibers in the motor cortex are distributed with approximately equal density through all cortical layers, while in somatic sensory cortex there is an increased density in layer V and a decreased density in layer IV (Houser et al 1985). There is general agreement that ChAT neurons are intrinsic to the neocortex. These cells are bipolar and are found in layers II–VI, with a slightly higher density in layers II and III (Eckenstein & Thoenen 1983, Houser et al 1983, 1985, Levey et al 1984), but lesion and biochemical data indicate that such intrinsic innervation probably constitutes a small fraction of the ACh innervation of neocortex. Double-labeling studies indicate that cortically projecting ACh neurons exhibit only limited collateralization, and a coarse topographic relationship exists between the locations of cells of origin and the locus of cortical termination (Bigl et al 1982, McKinney et al 1983). Perhaps source cells receive input from those cortical regions to which they project (Saper 1984). ChAT synapses have been described as predominantly symmetrical and occurring on dendritic shafts and spines in the cingulate and entorhinal cortices (Wainer et al 1984b) as well as in motor and somatosensory cortices (Houser et al 1985).

In area 17 of the cat (Bear et al 1985) AChE histochemistry reveals heavily stained pyramidal cells in layer V, as well as a network of fibers with striking laminar variations in density. Experimental manipulations indicate that the reactive fibers arise from the ipsilateral basal forebrain: Undercutting eliminates virtually all AChE fibers; injection of HRP into area 17 yields retrogradely labeled neurons in the basal forebrain; and basal forebrain lesions substantially reduce the density of AChE fibers. The projection appears to be strictly ipsilateral. These observations on innervation patterns (see Figure 3) have been verified with ChAT monoclonal antibodies (Stichel & Singer 1985). No ChAT+ neuronal profiles are observed in area 17, of the cat (Stichel & Singer 1985).

It has not yet proved possible to reliably visualize ChAT-immunoreactive fibers in the monkey cortex. Cortical fibers have been visualized with AChE histochemistry (Mesulam et al 1984) and AChE immunohistochemistry (Hedreen et al 1984). With histochemistry, regional specializations are evident in terms of laminar distribution and density of fibers. Primary visual, auditory, and somatosensory cortices contain a distinctive band of fine processes in layer IV. Motor and premotor cortices exhibit prominent, radially oriented fibers in deep layers. Association cortices contain the lowest density of reactive fibers of all cortical areas (Mesulam et al 1984). With AChE immunohistochemistry (Hedreen et al 1984), area 17 of the monkey (see Figure 4) exhibits enhanced fiber density in laminae in which lateral geniculate afferents terminate. Some moderately stained neuronal profiles are evident in layer VIB. Regional heterogeneity is also evident in monkeys when regional concentrations of ChAT are determined (Lehmann et al 1984).

In monkey, cortical ACh fibers appear to arise from the nucleus basalis of Meynert and, to a lesser extent, neurons within the diagonal band of Broca. Different subdivisions of the nucleus basalis project preferentially to major regions of the cortical mantle (Mesulam et al 1983). The topographic organization of this projection is more readily evident in primate than in rat (Rye et al 1984). In monkey, cortically projecting ACh neurons appear to receive input from a limited set of cortical and subcortical sites (Mesulam & Mufson 1984, Russchen et al 1985). Cortical input arises from prepyriform, orbitofrontal, anterior insular, temporal pole, entorhinal, and medial temporal areas, while subcortical afferents arise from septal nuclei, amygdala, nucleus accumbens–ventral pallidum complex, and the hypothalamus.

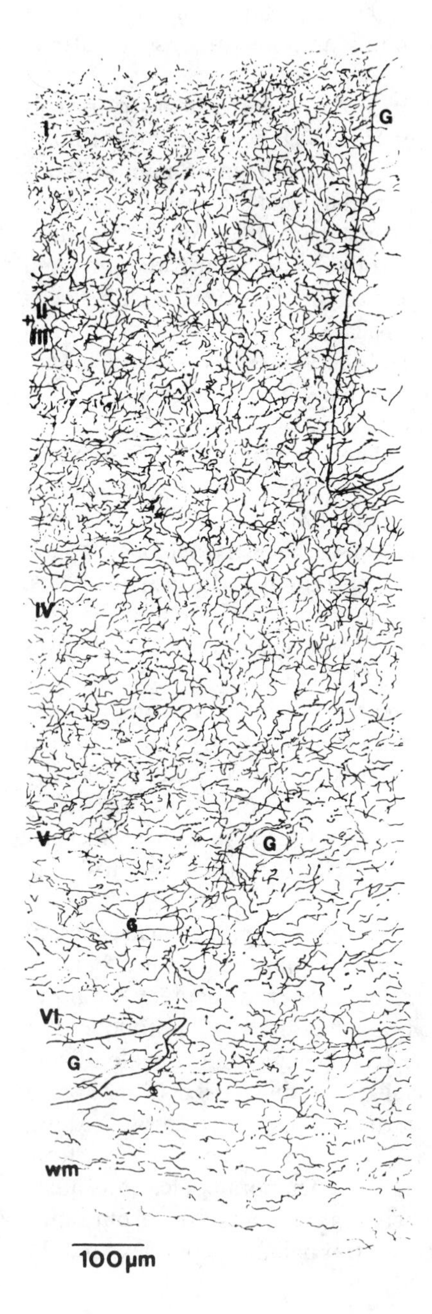

Figure 3 Camera lucida drawing of ChAT-immunoreactive fibers in a coronal section through area 17 of an adult cat. G = blood vessel. Courtesy of C. C. Stichel and W. Singer.

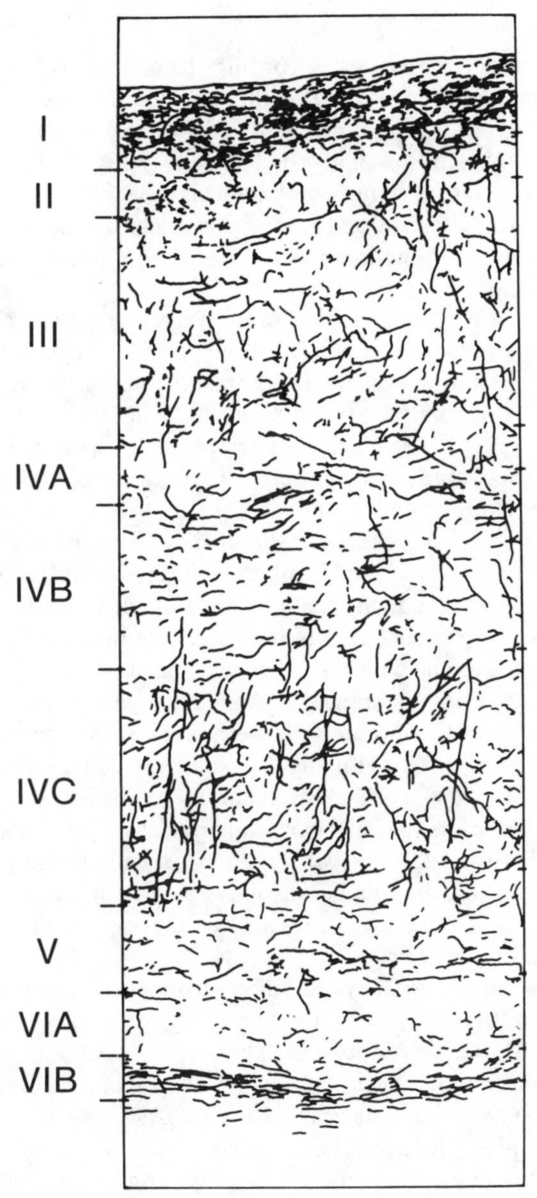

Figure 4 Drawing of AChE-immunoreactive fibers in an 8 μm section through area 17 of cynomolgus monkey. Laminar boundaries in the Broadmann system are indicated along the left-hand edge. From Hedreen et al (1984). Reprinted with permission.

Physiology

ACTIVITY OF SOURCE NEURONS Recordings from nucleus basalis neurons in behaving monkeys have revealed that the neurons are most active in situations involving the sight and taste of food rewards (DeLong 1971, Burton et al 1975, Rolls et al 1979). The desirability of the food reward and the animal's state of hunger both influence the intensity of activity. In the operant paradigms utilized in these studies, the neurons were not activated reliably as a function of simple sensory or motor variables.

CONDUCTION PROPERTIES OF AFFERENT AXONS Cortically projecting nucleus basalis neurons have been identified during single cell recordings in the nucleus basalis region by antidromic activation from frontal and parietal cortices in rats (Aston-Jones et al 1984a, 1985a) and monkeys (Aston-Jones et al 1984b). Such cells are physiologically heterogeneous, exhibiting a variety of rates and patterns of spontaneous discharge. A wide range of conduction latencies is also observed (1–26 msec for rat frontal cortex). One outstanding characteristic of these neurons in both rat and monkey is the tendency to exhibit multiple, discrete antidromic latencies, as a function of stimulus intensity and stimulation depth within the cortex. These authors interpret such results as evidence for pronounced branching of nucleus basalis axons within local cortical terminal fields. Conversely, there was no evidence of branching from a single neuron to innervate different cortical fields, since no cells were found to be driven from both frontal and parietal cortices. Finally, by examining conduction velocity as a function of depth of cortical stimulation, Aston-Jones et al (1985a) were able to determine that intracortically nucleus basalis fibers conduct impulses at about 0.3–0.8 m/s, whereas subcortically the impulse travels at 1.8–3.4 m/s. Thus, nucleus basalis fibers may be myelinated subcortically, but they lose their myelin sheaths as they approach their targets within cortical gray matter.

The nucleus basalis–ACh system differs from other extrathalamic cortical afferents in that these neurons form a physiologically heterogeneous population, with wide variations in spontaneous discharge, spike waveform, and conduction latencies to cortex. The functional significance of this heterogeneity remains unclear. However, Aston-Jones et al (1985a) note that the physiologically homogeneous noradrenaline-LC neurons appear to have more divergent efferent projections, such that individual neurons are found to project to widely separated brain areas. They speculate that the more restricted terminal fields of nucleus basalis–ACh neurons may correspond to their physiologic heterogeneity, such that more restricted target areas are differentially controlled by individual neurons to a greater extent than in the LC system.

NUCLEUS BASALIS–ACh EFFECTS ON CORTICAL NEURONAL ACTIVITY ACh has usually been found to have excitatory effects on cortical neurons when iontophoretically applied (e.g. Krnjevic & Phillis 1963, Krnjevic et al 1971, Spehlmann 1969, Spehlmann et al 1971, Foote et al 1975). It is of interest to note that the application of ACh to auditory cortex neurons in the unanesthetized monkey results in a large increase in spontaneous activity without a corresponding increase in stimulus-elicited activity (Foote et al 1975). Thus, the net effect of ACh is opposite that of noradrenaline, that is the signal-to-noise ratio of sensory responses is decreased. However, Inoue et al (1983), recording from monkey dorsolateral prefrontal cortex during bar-press feeding behavior, observed that the continuous iontophoretic application of ACh enhanced phasic neuronal responses to various aspects of the task. Scopolamine diminished task-related activity. Electrical stimulation of the nucleus basalis produced driven activity in cortical neurons, which was abolished by iontophoretic application of atropine.

Hypothesis

As noted by Mesulam & Mufson (1984), in monkey the subcortical afferents to nucleus basalis–ACh neurons projecting to neocortex are limbic and paralimbic. The information characteristic of these structures may be transmitted to neocortex by this ACh afferent system. Also, because the cortical regions projecting to the nucleus basalis complex are much more restricted than those regions receiving input from it, the cortical structures afferent to the nucleus basalis can control the ACh input to themselves and to other cortical structures. These authors speculate that this afferent system may be responsible for conveying information concerning relationships between complex environmental events and the internal milieu to many cortical areas involved in many different, specific functions. The possible heterogeneity of function may correspond to the physiological heterogeneity of the neurons described by Aston-Jones et al (1985a).

EXTRATHALAMIC MODULATION OF CORTICAL FUNCTION

What are the implications of our knowledge of the cellular anatomy and physiology of these extrathalamic systems for speculations concerning their roles in normal and abnormal brain function?

1. The source neurons for these systems generally reside outside the major sensory and motor pathways of the brain. However, the efferent pathways of each system innervate primary and secondary sensory and motor structures. Thus, each system probably plays a role in influencing such activities, probably by imposing state-dependent effects onto these highly topographic systems. The dopaminergic and ACh systems may well exert these influences with a greater degree of topographic specificity than do the noradrenergic and serotonergic systems.

2. The electrophysiologic effects of these putative transmitters on postsynaptic neurons also indicate a modulatory role. The present data suggest that these effects constitute well-defined alterations in the electrophysiologic properties of target cells that result in specific alterations of their operating characteristics. For example, the noradrenergic system appears to enhance the signal-to-noise characteristics of sensory neurons in many brain regions.

3. Electrophysiologic studies of the activity of extrathalamic source neurons indicate that noradrenaline neurons exert their effects in both a tonic and a phasic fashion; they are more active during waking but also exhibit bursts of activity during episodic increases in attentiveness during waking. Serotonin neurons appear to be more responsive to the sleep-wake cycle per se and might well initiate and maintain more tonic effects of behavioral state on target neuron function. The activity of dopamine and ACh neurons has been less extensively studied in behaving animals. Dopamine neurons appear to be involved in orienting behaviors, while ACh neurons are most active during relatively specific conditions involving motivated, or perhaps emotional, behaviors.

4. Although all three monoamines innervate the neocortex, each of the four clearly has preferred regions and laminae of termination. This suggests that their effects on the neocortex are not generalized excitation or inhibition but rather region-specific enhancement or diminution of activity in limited neuronal ensembles during certain stages of information processing.

5. Finally, because each monoamine innervates diverse functional systems, one would not expect a simple correlation of "one transmitter–one behavior." In both normal and abnormal states, each of these transmitters must influence a variety of behaviors.

The studies aimed at determining the behavioral functions of extrathalamic neocortical afferents have been limited in number (e.g. Brozoski et al 1979, Arnsten & Goldman-Rakic 1985). Such studies must be viewed as initial attempts to approach a very complex problem, but it is of interest that limited lesions and pharmacological manipulations of these systems can produce profound impairments in carefully controlled behavioral paradigms.

ACKNOWLEDGMENTS

Gary Aston-Jones provided essential assistance in evaluating available data on axonal conduction properties. The work reported here from our laboratories was supported by USPHS Grants MH40008 (S.L.F.), NS21384 (S.L.F.), AA06420 (S.L.F., J.H.M.), AG015131 (J.H.M.), and the MacArthur Foundation (S.L.F., J.H.M.).

Literature Cited

Aghajanian, G. K. 1981. The modulatory role of serotonin at multiple receptors. In *Serotonin Neurotransmission and Behavior*, ed. B. J. Jacobs, A. Gelperin, pp. 156–85. Cambridge: MIT Press

Aou, S., Oomura, Y., Nishino, H., Inokuchi, A., Mizuno, Y. 1983. Influence of catecholamines on reward-related neuronal activity in monkey orbitofrontal cortex. *Brain Res.* 267: 165–70

Arnsten, A. F. T., Goldman-Rakic, P. S. 1985. Alpha-2-adrenergic mechanisms in prefrontal cortex associated with cognitive decline in aged nonhuman primates. *Science* 205: 1273–76

Aston-Jones, G., Bloom, F. E. 1981a. Activity of norepinephrine-containing locus coeruleus neurons in behaving rats anticipates fluctuations in the sleep-waking cycle. *J. Neurosci.* 1: 876–86

Aston-Jones, G., Bloom, F. E. 1981b. Norepinephrine-containing locus coeruleus neurons in behaving rats exhibit pronounced responses to non-noxious environmental stimuli. *J. Neurosci.* 1: 887–900

Aston-Jones, G., Foote, S. L., Bloom, F. E. 1984c. Anatomy and physiology of locus coeruleus neurons: Functional implications. In *Norepinephrine*, ed. M. G. Ziegler, C. R. Lake, pp. 92–116. Baltimore: Williams & Wilkins

Aston-Jones, G., Foote, S. L., Segal, M. 1985b. Impulse conduction properties of noradrenergic locus coeruleus axons projecting to monkey cerebrocortex. *Neuroscience* 15: 765–77

Aston-Jones, G., Rogers, J., Grant, S., Ennis, M., Shaver, R., Bartus, R. 1984b. Physiology of cortically projecting neurons in monkey nucleus basalis of Meynert. *Soc. Neurosci. Abstr.* 10: 808

Aston-Jones, G., Shaver, R., Dinan, T. 1984a. Cortically projecting nucleus basalis neurons in rat are physiologically heterogeneous. *Neurosci. Lett.* 46: 19–24

Aston-Jones, G., Shaver, R., Dinan, T. 1985a. Nucleus basalis neurons exhibit axonal branching with decreased impulse conduction velocity in rat cerebrocortex. *Brain Res.* 325: 271–85

Bear, M. F., Carnes, K. M., Ebner, F. F. 1985. An investigation of cholinergic circuitry in cat striate cortex using acetylcholinesterase histochemistry. *J. Comp. Neurol.* 234: 411–30

Berger, B., Verney, C., Alvarez, C., Vigny, A., Helle, K. B. 1985. New dopaminergic terminal fields in the motor, visual (area 18b) and retrosplenial cortex in the young and adult rat. Immunocytochemical and catecholamine histochemical analyses. *Neuroscience* 15: 983–98

Bigl, V., Woolf, N. J., Butcher, L. L. 1982. Cholinergic projections from the basal forebrain to frontal, parietal, temporal, occipital, and cingulate cortices: A combined fluorescent tracer and acetylcholinesterase analysis. *Brain Res. Bull.* 8: 727–49

Bjorklund, A., Divac, I., Lindvall, O. 1978. Regional distribution of catecholamines in monkey cerebral cortex, evidence for a dopaminergic innervation of the primate prefrontal cortex. *Neurosci. Lett.* 7: 115–19

Bloom, F. E. 1979. Chemical integrative processes in the central nervous system. In *The Neurosciences: Fourth Study Program*, ed. F. O. Schmitt, F. G. Worden, pp. 51–58. Cambridge: MIT Press

Bloom, F. E., Hoffer, B. J., Siggins, G. R., Barker, J. L., Nicoll, R. A. 1972. Effects of serotonin on central neurones: Microiontophoretic administration. *Fed. Proc.* 31: 97–106

Brown, R. M., Crane, A. M., Goldman, P. S.

1979. Regional distribution of monamines in the cerebral cortex and subcortical structures of the rhesus monkey: Concentrations and in vivo synthesis rates. *Brain Res.* 168: 133–50

Brown, R. M., Goldman, P. S. 1977. Catecholamines in neocortex of rhesus monkeys: Regional distribution and ontogenetic development. *Brain Res.* 124: 576–80

Brozoski, T. J., Brown, R. M., Rosvold, H. E., Goldman, P. S. 1979. Cognitive deficit caused by regional depletion of dopamine in prefrontal cortex of rhesus monkey. *Science* 205: 929–32

Bunney, B. S., Aghajanian, G. K. 1976. Dopamine and norepinephrine innervated cells in the rat prefrontal cortex: Pharmacological differentiation using microiontophoretic techniques. *Life Sci.* 19: 1783–92

Burton, M. J., Mora, F., Rolls, E. T. 1975. Visual and taste neurones in the lateral hypothalamus and substantia innominata: Modulation of responsiveness by hunger. *J. Physiol.* 252: 50–51

DeLong, M. R., Crutcher, M. D., Geogopoulos, A. P. 1983. Relations between movement and single cell discharge in the substantia nigra of the behaving monkey. *J. Neurosci.* 3: 1599–1606

DeLong, M. R. 1971. Activity of pallidal neurons during movement. *J. Neurophysiol.* 34: 414–27

Dismukes, R. K. 1979. New concepts of molecular communication among neurons. *Behav. Brain Sci.* 2: 409–48

Eckenstein, F., Thoenen, H. 1983. Cholinergic neurons in the rat cerebral cortex demonstrated by immunohistochemical localization of choline acetyltransferase. *Neurosci. Lett.* 36: 211–15

Fallon, J. H., Loughlin, S. E. 1982. Monoamine innervation of the forebrain: Collateralization. *Brain Res. Bull.* 9: 295–307

Ferron, A., Thierry, A. M., LeDouarin, C., Glowinski, J. 1984. Inhibitory influence of the mesocortical dopaminergic system on spontaneous activity or excitatory response induced from the thalamic mediodorsal nucleus in the rat medial prefrontal cortex. *Brain Res.* 302: 257–65

Foote, S. L., Aston-Jones, G., Bloom, F. E. 1980. Impulse activity of locus coeruleus neurons in awake rats and monkeys is a function of sensory stimulation and arousal. *Proc. Natl. Acad. Sci.* 77: 3033–37

Foote, S. L., Bloom, F. E., Aston-Jones, G. 1983. The nucleus locus coeruleus: New evidence of anatomical and physiological specificity. *Physiol. Rev.* 63: 844–914

Foote, S. L., Freedman, R., Oliver, A. P. 1975. Effects of putative neurotransmitters on neuronal activity in monkey auditory cortex. *Brain Res.* 86: 229–42

Foote, S. L., Morrison, J. H. 1984. Postnatal development of laminar innervation patterns by monoaminergic fibers in *Macaca fascicularis* primary visual cortex. *J. Neurosci.* 4: 2667–80

Foote, S. L., Morrison, J. H. 1987. Development of the noradrenergic, serotonergic, and dopaminergic innervation of neocortex. *Curr. Top. Dev. Biol.* In press

Fuxe, K., Hökfelt, T., Johansson, O., Jonsson, G., Lidbrink, P., Ljungdahl, A. 1974. The origin of the dopamine nerve terminals in limbic and frontal cortex. Evidence for meso-cortico dopamine neurons. *Brain Res.* 82: 349–55

Hedreen, J. C., Uhl, G. R., Bacon, S. J., Fambrough, D. M., Price, D. L. 1984. Acetylcholinesterase-immunoreactive axonal network in monkey visual cortex. *J. Comp. Neurol.* 226: 246–54

Hökfelt, T., Johansson, O., Fuxe, K., Goldstein, M., Park, D. 1977. Immunohistochemical studies on the localization and distribution of monoamine neuron systems in the rat brain. II. Tyrosine hydroxylase in the telencephalon. *Med. Biol.* 55: 21–40

Hökfelt, T., Ljungdahl, A., Fuxe, K., Johansson, O. 1974. Dopamine nerve terminals in the rat limbic cortex: Aspects of the dopamine hypothesis of schizophrenia. *Science* 184: 177–79

Houser, C. R., Crawford, G. D., Barber, R. P., Salvaterra, P. M., Vaughn, J. E. 1983. Organization and morphological characteristics of cholinergic neurons: An immunocytochemical study with a monoclonal antibody to choline acetyltransferase. *Brain Res.* 266: 97–119

Houser, C. R., Crawford, G. D., Salvaterra, P. M., Vaughn, J. E. 1985. Immunocytochemical localization of choline acetyltransferase in rat cerebral cortex: A study of cholinergic neurons and synapses. *J. Comp. Neurol.* 234: 17–34

Inoue, M., Oomura, Y., Nishino, H., Aou, S., Sikdar, S. K., Hynes, M., Mizuno, Y., Katabuchi, T. 1983. Cholinergic role in monkey dorsolateral prefrontal cortex during bar-press feeding behavior. *Brain Res.* 278: 185–94

Jacobs, B. L., Heym, J., Steinfels, G. F. 1984. Physiological and behavioral analysis of raphe unit activity. In *Handb. Psychopharmacol.* 18: 343–95

Jones, R. S. G., Olpe, H. R. 1984. Activation of the noradrenergic projection from locus coeruleus reduces the excitatory responses of anterior cingulate cortical neurones to substance P. *Neuroscience* 13: 819–25

Kosofsky, B. E., Molliver, M. E., Morrison,

J. H., Foote, S. L. 1984. The serotonin and norepinephrine innervation of primary visual cortex in the old world monkey (*Macaca fascicularis*). *J. Comp. Neurol.* 230: 168–78

Krnjevik, K., Phillis, J. W. 1963. Acetylcholine-sensitive cells in the cerebral cortex. *J. Physiol.* 166: 296–327

Krnjevik, K., Pumain, R., Renaud, L. 1971. The mechanism of excitation by acetylcholine in the cerebral cortex. *J. Physiol.* 215: 247–68

Lehmann, J., Nagy, J. I., Atmadja, S., Fibiger, H. C. 1980. The nucleus basalis magnocellularis: The origin of a cholinergic projection to the neocortex of the rat. *Neuroscience* 5: 1161–74

Lehmann, J., Struble, R. G., Antuono, P. G., Coyle, J. T., Cork, L. C., Price, D. L. 1984. Regional heterogeneity of choline acetyltransferase activity in primate neocortex. *Brain Res.* 322: 361–64

Levey, A. I., Wainer, B. H., Rye, D. B., Mufson, E. J., Mesulam, M. M. 1984. Choline acetyltransferase-immunoreactive neurons intrinsic to rodent cortex and distinction from acetylcholinesterase-positive neurons. *Neuroscience* 13: 341–53

Levitt, P., Rakic, P., Goldman-Rakic, P. 1984. Region-specific distribution of catecholamine afferents in primate cerebral cortex: Fluorescence histochemical analysis. *J. Comp. Neurol.* 227: 23–36

Lewis, D. A., Campbell, M. J., Foote, S. L., Goldstein, M., Morrison, J. H. 1987. The distribution of tyrosine hydroxylase-immunoreactive fibers in primate neocortex is widespread but regionally specific. *J. Neurosci.* In press

Lidov, H. G. W., Grzanna, R., Molliver, M. E. 1980. The serotonin innervation of the cerebral cortex in the rat—An immunohistochemical analysis. *Neuroscience* 5: 207–27

Lindvall, O., Bjorklund, A. 1984. General organization of cortical monoamine systems. *Neurol. Neurobiol.* 10: 9–40

Lindvall, O., Bjorklund, A., Divac, I. 1978. Organization of catecholamine neurons projecting to the frontal cortex in the rat. *Brain Res.* 142: 1–24

Lindvall, O., Bjorklund, A., Moore, R. Y., Stenevi, U. 1984. Mesencephalic dopamine neurons projecting to neocortex. *Brain Res.* 81: 325–31

Loughlin, S. E., Foote, S. L., Fallon, J. H. 1982. Locus coeruleus projections to cortex: Topography, morphology, and collateralization. *Brain Res. Bull.* 9: 287–94

McCall, R. B., Aghajanian, G. K. 1979. Serotonergic facilitation of facial motoneuron excitation. *Brain Res.* 169: 11–27

McCall, R. B., Aghajanian, G. K. 1980. Pharmacological characterization of serotonin receptors in the facial motor nucleus: A microiontophoretic study. *Eur. J. Pharmacol.* 65: 175–83

McKinney, M., Coyle, J. T., Hedreen, J. C. 1983. Topographic analysis of the innervation of the rat neocortex and hippocampus by the basal forebrain cholinergic system. *J. Comp. Neurol.* 217: 103–21

Mesulam, M. M., Mufson, E. J., Levey, A., Wainer, B. H. 1983. Cholinergic innervation of cortex by the basal forebrain: Cytochemistry and cortical connections of the septal area, diagonal band nuclei, nucleus basalis (substantia innominata), and hypothalamus in the rhesus monkey. *J. Comp. Neurol.* 214: 170–97

Mesulam, M. M., Mufson, E. J. 1984. Neural inputs into the nucleus basalis of the substantia innominata (Ch4) in the rhesus monkey. *Brain* 107: 253–74

Mesulam, M. M., Rosen, A. D., Mufson, E. J. 1984. Regional variations in cortical cholinergic innervation: Chemoarchitectonics of acetylcholinesterase-containing fibers in the macaque brain. *Brain Res.* 311: 245–58

Miller, J. D., Farber, J., Gatz, P., Roffwarg, H., German, D. C. 1983. Activity of mesencephalic dopamine and non-dopamine neurons across stages of sleep and waking in the rat. *Brain Res.* 273: 133–41

Moises, H. C., Waterhouse, B. D., Woodward, D. J. 1983. Locus coeruleus stimulation potentiates local inhibitory processes in rat cerebellum. *Brain Res. Bull.* 10: 795–804

Moises, H. C., Woodward, D. J. 1980. Potentiation of GABA inhibitory action in cerebellum by locus coeruleus stimulation. *Brain Res.* 182: 327–44

Morrison, J. H., Foote, S. L. 1986. Noradrenergic and serotonergic innervation of cortical and thalamic visual structures in old and new world monkeys. *J. Comp. Neurol.* 243: 117–38

Morrison, J. H., Foote, S. L., Bloom, F. E. 1984. Laminar, regional, developmental, and functional specificity of monoaminergic innervation patterns in monkey cortex. *Neurol. Neurobiol.* 10: 61–75

Morrison, J. H., Foote, S. L., Molliver, M. E., Bloom, F. E., Lidov, H. G. 1982. Noradrenergic and serotonergic fibers innervate complementary layers in monkey primary visual cortex: An immunohistochemical study. *Proc. Natl. Acad. Sci.* 79: 2401–5

Morrison, J. H., Foote, S. L., O'Connor, D., Bloom, F. E. 1982b. Laminar, tangential and regional organization of the noradrenergic innervation of monkey cortex: Dopamine-hydroxylase immuno-

histochemistry. *Brain Res. Bull.* 9: 309–19
Morrison, J. H., Grzanna, R., Molliver, M. E., Coyle, J. T. 1978. The distribution and orientation of noradrenergic fibers in neocortex of the rat: An immunofluorescence study. *J. Comp. Neurol.* 181: 17–40
Morrison, J. H., Grzanna, R., Molliver, M. E. 1979. Noradrenergic innervation patterns in three regions of medial cortex: An immunofluorescence characterization. *Brain Res. Bull.* 4: 849–57
Morrison, J. H., Molliver, M. E., Grzanna, R., Coyle, J. T. 1981. The intra-cortical trajectory of the coeruleo-cortical projection in the rat: A tangentially organized cortical afferent. *Neuroscience* 6: 139–58
Nelson, C. N., Hoffer, B. J., Chu, N.-S., Bloom, F. E. 1973. Cytochemical and pharmacological studies on polysensory neurons in the primate frontal cortex. *Brain Res.* 62: 115–33
O'Hearn, E., Molliver, M. E. 1984. Organization of raphe-cortical projections in rat: A quantitative retrograde study. *Brain Res. Bull.* 13: 709–26
Olschowka, J. A., Molliver, M. E., Grzanna, R., Rice, F. L., Coyle, J. T. 1981. Ultrastructural demonstration of noreadrenergic synapses in the central nervous system by dopamine-β-hydroxylase immunocytochemistry. *J. Histochem. Cytochem.* 29: 271–80
Porrino, L. J., Goldman-Rakic, P. S. 1982. Brainstem innervation of prefrontal and anterior cingulate cortex in the rhesus monkey revealed by retrograde transport of HRP. *J. Comp. Neurol.* 205: 63–76
Rolls, E. T., Sanghera, M. K., Roper-Hall, A. 1979. The latency of activation of neurones in the lateral hypothalamus and substantia innominata during feeding in the monkey. *Brain Res.* 164: 121–35
Russchen, F. T., Amaral, D. G., Price, J. L. 1985. The afferent connections of the substantia innominata in the monkey, macaca fascicularis. *J. Comp. Neurol.* 242: 1–27
Rye, D. B., Wainer, B. H., Mesulam, M. M., Mufson, E. J., Saper, C. B. 1984. Cortical projections arising from the basal forebrain: A study of cholinergic and noncholinergic components employing combined retrograde tracing and immunohistochemical localization of choline acetyltransferase. *Neuroscience* 13: 627–43
Saper, C. B. 1984. Organization of cerebral cortical afferent systems in the rat. II. Magnocellular basal nucleus. *J. Comp. Neurol.* 222: 313–42
Saper, C. B. 1985. Organization of cerebral cortical afferent systems in the rat. II. Hypothalamocortical projections. *J. Comp. Neurol.* 237: 21–46
Saper, C. B., Akil, H., Watson, S. J. 1986. Lateral hypothalamic innervation of the cerebral cortex: Immunoreactive staining for a peptide resembling but immunochemically distinct from pituitary/arcuate α-melanocyte stimulating hormone. *Brain Res. Bull.* 16: 107–20
Spehlmann, R. 1969. Acetylcholine facilitation, atropine block of synaptic excitation of cortical neurons. *Science* 165: 404–5
Spehlmann, R., Daniels, J. C., Smathers, C. C. 1971. Acetylcholine and the synaptic transmission of specific impulses to the visual cortex. *Brain* 94: 125–38
Steinfels, G. F., Heym, J., Strecker, R. E., Jacobs, B. L. 1983. Behavioral correlates of dopaminergic unit activity in freely moving cats. *Brain Res.* 258: 217–28
Stichel, C. C., Singer, W. 1985. Organization and morphological characteristics of choline acetyltransferase-containing fibers in the visual thalamus and striate cortex of the cat. *Neurosci. Lett.* 53: 155–60
Strecker, R. E., Jacobs, B. L. 1985. Substantia nigra dopaminergic unit activity in behaving cats: Effect of arousal on spontaneous discharge and sensory evoked activity. *Brain Res.* 361: 339–50
Swanson, L. W. 1982. The projections of the ventral tegmental area and adjacent regions: A combined fluorescent retrograde tracer and immunofluorescence study in the rat. *Brain Res. Bull.* 9: 321–53
Takeuchi, Y., Sano, Y. 1983. Immunohistochemical demonstration of serotonin nerve fibers in the neocortex of the monkey (*Macaca fuscata*). *Anat. Embryol.* 166: 155–68
Thierry, A. M., Stinus, L., Blanc, G., Glowinski, J. 1973. Some evidence for the existence of dopaminergic neurons in the rat cortex. *Brain Res.* 50: 230–34
Tork, I., Turner, S. 1981. Histochemical evidence for a catecholaminergic (presumably dopaminergic) projection from the ventral mesencephalic tegmentum to visual cortex in the cat. *Neurosci. Lett.* 24: 215–19
Trulson, M. E., Preussler, D. W., Howell, G. A. 1981. Activity of substantia nigra units across the sleep-waking cycle in freely moving cats. *Neurosci. Lett.* 26: 183–88
Wainer, B. H., Levey, A. I., Mufson, E. J., Mesulam, M. M. 1984a. Cholinergic systems in mammalian brain identified with antibodies against choline acetyltransferase. *Neurochem. Intl.* 6: 163–82
Wainer, B. H., Bolam, J. P., Freund, T. F., Henderson, Z., Totterdell, S., Smith, A. D. 1984b. Cholinergic synapses in the rat brain: A correlated light and electron

microscopic immunohistochemical study employing a monoclonal antibody against choline acetyltransferase. *Brain Res.* 308: 69–76

Wang, R. Y., Aghajanian, G. E. 1977. Inhibition of neurons in the amygdala by dorsal raphe stimulation: Mediation through a direct serotonergic pathway. *Brain Res.* 120: 85–102

Waterhouse, B. D., Moises, H. C., Woodward, D. J. 1980. Noradrenergic modulation of somatosensory cortical neuronal responses to iontophoretically applied putative neurotransmitters. *Exp. Neurol.* 69: 30–49

Waterhouse, B. D., Moises, H. C., Woodward, D. J. 1981. Alpha-receptor-mediated facilitation of somatosensory cortical neuronal responses to excitatory synaptic inputs and iontophoretically applied aceylcholine. *Neuropharmacology* 20: 907–20

Woodward, D. J., Moises, H. C., Waterhouse, B. D., Hoffer, B. J., Freedman, R. 1979. Modulatory actions of norepinephrine in the central nervous system. *Fed. Proc.* 38: 2109–16

Yeh, H. H., Moises, H. C., Waterhouse, B. D., Woodward, D. J. 1981. Modulatory interactions between norepinephrine and taurine, beta-alanine, gamma-aminobutyric acid and muscimol, applied iontophoretically to cerebellar Prukinje cells. *Neuropharmacology* 20: 549–60

Ann. Rev. Neurosci. 1987. 10: 97–129

VISUAL MOTION PROCESSING AND SENSORY-MOTOR INTEGRATION FOR SMOOTH PURSUIT EYE MOVEMENTS

S. G. Lisberger, E. J. Morris, and L. Tychsen

Division of Neurobiology and Department of Physiology, University of California at San Francisco, San Francisco, California 94143

INTRODUCTION

Much of our motor activity is guided by what we see, with a final goal of grasping or tracking an object. Once intent has been established, the generation of visually guided movement involves processing sensory stimuli and transforming them into commands for activation of the relevant musculature. Because eye movements are simpler than other movements in many ways, the oculomotor system provides an ideal opportunity for investigation of the brain mechanisms subserving visually guided movement. The mechanics of the eyeball are straightforward (Robinson 1964), and the eyeball is not subject to sudden changes in load. In addition, the study of the eye movements is unencumbered by the problems of complex kinematics and dynamics that confound the study of limb motion (e.g. Atkeson & Hollerbach 1985).

Much is now known about the processing of visual inputs and their use in generating smooth pursuit eye movements. The important neural pathways are becoming well known, and a substantial body of psychophysical experiments have provided detailed information about the properties of pursuit. As a result, we know the properties of visual inputs that drive pursuit and we can correlate those properties with neural pathways that may transmit inputs for pursuit. Moreover, we have identified transformations performed in the brain to convert visual inputs into motor

0147–006X/87/0301–0097$02.00

commands. Finally, we can correlate this information with the firing properties and connections of brain cells in structures that are necessary for normal pursuit.

It now seems to be within our grasp to integrate anatomical, physiological, and behavioral results and to develop models that (*a*) include the elements, signals, and transformations identified by biological experimentation, and (*b*) when simulated on the computer, reproduce existing data on the performance of smooth pursuit. In most of our review, we focus on experiments that have provided the facts on which models of pursuit now can be based. In the final section, we discuss a model of pursuit as a means of synthesizing the wealth of recent data.

GENERAL PROPERTIES OF SMOOTH PURSUIT

Pursuit eye movements are most highly developed in primates. Pursuit occurs in response to a moving visual stimulus, and the resulting smooth eye rotation keeps the fovea pointed at the stimulus. Pursuit is different from other kinds of smooth ocular following, in that it allows primates to track a small object accurately, even as it moves across a stationary, patterned background (Collewijn & Tamminga 1984, Kowler et al 1978). This ability distinguishes primates from species such as cats, rabbits, and fish, all of which make smooth eye movements when the entire visual scene is moved, and none of which has strong pursuit systems. Pursuit has been investigated in two species of primates, humans and macaque monkeys. The performance of the two species is qualitatively similar, although monkeys generally respond more strongly to a given visual stimulus (compare Lisberger & Westbrook 1985 with Tychsen & Lisberger 1986a).

Normally, a smoothly moving small target evokes a combination of smooth and saccadic eye movements, as shown in Figure 1A. Although the saccades play an important role in maintaining accurate tracking, we define pursuit as the smooth component of the response. Even early investigators recognized the basic difference between pursuit and saccades (Dodge 1903, Westheimer 1954), and our rather conservative definition of pursuit is now supported by abundant evidence that the smooth and saccadic parts of the response are generated by separate neural systems. Smooth pursuit and saccades are affected differentially by lesions of specific regions of the cerebral cortex (Newsome et al 1985a,b) and cerebellum (Zee et al 1981, Optican & Robinson 1980), have different latencies (Rashbass 1961, Robinson 1965, Fuchs 1967a), and respond to different aspects of the visual stimulus (Rashbass 1961). In addition, single unit recordings in behaving animals have identified several classes of cells that modulate their discharge only in relation to saccades or to pursuit, but not both.

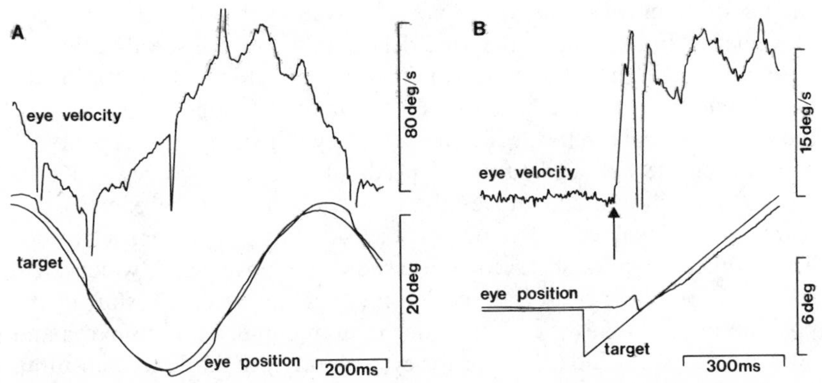

Figure 1 Pursuit eye movements of a monkey in response to sinusoidal target motion at 0.6 Hz, peak-to-peak amplitude 20 deg (*panel A*) and step-ramp target motion at 30 deg/s (*panel B*). The eye velocity records were obtained by electronic differentiation of the eye position records. The rapid deflections in eye velocity are associated with saccadic eye movements. In *panel B*, the *upward arrow* indicates the initiation of pursuit, which preceded the first saccade.

Properties of the Eye Movement

Most investigators have found that the onset of smooth target motion evokes pursuit after latencies of 80 to 130 ms. Pursuit latency can be as short as 65 ms for a small target (Lisberger & Westbrook 1985) and depends on a number of visual parameters, including the target's luminance, size, and initial position in the visual field (Lisberger & Westbrook 1985, Tychsen & Lisberger 1986a).

Pursuit eye movements are most effective when target speed is relatively slow. For ramp target motion, smooth eye velocity is usually unable to keep up with target motion at speeds in excess of 30 deg/s (Robinson 1965, Fuchs 1967a). Thus, pursuit ensures optimal vision only when the target is moving slowly, since visual acuity starts to decrease when the velocity of retinal image slip exceeds 3 deg/s (Westheimer & McKee 1975). For sine wave target motion, smooth pursuit is excellent when the amplitude and frequency of target motion are low (i.e. less than 1.0 Hz). When the amplitude or frequency is increased, smooth eye velocity becomes smaller than and lags increasingly behind target velocity (e.g. Fuchs 1967b, Collewijn & Tamminga 1984, Yasui & Young 1984, Lisberger et al 1981a). Because the smooth eye movement is not able to keep up with the target, the tracking is increasingly punctuated with saccades. Although targets moving that fast cannot be tracked accurately, pursuit eye velocity as high as 180 deg/s has been reported in humans (Lisberger et al 1981a).

Pursuit is specialized for tracking slowly moving targets and is inadequate for other functions, which are subserved by the other classes of eye movements. For example, the long delays and slow eye speeds achieved by pursuit make it too slow to stabilize the retinal images of stationary objects effectively during head turns. This function is subserved largely by the vestibulo-ocular reflex, which has a latency of 14 msec (Lisberger 1984) and is accurate for head turns at speeds in excess of 300 deg/s (Keller 1978). Similarly, pursuit moves the eyes too slowly to be effective for scanning the visual scene or to point the eyes at stationary eccentric objects. These functions are subserved by saccadic eye movements, which have relatively long latencies of about 200 ms, but can execute a shift in the position of gaze that (*a*) covers 20 deg in about 40 msec in monkeys and 70 msec in humans and (*b*) brings the eye accurately to the desired position. The reader is referred to several reviews of the neural subsystems subserving the vestibulo-ocular reflex (Miles & Lisberger 1981, Ito 1982) or saccades (Fuchs et al 1985).

Operational Definition of Pursuit

THE ADEQUATE STIMULUS FOR PURSUIT Under normal viewing conditions outside the laboratory, pursuit is generated only when there is a moving target. In most individuals, voluntary attempts to "move the eye from left to right as smoothly as possible" produce a staircase of small saccadic eye movements. In the intersaccadic intervals, which last about 200 msec each, the eyes are stationary. These anecdotal observations show that a visual stimulus is necessary to elicit pursuit, but Rashbass (1961) was the first to show that pursuit is a response to the *motion* of the target's images across the retina. He used the "step-ramp" target motion shown in Figure 1B to generate a situation in which target position and target motion called for eye movements in opposite directions. This stimulus evoked pursuit that took the eyes in the direction of target motion at latencies of about 100 ms. Saccadic eye movements had longer latencies and took the eyes in the direction of target position.

OTHER STIMULI FOR PURSUIT? The clear distinction between the adequate stimuli for saccades and pursuit eye movements has been challenged in recent years. Pola & Wyatt (1979) have shown that stationary stimuli can evoke pursuit if the target is artificially stabilized on the retina, slightly eccentric from the fovea. However, this kind of observation should not be taken as evidence that a stationary eccentric target normally elicits pursuit. Human subjects have proven to be extremely flexible in the use of their oculomotor systems, and one study (Cushman et al 1984) reports that they can learn to make a variety of smooth eye movements under conditions of image stabilization. Thus, some caution must be exercised in interpreting the results of experiments that use prolonged periods of image stabilization.

Although visual inputs provide the main stimulus for pursuit, cognitive factors also play an important role. First, activation of the pursuit system requires that a moving target be selected—targets can be ignored easily (Kowler et al 1984). Second, the pursuit system is capable of prediction, so that tracking is much more accurate when target motion is periodic than when target motion is unpredictable (Stark et al 1962, Michael & Melvill Jones 1966). Third, humans can make smooth eye movements that anticipate a change in the position or velocity of a target (Kowler & Steinman 1979). Although low in velocity (less than 1.0 deg/s), these drifts may be related to pursuit eye movements. Finally, humans can take advantage of "global motion" to pursue targets that are perceived but not actually visible, such as the center of a rolling wheel that is marked only by several small lamps attached to its rim (Steinbach 1976).

PURSUIT VS THE OPTOKINETIC RESPONSE In our review, we concentrate on the neural subsystem that generates smooth pursuit of small objects as they move across a structured background. Smooth eye movements can also be evoked by large or full field, moving patterns. These eye movements may share some of the neural mechanisms used by pursuit, so we briefly compare the smooth eye movements evoked by large and small targets.

A horizontally moving, full field, vertically-striped drum generates optokinetic nystagmus with slow phases in the direction of stimulus motion and quick phases (saccades) in the opposite direction. The smooth part of the eye movements has two components (Cohen et al 1977, Lisberger et al 1981b). One, the "slow component," is clearly different from pursuit. It takes seconds to reach its full amplitude, appears to be generated by neural networks in the brainstem, and is seen in species that do not pursue small objects. The other, "fast component," shows several properties that parallel pursuit. It takes only a few hundred milliseconds to reach its full amplitude and is largest in primates, which have excellent pursuit. Lesions of the primary visual cortex (Zee et al 1986) or the flocculus of the cerebellum (Zee et al 1981, Waespe et al 1983) cause large deficits in both the fast component of the optokinetic response and pursuit, so we assume that similar brain pathways are used. However, it seems likely that different afferent neurons are involved. Many neurons in the visual cortex respond to the motion of a small target but do not respond well to the motion of the full field stimuli that are required to generate an optokinetic response (Allman et al 1985, Tanaka et al 1986).

ANALYZING PURSUIT AS A NEGATIVE FEEDBACK CONTROL SYSTEM

Because of the backgrounds of the original investigators in the area, and because of the physical configuration of the pursuit system, it has become popular to view it as a negative feedback control system whose function is to minimize the retinal slip of the images from small objects. This approach to pursuit has provided many concepts that are useful, but, as Steinman (1986) has also pointed out, it has limited progress to some extent. We now discuss the control theory approach, extract the contributions it has made, and outline a modified approach that we have used to study pursuit.

Pursuit as a Control System

NORMAL MODE OF OPERATION The pursuit system is configured as a negative feedback control system like the one pictured in Figure 2A. The afferent pathways sense image motion across the retina and the efferent pathways move the eyeball (and retina) in an effort to match eye and target motion. In normal operation, accurate pursuit is achieved by taking advantage of the negative feedback. Under these circumstances, the pursuit system operates in "closed loop" mode, since the feedback loop shown by the *dashed line* in Figure 2A attempts to control and in fact minimize the difference between eye and target motion.

The negative feedback configuration of the system makes it imperative for us to distinguish between the motion of the tracking target and the resulting image motion across the retina. When the eyes are stationary, target motion and image motion are equivalent (although opposite in direction because the lens reverses images on the retina). During pursuit, however, target motion and image motion can be quite different, and image motion is target motion minus eye motion. We use the term "retinal error" to refer to any difference between eye and target position or motion.

INTERNAL COMMAND SIGNAL FOR PURSUIT Although visual inputs must be relied on to establish and maintain accurate pursuit, early investigators realized that there would be practical problems with a control system based on the simple configuration of Figure 2A. First, the system could never maintain perfect tracking, since that would reduce the retinal error driving the system to zero, and therefore transiently reduce eye velocity to zero. Second, with the 100 ms delay, the system was in danger of being unstable.

Young et al (1968) solved this apparent dilemma by suggesting that pursuit was driven by a central neural signal that encoded the velocity of the target with respect to the world. Then, the system could achieve and maintain accurate and stable performance, and it could also achieve perfect tracking. Robinson (1971) recognized that the brain did not receive a signal related to target velocity from any individual sensory input. He proposed that a target velocity signal was reconstructed by adding (*a*) a copy of the pursuit command encoding smooth eye velocity with respect to the head (corollary discharge) and (*b*) a visual input related to retinal velocity error (target velocity with respect to the eye). As shown in Figure 2B, the target velocity model retains the negative feedback pathway for visual inputs and adds internal positive feedback of the motor command for pursuit eye velocity.

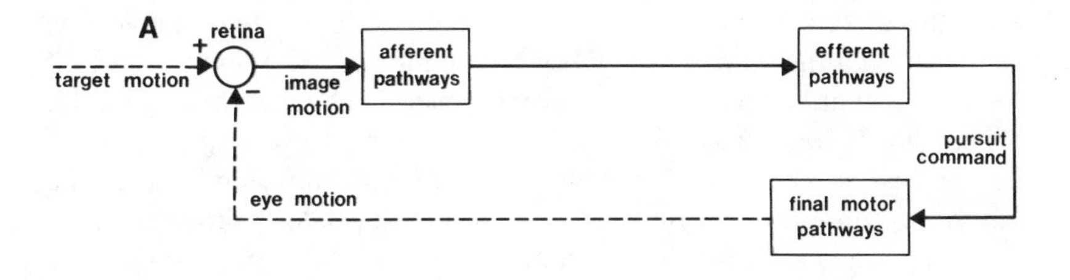

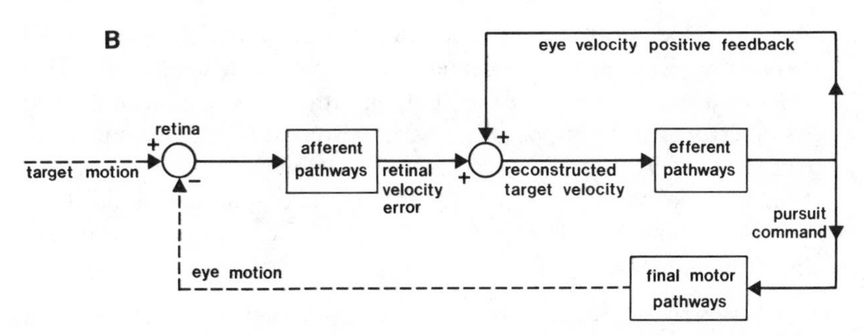

Figure 2 Two models of the pursuit system. *Panel A* shows a simple negative feedback control system in which image motion provides the central command to the efferent pathways for pursuit. *Panel B* shows a modification that includes a positive feedback pathway for the pursuit command for eye velocity. The mathematical addition of the positive feedback of eye velocity and the visual inputs signaling retinal velocity error (target velocity minus eye velocity) provides a reconstructed target velocity signal, which serves as the central command to the efferent pathways. Here and in Figure 5, the *solid lines* indicate the flow of neural signals and the *dashed lines* represent physical events. The *small circles* represent summing junctions that perform mathematical addition and subtraction of their inputs. Thus, the circle labeled "retina" compares target motion and eye motion, and its output is image motion, or target motion with respect to the eye.

OPENING THE VISUAL FEEDBACK LOOP FOR ANALYSIS OF PURSUIT The performance of a negative feedback system will be relatively independent of the properties of its internal elements. One advantage of such systems is that they allow the engineer (and the brain) to achieve reliable performance with little concern for the properties of the internal elements. Studies of eye movement in normal tracking conditions (closed loop) are useful for describing the performance of the pursuit system, but they characterize largely the existence and function of the negative feedback loop. To understand how pursuit is generated, we must use methods that characterize the properties of the brain pathways themselves. Although it is possible, this task is difficult with measures of steady-state, closed loop performance.

The properties of the brain pathways that generate pursuit can be evaluated directly by "opening the visual feedback loop." In principle, this entails cutting the *dashed lines* labeled "eye motion" in Figure 2, so that the subject cannot alter the retinal events generated by the tracking target. When the feedback loop is opened, the motion of the target uniquely determines the visual inputs for pursuit, and the relationship between the output (eye motion) and input (target motion) provides a direct estimate of the transformations that are going on in the brain. In practice, it is seldom possible to prevent eye motion altogether, so a number of equivalent methods have been used to achieve "open loop" stimulus conditions.

1. Ask the subject to track a foveal after-image, so that there can be no retinal slip (Kommerell & Taumer 1972, Heywood & Churcher 1971).
2. Use electronics to move the target along with the eye so that controlled visual errors can be imposed but never corrected (e.g. Robinson 1965, Collewijn 1969).
3. Present visual stimuli to an eye that has been immobilized surgically or by disease, and measure the movements of the other, moveable eye (Koerner & Schiller 1972, Leigh et al 1982).

THE TRANSIENT RESPONSE OF THE PURSUIT SYSTEM IS OPEN LOOP Although useful, the techniques enumerated above are highly artificial, in that any efforts to correct retinal errors are futile. A more natural way to open the visual feedback loop is to exploit the built-in latency of the visual feedback. Whenever there is a sudden change in the motion of a tracking target, the 100 ms latency of pursuit creates a short interval in which the system operates without feedback. The first 100 ms of the retinal error affects eye velocity in the subsequent 100 ms, in a first attempt to correct the original error. If the attempt is incorrect—either too big or too small—there will still be large retinal errors in the second 100 ms after the original error, but these cannot be corrected until the third 100 ms. Thus, the first 100 ms of the transient response is really an open loop response (Lisberger & Westbrook 1985), and it provides a fourth technique to reveal the properties of the neural pathways subserving pursuit in normal and lesioned subjects.

The transient response of the pursuit system, in addition to being a useful analytical tool, has a functional importance that is emphasized by the fact that pursuit is subject to adaptive plasticity. If a subject is forced to view through an eye that has peripheral muscle weakness, the pursuit system compensates over a time course of several days by increasing the magnitude of the initial response to a given ramp target motion (Optican et al 1985). Normally, the capacity to undergo adaptive changes has been associated with systems like the vestibulo-ocular reflex or saccadic eye movements, both of which operate open loop. Negative feedback control systems like pursuit depend on the feedback loop to achieve excellent steady-state performance, and are not thought to require adaptive control. The existence of an adaptive mechanism for pursuit suggests that an appropriate transient response is important behaviorally, perhaps serving to achieve accurate pursuit as quickly as possible.

Placing Biological Constraints on the Control System

We agree with the goal of the control theory approach, which is to develop a mathematical model that describes the pursuit system. We also favor the use of quantitative methods for both the development and testing of models, and we support the point of view that a block diagram is not really a model until it has been tested with computer simulations or pencil-and-paper computations. However, we believe that it is important to start by identifying the biological properties of the elements that are within the system, and the discussion that follows focuses on experiments that were designed specifically for that purpose. A model can then be built from the identified elements and evaluated on the basis of its transient performance as well as its open loop and closed loop behavior.

One premise of our approach has been to avoid simplifying assumptions about the inputs and outputs of the pursuit system. Instead we have determined the relevant variables for analysis: What retinal events are sensed, and what effect do they have on eye movement? We have distinguished between the initiation and maintenance of pursuit because the retinal events differ substantially between these two phases. In addition, we have divided the overall system into afferent and efferent limbs. Although this division can be fuzzy at the sensory motor interface, it has provided structure for designing experiments to evaluate one limb or the other and has allowed us to assign functions to neural structures that are clearly sensory or motor. It has also allowed us to define the properties of the sensory motor interface, which turn out to be quite simple. In the sections that follow, we evaluate the efferent limb, afferent limb, and sensory motor interface separately before showing our initial attempts to unify the system through mathematical modeling.

THE EFFERENT LIMB OF PURSUIT

Behavioral Properties Ascribed to the Efferent Limb

In this section, we describe the features of pursuit behavior that are subserved by the motor circuitry. We pay special attention to two features that are related: (*a*) The efferent limb of the pursuit system provides short-term "velocity memory"; (*b*) visual inputs drive eye *acceleration*.

VELOCITY MEMORY IN PURSUIT Psychophysical experiments have revealed that eye velocity is automatically sustained, even if retinal errors are prevented by stabilizing the target with respect to the moving eye (Morris & Lisberger 1983). It is not possible to account for the automatic maintenance of pursuit by the inertia of the eyeball and extraocular muscles. Left to its own passive properties, the eye would be expected to decelerate to zero velocity with a time constant of about 200 ms (Robinson 1964). Therefore, the maintenance of pursuit in the absence of retinal errors must be attributed to brain pathways. We suggest that the pursuit system employs a neural velocity memory that keeps the eyes moving at their current speed, unless visual inputs provide another command.

VISUAL ERRORS CAUSE EYE ACCELERATION Since velocity memory maintains ongoing pursuit, any retinal errors should cause *changes* in eye velocity, which would be measured as eye accelerations. Several studies have confirmed this prediction. Lisberger et al (1981a) recorded the eye movements evoked by sinusoidal target motion over a wide range of temporal frequencies and peak-to-peak amplitudes. Their data analysis revealed a single unifying graph that related visual inputs to pursuit motor outputs. Peak eye *acceleration* and not peak eye velocity was a consistent function of retinal velocity errors. In more recent studies, we have found that open loop presentation of constant retinal velocity errors causes constant eye accelerations (Lisberger & Westbrook 1985, Tychsen & Lisberger 1986a). The eye accelerations persist for the duration of the retinal errors, and their magnitude is a monotonic function of the magnitude of the retinal velocity error.

PURSUIT VS OPTOKINETIC VELOCITY MEMORY Velocity memory has also been invoked by Cohen et al (1977) to account for the persistent "optokinetic after-nystagmus (OKAN)" that endures for up to one minute or more after the end of a prolonged optokinetic stimulus. However, two facts argue that the mechanisms underlying pursuit and optokinetic velocity memory are different. First, continuous nystagmoid pursuit of the sawtooth motion of a small spot produces only a small after-nystagmus in humans (Muratore & Zee 1979) and no after-response in monkeys (Lisberger et al 1981b). Second, the mechanism underlying OKAN takes at least 15 seconds and sometimes up to one minute to reach its maximum output, while the velocity memory for pursuit appears to be fully activated within several hundred milliseconds.

Neural Pathways in the Efferent Limb of Pursuit

NEURAL BASIS FOR PURSUIT VELOCITY MEMORY Anatomical and physiological data strongly suggest that velocity memory is subserved at least partly by positive feedback of the command for pursuit eye velocity through the flocculus of the cerebellum. The flocculus receives inputs from the oculomotor areas of the brainstem and completes the loop by sending its outputs back to the same oculomotor areas.

Eye velocity inputs are transmitted to the flocculus over mossy fiber afferents that signal the current eye position and velocity. Although the exact origin of these inputs remains uncertain, the firing properties of the mossy fibers (Lisberger & Fuchs 1978b, Miles et al 1980) and anatomical studies of afferents to the flocculus (Langer et al 1985a) suggest that many originate in the nucleus prepositus hypoglossi. The predominant class of Purkinje cells, which provide the output from the flocculus, emits simple spikes at a rate that is correlated with pursuit eye velocity during steady-state pursuit (Miles & Fuller 1975, Lisberger & Fuchs 1978a, Noda & Suzuki 1979). Purkinje cell activity is really driven by pursuit eye velocity and is not due to correlated visual events: Firing rate was still related to eye velocity when open loop conditions were used to stabilize the target's image on the fovea during pursuit of ramp target motion (Stone & Lisberger 1985).

The output of the flocculus has relatively direct access to the motor and premotor circuits in the brainstem. First, stimulation of the flocculus evokes a twitch of eye velocity with a latency of 10.5 ms (S. G. Lisberger and T. A. Pavelko, unpublished observations). Second, Purkinje cells in the flocculus project to the vestibular nuclei (Langer et al 1985b), where they contact interneurons in vestibulo-ocular reflex pathways (e.g. Baker et al 1972). The cells that receive monosynaptic inhibition from the flocculus are near the lateral border of the medial vestibular nucleus and discharge in relation to eye movement (Lisberger & Pavelko 1984). Their projections are not known, but the positive feedback loop would be completed if the cells that receive inhibition from the flocculus had access to the cells that provide mossy fiber eye velocity inputs to the flocculus.

Of the Purkinje cells in the flocculus that discharge in relation to horizontal pursuit, almost all show increased firing for eye velocity toward the side of the recording (ipsilateral). This, along with the fact that electrical stimulation of the flocculus evokes ipsilateral smooth eye movement (Ron & Robinson 1973), suggests that increases in Purkinje cell firing cause ipsilateral eye velocity. The cells that receive inhibition from the flocculus are in the ipsilateral brainstem, and most show increased firing for contralateral eye movement (Lisberger & Pavelko 1984). The change in the preferred direction of firing across the Purkinje cell synapse does not affect our interpretation of the role of the flocculus. We assume that increases in floccular output cause ipsilateral eye motion by reducing the firing of brain stem cells that cause contralateral eye motion.

OTHER BRAIN AREAS IN THE EFFERENT LIMB OF PURSUIT In the brainstem, Eckmiller & Mackeben (1978) have identified a group of cells near the Abducens nucleus that discharge in relation to horizontal pursuit but not other kinds of horizontal eye movement. Chubb & Fuchs (1982) found a similar group of cells in the Y-group of the vestibular nuclei and the dentate nucleus of the cerebellum, for pursuit eye movements in the vertical plane. The connections and role of these cells are unknown. In addition, nothing is known about the physiological function of the basal interstitial nucleus of the cerebellum, which receives a projection from the flocculus (Langer 1985) and is probably involved in pursuit. Finally, the flocculus is not the only cerebellar structure that is involved in pursuit. Complete cerebellectomy produces a total loss of pursuit eye movements (Westheimer & Blair 1973), while flocculectomy produces only a partial loss. The remaining cerebellar influence on pursuit may be mediated by Purkinje cells in lobules VI and VII of the cerebellar vermis that discharge in relation to target velocity (Kase et al 1979, Suzuki et al 1981).

THE AFFERENT LIMB OF PURSUIT

Identification of the pathways that provide visual inputs for pursuit requires two kinds of information. First, we must catalog the kinds of signals that are available by studying the properties of visual pathways in the relevant parts of the brain. Second, we must measure eye movements in ways that reveal the properties of the signals that are actually used.

Motion Processing for the Initiation of Pursuit

INTRODUCTION TO THE CORTICAL MOTION PATHWAYS The striate and extrastriate cortex appear to contain two parallel pathways, each specialized for processing a different kind of visual information (Ungerleider & Mishkin 1982, Van Essen & Maunsell 1983, Maunsell & Newsome 1987, this volume). One of the pathways appears to be specialized for processing moving images and includes at least the primary visual cortex (area V1), the middle temporal visual area (MT), the medial superior temporal visual area (MST), and the posterior parietal cortex (area 7a). Much of the experimental work has focused on MT, and so our discussion emphasizes the role of MT in pursuit. About 80% of the cells in MT are sensitive to motion and selective for the direction of motion (e.g. Dubner & Zeki 1971, Maunsell & Van Essen 1983). All directions of motion are represented about equally. MT is organized topographically and contains a map of the contralateral visual field (Van Essen et al 1981).

THE VISUAL CORTEX PROVIDES INPUTS FOR PURSUIT Lesions of the visual cortex cause deficits in pursuit, and studies using this method have been instrumental in defining the visual areas that are involved in pursuit. Here, we concentrate on several recent lesion experiments that have used the step-ramp target motion pioneered by Rashbass (1961) to analyze the nature of the deficits more carefully.

Because it is an essential link in the overall logic, we begin by reviewing briefly the experimental conditions under which these experiments are done. Eye movements are measured in individual trials that begin with the monkey fixating a spot of light that is stationary in front of him. At an unpredictable time, the spot moves in a step to a new position and begins to move in a ramp of constant velocity. In the 100 ms before the eyes begin to track the spot, the retinal position and velocity of the stimulus are determined solely by the amplitude of the target step and the velocity of the target ramp. Thus, varying the amplitude of the step in different trials provides a means to test the integrity of pursuit for targets moving across different parts of the visual field. As we discussed above, the transient response evoked by step-ramp target motion provides a good estimate of the open loop performance of the pursuit pathways.

Lesions that cause visual field defects Pursuit is affected by lesions of the striate cortex or MT, but the deficits can be observed only if the pursuit stimulus moves across the area of the visual field whose cortical representation was destroyed. After ablation of the striate cortex of one hemisphere, monkeys can pursue targets moving in either direction, but they do so by positioning their eyes so that the target remains in the hemifield ipsilateral to the lesioned cortex. If the moving target is presented as part of a step-ramp trajectory into the contralateral hemifield, the monkey does not track it (Segraves et al 1986).

Newsome et al (1985a) found that small chemical lesions of MT produce deficits for targets that move across a small region of the visual field. These investigators recorded multiple unit activity at the site of the lesion just before they injected ibotenic acid. They then used step-ramp target motion to show that the defective region of the visual field corresponded with the receptive field locations of cells recorded at the site of the lesion. With the step-ramp stimulus, the deficits were limited to the initiation of pursuit. Steady-state tracking was normal once the monkey had used saccades to position the target near the fovea, in an intact part of the visual field. In addition, saccades to stationary targets were normal after lesions of MT but deficient after lesions of the primary visual cortex (Newsome et al 1985b), so the defect after a lesion of MT cannot be interpreted as a total blind spot. Thus, lesions of MT cause a visual field defect in the use of moving targets for the initiation of pursuit.

Pursuit receives multiple inputs from the motion pathways Anatomical studies have emphasized the hierarchical organization of the motion pathways, where V1 is the lowest level and MT, MST, and the parietal cortex are increasingly higher levels (Maunsell & Newsome 1987, this volume). An analogous hierarchy can be seen in the effects of lesions and in the discharge properties of cells in these areas. Lesions of MT cause a straightforward visual field defect for the initiation of pursuit; cells in MT respond exclusively to the passive visual stimulus, even in awake monkeys (Newsome & Wurtz 1981). Lesions of MST or foveal MT cause a combination of (*a*) the visual field defect seen following lesions of MT and (*b*) a deficit in steady-state pursuit toward the side of the lesion (Dursteler et al 1986). Lesions of the parietal cortex cause just the deficit in steady-state pursuit toward the side of the lesion (Lynch & McLaren 1982). Cells in these higher areas discharge in relation to a combination of the passive visual stimulation and the active eye movements generated by pursuit (Wurtz & Newsome 1985, Kawano et al 1984, Sakata et al 1983, Mountcastle et al 1975).

The pursuit system receives inputs from all levels in the motion processing hierarchy. V1, MT, MST and area 7a all project to the pontine nuclei (Glickstein et al 1980, Brodal 1978), which relay cortical inputs to the areas of the cerebellum that are concerned with pursuit (Brodal 1979, Langer et al 1985a). The multiple projections from the motion pathways to the brainstem may explain the rapid recovery of pursuit following small lesions of MT (Newsome et al 1985a), MST (Dursteler et al 1986), or area 7a (Lynch & McLaren 1982).

PARALLELS BETWEEN CORTICAL AND PURSUIT MOTION PROCESSING Because pursuit is in the direction of target motion, we assume that the visual inputs for the initiation of pursuit arise in cells that are selective for the direction of motion. The lesion data cited above, along with the fact that small moving spots are adequate stimuli for cells in MT and for pursuit, argue that MT is a major source of visual inputs for pursuit. The argument is strengthened by recent experiments showing parallels between the properties of optimal visual stimuli for activating cells in MT and the properties of visual motion that initiate pursuit.

Two components in the initiation of pursuit In humans (Tychsen & Lisberger 1986a) and monkeys (Lisberger & Westbrook 1985), the strength of pursuit initiation depends strongly on the velocity of the target motion, the position of the moving target in the visual field, the direction of target motion relative to the position of fixation, and the relative contrast of target and background. Parameters such as target size, shape, and intensity were less important. Other labs have found that some of the parameters mentioned above also affect steady-state pursuit, although the effects were generally more subtle (Barnes 1983, Winterson & Steinman 1978, Haegerstrom-Portnoy & Brown 1979).

Our experiments also revealed that visual parameters had differential effects, depending on whether we measured eye movement in the first 40 ms of pursuit or later in the first 100 ms. The differences suggest that there are at least two parallel visual inputs that affect eye movement early and late, creating two components in the initiation of pursuit. Figure 3 summarizes the properties of the two components; the late component is shown in the *top panels* and the early component in the *bottom panels*.

Contribution of MT to the late component of pursuit? We focus here on two properties of the discharge of cells in MT that find direct analogues in the initiation of pursuit. We take these parallels as further evidence that MT provides inputs for the initiation of pursuit.

The population of cells in MT emphasizes the central area of the visual field (Van Essen et al 1981), as does the magnitude of the late component of pursuit initiation (Lisberger & Westbrook 1985). In our behavioral experiments, the magnitude of the late component of pursuit was largest when the target moved through the central 5 deg of the visual field. Eye acceleration decreased smoothly but dramatically as the target was presented at more eccentric positions (Figure 3A). We attribute the pronounced weighting of central visual field in the initiation of pursuit to the central magnification in MT and most cortical maps. Stimuli moving across the central visual field should activate more cells in the first 100 ms of motion than do stimuli moving across more peripheral parts of the visual field.

The velocity selectivity of the population of cells in MT (Figure 7 of Maunsell & Van Essen 1983) is the same as the velocity selectivity of the late component of pursuit under comparable visual conditions. In the behavioral experiments, eye acceleration increased as a function of target velocity up to 30 deg/s, was maximal for target velocities between 30 and 75 deg/s, and decreased for higher target velocities. We suspect that the best velocity for pursuit initiation would represent that stimulus velocity for which the greatest number of cells in the afferent pathways responded the most vigorously. The correspondence between the properties of pursuit (*solid line* in Figure 3B) and the responses of the population of cells in MT (*dashed line* in Figure 3B) would argue that the velocity selectivity of pursuit results from inputs transmitted through MT.

The two basic properties of pursuit initiation outlined above—emphasis of central visual field and velocity selectivity—represent properties of the afferent limb rather than limitations in the motor system. First, eye acceleration was always well below the maxima that can be achieved by the oculomotor system, or even during pursuit of periodic target motion, so it is unlikely that our results represent any kind of eye acceleration

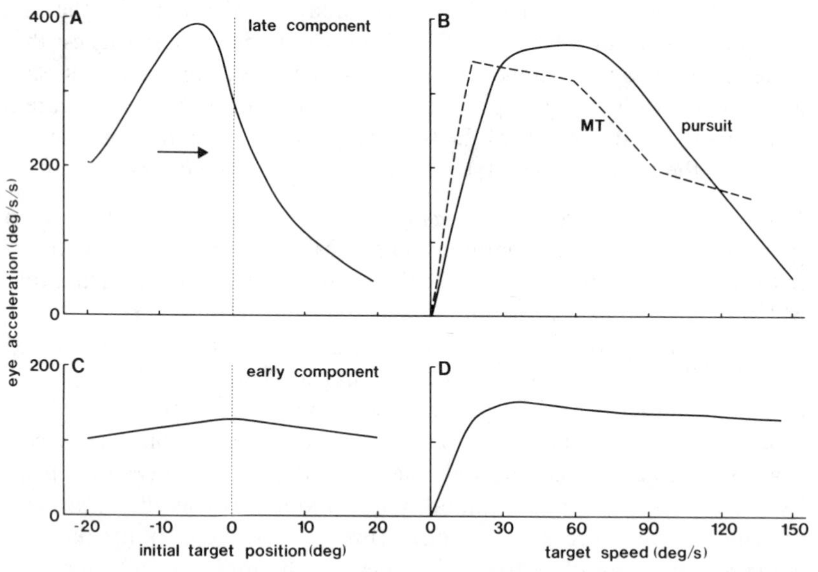

Figure 3 Properties of the early (*C*, *D*) and late (*A*, *B*) components in the initiation of pursuit. *Panels A and C* show the magnitude of each component (instantaneous eye acceleration) as a function of the moving target's initial position along the horizontal meridian. The *vertical dashed line* represents straight ahead gaze. As indicated by the *arrow* in *panel A*, these curves are for pursuit of rightward target motion. Negative values of initial position mean that the target started in the left visual field. *Panels B and D* show the magnitude of each component as a function of target speed. In *panel B*, the *solid line* represents the speed selectivity for the initiation of pursuit and the *dashed line* represents the response of the population of cells in macaque area MT (from Figure 7 of Maunsell & Van Essen 1983).

saturation. Second, the curves in Figure 3 were dependent strictly on the visual properties of the stimulus, and eye acceleration could always be improved by bringing the stimulus closer to the best velocity or the best position in the visual field.

Motion processing for the early component of pursuit Although MT appears to be a major part of the input pathway for the late component of pursuit initiation, it seems less likely to contribute to the early component. The first 40 ms of pursuit are distinguished by the fact that eye acceleration is little affected by the visual properties of the stimulus. The early component behaves as though its inputs come from cells that are selective for the direction of motion but are otherwise totally unselective. Most striking is the lack of emphasis of the central visual field.

Available data do not identify a visual pathway that could be transmitting the visual inputs for this part of the response, but two possibilities bear mentioning. First, Cohen et al (1981) noted that the cortico-pontine projection deemphasizes the central visual field. This suggests that one portion of the cortico-pontine system may be driving the early component of pursuit, while the part of the pathway that originates in MT may be driving the late component. Although the cortico-pontine projection from the primary visual cortex (V1) is relatively small, the fact that the early component has the shortest latencies raises the possibility that it arises directly from V1. Second, there are direct pathways from the retina through the brainstem to parts of the cerebellum that are involved in pursuit. These pathways, through the accessory optic system, are important in generating the optokinetic response in species that lack a pursuit system (Simpson 1984), and it is possible that they also respond to small targets at latencies short enough to drive the early component of pursuit (Hoffman & Distler 1985).

The relative unselectivity of the early component suggests that it functions just to start the eyes in the correct direction at the earliest possible time. In fact, the amplitude of the early component is relatively small. The late component provides about 75% of the initial eye acceleration of pursuit, and can also provide high fidelity information about target speed.

PURSUIT RECEIVES INPUTS FROM A GENERAL PURPOSE MOTION SYSTEM Until now, we have concentrated on the evidence that cortical motion processing pathways contribute visual inputs for the initiation of pursuit eye movements. With that role well established, we now can ask whether pursuit receives visual inputs from its own private motion system, or whether the pursuit motor system is just one target of a general-purpose cortical motion processing system. If the latter alternative obtains, there should be clear correlations between the properties of pursuit and of other behaviors that require information about visual motion, such as the perception of speed.

There is a striking correlation in the pursuit and perceptual responses of adult humans who had esotropic strabismus (crossed eyes) with onset in infancy (Tychsen & Lisberger 1986b). During monocular viewing, these subjects had severe nasal-temporal asymmetries in their responses to moving targets. When the task was the initiation of pursuit, the subjects exhibited normal magnitude responses when the targets moved nasally but weak or no pursuit when the targets moved temporally. When the task was to identify the perceived speed of a moving target (McKee & Welch 1985), the same subjects reported that nasally directed target motion was faster than temporally directed motion, even though the speed was actually the same for both directions. The correspondence of the perceptual and pursuit deficits in strabismics suggests that they have a maldevelopment of the cortical motion processing pathways, and that these pathways provide inputs to the motor and perceptual systems.

Motion Processing for the Maintenance of Pursuit

Pursuit must use a set of signals during the maintenance of pursuit different from that used to initiate pursuit. Eye velocities are nonzero and can be quite large during pursuit, while retinal errors can be quite small and occur on or near the fovea.

The velocity memory we discussed above provides one of the most important signals used during pursuit maintenance. It frees the visual system from the rather mundane task of keeping the eyes going, and allows visual inputs to fine-tune the pursuit response and make adjustments for changes in target direction and speed. In principle, velocity memory could maintain pursuit of a ramp target motion without any visual inputs, but one observation argues that retinal errors also play a role. In about half of a subject's attempts to track a target moving at constant velocity, the eye velocity records reveal substantial oscillations around the desired velocity (see Figure 1B). The oscillations cease if open loop conditions are used to stabilize the target with respect to the eye (E. J. Morris and S. G. Lisberger, unpublished observations). This observation shows that the oscillations are driven by visual inputs, and implies that visual inputs play an important role in the maintenance of pursuit.

To determine what visual signals are used during the maintenance of pursuit, it is necessary first to document what errors are present during normal, closed loop tracking. The errors that occur can then be imposed in open loop conditions to determine whether they affect pursuit. Finally, we must show that the retinal errors that are effective in open loop conditions are actually used in normal, closed loop pursuit.

RETINAL ERRORS THAT ARE EFFECTIVE DURING PURSUIT There was good agreement between the errors experienced in normal tracking and those effective in open loop conditions (Morris & Lisberger 1985). Monkeys responded to retinal position errors (RPEs) of 0.2 to 1.5 deg, retinal velocity errors (RVEs) of 0.25 to 4.0 deg/s, and retinal acceleration errors (RAEs) of 20 to 90 deg/s during the maintenance of pursuit. However, the relative importance of various kinds of errors was different in individual monkeys, varied as a function of the visual conditions for a given monkey, and could change quite dramatically over a time course of months for an individual monkey. All monkeys responded to retinal velocity errors, all but one of the monkeys we have studied responded to retinal acceleration errors, and three out of six responded to retinal position errors.

Especially the use of position errors appeared to depend on behavioral state. Retinal position errors caused significant eye accelerations only if imposed during pursuit (Morris & Lisberger 1983). They evoked only saccades when imposed in open loop conditions during fixation of a stationary target. When position errors did affect eye acceleration, the responses were usually largest in the first month of a monkey's participation in experiments. We conclude that the monkey pursuit system routinely responds to visual inputs that signal retinal acceleration error and retinal velocity error, and that it can respond to inputs that signal retinal position error under the appropriate behavioral conditions. This may explain why some human subjects make smooth eye movements in response to steps of target position (Carl & Gellman 1985) and can use small, prolonged position errors to initiate pursuit (Pola & Wyatt 1979).

RETINAL ERRORS THAT ARE USED DURING PURSUIT Caution must be exercised in the interpretation of experiments using open loop conditions, since the monkey's inability to correct the imposed errors may interfere with his normal tracking strategy. To reveal what errors are normally used, we cross-correlated eye acceleration and retinal error during closed-loop pursuit of ramp target motion; this revealed the functions relating eye movement to visual inputs for closed loop tracking (Morris & Lisberger 1985). The strongest correlation was obtained when the retinal position, velocity, and acceleration errors were paired with the eye acceleration 90 ms later, implying that the latency for all three kinds of error is about 90 ms. We found that the visual signals that are used during closed loop pursuit (*bottom panels* in Figure 4) are the same as those that are effective in open loop conditions on the same day (*top panels* in Figure 4).

VISUAL ERRORS DRIVE OSCILLATIONS OF EYE VELOCITY During pursuit, eye velocity often oscillates around target velocity. The period of the oscillations varied among our monkeys, over the range 220 to 390 ms. Similar periods, ranging from 300 to 400 ms, were found by other laboratories (Robinson 1965, Fuchs 1967a, Optican et al 1985). As mentioned above, the oscillations are visually driven, since they disappear when the target is artificially stabilized with respect to the moving eye.

The 90 ms delay in the visual inputs to the pursuit system should cause the system to oscillate, and, if the visual inputs signal only retinal velocity error, the period of the oscillation should be approximately four times the latency of the visual input (Robinson 1965). However, the calculation changes if, as we have found, pursuit receives inputs about both retinal velocity error and retinal acceleration error. The predicted period of oscillation shortens as the sensitivity to retinal acceleration error increases. In our data, the actual period of oscillation in each monkey agreed well with that predicted by his sensitivity to retinal acceleration and velocity errors during the maintenance of pursuit (Morris & Lisberger 1985). This tells us that that the oscillation can be explained as a direct consequence of the visual signals that are employed to maintain pursuit. Also, the relation between sensitivity to retinal acceleration error and the period of oscillation

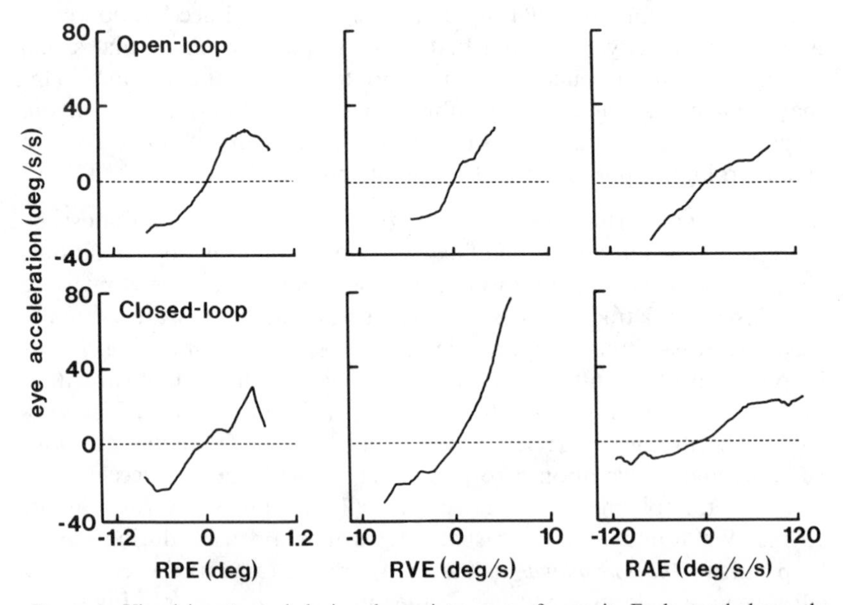

Figure 4 Visual inputs used during the maintenance of pursuit. Each panel shows the relationship between eye acceleration and retinal position, velocity, or acceleration errors (RPE, RVE, and RAE). The *upper panels* show the relationships obtained when small controlled errors were imposed in open loop conditions. The *lower panels* show the relationships for the same monkey on the same day during normal, closed loop tracking.

strengthens the argument that visual inputs related to both retinal acceleration and velocity errors are used by pursuit.

THE SENSORY MOTOR INTERFACE FOR PURSUIT

Transformations Done by the Sensory Motor Interface

First, the sensory motor interface must combine visual inputs from a large number of input cells and cortical areas. In the visual cortex, each individual cell provides information about motion at a narrow range of speeds across a small part of the visual field. In the pursuit motor pathways, each cell responds in proportion to smooth eye velocity, without regard to the receptive fields of the visual input cells that were activated by the moving target. Therefore, the sensory motor interface need not retain the retinotopy of the visual cortex or the speed selectivity of individual cortical cells. It can merely add the inputs from direction-selective cells with different velocity selectivities and different receptive field locations without special transformations. If the interface were merely a summing junction, the motor output would reflect the visual properties of the population of input cells weighted by the relative strengths of their synapses. The fact that pursuit initiation shows an emphasis of central retina and a broadly tuned velocity selectivity related to that in MT argues that this is how the sensory motor interface could work.

Second, the interface must decode the inputs from cells with different preferred directions of image motion and encode outputs in the directions of action of the six extraocular muscles. No information is available to indicate how well the direction transformation is performed, or to suggest how it is done. The obvious approach would be to give the appropriate weighting to each connection between a sensory cell and the motor pathways, so that visual motion in a given direction causes the correct proportion of inputs to each extraocular muscle.

The sensory motor interface proposed above is much simpler than that for saccades. The appearance of a target for a saccade will activate a subset of cells in neural maps of the visual field, with the location of the cells determined by the location of the target in the visual field. To generate a saccade, the retinotopic inputs must be transformed so that all extraocular motoneurons are driven by a burst of spikes. The duration of the burst will determine the duration of the saccade, and the intraburst firing rate will determine saccade velocity. This "spatio-temporal" transformation requires a complicated sensory motor interface. A substantial body of work now argues that the saccadic system employs internal feedback loops to drive the eye until an internally monitored "motor error" is corrected (Zee et al 1976, Fuchs et al 1985). However, the pursuit system does not seem to need spatio-temporal transformation, even to process small retinal position errors, and we do not see any reason to invoke such a system now.

Cortico-ponto-cerebellar Pathways

DORSOLATERAL PONTINE NUCLEUS Anatomical and physiological data all point to a number of pontine nuclei as the sites of sensory motor integration for pursuit. Most attention has been devoted to the dorsolateral pontine nucleus. Although it is but one of the pontine nuclei that receives cortical visual inputs, the dorsolateral pontine nucleus has anatomical connections that are ideal for relaying cortical visual inputs to the pursuit motor pathways. It receives inputs from MT (Fries 1981) as well as other parts of the cortical motion pathways (Brodal 1978, Glickstein et al 1980), and projects abundantly to the flocculus and the vermis of the cerebellum (Brodal 1979, Langer et al 1985a).

Cells in the dorsolateral pontine nucleus encode all of the different kinds of visual signals that our behavioral studies suggested were used by pursuit (Suzuki & Keller 1984, Suzuki et al 1985). These include signals that are appropriate for driving the early and late components of pursuit initiation as well as signals related to RPE, RVE, and RAE. Lesions of the dorsolateral pontine nucleus cause deficits in pursuit (Suzuki et al 1986). The direction of the deficient pursuit depended on the site of the lesion, and coincided with the "on direction" for the cells recorded at the site of the lesion. Electrical stimulation of the dorsolateral pontine nucleus evokes smooth eye movements (May et al 1985). The stimulation is most effective if delivered during pursuit, implying that there is a switch or gate between the dorsolateral pontine nucleus and the final oculomotor pathways.

Why should the dorsolateral pontine nucleus be considered part of the sensory motor interface rather than part of the afferent limb of the pursuit system? At least one of the transformations we discussed above has occurred by the time the signals leave the dorsolateral pontine nucleus. Visual responses are no longer encoded in small receptive fields organized retinotopically within the nucleus. Instead, many of the cells have large receptive fields and are sensitive to the motion of small targets over a wide range of the visual field. As expected from our studies of the initiation of pursuit (Lisberger & Westbrook 1985), some of the cells are differentially sensitive to motion across the central visual field (Suzuki & Keller 1984). Thus, the dorsolateral pontine nucleus seems to receive convergent inputs from a large number of cortical neurons, and it transmits the kind of visual information that is used by the pursuit system.

VISUAL RESPONSES IN THE CEREBELLUM We discussed the flocculus of the cerebellum as a key structure for providing velocity memory in the efferent limb of the pursuit system. However, recent work has shown that visual signals are also transmitted through the flocculus. Several labs have found mossy fiber afferents in the monkey flocculus that responded to retinal slip (Miles & Fuller 1975, Miles et al 1980, Noda 1981). In addition, Warabi et al (1979) found some Purkinje cells that showed weak "simple spike" responses to visual inputs. We have now found that most of the Purkinje cells in the flocculus emit a visually driven pulse of simple-spike firing about 100 ms after the start of a step-ramp target motion at 30 deg/s (Stone & Lisberger 1985). It seems likely that these visual inputs are transmitted to the flocculus from the dorsolateral pontine nucleus and other visual areas of the pons that are afferent to the flocculus (Langer et al 1985a). We also suspect that the visual output from the flocculus plays an important role in the initiation of pursuit.

Purkinje cells in the vermis of the cerebellum also discharge in relation to pursuit eye movements and visual inputs (Kase et al 1979, Suzuki et al 1981). The firing of these cells encodes the velocity of the target, but detailed analysis of the signals that contribute to their firing has not yet been conducted. The vermis receives visual inputs from the same pontine nuclei that project to the flocculus (Brodal 1979), and electrical stimulation of some parts of the vermis causes smooth eye movements (Ron & Robinson 1973). Thus available data indicate that the cerebellar vermis is involved in pursuit, but we cannot evaluate its exact functional role yet.

A MATHEMATICAL MODEL OF PURSUIT

To complete the description of pursuit, we would like to be able to synthesize the biological observations in the form of a mathematical model. This requires that we (*a*) put together all the pieces discovered by physiological experiments with the constraints specified by anatomical observations, and (*b*) simulate the model on the computer to verify that it performs the same way as does the biological system. As yet, no model of the pursuit system fulfills the two criteria given above. However, mathematical modeling is also a process by which the experimental biologist can test ideas about how the unified system works. From this viewpoint, the failures of any given model are as important as its successes, and modeling is a valuable approach even when biological information about a system is still incomplete, as it is about pursuit.

A Model that Simulates Pursuit on a Millisecond Time Scale

We present a model that has been developed based on our behavioral studies of the maintenance of pursuit (Morris & Lisberger 1985). The model shows that our observations on the visual inputs used during the maintenance of pursuit can explain the monkey's normal smooth eye movement performance.

PSYCHOPHYSICAL BASIS FOR THE MODEL The model is shown in Figure 5A. It includes inputs from three kinds of visual signals, all with 90 ms delays: retinal position, velocity, and acceleration errors (RPE, RVE, RAE). In attempting to simulate the performance of each monkey, the functions relating eye acceleration to each kind of error were taken from the cross-correlation analysis of closed loop tracking in that monkey (e.g. *lower panels* of Figure 4). At the summing junction, the eye acceleration commands from the three kinds of errors are added linearly. The summed visual inputs are then fed into the efferent pathways, which act as velocity memory. We have simulated velocity memory as a single mathematical integration, since the postulated eye velocity positive feedback through the flocculus will act as a mathematical integrator when the delay around the feedback loop is much less than the 90 ms delay for visual inputs (Lisberger et al 1981a). Although derived from psychophysical data, one feature of our model is that its structure (Figure 5A) parallels the anatomical connections of pathways that are involved in pursuit (Figure 5B).

PERFORMANCE OF THE MODEL Figure 6 shows that the model is sensitive to its internal parameters, and, with the correct parameters, accounts for the performance of the monkey. The *top panel* shows that the details of the monkey's eye velocity during one behavioral trial are reproduced in the performance of the model. The *lower panel* shows a quantitative assessment of the performance of the model. Performance has been averaged over many trials and expressed so that a value of 0% on the y-axis corresponds to a perfect prediction of the monkey's eye velocity, while higher values represented poorer predictions.[1] The *trace* labeled "optimum" is the performance of the model when it was used to simulate the pursuit of one monkey using his own functions relating eye acceleration to RPE, RVE, and RAE. The *trace* labeled "other" is the performance of the model when it again was used to simulate the pursuit of the same monkey, but now using the functions for a different monkey.

In one sense, the "excellent" performance of our model is not surprising,

[1] The mean squared error was computed at each time point as the average over many trials of the square of the difference between the model's prediction and the monkey's eye velocity. The "prediction error" was then computed by normalizing: The mean squared error of the prediction was divided by the mean squared error that would have been computed if eye velocity had remained constant from time zero. A model that was unresponsive to visual errors would have a prediction error of 100% and a model that was counterproductive could produce a prediction error greater than 100%.

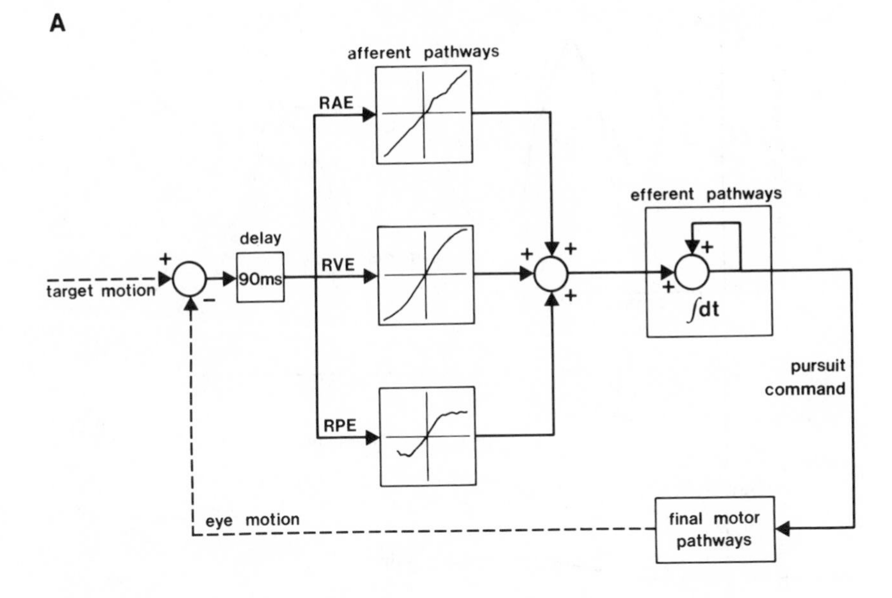

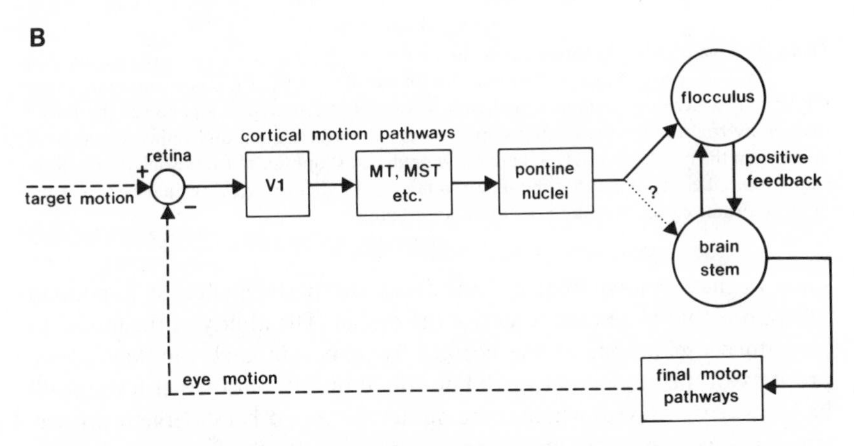

Figure 5 A computer model that simulates pursuit on a millisecond time scale and its relation to the pathways subserving pursuit. The model in *panel A* has the same basic structure as those in Figure 2. Retinal inputs are processed through a 90 ms delay and retinal acceleration, velocity, and position errors (RAE, RVE, and RPE) are transformed according to the relationships obtained in psychophysical experiments (see Figure 4). The efferent pathways contain eye velocity positive feedback, and perform a mathematical integration. In *panel B*, the "cortical motion pathways" correspond to the afferent limb of pursuit, the "pontine nuclei" represent the sensory motor interface, and the positive feedback between the "flocculus" and the "brainstem" corresponds to the efferent pathways.

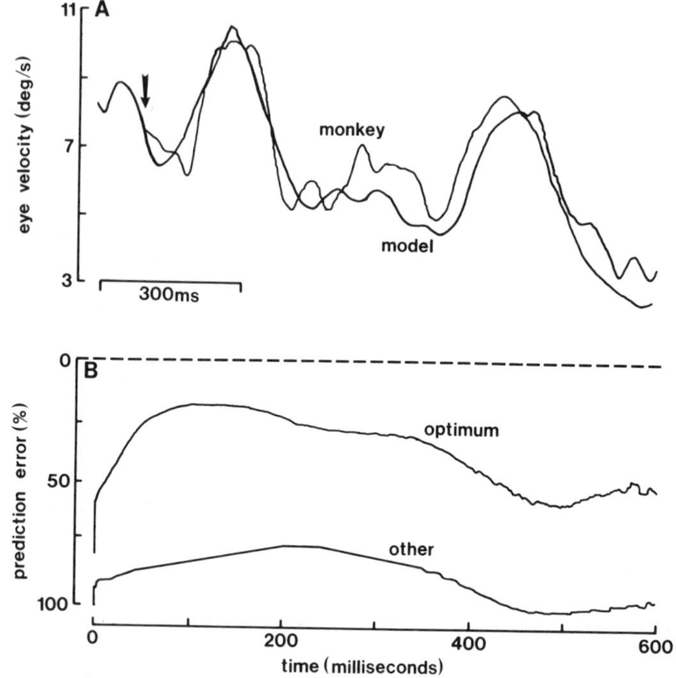

Figure 6 Quantitative performance of the model. *A*. Comparison of the eye velocity of a monkey and the performance of the model for an individual trial. The *downward arrow* shows the time at which the attempted prediction began. *B*. Average performance of the model over many trials, when the relationships between eye acceleration and retinal errors were taken from the monkey whose data are being predicted ("optimum") and when the relationships were taken from another monkey ("other"). "Prediction error" would be 0% if the model reproduced the monkey's performance exactly.

since we have shown that it reproduces the performance of pursuit in conditions that were used to derive the model. The ability of the model to reproduce oscillations at the correct frequency in each monkey is one property of the model that we did not build into it, but further testing will be necessary to reveal whether the model has any other emergent properties. The success of the model in reproducing the performance of each monkey argues that it contains the elements that are the most important in determining the performance of the pursuit system. If one of the elements, say the contribution of RAE, is left out, the model does not perform as well (Morris & Lisberger 1985). In fact, our inability to construct a satisfactory model with just RVE and RPE led us to conduct experiments designed to determine whether RAE contributed to the maintenance of pursuit.

WHAT IS THE CENTRAL COMMAND FOR PURSUIT? Since the early work of Young et al (1968), it has been thought that the central command for pursuit was a neural correlate of target velocity (Figure 2B). This view was put forward in its purest form by Young (1977), who suggested that the pursuit system uses a "target velocity predictor." He suggested that the predictor receives inputs from all the senses and combines them with eye velocity feedback to provide a reconstructed signal related to target velocity.

Our model is closely related to Young's since it also uses a combination of positive feedback (velocity memory) and visual error signals to drive pursuit. In fact, the performance of the two models would be similar in many regards, and it might be difficult to find any single experiment that would contradict one model or the other. But our model is different in that we make no explicit effort to reconstruct a target velocity signal as a command for pursuit. Our model has the additional advantage that its internal elements were identified by observing pursuit performance. It reproduces the pursuit performance of the subject, and it does so with elements that have been identified in psychophysical experiments.

As we explored the performance of our model, we discovered that it always performed quite well when we allowed it to use feedback to correct its own input. The closed loop, negative feedback configuration of the model (and the pursuit system) guaranteed excellent performance most of the time, so that one monkey's closed loop pursuit could be predicted by the model using internal elements derived from another monkey's data. Only when we prevented feedback (open loop conditions) did the performance of the model show the differences between monkeys that we have illustrated in Figure 6B.

The good closed loop performance of the model using functions derived from any monkey allayed our initial concern about the fact that we obtained different functions relating eye acceleration to retinal errors in different monkeys. It appears that the brain's pursuit system does not seek out specially designed input signals (such as target velocity) in the way we might in designing a pursuit control system. Rather, it manages with the signals that are available. The negative feedback loop ensures that the pursuit system will perform well in all subjects, at least in the steady-state, in spite of large individual variations in the input signals derived from the afferent pathways.

Limitations of the Model

FEATURES OF PURSUIT THAT ARE NOT REPRODUCED BY THE MODEL Our model is based on observations made during pursuit maintenance, and we have shown that it reproduces with high fidelity the details of each monkey's steady-state performance. However, some elements will have to be added to the model before it can reproduce the eye velocity recorded during the initiation of pursuit. Thus, our modeling effort is preliminary, but we take its success in reproducing steady-state pursuit as evidence that it will be possible to build more complete models that faithfully reflect biological observations.

STRUCTURES INVOLVED IN PURSUIT BUT NOT INCLUDED IN THE MODEL Like the amateur automobile mechanic, we have built a machine and the result works even though we have a few parts left over that must serve some function. The "spare parts" include at least cortical area MST, the posterior parietal cortex, and the cerebellar vermis. In addition, recent ablation studies have suggested that the frontal eye fields may play a role in pursuit (Keating et al 1985, Lynch & Allison 1985). Two kinds of studies will be helpful in establishing the roles of these structures. First, small chemical lesions in physiologically identified areas of the cerebral cortex will help to establish which of these areas are really involved in pursuit. Second, recordings from neurons in these areas during step-ramp target motion will help to establish the latency at which these structures could contribute to pursuit, and to determine whether these areas are involved in the initiation of pursuit, the maintenance of pursuit, neither, or both. As we discussed above, MST and the parietal cortex both contain cells that respond to a complicated combination of visual inputs and pursuit eye movements. Perhaps these areas regulate the more cognitive aspects of pursuit, such as target selection, prediction and smooth eye movements based on expectations.

SUMMARY

The function of smooth pursuit is to keep the fovea pointed at a small visual target that moves smoothly across a patterned background. Chemical lesions, single cell recordings, and behavioral measures have shown that the cortical motion processing pathways form the afferent limb for pursuit. Important areas include at least the striate cortex and the middle temporal visual area, and probably the medial superior temporal visual area and the posterior parietal cortex. We argue that the visual inputs are transmitted through a simple sensory motor interface in the pons, to the efferent limb in the brain stem and cerebellum. The efferent limb uses neural velocity memory to maintain pursuit automatically. We present evidence that the velocity memory is provided, at least in part, by eye velocity positive feedback between the flocculus of the cerebellum and the brain stem. Finally, we use a computer model to show how the maintenance of pursuit can be simulated on a millisecond time scale. The structure and internal elements of the model are based on the biological experiments reviewed in our paper.

In the past five years, progress on the neural basis of pursuit eye movements has been rapid. Several areas of research have made substantial contributions, by using combinations of new and conventional methods. Many of the pathways that contribute to pursuit have been identified, and their physiological activity and functions are becoming understood. Continuing progress promises to yield an understanding of one specific form of visually guided movement, at the level of neuronal circuits and behavior, in the primate.

ACKNOWLEDGMENTS

Our paper is dedicated to the late Edward V. Evarts, who provided invaluable encouragement and support to the senior author. We are also grateful to Drs. Bill Newsome and Bob Wurtz for their contributions at the outset of our work on pursuit, and we thank Dr. Newsome and especially Dr. Lee Stone for their critical comments on earlier versions of the review. The work from our laboratory was supported by NIH grant EY 03878, NSF grant BNS 8314050, a fellowship from the Alfred P. Sloan Foundation, and a Scholars Award from the McKnight Foundation, all to S.G.L.

Literature Cited

Allman, J., Miezin, F., McGuinness, E. 1985. Stimulus specific responses from beyond the classical receptive field: Neurophysiological mechanisms for local-global comparisons in visual neurons. *Ann. Rev. Neurosci.* 8: 407–30

Atkeson, C. G., Hollerbach, J. M. 1985. Kinematic features of unrestrained vertical arm movements. *J. Neurosci.* 9: 2318–30

Baker, R., Precht, W., Llinas, R. 1972. Cerebellar modulatory action on the vestibulo-trochlear pathway in the cat. *Exp. Brain Res.* 15: 364–85

Barnes, G. R. 1983. The effects of retinal target location on suppression of the vestibulo-ocular reflex. *Exp. Brain Res.* 49: 257–68

Brodal, P. 1978. The corticopontine projection in the rhesus monkey: Origin and principles of organization. *Brain* 101: 251–83

Brodal, P. 1979. The pontocerebellar projection in the rhesus monkey: An experimental study with retrograde axonal transport of horseradish peroxidase. *Neuroscience* 4: 193–208

Carl, J. R., Gellman, R. S. 1985. Human smooth pursuit: The response to conflicting velocity and position stimuli. *Soc. Neurosci. Abstr.* 11: 78

Chubb, M. C., Fuchs, A. F. 1982. Contribution of y group of vestibular nuclei and dentate nucleus of cerebellum to generation of vertical smooth eye movements. *J. Neurophysiol.* 48: 75–79.

Cohen, B., Matsuo, V., Raphan, T. 1977. Quantitative analysis of the velocity characteristics of optokinetic nystagmus and optokinetic after-nystagmus. *J. Physiol.* 270: 321–44

Cohen, J. L., Robinson, F., May, J., Glickstein, M. 1981. Corticopontine projections of the lateral suprasylvian cortex: De-emphasis of the central visual field. *Brain Res.* 219: 239–48

Collewijn, H. 1969. Optokinetic eye movements in the rabbit: Input-output relationships. *Vision Res.* 9: 117–32

Collewijn, H., Tamminga, E. P. 1984. Human smooth and saccadic eye movements during voluntary pursuit of different target motions on different backgrounds. *J. Physiol.* 351: 217–50

Cushman, W. B., Tangney, J. F., Steinman, R. M., Ferguson, J. L. 1984. Characteristics of smooth eye movements with stabilized targets. *Vision Res.* 24: 1003–9

Dodge, R. 1903. Five types of eye movements in the horizontal meridian plane of the field of regard. *Am. J. Physiol.* 8: 307–29

Dubner, R., Zeki, S. M. 1971. Response properties and receptive fields of cells in an anatomically defined region of the superior temporal sulcus in the monkey. *Brain Res.* 35: 239–48

Dursteler, M. R., Wurtz, R. H., Newsome, W. T. 1987. Directional and retinotopic deficits following lesions of the foveal representation within the superior temporal sulcus of the macaque monkey. *J. Neurophysiol.* Submitted

Eckmiller, R., Mackeben, M. 1978. Pursuit eye movements and their neural control in the monkey. *Pfluegers Arch.* 377: 15–23

Fries, W. 1981. The projections from striate and prestriate visual cortex onto the pontine nuclei in the macaque monkey. *Soc. Neurosci. Abstr.* 7: 762

Fuchs, A. F. 1967a. Saccadic and smooth pursuit eye movements in the monkey. *J. Physiol.* 191: 609–31

Fuchs, A. F. 1967b. Periodic eye tracking in the monkey. *J. Physiol.* 193: 161–71

Fuchs, A. F., Kaneko, C. R. S., Scudder, C. A. 1985. Brainstem control of saccadic eye movements. *Ann. Rev. Neurosci.* 8: 307–37

Glickstein, M., Cohen, J. L., Dixon, B., Gibson, A., Hollins, M., Labossiere, E., Robinson, F. 1980. Corticopontine visual projections in macaque monkeys. *J. Comp. Neurol.* 190: 209–29

Haegerstrom-Portnoy, G., Brown, B. 1979. Contrast effects on smooth pursuit eye movement velocity. *Vision Res.* 19: 169–74

Heywood, S., Churcher, J. 1971. Eye movements and the afterimage. I. Tracking the afterimage. *Vision Res.* 11: 1163–68

Hoffman, K.-P., Distler, C. 1986. The role of direction selective cells in the nucleus of the optic track of cat and monkey during optokinetic nystagmus. In *Adaptive Processes in Visual and Oculomotor Systems*, ed. E. L. Keller, D. S. Zee. Oxford: Pergamon

Ito, M. 1982. Cerebellar control of the vestibulo-ocular reflex—around the flocculus hypothesis. *Ann. Rev. Neurosci.* 5: 275–96

Kase, M., Noda, H., Suzuki, D., Miller, D. C. 1979. Target velocity signals of visual tracking in vermal Purkinje cells of the monkey. *Science* 205: 717–20

Kawano, K., Sasaki, M., Yamashita, M. 1984. Response properties of neurons in posterior parietal cortex of monkey during visual-vestibular stimulation. I. Visual tracking neurons. *J. Neurophysiol.* 51: 340–51

Keating, E. G., Gooley, S. G., Kenney, D. V. 1985. Impaired tracking and loss of predictive eye movements after removal of the frontal eye fields. *Soc. Neurosci. Abstr.* 11: 472

Keller, E. L. 1978. Gain of the vestibulo-ocular reflex in monkey at high rotational frequencies. *Vision Res.* 20: 535–38

Koerner, F., Schiller, P. H. 1972. The optokinetic response under open and closed loop conditions in the monkey. *Exp. Brain Res.* 14: 318–30

Kommerell, G., Taumer, R. 1972. Investigations of the eye tracking system through stabilized retinal images. *Bibl. Ophthal.* 82: 288–97

Kowler, E., Murphy, B. J., Steinman, R. M. 1978. Velocity matching during smooth pursuit of different targets on different backgrounds. *Vision Res.* 18: 603–5

Kowler, E., Steinman, R. M. 1979. The effect of expectations on slow oculomotor control. II. Single target displacements. *Vision Res.* 19: 633–46

Kowler, E., van der Steen, J., Tamminga, E. P., Collewijn, H. 1984. Voluntary selection of the target for smooth eye movement in the presence of superimposed, full-field stationary and moving stimuli. *Vision Res.* 12: 1789–98

Leigh, R. J., Newman, S. A., Zee, D. S., Miller, N. R. 1982. Visual following during stimulation of an immobile eye (the open loop condition). *Vision Res.* 22: 1193–97

Langer, T. 1985. Basal interstitial nucleus of the cerebellum: A deep cerebellar nucleus related to the flocculus. *J. Comp. Neurol.* 235: 38–47

Langer, T., Fuchs, A. F., Scudder, C. A., Chubb, M. C. 1985a. Afferents to the flocculus of the cerebellum in the rhesus macaque as revealed by retrograde transport of horseradish peroxidase. *J. Comp. Neurol.* 235: 1–25

Langer, T., Fuchs, A. F., Chubb, M. C., Scudder, C. A., Lisberger, S. G. 1985b. Floccular efferents in the rhesus monkey as revealed by autoradiography and horseradish peroxidase. *J. Comp. Neurol.* 235: 26–37

Lisberger, S. G. 1984. The latency of pathways containing the site of motor learning in the monkey vestibulo-ocular reflex. *Science* 225: 74–76

Lisberger, S. G., Fuchs, A. F. 1978a. Role of primate flocculus during rapid behavioral modification of vestibulo-ocular reflex. I. Purkinje cell activity during visually guided horizontal smooth-pursuit eye movements and passive head rotation. *J. Neurophysiol.* 41: 733–63

Lisberger, S. G., Fuchs, A. F. 1978b. Role of primate flocculus during rapid behavioral modification of vestibulo-ocular reflex. II. Mossy fiber firing patterns during horizontal head rotation and eye movement. *J. Neurophysiol.* 41: 764–77

Lisberger, S. G., Evinger, C., Johanson, G. W., Fuchs, A. F. 1981a. Relationship between eye acceleration and retinal image velocity during foveal smooth pursuit eye movements in man and monkey. *J. Neurophysiol.* 46: 229–49

Lisberger, S. G., Miles, F. A., Optican, L. M., Eighmy, B. B. 1981b. Optokinetic response in monkey: Underlying mechanisms and their sensitivity to long-term adaptive changes in vestibulo-ocular reflex. *J. Neurophysiol.* 45: 869–90

Lisberger, S. G., Pavelko, T. A. 1984. Functional properties of brainstem cells inhibited from the cerebellar flocculus in monkey. *Neurosci. Abstr.* 10: 988

Lisberger, S. G., Westbrook, L. E. 1985. Properties of visual inputs that initiate horizontal smooth pursuit eye movements in monkeys. *J. Neurosci.* 5: 1662–73

Lynch, J. C., Allison, J. C. 1985. A quantitative study of visual pursuit deficits following lesions of the frontal eye fields in rhesus monkeys. *Soc. Neurosci. Abstr.* 11: 473

Lynch, J. C., McLaren, J. W. 1982. The contribution of parieto-occipital association cortex to the control of slow eye movements. In *Functional Basis of Ocular Motility Disorders*, ed. D. S. Zee, E. L. Keller, pp. 501–10. Oxford: Pergamon

Maunsell, J. H. R., Newsome, W. T. 1987. Visual processing in monkey extrastriate cortex. *Ann. Rev. Neurosci.* 10: 363–401

Maunsell, J. H. R., Van Essen, D. C. 1983. Functional properties of neurons in middle temporal visual area of the macaque monkey. I. Selectivity for stimulus direction, speed, and orientation. *J. Neurophysiol.* 49: 1127–47

May, J. G., Keller, E. L., Crandall, W. F. 1985. Changes in eye velocity during smooth pursuit tracking induced by microstimulation in the dorsolateral pontine nucleus of the macaque. *Soc. Neurosci. Abstr.* 11: 79

McKee, S. M., Welch, L. 1985. Sequential recruitment in the discrimination of velocity. *J. Opt. Soc. Am. A* 2: 243–51

Michael, J. A., Melvill Jones, G. 1966. Dependence of visual tracking capability upon stimulus predictability. *Vision Res.* 6: 707–16

Miles, F. A., Fuller, J. H. 1975. Visual tracking and the primate flocculus. *Science* 189: 1000–2

Miles, F. A., Fuller, J. H., Braitman, D. J., Dow, B. M. 1980. Long-term adaptive changes in primate vestibuloocular reflex. III. Electrophysiological observations in flocculus of normal monkeys. *J. Neurophysiol.* 43: 1437–76

Miles, F. A., Lisberger, S. G. 1981. Plasticity in the vestibulo-ocular reflex: A new hypothesis. *Ann. Rev. Neurosci.* 4: 273–99

Morris, E. J., Lisberger, S. G. 1983. Signals used to maintain smooth pursuit eye movements in monkeys: Effects of small retinal position and velocity errors. *Soc. Neurosci. Abstr.* 9: 866

Morris, E. J., Lisberger, S. G. 1985. A computer model that predicts monkey smooth pursuit eye movements on a millisecond timescale. *Soc. Neurosci. Abstr.* 11: 79

Mountcastle, V. B., Lynch, J. C., Georgopoulos, A., Sakata, H., Acuna, C. 1975. Posterior parietal association cortex of the monkey: Command functions for operations within extrapersonal space. *J. Neurophysiol.* 38: 871–908

Muratore, R., Zee, D. S. 1979. Pursuit afternystagmus. *Vision Res.* 19: 1057–59

Newsome, W. T., Wurtz, R. H. 1981. Response properties of single neurons in the middle temporal visual area (MT) of alert macaque monkeys. *Soc. Neurosci. Abstr.* 7: 832

Newsome, W. T., Wurtz, R. H., Dursteler, M. R., Mikami, A. 1985a. Deficits in visual motion processing following ibotenic acid lesions of the middle temporal visual area of the macaque monkey. *J. Neurosci.* 5: 825–40

Newsome, W. T., Wurtz, R. H., Dursteler, M. R., Mikami, A. 1985b. Punctate chemical lesions of striate cortex in the macaque monkey: Effect on visually guided saccades. *Exp. Brain Res.* 58: 392–99

Noda, H. 1981. Visual mossy fiber inputs to the flocculus of the monkey. *Ann. NY Acad. Sci.* 374: 465–75

Noda, H., Suzuki, D. A. 1979. The role of the flocculus of the monkey in fixation and smooth pursuit eye movements. *J. Physiol.* 294: 335–48

Optican, L. M., Robinson, D. A. 1980. Cerebellar-dependent adaptive control of primate saccadic system. *J. Neurophysiol.* 44: 1058–76

Optican, L. M., Zee, D. S., Chu, F. C. 1985. Adaptive response to ocular muscle weakness in human pursuit and saccadic eye movements. *J. Neurophysiol.* 54: 110–22

Pola, J., Wyatt, H. J. 1979. Target position and velocity: The stimuli for smooth pursuit eye movements. *Vision Res.* 20: 523–34

Rashbass, C. 1961. The relationship between saccadic and smooth tracking eye movements. *J. Physiol.* 159: 326–38

Robinson, D. A. 1964. The mechanics of human saccadic eye movement. *J. Physiol.* 174: 245–64

Robinson, D. A. 1965. The mechanics of human smooth pursuit eye movements. *J. Physiol.* 180: 569–91

Robinson, D. A. 1971. Models of oculomotor neural organization. In *The Control of Eye Movements*, ed. P. Bach-y-Rita, C. C. Collins, pp. 519–38. New York: Academic

Ron, S., Robinson, D. A. 1973. Eye movements evoked by cerebellar stimulation in the alert monkey. *J. Neurophysiol.* 36: 1004–22

Sakata, H., Shibutani, H., Kawano, K. 1983. Functional properties of visual tracking neurons in the posterior parietal association cortex of the monkey. *J. Neurophysiol.* 49: 1364–80

Segraves, M. A., Goldberg, M. E., Deng, S.-Y., Bruce, C. J., Ungerleider, L. G., Mishkin, M. 1986. No notion of motion: Monkeys with unilateral striate lesions have long-term deficits in the utilization of stimulus velocity information by the oculomotor system. See Hoffman & Distler 1986

Simpson, J. I. 1984. The accessory optic system. *Ann. Rev. Neurosci.* 7: 13–41

Stark, L., Vossius, G., Young, L. R. 1962. Predictive control of eye tracking movements. *IRE Trans. Hum. Factors Electron.* 3: 52–56

Steinbach, M. J. 1976. Pursuing the perceptual rather than the retinal stimulus. *Vision Res.* 16: 1371–76

Steinman, R. M. 1986. The need for an eclectic, rather than systems, approach to the study of the primate oculomotor system. *Vision Res.* 26: 101–12

Stone, L. S., Lisberger, S. G. 1985. Visual responses in "gaze velocity" Purkinje cells in the primate cerebellar flocculus. *Soc. Neurosci. Abstr.* 11: 1034

Suzuki, D. A., Keller, E. L. 1984. Visual signals in the dorsolateral pontine nucleus of the alert monkey: Their relationship to smooth pursuit eye movements. *Exp. Brain Res.* 53: 473–78

Suzuki, D. A., Keller, E. L., Yee, R. D. 1985. Smooth pursuit eye movement related visual and visuo-motor responses in dorsolateral pontine nucleus of alert monkey. *Soc. Neurosci. Abstr.* 11: 473

Suzuki, D. A., May, J. G., Keller, E. L. 1984. Smooth pursuit eye movement deficits with pharmacological lesions in monkey dorsolateral pontine nucleus. *Soc. Neurosci. Abstr.* 10: 58

Suzuki, D. A., Noda, H., Kase, M. 1981. Visual and pursuit eye movement-related activity in posterior vermis of monkey cerebellum. *J. Neurophysiol.* 46: 1120–39

Tanaka, K., Hikosaka, K., Saito, H., Yukie, M., Fukada, Y., Iwai, E. 1986. Analysis of local and wide-field movements in the superior temporal visual areas of the macaque monkey. *J. Neurosci.* 6: 134–44

Tychsen, L., Lisberger, S. G. 1986a. Visual motion processing for initiation of smooth pursuit eye movements in humans. *J. Neurophysiol.* 6: In press

Tychsen, L., Lisberger, S. G. 1986b. Maldevelopment of visual motion processing humans who had strabismus with onset in infancy. *J. Neurosci.* 6: 2495–2508

Ungerleider, L. G., Mishkin, M. 1982. Two cortical visual systems. In *Analysis of Visual Behavior*, ed. D. J. Ingle, M. A. Goodale, R. J. W. Mansfield, pp. 549–86. Cambridge: MIT Press

Van Essen, D. C., Maunsell, J. H. R. 1983. Hierarchical organization and functional streams in the visual cortex. *Trends. Neurosci.* 6: 370–75

Van Essen, D. C., Maunsell, J. H. R., Bixby, J. L. 1981. The middle temporal visual area in the macaque: Myeloarchitecture, connections, functional properties and topographic organization. *J. Comp. Neurol.* 199: 293–326

Waespe, W., Cohen, B., Raphan, T. 1983. Role of the flocculus and paraflocculus in optokinetic nystagmus and visual-vestibular interactions: Effects of lesions. *Exp. Brain Res.* 50: 9–33

Warabi, T., Noda, H., Ishi, N. 1979. Effect of retinal image motion upon flocculus Purkinje cell activity during smooth pursuit eye movements. *Soc. Neurosci. Abstr.* 5: 108

Westheimer, G. 1954. Eye movement responses to a horizontally moving visual stimulus. *A.M.A. Arch. Ophthal.* 52: 932–41

Westheimer, G., Blair, S. M. 1973. Oculomotor defects in cerebellectomized monkeys. *Invest. Ophthalmol.* 12: 618–21

Westheimer, G., McKee, S. M. 1975. Visual acuity in the presence of retinal image motion. *J. Opt. Soc. Am.* 65: 847–50

Winterson, B. J., Steinman, R. M. 1978. The effect of luminance on human smooth pursuit of perifoveal and foveal targets. *Vision Res.* 18: 1165–72

Wurtz, R. H., Newsome, W. T. 1985. Divergent signals encoded by neurons in extrastriate areas MT and MST during smooth pursuit eye movements. *Soc. Neurosci. Abstr.* 11: 1246

Yasui, S., Young, L. R. 1984. On the predictive control of foveal eye tracking and slow phases of optokinetic and vestibular nystagmus. *J. Physiol.* 347: 17–33

Young, L. R. 1977. Pursuit eye movement—what is being pursued? In *Control of Gaze by Brain Stem Neurons. Dev. Neurosci.* 1: 29–36

Young, L. R., Forster, J. D., van Houtte, N. 1968. A revised stochastic sampled model for eye tracking movements. *4th Ann. NASA-Univ. Conf. Manual Control*, Univ. Mich., Ann Arbor

Zee, D. S., Optican, L. M., Cook, J. D., Robinson, D. A., Engel, W. K. 1976. Slow saccades in spinocerebellar degeneration. *Arch. Neurol.* 33: 243–51

Zee, D. S., Tusa, R. J., Butler, P. H., Herdman, S. J., Gucer, G. 1986. The acute and chronic effects of bilateral occipital lobectomies on eye movements in monkey. See Hoffman & Distler 1986

Zee, D. S., Yamazaki, A., Butler, P. H., Gucer, G. 1981. Effects of ablation of flocculus and paraflocculus on eye movements in primate. *J. Neurophysiol.* 46: 878–99

Ann. Rev. Neurosci. 1987. 10 : 131–61

LONG-TERM POTENTIATION

T. J. Teyler and P. DiScenna

Neurobiology Department, Northeastern Ohio Universities College of Medicine, Rootstown, Ohio 44272

PHENOMENOLOGY

History and Definition

In 1973, two reports appeared in the *Journal of Physiology* that described a long-term potentiation (LTP) of synaptic transmission at a monosynaptic junction in the mammalian CNS. These two papers, one dealing with the anesthetized rabbit preparation (Bliss & Lømo 1973) and the other dealing with the unanesthetized rabbit (Bliss & Gardner-Medwin 1973), were of considerable interest because they marked the first demonstration of a neurophysiological alteration in the mammalian brain possessing a considerable time-course. Prior to these observations, neurophysiological phenomena were known to be of relatively modest temporal extent.

The experiments of Bliss and colleagues showed that in the perforant path to dentate gyrus synapse, substantial increases in synaptic efficacy were observed following tetanic stimulation of afferent fibers. Synaptic efficacy refers to the postsynaptic response to a constant afferent volley. The changes (on the order of a 50% increase in the amplitude of the response) lasted for at least ten hours in the anesthetized preparation and up to 16 weeks in the unanesthetized preparation following a series of tetanic stimuli. These results attracted great interest because of the possibility that the phenomenon might underlie some aspects of memory storage. A critical question—not yet completely answered—was stated by Bliss & Lømo: "Whether or not the intact animal makes use in real life of a property which has been revealed by synchronous, repetitive volleys to a population of fibers the normal and pattern of activity along which are unknown, is another matter" (Bliss & Lømo 1973, p. 355). As we shall see, the available evidence suggests that LTP does occur in conjunction with behavioral learning—although whether it is necessary and sufficient is far from certain.

0147–006X/87/0301–0131$02.00

Long-term potentiation is defined as a stable, relatively long lasting increase in the magnitude of a post-synaptic response to a constant afferent volley following brief tetanic stimulation of the same afferents. LTP is thus an increase in synaptic efficacy, at monosynaptic junctions, occurring as the result of an afferent fiber tetanization. Initially observed in the hippocampus, LTP has now been documented in a variety of brain structures and in a variety of species. LTP is also known as LTE (long-term enhancement) and LLP (long-lasting potentiation). In this review we consider the phenomenology of LTP, its hypothesized mechanism(s) of action, and the implications that arise from considering LTP as a candidate mnemonic device in the brain.

Preparations

LTP was initially observed in the hippocampus of both anesthetized and unanesthetized preparations (Bliss & Lømo 1973, Bliss & Gardner-Medwin 1973, Douglas & Goddard 1975). Subsequently, LTP was observed in slices of hippocampus in area CA1 (Schwartzkroin & Wester 1975, Lynch et al 1977, Andersen et al 1977), CA3 (Alger & Teyler 1976, Yamamoto & Chujo 1978), and dentate gyrus (Alger & Teyler 1976) with few differences noted between in vivo and in vitro preparations (Teyler et al 1977). The hippocampus remained the structure of choice for studying LTP in the mammalian brain.

Both in vivo and in vitro preparations continue to be utilized today, with the choice determined by the nature of the experiment. In the future, we will probably see the combined use of both approaches—the utility of which was recently demonstrated by Disterhoft et al (1986), who obtained evidence that behavioral eyelid conditioning of rabbits subsequently results in a reduction of Ca^{2+}-dependent K^{+} current of hippocampal CA1 cells studied in vitro. The demonstration that neuronal changes induced in vivo can be preserved and studied in vitro is important and may permit an extension of the advantages inherent in the brain slice preparation to behavioral studies.

Distribution of LTP

The analysis of other monosynaptic junctions in the CNS began to yield data suggesting that LTP was not limited to the trisynaptic circuitry of the hippocampus. LTP has been established in the sprouted, crossed entorhinal projection to dentate gyrus (Wilson et al 1979); in the septal projections to CA1 (Racine et al 1983) and dentate gyrus (McNaughton & Miller 1984); the perforant path projection to ipsilateral CA1 (Doller & Weight 1984); and in the commissural projection to area CA1 (Buzsáki 1980). It is interesting that the magnitude of the induced LTP was in many cases considerably less than that recorded within the trisynaptic pathway. The reasons for this difference are unclear but may relate to the innervation density of these pathways and the coactivation requirement (discussed below). Inhibitory, feedforward interneurons of the CA1 area have been found to exhibit LTP in anesthetized rat (Buzsáki & Eidelberg 1982), as evidenced by extracellular recording techniques, thus raising the possibility that IPSPs can also exhibit LTP, a conclusion recently supported by Patneau & Stripling (1985) in work with the pyriform cortex.

Recent studies have shown that LTP can be observed across a wide variety of brain areas and species. In mammalian systems LTP has been demonstrated in the subiculum and septum (Racine et al 1983) and medial geniculate (Gerren & Weinberger 1983). The role of cerebellar circuitry in learning and memory, which figures prominently in the theories of Marr (1969) and Albus (1971), has come under increased scrutiny and interest of late (Thompson 1983, Thompson et al 1983). Given the role that cerebellar circuits play in certain forms of motor learning, it is important to know whether LTP is present at cerebellar synapses. Enduring changes in synaptic efficacy are seen in cerebellar cortex (Ito 1983) and deep cerebellar nuclei (Racine et al 1986). Interestingly, the output of potentiated cerebellar Purkinje cells displays an enduring response depression (Y. Miyashita, personal communication).

Neocortical LTP has been reported for striate cortex (Komatsu et al 1981, Lee 1982) and somatosensory cortex (Voronin 1985) following stimulation of subjacent white matter, and for entorhinal cortex (Racine et al 1983) following stimulation of the amygdala. However, relative to the hippocampus, cortical LTP is difficult to study given the heterogeneous nature of the neuronal elements within cortical tissue. A better definition of the synapses involved is provided by LTP in pyriform cortex following stimulation of the olfactory bulb (Stripling et al 1984, Stripling & Patneau 1985, Patneau & Stripling 1985), the callosal fibers of the neocortex (Wilson 1984), and the claustrum projections to entorhinal cortex (Wilhite et al 1986). Apart from the obvious interest in cortical LTP (Eccles 1983, Teyler & DiScenna 1984, 1985), the study of Komatsu et al (1981) is particularly provocative in that the ability to induce neocortical LTP was correlated with a critical period for visual plasticity.

LTP has also been observed in muscarinic systems in autonomic ganglia (Libet et al 1975) and in nicotinic systems in the superior cervical ganglion of rat (Brown & McAfee 1982) and the sympathetic ganglia of bullfrog (Koyano et al 1985). These latter preparations are of interest in that apparently presynaptic mechanisms leading to increased neurotransmitter release are responsible for the LTP observed (Briggs et al 1985). A quantal analysis in bullfrog sympathetic ganglia indicated that the response enhancement was associated with an increase in quantal content and/or quantal size (Koyano et al 1985). Consistent with this proposed presynaptic locus of effect, pharmacologic blockade of nicotinic, muscarinic, and adrenergic receptors failed to block LTP development (Briggs et al 1985). Both groups of investigators have implicated presynaptic Ca^{2+} levels in LTP in these preparations.

Several nonmammalian vertebrate systems demonstrate LTP. The cholinergic retinal input to goldfish optic tectum was studied by Lewis & Teyler (1986a), who found a gradual response enhancement following low-frequency stimulation of optic nerve fibers. Larson & Lynch (1985) examined an isolated lizard cortex preparation and found that the apical dendrites of medial cortex neurons (but not basilar dendrites) support LTP. It has been suggested that reptile medial cortex is homologous to parts of mammalian hippocampus (Northcutt 1981). However, pyramidal cells in rodent hippocampus show potentiation of afferents to both apical and basilar dendrites. Thus, the lizard preparation may provide information regarding the mechanism(s) of LTP (comparing the properties of apical vs basilar dendrites, as suggested by Larson & Lynch) as well as the beginning of a phylogenetic analysis of LTP distribution.

LTP is not limited to vertebrate systems, as shown by Walters & Byrne (1985) in *Aplysia* and, perhaps surprisingly, in the crayfish neuromuscular junction. In these phasic motoneurons, an enhancement of junctional transmission is seen following motor nerve stimulation (Lnenicka & Atwood 1985), an effect attributed to an increase in quantal content (not size) following tetanization (Baxter et al 1985).

In the only human study, electrical stimulation of the hippocampal gyrus yields response changes associated with learning and memory tasks that may reflect the operation of an underlying process such as LTP (Babb 1982). These experiments, while provocative, must remain equivocal due to the necessary limitations in experimental design.

The ontogenetic development of LTP has been examined in the rodent hippocampus, where it is found to develop between postnatal days 7 and 10 for area CA1 (Harris & Teyler 1984, Baudry et al 1981) and somewhat later (7–28 days) in the dentate gyrus (Wilson 1984, Duffy & Teyler 1978). The magnitude and incidence of LTP shows a developmental gradient. Prior to the appearance of LTP, synapses have established functional connectivity and are capable of displaying such phenomena as post-tetanic potentiation (PTP). Similar findings have been reported for neocortical transcortical LTP (Wilson & Racine 1983).

LTP appears to be a widespread phenomenon; however, it is not yet established that these diverse examples of enhanced synaptic activity represent the same phenomenon. In many experiments the parameters required to initiate LTP (as well as the phenomenon of LTP itself) are different from those required in the trisynaptic circuitry of the hippocampus. In some cases (the medial geniculate nucleus, for example) the magnitude of the induced LTP is considerably less than that seen in the hippocampus and its duration is much shorter; in others (the goldfish optic tectum, for example) the stimulation parameters required to elicit LTP are outside the range effective in hippocampus. The question to be resolved is whether all of these forms of plasticity represent a unitary phenomenon or whether different phenomena are being grouped under the heading of LTP. Ultimately the answer to this question will rely upon determining the mechanisms underlying all forms of LTP.

The vast majority of experiments with LTP have been done in the rat, guinea pig, and rabbit, with emphasis on hippocampal circuitry. There are large gaps in our knowledge of the distribution of LTP within different regions of rodent brain, across developmental stages, and across phyla.

Parameters and Necessary Conditions

The induction of LTP (we limit our consideration here to hippocampus/dentate gyrus) begins with the delivery of tetanic stimulation to afferent fibers. Depending upon the intensity, frequency, and pattern of afferent activation, LTP can be established in a gradual or all-or-none fashion. At higher stimulus intensities and higher tetanic frequencies, LTP reaches an asymptotic level after one or a few tetanizations.

Following termination of the tetanus it is possible to record a substantial post-tetanic potentiation (PTP) from the effected synapses (Bliss & Lømo 1973, Deadwyler et al 1975, McNaughton et al 1978, Harris & Teyler 1984). PTP is a short-term increase in synaptic efficacy (e.g. time constant = 90 sec at the perforant path-granule cell synapse; McNaughton 1982). PTP and LTP appear to be subserved by different neural mechanisms (McNaughton 1982, Stanton & Sarvey 1984). Following the PTP one of three things may happen: (*a*) the response may increase to form the relatively long-lasting change known as LTP; (*b*) a brief heterosynaptic depression may be seen prior to the development of LTP (such depression has been observed to peak several minutes after the offset of the tetanus); (*c*) LTP may fail to develop (Duffy & Teyler 1978). Various laboratories have reported differing incidence of LTP failure ranging from 10% to more than 60%. The significance of this finding is difficult to evaluate due to the unknown reason(s) for the failure of LTP (Yamamoto & Sawada 1981). Although it has not been systematically studied, LTP failure does not appear to be related to deteriorated physiological conditions. Since LTP reaches an asymptote following the delivery of tetanic stimulation, it is possible that those preparations failing to exhibit LTP have previously been potentiated and thus are incapable of changing further. This is a difficult proposition to prove.

LTP can be induced by a wide range of tetanus frequencies and patterns. Tetanus frequencies supporting LTP range from 0.2 to 400 Hz (Douglas 1977, Skelton et al 1983). Usually a total of 100–200 pulses are delivered. At lower frequencies of stimulation (below 50 Hz), a heterosynaptic response depression can be seen following the termination of the tetanus (Alger et al 1978). The heterosynaptic depression following low frequency tetanus may be caused by the enhanced ability of a low frequency stimulus to activate inhibitory interneurons projecting on the target cells. At higher frequencies of stimulation (between 100 and 400 Hz), such heterosynaptic depression is rarely observed (McNaughton et al 1978, Douglas 1977, but see Abraham & Goddard 1983). Often the tentanus is delivered as a package of short bursts (8 to 10 stimuli) of 400 Hz stimulation repeated every few seconds or minutes. This technique produces reliable LTP, avoids a potential problem with afterdischarges, which sometimes accompany long trains of stimuli, and provides a more physiologic stimulus (Douglas 1977).

The duration of LTP is considerable by neurophysiological standards, but insufficient when considered in behavioral terms. The original descriptions of LTP in intact rabbits indicated a duration of about three days following a single tetanus (Bliss & Gardner-Medwin 1973). However, they also showed that the duration could be extended considerably by repeating the tetanic stimulus. Recent studies on rodents have demonstrated decay rates ranging from 3 to 13 days (Barnes 1979, Racine et al 1983, de Jonge & Racine 1985). The effect of delivering the tetanic stimulation as multiple exposures over time may prolong the duration of LTP (Barnes 1979, Barnes & McNaughton 1980, Reymann et al 1985), although the experiment of de Jonge & Racine (1985) argues otherwise. This question is important in terms of the role of in vivo afferent activity in inducing LTP in behavioral settings.

Coactivation

In a general sense, "coactivation" refers to the repeated demonstration that as the intensity of afferent drive during the tetanus is increased, the probability that LTP will occur increases and the magnitude of LTP increases (Bliss & Lømo 1973, Schwartzkroin & Wester 1975, McNaughton et al 1978, Yamamoto & Sawada 1981, Lee 1983). Coactivation (also termed "cooperativity") is also implied by the demonstration of a stimulus intensity threshold for the production of LTP at the perforant path–dentate granule cell (PP-DGC) synapses. This LTP threshold is at or near the stimulus intensity required to elicit a population spike (McNaughton et al 1978). The population spike, per se, does not appear to be necessary for LTP (McNaughton et al 1978, Douglas et al 1982). Rather, it appears that a minimum number of afferent fibers or synaptic contacts must be coactive in order for LTP to occur. Perhaps this reflects a requirement for a given level of depolarization by some integrating component. It is of interest to know whether this threshold exists at the intracellular level.

The question of an LTP threshold has not been systematically examined outside of the PP-DGC system. There are conflicting reports regarding area CA1 as to whether orthodromic stimulation must produce spike activity in order to induce LTP. Two intracellular studies suggest that spike activity is not necessary (Wigström et al 1982, Haas & Rose 1984), whereas an extracellular investigation suggests that it is essential (Scharfman & Sarvey 1985).

A third line of evidence for the coactivation phenomenon comes from the associative LTP paradigm. In associative LTP, the concurrent activation of independent weak and strong synaptic inputs produces LTP of the weak input, where independent tetanization of the weak input would not result in LTP. This has been demonstrated in the ipsilateral and contralateral entorhinal projections to dentate (Levy & Steward 1979), the septal and entorhinal projections to dentate (Robinson & Racine 1982), converging pathways in motor cortex (Baranyi & Feher 1981), two independent Schaffer collateral inputs to CA1 (Barrionuevo & Brown 1983), and converging inputs into cat autonomic (stellate) ganglia (Mochida & Libet 1985). Barrionuevo & Brown (1983) reported associative LTP at extra- and intracellular levels.

It has been observed that the hippocampal synapses possess nonlinear properties (Robinson & Racine 1982, McNaughton et al 1981, Sclabassi et al 1985). The sum of individual pathway effects when exposed to the associative LTP paradigm shows that the final level of LTP is more than the sum of its separate effects and becomes more linear (Berger et al 1984). The implication of this phenomenon has been considered (McNaughton et al 1978, 1981).

The observation that associative LTP is more than the sum of its independent effects suggests that associative LTP could provide an associative memory mechanism in the CNS (Levy & Steward 1979, McNaughton & Barnes 1977). If so, the timing requirements for the induction of associative LTP might be expected to be governed by rules similar to those known to control associative learning. This aspect was initially investigated by Levy & Steward (1983), with the ipsilateral and contralateral entorhinal projections to dentate. They showed that if weak contralateral stimulation preceded the strong ipsilateral stimulation by 20 msec, associative LTP of the weak input could occur. If, however, the interstimulus interval between the weak and strong pathway tetani was increased to 200 msec, not LTP, but rather a long-lasting depression was noted (Levy & Steward 1983). The morphological consequences of associative potentiation/depression are described in Desmond & Levy (1983).

Kelso & Brown (1986) showed that pairing independent weak and strong Schaffer collateral projections (the tetanus to the weak pathway precedes the tetanus to the strong pathway by 200 msec) produced associative LTP in the weak pathway. This design is identical to that employed in forward classical conditioning in many behavioral experiments. Utilization of a backward design (stimulation of the strong pathway occurs 600 msec prior to stimulation of the weak pathway) failed to produce LTP. These results are interesting in terms of the strong temporal parallels to the classical conditioning literature (Burger & Levy 1985). They are different from the results of Levy & Steward (1983), who showed that a 200 msec interval failed to produce LTP, yielding depression instead. That the two experiments were done in CA1 and dentate, respectively, perhaps accounts for their contrasting results.

Interest in coactivation and associative LTP (which probably represents the same phenomenon) derives primarily from the potential for such a device to form associations between coactive elements on a target neuron. While much remains to be learned regarding the rules governing the nature of the patterned input, as well as the outcome of the associative LTP depending on the nature of the elements activated, further understanding of this process will be important for understanding how the CNS encodes near contiguous events. Theoretically, these observations may require a modification of the classical "Hebb Synapse" (Hebb 1949). Hebb proposed that correlated afferent activity and postsynaptic activation leads to an alteration in synaptic efficacy (see also Wigström & Gustafsson 1985). Since it has been demonstrated that LTP can be induced in the absence of action potentials in the postsynaptic cell, it may be necessary to modify the notion of the "Hebb Synapse" to a system that requires coactivity among converging presynaptic elements onto a common postsynaptic membrane.

The phenomenon of coactivation suggests the presence of a mechanism for integrating activity over some given space and time, capable of triggering a series of events controlling the occurrence and magnitude of LTP. Classically, this would indicate a postsynaptic component of LTP; however, the possible involvement of presynaptic or extraneuronal integration has not been eliminated. The associative LTP paradigm may provide information regarding the mechanism of LTP, by revealing the limits of the temporal and spatial constraints on the coactive inputs.

Input Specificity

LTP displays input specificity such that only inputs that are simultaneously active (or nearly so) can be potentiated (Andersen et al 1977, Lynch et al 1977). This is also termed "homosynaptic potentiation." LTP appears to be a local process and not a generalized increase in neuronal excitability. Intracellular studies of LTP do not reveal consistent changes in membrane potential or input resistance (Andersen et al 1980, Barrionuevo & Brown 1983).

The response of area CA3 pyramidal cells to mossy fiber tetanization, however, demonstrates heterosynaptic LTP (increased response to other converging, but independent, nontetanized inputs) as well as homosynaptic LTP (Yamamoto & Chujo 1978, Misgeld et al 1979). This puzzling exception to input specificity has been recently evaluated. Higashima & Yamamoto (1985) analyzed the mossy fiber–CA3 evoked response in terms of an early component (due to the mossy fiber–CA3 synaptic activation) and a late component (presumably polysynaptic). They reported that only the late component supports heterosynaptic potentiation, while the monosynaptic activation maintains input specificity. The authors suggest a number of hypotheses to explain the heterosynaptic LTP, including the possibility of increased electrotonic coupling.

Another exception to input specificity was recently reported by Abraham, Bliss & Goddard (1985). LTP is usually defined as an increase in synaptic efficacy. However, a number of studies have demonstrated LTP of the population spike (an extracellular measure of summed action potentials) beyond that which could be explained by potentiation of the population excitatory post-synaptic potential (EPSP) (Bliss & Lømo 1973, Wigström & Swann 1980, Andersen et al 1980, Wilson et al 1981). A population EPSP of a given size produces a larger population spike after LTP, relative to pretetanus controls. Therefore, the EPSP/spike relationship appears to have shifted. EPSP/spike potentiation has not been systematically investigated at the intracellular level.

Abraham et al (1985) demonstrated homosynaptic and heterosynaptic EPSP/spike LTP between two converging, but independent, PP-DGC inputs (medial and lateral PP). The apparent maintenance of control level responses in the nontetanized pathway is reported to consist of a generalized heterosynaptic depression of the EPSP, coupled with a heterosynaptic EPSP/spike LTP. LTP of the population EPSP, however, remains homosynaptic. These findings again indicate that LTP is a complex process that is not completely characterized.

It follows from the concept of input specificity that some of the mechanisms subserving LTP of the population EPSP should be located near the activated synaptic membrane. If not, there must be a process for marking tetanized synapses.

Summary

Input specificity suggests localized changes, most easily explained by synaptic alterations. Coactivation suggests the existence of a postsynaptic mechanism for integrating some aspect of synaptic activation. The long-lasting nature of LTP seems to require conformational changes in neuronal architecture and/or de novo protein synthesis. While these inferences may be superficial, they suggest that LTP is a complex process. Our understanding of its mechanisms awaits a more complete understanding of neurotransmission, neuromodulation, local circuits, intrinsic currents, and protein synthesis and modification. In spite of this problem, we discuss the changes that are associated with LTP, as well as the modulators of LTP, in an attempt to circumscribe the potential neural mechanisms subserving LTP.

MECHANISM

Neural Transmission

Andersen et al (1980) demonstrated that the presynaptic fiber volley does not show any consistent change after LTP induction. It is inferred that LTP is not caused by the recruitment of fibers during the high frequency stimulation.

Sastry (1982) showed that the presynaptic terminals in CA1 are less excitable post-LTP. This effect may be due to presynaptic hyperpolarization. If so, it would be compatible with a number of studies indicating increased neurotransmitter release associated with LTP. Increased release of preloaded *d*-[3H]-aspartate by stimulation of Schaffer collaterals was demonstrated in area CA1 (slice preparation; Skrede & Malthe-Sørenssen 1981). Dolphin et al (1982) reported increased release of newly synthesized [3H]-glutamate in the dentate gyrus (preloaded with a precursor, [3H]-glutamine) associated with LTP of the PP-DGC synapses. These studies strongly suggest a presynaptic component of LTP. However, at present it is not possible to understand coactivation or EPSP/spike LTP from a strictly presynaptic mechanism. Neither is it yet possible to ascertain the source of the increased release. It is assumed to arise from the tetanized terminals.

Synaptic transmission is required for LTP induction. This has been demonstrated with the use of neurotransmitter antagonists and manipulation of Ca^{2+} concentration. Since glutamate is believed to be the neurotransmitter at the principal trisynaptic junctions of the hippocampus (White et al 1977), Dunwiddie et al (1978) tentanized perforant path fibers while bathing the dentate gyrus slice in APB (a glutamate antagonist). No LTP was later observed after washout of the APB, thus suggesting that glutamate neurotransmission is necessary for LTP in dentate. Similar results were obtained by Krug et al (1982) utilizing GDEE (another glutamate antagonist). Dolphin (1983a), however, using the amino acid antagonist gamma-D-glutamylglycine (gamma-DGG) in dentate gyrus, showed that LTP appeared to be blocked in its presence, but upon clearance LTP was observed. Dolphin concluded that gamma-DGG sensitive glutamate receptors are not involved in LTP and suggested that presynaptic changes may be implicated. Table 1 compares these, and other, manipulations as they affect LTP.

Recently, considerable interest has focused on the activity of an amino acid receptor subtype, classified as the *N*-methyl-D-aspartate (NMDA) receptor (Collingridge 1985). A specific NMDA antagonist, D-2-amino-5-phospho-novalerate (APV), does not affect normal transmission (which appears to be gated through kainate/quisqualate receptors), but it does block LTP (CA1, slice preparation; Collingridge et al 1983, Wigström & Gustafsson 1984, Harris et al 1984). APV also blocks LTP in the intact dentate gyrus (Morris et al 1986) and the LTP-dependent, increased glutamate release from dentate gyrus (Lynch et al 1985b).

These results point to a receptor that is sensitive to some aspect of an intense, high frequency stimulus, or the consequences thereof. During normal activity, NMDA receptors appear to be blocked by a voltage dependent, Mg^{2+}-mediated blockade (Nowak et al 1984). Collingridge suggests that the depolarization produced by the high frequency stimulus might be sufficient to open NMDA receptor channels. This additional depolarization would in turn open more NMDA receptors, producing a cascade effect and somehow initiating the process of LTP. The activity of the NMDA receptors would have to be coupled to a mechanism for maintaining the synaptic enhancement through the kainate/quisqualate receptors, since NMDA receptors are thought to be involved in the initiation but not the maintenance of LTP (Collingridge 1985).

Obviously, the site of this activation (presynaptic and/or postsynaptic) needs to be determined, as well as the mechanism to which NMDA activity is coupled to maintain LTP. Perhaps examination of PTP in the presence of APV could provide a better understanding of this effect.

Calcium

Given the central role of calcium in synaptic processes, several laboratories have studied the role of calcium-calmodulin systems in LTP. Dunwiddie

Table 1 The effect on LTP of manipulations of synaptic transmission and neuromodulation

Manipulation	Area	Effect on LTP +	Effect on LTP −	Comment	References
Block transmission					
APB (Glu antagonist)	DG		X[a]	tetanus in presence of APB, test without; no LTP	Dunwiddie et al (1978)
GDEE (Glu antagonist)	DG		X	tetanus in presence of GDEE, test without; no LTP	Krug et al (1982)
Gamma-DGG (amino acid antagonist)	DG	no effect		tetanus in presence of DGG, test without; LTP, suggests masking effect	Dolphin (1983a)
APV (NMDA antagonist)	CA1		X	tetanus in presence of APV, test without; no LTP	Collingridge et al (1983) Wigström & Gustafsson (1984) Harris et al (1984)
Manipulate calcium					
Ca^{2+} free media	CA1		X	tetanus in ↓ Ca^{2+} media, test in normal; no LTP	Dunwiddie et al (1978) Wigström et al (1979)
EGTA	CA1		X	Ca^{2+} chelator blocked LTP	Lynch et al (1983)
2-chloroadenosine	DG		X	affects both pre/postsynaptic	Dolphin (1983b)
Trifluoperazine	CA1		X	calmodulin antagonist blockage of LTP and Ca^{2+} induced LTP	Mody et al (1984)
Pimozide	CA1		X		Mody et al (1984)
Ca^{2+} uptake	CA1			uptake of labeled Ca^{2+} ↑ after LTP	Baimbridge & Miller (1981)
Increase Ca^{2+}	CA1	X		↑ Ca^{2+} (4 mM) yields LTP-like effect without tetanus	Turner et al (1982)
Trifluoperazine	CA1		X	blockage by this phenothiazine neuroleptic	Finn et al (1980)

Catecholamines					
AMPT	DG	X		CA depletion by AMPT ↑ pop spike only, not EPSP	Krug et al (1983)
Haloperidol	DG	X		CA block by Haloperidol ↑ pop spike, not EPSP	Krug et al (1983)
Reserpine, 6-OHDA	DG		X	↓ EPSP, normal spike shortly after tetanus	Bliss et al (1983)
Amphetamine	DG	X		dose-dependent effect, parallels to behavioral effect noted	Gold et al (1984)
Epinephrine	DG	X			Gold et al (1984)
Reserpine	DG	X		short-term : ↑ EPSP, ↓ spike	Robinson & Racine (1985)
	DG	no effect		normal LTP 24 hr later	
PCPA ; 5, 7-DHT	DG	no effect		5-HT depletion without effect	Stanton & Sarvey (1985a)
	CA1	no effect			
NE (25 μM)	CA1	no effect		suggests NE not involved	Dunwiddie et al (1982)
Amphetamine, reserpine	CA1	no effect		CA facilitators without effect	Dunwiddie et al (1982)
Isoproteronol (beta agonist)	CA1		X	suggests beta receptor involvement	Dunwiddie et al (1982)
Neuroleptics	CA1		X	suggests effect related to neuroleptic actions on calmodulin	Dunwiddie et al (1982)
6-OHDA	CA1		X	LTP restored with forskolin	Stanton & Sarvey (1985a)
NE (10 μM)	CA3	X		↑ NE results in ↑ LTP duration	Hopkins & Johnston (1983)
NE (1–10 μM)	CA3	X		↑ NE during test only ↑ LTP duration	Hopkins & Johnston (1983, 1984)
Propanolol (beta antagonist)	CA3		X	alpha antagonist without effect	Hopkins & Johnston (1983, 1984)
Opiates					
Naloxone	CA3		X	proposed action with GABA inhibition	Martin (1983)
Naloxone	CA1	no effect		in vivo	Linseman & Corringall (1981)
PCP	CA1		X	other opiates without effect	Stringer et al (1983)
Naloxone	CA1	no effect			Stringer et al (1983)
opiates (sigma)	CA1		X		Stringer et al (1983)
PCP	CA1		X	in vivo	Stringer & Guyenet (1983)
Ketamine	CA1		X		Stringer & Guyenet (1983)

Table 1 (*continued*)

Manipulation	Area	Effect on LTP +	Effect on LTP −	Comment	References
Protein synthesis					
Protein synthesis	CA1			specific protein synthesis increased with LTP	Browning et al (1979)
Cyclohexamide	CA1		X	differential effect noted among protein synthesis inhibitors with Ani being ineffective	Stanton & Sarvey (1984)
Emetine, puromycin			X		
Anisomycin		no effect			
Protein synthesis/secretion	DG/CA1			secretion of synthesized proteins after LTP	Duffy et al (1981)
Anisomycin		no effect		Pr^- synthesis inhibition in blocking LTP by Ani ineffective	Swanson et al (1982)
Anisomycine	DG		X	no effect until 3–4 hr post-tetanus, then ↓ LTP	Krug et al (1984)
Cyclohexamide	DG		X	reversible block	Steward & Brassel (1982)
Other					
TTX, GABA, pentobarb			X	LTP blocked, PTP unaffected by these postsynaptic spike blockers	Scharfman & Sarvey (1985)
Insecticides	DG	X		dieldrin, lindane, acute	Woolley et al (1984)
Heavy metal	DG	X		trimethyltin, acute	Woolley et al (1984)
Adrenalectomy	DG		X	corticosterone replacement restores LTP	Dana et al (1982)
Adrenalectomy	CA1		X	lower magnitude LTP, corticosterone enhances LTP magnitude	Nowicky et al (1983)
Stress	CA1		X	stressed animals failed to show LTP	Foy et al (1985)
ETOH	CA1		X	chronic ETOH reversibly blocks LTP	Durand & Carlen (1984)
Antibody to S100	CA1		X	also erases established LTP	Lewis et al (1986)
Circadian rhythms	DG/CA1	biphasic		DG spike LTP ↑ at night, CA1 at day	Harris & Teyler (1983)

[a] The effect of various manipulations is listed either as a facilitation (+) of LTP magnitude, a complete or partial block of LTP (−), or as no effect on LTP, that is, LTP magnitude and other parameters are normal.

et al (1978) and Wigström et al (1979) showed that LTP was blocked under low calcium conditions, indicating that calcium plays a critical role in LTP production. The experimental paradigm involves delivering the tetanic stimulus in a low calcium environment (all experiments done in vitro) and testing for LTP later in a normal calcium environment. These observations were extended by the experiments of Baimbridge & Miller (1981), who demonstrated that the uptake and retention of labeled calcium is increased after LTP. Kuhnt et al (1985) demonstrated a large increase in dentritic calcium deposits with the induction of LTP. These results suggest an additional role of calcium in LTP beyond that of normal synaptic transmission. This interpretation is supported by the experiments of Lynch et al (1983) showing that intracellular injection of the calcium chelator, EGTA, into area CA1 pyramidal cells (hippocampal slice) blocks the appearance of LTP. Similarly, in dentate gyrus, Dolphin (1983b) found that the utilization of 2-chloradenosine (an adenosine blocker that may inhibit calcium influx) blocks LTP production.

Considering the ubiquitous role of calcium in neuronal regulation, it is not surprising that manipulations of calcium levels would influence LTP. This was supported by the experiments of Finn et al (1980), who showed that utilization of the calmodulin-acting neuroleptic trifluoperazine (TFP) blocks LTP in area CA1. The most startling result is that of Turner et al (1982), who demonstrated that merely incubating a hippocampal slice for five to ten minutes in medium containing elevated calcium levels (4 mM) is sufficient to produce LTP that is identical in many ways to that produced by afferent activation. Mody et al (1984) found that TFP and pimozide also block this calcium-induced LTP and suggested that calmodulin-mediated mechanisms are involved in the production of LTP.

The Baudry-Lynch model of LTP (Baudry et al 1980, Lynch & Baudry 1984) is based on a long series of experiments indicating a calcium mediated increase in receptor binding. Simply, Lynch & Baudry suggest that tetanic stimulation causes an increase in postsynaptic calcium concentration, which triggers a series of events leading to an increase in the number (but not affinity) of glutamate receptors. Recently, Sastry & Goh (1984) and M. A. Lynch, Errington & Bliss (1985a) reported that they were unable to replicate an LTP-induced increase in receptor binding. In addition, leupeptin, a potent inhibitor of a class of calcium-activated proteinases [which Lynch & Baudry (1984) suggest is involved in the uncovering of glutamate receptors] does not appear to block LTP (Sastry 1985). Significant differences in the methods of inducing LTP and/or the experimental preparations could possibly account for some of these differences. Further experiments will be required to clarify the association of LTP and increased glutamate receptor binding.

Catecholamines

Numerous studies have investigated the modulatory role of catecholamines in LTP production. Catecholamine depletion by AMPT was studied by Krug et al (1983), who showed that the magnitude of dentate LTP (measured 24 hr later) was increased but only for the population spike and observed no effect on the EPSP. Both response measures were increased for a period up to 5 hr after tetanus in the depleted animals. A similar effect was seen as a result of haloperidol blockage on catecholamine activity. Krug and colleagues interpret this result as an indirect modulatory effect on the homosynaptic mechanism of LTP.

The opposite effects were reported by Robinson & Racine (1985), who depleted catecholamines with reserpine. Shortly after (15 min) the delivery of a tetanus to the perforant path, depleted animals showed an elevated LTP of the field EPSP and a reduced LTP of the population spike. This effect, however, was not maintained and no differences were noted between depleted and control animals one week later. Bliss and colleagues (1983) also noted a short-term effect of catecholamine depletion (via reserpine or 6-hydroxydopamine), but in contrast to Robinson & Racine's results, the field EPSP was depressed relative to controls whereas no difference was seen in the population spike.

Stanton & Sarvey (1985a) have shown that 6-hydroxydopamine depletion of norepinephrine reduces population spike LTP in the dentate gyrus (but not CA1). LTP could be restored in the dentate gyrus of the depleted animals by treatment with the adenylate cyclase stimulant, forskolin. Stanton & Sarvey (1985a) found that the beta-adrenergic receptor antagonists, propranolol and metoprolol, were effective in blocking LTP and suggested that noradrenergic stimulation of cAMP production plays a role in the production of LTP in the dentate gyrus since both tetanic stimulation and norepinephrine elicit increased levels of cAMP in hippocampal slices (Stanton & Sarvey 1985b). The serotonin depleting agents, PCPA and 5,7-DHT, were without effect in either area CA1 or the dentate gyrus (Stanton & Sarvey 1985a).

The effect of adding norepinephrine (10 μM) to the slice medium was shown to facilitate CA3 LTP, particularly in terms of increasing its duration (Hopkins & Johnston 1983). Hopkins & Johnston (1983) showed that raising the concentration of norepinephrine (1–10 μM) only during the tetanic stimulation has the same effect. The beta-adrenergic receptor was implicated in LTP because of the effect of propanolol on blocking both LTP induction and the aforementioned noradrenergic effect (Hopkins & Johnston 1984). They report that alpha-adrenergic antagonists are ineffective in manipulating LTP. Similarly, Gold et al (1984) have shown that peripheral administration of both amphetamine and epinephrine facilitate the production of LTP in the dentate and do so in a dose-dependent manner, similar to the effect of these agents in behavioral domains. In contrast to the above, Dunwiddie et al (1982) find that norepinephrine is not involved in LTP production in CA1. They see no effect of the addition of norepinephrine (25 μM), amphetamine, reserpine, or adrenergic antagonists. Dunwiddie and colleagues report a partial blockade of LTP as a result of the utilization of isoproteronol (a beta-adrenergic agonist) and a block of LTP with the use of various neuroleptics, which, as mentioned above, may act via their action on calmodulin.

Opiates

The activity of opiates in modulating hippocampal LTP has been examined by several laboratories. Martin (1983) demonstrated that LTP in area CA3 can be blocked with nanomolar concentrations of naloxone. Martin suggests that the mechanism of action involves the release from opiate depression of GABA interneuron activity. However, Stringer et al (1983) report that naloxone does not affect CA1 LTP, but that PCP and certain opiates (sigma opiates) blocked the development of LTP. Similarly, Linseman & Corrigall (1981) report no effect of naloxone on LTP in area CA1, in vivo. This report follows an earlier in vivo study (Stringer & Guyenet 1983) wherein PCP and ketamine were effective in blocking LTP in area CA1. A recent study (Thomson & Lodge 1985) suggests that sigma opiates block the NMDA receptor. Collingridge (1985) discusses a similar effect of PCP and ketamine.

Inhibition

Bliss & Lømo (1973) originally suggested a shift in tonic inhibition as one of four possible mechanisms for synaptic enhancement. It has been reported that LTP in area CA3 is associated with a decrease in intracellular IPSP amplitude (Yamamoto & Chujo 1978, Misgeld et al 1979). However, whether the IPSP is simply being masked by the concomitant increased depolarization is unclear.

Haas & Rose (1982) studied the effects of LTP on inhibitory circuitry (area CA1, hippocampal slice), using both extracellular and intracellular recording techniques. They concluded that these systems were not directly involved in LTP induction in CA1 pyramidal cells. A further report (Haas & Rose 1984) suggested a contribution from intrinsic inhibitory mechanisms. The use of CeCl electrodes to block potassium currents prevented LTP induction in CA1 pyramidal cells. The authors propose that intrinsic inhibitory currents may be involved in LTP, perhaps altered by neuromodulators released during intense synaptic activation. A recent test of this hypothesis (Haas & Greene 1985) showed that one of the potassium currents, the transient A current, is not involved in LTP.

LTP can be produced in the absence of GABA-mediated inhibition. Wigström & Gustafsson (1983a,b) and Hendricks & Teyler (1983) show that LTP induction is, in fact, facilitated by the addition of picrotoxin or bicuculline, GABA antagonists (hippocampal slice; CA1, dentate). The facilitation is particularly prominent in terms of a decrease in the number of stimuli required to induce LTP.

Douglas et al (1982) coactivated perforant path and commissural inputs to dentate granule cells. Single pulses to the commissural fibers block the population spike evoked by PP tetanization; however, they do not block perforant path LTP. Co-tetanization of commissural input and perforant path is required to block PP-LTP. Therefore, blocking the action potential via inhibitory circuitry does not affect LTP in the PP-DGC system. Tetanization of that same input, however, appears to modify the conditions for the production of LTP. These findings suggest that local inhibition may not be directly involved in the generation of LTP. Rather, it may bias the level of activity required to initiate LTP (Douglas et al 1982, Wigström & Gustafsson 1985).

Protein Synthesis

The duration of LTP suggests long-term changes in neuronal structure. A number of investigators have reported such changes in synaptic contacts and spine dimensions at the electron microscopic level following high frequency stimulation of the hippocampus (Fifkova & Van Harreveld 1977, Lee et al 1980, Desmond & Levy 1983). These factors gave rise to a number of studies on the role of protein synthesis in LTP.

Following high frequency stimulation, changes in the phosphorylation level of a number of proteins have been reported (Browning et al 1979, Bär et al 1980, Routtenberg et al 1985). Differences between labs in terms of preparation, stimulation parameters, sampling times, and biochemical assays have no doubt led to some inconsistent results (Hoch et al 1984). Duffy et al (1981) showed that LTP was associated with specific increases in the synthesis and secretion of newly synthesized proteins in the hippocampal slice.

The protein synthesis inhibitors cyclohexamide (dentate gyrus, intact preparation; Steward & Brassel 1983) emetine, and puromycin (CA1, hippocampal slice; Stanton & Sarvey 1984) are effective in blocking LTP. The effects of anisomycin appear to be more complex. In the hippocampal slice, Stanton & Sarvey (1984) could not detect an effect on LTP. Krug et al (1984) using a chronically implanted preparation (PP-DGC system), also showed that anisomycin does not affect the early stages of LTP (up to 3 hr); however, at 3–4 hr post-tetanus, the response declined back to control levels and remained stable for up to 7 d. Although these results may be confounded by a long-latency, anisomycin-induced depression, this dissociation deserves further attention. The possibility that LTP depends upon the synthesis or replenishment of several classes of new proteins for its expression is interesting although the kinetics of such a requirement are made difficult by the rapid appearance of LTP.

LTP Modulators

A number of other observations have been made regarding modulators of LTP. Scharfman & Sarvey (1985) demonstrated that the agents TTX, GABA, and pentobarbital (all of which are postsynaptic action potential blockers) block the induction of LTP when the tetanic stimulation is delivered in their presence and the tissue tested later in their absence. Dana et al (1982) demonstrated that depletion of corticosterone and other adrenal steroid hormones by adrenalectomy abolishes LTP in dentate gyrus of intact rats. Replacement treatment with corticosterone restores the ability of the dentate gyrus to display LTP. Similarly, Nowicky and colleagues (1983) showed in CA1 that adrenalectomy reduces LTP magnitude, which can be restored with corticosterone. Consistent with the hypothesized role of adrenal steroids in LTP modulation, Foy et al (1985) reported that animals receiving an intense stressor (30 min of tail shock) prior to the preparation of hippocampal slices failed to generate significant LTP and, further, that the LTP achieved by individual slices was inversely correlated with plasma corticosterone levels. Lewis & Teyler (1986b) have demonstrated that antibody to the brain specific protein, S100, is capable of blocking the induction of LTP in area CA1 and, if administered after LTP has been induced, is capable of erasing the LTP and restoring a baseline response. Circadian rhythms modulate LTP with dentate gyrus LTP increased during the dark cycle and CA1 LTP enhanced during the light cycle (Harris & Teyler 1983).

Exogenous agents have also been demonstrated to modulate LTP. Delta-9-tetrahydrocannabinol (a major psychoactive agent in marijuana) has been reported to reduce the duration of LTP in hippocampal slices (Nowicky et al 1985) at a dose of 10 pM. Chronic alcohol administration has been shown to lead to a reversible blockage of LTP in area CA1 (Durand & Carlen 1984). Enrichment of CA1 synaptic membrane with GMI ganglioside increased the magnitude of LTP without affecting baseline physiology (Wieraszko & Seifert 1985). Finally, some insecticides (dieldrin, Lindane) and heavy metals (trimethyltin) produce a response facilitation in the dentate gyrus that is quite similar to LTP, although perhaps induced by the convulsant activity of these agents as the authors suggest (Woolley et al 1984). The mechanism(s) by which these endogenous and exogenous agents modulate LTP remain unknown.

Summary

The neural substrates of LTP remain unknown. Clearly LTP has not yet been completely characterized. A number of experiments need to be completed and inconsistencies carefully examined. Conceptually, the focus of interest is whether LTP is a presynaptic or a postsynaptic process. This question has impact on the potential relationship of LTP to learning and memory, and the models of Hebb, Marr, and others.

Studies of the peripheral nervous system indicate that LTP is primarily a presynaptic process (Dolphin 1985). However, it is not understood how these forms of LTP are related to hippocampal LTP. The indications of increased neurotransmitter release in the hippocampus (Skrede & Malthe-Sørenssen 1981, Dolphin et al 1982) suggest that the presynaptic terminal is the source of a component of the LTP mechanism. In addition, a recent report suggests that there may be presynaptic NMDA receptors (Lynch et al 1985b). The LTP block produced by intracellular EGTA injection (Lynch et al 1983) argues for a postsynaptic component of LTP. Cooperativity and EPSP/spike potentiation, as well as the influence of inhibitory circuitry on LTP, are easier to integrate into a postsynaptic model.

A number of hypotheses have been proposed to integrate the apparent presynaptic and postsynaptic contributions to LTP. Eccles (1983) proposes that postsynaptic changes, secondary to tetanus-induced calcium influx, induce the increased release of neurotransmitter by means of a trophic factor. Conversely, Krug et al (1984) have suggested that the primary event is increased neurotransmitter release, which in turn induces long-term structural changes in the postsynaptic neuron.

BEHAVIORAL CONSIDERATIONS

Behavioral LTP

Much of the interest in LTP research stems from the possibility that LTP may be involved in memory storage. However, that LTP is produced in association with behavioral learning in intact animals and is necessary and sufficient for the behavioral learning must be convincingly demonstrated. Although LTP has been known for over a decade, research into its behavioral correlates is recent. Among the first studies of the behavioral correlates of LTP was that of Barnes (1979), who studied the behavioral and electrophysiological properties of young and old rats. She demonstrated a relationship between the longevity of electrically induced LTP and the level of retention on a behavioral task.

In an attempt to determine whether LTP occurs during the normal activity of an animal, as opposed to only during laboratory learning tasks, the PP-DGC system was studied in chronically implanted rats who were placed in a novel, complex environment (Sharp et al 1985). No tetanic stimuli were delivered, only a single daily probe stimulus. Across several days in the novel environment an LTP-like change was observed, which gradually subsided, only to be reinstated upon the rats' transfer to a second, different complex environment. An important control used in this study was to place animals in restricted environments for one month so that any prior behavioral LTP might decay to baseline before the experimental sessions.

In a more direct test of the behavioral correlates of LTP, Thompson and colleagues implanted rabbits with perforant path stimulating electrodes and dentate gyrus recording electrodes. The behaving rabbits were trained on a classical conditioned eyeblink (nictitating membrane) response (Thompson et al 1983). This preparation allows any behaviorally induced change in synaptic efficacy at the dentate gyrus to be reflected electrophysiologically. Since no tetanic stimuli were presented, any change in synaptic efficacy must be elicited by naturally occurring patterns of afferent activity. The results indicated that dentate synaptic efficacy increased approximately in parallel with the acquisition of the behavioral response (Weisz et al 1984). Similar results showing that behavioral training results in LTP-like changes at monosynaptic junctions in hippocampus have been reported by Laroche & Bloch (1982) in a tone-footshock conditioning paradigm and by Ruthrich et al (1982) in a brightness discrimination task.

In a continuing series of experiments with the nictitating membrane response of rabbit, it has been demonstrated that the induction of an LTP-like mechanism in rabbit hippocampus after conditioning alters the original distribution of tritiated glutamate binding (Mamounas et al 1984). This result is similar to that seen in rat hippocampal slices following the induction of LTP by tetanic stimulation wherein the number of tritiated glutamate binding sites in the potentiated region increases (Lynch & Baudry 1984). In the rabbit preparation, the maximal number of tritiated glutamate binding sites in the hippocampus is increased only in animals receiving paired presentations of conditioned and unconditioned stimuli and is not observed in naive or unpaired control animals. These results provide additional evidence that an LTP-like mechanism is occurring during this behavioral learning paradigm.

Utilizing a tone-foot shock conditioning paradigm, Laroche et al (1983) showed that stimulation of the reticular formation in association with the learning of the behavioral task influenced the magnitude of LTP induced, at the end of the behavioral experience, with electrical stimulation. The authors suggest that during the establishment of associative LTP or associative learning, modulation by reticular formation activity can act to enhance the properties of LTP. Similarly, post-trial stimulation of the reticular formation also facilitates LTP, with a temporal gradient comparable to that seen in behavioral studies (Bloch & Laroche 1985).

These results collectively provide some evidence that behavioral processes give rise to LTP-like phenomena that are similar in many respects to LTP induced by electrical stimulation. However, as we (Teyler & DiScenna 1984) have pointed out, these observations raise additional questions regarding the underlying phenomenon of LTP. For example, it is unclear why participation in a variety of behavioral tasks should be reflected in alterations of hippocampal efficacy at electrode locations apparently chosen at random. LTP is a relatively enduring phenomenon, yet behavioral responding can be altered quite rapidly (extinction) given changes in the contingency of the training task (non-reinforcement). It is important to know the fate of behavioral LTP in animals subjected to extinction training.

LTP Effects on Learning

If LTP has a relationship to the neuronal mechanisms underlying behavioral learning, then the previous establishment of LTP in circuits responsible for behavioral responding should alter the behavior in some way. Clearly one could hypothesize that the induction of LTP would either facilitate or disrupt subsequent behavior depending upon the effect of the induced LTP on the signal processing characteristics of the potentiated circuit. Berger (1984) investigated this question in the rabbit nictitating membrane conditioning paradigm. Electrical stimulation was used to produce LTP in the dentate gyrus of untrained rabbits. Animals were then trained on the eyelid response. Animals receiving LTP treatments prior to training subsequently showed accelerated behavioral learning in this paradigm. Ott et al (1982) have shown that perforant path stimulation, which induces LTP, can also be used successfully as a conditioned stimulus in a shuttlebox avoidance paradigm. Barnes (1979) found, conversely, that the induction of LTP in old rats interfered with subsequent learning of a T-maze spontaneous alternation. Interestingly, mature adult rats did not display this interference.

Reymann et al (1982) showed that 15 or 100 Hz electrical stimulation (but not 1.7 Hz) of the perforant path can serve as a conditioned stimulus (CS) in a two-way shuttle avoidance learning task. In a conceptually related experiment (Skelton et al 1985), single pulse electrical stimulation of the perforant path was used as the discriminative stimulus in a food-reward operant chamber. The rate of behavioral learning was related to the magnitude of the stimulus-evoked dentate response. Transection of the perforant path impaired behavioral learning, suggesting that this limbic circuit is directly involved in the learning. Tetanic stimulation of the perforant path in naive animals both enhanced the stimulus-evoked dentate activity and facilitated subsequent behavioral learning, thus demonstrating that LTP can alter both synaptic and behavioral responses.

Morris et al (1986) reported that APV (NMDA receptor antagonist) blocked LTP at the PP-DGC synapses in the rat. It also impaired their performance on a spatial location task, but had no significant effect on a nonspatial, visual discrimination task. This indicates that LTP, if it is directly involved in behavioral learning, does not subserve all forms of learning.

Clearly, many more experiments are needed in the area of behavioral LTP in order to be confident of the nature of the relationship between the behavioral learning and LTP. Important questions relate to establishing that LTP is necessary and sufficient for the neuronal mechanism underlying behavioral memory. A full appreciation of the answer to this question must await our understanding of the mechanism of action of LTP as well as an understanding of brain distribution of LTP.

IMPLICATIONS AND CONCLUSIONS

The existence of a neural phenomenon that persists for considerable lengths of time and is seen to occur both as a result of electrical stimulation of afferents and in conjunction with behavioral learning has led many to consider the hypothesis that LTP underlies long-term information storage in the brain (Goddard 1980, Andersen et al 1980, Swanson et al 1982, Eccles 1983, Voronin 1983, Lynch & Baudry 1984, Teyler & DiScenna 1984, 1985). LTP is attractive as a memory model because of its lengthy time-course. But is it sufficiently long-lasting to account for the duration of behavioral memory? If LTP is not permanent, what is permanent? And what role does LTP play in this process? The mechanism underlying LTP has yet to be established. Is the same mechanism responsible for LTP in all regions of brain? In all species? Does behavioral LTP result in the same mechanistic changes? Are behavioral LTP and electrically induced LTP identical? Do different experiences result in different patterns of LTP distribution in the brain? What is the nature and significance of coactivation? Thus, acceptance of the hypothesis that LTP does underlie memory is premature at present. As can be seen from this review, LTP is incompletely understood both mechanistically and conceptually.

Unstated in most accounts of LTP is the nature of the information that is being represented by synapses undergoing LTP. Given the presently known distribution of LTP in the brain and the prominence of LTP in structures of the limbic forebrain long associated with learning and memory, a logical choice is the encoding of episodic or procedural events. However, many aspects of normal brain function have recourse to stored information, and neuroendocrine, circadian, and homeostatic information also may be encoded by LTP. LTP may be a general synaptic mechanism for modifiable synapses, to be used in whatever brain circuit requires this feature.

Our present ignorance hinders our understanding of LTP in several areas. Foremost among these areas is to strengthen our knowledge of the links between LTP and behavioral processes. Although a good beginning has been made, much remains in question. If LTP is a substrate for memory it must be shown that it is both necessary and sufficient for behavioral learning and memory. This can best be accomplished in model systems wherein both the behavioral variables are well controlled and the brain circuit understood. These efforts will further our understanding of whether one or several varieties of LTP exist and to specify the neuronal circuits and elements involved. Although the hippocampus has been the focus of most of the work cited in this review, one must recognize that humans suffering from destruction of the hippocampus do not loose access to prior memories (Squire 1982). While not negating a role for the hippocampus in human memory, such evidence indicates that other structures are also involved and should be examined with respect to LTP. Finally, if LTP remains a candidate mnemonic device, attention will have to be given to the nature of the memory stored, how it is reaccessed and how it is used to influence ongoing behavior.

NOTE ADDED IN PROOF This review was completed in January 1986. Since then the activity in many of the areas covered in this review has been significant. Among the most interesting work are studies on conductance changes accompanying LTP (Barrionuevo et al 1986), conditions for LTP induction (Gustafsson & Wigström 1986, Larson & Lynch 1986, Malinow & Miller 1986, Sastry et al 1986), the involvement of phorbol esters and protein kinase C (Malenka et al 1986, Akers et al 1986), and the behavioral effects of LTP (McNaughton et al 1986).

ACKNOWLEDGMENT

Preparation of this review has been partially supported by grants from NIH (DA 03755), EPA (CR 813394), and ONR (86-188). We are grateful to Dr. Philip Schwartzkroin for his suggestions regarding the scope and organization of this review.

Literature Cited

Abraham, W. C., Bliss, T. V. P., Goddard, G. V. 1985. Heterosynaptic changes accompany long-term, but not short-term potentiation of the perforant path in the anesthetized rat. *J. Physiol.* 363: 335–49

Abraham, W. C., Goddard, G. V. 1983. Asymmetric relationships between homosynaptic long-term potentiation and heterosynaptic long-term depression. *Nature* 305: 717–19

Akers, R. F., Lovinger, D. M., Colley, P. A., Linden, D. J., Routtenberg, A. 1986. Translocation of protein kinase C activity may mediate hippocampal long-term potentiation. *Science* 231: 587–89

Albus, J. S. 1971. A theory of cerebellar function. *Math. Biosci.* 10: 25–61

Alger, B. E., Megela, A. L., Teyler, T. J. 1978. Transient heterosynaptic depression in the hippocampal slice. *Brain Res. Bull.* 3: 181–84

Alger, B. E., Teyler, T. J. 1976. Long-term and short-term plasticity in the CA1, CA3 and dentate region of the rat hippocampal slice. *Brain Res.* 110: 463–80

Andersen, P., Sundberg, S. H., Sveen, O., Wigström, H. 1977. Specific long-lasting potentiation of synaptic transmission in hippocampal slices. *Nature* 266: 736–37

Andersen, P., Sundberg, S. H., Sveen, O., Swann, J. W., Wigström, H. 1980. Possible mechanisms for long-lasting potentiation of synaptic transmission in hippocampal slices from guinea-pigs. *J. Physiol.* 302: 463–82

Babb, T. L. 1982. Short-term and long-term modifications of neurons and evoked potentials in the human hippocampal formation. *Hippocampal Long-Term Potentiation: Mechanisms and Implications for Memory*, *Neurosci. Res. Progr. Bull.* 20: 729–39

Baimbridge, K. G., Miller, J. J. 1981. Calcium uptake and retention during long-term potentiation of neuronal activity in the rat hippocampal slice preparation. *Brain Res.* 221: 299–305

Bär, P. R., Schotman, P., Gispen, W. H., Tielen, A. M., Lopes da Silva, F. H. 1980. Change in synaptic membrane phosphorylation after tetanic stimulation in the dentate area of the rat hippocampal slice. *Brain Res.* 198: 478–84

Baranyi, A., Feher, O. 1981. Synaptic facilitation requires paired activation of convergent pathways in the neocortex. *Nature* 290: 413–15

Barnes, C. A. 1979. Memory deficits associated with senescense: A neurophysiological and behavioral study in the rat. *J. Comp. Physiol. Psychol.* 93: 74–104

Barnes, C. A., McNaughton, B. L. 1980. Spatial memory and hippocampal synaptic plasticity in senescent and middle-aged rats. In *The Psychobiology of Aging: Problems and Perspectives*, ed. D. G. Stein, pp. 253–72. Amsterdam: Elsevier

Barrionuevo, G., Brown, T. H. 1983. Associative long-term potentiation in hippocampal slices. *Proc. Nat. Acad. Sci. USA* 80: 7347–51

Barrionuevo, G., Kelso, S. R., Johnston, D., Brown, T. H. 1986. Conductance mechanism responsible for long-term potentiation in monosynaptic and isolated excitatory synaptic inputs to hippocampus. *J. Neurophysiol.* 55: 540–50

Baudry, M., Arst, O., Oliver, M., Lynch, G. 1981. Development of glutamate binding sites and their regulation by calcium in rat hippocampus. *Dev. Brain Res.* 1: 37–48

Baudry, M., Oliver, M., Creager, R., Wieraszko, A., Lynch, G. 1980. Increase in glutamate receptors following repetitive electrical stimulation in hippocampal slices. *Life Sci.* 27: 325–30

Baxter, D. A., Bittner, G. D., Brown, T. H. 1985. Quantal analysis of long-term synaptic potentiàtion. *Proc. Natl. Acad. Sci. USA* 82: 5978–82

Berger, T. 1984. Long-term potentiation of hippocampal synaptic transmission affects rate of behavioral learning. *Science* 224: 627–30

Berger, T. W., Balzer, J. R., Eriksson, J. L., Sclabassi, R. J. 1984. Long-term potentiation alters non-linear characteristics of hippocampal perforant path-dentate synaptic transmission. *Soc. Neurosci. Abstr.* 10: A305.5

Bliss, T. V. P., Gardner-Medwin, A. R. 1973. Long-lasting potentiation of synaptic transmission in the dentate area of the unanaesthetized rabbit following stimulation of the perforant path. *J. Physiol.* 232: 357–74

Bliss, T. V. P., Goddard, G. V., Riives, M. 1983. Reduction of long-term potentiation in the dentate gyrus of the rat following selective depletion of monoamines. *J. Physiol.* 334: 475–91

Bliss, T. V. P., Lømo, T. 1973. Long-lasting potentiation of synaptic transmission in the dentate area of the anaesthetized rabbit following stimulation of the perforant path. *J. Physiol.* 232: 331–56

Bloch, V., Laroche, S. 1985. Enhancement of long-term potentiation in the rat dentate gyrus by post-trial stimulation of the reticular formation. *J. Physiol.* 360: 215–31

Briggs, C. A., McAfee, D. A., McCaman, R.

E. 1985. Long-term potentiation of synaptic acetylcholine release in the superior cervical ganglion of the rat. *J. Physiol.* 363: 181–90

Brown, T. H., McAfee, D. A. 1982. Long-term synaptic potentiation in the superior cervical ganglion. *Science* 215: 1411–13

Browning, M., Dunwiddie, T., Bennett, W., Gispen, W., Lynch, G. 1979. Synaptic phosphoproteins: Specific changes after repetitive stimulation of the hippocampal slice. *Science* 203: 60–62

Burger, B., Levy, W. B. 1985. Long-term associative potentiation/depression as an analogue of classical conditioning. *Soc. Neurosci. Abstr.* 11: A245.3

Buzsáki, G. 1980. Long-term potentiation of the commissural path-CA1 pyramidal cell synapse in the hippocampus of the freely moving rat. *Neurosci. Lett.* 19: 293–96

Buzsáki, G., Eidelberg, E. 1982. Direct afferent excitation and long-term potentiation of hippocampal interneurons. *J. Neurophysiol.* 48: 597–607

Collingridge, G. 1985. Long-term potentiation in the hippocampus: Mechanisms of initiation and modulation by neurotransmitters. *Trends Pharmacol. Sci.* 6: 407–11

Collingridge, G. L., Kehl, S. J., McLennan, H. 1983. Excitatory amino acids in synaptic transmission in the Schaffer-commissural pathway of the rat hippocampus. *J. Physiol.* 334: 33–46

Dana, R. C., Gerren, R. A., Sternberg, D. B., Martinez, J. L., Hall, J., Stansbury, N. A. Weinberger, N. M. 1982. Long-term potentiation is impaired by adrenalectomy and restored by corticosterone. *Soc. Neurosci. Abstr.* 8: A86.11

Deadwyler, S. A., Dudek, F. E., Cotman, C. W., Lynch, G. 1975. Intracellular responses of rat dentate gyrus granule cells *in vitro*: Posttetanic potentiation to perforant path stimulation. *Brain Res.* 88: 59–65

de Jonge, M., Racine, R. J. 1985. The effects of repeated induction of long-term potentiation in the dentate gyrus. *Brain Res.* 328: 181–85

Desmond, N. L., Levy, W. B. 1983. Synaptic correlates of associative potentiation/depression: An ultrastructural study in the hippocampus. *Brain Res.* 265: 21–30

Disterhoft, J. F., Coulter, D. A., Alkon, D. L. 1986. Conditioning-specific membrane changes of rabbit hippocampal neurons measured in vitro. *Proc. Natl. Acad. Sci. USA* 83: 2733–37

Doller, H. J., Weight, F. F. 1985. Perforant pathway-evoked long-term potentiation of CA1 neurons in the hippocampal slice preparation. *Brain Res.* 333: 305–10

Dolphin, A. C. 1983a. The excitatory amino-acid antagonist gamma-D-glutamyl-glycine masks rather than prevents long-term potentiation of the perforant path. *Neuroscience* 2: 377–83

Dolphin, A. C. 1983b. The adenosine agonist 2-chloroadenosine inhibits the induction of long-term potentiation of the perforant path. *Neurosci. Lett.* 39: 83–89

Dolphin, A. C. 1985. Long-term potentiation at peripheral synapses. *Trends Neurosci.* 8: 376–78

Dolphin, A. C., Errington, M. L., Bliss, T. V. P. 1982. Long-term potentiation of the perforant path *in vivo* is associated with increased glutamate release. *Nature* 297: 496–98

Douglas, R. M. 1977. Long-lasting synaptic potentiation in the rat dentate gyrus following brief high frequency stimulation. *Brain Res.* 126: 361–65

Douglas, R. M., Goddard, G. V. 1975. Long-term potentiation of the perforant path-granule cell synapse in the rat hippocampus. *Brain Res.* 86: 205–15

Douglas, R. M., Goddard, G. V., Riives, M. 1982. Inhibitory modulation of long-term potentiation: Evidence for a postsynaptic locus of control. *Brain Res.* 240: 259–72

Duffy, C. J., Teyler, T. J. 1978. Development of potentiation in the dentate gyrus of rat: Physiology and anatomy. *Brain Res. Bull.* 3: 425–30

Duffy, C. J., Teyler, T. J., Shashoua, V. E. 1981. Long-term potentiation in the hippocampal slice: Evidence for stimulated secretion of newly synthesized proteins. *Science* 212: 1148–51

Dunwiddie, T., Madison, D., Lynch, G. 1978. Synaptic transmission is required for initiation of long-term potentiation. *Brain Res.* 150: 413–17

Dunwiddie, T. V., Roberson, N. L., Worth, T. 1982. Modulation of long-term potentiation: Effects of adrenergic and neuroleptic drugs. *Pharm. Biochem. Behav.* 17: 1257–64

Durand, D., Carlen, P. L. 1984. Impairment of long-term potentiation in rat hippocampus following chronic ethanol treatment. *Brain Res.* 308: 325–32

Eccles, J. C. 1983. Calcium in long-term potentiation as a model for memory. *Neuroscience* 4: 1071–81

Fifkova, E., Van Harreveld, A. 1977. Long-lasting morphological changes in dendritic spines of dentate granular cells following stimulation of the entorhinal area. *J. Neurocytol.* 6: 211–30

Finn, R. C., Browning, M., Lynch, G. 1980. Trifluoperazine inhibits long-term potentiation and the phosphorylation of a

40,000 dalton protein. *Neurosci. Lett.* 19: 103–8
Foy, M. R., Stanton, M. E., Levine, S., Thompson, R. F. 1985. Stress impairs long-term potentiation in rodent hippocampus. *Soc. Neurosci. Abstr.* 11: A225.18
Gerren, R. A., Weinberger, N. M. 1983. Long-term potentiation in the magnocellular medial geniculate nucleus of the anesthetized cat. *Brain Res.* 265: 138–42
Goddard, G. V. 1980. Component properties of the memory machine: Hebb revisited. In *The Nature of Thought: Essays in Honour of D. O. Hebb*, ed. P. W. Jusczyk, R. M. Klein, pp. 231–47. London: Erlbaum
Gold, P. E., Delanoy, R. L., Merrin, J. 1984. Modulation of long-term potentiation by peripherally administered amphetamine and epinephrine. *Brain Res.* 305: 103–7
Gustafsson, B., Wigström, H. 1986. Hippocampal long-lasting potentiation produced by pairing single volleys and brief conditioning tetani evoked in separate afferents. *J. Neurosci.* 6: 1575–82
Haas, H. L., Greene, R. W. 1985. Long-term potentiation and 4-aminopyridine. *Cell Molec. Neurobiol.* 5: 297–300
Haas, H. L., Rose, G. 1982. Long-term potentiation of excitatory synaptic transmission in the rat hippocampus: The role of inhibitory processes. *J. Physiol.* 329: 541–52
Haas, H. L., Rose, G. 1984. The role of inhibitory mechanisms in hippocampal long-term potentiation. *Neurosci. Lett.* 47: 301–6
Harris, E. W., Ganong, A. H., Cotman, C. W. 1984. Long-term potentiation in the hippocampus involves activation of *N*-methyl-D-aspartate receptors. *Brain Res.* 323: 132–37
Harris, K. M., Teyler, T. J. 1983. Age differences in a circadian influence on hippocampal LTP. *Brain Res.* 261: 69–73
Harris, K. M., Teyler, T. J. 1984. Developmental onset of long-term potentiation in area CA1 of the rat hippocampus. *J. Physiol.* 346: 27–48
Hebb, D. O. 1949. *The Organization of Behavior*. New York: Wiley
Hendricks, C., Teyler, T. J. 1983. Effects of blocking GABA on hippocampal CA1 inhibition and plasticity. *Soc. Neurosci. Abstr.* 9: A248.3
Higashima, M., Yamamoto, C. 1985. Two components of long-term potentiation in mossy fiber-induced excitation in hippocampus. *Exp. Neurol.* 90: 529–39
Hoch, D. B., Dingledine, R. J., Wilson, J. E. 1984. Long-term potentiation in the hippocampal slice: Possible involvement of pyruvate dehydrogenase. *Brain Res.* 302: 125–34
Hopkins, W. F., Johnston, D. 1983. Beta-adrenergic receptor regulation of long-term potentiation in the hippocampus. *Soc. Neurosci. Abstr.* 9: A248.12
Hopkins, W. F., Johnston, D. 1984. Frequency-dependent noradrenergic modulation of long-term potentiation in the hippocampus. *Science* 226: 350–52
Ito, M. 1983. Evidence for synaptic plasticity in the cerebellar cortex. *Acta Morph. Acad. Sci. Hung.* 31: 213–18
Kelso, S. R., Brown, T. H. 1986. Differential conditioning of associative synaptic enhancement in hippocampal brain slices. *Science* 232: 85–87
Komatsu, Y., Toyama, K., Maeda, J., Sakaguchi, H. 1981. Long-term potentiation investigated in a slice preparation of striate cortex of young kittens. *Neurosci. Lett.* 26: 269–74
Koyano, K., Kuba, K., Minota, S. 1985. Long-term potentiation of transmitter release induced by repetitive presynaptic activities in bullfrog sympathetic ganglia. *J. Physiol.* 359: 219–33
Krug, M., Brodemann, R., Ott, T. 1982. Blockade of long-term potentiation in the dentate gyrus of freely moving rats by the glutamic acid antagonist GDEE. *Brain Res.* 249: 57–62
Krug, M., Chepkova, A. N., Geyer, C., Ott, T. 1983. Aminergic blockade modulates long-term potentiation in the dentate gyrus of freely moving rats. *Brain Res. Bull.* 11: 1–6
Krug, M., Lössner, B., Ott, T. 1984. Anisomycin blocks the late phase of long-term potentiation in the dentate gyrus of freely moving rats. *Brain Res. Bull.* 13: 39–42
Kuhnt, U., Mihaly, A., Joó, F. 1985. Increased binding of calcium in the hippocampus during long-term potentiation. *Neurosci. Lett.* 53: 149–54
Laroche, S., Bergis, O. E., Bloch, V. 1983. Posttrial reticular facilitation of dentate multiunit conditioning is followed by an increased long-term potentiation. *Soc. Neurosci. Abstr.* 9: A191.1
Laroche, S., Bloch, V. 1982. Conditioning of hippocampal cells and long-term potentiation: An approach to mechanisms of posttrial memory formation. In *Neuronal Plasticity and Memory Formation*, ed C. Ajmone Marsan, H. Matthies, pp. 575–87. New York: Raven
Larson, J. R., Lynch, G. 1985. Long-term potentiation in lizard cerebral cortex. *Soc. Neurosci. Abstr.* 11: A225.2
Larson, J., Lynch, G. 1986. Induction of synaptic potentiation in hippocampus by patterned stimulation involves two events. *Science* 232: 985–88

Lee, K. S. 1982. Sustained enhancement of evoked potentials following brief, high frequency stimulation of the cerebral cortex *in vitro*. *Brain Res.* 239: 617–23

Lee, K. S. 1983. Cooperativity among afferents for the induction of long-term potentiation in the CA1 region of the hippocampus. *J. Neurosci.* 3: 1369–72

Lee, K. S., Schottler, F., Oliver, M., Lynch, G. 1980. Brief bursts of high-frequency stimulation produces two types of structural change in rat hippocampus. *J. Neurophysiol.* 44: 247–58

Levy, W. B., Steward, O. 1979. Synapses as associative memory elements in the hippocampal formation. *Brain Res.* 175: 233–45

Levy, W. B., Steward, O. 1983. Temporal contiguity requirements for long-term associative potentiation/depression in the hippocampus. *Neuroscience* 8: 791–97

Lewis, D., Teyler, T. J. 1986a. Long-term potentiation in the goldfish optic tectum. *Brain Res.* 375: 246–50

Lewis, D., Teyler, T. J. 1986b. Anti-S-100 serum blocks long-term potentiation in the hippocampal slice. *Brain Res.* In press

Libet, B., Kobayashi, H., Tanaka, T. 1975. Synaptic coupling into the production and storage of a neuronal memory trace. *Nature* 258: 155–57

Linseman, M. A., Corrigall, W. A. 1981. Are endogenous opiates involved in potentiation of field potentials in the hippocampus of the rat? *Neurosci. Lett.* 27: 319–24

Lnenicka, G. A., Atwood, H. L. 1985. Long-term facilitation and long-term adaptation at synapses of a crayfish phasic motoneuron. *J. Neurobiol.* 16: 97–110

Lynch, G., Baudry, M. 1984. The biochemistry of memory: A new and specific hypothesis. *Science* 224: 1057–63

Lynch, G. S., Dunwiddie, T., Gribkoff, V. 1977. Heterosynaptic depression: A postsynaptic correlate of long-term potentiation. *Nature* 266: 736–37

Lynch, G., Larson, J., Kelso, S., Barrionuevo, G., Schottler, F. 1983. Intracellular injections of EGTA block induction of hippocampal long-term potentiation. *Nature* 305: 719–21

Lynch, M. A., Errington, M. L., Bliss, T. V. P. 1985a. Long-term potentiation of synaptic transmission in the dentate gyrus: Increased release of [^{14}C]glutamate without increase in receptor binding. *Neurosci. Lett.* 62: 123–29

Lynch, M. A., Errington, M. L., Bliss, T. V. P. 1985b. Long-term potentiation and the sustained increase in glutamate release which follows tetanic stimulation of the perforant path are both blocked by D(-)aminophosphonovaleric acid. *Soc. Neurosci. Abstr.* 11: 245.2

Malenka, R. C., Madison, D. V., Nicoll, R. A. 1986. Potentiation of synaptic transmission in the hippocampus by phorbol esters. *Nature* 321: 175–77

Malinow, R., Miller, J. P. 1986. Postsynaptic hyperpolarization during conditioning reversibly blocks induction of long-term potentiation. *Nature* 320: 529–30

Mamounas, L. A., Thompson, R. F., Lynch, G., Baudry, M. 1984. Classical conditioning of the rabbit eyelid response increases glutamate receptor binding in hippocampal synaptic membranes. *Proc. Natl. Acad. Sci. USA* 81: 2548–52

Marr, D. 1969. A theory of cerebellar cortex. *J. Physiol.* 202: 437–70

Martin, M. R. 1983. Naloxone and long-term potentiation of hippocampal CA3 field potentials *in vitro*. *Neuropeptides* 4: 45–50

McNaughton, B. L. 1982. Long-term synaptic enhancement and short-term potentiation in rat fascia dentata act through different mechanisms. *J. Physiol.* 324: 249–62

McNaughton, B. L., Barnes, C. A. 1977. Physiological identification and analysis of dentate granule cell responses to stimulation of the medial and lateral perforant pathways in the rat. *J. Comp. Neurol.* 175: 439–54

McNaughton, B. L., Barnes, C. A., Andersen, P. 1981. Synaptic efficacy and EPSP summation in granule cells of rat fascia dentata studied *in vitro*. *J. Neurophysiol.* 46: 952–66

McNaughton, B. L., Barnes, C. A., Rao, G., Baldwin, J., Rasmussen, M. 1986. Long-term enhancement of hippocampal synaptic transmission and the acquisition of spatial information. *J. Neurosci.* 6: 563–71

McNaughton, B. L., Douglas, R. M., Goddard, G. V. 1978. Synaptic enhancement in fascia dentata: Cooperativity among coactive afferents. *Brain Res.* 157: 277–93

McNaughton, N., Miller, J. J. 1984. Medial septal projections to the dentate gyrus of the rat: Electrophysiological analysis of distribution and plasticity. *Exp. Brain Res.* 56: 243–56

Misgeld, U., Sarvey, J. M., Klee, M. R. 1979. Heterosynaptic postactivation potentiation in hippocampal CA3 neurons: Long-term changes of the post-synaptic potentials. *Exp. Brain Res.* 37: 217–29

Mochida, S., Libet, B. 1985. Synaptic long-term enhancement (LTE) induced by a heterosynaptic neural input. *Brain Res.* 329: 360–63

Mody, I., Baimbridge, K. G., Miller, J. J. 1984. Blockade of tetanic- and calcium-induced long-term potentiation in the hippocampal slice preparation by neuroleptics. *Neuropharmacology* 23: 625–31

Morris, R. G. M., Anderson, E., Lynch, G. S., Baudry, M. 1986. Selective impairment of learning and blockade of long-term potentiation by an *N*-methyl-D-aspartate receptor antagonist, AP5. *Nature* 319: 774–76

Northcutt, R. G. 1981. Evolution of the telencephalon in nonmammals. *Ann. Rev. Neurosci.* 4: 301–50

Nowak, L., Bregestovski, P., Ascher, P., Herbet, A., Prochiantz, A. 1984. Magnesium gates glutamate-activated channels in mouse central neurones. *Nature* 307: 462–64

Nowicky, A. V., Vardaris, R. M., Teyler, T. J. 1983. Corticosterone and long-term potentiation in the hippocampus. *Soc. Neurosci. Abstr.* 9: A350.14

Nowicky, A. V., Vardaris, R. M., Teyler, T. J. 1985. The modulation of long-term potentiation by delta-9-tetrahydrocannabinol in the *in vitro* rat hippocampal slice. *Soc. Neurosci. Abstr.* 11: A117.1

Ott, T., Ruthrich, H., Reymann, K., Lindenau, L., Matthies, H. 1982. Direct evidence for the participation of changes in synaptic efficacy in the development of behavioral plasticity. See Laroche & Bloch 1982, pp. 441–52

Patneau, D. K., Stripling, J. S. 1985. Functional correlates of selective long-term potentiation in the pyriform cortex. *Soc. Neurosci. Abstr.* 11: A225.11

Racine, R. J., Milgram, N. W., Hafner, S. 1983. Long-term potentiation phenomena in the rat limbic forebrain. *Brain Res.* 260: 217–31

Racine, R. J., Wilson, D. A., Gingell, R., Sunderland, D. 1986. Long-term potentiation in the interpositus and vestibular nuclei in the rat. *Exp. Brain Res.* 63: 158–62

Reymann, K. G., Malisch, R., Schulzeck, K., Brodemann, R., Ott, T., Matthies, H. 1985. The duration of long-term potentiation in the CA1 region of the hippocampal slice preparation. *Brain Res. Bull.* 15: 249–55

Reymann, K., Ruthrich, H., Lindenau, L., Ott, T., Matthies, H. 1982. Monosynaptic activation of the hippocampus as a conditioned stimulus: Behavioral effects. *Physiol. Behav.* 29: 1007–12

Robinson, G. B., Racine, R. J. 1982. Heterosynaptic interactions between septal and entorhinal inputs to the dentate gyrus: Long-term potentiation effects. *Brain Res.* 249: 162–66

Robinson, G. B., Racine, R. J. 1985. Long-term potentiation in the dentate gyrus: Effects of noradrenaline depletion in the awake rat. *Brain Res.* 325: 71–78

Routtenberg, A., Lovinger, D. M., Steward, O. 1985. Selective increase in phosphorylation of a 47-KDa protein (F1) directly related to long-term potentiation. *Behav. Neurol. Biol.* 43: 3–11

Ruthrich, H., Matthies, H., Ott, T. 1982. Long-term changes in synaptic excitability of hippocampal cell populations as a result of training. See Laroche & Bloch 1982, pp. 589–94

Sastry, B. R. 1982. Presynaptic change associated with long-term potentiation in hippocampus. *Life Sci.* 30: 2003–8

Sastry, B. R. 1985. Leupeptin does not block the induction of long-lasting potentiation in hippocampal CA1 neurones. *Br. J. Pharmacol.* 86: 589P

Sastry, B. R., Goh, J. W. 1984. Long-lasting potentiation in hippocampus is not due to an increase in glutamate receptors. *Life Sci.* 34: 1497–1501

Sastry, B. R., Goh, J. W., Auyeung, A. 1986. Associative induction of posttetanic and long-term potentiation in CA1 neurons of rat hippocampus. *Science* 232: 988–90

Scharfman, H. E., Sarvey, J. M. 1985. Postsynaptic firing during repetitive stimulation is required for long-term potentiation in hippocampus. *Brain Res.* 331: 267–74

Schwartzkroin, P. A., Wester, K. 1975. Long-lasting facilitation of a synaptic potential following tetanization in the *in vitro* hippocampal slice. *Brain Res.* 89: 107–19

Sclabassi, R. J., Eriksson, J. L., Berger, T. W. 1985. Nonlinear characteristics of hippocampal perforant path-dentate synaptic transmission are different for synaptic and action potential currents. *Soc. Neurosci. Abstr.* 10: A305.4

Sharp, P. E., McNaughton, B. L., Barnes, C. A. 1985. Enhancement of hippocampal field potentials in rats exposed to a novel, complex environment. *Brain Res.* 339: 361–65

Skelton, R. W., Miller, J. J., Phillips, A. G. 1983. Low-frequency stimulation of the perforant path produces long-term potentiation in the dentate gyrus of unanesthetized rats. *Can. J. Physiol. Pharmacol.* 61: 1156–61

Skelton, R. W., Miller, J. J., Phillips, A. G. 1985. Long-term potentiation facilitates behavioral responding to single-pulse stimulation of perforant path. *Behav. Neurosci.* 99: 603–20

Skrede, K. K., Malthe-Sørenssen, D. 1981. Increased resting and evoked release of transmitter following repetitive electrical tetanization in hippocampus: A biochemical correlate to long-lasting synaptic potentiation. *Brain Res.* 208: 436–41

Squire, L. 1982. The neuropsychology of human memory. *Ann. Rev. Neurosci.* 5: 241–73

Stanton, P. K., Sarvey, J. M. 1984. Blockade of long-term potentiation in rat hippocampal CA1 region by inhibitors of protein synthesis. *J. Neurosci.* 4: 3080–88

Stanton, P. K., Sarvey, J. M. 1985a. Depletion of norepinephrine, but not serotonin, reduces long-term potentiation in the dentate of rat hippocampal slices. *J. Neurosci.* 5: 2169–76

Stanton, P. K., Sarvey, J. M. 1985b. The effect of high-frequency electrical stimulation and norepinephrine on cyclic AMP levels in normal versus norepinephrine-depleted rat hippocampal slices. *Brain Res.* 358: 343–48

Steward, O., Brassel, S. 1983. Intrahippocampal injections of cyclohexamide reversibly block long-term potentiation. *Soc. Neurosci. Abstr.* 9: A248.7

Stringer, J. L., Greenfield, L. J., Hackett, J. T., Guyenet, P. G. 1983. Blockade of long-term potentiation by phencyclidine and sigma opiates in the hippocampus *in vivo* and *in vitro*. *Brain Res.* 280: 127–38

Stringer, J. L., Guyenet, P. G. 1983. Elimination of long-term potentiation in the hippocampus by phencyclidine and ketamine. *Brain Res.* 258: 159–64

Stripling, J. S., Patneau, D. K. 1985. Selective long-term potentiation in the pyriform cortex. *Soc. Neurosci. Abstr.* 11: A225.10

Stripling, J. S., Patneau, D. K., Granlich, C. A. 1984. Long-term changes in the pyriform cortex evoked potential produced by stimulation of the olfactory bulb. *Soc. Neurosci. Abstr.* 10: A26.6

Swanson, L. W., Teyler, T. J., Thompson, R. F. 1982. Hippocampal long-term potentiation: Mechanisms and implications for memory. *Neurosci. Res. Progr. Bull.* 20: 613–769

Teyler, T. J., Alger, B. E., Bergman, T., Livingston, K. 1977. A comparison of long-term potentiation in the *in vitro* and *in vivo* hippocampal preparations. *Behav. Biol.* 19: 24–34

Teyler, T. J., DiScenna, P. 1984. Long-term potentiation as a candidate mnemonic device. *Brain Res. Rev.* 7: 15–28

Teyler, T. J., DiScenna, P. 1985. The role of hippocampus in memory: A hypothesis. *Neurosci. Biobehav. Rev.* 9: 377–89

Thompson, R. F. 1983. Neuronal substrates of simple associative learning: Classical conditioning. *Trends Neurosci.* 6: 270–75

Thompson, R. F., Berger, T. W., Madden, J. 1983. Cellular processes of learning and memory in the mammalian CNS. *Ann. Rev. Neurosci.* 6: 447–91

Thomson, A. M., Lodge, D. 1985. Selective blockade of an excitatory synapse in rat cerebral cortex by the sigma opiate cyclazocine: An intracellular, *in vitro* study. *Neurosci. Lett.* 54: 21–26

Turner, R. W., Baimbridge, K. G., Miller, J. J. 1982. Calcium-induced long-term potentiation in the hippocampus. *Neuroscience* 7: 1411–16

Voronin, L. L. 1983. Long-term potentiation in the hippocampus. *Neuroscience* 4: 1051–69

Voronin, L. L. 1985. Synaptic plasticity at archicortical and neocortical levels. *Neirofiziologiya* 16: 651–65

Walters, E. T., Byrne, J. H. 1985. Long-term enhancement produced by activity-dependent modulation of *Aplysia* sensory neurons. *J. Neurosci.* 5: 662–72

Weisz, D. J., Clark, G. A., Thompson, R. F. 1984. Increased responsivity of dentate granule cells during nictitating membrane response conditioning in rabbit. *Behav. Brain Res.* 12: 145–54

White, W. F., Nadler, J. V., Hamberger, A., Cotman, C. W. 1977. Glutamate as transmitter of hippocampal perforant path. *Nature* 270: 356–57

Wieraszko, A., Seifert, W. 1985. The role of monosialoganglioside GM1 in synaptic plasticity: An *in vitro* study on rat hippocampal slices. *Brain Res.* 345: 159–64

Wigström, H., Gustafsson, B. 1983a. Facilitated induction of hippocampal long-lasting potentiation during blockade of inhibition. *Nature* 301: 603–4

Wigström, H., Gustafsson, B. 1983b. Large long-lasting potentiation in the dentate gyrus *in vitro* during blockade of inhibition. Brain Res 275: 153–58

Wigström, H., Gustafsson, B. 1984. A possible correlate of the postsynaptic condition for long-lasting potentiation in the guinea pig hippocampus *in vitro*. *Neurosci. Lett.* 44: 327–32

Wigström, H., Gustafsson, B. 1985. On long-lasting potentiation in the hippocampus: A proposed mechanism for its dependence on coincident pre- and postsynaptic activity. *Acta Physiol. Scand.* 123: 519–22

Wigström, H., McNaughton, B. L., Barnes, C. A. 1982. Long-term synaptic enhancement in hippocampus is not regulated by postsynaptic membrane potential. *Brain Res.* 233: 195–99

Wigström, H., Swann, J. W. 1980. Strontium supports synaptic transmission and long-lasting potentiation in the hippocampus. *Brain Res.* 194: 181–91

Wigström, H., Swann, J. W., Andersen, P. 1979. Calcium dependency of synaptic long-lasting potentiation in the hippocampal slice. *Acta Physiol. Scand.* 105: 126–28

Wilhite, B. L., Teyler, T. J., Hendricks, C. 1986. Functional relations of the rodent claustral-entorhinal-hippocampal system. *Brain Res.* 365: 54–60

Wilson, D. A. 1984. A comparison of the postnatal development of post-activation potentiation in the neocortex and dentate gyrus of the rat. *Dev. Brain Res.* 16: 61–68

Wilson, D. A., Racine, R. J. 1983. The postnatal development of post-activation potentiation in the rat neocortex. *Dev. Brain Res.* 7: 271–76

Wilson, R., Levy, W. B., Steward, O. 1979. Functional effects of lesion-induced plasticity: Long-term potentiation in the normal and lesion-induced temporodentate circuits. *Brain Res.* 176: 65–78

Wilson, R. C., Levy, W. B., Steward, O. 1981. Changes in translation of synaptic excitation to dentate granule cell discharge accompanying long-term potentiation. II. An evaluation of mechanisms utilizing dentate gyrus dually innervated by surviving ipsilateral and sprouted crossed temporodentate inputs. *J. Neurophysiol.* 46: 324–38

Woolley, D., Zimmer, L., Hasan, Z., Swanson, K. 1984. Do some insecticides and heavy metals produce long-term potentiation in the limbic system. In *Cellular and Molecular Neurotoxicology*, ed. T. Narahashi, pp. 45–69. New York: Raven

Yamamoto, C., Chujo, T. 1978. Long-term potentiation in thin hippocampal sections studied by intracellular and extracellular recordings. *Exp. Neurol.* 58: 242–50

Yamamoto, C., Sawada, S. 1981. Important factors in induction of long-term potentiation in thin hippocampal sections. *Exp. Neurol.* 74: 122–30

Ann. Rev. Neurosci. 1987. 10 : 163–94

MOLECULAR BIOLOGY OF VISUAL PIGMENTS

Jeremy Nathans

Department of Biochemistry, Stanford University School of Medicine, Stanford, California 94305

INTRODUCTION

We have witnessed exciting advances in the study of vision over the past several years. In particular, our understanding of the biochemical events responsible for visual transduction has grown enormously. By visual transduction we mean those events within the retinal photoreceptor cell that convert a photon stimulus into a neural signal. At center stage are the visual pigments. These light-absorbing molecules initiate an intracellular enzyme cascade that eventually leads either to decreasing (vertebrates) or increasing (invertebrates) plasma membrane cation conductance. I focus in this review on the structure and function of the visual pigments, with emphasis on vertebrates. I have tried throughout to highlight interesting areas in which our knowledge is still incomplete. Recent reviews of this and related topics include Birge 1981, Hargrave 1982, Chabre 1985, Lamb 1986, and Stryer 1986.

PHOTORECEPTOR STRUCTURE

Retinal photoreceptors—rods and cones—are highly specialized cells. They each have a sensory organelle, the outer segment, which contains the biochemical machinery responsible for visual transduction; the biosynthetic machinery resides in the inner segment. The outer segments are filled with stacks of flattened membrane saccules (discs), which are derived from the plasma membrane. In cones, but not in rods, the discs retain a connection with the plasma membrane. Outer segment biosynthesis is an ongoing process (Young 1971): in rhesus monkeys, approximately 90 discs per day are added at the base of the rod outer segment. Ten days later,

0147–006X/87/0301–0163$02.00

after moving through the entire length of the outer segment, these discs are phagocytosed at its distal end by pigment epithelial cells. How the outer segment proteins are assembled and segregated away from other cellular constituents is still a mystery. The visual pigment, which represents 80% of all protein and at least 95% of membrane protein of the outer segment, is probably targeted first to the plasma membrane and then carried along during membrane invagination to produce the outer segment disc. A recent intriguing observation is that tunicamycin inhibition of rhodopsin glycosylation disrupts orderly disc biogenesis (Fliesler & Basinger 1985, Fliesler et al 1985). Newly synthesized nonglycosylated opsin (the apoprotein of rhodopsin) accumulates in a heterogeneous mass of vesicles at the base of the outer segment. The question of inner vs outer segment protein segregation is especially acute with respect to cytosolic components of the transduction machinery, since small molecules (e.g. ATP) can certainly pass freely between inner and outer segments. Are these proteins carried into the outer segment by a transient membrane or cytoskeletal attachment? If so, could this mechanism alone account for the absence in outer segments of other soluble cellular proteins?

VISUAL PIGMENTS

Protein Structure

All visual pigments are built upon a common plan. An apoprotein, opsin, is joined covalently to 11-*cis* retinal via a protonated Schiff's base linkage between the *e*-amino group of a lysine and the aldehyde of retinal (Wald 1968). Photon absorption causes retinal to isomerize from 11-*cis* to all-*trans* (Hubbard & Kropf 1958). This movement of approximately 0.7 nm drives a series of conformational changes in the attached apoprotein. In vertebrates, all-*trans* retinal dissociates from opsin on a time scale of minutes. Free all-*trans* retinal is enzymatically converted back to 11-*cis* and is eventually reattached to opsin to generate the native photopigment. In invertebrates, all-*trans* retinal is photochemically converted to 11-*cis* without dissociation from opsin (Hamdorf 1979).

Because of the ready availability of cattle eyes, most biochemical studies have concentrated on bovine rhodopsin. The complete amino acid sequence of this protein has been determined (Ovchinnikov et al 1983, Hargrave et al 1983), as has that of ovine rhodopsin (Pappin et al 1984). Bovine rhodopsin is 348 amino acids in length and displays alternating stretches of hydrophobic and hydrophylic amino acids. Its transmembrane topology has been mapped by limited proteolysis and by derivatization with radiolabeled hydrophilic chemical probes (Ovchinnikov 1982, Martynov et al 1983, Mullen & Akhtar 1983, Barclay & Findlay 1984). These experiments use (*a*) intact outer segment disc membranes, in which case only cytosolic regions are cleaved or derivatized, and (*b*) frozen and thawed membranes, which expose both cytosolic and luminal faces. This analysis reveals that the amino-terminus faces the lumen of the rod outer segment discs (topologically equivalent to the outside of the cell) and the carboxy-terminus faces the cytosol. Therefore, opsin must traverse the membrane an odd number of times. The experimental data also constrain the polypeptide to cross the membrane from lumen to cytoplasm between residues 30 and 67, from cytoplasm to lumen between residues 146 and 186, and from lumen to cytoplasm again from residues 186 to 236. That these are not the only transmembrane crossings is suggested by the plot shown in Figure 1. Each point on the curve represents the free energy of transfer of that residue plus the next 20 from a polar to a nonpolar environment (Engelman et al 1982). If the transfer is thermodynamically favorable (a large negative ΔG), then that segment is considered likely to span the bilayer. The seven horizontal bars in Figure 1 indicate the most likely transmembrane segments. Also shown in Figure 1 is a schematic view of rhodopsin in the membrane. In this schematic, the chain is represented as a two-dimensional object. In reality, the seven membrane-spanning segments probably form a tight bundle around a central retinal binding pocket.

The model formulated above is supported by spectroscopic and neutron diffraction studies. It was observed 15 years ago that isolated rod outer segments oriented their long axes along the lines of a homogeneous magnetic field (Chalazonitis et al 1970, Chabre 1978), a phenomenon that almost certainly reflects a preferential arrangement of alpha-helices perpendicular to the plane of the disc membrane (Worcester 1978). This interpretation has been confirmed by measuring dichroic infrared absorbance due to peptide bond vibrations in magnetically oriented rod outer segments: those absorption bands characteristic of alpha helices display the predicted dichroism (Michel-Villaz et al 1979). Similar results obtained with multilamellar films of rod outer segment membranes predict that the helices have an average tilt of 40° or less (Rothschild et al 1980b). Moreover, in oriented rod outer segments, those amide protons that give rise to dichroic absorption are also unable to exchange with D_2O, as measured by shifts in infrared absorption. Conversely, those amide protons that do not display dichroic absorption can exchange with D_2O. Exchangeable protons derive from parts of the polypeptide chain that are exposed to the aqueous environment, and nonexchangeable protons derive from parts of the chain that are either membrane embedded or folded within the protein interior. These data, together with previous estimates from circular dichroism studies that 50% of rhodopsin is alpha-helical (Shichi

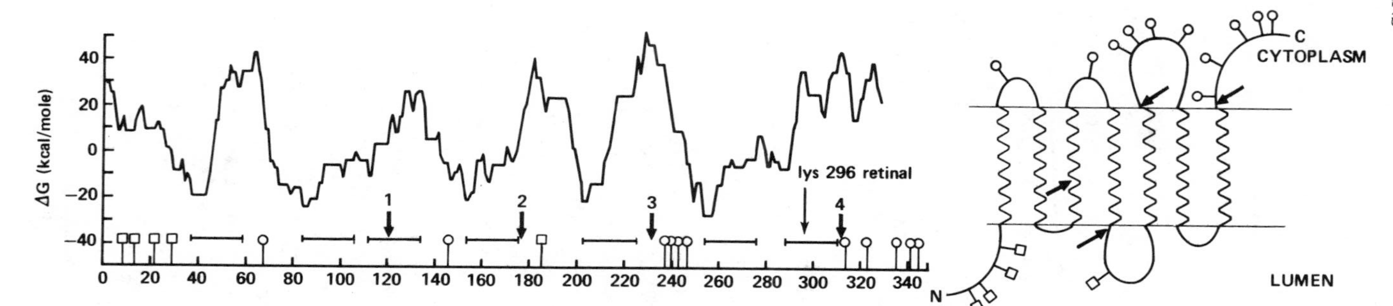

Figure 1 Predicted transmembrane topography of bovine rhodopsin. The *horizontal axis* marks the position along the polypeptide chain. *Squares* and *circles* indicate sites of accessibility to hydrophilic probes from luminal and cytosolic faces, respectively (Martynov et al 1983, Mullen & Akhtar 1983, Barclay & Findlay 1984). The *vertical axis* and the *plot* show the free energy of transfer of successive segments of 21 amino acids from a polar to a nonpolar environment (Engleman et al 1982). *Horizontal bars* demarcate the seven predicted transmembrane segments, and *numbered arrows* mark intron positions in the human and bovine rhodopsin genes. To the right is a schematic of rhodopsin in the membrane. (After Nathans & Hogness 1983.)

& Shelton 1974), suggest that approximately half of the polypeptide consists of alpha-helical rods embedded in the membrane. This model is also consistent with neutron and x-ray scattering studies indicating that approximately 50% of rhodopsin's mass lies in the bilayer (Saibil et al 1976, Sardet et al 1976, Osborne et al 1978).

Further constraints on the structure come from (*a*) the location of 11-*cis* retinal near the center of the bilayer (Thomas & Stryer 1982) and (*b*) the covalent attachment of 11-*cis* retinal to lysine^{296} (Bownds 1967, Wang et al 1980, Mullen & Akhtar 1981, Findlay et al 1981). The seventh transmembrane segment is therefore centered at this position (see Figure 1). Our only direct glimpse of rhodopsin comes from electron micrographs of negatively stained two-dimensional crystals of frog rhodopsin (Corless et al 1982). At 2.2 nm resolution, a pair of rhodopsin molecules appears as a slightly S-shaped ellipse when viewed perpendicular to the membrane. Taken together, this combination of experimental approaches gives us a model of rhodopsin's structure that is highly constrained.

Gene Structure

Recent molecular genetic experiments have shed light on the structure and evolution of visual pigments. Genes encoding bovine (Nathans & Hogness 1983), human (Nathans & Hogness 1984), and *Drosophila* (O'Tousa et al 1985, Zuker et al 1985) rhodopsins have been isolated and sequenced. The genes encoding the three human cone pigments have also been isolated and sequenced (Nathans et al 1986a) and are discussed separately at the end of this article. The bovine and human rhodopsin genes are highly homologous. They are both interrupted by four introns that lie at exactly analogous locations, and the deduced amino acid sequence of human rhodopsin is identical to the sequence of bovine rhodopsin at 94% of its residues. The three cytosolic loops and adjacent 10–12 transmembrane residues are identical in these two proteins; these regions probably constitute the enzymatic face of light-activated rhodopsin (see below). Interestingly, three of the four introns are located just beyond those sequences that encode three of the seven predicted transmembrane segments (Figure 1). In each case the putative transmembrane segment consists of a stretch of hydrophobic amino acids followed by a single positively charged amino acid. The positively charged amino acid might plausibly interact with negatively charged phospholipid head groups and thus serve to anchor the transmembrane segment. A similar pattern is observed in the sequences that encode membrane-anchoring segments of surface immunoglobulins and histocompatibility antigens (Figure 2). This observation is consistent with the general notion that introns divide genes into segments that encode structural or functional domains.

The amino acid sequence of the major *Drosophila* rhodopsin, deduced from its gene sequence, is 22% identical to bovine rhodopsin. To date, this is the only invertebrate—indeed, the only nonmammalian—visual pigment whose sequence has been determined. It firmly establishes the common ancestry of vertebrate and invertebrate visual pigments despite the anatomic dissimilarity of their photoreceptors. These sequence homologies, together with the very similar hydropathy profiles of vertebrate and invertebrate pigments, suggest that *Drosophila* rhodopsin adopts transmembrane and tertiary conformations similar to that of the mammalian protein. The *Drosophila* gene is also interrupted by four introns: The first intron has no homologue among the mammalian genes, the second intron resides at a position exactly analogous to that of the second mammalian intron, and the third and fourth introns are located several codons before the corresponding positions of the third and fourth mammalian introns. The cloned *Drosophila* gene corresponds to *ninaE*, a locus previously defined genetically. *ninaE* has a gene dosage effect on visual pigment content in the R1–R6 photoreceptor cells (Scavarda et al 1983), and its product is required for normal phototransduction. Several *ninaE* mutations are associated with disruptions of the *Drosophila* rhodopsin gene (O'Tousa et al 1985).

```
RHODOPSIN EXON 2                                                        ↱ INTRON
CATCATGGGCGTCGCCTTCACCTGGGTCATGGCTCTGGCCTGTGCCGCGCCCCCCTCGTCGGCTGGTCCAGGTAACGGCCC
 IleMetGlyValAlaPheThrTrpValMetAlaLeuAlaCysAlaAlaProProLeuValGlyTrpSerArg
                                                                       +
RHODOPSIN EXON 3                                                        ↱ INTRON
GTGGTCCACTTCATCATCCCCCTGATTGTCATATTCTTCTGCTACGGGCAGCTGGTGTTCACCGTCAAGGAGGTAGGGCCTT
ValValHisPheIleIleProLeuIleValIlePhePheCysTyrGlyGlnLeuValPheThrValLysGlu
                                                                    +
RHODOPSIN EXON 4                                                        ↱ INTRON
ACCATCCCGGCTTTCTTTGCCAAGACTTCTGCCGTCTACAACCCCGTCATCTACATCATGATGAACAAGCAGGTGCCTGCTG
ThrIleProAlaPhePheAlaLysThrSerAlaValTyrAsnProValIleTyrIleMetMetAsnLysGln
                                                                    +
IgG-1 EXON 7                                                            ↱ INTRON
ACGACCATCACCATCTTCATCAGCCTCTTCCTGCTCAGCGTGTGCTACAGCGCTGCTGTCACACTCTTCAAGGTCAGCCATA
ThrThrIleThrIlePheIleSerLeuPheLeuLeuSerValCysTyrSerAlaAlaValThrLeuPheLys
                                                                       +
H-2 K^D EXON 5                                                          ↱ INTRON
TGGTTGTCCTTGGAGCTGCAATAGTCACTGGAGCTGTGGTGGCTTTTGTGATGAAGATGAGAAGGAACACAGGTTAGTTGTG
LeuValValLeuGlyAlaAlaIleValThrGlyAlaValValAlaPheValMetLysMetArgArgAsnThr
                                                        +     +  +
```

Figure 2 Nucleotide and amino acid sequences at intron-exon junctions. The *upper three sequences* are from the 3′-ends of exons 2, 3, and 4 of the bovine rhodopsin gene (Nathans & Hogness 1983); the *lower two sequences* are from the 3′-ends of exons encoding mouse immunoglobulin and histocompatiability antigen membrane anchoring segments (Kvist et al 1983, Yamawaki-Kataoka et al 1982).

Energetics and Photochemistry

A detailed review of the photochemistry of visual pigments is beyond the scope of this article, but may be found in Honig 1978 and Birge 1981. Photon absorption triggers a sequence of conformational changes that are defined by their absorption spectra and lifetimes. Visual pigments containing either 11-*cis* or 9-*cis* retinal (the latter is an artificial laboratory construct) yield bathorhodopsin as the first intermediate (Yoshizawa & Wald 1963). This suggests that bathorhodopsin contains all-*trans* retinal. As judged by picosecond resonance Raman spectra, *cis-trans* isomerization occurs within 30 psec (Hayward et al 1981). The initial all-*trans* retinal product displays vibrational lines not displayed by retinal in solution and is presumed to be in a strained configuration.

Calorimetric measurements show that the energy (enthalpy) stored in bovine bathorhodopsin amounts to 35 kcal/mole, 60% of the available photon energy (57 kcal/mole for 500 nm light) (Cooper 1979, 1981). Because there is no detectable thermal decay of bathorhodopsin back to rhodopsin, this reverse pathway presents a barrier of at least 10 kcal/mole. Therefore, the rhodopsin-to-bathorhodopsin transition presents a barrier of at least $35+10 = 45$ kcal/mole. This insures a very low rate of spontaneous thermal isomerization. In contrast, retinal isomerization in solution is restrained by an energy barrier of only 25 kcal/mole (Hubbard 1966).

The energy stored in bathorhodopsin is harnessed to drive a series of conformational changes: bathorhodopsin $\rightarrow$ lumirhodopsin $\rightarrow$ metarhodopsin I $\rightarrow$ metarhodopsin II. As discussed below, metarhodopsin II, but not metarhodopsin I or III, binds to and activates a peripheral membrane protein, transducin. The metarhodopsin I $\rightarrow$ II transition probably involves a large protein conformational change. The Schiff's base nitrogen linking retinal to lysine^{296} loses its proton during this transition (Doukas et al 1978, Longstaff et al 1986). Nonspecific steric constraints such as delipidation (Applebury et al 1974), dehydration of the rod outer segment membrane stacks (Rothschild et al 1980a), or increased viscosity in reconstituted artifical lipid membranes (O'Brien et al 1977) inhibit this transition. Under these conditions rhodopsin remains in the metarhodopsin I state. In the first two cases, when the steric constraint is relieved by adding detergent or increasing humidity, the metarhodopsin I $\rightarrow$ II transition proceeds. These observations may reflect a rearrangement of membrane-embedded alpha-helices during this transition.

Detailed models have been proposed to account for the photochemical properties of visual pigments. Of central interest are the mechanisms by which opsin tunes the absorption spectrum of 11-*cis* retinal. A protonated

Schiff's base in solution (with a counterion) absorbs maximally at 440 nm. Bovine rhodopsin absorbs maximally at 500 nm, and the absorption maxima of cone pigments based upon 11-*cis* retinal range from 432 to 575 nm (Liebman 1971). Thirty years ago, Kropf & Hubbard (1958) noted that molecular orbital calculations predict a redistribution of positive charge from the protonated Schiff's base nitrogen to the retinal pi-electron system in the photoexcited state (Figure 3, structure B). The experimentally measured change in dipole moment is 12 debye, an extremely large value (Mathies & Stryer 1976). In the excited state, resonance structures such as B in Figure 3 enhance pi-electron delocalization and decrease the distinction between single and double bonds. Isomerization about the 11–12 double bond is promoted by this increase in its single bond character. To the extent that opsin stabilizes excited state structures and promotes pi-electron delocalization, it lowers the photon energy required for excitation. This model predicts that positively or negatively charged amino acids near the pi-electron system should preferentially destabilize or stabilize, respectively, the excited state, whereas positively or negatively charged amino acids near the Schiff's base nitrogen should preferentially destabilize or stabilize, respectively, the ground state.

By way of comparison, cyanine dyes with six double bonds and six single bonds (Figure 3, structures D and E) absorb maximally far in the red, at

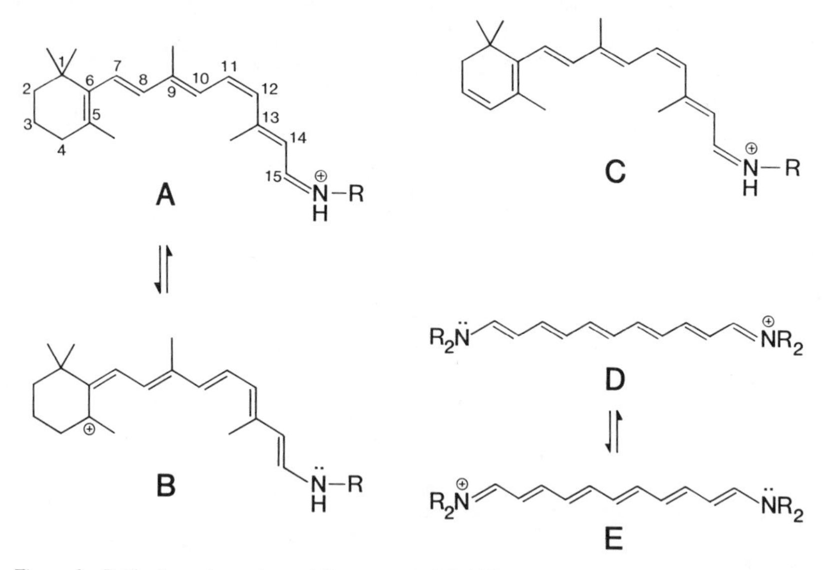

Figure 3 Retinals and cyanines. (*A*) protonated Schiff base of 11-*cis* retinal; (*B*) alternate resonance structure of *A*; (*C*) protonated Schiff base of 11-*cis* dehydroretinal; (*D, E*) resonance structures of a cyanine.

734 nm (Figure 4; Suzuki 1967). This appears to result from the highly delocalized pi-electron density and the near equivalence of bonds; the cyanine is a hybrid between the two resonance structures D and E in Figure 3. The model predicts that the chromophores of long-wavelength visual pigments resemble cyanines, whereas the chromophores of short-wavelength pigments retain the properties of polyenes, which have strong single bond/double bond alternation along the chain.

The calculations of Honig and co-workers show that the effects of charged amino acids are sufficient to explain the observed pigment spectra (Honig et al 1976). Two other long-standing observations are also predicted by these calculations and by the model discussed above:

1. The visual pigment red shift that accompanies replacement of 11-*cis* retinal (Figure 3, structure A) with 11-*cis* 3,4-dehydroretinal (Figure 3, structure C) increases linearly with wavelength of maximal absorption (6 nm for a pigment absorbing at 432 nm, and 45 nm for a pigment absorbing at 575 nm). This fits nicely with the observation that cyanine absorption maxima are strongly red-shifted by each additional double bond, whereas polyenes (like retinal) display a far smaller red-shift per additional double bond (Figure 4).

2. All visual pigment absorption spectra, when plotted on a frequency scale, have bandwidths approximately proportional to frequency of maximal absorption, i.e. long-wavelength pigments have narrow bandwidths and short-wavelength pigments have broad bandwidths. Narrow or broad bandwidth reflects small or large changes in bond length, respectively, during the transition from ground to excited states. This is consistent with a cyanine-like character of long wavelength pigments: cyanine bonds undergo only a small change in length upon photoexcitation, and as a consequence cyanine absorption spectra are narrow (Greenberg et al 1975). Because of their high degree of alternation, polyene bonds undergo large

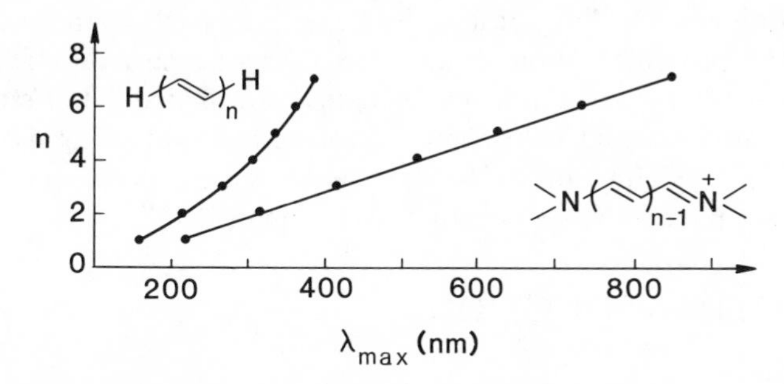

Figure 4 Polyene (*left*) and cyanine (*right*) λ_{max} as a function of number of double bonds (*n*). (After Suzuki 1967.)

changes in length upon photoexcitation; as a consequence, polyene absorption spectra are broad.

Another means of wavelength regulation that accounts for observation 1 has been proposed (Blatz & Liebman 1973). Twisting about the C_6–C_7 single bond would, as a consequence of isolating the beta-ionone ring pi-electrons, decrease λ_{max} and diminish the red-shift caused by 3,4 desaturation.

For bovine rhodopsin ($\lambda_{max} = 500$ nm; photon energy = 57 kcal/mole), the apoprotein causes a decrease in photon energy of 7.8 kcal/mole compared to absorption by a protonated Schiff's base of retinal ($\lambda_{max} = 440$ nm; photon energy = 64.8 kcal/mole). Electrostatic effects near the beta-ionone ring contribute approximately 2.5 kcal/mole because reconstitution of bovine opsin with retinal derivatives saturated at the C5–C_6, C_7–C_8, or C_9–C_{10} double bonds show red shifts corresponding approximately to 2.5 kcal/mole less than the red shift shown by 11-*cis* retinal (Arnaboldi et al 1979). Moreover, retinal derivatives saturated at the C_{11}–C_{12} double bond show relatively large red-shifts upon reconstitution with bovine opsin. From these data, Honig and co-workers propose that a single, negatively charged residue in the vicinity of C_{12} and C_{14} accounts for most of bovine rhodopsin's spectral shift (Honig et al 1979a).

Charged residues are also proposed to play a role in energy storage and the spectral red-shift that follows photoexcitation (Honig et al 1979b). Bathorhodopsin, the first spectral intermediate, is so named because it is red-shifted with respect to the parental pigment; for bovine rhodopsin there is a red shift of 38 nm. (From Greek: bathos = deep; bathorhodopsin = deeper red absorbing than rhodopsin.) As discussed above, bathorhodopsin also lies 35 kcal/mole above the energy of the ground state. Both properties may be simply explained by supposing that *cis–trans* isomerization moves the protonated Schiff's base nitrogen further from its counterion, presumably an aspartate or glutamate. This would favor delocalization of the Shiff's base positive charge throughout the pi-electron system and promote a more cyanine-like electronic structure, i.e. a red-shift. Indeed, large red-shifts are seen when protonated Schiff's bases in solution are neutralized with counterions of increasing atomic radius (Blatz et al 1972). This effect may also play a role in wavelength regulation of the primary pigment spectrum (Blatz & Liebman 1973).

PHOTOTRANSDUCTION

Excitation

How does rhodopsin activation lead to a change in membrane conductance? The answer (at least for vertebrates) has come from a recent convergence of biochemical and electrophysiological experiments (Lamb 1986, Stryer 1986).

Some years ago George Wald proposed that visual transduction might proceed via an enzyme cascade in which one light-activated enzyme (rhodopsin) converted many substrate proenzymes to an active form (Wald 1956, 1965). These newly activated enzymes could then activate other proenzymes, and so on. By analogy with the blood-clotting enzyme cascade, Wald reasoned that such a system would provide the amplification necessary for reliable single photon detection. A similar proposal arose from electrophysiological experiments in which the time course of the Limulus photoresponse was modeled by a chain of events mathematically equivalent to an enzyme cascade (Fourtes & Hodgkin 1964, Borsellino et al 1965). Thirty years after Wald's initial proposal, we now know that photoexcited rhodopsin possesses an enzymatic activity that dark rhodopsin lacks. Phoexcited rhodopsin catalyzes release of GDP and uptake of GTP by a peripheral membrane protein, transducin (also called G-protein) (Fung & Stryer 1980). In the dark, transducin binds GDP and exists as a complex of three subunits, α, β, and γ. Photoexcited rhodopsin stabilizes transducin in a form that permits release of GDP into the cytosol (Bennett & Dupont 1985). GTP then binds to the vacant nucleotide binding site, and transducin is released from photoexcited rhodopsin in a dissociated form : transducin$_{\alpha}$-GTP and transducin$_{\beta\gamma}$. Photoexcited rhodopsin is recycled to catalyze further rounds of GTP–GDP exchange.

Metarhodopsin II is the bleaching intermediate that interacts with transducin. In the presence of stoichiometric quantities of transducin and in the absence of GTP, the metarhodopsin I$\rightleftharpoons$II equilibrium is shifted toward metarhodopsin II and the metarhodopsin II $\rightarrow$ III decay is retarded (Pfister et al 1983, Bennett et al 1982, Kuhn 1984, Chabre 1985). When GTP is added these effects are not seen. This shows that transducin binds to and stabilizes metarhodopsin II until it is released by GTP uptake.

At 20°C, toad rod plasma membrane hyperpolarization begins within hundreds of milliseconds following absorption of a single photon and peaks at approximately two seconds (Baylor et al 1979). Therefore, if GTP uptake by transducin is a causal link in phototransduction, it should begin within tens of milliseconds and be complete within several seconds. This reaction can be conveniently monitored on a rapid time scale because it produces characteristic far-red light-scattering changes (Kuhn 1984, Chabre 1985). The time course of these scattering signals show that GTP–GDP exchange is well under way within 100 msec following a modest flash of light. In magnetically oriented rod outer segments, the scattering signal produced by release of T_{α}-GTP from the membrane has an amplitude corresponding to a turnover of approximately 1 transducin per msec per photoexcited rhodopsin (Vuong et al 1984). This demonstrates that the reaction is sufficiently rapid to be an excitatory intermediate in phototransduction.

In the next step transducin$_{\alpha}$-GTP stoichiometrically activates a hitherto inactive cGMP phosphodiesterase (Wheeler & Bitensky 1977, Liebman & Pugh 1979, Fung et al 1981). In the dark, phosphodiesterase is maintained in an inactive form by the action of a small inhibitory subunit (Hurley & Stryer 1982). Transducin$_{\alpha}$-GTP relieves the inhibitory constraint and thereby initiates cGMP hydrolysis. This reaction can be conveniently monitored in vitro because cGMP hydrolysis liberates a proton; the resultant drop in pH is measured with a pH sensitive electrode (Yee & Liebman 1979). The kinetics of cGMP hydrolysis have recently been examined in a photoreceptor preparation consisting of intact outer segments attached to the mitochondria-rich part of the inner segment (Cote et al 1984). At low levels of bleaching, cGMP hydrolysis is complete within 50 msec and precedes the light-induced decrease in plasma membrane conductance.

The light-induced drop in cGMP concentration is sensed directly by the plasma membrane. Recent recordings from excised rod outer segment plasma membrane patches show that cGMP causes the cation channels to open (Fesenko et al 1985, Nakatani & Yau 1985). Other nucleotides (ATP, GTP, cAMP) have no effect on the conductance. The simplest interpretation is that cGMP binds directly to the channels. In the dark, cGMP holds the channels open and sodium, and to a lesser extent calcium, rush inward. In the light, a decrease in cGMP concentration causes the channels to close. The result is a graded hyperpolarization that is transmitted passively to the synaptic region of the cell, where it modulates neurotransmitter release.

Recovery

The transducin/cGMP phosphodiesterase cascade provides a large and rapid signal amplification. During its lifetime of several hundred milliseconds, a single photoexcited rhodopsin activates 100–500 transducins, and each transducin-activated phosphodiesterase hydrolyzes cGMP at a rate of approximately 3700 per sec (Stryer 1986). The system is also engineered for a rapid recovery. This is important because temporal resolution depends on the speed of both signal production and signal termination.

How is photoexcited rhodopsin turned off? Kuhn has shown that it is a substrate for the combined action of (*a*) a rhodopsin kinase (a more accurate name would be photoexcited rhodopsin kinase), which transfers from one to nine phosphates to the C-terminus and third cytoplasmic loop of photoexcited rhodopsin (Sale et al 1978, Hargrave 1982, Thompson & Findlay 1984, McDowell et al 1985), and (*b*) a 48 K protein [also called

S-antigen (Pfister et al 1985)] that binds to phosphorylated rhodopsin and sterically blocks transducin activation (Wilden & Kuhn 1982, Kuhn et al 1984, Kuhn 1984). The best kinetic experiments to date show that at low bleaches, radioactive phosphate transfer from gamma-^{32}P-ATP to photoexcited rhodopsin is complete within two seconds, the resolution of the technique (Sitaramayya & Liebman 1983). It therefore seems likely that phosphorylation is sufficiently rapid to mediate photoexcited rhodopsin turn-off. Previous workers had observed a much slower time course of phosphorylation, probably because they were using saturating substrate (photoexcited rhodopsin) concentrations. Indeed, the kinase is present at only one part per 1000 rhodopsins and is easily lost during preparation of rod outer segments. A second technique that has been interpreted as an indirect assay of photoexcited rhodopsin turn-off is the measurement of ATP-dependent turn-off of phosphodiesterase (Liebman & Pugh 1980). Although the interpretation of these experiments is complicated by the possibility that other ATP-dependent processes may be occurring, it is clear that ATP mediates a rapid quench of phosphodiesterase activity (beginning in less than 100 msec). In a still obscure process, phosphorylated photoexcited rhodopsin is eventually recycled back to the dark state.

Transducin has a built-in mechanism for returning to the dark state. It has an intrinsic GTPase activity that hydrolyzes GTP to GDP, causing transducin$_{\alpha}$ to reassociate with transducin$_{\beta\gamma}$ and returning phosphodiesterase to its inactive state. Interestingly, gamma-^{32}P release from transducin-GTP hydrolysis measured in vitro occurs on a time scale of tens of seconds—far too slow to account for observed recovery times (Navon & Fung 1984). One simple solution to this problem would be to suppose that GTP hydrolysis and transducin inactivation are rapid, but release of inorganic phosphate from transducin is slow (P. A. Liebman, personal communication). A second solution would be to suppose that an as yet undiscovered active mechanism turns off transducin-GTP-activated phosphodiesterase.

Reproducibility

Recordings of membrane current in single toad rod outer segments show a highly reproducible signal amplitude, duration, and shape following absorption of a single photon (Baylor et al 1979). This fidelity is in part a consequence of the large number of second messengers used in signal transduction. Since statistical fluctuations in transmitter numbers and lifetimes (N) are proportional to $(N)^{1/2}$, the fractional fluctuation is proportional to $(N)^{1/2}/(N) = (N)^{-1/2}$.

Because one photon photoexcites only a single rhodopsin molecule, the first stage in transduction is not subject to this "law of large numbers."

Yet the data clearly indicate that different single photon events produce approximately the same signal. Could this reproducibility be a consequence of a narrow range of photoexcited rhodopsin lifetimes? The early photochemical events that control its production are so rapid (milliseconds) that on a physiological time scale photoexcited rhodopsin appears nearly instantaneously following photon absorption. However, deactivation may not be nearly so synchronous, and therefore it presents a potential problem. Suppose, in the simplest case, that a single encounter between photoexcited rhodopsin and rhodopsin kinase is sufficient to convert photoexcited rhodopsin into a form that binds 48 K protein. Its turn-off would then be a two step process: a photoexcited rhodopsin–rhodopsin kinase encounter followed by a phosphorylated photoexcited rhodopsin–48 K protein encounter. However, the 48 K protein is quite abundant (approximately 10% of rod outer segment protein) and rhodopsin kinase is rare (approximately 0.1% of rod outer segment protein). If the ratio of their encounter frequencies is at all comparable to the ratio of their concentrations, then turn-off kinetics will be dominated by the slower of these two processes and will appear to be essentially a one step process. A one step process produces an exponential curve of probability of occurrence as a function of time. If the activity of photoexcited rhodopsin is proportional to its lifetime, then a one step turn-off process would result in a far greater dispersion of activity than the single photon data permit.

One solution to this quandary is to suppose that multiple encounters between photoexcited rhodopsin and kinase are necessary to produce a degree of phosphorylation sufficient to promote 48 K protein binding. Several independent sequential steps with similar time constants would result in a narrower distribution of turn-off probabilities vs time. A second solution would be to suppose that the signal from a single photoexcited rhodopsin is quantized by rod outer segment architecture: the partitioning of the outer segment membrane into topologically separate domains (discs) might constrain the single photon signal. This would occur if a single photoexcited rhodopsin lives long enough to activate all or nearly all of the phosphodiesterase molecules associated with the disc in which it resides, but is unable to activate those associated with other discs. (More realistically, it might activate all of the phorphodiesterases on the nearer side of the adjacent disc as well, since transducin$_\alpha$-GTP is soluble). If we start with the estimate of 20,000 rhodopsins per frog rod outer segment disc, there should be approximately 200 phosphodiesterases per disc (Liebman & Pugh 1979). This is in fact approximately the number of phosphodiesterases activated per photoexcited rhodopsins in the limit of low light levels and is therefore consistent with the above hypothesis. A third solution is suggested by the recent observation that rhodopsin kinase is inhibited by cGMP (Shuster & Farber 1984). This suggests a negative feedback loop in which rhodopsin turn-off occurs only after a sufficiently large drop in cGMP concentration. Of course, a fourth solution to the problem is to suppose that a completely different mechanism, perhaps acting upon phosphodiesterase, controls the kinetics of turn-off.

Absolute Sensitivity and Dark Noise

How reliable is visual transduction? This question can be divided into two parts. First, what is the probability that photon capture will generate a signal? Second, how often is a signal generated in the absence of a photon? Dartnall has shown that for rhodopsin in detergent solution the probability of isomerization following photon absorption is 0.67 (Dartnall 1972). This is an intrinsic property, the quantum efficiency. Baylor and co-workers have arrived at a slightly lower figure, 0.5, for the probability of signal production in the toad rod outer segment following photon absorption (Baylor et al 1979). Since the rod outer segment captures approximately 50% of photons at the wavelength of maximal absorption that strike the retina (Bowmaker & Dartnall 1980), the overall detection efficiency is approximately 25%. This is comparable to the sensitivities of the best photomultiplier tubes.

Reliable detection of a single photolyzed rhodopsin requires that the remaining rhodopsins (10^8 in a primate rod and 3×10^9 in a toad rod) remain "silent." In the absence of light, recordings of toad rod outer segment membrane current are interrupted on average once per 50 sec (at 20°C) by electrical events resembling single photon responses (Baylor 1980). These events are Poisson distributed and show a temperature dependence corresponding to an energy (enthalpy) of activation of 22 kcal/mole. A component of membrane noise in dogfish bipolar cells that may reflect the same events shows an energy (enthalpy) of activation of 36 kcal/mole (Ashmore & Falk 1977). These characteristics strongly suggest that the events arise from thermal activation (isomerization) of single rhodopsins. If we extrapolate to 37°C, then the half time of thermal isomerization is 400 years. Recent recordings of dark noise in primate rods at 36°C confirm this calculation (Baylor et al 1984).

A different sort of dark noise has recently been observed in toad (Lamb 1980) and primate rods (Baylor et al 1984) following strong flashes of light. In toad rods it resembles a summation of single photon-like events. In primate rods it consists of one or several discrete step changes in membrane current of 1.2 ± 0.4 pA per step. An important characteristic of the steps is that they are both ascending and descending, and, as a result, they cannot merely derive from a delayed turning off of light-activated components. In both toad and primate rods the noise appears to correspond to an event whose probability of occurrence falls approximately exponentially with increasing time after the flash. The time over which these steps are seen can last for tens of seconds, and individual steps sometimes last for a large fraction of that time. They are seen most readily during recovery from strong flashes because the number of steps is roughly propotional to the number of absorbed photons. Within a factor of three, an average of one step is seen per 800 absorbed photons. These electrophysiological findings may explain the well-known phenomenon of "after-images" following a bright stimulus (Barlow 1964).

Are the post-stimulus events due to some form of activated rhodopsin? In toad rods the power spectrum of these events is nearly coincident with that calculated for randomly occurring single photon responses. In primates, the step amplitude (1.2 pA) is tantalizingly close to the single photon response amplitude (0.7 pA). Indeed, this correspondence may be even better if the single photon response amplitude is attenuated to some extent by rapid turn-off of photoexcited rhodopsin. Perhaps in both systems the events result from rare errors in recycling of photolyzed rhodopsin.

Homologies with Other Systems

Phosphodiesterase activation by photoexcited rhodopsin bears a striking resemblance to adenylate cyclase activation by hormone-receptor complexes. In the β-adrenergic system, for example, hormone binding causes the receptor to assume a conformation that catalyzes a GTP–GDP exchange reaction analogous to that catalyzed by photoexcited rhodopsin (Schramm & Selinger 1984). The hormone system has an additional twist: Within the same cell there exist at least two types of transducin analogues, G_s and G_i (Gilman 1984). These correspond to two or possibly more opposing receptor classes. Stimulatory receptors activate adenylate cyclase by activating G_s, and inhibitory receptors inhibit adenylate cyclase by activating G_i. Following GTP uptake, G_s (or G_i) dissociates into $G_{s\alpha}$-GTP (or $G_{i\alpha}$-GTP) and $G_{\beta\gamma}$ (the β and γ subunits are the same for G_s and G_i). $G_{s\alpha}$-GTP stoichiometrically activates adenylate cyclase in much the same way that transducin$_\alpha$-GTP activates phosphodiesterase. In contrast, the inhibitory pathway appears to act indirectly. By increasing the free $G_{\beta\gamma}$ concentration, the inhibitory pathways drives free $G_{s\alpha}$-GTP into an inert $G_{s\alpha\beta\gamma}$ complex.

How homologous are the visual and hormone systems? Bitensky and co-workers (1982) have demonstrated functional interchange of some components. In preparations of rat brain synaptic membranes they observe that (*a*) photoactivated rhodopsin, but not dark rhodopsin, stimulates adenylate cyclase, (*b*) transducin augments isoproteranol activation of adenylate cyclase, and (*c*) the small inhibitory subunit of rod outer segment phosphodiesterase reversably inhibits adenylate cyclase. Recent experiments with purified components show that G_i is a good photoexcited rhodopsin substrate but that G_s is much poorer (Kanaho et al 1984, Tsai et al 1984, Cerione et al 1985).

Another criterion of similarity is the ability of transducin to serve as a substrate for ADP ribosylation by cholera and pertussis toxins (Abood et al 1982, Van Dop et al 1984, Navon & Fund 1984). The normal substrates are G_s and G_i, respectively. Amino acid sequencing and DNA cloning experiments have revealed homology between $G_{s\alpha}$ and transducin (as well as between G_o, *ras*, and EF-Tu; see below; Hurley et al 1984, Harris et al 1985, Lochrie et al 1985, Medynski et al 1985, Tanabe et al 1985, Yatsunami & Khorana 1985), and proteolytic mapping studies indicate that transducin$_\beta$ and G_β may be nearly identical (Manning & Gilman 1983). Partially purified preparations of β-adrenergic receptor kinase, an enzyme that phosphorylates only the active form of the receptor, also phosphorylates photoexcited rhodopsin but not dark rhodopsin (Benovic et al 1986). Finally, the recently determined amino acid sequence of the β-adrenergic receptor shows three small regions of sequence homology with rhodopsin and a pattern of alternating hydrophobicity and hydrophilicity strikingly similar to that of rhodopsin (Dixon et al 1986).

Signal coupling via a GTP–GDP exchange protein is even more widespread than the above discussion indicates. Table 1 lists some of the systems in which it has been implicated. By extension, each has a catalyst analogous to rhodopsin. Hormone receptors must communicate a conformational change from a binding pocket in contact with the external environment to a catalytic site facing the cytosol. This distance, comparable to the width of a lipid bilayer, suggests that there are long-range changes in protein conformation. The conformational states that photoactivated rhodopsin assumes may be directly relevant to these other systems.

COLOR VISION

In most vertebrates the visual pigments responsible for color vision reside in the cones, which usually represent only a tiny minority of the photoreceptor cells. As a consequence, these pigments have been purified from only a few animals, notably chickens, that have a large number of cones (Fager & Fager 1982, Matsumoto & Yoshizawa 1982). Much of what we know about cone pigments and cone function comes from microspectrophotometric and electrophysiological experiments. In humans, and to a lesser extent in other primates, psychophysical experiments have

Table 1 Some systems in which G proteins or their homologues have been implicated

System	Catalyst	Transducer	Amplified reaction	Ref.
Vision	rhodopsin	transducin/G	cGMP hydrolysis	Stryer 1986
ACTH, beta-adrenergic, gonadotropin hormone systems	ACTH, beta-adrenergic, gonadotropin receptors	G_s, G_i	cAMP synthesis	Gilman 1984
Eukaryotic translation initiation	eIF-2B	eiF-2	association of 80S initiation complex	Safer 1983
Leukocyte chemotaxis	? *N*-formylated peptide receptor	G-like protein	? release of inositol triphosphate and diacyl glycerol	Joseph 1985
Olfaction	? olfactory receptor	G-like protein	cAMP synthesis	Pace et al 1985
Memory, learning	serotonin receptor	55 K G protein	cAMP synthesis	Schwartz et al 1983
Control of cell growth	?	*ras* p21	?	Weinberg 1985

been a major source of information. Color vision is widespread; many fish, birds, amphibia, reptiles, mammals, and insects have color vision. For a thorough treatment of this subject, the reader is referred to Boynton 1979, Jacobs 1981, and Lythgoe 1979.

Central to all color vision systems is the principle of univariance. It states that visual pigment excitation is a stereotyped event, independent of the wavelength of the absorbed photon; wavelength merely determines the probability of absorption as expressed by the spectral absorption curve. As a consequence, color vision requires a comparison of the ouputs of at least two photoreceptor cells that differ with respect to their pigment absorption spectra. The relative extents of excitation of the different cells provide information about the wavelength composition of the stimulus, but do not uniquely define it.

In evolutionary terms, color vision can be imagined to have evolved in four steps:

1. duplication of a primordial visual pigment gene;
2. accumulation of DNA sequence changes in one of the duplicated genes that alter the spectral absorbance of the encoded pigment;
3. accumulation of DNA sequence changes that lead to expression of one of the duplicated genes in a set of photoreceptor cells different from that set in which the other gene is expressed;
4. development of second order neurons sensitive to differences in the extents of excitation of the two photoreceptor sets.

A fifth step has occurred in some species: colored oil droplets within single photoreceptor cells filter the incoming light and produce different action spectra (Lythgoe 1979, Jacobs 1981). If at first steps 2, 3, and 4 occur imperfectly, selective pressure can be expected to promote favorable refinements. This model tacitly assumes that differences in the apoprotein, rather than the chromophore, account for pigment diversity, an assumption in keeping with the finding in different species of a common 11-*cis* retinal chromophore in pigments that differ in their absorption maxima. The model also assumes that the transduction machinery will be conserved, at least early in evolution, between different photoreceptor cell types. Recent experiments support this notion: (*a*) injection of cGMP into salamander cones via a patch pipette causes a large increase in dark current that is suppressible by light (Cobbs et al 1985); (*b*) cGMP opens cation channels in excised membrane patches from catfish cones with very nearly the same concentration dependence as in frog and toad rods (Haynes & Yau 1985); and (*c*) human cones have a cGMP phosphodiesterase that is chromatographically distinct but immunologically cross-reactive with human rod cGMP phosphodiesterase (Hurwitz et al 1985). These data argue strongly that cones, like rods, respond to light by hydrolyzing cGMP.

Human Color Vision

Most humans can match any color by either combining three primary colors or combining two primary colors and adding the third primary to the given color. This is a direct consequence of our possessing three pigment types: our vision is trichromatic. The three pigments that mediate human color vision have absorption maxima at approximately 420–440 nm (the "blue pigment"), 530–535 nm (the "green pigment"), and 560–570 nm (the "red pigment"), as determined psychophysically (Boynton 1979) and by microspectrophotometry of single human cones (Bowmaker & Dartnall 1980, Bowmaker et al 1983).

Recent advances in recombinant DNA technology have given us a direct look at the amino acid sequences of the human cone pigments. The genes encoding the blue, green, and red cone pigments have recently been isolated from human DNA by using DNA encoding bovine rhodopsin as a hybridization probe (Nathans & Hogness 1983, 1984, Nathans et al 1986a). As judged by their similarity in structure and sequence, these genes are all derived from a common ancestral gene; indeed, the success of the cross-hybridization experiments depended upon this evolutionary mechanism.

As discussed above, the human and bovine rhodopsin genes are highly homologous. The cone pigment genes show far less homology to bovine (or human) rhodopsin. Their identification relies on the well-characterized genetics of inherited variation in human color vision ("color-blindness"). One of the cone pigment genes resides on the seventh chromosome, whereas the rhodopsin gene resides on the third chromosome, and all of the other cone pigment genes are X-linked. As discussed below, variations in the blue cone system are inherited in an autosomal fashion, and variations in the red and green cone systems are inherited in an X-linked fashion. Therefore, the seventh chromosome locus is presumed to be the blue pigment gene. The X-linked pigment genes are of two types; they were identified as red and green pigment genes because their structures are specifically altered in individuals with inherited variations in the red and green cone systems.

The genes encoding red and green pigments have been examined by Southern blotting of genomic DNA from different color-normal males. The genes show an unexpected variation: each X-chromosome has one red pigment gene and either one, two, or three green pigment genes. Because of their high degree of sequence homology and their propensity to vary in number, these genes are proposed to reside in a head-to-tail tandem array (Figure 5A). Unequal homologous recombination, as shown

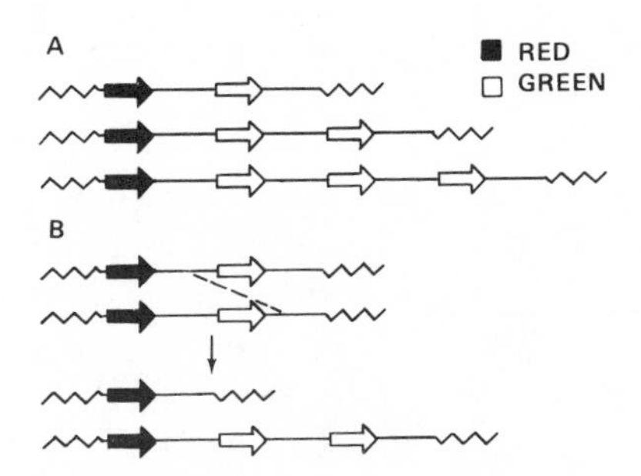

Figure 5 Arrangement and rearrangement of red and green pigment genes in color normal males. The *base* of each arrow corresponds to the 5′-end of the gene and the *tip* corresponds to the 3′-end. The *dashed line* indicates an unequal homologous recombination event. (After Nathans et al 1986a.)

in Figure 5B, produces variation in gene number. Importantly, the variation is asymmetric: only green pigment genes vary among color normals. To account for this observation and for the divergence in red and green pigment gene sequences just upstream of the 5′-end, the red pigment gene is proposed to lie at the edge of the tandem array as drawn in Figure 5. If the diverged 5′ flanking sequences constitute a part of the red pigment gene, then the complete duplication or loss of this gene cannot occur by unequal homologous recombination. In contrast, the green pigment genes, shown downstream of the red pigment gene, behave as though they are embedded entirely within duplicated sequences.

This arrangement may be a consequence of the first and third evolutionary steps in the development of color vision as outlined above. Immediately following gene duplication, the two protein products were probably expressed in the same photoreceptor cells. However, red and green pigment genes have since acquired a mechanism to insure expression of green pigment in one set of cones and red pigment in a second set of cones. This control may well be at the level of transcription. Juxtaposition of the red pigment gene 5′-end next to unique flanking DNA may provide the necessary sequence differences for cell-specific transcription of red and green pigment genes. A related aspect of red and green pigment gene structure is the finding that they both begin with a small exon that has no homologue in the rhodopsin gene or the blue pigment gene. It seems reasonable to suppose that at some point in evolution this small exon was added to an otherwise intact pigment gene to provide a new source of 5′-controlling sequences.

Cone Pigment Amino Acid Sequences

Pairwise comparisons of amino acid sequences for all of the human visual pigments are shown in Figure 6. The sequences were deduced by conceptual translation of cDNA and genomic DNA sequences that encode them. The red and green pigments are far more similar (96% identity) than are any

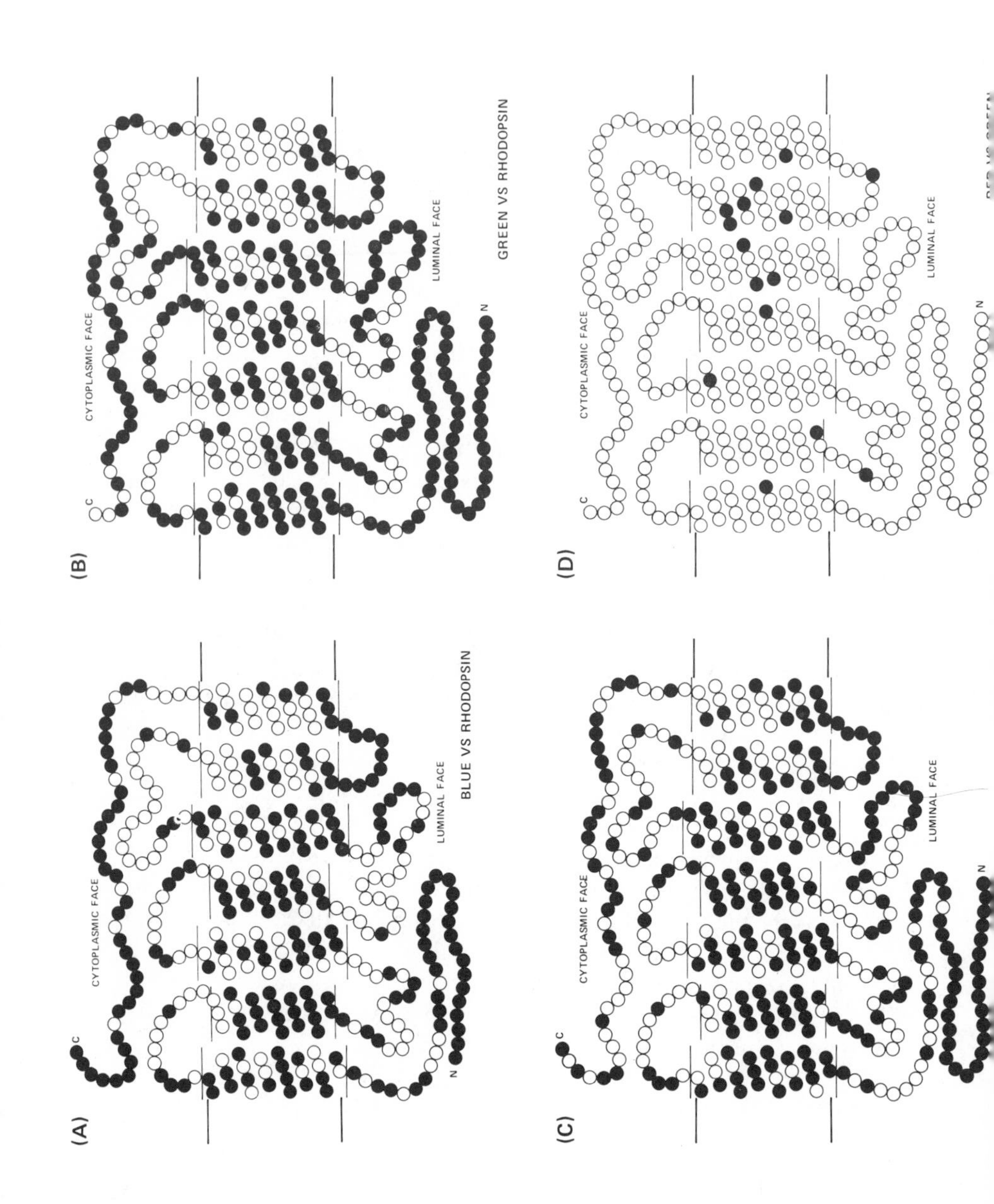
(A)
CYTOPLASMIC FACE
LUMINAL FACE
BLUE VS RHODOPSIN
C
N
(B)
CYTOPLASMIC FACE
LUMINAL FACE
GREEN VS RHODOPSIN
C
N
(C)
CYTOPLASMIC FACE
LUMINAL FACE
C
N
(D)
CYTOPLASMIC FACE
LUMINAL FACE
C
N

other pair of pigments. Every other pairwise comparison shows between 40% and 45% amino acid identity. This is in keeping with the observation that humans and Old World monkeys have both red and green pigments, presumably encoded in each species by two X-linked genes, whereas New World monkeys have only a single polymorphic long wavelength pigment encoded on the X-chromosome (Jacobs 1983, 1984). Most likely, a recent gene duplication has generated the human and Old World monkey red and green pigment genes.

The presence of multiple serines and threonines near the carboxy termini of the cone pigments suggests that a kinase, analogous to the one present in rods, turns off light-activated cone pigments. Curiously, this carboxy terminal region is more similar among the cone pigments than is the corresponding region of rhodopsin. This may bear upon the observation that primate (Nunn et al 1984) and salamander (Cobbs et al 1985) cone response amplitudes are at least 100-fold lower per photon than primate and salamander rod response amplitudes, respectively. Perhaps cone pigments are more rapidly turned off by phosphorylation than are rod pigments.

The cone pigment sequences provide some clues regarding the mechanism of spectral tuning. The charged residues that are predicted to reside within the bilayer, and hence are the most likely to be in contact with retinal, are identical in both red and green visual pigments. The blue pigment and rhodopsin differ from each other and from the red and green pigments in their intramembrane charge distributions. The net intramembrane charge of the pigments are the following: blue, +1; rhodopsin, 0; red and green, −1. If functional visual pigments can be expressed from the cloned DNAs, the role of these charged residues in spectral tuning can be tested by site directed mutagenesis. One can rationalize the subtlety of the red vs green pigment differences by recalling that the energy difference between photons of 530 nm and 560 nm is only 3 kcal/mole.

Inherited Variation in Human Color Vision

Approximately 8% of males and 0.5% of females differ from the majority on the proportions of the three primaries required to match a given color. These individuals fall into two classes: those who require only two primaries (dichromats) and those who require a different ratio of the three primaries (anomalous trichromats). Dichromats appear to have lost one of the three color sensitive systems; anomalous trichromats appear to have

---

Figure 6 Pairwise comparisons of human visual pigment amino acid sequences. Each *black dot* indicates an amino acid difference. (After Nathans et al 1986a.)

a change in the spectral sensitivity of one of the systems. Each class is further divided into three types according to whether the blue, green, or red cone systems are defective. Because defects involving the blue cone system are rare, most of our information comes from studying defects in the green and red cone systems. In particular, the psychophysical experiments described below involve only the red and green systems. [In this article the following simplified nomenclature is used: G^+, normal green system; G', anomalous green system (deuteranomaly); G'', extremely anomalous green system (extreme deuteranomaly); G^-, absent green system (deuteranopia); R^+, normal red system; R', anomalous red system (protanomaly); R'', extremely anomalous red system (extreme protanomaly); R^-, absent red system (protanopia).]

The heritability of variant color vision has been recognized for well over a century and accounts for the male/female bias in affected individuals. Defects in the green and red cone systems map to two tightly linked loci on the X-chromosome. Defects in the blue cone system are inherited in an autosomal fashion. The simplest hypothesis consistent with the data is that these loci correspond to the genes encoding each of the cone pigment apoproteins (Wald 1966, Piantanida 1976). Alternative models based upon defects in neural circuitry have also been proposed.

The hypothesis of identity of color variant loci and cone pigment genes is supported by the observation that the retinas of dichromats lack the affected pigment (Rushton 1963, 1965, Alpern & Wake 1977). In these experiments, a test light was directed into the pupil so as to fall on the fovea, the cone rich region of the human retina. By the measured differences between light entering and exiting the pupil, the presence or absence of each cone pigment can be assayed. These results have been beautifully confirmed by recent microspectrophotometric measurements of single cones in a dichromat retina (Dartnall et al 1983). Psychophysical measurements of photopic sensitivity also confirm this result (Wald 1966). Psychophysical experiments have also measured the sensitivities of anomalous mechanisms in both $G'R^+$ and G^+R' variants. Interestingly, in both cases the anomalous sensitivity lies in the interval between normal red and green cone sensitivities (Piantanida & Sperling 1973a, 1973b, Rushton et al 1973). These results account for the characteristic decrease in wavelength discrimination that many anomalous trichromats display; as the interval between the spectral absorbances of normal and anomalous cone sensitivities narrows, the difference in the respective cone outputs decreases. When the interval becomes very small, wavelength discrimination is poor. These shifted sensitivities can be simply explained either by shifts in pigment absorption spectra or by mixing of different pigments within a single cone. Either explanation might involve alterations in the genes encoding the visual pigment apoproteins.

The hypothesis of identity of red and green variant loci and red and green pigment genes has been tested by a direct analysis of these genes in color variant males (Nathans et al 1986b). The corresponding hypothesis regarding blue variation has not yet been tested. Nathans and co-workers examined the red and green pigment genes in 25 red or green color variant males; in 24 out of 25 the pattern of DNA fragments derived from these genes was clearly different from that seen among color normal males. The gene rearrangements are consistent with either unequal homologous recombination or, in some cases, gene conversion. The observed genotypes and phenotypes are summarized in Figure 7.

G^-R^+ genotypes are of two classes: (*a*) six out of nine subjects have a red pigment gene and no green pigment genes, and (*b*) three out of nine subjects have a red pigment gene and a hybrid gene in which the 5′-end derives from a green pigment gene and the 3′-end derives from a red pigment gene. It is presumed that the hybrid gene produces a red-like pigment in the "green" cones. The six G^+R^- subjects so far examined are

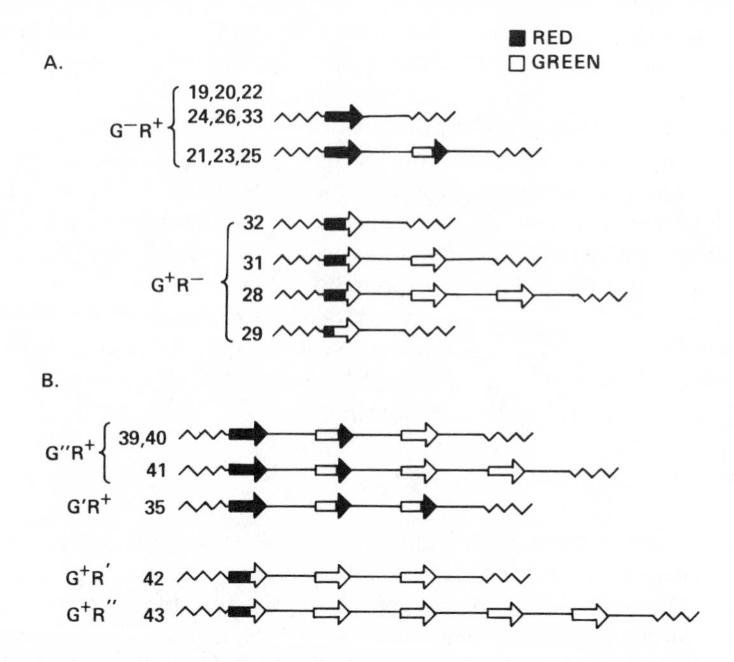

Figure 7 Arrangements of red and green pigment genes in color variant males. Each number refers to an individual whose phenotype is indicated on the left. (After Nathans et al 1986b.)

more complex. In each case the 5′ end of the red pigment gene is intact but in five out of six cases the 3′ end of the red gene is replaced by sequences derived from a green pigment gene. (The sixth case is ambiguous.) The hybrid genes are presumed to produce a green-like pigment in the "red" cones. In some of the G^+R^- subjects, one or more normal green pigment genes are also present. This is essentially the reverse of the second class of G^-R^+ genotypes. The data suggest that 3′ proximal sequences are responsible for the spectral differences between red and green pigments. Figure 7 shows the genotypes of four of these G^+R^- subjects.

Anomalous trichromats also possess hybrid genes (Figure 7). Two red anomalous subjects—one G^+R' and one G^+R''—possess a 5′ red–3′ green hybrid gene in place of the normal red pigment gene. Multiple green pigment genes are also present. In these subjects the hybrid gene is presumed to produce a pigment in the red cones that is neither red nor green in its spectral properties. The spectral properties of the hybrid pigment probably depend on the exact point of cross-over. All eight green anomalous subjects—three $G'R^+$ and five $G''R^+$—possess one or more hybrid genes with the reverse structure, namely 5′ green–3′ red. These are presumed to produce hybrid pigments in the "green" cones and thereby shift the sensitivity of these cones. These eight subjects all possess a red pigment gene and, in most cases, one or more green pigment genes. The genotypes of four green anomalous subjects are shown.

These data fit nicely with the observation that the anomalous green and red sensitivity curves lie between the normal green and red sensitivity curves. The spectral properties of pigments produced by hybrid genes might be partway between the properties of the normal red and green pigments. In contrast, point mutation, which probably occurs at a much lower frequency, would not be expected to preferentially shift an absorption spectrum in one direction. These models also explain qualitatively the observed excess in the population of $G'R^+$ and $G''R^+$ genotypes over G^+R' and G^+R'' genotypes. The reader is referred to Nathans et al 1986b for a full explanation of this phenomenon.

CONCLUSIONS AND PROSPECTS

A recent convergence of biochemical, electrophysiological, spectroscopic, and molecular genetic approaches has produced exciting advances in the study of vision. The molecular genetic approach is the most recent to appear on the scene; its potential has barely been tapped. The isolation of genes that encode visual pigments, as well as those encoding components of the transduction machinery, makes possible a variety of experiments. Expression of either natural or synthetic opsin coding sequences (Ferretti et al 1986) will facilitate detailed structure/function analyses using site-directed mutagenesis. The recent isolation of the major *Drosophila melanogaster* rhodopsin gene (O'Tousa et al 1985, Zuker et al 1985) has paved the way for an analysis of these proteins through use of the powerful tools of *Drosophila* genetics, which now include germ-line transformation with cloned DNA (Rubin & Spradling 1982).

ACKNOWLEDGMENTS

I thank Drs. Denis Baylor, David Hogness, Tim Kraft, Daniel Nathans, Julie Schnapf, Lubert Stryer, and Anita Zimmerman for valuable discussions. The author is a trainee of the Medical Scientist Training Program; his work was supported by a grant from the National Eye Institute to Dr. David Hogness.

Literature Cited

Abood, M. E., Hurley, J. B., Pappone, M. C., Bourne, H., Stryer, L. 1982. Functional homology between signal-coupling proteins: Cholera toxin inactivates the GTPase activity of transducin. *J. Biol. Chem.* 257: 10540–43

Alpern, M., Wake, T. 1977. Cone pigments in human deutan color vision defects. *J. Physiol.* 266: 595–612

Applebury, M. L., Zuckerman, D. M., Lamola, A. A., Jovin, T. M. 1974. Rhodopsin. Purification and recombination with phospholipids assayed by the metarhodopsin I–metarhodopsin II transition. *Biochemistry* 13: 3448–58

Arnaboldi, M., Motto, M. G., Tsujimoto, K., Balough-Nair, V., Nakanishi, K. 1979. Hydroretinals and hydrorhodopsins. *J. Am. Chem. Soc.* 101: 7082–84

Ashmore, J. F., Falk, G. 1977. Dark noise in retinal bipolar cells and stability of rhodopsin in rods. *Nature* 270: 69–71

Barclay, P. L., Findlay, J. B. C. 1984. Labelling of the cytoplasmic domains of ovine rhodopsin with hydrophilic chemical probes. *Biochem. J.* 220: 75–84

Barlow, H. B. 1964. Dark-adaptation: A new hypothesis. *Vision Res.* 4: 47–58

Baylor, D. A., Lamb, T. D., Yau, K.-W. 1979. Responses of retinal rods to single photons. *J. Physiol.* 288: 613–34

Baylor, D. A., Mathews, G., Yau, K.-W. 1980. Two components of electrical dark noise in toad retinal rod outer segments. *J. Physiol.* 309: 591–621

Baylor, D. A., Nunn, B. J., Schnapf, J. L. 1984. The photocurrent, noise and spectral sensitivity of rods of the monkey *Macaca Frascicularis*. *J. Physiol.* 357: 575–607

Bennett, N., Michel-Villaz, M., Kuhn, H. 1982. Light-induced interaction between rhodopsin and the GTP-binding protein. Metarhodopsin II is the major photoproduct involved. *Eur. J. Biochem.* 127: 97–103

Bennet, N., Dupont, Y. 1985. The G-protein of retinal rod outer segments (transducin): Mechanism of interaction with rhodopsin and nucleotides. *J. Biol. Chem.* 260: 4156–68

Benovic, J. L., Mayor, F., Somers, R. L., Caron, M. G., Lefkowitz, R. J. 1986. Light-dependent phosphorylation of rhodopsin by the β-adrenergic receptor kinase. *Nature* 321: 869–72

Bitensky, M. W., Wheeler, M. A., Rasenick, M. M., Yamazaki, A., Stein, P. J., Halliday, K. R., Wheeler, G. L. 1982. Functional exchange of components between light-activated photoreceptor phosphodiesterase and hormone-activated adenylate cyclase systems. *Proc. Natl. Acad. Sci. USA* 79: 3408–12

Birge, R. R. 1981. Photophysics of light energy transduction in rhodopsin and bacteriorhodopsin. *Ann. Rev. Biophys. Bioeng.* 10: 315–54

Blatz, P. E., Mohler, J. H., Navangul, H. V. 1972. Anion-induced wavelength regulation of absorption maxima of Schiff bases of retinal. *Biochemistry* 11: 848–55

Blatz, P. E., Liebman, P. A. 1973. Wavelength regulation in visual pigments. *Exp. Eye Res.* 17: 573–80

Borsellino, A., Fourtes, M. G. F., Smith, T.

G. 1965. Visual responses in *Limulus*. *Cold Spring Harbor Symp. Quant. Biol.* 30: 429–43

Bowmaker, J. K., Dartnall, H. J. A. 1980. Visual pigments of rods and cones in a human retina. *J. Physiol.* 298: 501–11

Bowmaker, J. K., Mollon, J. D., Jacobs, G. H. 1983. Microspectrophotometric results for old and new world primates. In *Colour Vision*, ed. J. D. Mollon, L. T. Sharpe, pp. 57–68. New York: Academic

Bownds, D. 1967. Site of attachment of retinal in rhodopsin. *Nature* 216: 1178–81

Boynton, R. M. 1979. *Human Color Vision*. New York: Holt, Rinehart & Winston. 438 pp.

Cerione, R. A., Staniszewski, C., Benovic, J. L., Lefkowitz, R. J., Caron, M. G., Gierschik, P., Somers, R., Spiegel, A. M., Codina, J., Birnbaumer, L. 1985. Specificity of the functional interactions of the beta-adrenergic receptor and rhodopsin with guanine nucleotide regulatory proteins reconstituted in phospholipid vesicles. *J. Biol. Chem.* 260: 1493–1500

Chalazonitis, N., Chagneux, R., Artavanitaki, A. 1970. Rotation des segments externes des photorecepteurs dan le champ magnetique constant. *CR Acad. Sci. Paris Ser. D* 271: 130–33

Chabre, M. 1978. Diamagnetic anisotropy and orientation of alpha-helix in frog rhodopsin and meta II intermediate. *Proc. Natl. Acad. Sci.* 75: 5471–74

Chabre, M. 1985. Trigger and amplification mechanisms in visual phototransduction. *Ann. Rev. Biophys. Biophys. Chem.* 14: 331–60

Cobbs, W. H., Barkdoll, A. E., Pugh, E. N. 1985. Cyclic GMP increases photocurrent and light sensitivity of retinal cones. *Nature* 317: 64–66

Cooper, A. 1979. Energy uptake in the first step of visual excitation. *Nature* 282: 531–33

Cooper, A. 1981. Rhodopsin photoenergetics: Lumirhodopsin and the complete energy profile. *FEBS Lett.* 123: 324–26

Corless, J. M., McCaslin, D. R., Scott, B. L. 1982. Two dimensional rhodopsin crystals from disc membranes of frog retinal rod outer segments. *Proc. Natl. Acad. Sci. USA* 79: 1116–20

Cote, R. H., Biernbaum, M. S., Nicol, G. D., Bownds, M. D. 1984. Light-induced decreases in cGMP concentration precede changes in membrane permeability in frog rod outer segments. *J. Biol. Chem.* 259: 9635–41

Dartnall, H. J. A. 1972. Photosensitivity. *Handb. Sensory Physiol.* 7(1): 122–45

Dartnall, H. J. A., Bowmaker, J. K., Mollon, J. D. 1983. Microspectrophotometry of human photoreceptors. See Bowmaker et al 1983, pp. 69–80

Dixon, R. A. F., Kobilka, B. K., Strader, D. J., Benovic, J. L., Dohlman, H. G., et al. 1986. Cloning of the gene and cDNA for mammalian β-adrenergic receptor and homology with rhodopsin. *Nature* 321: 75–79

Doukas, A. G., Aton, B., Callender, R. H., Ebrey, T. G. 1978. Resonance raman studies of bovine metarhodopsin I and metarhodopsin II. *Biochemistry* 17: 2430–35

Engelman, D. M., Goldman, A., Steitz, T. A. 1982. The identification of helical segments in the polypeptide chain of bacteriorhodopsin. *Meth. Enzymol.* 88: 81–88

Fager, L. Y., Fager, R. S. 1982. Chromatographic separation of rod and cone pigments from chicken retinas. *Meth. Enzymol.* 81: 160–66

Ferretti, L., Karnik, S. S., Khorana, H. G., Nassal, M., Oprian, D. 1986. Total synthesis of a gene for bovine rhodopsin. *Proc. Natl. Acad. Sci. USA* 83: 599–603

Fesenko, E. E., Kolesnikov, S. S., Lyubarsky, A. L. 1985. Induction by cyclic GMP of cationic conductance in plasma membrane of retinal rod outer segments. *Nature* 313: 310–13

Findlay, J. B. C., Brett, M., Pappin, D. J. C. 1981. Primary structure of C-terminal functional sites in ovine rhodopsin. *Nature* 293: 314–16

Fliesler, S. J., Basinger, S. F. 1985. Tunicamycin blocks the incorporation of opsin into retinal rod outer segments. *Proc. Natl. Acad. Sci. USA* 82: 1116–20

Fliesler, S. J., Rayborn, M. E., Hollyfield, J. G. 1985. Membrane morphogenesis in retinal rod outer segments: Inhibition by tunicamycin. *J. Cell Biol.* 100: 574–87

Fung, B. K.-K., Stryer, L. 1980. Photolyzed rhodopsin catalyzes the exchange of GTP for bound GDP in retinal rod outer segments. *Proc. Natl. Acad. Sci. USA* 77: 2500–4

Fung, B. K.-K., Hurley, J. B., Stryer, L. 1981. Flow of information in the light-triggered cyclic nucleotide cascade of vision. *Proc. Natl. Acad. Sci. USA* 78: 152–56

Fourtes, M. G. F., Hodgkin, A. L. 1964. Changes in time scale and sensitivity in the ommatidia of Limulus. *J. Physiol.* 172: 239–63

Gilman, A. G. 1984. G proteins and the dual control of adenylate cyclase. *Cell* 36: 577–79

Greenberg, A., Honig, B., Ebrey, T. G. 1975.

Wavelength regulation of the bandwidths of visual pigment spectra. *Nature* 257: 823–24

Hamdorf, K. 1979. The physiology of invertebrate visual pigments. *Handb. Sensory Physiol.* 7(6A): 146–224

Hargrave, P. A. 1982. Rhodopsin chemistry, structure, and topology. *Prog. Retinal Res.* 1: 1–51

Hargrave, P. A., McDowell, J. H., Curtis, D. R., Wang, J. K., Juszczak, E., Fong, S. L., Mohanna Rao, J. K., Argos, P. 1983. The structure of bovine rhodopsin. *Biophys. Struct. Mech.* 9: 235–44

Harris, B. A., Robishaw, J. D., Mumby, S. M., Gilman, A. G. 1985. Molecular cloning of complementary DNA for the alpha subunit of the G-protein that stimulates adenylate cyclase. *Science* 229: 1274–77

Haynes, L., Yau, K.-W. 1985. Cyclic GMP-sensitive conductance in outer segment membranes of catfish cones. *Nature* 317: 61–64

Hayward, G., Carlsen, W., Siegman, A., Stryer, L. 1981. Retinal chromophore of rhodopsin photoisomerizes within picoseconds. *Science* 211: 942–44

Honig, B., Greenberg, A. D., Dinur, U., Ebrey, T. G. 1976. Visual pigment spectra: Implications of the protonation of the retinal Schiff base. *Biochemistry* 15: 4593–99

Honig, B. 1978. Light energy transduction in visual pigments and bacteriorhodopsin. *Ann. Rev. Phys. Chem.* 29: 31–57

Honig, B., Dinur, U., Nakanishi, K., Balough-Nair, V., Gawinowicz, M. A., Arnaboldi, M., Motto, M. G. 1979a. An external point charge model for wavelength regulation in visual pigments. *J. Am. Chem. Soc.* 101: 7084–86

Honig, B., Ebrey, T., Callender, R., Dinur, U., Ottolenghi, M. 1979b. Photoisomerization, energy storage and charge separation: A model for light energy transduction in visual pigments and bacteriorhodopsin. *Proc. Natl. Acad. Sci. USA* 76: 2503–7

Hubbard, R. 1966. The stereoisomerization of 11-*cis* retinal. *J. Biol. Chem.* 241: 1814–18

Hubbard, R., Kropf, A. 1958. The action of light on rhodpsin. *Proc. Natl. Acad. Sci. USA* 44: 130–39

Hurley, J. B., Stryer, L. 1982. Purification and characterization of the gamma regulatory subunit of the cyclic GMP phosphodiesterase from retinal rod outer segments. *J. Biol. Chem.* 257: 11094–99

Hurley, J. B., Simon, M. I., Telow, D. B., Robishaw, J. D., Gilman, A. G. 1984. Homologies between signal transducing G-proteins and *ras* gene products. *Science* 266: 860–64

Hurwitz, R. L., Bunt-Milam, A. A., Chang, M. L., Beavo, J. A. 1985. cGMP phosphodiesterase in rod and cone outer segments of the retina. *J. Biol. Chem.* 260: 568–73

Jacobs, G. H. 1981. *Comparative Color Vision*. New York: Academic. 209 pp.

Jacobs, G. H. 1983. Within-species variation in visual capacity among squirrel monkeys (*Saimiri sciureus*): Sensitivity differences. *Vision Res.* 23: 239–48

Jacobs, G. H. 1984. Within-species variations in visual capacity among squirrel monkeys (*Saimiri sciureus*): Color vision. *Vision Res.* 24: 1267–77

Joseph, S. K. 1985. Receptor-stimulated phosphoinositide metabolism: A role for GTP-binding proteins? *Trends Biochem. Sci.* 10: 297–98

Kanaho, Y., Tsai, S.-C., Adamik, R., Hewlett, E. L., Moss, J., Vaughan, M. 1984. Rhodopsin-enhanced GTPase activity of the inhibitory GTP-binding protein of adenylate cyclase. *J. Biol.* 259: 7378–81

Kropf, A., Hubbard, R. 1958. The mechanism of bleaching rhodopsin. *Ann. NY Acad. Sci.* 74: 266–80

Kuhn, H. 1984. Interactions between photoexcited rhodopsin and light-activated enzymes in rods. *Prog. Retinal Res.* 3: 123–56

Kuhn, H., Hall, S. W., Wilden, U. 1984. Light-induced binding of 48-kDa protein to photoreceptor membranes is highly enhanced by phosphorylation of rhodopsin. *FEBS Lett.* 176: 473–78

Kvist, S., Roberts, L., Dobberstein, B. 1983. Mouse histocompatability genes: Structure and organization of a K^d gene. *EMBO J.* 2: 245–54

Lamb, T. 1980. Spontaneous quantal events induced in toad rods by pigment bleaching. *Nature* 87: 349–51

Lamb, T. 1986. Transduction in vertebrate photoreceptors: The role of cyclic GMP and calcium. *Trends Neurosci.* 9: 224–28

Liebman, P. A. 1971. Microspectrophotometry of photoreceptors. *Handb. Sensory Physiol.* 7(1): 482–528

Liebman, P. A., Pugh, E. N. 1980. ATP mediates rapid reversal of cGMP phosphodiesterase activation in visual photoreceptor membranes. *Nature* 287: 734–36

Liebman, P. A., Pugh, E. N. 1979. The control of phosphodiesterase in rod disc membranes: Kinetics, possible mechanisms and significance for vision. *Vision Res.* 19: 375–80

Lochrie, M., Hurley, J. B., Simon, M. 1985. Sequence of the alpha subunit of photoreceptor G protein: Homology between transducin, *ras*, and elongation factors. *Science* 228 : 96–99

Longstaff, C., Calhoun, R. D., Rando, R. R. 1986. Deprotonation of the Schiff base of rhodopsin is obligate in the activation of the G protein. *Proc. Natl. Acad. Sci. USA* 83 : 4209–13

Lythgoe, J. N. 1979. *The Ecology of Vision*. Oxford : Clarendon. 244 pp.

Manning, D. R., Gilman, A. G. 1983. The regulatory components of adenylate cyclase and transducin. A family of structurally homologous nucleotide-binding proteins. *J. Biol. Chem.* 258 : 7059–63

Martynov, V. I., Kostina, M. B., Feigina, M. Yu., Miroshnikov, A. I. 1983. A study of the molecular organization of visual rhodopsin in photoreceptor membranes by limited proteinolysis. *Bioorg. Khim.* 9 : 734–45

Mathies, R., Stryer, L. 1976. Retinal has a highly dipolar vertically excited singlet state : Implications for vision. *Proc. Natl. Acad. Sci. USA* 73 : 2169–73

Matsumoto, H., Yoshizawa, T. 1982. Preparation of chicken iodopsin. *Meth. Enzymol.* 81 : 154–60

McDowell, H. J., Curtis, D. R., Baker, U. A., Hargrave, P. A. 1985. Phosphorylation of rhodopsin : Localization of phosphorylated residues in the helix V–helix VI connecting loop. *Invest. Ophthalmol. Vis. Sci.* 26(3) : 291

Medynski, D. C., Sullivan, K., Smith, D., Van Dop, C., Chang, F.-H., Fung, B. K.-K., Seeburg, P., Bourne, H. 1985. Amino acid sequence of the alpha-subunit of transducin deduced from the cDNA sequence. *Proc. Natl. Acad. Sci. USA* 82 : 4311–15

Michel-Villaz, M., Saibil, H. R., Chabre, M. 1979. Orientation of rhodopsin alpha-helices in retinal rod outer segment membrane studied with infrared linear dichroism. *Proc. Natl. Acad. Sci. USA* 76 : 4405–8

Mullen, E., Akhtar, M. 1981. Topographic and active site studies on bovine rhodopsin. *FEBS Lett.* 132 : 261–64

Mullen, E., Akhtar, M. 1983. Structural studies on membrane-bound bovine rhodopsin. *Biochem. J.* 211 : 45–54

Nakatani, K., Yau, K.-W. 1985. cGMP opens the light-sensitive conductance in retinal rods. *Biophys. J.* 47 : 356a

Nathans, J., Hogness, D. S. 1983. Isolation, sequence analysis, and intron-exon arrangement of the gene encoding bovine rhodopsin. *Cell* 34 : 807–14

Nathans, J., Hogness, D. S. 1984. Isolation and nucleotide sequence of the gene encoding human rhodopsin. *Proc. Natl. Acad. Sci. USA* 81 : 4851–55

Nathans, J., Thomas, D., Hogness, D. S. 1986a. Molecular genetics of human color vision : The genes encoding blue, green, and red pigments. *Science* 232 : 193–202

Nathans, J., Piantanida, T. P., Eddy, R., Shows, T. B., Hogness, D. S. 1986b. Molecular genetics of inherited variation in human color vision. *Science* 232 : 203–10

Navon, S. E., Fung, B. K.-K. 1984. Characterization of transducin from bovine retinal rod outer segments. Mechanism and effects of cholera toxin catalyzed ADP ribosylation. *J. Biol. Chem.* 259 : 6686–93

Nunn, B. J., Schanpf, J. L., Baylor, D. A. 1984. Spectral sensitivity of single cones in the retina of *Macaca fascicularis*. *Nature* 309 : 264–66

O'Brien, D. F., Costa, L. F., Ott, R. A. 1977. Photochemical functionality of rhodopsin-phospholipid recombinant membranes. *Biochemistry* 16 : 1295–1303

Osborne, H. B., Sardet, C., Michel-Villaz, M., Chabre, M. 1978. Structural study of rhodopsin in detergent micelles by small angle neutron scattering. *J. Mol. Biol.* 123 : 177–206

O'Tousa, J., Baehr, W., Martin, R. L., Hirsch, J., Pak, W. L., Applebury, M. L. 1985. The *Drosophila ninaE* gene encodes an opsin. *Cell* 40 : 839–50

Ovchinnikov, Yu. A. 1982. Rhodopsin and bacteriorhodopsin : Structure-function relationships. *FEBS Lett.* 148 : 179–91

Ovchinnikov, Yu. A., Abdulaev, N. G., Feigina, M. Yu., Artamonov, I. D., Bogachuk, A. S., Zolotarev, A. S., Eganyan, E. R., Kostetskii, P. V. 1983. Visual rhodopsin III : Complete amino acid sequence and topography in the membrane. *Bioorg. Khim.* 9 : 1331–40

Pace, U., Hanski, E., Salomon, Y., Lancet, D. 1985. Odorant-sensitive adenylate cyclase may mediate olfactory reception. *Nature* 316 : 255–58

Pappin, D. J. C., Eliopoulis, E., Brett, M., Findlay, J. B. C. 1984. A structural model for ovine rhodopsin. *Int. J. Biol. Macromol.* 6 : 73–76

Pfister, C., Kuhn, H., Chabre, M. 1983. Interaction between photoexcited rhodopsin and peripheral enzymes in frog retinal rods. Influence on the postmetarhodopsin II decay and phosphorylation and rhodopsin. *Eur. J. Biochem.* 136 : 489–99

Pfister, C., Chabre, M., Plouet, J., Tuyen, V. V., DeKozak, Y., Faure, J. P., Kuhn, H. 1985. Retinal S antigen identified as the 48 K protein regulating light dependent phosphodiesterase in rods. *Science* 228 : 891–93

Piantanida, T. P. 1976. Polymorphism of human color vision. *Am. J. Optomet. Physiol. Optics* 53 : 647–57

Piantanida, T. P., Sperling, H. G. 1973a. Isolation of a third chromatic mechanism in the protanomalous observer. *Vision Res.* 13: 2033–47

Piantanida, T. P., Sperling, H. G. 1973b. Isolation of a third chromatic mechanism in the deuteranomalous observer. *Vision Res.* 13: 2049–58

Rothschild, K. J., Rosen, K. M., Clark, N. A. 1980a. Incorporation of photoreceptor membrane into a multilamellar film. *Biophys. J.* 31: 45–52

Rothschild, K. J., Sanches, R., Hsiao, T. L., Clark, N. A. 1980b. A spectroscopic study of rhodopsin alpha-helix orientation. *Biophys. J.* 31: 53–64

Rubin, G. M., Spradling, A. C. 1982. Genetic transformation of *Drosophila* with transposable element vectors. *Science* 218: 348–53

Rushton, W. A. H. 1963. A cone pigment in the protanope. *J. Physiol.* 168: 345–59

Rushton, W. A. H. 1965. A foveal pigment in the deuteranope. *J. Physiol.* 176: 24–37

Rushton, W. A. H., Powell, D. S., White, K. D. 1973. Pigments in anomalous trichromats. *Vision Res.* 13: 2017–31

Safer, B. 1983. 2B or not 2B: Regulation of the catalytic utilization of eIF-2. *Cell* 33: 7–8

Saibil, H., Chabre, M., Worcester, D. 1976. Neutron diffraction studies of retinal rod outer segment membranes. *Nature* 262: 266–70

Sale, G. J., Towner, D., Akhtar, M. 1978. Topography of the rhodopsin molecule. Identification of the domain phosphorylated. *Biochem. J.* 175: 421–30

Sardet, C., Tardieu, A., Luzzati, V. 1976. Shape and size of bovine rhodopsin: A small angle X-ray scattering study of a rhodopsin detergent complex. *J. Mol. Biol.* 105: 383–407

Scavarda, N. J., O'Tousa, J., Pak, W. L. 1983. *Drosophila* locus with gene dosage effect on rhodopsin. *Proc. Natl. Acad. Sci. USA* 80: 4441–45

Schramm, M., Selinger, Z. 1984. Message transmission: Receptor controlled adenylate cyclase system. *Science* 225: 1350–56

Schwartz, J. H., Bernier, L., Castellucci, V. F., Palazollo, M., Saitoh, T., Stapleton, A., Kandel, E. R. 1983. What molecular steps determine the time course of the memory of short-term sensitization in *Aplysia? Cold Spring Harbor Symp. Quant. Biol.* 48: 811–19

Shichi, H., Shelton, E. 1974. Assessment of physiological integrity of sonicated retinal rod membranes. *J. Supramolec. Struct.* 2: 7–16

Shuster, T. A., Farber, D. B. 1984. Phosphorylation in sealed rod outer segments: Effect of cyclic nucleotides. *Biochemistry* 23: 515–21

Sitaramayya, A., Liebman, P. A. 1983. Phosphorylation of rhodopsin and quenching of cyclic GMP phosphodiesterase activation by ATP at weak bleaches. *J. Biol. Chem.* 258: 12106–9

Stryer, L. 1986. Cyclic GMP cascade of vision. *Ann. Rev. Neurosci.* 9: 87–119

Suzuki, H. 1967. *Electronic Absorption Spectra and Geometry of Organic Molecules.* New York: Academic. 568 pp.

Tanabe, T., Nukada, T., Nishikawa, Y., Sugimoto, K., Suzuki, H., Takahashi, H., Noda, M., Haga, T., Ichiyama, A., Kongawa, K., Minamino, N., Matsuo, H., Numa, S. 1985. Primary structure of the alpha subunit of transducin and its relationship to *ras* proteins. *Nature* 315: 242–45

Thomas, D. D., Stryer, L. 1982. The transverse location of the retinal chromophore of rhodopsin in rod outer segment disc membranes. *J. Mol. Biol.* 154: 145–57

Thompson, P., Findlay, J. B. C. 1984. Phosphorylation of ovine rhodopsin. Identification of phosphorylated sites. *Biochem. J.* 220: 773–80

Tsai, S. C., Adamik, R., Kanaho, Y., Hewlett, E. L., Moss, J. 1984. Effects of guanylate nucleotides and rhodopsin on ADP-ribosylation of the inhibitory GTP-binding component of adenylate cyclase by pertussis toxin. *J. Biol. Chem.* 259: 15320–23

Van-Dop, C., Yamanaka, G., Steinberg, F., Sekura, R. D., Manclarck, C. R., Stryer, L., Bourne, H. 1984. ADP-ribosylation of transducin by pertussis toxin blocks the light-stimulated hydrolysis of GTP and cGMP in retinal photoreceptors. *J. Biol. Chem.* 259: 23–26

Vuong, T. M., Chabre, M., Stryer, L. 1984. Millisecond activation of transducin in the cyclic nucleotide cascade of vision. *Nature* 311: 659–61

Wald, G. 1956. The biochemistry of visual excitation. In *Enzymes: Units of Biological Structure and Function*, ed. O. Gaebler, pp. 355–67. New York: Academic

Wald, G. 1965. Visual excitation and blood clotting. *Science* 150: 1028–30

Wald, G. 1966. Defective color vision and its inheritance. *Proc. Natl. Acad. Sci. USA* 55: 1347–63

Wald, G. 1968. The molecular basis of visual excitation. *Nature* 219: 800–7

Wang, J. K., McDowell, J. H., Hargrave, P. A. 1980. Site of attachment of 11-*cis* retinal in bovine rhodopsin. *Biochemistry* 19: 5111–17

Weinberg, R. A. 1985. The action of onco-

genes in the cytoplasm and nucleus. *Science* 230: 770–76
Wheeler, G. L., Bitensky, M. W. 1977. A light-activated GTPase in vertebrate photoreceptors: Regulation of light activated cyclic GMP phosphodiesterase. *Proc. Natl. Acad. Sci. USA* 74: 4238–42
Wilden, U., Kuhn, H. 1982. Light-dependent phosphorylation of rhodopsin. Number of phosphorylation sites. *Biochemistry* 21: 3014–22
Worcester, D. L. 1978. Structural origins of diamagmetic anisotropy in proteins. *Proc. Natl. Acad. Sci. USA* 75: 5475–77
Yamawaki-Kataoka, Y., Nakai, S., Miyata, T., Honjo, T. 1982. Nucleotide sequence of gene segments encoding membrane domains of immunoglobulin gamma chains. *Proc. Natl. Acad. Sci. USA* 79: 2623–27
Yatsunami, K., Khorana, H. G. 1985. GTPase of bovine rod outer segments: The amino acid sequence of the alpha subunit as derived from cDNA sequence. *Proc. Natl. Acad. Sci. USA* 82: 4316–20
Yee, R., Liebman, P. A. 1979. Light activated phosphodiesterase of the rod outer segment: Kinetics and parameters of activation and deactivation. *J. Biol. Chem.* 253: 8902–9
Yoshizawa, T., Wald, G. 1963. Pre-lumirhodopsin and the bleaching of visual pigments. *Nature* 197: 1279–86
Young, R. W. 1971. The renewal of rod and cone outer segments in the rhesus monkey. *J. Cell Biol.* 49: 303–18
Zuker, C., Cowman, A. F., Rubin, G. M. 1985. Isolation and structure of a rhodopsin gene from *D. melanogaster*. *Cell* 40: 851–58

Ann. Rev. Neurosci. 1987. 10 : 195–236

MOLECULAR PROPERTIES OF THE MUSCARINIC ACETYLCHOLINE RECEPTOR

Neil M. Nathanson

Department of Pharmacology, University of Washington School of Medicine, Seattle, Washington 98195

INTRODUCTION

The action of acetylcholine (ACh)[1] as a neurotransmitter was first confirmed by Loewi's demonstration that stimulation of the vagus nerve released a substance that decreased beating rate when applied to an isolated heart (Loewi 1921). Prior to this ACh had been shown to act at two pharmacologically distinct receptors differentiated by the relative abilities of muscarine and nicotine to activate them (Dale 1914). Muscarinic acetylcholine receptors are present in neurons in the central and peripheral nervous systems, cardiac and smooth muscles, and a variety of exocrine glands. Muscarinic receptors play a key role in regulating the functions of the target organs of the autonomic nervous system and thus are important in the maintenanoce of the homeostasis of the organism. Activation of these receptors decreases the rate and force of contraction of the heart, constricts the airways, increases motility and secretions in the gastrointestinal tract, and increases secretions from salivary and sweat glands. In addition, the majority of cholinergic synapses in the central nervous system are muscarinic, and ACh is thought to play a key role in neural

[1] Abbreviations used: Ach, acetylcholine; CHAPS, 3-[(3-cholamidopropyl)-dimethylammonio]-1-propane sulfonate; DAG, diacylglycerol; IAP, islet-activating protein; IP_3, inositol trisphosphate; NMS, [^{3}H]*N*-methyl-scopolamine; PI, phosphatidylinositol; PIP_2, phosphatidylinositol 4,5-bis-phosphate; PKC, protein kinase C; PrBCM, [^{3}H] propylbenzilycholine mustard; PZ, pirenzepine; QNB, [^{3}H]quinuclidinyl benzilate; VIP, vasoactive intestinal peptide.

0147–006X/87/0301–0195$02.00

mechanisms underlying memory, learning, arousal, control of movement, and possibly the generation of epileptic foci.

Because of the availability of tissues with high concentrations of nicotinic ACh receptors and snake toxins with high affinity for the nicotinic receptor, studies of the nicotinic ACh receptors have yielded a vast body of information on the characterization, purification, reconstitution, molecular cloning, and localized mutagenesis of the receptor. While muscarinic ACh receptors have been the object of intense investigation by an ever-increasing number of investigators for the last dozen years, crucial aspects of their structure and function remain unknown. However, biochemical and molecular investigations have begun to elucidate the structure and mechanisms of action and regulation of the muscarinic receptor.

BIOCHEMICAL RESPONSES MEDIATED BY THE MUSCARINIC RECEPTOR

Activation of muscarinic receptors in a variety of cell types can lead to one or more measureable biochemical responses, such as increased levels of intracellular cGMP, decreased levels of cAMP, and increased turnover of certain membrane phospholipids. While it is clear that no single biochemical change mediates all physiological responses due to muscarinic ACh receptor activation, each of these biochemical responses may be involved in mediation of certain physiological actions of the receptor.

Inhibition of Adenylate Cyclase

Activation of the muscarinic ACh receptors in many cells frequently leads to a decrease in the rate of formation of cAMP due to inhibition of the activity of adenylate cyclase. In washed membrane preparations, inhibition of adenylate cyclase by muscarinic agonists requires the addition of GTP (Watanabe et al 1978, Jakobs et al 1979). The muscarinic receptor does not directly interact with the enzyme, but rather is coupled to adenylate cyclase via a GTP-binding protein, termed N_i or G_i. N_i is a protein consisting of three polypeptides, of molecular weights 41,000 (alpha), 35,000 (beta), and 9000 (gamma). N_i is a member of a structurally homologous family of GTP-binding proteins, which includes N_s, which couples stimulatory receptors to activation of adenylate cyclase; transducin, which couples rhodopsin to phosphodiesterase in the retina; and N_o (the "other" GTP-binding protein), which also interacts with the muscarinic receptor (Sternweis & Robishaw 1984, Florio & Sternweis 1985, Roof et al 1985).

A tool that has been extremely useful in the study of the role of the GTP-binding proteins in the coupling of the muscarinic ACh receptor to physiological responses is islet-activating protein (IAP), a toxin secreted by *Bordetella pertussis*. Treatment of intact cells or membranes with IAP results in the ADP-ribosylation of the alpha subunit of N_i (and N_o). This covalent modification of N_i inactivates the protein, resulting in the loss of the ability of muscarinic receptors to inhibit adenylate cyclase and to exhibit guanine nucleotide regulation of agonist binding to the receptor (see below). Thus, treatment of neural cells (Kurose et al 1983), rat cardiac cells (Hazeki & Ui 1981), or GH_3 pituitary cells (Schlegel et al 1985) with IAP abolishes muscarinic receptor mediated inhibition of adenylate cyclase. As described below, the ability of IAP treatment to inactivate N_i can also be used to test the role of the GTP-binding proteins in other muscarinic receptor mediated responses.

While inhibition of adenylate cyclase has been implicated in a number of physiological responses mediated by the muscarinic receptor, the clearest evidence for this is in cases in which the receptor inhibits responses mediated by stimulatory hormones or other activators known to act through stimulation of adenylate cyclase. For example, the release of prolactin and growth hormone from GH_3 pituitary cells can be evoked by vasoactive intestinal peptide, which stimulates adenylate cyclase, and by treatment with a phosphodiesterase inhibitor, which raises intracellular cAMP levels. Activation of muscarinic receptors inhibits both basal adenylate cyclase activity and secretagogue-stimulated activity, and there is a corresponding inhibition of hormone release in both cases (Wojcikiewicz et al 1984). The time required for maximal inhibition of adenylate cyclase precedes that of maximal inhibition of hormone secretion, as expected if cAMP is the second messenger (Wojcikiewicz et al 1984). Similarly, muscarinic receptors antagonize the relaxation of canine tracheal smooth muscle due to treatment with isoproterenol, prostaglandin E_2, or forskolin: The ability of the receptor to prevent the increase in intracellular cAMP (caused by isoproterenol, etc) correlates with its inhibition of the cAMP-dependent protein kinase. Although it has not been determined whether the decrease in intracellular cAMP in this case is due to inhibition of adenylate cyclase or stimulation of a phosphodiesterase, evidence suggests that muscarinic ACh receptors act on adenylate cyclase and not phosphodiesterase (Torphy et al 1985).

Because of the well documented effects of cAMP on cardiac contractility and the ability of cardiac muscarinic ACh receptors to inhibit both basal- and hormone-stimulated adenylate cyclase, probably at least some of the inhibitory effects of muscarinic receptors in the heart are mediated via inhibition of adenylate cyclase activity. Considerable electrophysiological evidence indicates that activation of cardiac β-adrenergic receptors increases the permeability of the cardiac muscle membrane to calcium through the activation of cAMP-dependent protein kinase and resultant phosphorylation of calcium channels (Kameyama et al 1985). Activation of the muscarinic receptors decreases the β-adrenergic receptor-induced increase in calcium permeability and decreases β-adrenergic stimulation of adenylate cyclase (Biegon & Pappano 1980). One type of calcium channel has recently been purified and shown to be a substrate for cAMP-dependent protein kinase in skeletal muscle membranes (Curtis & Catterall 1985); however, calcium channels in heart have not been directly shown to be phosphorylated in response to β-adrenergic agonists and that this is inhibited by muscarinic ACh receptor agonists.

Muscarinic inhibition of the β-adrenergic stimulated phosphorylation of some well-characterized proteins known to be involved in cardiac contractility has been demonstrated. Even in these cases, however, inhibition of adenylate cyclase activity may not be the sole mechanism involved. For example, C-protein is a protein in the thick filament whose phosphorylation correlates with the relaxation of twitch tension. Muscarinic receptors inhibit the phophorylation of C-protein induced not only by β-adrenergic agonists but also by cAMP derivatives (Hartzell & Titus 1982, Hartzell 1984). Electrophysiological experiments have also been able to demonstrate muscarinic receptor mediated inhibition of cardiac contractility without reduction of cAMP levels (Pappano et al 1982). These results thus demonstrate that muscarinic receptors may regulate cardiac contractility and protein phosphorylation not only via inhibition of cAMP synthesis but by cAMP-independent mechanisms as well.

Phosphatidylinositol Turnover

Hokin & Hokin (1954) demonstrated that acetylcholine acted at muscarinic receptors in the pancreas to increase the incorporation of $^{32}P_i$ into phosphatidylinositol (PI). Subsequent work has demonstrated that a large number of hormones and transmitters in many different cells and tissues are able to stimulate PI metabolism. PI in the inner leaflet of the plasma membrane is phosphorylated in two successive steps to yield phosphatidylinositol 4,5-bis-phosphate (PIP_2). Activation of the muscarinic receptors activates phophoplipase C, which cleaves PIP_2 to yield two potential intracellular second messengers, inositol trisphosphate (IP_3) and diacylglycerol (DAG) (Berridge & Irvine 1984). While the mechanism of coupling of the muscarinic receptors to phospholipase C is not as well characterized as the coupling to adenylate cyclase, evidence suggests that GTP-binding proteins may also couple receptors to phospholipase C (Cockcroft & Gompert 1985, Nakamura & Ui 1985). Clearly, however, muscarinic receptors are not coupled to phospholipase C by N_i or N_o, because treatment of cells with IAP does not affect PI turnover (Hughes et al 1984). The muscarinic receptor in cultured astrocytoma cells has been shown to interact with a GTP-binding protein distinct from N_i (Martin et al 1985b, Evans et al 1984) that is likely to be involved in coupling the muscarinic receptors in both these and other cell types to phospholipase C.

PHYSIOLOGICAL ROLES OF IP_3 IP_3 is an intracellular second messenger that promotes the release of calcium from intracellular stores (see Berridge & Irvine 1984). Streb et al (1983) first demonstrated that IP_3 caused the release of calcium from intracellular stores in permeabilized acinar cells, and that activation of muscarinic receptors and IP_3 appeared to release Ca^{2+} from the same intracellular stores. Mobilization of Ca^{2+} by muscarinic receptors has been demonstrated by a number of techniques. Increases in free intracellular Ca^{2+} levels have been detected with fluorescence measurements using quin-2 in chromaffin cells (Kao & Schneider 1985), neuroblastoma cells (Ohsako & Deguchi 1984), and PC12 cells (Vincentini et al 1985). Muscarinic receptor mediated decreases in membrane-associated calcium levels have been found by using fluorescence measurements with chlorotetracycline in pancreatic acinar cells (Chandler & Williams 1978) and chicken fibroblasts (Oettling et al 1985), and activation of muscarinic receptors causes increased efflux of $^{45}Ca^{2+}$ in astrocytoma (Masters et al 1984) and pancreatic acini (Dehaye et al 1984).

This mobilization of intracellular Ca^{2+} is probably involved in a number of responses. It probably mediates muscarinic receptor induced secretion of catecholamines from the adrenal medulla and of amylase from pancreas. Furthermore, increased levels of intracellular Ca^{2+} can decrease cAMP levels in some cells: in both thyroid and astrocytoma cells, muscarinic receptors lower cAMP levels not by inhibiting adenylate cyclase activity but by stimulating the activity of a calmodulin-regulated phosphodiesterase (Evans et al 1984, Cochaux et al 1985). In addition, injection of IP_3 into oocytes can mimic the activation by muscarinic receptors of a chloride current (Oron et al 1985), thus suggesting that IP_3 may also mediate some of the depolarizing actions of muscarinic receptors in the nervous system.

ROLE OF DIACYLGLYCEROL The other product of the breakdown of PIP_2 is DAG. DAG activates protein kinase C (PKC), a Ca^{2+}- and phospholipid-dependent enzyme, by increasing the apparent affinity of the kinase for Ca^{2+}, so that it can be activated at resting levels of intracellular calcium. Protein kinase C has been implicated as a regulator of a variety of cellular and physiological responses. The tumor-promoting phorbol esters act by substituting for endogenously generated DAG to activate PKC in intact cells and tissues and PKC has been shown to be involved (either alone or acting synergistically with increased levels of intracellular Ca^{2+} released by the action of IP_3) in release of hormones and transmitters, platelet activation, and regulation of a variety of receptor-mediated responses (see Nishizuka 1984 for review). Evidence shows that PKC may be involved both in short-term desensitization of muscarinic receptor function and agonist-induced long-term regulation of muscarinic receptor number in at least some cells (see below). PKC may also mediate the ability of muscarinic receptors to inhibit a K^+ channel in *Xenopus oocytes* (Dascal et al 1985).

Stimulation of cGMP Synthesis

Activation of muscarinic receptors in many cells causes increased levels of cGMP. The precise mechanism of coupling of activation of these receptors to stimulation of guanylate cyclase has not been unequivocally established. Both increased concentrations of intracellular calcium and increased concentrations of metabolites of arachidonic acid have been suggested as possible mediators. The increased level of cGMP in N1E-115 cells requires extracellular Ca^{2+}, and activation of voltage-insensitive calcium channels has been implicated in this response (Study et al 1978, El-Fakahany & Richelson 1983). Snider et al (1984) reported that muscarinic agonists did not elevate internal calcium levels in N1E-115 cells as measured with aequorin, and that carbachol-induced increases in cGMP levels and increases in arachidonic acid release could be blocked by the phospholipase A_2 inhibitor, quinacrine. It was therefore suggested that a metabolite of arachidonic acid may mediate the stimulation of guanylate cyclase evoked by the muscarinic receptor. This conclusion, however, is complicated because the concentrations of quinacrine used in this study are able to inhibit significantly ligand binding to the receptor (N. M. Nathanson, unpublished observations). Furthermore, Ohsako & Deguchi (1984) demonstrated that addition of carbachol to N1E-115 cells caused increased efflux of preloaded $^{45}Ca^{2+}$, release of membrane bound calcium, and increased levels of intracellular calcium as measured by quin-2 fluorescence. The time courses of these responses were similar to the time course for the increase in cGMP. Although these results do demonstrate a concomitant increase in intracellular levels of calcium and cGMP, they do not distinguish between elevated calcium as causing the increase in cGMP or vice versa. Further work is thus required to determine the relative contributions of calcium and arachidonic acid metabolites in the stimulation of cGMP synthesis.

A role for cGMP in the mediation of muscarinic responses has been suggested in a number of cell types. Thus, both cGMP and muscarinic stimulation have been shown to elicit such responses as inhibition of ACh release from hippocampal slices (Nördstrom & Bartfai 1981), hyperpolarization of neocortical neurons (Swartz & Woody 1979), depolarization of sympathetic neurons (Dun et al 1978), and hyperpolarization of N1E-115 cells (Wastek et al 1981). However, most reports have not determined whether the same ionic channels are regulated by both cGMP and the muscarinic ACh receptors. The importance of this is shown by the results of Dun et al (1978), who demonstrated that muscarinic receptors and cGMP depolarized sympathetic ganglia via different ionic mechanisms. Electrophysiological evidence supporting a role for cGMP in muscarinic action has been obtained using voltage clamp recording techniques in the *Xenopus oocyte*, where application of muscarinic agonists and intracellular injection of cGMP both produce hyperpolarization due to activation of a similar outward K^+ channel (Dascal et al 1984).

A novel role for cGMP has been implicated for muscarinic action in vascular smooth muscle. Activation of muscarinic receptors usually produces contraction in smooth muscle but causes vascular smooth muscle to relax. Furchgott & Zawadzki (1980) demonstrated that ACh only relaxed isolated smooth muscle when the endothelial cell layer in the vessel was intact. Removing or damaging the endothelial cells resulted in muscarinic receptor evoked contraction. Perfusion experiments have demonstrated that ACh acts on the endothelium cells to cause the release of a factor that produces relaxation of the smooth muscle (Griffith et al 1984). The released factor appears to produce relaxation by the elevation of cGMP levels and activation of a cGMP-dependent protein kinase in the muscle (Rapoport et al 1983).

ELECTROPHYSIOLOGICAL RESPONSES OF THE MUSCARINIC RECEPTOR

Activation of muscarinic receptors in various electrically excitable cells can lead to either depolarization or hyperpolarization, due to opening or closing of potassium, calcium, or chloride channels. Thus, muscarinic receptors produce hyperpolarization by activating potassium channels in sympathetic and parasympathetic neurons (Hartzell et al 1977, Dodd & Horn 1983), in heart (Giles & Noble 1976), and oocytes (Dascal et al 1984), but they produce depolarization by closing potassium channels in sympathetic neurons (Adams et al 1982), spinal cord neurons (Nowak & MacDonald 1983), hippocampal neurons (Cole & Nicoll 1984), and myenteric neurons (Morita et al 1982). Muscarinic agonists activate calcium-dependent K^+ channels in lacrimal glands (Trautmann & Marty 1984) and pancreatic acinar cells (Maruyama et al 1983) and close calcium-dependent K^+ channels in hippocampal neurons (Cole & Nicoll 1984) and myenteric neurons (North & Tokimasa 1983). Muscarinic receptors increase Cl^- permeability in oocytes (Oron et al 1985), decrease Ca^+ permeability in heart (Biegon & Pappano 1980), and activate a nonselective cation channel in smooth muscle (Benham et al 1985). In this section I describe two examples of muscarinic regulation of specific ion channel activity.

Regulation of Inwardly Rectifying Potassium Channels in the Heart

Activation of muscarinic receptors in the heart decreases beating rate due to an increase in the permeability of the cardiac muscle membrane to potassium ions. The channel activated by ACh has a high conductance for inward K^+ currents and a lower conductance for outward currents, so that it is an inward rectifying potassium channel (Giles & Noble 1976). The minimum latency in the response to ACh is relatively long—tens to hundreds of milliseconds (Hill-Smith & Purves 1978, Hartzell 1980). Because this is several orders of magnitude slower than the latency for the nicotinic acetylcholine receptor at the neuromuscular junction, and because the magnitude of the latency for the muscarinic receptor mediated hyperpolarization exhibits a relatively large dependence on temperature (Hartzell 1980), either intermediate steps or second messengers are probably involved in coupling muscarinic receptors to the channel. Nargeot et al (1982) studied the activation kinetics of a K^+ channel in the presence of a photoisomerizable muscarinic antagonist and concluded that the rate-limiting step was subsequent to binding of ligand to the muscarinic receptor. Attempts to identify a cytosolic second messenger in coupling muscarinic receptor to the K^+ channel have been unsuccessful. Electrophysiological experiments have demonstrated that neither cGMP nor cAMP is involved in this response (Nawrath 1977, Trautwein et al 1982, Nargeot et al 1983). Furthermore, Sakmann et al (1983) and Soejima & Noma (1984) reported that in "cell-attached" patch clamp recordings, ACh activated channels under the recording pipette only when it was applied inside the pipette but not when it was applied outside the pipette. If a diffusible cytosolic second messenger were involved, application of ACh outside the pipette should have been able to activate channels inside the pipette. In addition, Soejima & Noma (1984) also reported that ACh activation of K^+ channels could be observed in isolated membrane patches excised from cells. However, because they compared recordings from excised patches in the presence of ACh to cell-attached patches in the absence of ACh, their results did not unequivocally rule out a requirement for intracellular components.

Halvorsen & Nathanson (1984) reported that the appearance of negative chronotropic responses in developing embryonic chick heart mediated by muscarinic ACh receptors was correlated with functional and physical changes in the GTP-binding proteins (N_o and N_i) associated with the receptor; thus these proteins may play a role in coupling muscarinic receptors to physiological responses in the heart. Because cyclic nucleotides had been ruled out as possible second messengers in coupling the receptor to the K^+ channel, these results suggested that the GTP-binding proteins may directly couple the receptor to the channel. Treatment of cultured chick cardiac cells with IAP blocks muscarinic receptor mediated increases in potassium permeabiliity (Martin et al 1985a), and treatment of rats and chicks with IAP blocks muscarinic receptor mediated hyperpolarization of the atrium (Endoh et al 1985, Sorota et al 1985). Because the only two polypeptides covalently modified by IAP treatment in the embryonic chick heart are N_i and N_o (Halvorsen & Nathanson 1984, Martin et al 1985a), these results implicate N_i and/or N_o in coupling the channel to the receptor. Pfaffinger et al (1985) demonstrated that in whole cell voltage clamp experiments with embryonic chick atrial cells, which allow control of the intracellular contents while recording membrane currents, muscarinic receptor activation of the inward rectifying K^+ channel required the presence of intracellular GTP. Even in the presence of GTP, activation of the channel by the receptors was blocked by pretreatment with IAP. Breitwieser & Szabo (1985), using whole cell voltage clamp of bullfrog atrial cells, found that intracellular application of a nonhydrolyzable GTP analogue caused an agonist-induced persistant activation of the potassium channels. While the precise molecular details remain to be resolved, these results demonstrate that the muscarinic receptor interacts with the potassium channel through a GTP-binding protein such as N_i or N_o.

Regulation of M-Current Potassium Channels

Frog sympathetic neurons possess a time- and voltage-dependent potassium current activated by depolarization that can be decreased by muscarinic agonists and thus has been called an M-current (Adams et al 1982). Suppression of the M-current results in the excitatory postsynaptic potential produced by muscarinic receptor activation in these neurons. Regulation of the M-current potassium channel by muscarinic receptors has subsequently been demonstrated in a variety of cells, including spinal cord neurons (Nowak & MacDonald 1983) and hippocampal neurons (Cole & Nicoll 1984). Because the M-current can be regulated not only by muscarine but also by neuropeptides, a common intracellular pathway may be responsible for regulation of channel activity. Intracellular injections of cGMP or cAMP did not affect the M-current, and the mechanism of regulation of the channel by muscarinic and peptide receptors has been unclear (Adams et al 1982). The demonstration that GTP-binding proteins couple muscarinic receptors to K^+ channels in the heart raises the possibility that a similar mechanism may be involved in the regulation of the M-current potassium channel by acetylcholine and peptide receptors.

LIGAND BINDING AND BIOCHEMICAL STUDIES OF THE MUSCARINIC RECEPTOR

The development of radioligands with high affinity and specificity for the muscarinic receptor has played a key role in the tremendous increase over the last ten years in biochemical information available about the muscarinic receptor. The use of these ligands has permitted the identification, quantitation, and purification of the muscarinic receptor. Countless reports on the identification of muscarinic receptor binding sites in innumerable cells and tissues have been published. In this section I will summarize some of the key features of the ligand binding properties of the receptor.

Binding of Antagonists

The first radioligand binding studies of the muscarinic receptor were performed by Paton & Rang (1965), who demonstrated that the binding of tritiated atropine could be used to identify muscarinic receptors in guinea pig smooth muscle. The very low specific activity available (less than 1/100 of the activity of ligands now in common use) and the use of intact pieces of tissue instead of membrane homogenates (which increases nonspecific binding and complicates analysis of the binding kinetics), however, prevented detailed biochemical experimentation. Not until nearly a decade later did a number of investigators demonstrate the feasibility of using ligands of higher specific radioactivity to identify and characterize muscarinic receptors biochemically. The muscarinic antagonist [^{3}H]dexetimide was used to label muscarinic receptor in homogenates of bovine smooth muscle and brain (Beld & Ariëns 1974); the affinity alkylating antagonist, [^{3}H]propylbenzilylcholine mustard (PrBCM), was used to label muscarinic receptors in intact intestinal smooth muscle and in homogenates from mammalian brain (Burgen et al 1974a,b); and the high affinity antagonist, [^{3}H]quinuclidinyl benzilate (QNB), was used to identify muscarinic receptor in homogenates from rat brain and guinea pig ileum (Yamamura & Snyder 1974a,b). A good correlation was found between the concentrations of various agonists and antagonists required to compete for the binding of QNB and to produce physiological effects at muscarinic receptors in the intact ileum (Yamamura & Snyder 1974b). These results demonstrated that the sites labeled by QNB were in fact the physiologically active muscarinic receptors in this tissue.

The binding of a large number of antagonists to muscarinic receptors in rat cerebral cortex membrane homogenates was extensively studied by Hulme et al (1978). They found that [^{3}H]QNB, [^{3}H]atropine, [^{3}H]PrBCM, [^{3}H]*N*-methyl-scopolamine ([^{3}H]NMS), and [^{3}H]*N*-methylatropine all bound to an equal number of sites. The Hill coefficients for binding were close to unity, indicating that the antagonists bound to a single class of high affinity binding sites. The affinity binding constants determined by direct binding of the radioligands were identical to those determined by competition using unlabeled antagonists. Furthermore, the affinity constants of the 20 antagonists determined by competition binding studies agreed well with the physiological affinity constants determined by inhibition of contraction of isolated guinea pig ileum, further confirming the validity of the identification of these binding sites as the muscarinic receptor.

While the equilibrium binding of radiolabeled antagonists in physiological buffers can be well described by the binding of the drugs to a single class of high affinity sites, kinetic analysis indicates that the binding is more complex than expected for a simple bimolecular reaction. Semilogarithmic plots of the rate of formation of the ligand-receptor complex under pseudo–first order conditions (i.e. with the concentration of ligand much greater than that of the receptor) were biphasic, suggesting that the binding of antagonist to receptor was a multistep process (Galper et al 1977). Confirmation of a two-step process was provided by Järv et al (1979), who demonstrated that the pseudo–first order rate constants for association of two different radiolabeled antagonists with the muscarinic receptor exhibited a hyperbolic dependence on the concentration of antagonist. The simplest model to account for this was a rapid binding step followed by a slower isomerization of the receptor-antagonist complex. Schreiber et al (1985a) have described a general method for the analysis of ligand binding by competition kinetics that allows the calculation of rate constants for binding and isomerization of unlabeled ligands to the receptor.

Several additional complications regarding the binding of antagonists have been reported. For example, in contrast to the homogeneity of antagonist binding sites seen in buffers of physiological ionic strength, experiments performed in buffers of low ionic strength and/or in the presence of tris or choline ions can exhibit heterogeneity of antagonist binding. Guanine nucleotides, which regulate agonist binding but not antagonist binding in physiological bufferss (see below), also affect antagonist affinity in low ionic strength buffers (Hulme et al 1981, Burgisser et al 1982, Murphy & Sastre 1983). While the molecular basis for these differences is not clear, these results demonstrate the importance that seeminly minor differences in assay conditions can have in determining binding properties of the receptor.

In contrast to the equal number of binding sites for a variety of muscarinic ligands demonstrated by Hulme et al (1978), some studies have suggested differences in the binding of different radiolabeled antagonists. Thus, Lee & El-Fakahany (1985a) reported that the lipophobic ligand [^{3}H]NMS labeled 65% of the number of sites labeled by the lipophilic ligand [^{3}H]QNB. Furthermore, competition binding curves of quaternary ammonium antagonists with [^{3}H]QNB demonstrated heterogeneity of binding sites. Lee & El-Fakahany concluded that lipophobic antagonists may distinguish between different classes of muscarinic receptors. However, the results to date cannot exclude a trivial reason for the discrepancy between the results of Hulme et al and Lee & El-Fakahany, such as differences in the method of homogenization (which could result in trapping a fraction of the receptors inside vesicles, thereby rendering them inaccessible to NMS but still able to bind QNB) or differences in the radiochemical purity of the ligands.

Binding of Agonists

In contrast to the homogeneity of antagonist binding sites usually reported, agonist binding to the muscarinic ACh receptor in membrane homogenates exhibits pronounced deviation from the law of mass action, with Hill coefficients significantly less than unity, indicating that the binding cannot be adequately described by the interaction of ligand with a single class of binding sites. The binding of agonists to the muscarinic receptor is most commonly determined by competition of unlabeled agonists with radiolabeled antagonists for the binding to the receptor, although radiolabeled agonists can be used to label those receptor binding sites with relatively high affinity for agonist. The interaction of agonists with the muscarinic receptor from rat cerebral cortex was extensively studied by Birdsall et al (1978). Competition binding experiments using 21 different agonists indicated that the binding data could be explained by the presence of two binding sites with different affinities for agonist ("high" and "low") but with the same affinity for antagonists. Birdsall et al also examined the binding of a number of radiolabeled agonists to the muscarinic receptor. Direct binding of agonists labeled the high affinity but not low affinity sites, and in addition revealed the presence of a minor population of "superhigh" affinity sites. (Because of the practical limits on the concentrations of ligand possible to use in a binding assay and still detect specific binding, few low affinity sites are labeled by tritiated agonists.) This apparent heterogeneity of agonist binding sites did not appear to be due to negative cooperativity between interacting sites, as the agonist binding curves were unaffected by prior affinity alkylation of over 90% of the receptor sites with PrBCM. Direct confirmation of the existence of these three classes of sites was provided by an elegant series of alkylation and protection experiments. Membranes were incubated with a low concentration of agonist that should occupy primarily superhigh and high affinity sites and the remaining unoccupied sites were alkylated with PrBCM. After washing out the unbound agonist, binding studies indicated that PrBCM alkylated almost exclusively the low affinity sites, and that the remainder of the sites in fact corresponded primarily to the high and superhigh affinity sites.

Heterogeneity of agonist binding sites is observed for muscarinic receptors in virtually every tissue and cell type examined, although both the ratios and absolute values of high and low affinity states vary from tissue to tissue. Such differences have also been seen by using unlabeled agonist-radiolabeled antagonist competition experiments in tissue sections by autoradiography (Wamsley et al 1980), and by the use of radiolabeled agonists either by binding studies with membrane homogenates (Kellar et al 1985, Gurwitz et al 1985) or by autoradiography of tissue sections (Yamamura et al 1985). Kinetic studies indicate that, in contrast to the isomerization of receptor-antagonist complexes noted above, agonists do not induce isomerization or interconversions between high and low affinity states (Järv et al 1980, Schreiber et al 1985b). Selective alkylation of either high or low affinity sites in cerebral cortex has been demonstrated by Ehlert & Jenden (1985), who used an alkylating derivative of the agonist, oxotremorine, which exhibited kinetic differences in its interactions between high and low affinity states. The selective alkylations performed by Birdsall et al (1978) and Ehlert & Jenden (1985) further suggest that the different agonist affinity states in cerebral cortex are not freely interconvertable.

Conversions between high and low agonist affinity states can be induced by divalent cations and by guanine nucleotides. Addition of guanine nucleotides decreases the apparent affinity of the muscarinic receptors for agonists, although the extent of this effect varies from tissue to tissue. Guanine nucleotides have large effects on agonist binding to muscarinic receptors in the heart (Berrie et al 1979) and in cerebellum and brainstem (Korn et al 1983), but have relatively little effect on muscarinic receptors in retina, hippocampus, or cerebral cortex (Korn et al 1983, Dunlap & Brown 1984). The regulation of agonist binding by guanine nucleotides is due to the interaction of the receptor with the guanine nucleotide regulatory proteins, N_i and N_o. Isolated receptors in the absence of the N-proteins exhibit a low affinity for agonist, whereas the receptor-N-protein complex in the absence of nucleotide has a high affinity for agonist (Florio & Sternweis 1985). Guanine nucleotides decrease the apparent heterogeneity of agonist binding sites by decreasing the fraction of high affinity sites and/or the ratio of the affinities for the high and low affinity sites (Halvorsen & Nathanson 1981, Burgisser et al 1982, Waelbroeck et al 1982, Halvorsen et al 1983, Nathanson 1983). The regulation of agonist binding by guanine nucleotides is well-described by a ternary complex model for the interaction among ligand receptor, and N-protein (Ehlert 1985), similar to the ternary complex model evoked for the interaction of β-adrenergic agonists with the β-adrenergic receptor and the stimulatory guanine nucleotide binding protein, N_s (De Lean et al 1980). Treatment of cardiac and neural cells with IAP, which ADP-ribosylates N_i and N_o, blocks the interaction of the receptor with these proteins, and shifts the receptors to a uniform population of low affinity states with no further effect of guanine nucleotides (Kurose et al 1983, Martin et al 1985a). In astrocyoma cells, treatment with concentrations of IAP that completely ADP-ribosylate N_i does not affect guanine nucleotide regulation of agonist binding, thus indicating that the muscarinic receptor in these cells interacts with an unidentified GTP-binding protein (Hughes et al 1984, Martin et al 1985b).

The interaction of the muscarinic receptor with the N-proteins involves an essential sulfhydryl group and can be blocked by pretreatment of membranes with the sulfhydryl alkylating reagent, *N*-ethyl maleimide (Harden et al 1982). Under conditions in which the amount of N-protein may be limiting, detection of guanine nucleotide effects on agonist binding can require the presence of a sulfhydryl reducing agent (Halvorsen & Nathanson 1984). Divalent cations can also regulate agonist affinity and sensitivity to guanine nucleotides (Hulme et al 1983). Guanine nucleotide regulation of agonist binding to muscarinic receptors in rat brainstem membranes has been shown to be eliminated after a 30 minute incubation at 37° by the action of an endogenous protease (Aronstam & Greenbaum 1984). It is not known to what extent the different magnitudes of guanine nucleotide regulation observed in different tissues arise from artifactual conditions such as proteolysis or oxidation of sulfhydryl groups. Although the heterogeneity of agonist binding in the heart can be adequately explained by the interaction of the muscarinic receptor with N_i and N_o, the multiple affinity states for agonist in the CNS may be due at least in part to the existence of distinct subtypes of the receptor (see below).

A novel study of the agonist binding properties of the muscarinic receptor was performed by Zarbin et al (1982), who used competition-binding autoradiographic experiments to examine axonal transport of muscarinic receptors in the rat vagus nerves. Anterogradely transported muscarinic receptors exhibited high affinity agonist binding that was sensitive to guanine nucleotides, while retrogradely transported receptors exhibited primarily a low affinity for agonist that was not regulated by guanine nucleotides. These results raise the intriguing possibility that new muscarinic receptors may undergo transport to the nerve terminal associated with their GTP-binding coupling proteins, whereas the receptors that have been removed and are returning to the soma become uncoupled.

A large number of compounds have been reported to have allosteric effects on the binding of muscarinic ligands to the receptor. The nicotinic neuromuscular blockers, gallamine and pancuronium, decrease the binding of muscarinic agonists and antagonists to the muscarinic receptor in a negatively cooperative manner (Stockton et al 1983, Dunlap & Brown 1983). The nicotinic blockers both decrease the rate of association and increase the rate of dissociation of muscarinic antagonists from the muscarinic receptor. Gallamine does not protect the receptor from affinity alkylation by PrBCM. These results indicate that gallamine and other neuromuscular blockers can interact at a second site on the muscarinic receptor (or on another protein that it interacts with) to regulate ligand binding. The effects of gallamine are much greater on ligand binding to muscarinic receptors in heart than in other tissues, consistent with the known physiological actions of gallamine. Pancuronium and gallamine antagonize the muscarinic receptor mediated inhibition of cAMP accumulation in intact atria (Dunlap & Brown 1983), although it has recently been claimed that no effect of gallamine was observed on muscarinic receptor mediated inhibition of adenylate cyclase activity in membrane homogenates (Jagadesh & Sulakhe 1985).

The effects of other compounds on muscarinic receptor binding have been noted and interpreted to indicate a possible interaction of the receptor with other macromolecules. For example, the steroid sex hormones, β-estradiol and progesterone, and the antiestrogen drug, clomiphene, have allosteric effects on muscarinic receptor binding, and this could be due to an interaction between the muscarinic and estrogen receptors (Sokolovsky et al 1981, Ben-Baruch et al 1982). Drugs that can act on the voltage-dependent sodium channel, such as quinidine, lidocaine, and bretylium, and the alkaloid activator, batrachotoxin, are able to affect muscarinic ACh receptor-ligand binding, and the possibility that this may reflect an interaction between the muscarinic receptor and the sodium channel has been raised (Cohen-Armon et al 1985a,b). The potassium channel blocker 4-aminopyridine alters ligand binding to the muscarinic receptor, and an interaction between the muscarinic receptors and K^+ channels has been suggested (Lai et al 1985). In all these cases, however, both the molecular mechanisms and the physiological significance of the effects of these drugs on the muscarinic receptor remain to be determined.

It has been suggested that the muscarinic agonists are able to regulate the binding of β-adrenergic agonists to the β-adrenergic receptor and that β-adrenergic agonists are able to regulate binding of muscarinic agonists to the muscarinic receptor. Watanabe et al (1978) reported that methacholine blocked the ability of GTP to decrease the apparent affinity of isoproterenol for the cardiac β-adrenergic receptor, and Rosenberger et al (1980) found that isoproterenol decreased the apparent affinity of the muscarinic agonist, oxotremorine, for cardiac muscarinic receptors. These results, if confirmed by others, might suggest that the muscarinic and β-adrenergic receptors are able to interact via their GTP-binding proteins. The prize for the largest effect reported on ligand binding to the muscarinic receptor should be awarded to Lundberg et al (1982), who demonstrated that vasoactive intestinal peptide (VIP) increased the apparent affinity of carbachol to muscarinic receptors in the feline submandibular salivary gland by 100,000-fold. Although this result may be consistent with the observation that VIP potentiates ACh-induced secretion, further work is required to confirm that VIP and muscarinic receptors interact.

Because the muscarinic receptor may possess a regulatory site that allosterically affects ligand binding to the receptor, several groups have tested for the presence in tissues of endogenous regulators of muscarinic receptor–ligand interactions. A soluble, nonpeptide low molecular weight inhibitor of [^{3}H]QNB binding was found in calf brain and thymus (Diaz-Arrastia et al 1985) and in embryonic chick heart and brain (Creazzo & Hartzell 1985). Although the identity and physiological significance of this is unknown, the presence of this compound could result in significant differences when the results of ligand binding to muscarinic receptor in unfractionated homogenates and purified membrane preparations are compared.

Muscarinic Receptor Subtypes

The heterogeneity of both antagonist and agonist binding observed under some conditions and the plethora of biochemical and physiological responses evoked by the muscarinic receptor raises the question of whether more than one type of receptor exists. Hammer et al (1980) first clearly demonstrated that the ligand binding properties of muscarinic receptors could be differentiated by the novel antagonist, pirenzepine (PZ). In contrast to the homogeneous binding seen for classical muscarinic antagonists, competition curves between pirenzepine and [^{3}H]NMS indicated the presence of multiple affinity states for pirenzepine in most tissues. Cardiac and smooth muscle had a single class of binding sites with a relatively low affinity, while different glands and regions of the nervous system possessed varying proportions of an additional site, with approximately a 50-fold higher affinity. Hammer & Giachetti (1982) compared the functional and biochemical properties of muscarinic receptors in atria and sympathetic ganglia. The classical antagonist, atropine, bound to both receptors with similar affinities. In contrast, whereas pirenzepine bound to a single class of binding sites in the atria with low affinity, it bound to two classes of sites in the ganglia. Physiological studies demonstrated that while atropine was equipotent in blocking the effects of cholinergic agonists in each tissue, a nearly 100-fold difference in the concentrations of pirenzepine was required to block muscarinic receptors in the atria compared to the ganglia. Similar functional differences in sensitivity for blockade by pirenzepine and activation by the novel agonist, McN-A-343, for muscarinic receptors in sympathetic ganglia and vascular smooth muscle have been reported by Wess et al (1984). These results show that these receptors in different tissues can exhibit vastly different affinities for pirezepine. Receptors with relatively high affinity for pirenzepine and McN-A-343 have come to be called M_1-muscarinic receptors, and those with relatively low affinity, M_2-receptors.

Binding studies using [^{3}H]PZ have directly demonstrated that high affinity binding sites for pirezepine are a subset of the available muscarinic receptor binding sites (Watson et al 1983, Luthin & Wolfe 1984a). Consistent with the functional and competition binding studies described above, relatively few of the [^{3}H]QNB binding sites in rat heart, ileum, and cerebellum had high affinity for [^{3}H]PZ, while the density of [^{3}H]PZ binding sites in cortex, hippocampus, and cerebellum was approximately half that of [^{3}H]QNB. Luthin & Wolfe (1984b) demonstrated that these differences in the binding densities for the two ligands were not due to differences in the rate of isomerization of the receptor-ligand complexes.

A number of studies have suggested relationships among the various classes of agonist affinity binding sites and pirenzepine binding sites. Based on the effects of the sulfhydryl agent, *p*-chloromercuribenzoate, on the ligand binding properties of the muscarinic receptor, Birdsall et al (1983) concluded that the low affinity agonist binding site was related to the high affinity pirenzepine binding site and the high affinity agonist binding site was related to the low affinity pirenzepine binding site. Similarly, Kellar et al (1985) examined the binding of [^{3}H]acetylcholine and suggested that most of the M_2 sites (i.e. those with a low affinity for pirenzepine) had a high affinity for ACh, whereas the majority of M_1 sites had a low affinity for ACh. However, Rodrigues de Miranda et al (1985) have reported that in rat nasal mucosa the muscarinic receptors are predominantly in a low affinity agonist state yet also exhibit a low affinity for pirenzepine. As described below, structural and biochemical studies have not yet unequivocally determined whether the heterogeneity of pirenzepine affinity states results from the existence of more than one type of polypeptide with different ligand binding sites or from the association of a single polypeptide with a variety of different polypeptides that can modify the ligand binding properties of the receptor.

Correlation of Physiological Responses with Different Ligand Binding Sites

Because of the heterogeneity of muscarinic ligand binding sites, there have been a number of attempts to correlate various biochemical responses with the different agonist and pirenzepine affinity states. In general, the biochemical and physiological responses mediated by the muscarinic receptor are correlated with occupancy of the low affinity agonist binding site. Strange et al (1977) first demonstrated that the concentrations of agonists required for stimulation of guanylate cyclase activity in N1E-115 neuroblastoma cells were very similar to those calculated to be required for binding to the low affinity site in membranes. Birdsall et al (1978) showed that, after elimination of spare receptors, occupancy of the low affinity site correlated well with contraction of smooth muscle. In a similar fashion, Halvorsen & Nathanson (1981) demonstrated that the negative chronotropic response of embryonic chick atria was correlated with occupancy of the low affinity site. In cultured cardiac cells, which are either grown under conditions in which they do not possess spare receptors (Galper et al 1982a) or are treated with PrBCM to eliminate spare receptors (Hunter & Nathanson 1985), the concentrations of carbachol required for increasing membrane permeability to potassium ions are also similar to those required for occupancy of the low affinity site.

A number of studies have demonstrated that there are relatively few spare receptors for muscarinic receptor mediated stimulation of inositol phospholipid turnover, and that occupancy of the low affinity agonist site correlates well with this response (Fisher et al 1983, Jacobson et al 1985). Dehaye et al (1984) found that the low affinity site in pancreatic acinar cells also mediated muscarinic receptor induced Ca^{2+} mobilization and amylase secretion. In contrast to these conclusions, Brown & Brown (1984) and Oettling et al (1985) compared ligand binding to intact cells with muscarinic receptor mediated PI turnover in cultured cardiac cells and muscarinic receptor mediated calcium mobilization in fibroblasts and concluded that the high affinity agonist sites mediated these responses. As described below, because of the radioligand used in these studies, their "high affinity" form actually corresponds to the low affinity state normally seen in membrane homogenates, and their "low affinity" form most likely corresponds to an internalized form of the receptor. These studies are thus also consistent with the correlation between occupancy of the low affinity state and response.

Several studies have suggested that occupancy of the high affinity agonist state correlates with muscarinic receptor mediated inhibition of adenylate cyclase (Brown & Brown 1984, McKinney et al 1984). However, these studies ignored the possibility that there may be spare receptors for this response as there are for contraction of smooth muscle and inhibition of the heart beat. In fact, Delhaye et al (1984) demonstrated that there was a significant spare receptor reserve for both muscarinic receptor mediated inhibition of adenylate cyclase and the negative inotropic effect in the human heart, and that when this was taken into account, occupancy of the low affinity agonist site again correlated with the response. Ehlert (1985) has demonstrated that the presence of high and low affinity states and their regulation by guanine nucleotides in the heart can be well described mathematically by a ternary complex model linking agonist, receptor, and GTP-binding protein. Ehlert further showed that the ability of various agonists to inhibit adenylate cyclase activity was strongly correlated with their ability to stabilize the ternary complex. The results of Ehlert (1985) indicate that, when binding is analyzed with a two-site model for agonist binding, the correlation of physiological responses with occupancy of the low affinity site is a reflection of this ternary complex, which results from the requirement of a GTP-binding protein in muscarinic receptor mediated responses and the low affinity of the muscarinic receptor for agonist in the presence of guanine nucleotides.

A number of studies have suggested that high and low pirenzepine affinity receptors (M_1 and M_2) are coupled to different physiological responses. Gil & Wolfe (1984) found that the concentrations of pirenzepine required to antagonize muscarinic receptor mediated phosphoinositide turnover in rat brain and parotid gland were an order of magnitude lower than those required to antagonize muscarinic receptor mediated inhibition of adenylate cyclase in rat brain and heart. McKinney et al (1985) studied muscarinic receptor mediated inhibition of adenylate cyclase and stimulation of guanylate cyclase in the clonal neuroblastoma cell line, N1E-115, and found that much higher concentrations of pirenzepine were required to antagonize inhibition of adenylate cyclase than stimulation of guanylate cyclase. Because of the nature of these experiments, the presence or absence of spare receptors does not complicate their interpretation. Thus, different muscarinic receptors mediating different responses can have different affinities for pirenzepine. However, the receptors mediating the same physiological response can also have different affinities for pirenzepine in different tissues. Brown et al (1985) have demonstrated that in contrast to the above results, receptors mediating inhibition of adenylate cyclase in chick heart have a high affinity for pirenzepine, and receptors mediating phosphoinositide breakdown in both chick heart and 1321N1 astrocytoma cells have low affinity for pirenzepine. Thus, the different affinity states for pirenzepine do not appear to be correlated in an obligatory fashion with different muscarinic receptor mediated biochemical responses.

Ligand Binding in Intact Cells and Tissues

While most ligand binding studies have used membrane homogenates, the binding of muscarinic agonists and antagonists to receptors on intact cells and tissues has also been investigated. Early studies by Paton & Rang (1965) and Burgen et al (1974a) reported the binding of radioligands to intact intestinal smooth muscle. Hartzell (1980) used autoradiography with [^{3}H]QNB in frog sinus venosus to show that cardiac muscarinic receptors were diffusely distributed over the entire surface of the muscle. James & Klein (1982) used autoradiography to demonstrate that [^{3}H]QNB binding sites on retinal neurons were confined to dentritic processes. Muscarinic antagonists labeled with ^{123}I and ^{13}C have also been employed to identify muscarinic receptors in the brain and heart of living subjects by using positron and single-photon emission tomography (Eckelman et al 1984, Syrota et al 1985).

Galper et al (1982a) demonstrated that the binding to intact cardiac cells of the lipophilic ligand [^{3}H]QNB and the lipophobic ligand [^{3}H]NMS could be used to differentiate total cellular muscarinic receptors (i.e. those on the cell surface plus internalized receptors) from those only on the cell surface. The number of binding sites for [^{3}H]NMS on intact cardiac (Galper et al 1982b, Nathanson 1983, Siegal & Fischbach 1984), neuroblastoma (Feigenbaum & El-Fakahany 1985, Liles & Nathanson 1986), and astrocytoma cells (Harden et al 1985) are similar (70–100%) to the number of sites for [^{3}H]QNB, thus indicating that there is normally not a significant reservoir of internalized receptors.

Binding studies of antagonists to intact cells are well described by the binding to a single class of sites with affinities similar to those seen in homogenates (Galper et al 1982b, Nathanson 1983). In contrast to the pronounced heterogeneity of agonist binding to muscarinic receptors in membrane homogenates, competition curves between carbachol and [^{3}H]NMS to muscarinic receptors on intact cardiac cells can be well described by the binding of agonist to a single class of sites (Nathanson 1983). The affinity of the binding to intact cells is identical to the low affinity agonist binding site seen in membrane homogenates in the presence of guanine nucleotides. Kinetic studies of the binding of [^{3}H]NMS in the absence and presence of agonist showed that there was no transient (>30 sec) higher affinity form that subsequently was converted to lower affinity. In addition, in contrast to the decrease in the affinity of muscarinic receptor in membrane homogenates for agonist after IAP treatment, treatment with IAP does not further decrease the affinity of muscarinic receptor on intact cells for agonist (Martin et al 1985a). These results are consistent with the interaction of the muscarinic receptor with the GTP-binding proteins and endogenous GTP in the intact cell to form the low affinity agonist state.

In contrast to the single class of binding sites ($K_D \sim 10^{-5}$ M) for carbachol seen in competition binding experiments with [^{3}H]NMS, competition curves between [^{3}H]QNB and carbachol with intact cardiac cells and fibroblasts yield a second site with a $K_D > 10^{-3}$ M (Halvorsen et al 1983, Brown & Brown 1984, Oettling et al 1985). Because both NMS and carbachol have quaternary ammonium groups and should therefore be lipophobic while QNB is lipophilic, it is likely that this additional extremely low affinity site seen in [^{3}H]QNB/carbachol competition experiments represents internalized receptors that are accessible to QNB but not to carbachol. A somewhat different result was found by McKinney et al (1984) and Lee & El-Fakahany (1985b), who performed [^{3}H]QNB/agonist competition curves with intact N1E-115 and rat brain cells at 15° to prevent agonist-induced internalization. They found multiple classes of agonist binding sites with affinities similar to those seen in homogenates. Because low temperature could decrease the cooperative interactions between muscarinic receptors and the GTP-binding proteins required to shift receptors to the low affinity form, these results do not necessarily indicate that there is a difference between the binding properties of muscarinic receptors on intact cardiac and neuronal cells.

STRUCTURAL STUDIES AND PURIFICATION OF THE MUSCARINIC RECEPTOR

Identification of the Ligand Binding Polypeptide by Affinity Alkylation

Birdsall et al (1979) and Ruess & Liefländer (1979) first used the affinity alkylating antagonist, [^{3}H]PrBCM, to identify the polypeptide containing the ACh-binding site. They found that after labeling muscarinic receptors in rat, guinea pig, and frog brain, and in guinea pig intestinal smooth muscle, a single polypeptide of 75–83 kD was identified by SDS gel electrophoresis. Porcine atrial muscarinic receptors were identified by labeling both with [^{3}H]PrBCM (Petersen & Schimerlik 1984) and a photoaffinity analogue of atropine (Cremo & Schimerlik 1984) and shown to be a single polypeptide with molecular weight 70–80 kD. Venter (1983) also demonstrated that muscarinic receptors from a variety of species and tissues labeled with [^{3}H]PrBCM yielded a single polypeptide of 80 kD. Treatment of labeled membranes with either trypsin or papain yielded identical partial proteolytic maps for receptors from different sources. This suggested significant homologies between muscarinic receptors from different tissues and species.

Venter (1983) also reported that partial digestion of [^{3}H]PrBCM-labeled membranes initially yielded a water-soluble labeled fragment of 40 kD (as well as smaller fragments), suggesting that this portion of the receptor is extracellular and contains the ligand binding site. In contrast to these results, Hootman & Picado-Leonard (1985) have reported that digestion of pancreatic acinar cells with papain resulted in the decrease in the molecular weight of the muscarinic receptors from 88,000 to 46,000, but that the digested receptor remained associated with the membrane and was able to induce amylase secretion. The muscarinic receptor in cultured heart cells is rapidly converted from 70–80 kD to 40–50 kD by an endogenous protease when cells are homogenized, but the smaller receptors also remain associated with the cell membrane, exhibit guanyl nucleotide regulation of agonist binding, and are able to inhibit adenylate cyclase (D. H. Hunter and N. M. Nathanson, unpublished observations). The topology of the muscarinic receptor in the membrane is thus not yet resolved. Extensive digestion of [^{3}H]PrBCM-labeled bovine brain membranes with trypsin does release a proteolytic fragment of molecular weight of $\sim$5000 that can be immunoprecipitated by a monoclonal antibody specific for PrBCM (M. W. Gainer and N. M. Nathanson 1986 and unpublished observations).

Several studies have reported affinity labeling of multiple polypeptides in addition to that noted above. Avissar et al (1983) used a photoaffinity antagonist to covalently label muscarinic receptors in various brain regions and in heart. They found polypeptides of 40 kD and 160 kD in addition to 86 kD. They suggested that the receptor may exist in interconvertable forms that are related to the high and low affinity states of the receptor. Birdsall et al (1979), however, reported that agonist protection followed by selective alkylation of high or low affinity sites showed identical electrophoretic profiles. Dadi & Morris (1984) have demonstrated that the muscarinic receptors from rat forebrain yielded two [^{3}H]PrBCM-labeled polypeptides of 68 and 73 kD. Even in the presence of 2-mercaptoethanol, intermolecular disulfide bond formation could generate higher molecular weight aggregates unless membrane sulfhydryl groups were carboxyamidated with iodoacetamide. Hootman et al (1985) reported that labeling of pancreatic, lacrimal, and parotid acinar cells and lymphocyte and neural cells with [^{3}H]PrBCM yielded a number of polypeptides between 118 kD and 63 kD. It is not known whether these different-sized polypeptides are due to differences in biosynthetic or degradative processes or functional differences between muscarinic receptors in different tissues. Large et al (1985) reported that the molecular weight of the muscarinic receptor in developing chick retina decreased from 86,000 to 72,000 concurrent with synaptogenesis. The larger form had a faster turn-over rate than the smaller form; it is not known whether the smaller form is derived from the larger by degradation or other covalent modifications.

A number of biophysical studies have also yielded information about the molecular weight of the muscarinic receptors. Target size analysis of radiation inactivation experiments have shown that the minimum molecular mass of the [^{3}H]QNB binding component in membranes is 76–80 kD (Uchida et al 1982, Venter 1983). Guanine nucleotide regulation of agonist binding was inactivated with a target size of 179 kD (Uchida et al 1982), consistent with the involvement of the GTP-binding proteins in addition to the 80 kD ACh-binding polypeptide. In an analogous study, Shirakawa & Tanaka (1985) reported that [^{3}H]QNB binding had a target size of 91 kD and [^{3}H]PZ binding had a target size of 157 kD, suggesting that high affinity binding of pirenzepine required an additional component to the ACh-binding polypeptide.

Solubilization of Muscarinic Receptors

The reports by several groups that muscarinic ACh receptors could be solubilized by high salt treatment (Alberts & Bartfai 1976, Carson et al 1977) have not been confirmed by others (Hurko 1978, Aronstam et al 1978). Beld & Ariëns (1974) reported that prelabeled muscarinic receptor–[^{3}H]benzetimide complexes could be solubilized with digitonin. Hurko (1978), Gorissen et al (1978), and Aronstam et al (1978) demonstrated that digitonin could also be used to solubilize unlabeled receptors with retention of the ability to bind muscarinic ligands. A mixed detergent system of digitonin-cholate (Cremo & Schimerlik 1984) and cholate-salt solutions (Carson 1982) has also been used to obtain solubilized preparations of muscarinic receptors. Most reports with those detergents indicate that these solubilized preparations of muscarinic receptors are able to bind muscarinic antagonists with high affinity and specificity. In contrast to the heterogeneity of agonist binding sites seen in membrane preparations, however, the binding of agonists can be well described by the interaction with a single class of low affinity sites that are insensitive to regulation by guanine nucleotides, although solubilization of the receptor in the presence of agonist could stabilize the receptor-GTP binding protein complex (Harden et al 1983).

Other studies have demonstrated complex agonist binding properties after solubilization with digitonin. Baron & Abood (1984) and Luthin & Wolfe (1985) reported heterogeneity of agonist binding for muscarinic receptors solubilized from bovine and rat brain. Wenger et al (1985) reported that muscarinic receptors from different regions of the central nervous system retained their regional differences in agonist binding site heterogeneity after solubilization. Berrie et al (1984) found that solubilization of cardiac muscarinic receptors with digitonin in the presence of magnesium yielded preparations that exhibited guanyl nucleotide-sensitive agonist binding. Although Luthin & Wolfe (1985) found only a single class of low affinity pirenzepine sites in digitonin-solubilized rat cerebral cortex preparations, Berrie et al (1985) have reported that both high and low affinity binding sites for pirenzepine could be solubilized with digitonin. The zwitterionic detergent 3-[(3-cholamidopropyl)-dimethylammonio]-1-propane sulfonate (CHAPS) has also been used to obtain preparations of muscarinic receptor that retain heterogeneity of agonist binding and sensitivity to guanine nucleotides (Gavish & Sokolovsky 1982, Kuno et al 1983). However, because high concentrations of CHAPS inhibit ligand binding to the receptor, the concentration of detergent was decreased below the critical micellar concentration during the binding assay in these studies. The receptor in these preparations may thus no longer be truly soluble under these assay conditions.

Purification and Reconstitution of the Muscarinic Receptor

Andre et al (1983) described the purification of muscarinic receptors from rat forebrain by using affinity chromatography on dexetimide-agarose. The preparation contained a major polypeptide of 70 kD that comigrated with the receptor in [^{3}H]PrBCM-labeled membranes. Ligand binding activity could not, however, be demonstrated in the preparation. Peterson et al (1984) used a combination of lectin, ion-exchange, and affinity chromatography to obtain highly purified preparations from porcine atria that retain ligand binding. The preparation contained two main polypeptides, a diffuse band of molecular weight 78,000, which could be affinity-alkylated with [^{3}H]PrBCM, and a second polypeptide of molecular weight 15,000, which was present in less than stochiometric amounts. This preparation exhibited both high affinity antagonist binding and heterogeneity of agonist binding similar to the membrane-bound receptor (the latter is somewhat surprising as the solubilized starting material exhibited a single class of low affinity sites). Haga & Haga (1985) purified the muscarinic receptor from porcine brain by using affinity and gel permeation high pressure liquid chromatography. This preparation consisted primarily of a diffuse polypeptide of 70 kD, which could be affinity-alkylated by [^{3}H]PrBCM, as well as low amounts of a smaller polypeptide of 14 kD. This preparation exhibited high affinity antagonist binding; agonists appeared to bind to a homogeneous population of low affinity sites.

A partial functional reconstitution of the muscarinic receptor was reported by Shreeve et al (1984), who suggested that crude preparations of detergent solubilized cardiac muscarinic receptor preparations could be reconstituted into erythrocyte membranes by dialysis; guanine nucleotide regulation of agonist binding reappeared with reconstitution. The receptors in this preparation did not, however, regain the heterogeneity of agonist binding seen in native membranes, and the effects of guanine nucleotides on agonist affinity were also significantly less than in cardiac membranes. Florio & Sternweis (1985) resolved bovine brain muscarinic receptor from GTP-binding proteins and demonstrated that reconstitution of the resolved receptor into liposomes with either purified N_i or purified N_o restored heterogeneity of agonist binding and guanine nucleotide regulation of agonist binding. Haga et al (1985) showed that purified N_i and purified muscarinic receptor could be reconstituted into liposomes to yield receptor activation of the GTPase activity of N_i. The availability of purified functional preparations of muscarinic receptors and coupling proteins should facilitate studies on the molecular mechanisms involved in receptor action.

Immunology of the Muscarinic Receptor

Two laboratories to date have reported the isolation of monoclonal antibodies specific for the muscarinic receptor. Venter et al (1984) reported the isolation of several antibodies that recognized the receptor. Some of these antibodies recognized the alpha$_1$-adrenergic receptors as well, and the existence of structural homologies between the two receptors was suggested. However, these antibodies were isolated after immunization with a crude immunogen, and immunoblots demonstrating the specificity of these antibodies were not presented. The published data thus cannot exclude the possibility that these antibodies are not specific for the muscarinic receptor itself but may recognize epitopes (such as carbohydrate) present on many proteins.

Two monoclonal antibodies raised after immunization with purified preparations of muscarinic receptors have been described. One antibody recognized denatured receptors and reacted only with a single polypeptide of 70 kD on immunoblots of crude brain membranes. The other antibody recognized native receptors and could immunoprecipitate a polypeptide of 70 kD from solution. Both antibodies exhibited agonist-like activity in intact tissues that could be blocked by atropine. These results suggest that these antibodies exhibit a high degree of specificity for the muscarinic receptor (Lieber et al 1984).

REGULATION AND DEVELOPMENT OF RECEPTOR NUMBER AND FUNCTION

Regulation of Muscarinic Receptor Number and Function in Cell Culture

REGULATION BY EXPOSURE TO AGONIST Like many other hormone and neurotransmitter receptors, the number of muscarinic receptors on cultured cells is decreased by long-term exposure (several hours) to agonists. This agonist-induced decrease in receptor number ("down-regulation") has been postulated to represent a mechanism for the regulation of sensitivity to receptor activation in response to varying levels of physiological activity in vivo. Decreases in receptor number in cell culture have been reported after prolonged agonist exposure of neuronal cell lines (Klein et al 1979, Taylor et al 1979, Shifrin & Klein 1980), heart cells (Galper & Smith 1980), neurons (Siman & Klein 1979, Burgoyne & Pearce 1981), smooth muscle (Takeyasu et al 1981), thyroid cells (Champion & Mauchamp 1982), and AtT-20 pituitary cells (Heisler et al 1985). The decrease in receptor number occurs over the course of several hours and is thus distinct from short-term desensitization of receptor function, which occurs on a time scale of minutes and is not accompanied by a decrease in total cellular receptor number (Taylor et al 1979). The long-term decrease in muscarinic receptor number is accompanied by a decreased physiological sensitivity to cholinergic stimulation. Thus, neuronal cells show a concomitant decrease in receptor-mediated inhibition of adenylate cyclase (Nathanson et al 1978), stimulation of guanylate cyclase (Taylor et al 1979), and stimulation of phosphatidylinositol turnover (Siman & Klein 1981).

Binding of an antagonist to the receptor will not induce a decrease in receptor number and can block the decrease due to simultaneous agonist addition. Thus, simple occupancy of the ligand binding site without receptor activation is not sufficient to promote loss of the muscarinic receptor. In neuronal cells, the apparent half-life of the receptor decreased from 11 to 3 hours after agonist addition (Klein et al 1979). Following removal of agonist, a gradual recovery of receptor number occurs, and this recovery can be prevented by inhibition of de novo protein synthesis (Klein et al 1979, Taylor et al 1979) or by inhibition of protein glycosylation (Liles & Nathanson 1986). Galper et al (1982b) examined the binding of the lipophilic ligand [^{3}H]QNB and the lipophobic ligand [^{3}H]NMS to intact cardiac cells after agonist exposure. Due to the different hydrophobicities of these two ligands, the binding of [^{3}H]NMS to intact cells should label only receptors exposed on the cell surface, whereas [^{3}H]QNB should label the total receptor population. Agonists caused a rapid disappearance of the muscarinic receptor from the cell surface, so that the receptors were no longer available to bind [^{3}H]NMS but could still bind [^{3}H]QNB; [^{3}H]QNB binding sites were lost from intact cells with a slower time course, similar to the loss of binding sites seen in membrane homogenates. Maloteaux et al (1983) and Feigenbaum & El-Fakahany (1985) demonstrated that the rapid loss of [^{3}H]NMS sites from intact neuronal cells was also rapidly reversible and was temperature sensitive. Harden et al (1985) demonstrated that the agonist-induced conversion of the muscarinic receptor to a form that was not accessible to [^{3}H]NMS was accompanied by an alteration in the apparent subcellular localization of the receptor from the plasma membrane to a form that sedimented in sucrose gradients with lighter membrane vesicles. These results provide strong evidence for the hypothesis that agonist-induced decreases in receptor number result from a rapid and initially reversible internalization of the receptor from the cell surface, followed by a slower degradation of the receptor.

REGULATION BY ELECTRICAL ACTIVITY Membrane depolarization has been shown to regulate the expression of a number of important macromolecules in excitable cells. For example, increased electrical activity has pronounced effects on the synthesis of nicotinic acetylcholine receptors (Reiness & Hall 1977). Luqmani et al (1979) demonstrated that depolarization of brain synaptosomes either by electrical stimulation or with veratrine decreased muscarinic receptor number by up to 50%. This decrease was blocked by the sodium channel blocker, tetrodotroxin. Depolarization with high potassium medium did not decrease receptor number, suggesting that sodium entry might be required. Milligan & Strange (1985) studied the effects of sodium channel activators on muscarinic receptors in N1E-115 cells. They reported that incubation of the cells with these drugs decreased the amount of [^{3}H]NMS bound to the cells not by decreasing the actual number of receptors but merely by competing with the radioligand for binding to the receptor. No effect on receptor number was found after incubation with these drugs for up to 6 hours, so Milligan & Strange concluded that there was no effect of electrical activity on receptor number in these cells. Further work has shown that depolarization of N1E-115 cells with either veratridine or high potassium medium caused an 80–100% increase in both total and cell surface number over a 12-hour period, with a 6–8 hour lag before receptor number began to increase. Inhibitors of either protein synthesis or protein glycosylation did not block the increase in receptor number, suggesting that the increase was due not to an increase in the rate of synthesis but rather a decrease in the rate of degradation of the receptor (W. C. Liles and N. M. Nathanson, unpublished observations). The cellular mechanism mediating this effect remains to be determined.

POSSIBLE ROLES OF PHOSPHORYLATION IN RECEPTOR REGULATION Burgoyne (1983 and references therein) has suggested that protein phosphorylation may regulate muscarinic receptor number and function. He reported that incubation of rat bovine cortex membranes for 5 minutes under "phosphorylating conditions" (i.e. in the presence of ATP) decreased [^{3}H]QNB binding by 20–40%. The remaining receptor binding sites also exhibited altered agonist binding properties after this treatment. Because the maximal loss of binding sites required the addition of both cAMP and calmodulin, it was suggested that both cAMP-dependent and calmodulin-dependent protein kinases were involved in receptor down-regulation. The relationship of agonist-induced decreases in receptor number to the loss of binding sites observed here is not clear. It was reported that a polypeptide in crude membrane fractions with an electrophoretic mobility similar to the muscarinic receptor was phosphorylated by incubation of neurons with muscarinic agonists (Burgoyne & Pearce 1981). Because of the very low density of the receptor in these cells, however, it seems unlikely that phosphorylation of the receptor could be detected by SDS gel electrophoresis of crude membrane fractions.

Some evidence suggests a role for the calcium- and phospholipid-dependent enzyme, protein kinase C (PKC), in the regulation of muscarinic receptor number and function. Treatment of N1E-115 cells with the tumor-promoting phorbol esters, which can activate PKC in intact cells, causes a rapid initial internalization and subsequent degradation of the receptor with the same time course as agonist treatment (Liles et al 1986a). Muscarinic agonists cause a rapid activation of PKC that occurs prior to agonist-mediated receptor internalization and persists during subsequent receptor degradation. Because of analogous results seen in other systems (Leeb-Lundberg et al 1985), these results suggest that PKC-mediated phosphorylation of the muscarinic receptor may be involved in agonist-induced receptor loss in these cells. Direct demonstration of similar time courses of receptor loss and phosphorylation are required for confirmation of this hypothesis.

A different role for PKC has been suggested in other cell types. The tumor-promoting phorbol esters can block muscarinic receptor stimulation of phosphoinositide turnover and calcium mobilization in hippocampal slices (Labraca et al 1985), pheochromotcytoma cells (Vincentini et al 1985), adrenal medullary cells (Misbahuddin et al 1985), and astrocytoma cells (Orellana et al 1985). In contrast to the loss of muscarinic receptors seen after phorbol ester treatment of N1E-115 cells, the blockade of receptor function in these studies was not accompanied by changes in either cell surface or total receptor number. Because phorbol esters have been shown to promote the PKC mediated phosphorylation and subsequent inactivation of N_i in lymphoma cells and platelets (Katada et al 1985), it is possible that these effects on muscarinic receptor function are due to analagous phosphorylation and inactivation of the coupling proteins involved in these responses.

Regulation of Muscarinic Receptors in Vivo

REGULATION BY AGONIST AND ANTAGONIST ADMINISTRATION The number of muscarinic receptors can be altered in vivo by a variety of pharmacological treatments. Direct administration of muscarinic agonists decreases receptor number and physiological responsiveness in heart, brain, and spinal cord (Halvorsen & Nathanson 1981, Marks et al 1981, Meyer et al 1982, Taylor et al 1982). Chronic administration of acetylcholinesterase inhibitors results in decreased numbers and physiological responsiveness of receptors in brain, ileum, and rat submandibular gland (Gazit et al 1979, Dawson & Jarrott 1981, Olianas et al 1984, Costa & Murphy 1985). This decrease in receptor number is presumably the result of increased levels of endogenous ACh-promoting-agonist-induced down-regulation of the receptor. Conversely, increased numbers of muscarinic receptors are found after administration of muscarinic antagonists, presumably by preventing synaptically released ACh from interacting with the receptor (Ben-Barak & Dudai 1980, Westlind et al 1981).

The regulation of receptor number by direct administration of agonists in vivo exhibits characteristics similar to down-regulation in cell culture. Kinetic studies and experiments with partial agonists on the loss of muscarinic receptors from the embryonic chick heart indicate that the number of receptors remaining after treatment with various concentrations of agonist is a reflection of the fraction of receptors occupied and activated by that particular concentration of agonist (Halvorsen & Nathanson 1981, Nathanson et al 1984). Receptor number gradually returns to control values if further receptor-agonist interactions are blocked by administration of a muscarinic antagonist (Halvorsen & Nathanson 1981, Meyer et al 1982). This recovery of receptors can be blocked by concomitant administration of protein synthesis inhibitors (Hunter & Nathanson 1984), a finding consistent with experiments in cell culture indicating that the receptors that reappear following agonist-induced decreases in receptor number represent newly synthesized receptors. These newly synthesized receptors in the embryonic chick heart exhibit a diminished physiological responsiveness, suggesting that the muscarinic receptor may be initially synthesized in an immature form and slowly converted to a more physiologically functional form (Hunter & Nathanson 1984).

REGULATION BY INNERVATION The regulation of receptor number by administration of cholinergic drugs suggests that innervation and denervation should also affect receptor number. Lesions of the cholinergic input to various regions of the central nervous system have been reported to result in no change (Kamiya et al 1981), increases (Westlind et al 1981), or decreases (de Belleroche et al 1985) in muscarinic receptor number. Both no change (Burt 1978) and increases (Taniguchi et al 1983) in receptor number have been reported after denervation of sympathetic ganglia. These binding studies on tissue homogenates are difficult to interpret, however. As others have noted previously (Burt 1978), the relative contributions of presynaptic, postsynaptic, and nonneuronal cells cannot be distinguished, and possible changes in the distribution of receptors are also ignored. Ultimate resolution of these problems will probably require ultrastructural localization with the electron microscope.

Many smooth muscles exhibit supersensitivity to ACh after denervation, but the possible role of alterations in muscarinic receptor number is not clear. Increases in receptors after denervation were reported in the guinea pig vas deferens (Hata et al 1980) but not in cat iris (Sachs et al 1979). Electrophysiological studies indicate that changes in resting potential and electrical excitability of the membrane can play an important role in the development of supersensitivity after denervation of at least some smooth muscles (Gerthoffer et al 1979). Denervation of the cholinergic sympathetic nerves innervating the sweat glands results in loss of physiological responsiveness to muscarinic agonists; it is not yet known whether this is due to a decrease in receptor number or a physiological uncoupling of the receptor (Kennedy et al 1984).

A novel example of the regulation of muscarinic receptors by innervation is seen in the expansor secondarium, a muscle in the chick wing that receives only adrenergic innervation. Immediately after hatching, both muscarinic and adrenergic agonists elicit contraction. The physiological sensitivity to ACh but not norepinephrine decreases over the next several weeks. The developmental decrease in muscarinic responsiveness could be prevented by denervation of the muscle (Kuromi & Hagihara 1976). These changes in cholinergic sensitivity are due to changes in the number of receptors present in the muscle (Bennett et al 1982, Rush et al 1982). Organ culture experiments indicate that muscle contraction plays a role in the effect of innervation on receptor number (Gonoi 1980), in a manner that may be analagous to the regulation of nicotinic AChR synthesis in skeletal muscle (Reiness & Hall 1977).

REGULATION BY ELECTRICAL ACTIVITY Decreased numbers of muscarinic receptors in the central nervous system have been reported following the induction of kindling, a model for epilepsy in which the repeated administration of initially subconvulsive electrical stimuli eventually induces seizure activity. This decrease in receptor number has been suggested to be an agonist-independent phenomenon secondary to the increased electrical activity that occurs during seizures (Savage et al 1985, and references therein). Decreased numbers of receptors in the central nervous system in the mouse mutant, *totterer*, an animal model of epilepsy, have been observed with a developmental time course that suggests the changes in receptor number are also secondary to increased neuronal activity (Liles et al 1986b). Direct electrical or pharmacological stimulation can decrease muscarinic receptor number in the central nervous system (Abdul-Ghani et al 1981).

DEVELOPMENTAL CHANGES IN CARDIAC MUSCARINIC RECEPTOR Muscarinic receptors in the chick heart undergo a number of physiological changes during the course of embryonic development. Atria from 3–4 day embryos exhibit a reduced negative chronotropic response (heart rate slowing) to muscarinic agonists than atria from older embryos, even though the density of receptor binding sites is similar in younger and more responsive older hearts (Pappano & Skowronek 1974, Galper et al 1977). Muscarinic receptors at both 3–4 days and older exhibit similar sensitivities for inhibition of adenylate cyclase and stimulation of phosphoinositide turnover (Halvorsen & Nathanson 1984, Orellana & Brown 1985), suggesting that the receptor itself is not altered at 3–4 days. Guanine nucleotide sensitivity of agonist binding to the muscarinic receptor and receptor-independent guanine nucleotide inhibition of adenylate cyclase activity were greatly decreased in younger atria, and the electrophoretic mobility of the GTP-regulatory proteins associated with the receptor also changed during this time (Halvorsen & Nathanson 1984). These results indicate both functional and physical changes in these GTP-binding proteins that occur with the same time course during embryonic development as the onset of the muscarinic receptor negative chronotropic response, and suggest that these changes may contribute to the onset of physiological responsiveness. Because changes in cyclic nucleotides are not involved in coupling receptors to potassium channels in the heart, these results also suggested that the GTP-binding proteins may directly link muscarinic receptors to K^+ channels (see above).

Several other properties of the receptors change during development. Muscarinic receptors in the ventricle inhibit β-adrenergic stimulation of cardiac contractility both before and after hatching, but have direct effects on contractility in the absence of β-adrenergic stimulation only after hatching (Biegon & Pappano 1980). The negative chronotropic response to muscarinic agonists rapidly desensitizes in atria younger than embryonic day 12 but does not desensitize at older ages (Pappano & Skowronek 1974). A significant increase in receptors occurs in atria (but not in ventricles) between embryonic days 12–15, which may be related to the onset of functional cholinergic innervation (Kirby & Aronstam 1983). The amount of at least some of the subunits of the GTP-regulatory proteins are similarly regulated, raising the possibility that both the muscarinic receptor and its coupling proteins may be regulated in a coordinate fashion during development and innervation of the heart (C. W. Luetje and N. M.

Nathanson, unpublished observations). The number of receptors in the embryonic chick ventricle has been reported to increase around the time of hatching (Hosey et al 1985, Sullivan et al 1985). In all these cases, the molecular mechanisms responsible for the developmental changes remains to be determined.

CONCLUSIONS

Great progress has been made in the last several years in purifying both the muscarinic receptor and the coupling proteins required for its action, in beginning studies on the functional reconstitution of the receptor, and in elucidating new mechanisms for the regulation of ion channel function by the muscarinic receptor. Many unanswered questions remain regarding the molecular and cellular mechanisms responsible for the heterogeneity of ligand binding sites, different coupling mechanisms for the various physiological and biochemical responses mediated by the muscarinic receptor, desensitization of receptor function, and regulation of receptor number and function by pharmacological and physiological factors. The use of the full range of physiological, biochemical, immunological, and molecular biological techniques available to the neuroscientist will continue to yield new and exciting information about this important macromolecule.

ACKNOWLEDGMENTS

Research in the author's laboratory has been supported by a Grant-in-Aid and an Established Investigator Award from the American Heart Association, by the National Institutes of Health (HL 30639), and by the US Army Research Office.

Literature Cited

Abdul-Ghani, A. S., Boyar, M. M., Coutinho-Netto, J., Bradford, H. F., Bernie, C. P., Hulme, E. C., Birdsall, N. J. M. 1981. Effect of Tityus toxin and sensory stimulation on muscarinic cholinergic receptors *in vivo*. *Biochem. Pharmacol.* 30: 2713–14

Adams, P. R., Brown, D. A., Constanti, A. 1982. Pharmacological inhibition of the M-current. *J. Physiol.* 332: 223–62

Alberts, P., Bartfai, T. 1976. Muscarinic acetylcholine receptor from rat brain. *J. Biol. Chem.* 251: 1543–47

Andre, C., DeBacker, J. P., Guillet, J. C., Vanderheyden, P., Vauquelin, G., Strosberg, A. D. 1983. Purification of muscarinic acetylcholine receptors by affinity chromatography. *EMBO J.* 2: 499–504

Aronstam, R. S., Greenbaum, L. M. 1984. Guanine nucleotide sensitivity of muscarinic acetylcholine receptors from rat brainstem is eliminated by endogenous proteolytic activity. *Neurosci. Lett.* 47: 131–37

Aronstam, R. S., Schuessler, D. C. Jr., Eldefrawi, M. E. 1978. Solubilization of muscarinic acetylcholine receptors of bovine brain. *Life Sci.* 23: Sci. 23: 1377–82

Avissar, S., Amitai, G., Sokolovsky, M. 1983. Oligomeric structure of muscarinic receptors is shown by photoaffinity labeling: Subunit assembly may explain high- and low-affinity agonist states. *Proc. Natl. Acad. Sci. USA* 80: 156–59

Baron, B., Abood, L. G. 1984. Solubilization

and characterization of muscarinic receptors from bovine brain. *Life Sci.* 35 : 2407–14

Beld, A. J., Ariëns, E. J. 1974. Stereospecific binding as a tool in attempts to localize and isolate muscarinic receptors. Part II. Binding of (+)-benzetimide, (−)-benzetimide and atropine to a fraction from bovine tracheal smooth muscle and to bovine caudate nucleus. *Eur. J. Pharmacol.* 25 : 203–9

Ben-Barak, J., Dudai, Y. 1980. Scopolamine induces an increase in muscarinic receptor level in rat hippocampus. *Brain Res.* 193 : 309–13

Ben-Baruch, G., Schreiber, G., Sokolovsky, M. 1982. Cooperativity pattern in the interaction of the antiestrogen drug clomiphene with the muscarinic receptors. *Molec. Pharmacol.* 21 : 287–93

Benham, C. D., Bolton, T. B., Lang, R. J. 1985. Acetylcholine activates an inward current in single mammalian smooth muscle cells. *Nature* 316 : 345–46

Bennett, T., Lot, T. Y., Strange, P. G. 1982. The effects of noradrenergic denervation on muscarinic receptors of smooth muscle. *Br. J. Pharmacol.* 76 : 177–83

Berridge, M. J., Irvine, R. F. 1984. Inositol trisphosphate, a novel second messenger in cellular signal transduction. *Nature* 312 : 315–21

Berrie, C. P., Birdsall, J. M., Burgen, A. S. V., Hulme, E. C. 1979. Guanine nucleotides modulate muscarinic receptor binding in the heart. *Biochem. Biophys. Res. Commun.* 87 : 1000–5

Berrie, C. P., Birdsall, N. J. M., Hulme, E. C., Keen, M., Stockton, M. J. 1984. Solubilization and characterization of guanine nucleotide-sensitive muscarinic agonist binding sites from rat myocardium. *Br. J. Pharmacol.* 82 : 853–61

Berrie, C. P., Birdsall, N. J. M., Hulme, E. C., Keen, M., Stockton, J. M. 1985. Solubilization and characterization of high and low affinity pirenzepine binding sites from rat cerebral cortex. *Br. J. Pharmacol.* 85 : 697–703

Biegon, R. L., Pappano, A. J. 1980. Dual mechanism for inhibition of calcium-dependent action potentials by acetylcholine in avian ventricular muscle : Relationship to cyclic AMP. *Cir. Res.* 46 : 353–62

Birdsall, N. J. M., Burgen, A. S. V., Hulme, E. C. 1978. The binding of agonists to brain muscarinic receptors. *Molec. Pharmacol.* 14 : 723–36

Birdsall, N. J. M., Burgen, A. S. V., Hulme, E. C. 1979. A study of the muscarinic receptor by gel electrophoresis. *Br. J. Pharmacol.* 66 : 337–42

Birdsall, N. J. M., Burgen, A. S. V., Hulme, E. C., Wong, E. H. F. 1983. The effects of *p*-chloromercuribenzoate on muscarinic receptors in the cerebral cortex. *Br. J. Pharmacol.* 80 : 187–96

Breitwieser, G. E., Szabo, G. 1985. Uncoupling of cardiac muscarinic and β-adrenergic receptors from ion channels by a guanine nucleotide analogue. *Nature* 317 : 538–40

Brown, J. H., Brown, S. L. 1984. Agonists differentiate muscarinic receptors that inhibit cyclic AMP formation from those that stimulate phosphoinositide metabolism. *J. Biol. Chem.* 259 : 3777–81

Brown, J. H., Goldstein, D., Masters, S. B. 1985. The putative M_1 muscarinic receptor does not regulate phosphoinositide hydrolysis. *Molec. Pharmacol.* 27 : 525–31

Burgen, A. S. V., Hiley, C. R., Young, J. M. 1974a. The binding of [^{3}H]-propylbenzilylcholine mustard by longitudinal muscle strips from guinea-pig small intestine. *Br. J. Pharmacol.* 50 : 145–51

Burgen, A. S. V., Hiley, C. R., Young, J. M. 1974b. The properties of muscarinic receptors in mammalian cerebral cortex. *Br. J. Pharmacol.* 51 : 279–85

Burgisser, E., De Lean, A., Lefkowitz, R. J. 1982. Reciprocal modulation of agonist and antagonist binding to muscarinic cholinergic receptor by guanine nucleotide. *Proc. Natl. Acad. Sci. USA* 79 : 1732–36

Burgoyne, R. D. 1983. Regulation of the muscarinic acetylcholine receptor : Effects of phosphorylating conditions on agonist and antagonist binding. *J. Neurochem.* 40 : 324–31

Burgoyne, R. D., Pearce, B. 1981. Muscarinic acetylcholine receptor regulation and protein phosphorylation in primary cultures of rat cerebellum. *Dev. Brain Res.* 2 : 55–63

Burt, D. 1978. Muscarinic receptor binding in rat sympathetic ganglia is unaffected by denervation. *Brain Res.* 143 : 573–79

Carson, S. 1982. Cholate-salt solubilization of bovine brain muscarinic receptors. *Biochem. Pharmacol.* 31 : 1806–9

Carson, S., Godwin, S., Massoulie, J. Kato, G. 1977. Solubilization of atropine-binding material from brain. *Nature* 266 : 176–78

Champion, S., Mauchamp, J. 1982. Muscarinic cholinergic receptors on cultured thyroid cells. II. Carbachol-induced desensitization. *Molec. Pharmacol.* 21 : 66–72

Chandler, D. E., Williams, J. A. 1978. Intracellular divalent cation release in pancreatic acinar cells during stimulus-secretion coupling. *J. Cell Biol.* 76 : 371–85

Cochaux, P., Van Sande, J., Dumont, J. E. 1985. Islet-activating protein discriminates between different inhibitors of thyroidal cyclic AMP system. *FEBS Lett.* 179: 303–6

Cockcroft, S., Gomperts, B. D. 1985. Role of guanine nucleotide binding protein in the activation of polyphosphoinositide phosphodiesterase. *Nature* 314: 534–36

Cohen-Armon, M., Henis, Y. I., Kloog, Y., Sokolovsky, M. 1985a. Interactions of quinidine and lidocaine with rat brain and heart muscarinic receptors. *Biochem. Biophys. Res. Commun.* 127: 326–32

Cohen-Armon, M., Kloog, Y., Henis, Y. I., Sokolovsky, M. 1985b. Batrachotoxin changes the properties of the muscarinic receptor in rat brain and heart: Possible interaction(s) between muscarinic receptors and sodium channels. *Proc. Natl. Acad. Sci. USA* 82: 3524–27

Cole, A. E., Nicoll, R. A. 1984. Characterization of a slow cholinergic postsynaptic potential recorded *in vitro* from rat hippocampal pyramidal cells. *J. Physiol.* 352: 173–88

Costa, L. G., Murphy, S. D. 1985. Characterization of muscarinic cholinergic receptors in the submandibular gland of the rat. *J. Autonom. Nerv. Syst.* 13: 287–301

Creazzo, T. L., Hartzell, H. C. 1985. Reduction of muscarinic acetylcholine receptor number and affinity by an endogenous substance. *J. Neurochem.* 45: 710–18

Cremo, C. R., Schimerlik, M. I. 1984. Photoaffinity labeling of the solubilized, partially purified muscarinic acetylcholine receptor from porcine atria by *p*-azidoatropine methyl iodide. *Biochemistry* 23: 3494–3501

Curtis, B. M., Catterall, W. A. 1985. Phosphorylation of the calcium antagonist receptor of the voltage-sensitive calcium channel by cAMP-dependent protein kinase. *Proc. Natl. Acad. Sci. USA* 82: 2528–32

Dadi, H. K., Morris, R. J. 1984. Muscarinic cholinergic receptor of rat brain. *Eur. J. Biochem.* 144: 617–28

Dale, H. H. 1914. The action of certain esters and esters of choline, and their relation to muscarine. *J. Pharmacol. Exp. Ther.* 6: 147–90

Dascal, N., Ilana, L., Gillo, B., Lester, H. A., Lass, Y. 1985. Acetylcholine and phorbol esters inhibit potassium currents evoked by adenosine and cAMP in *Xenopua* oocytes. *Proc. Natl. Acad. Sci. USA* 82: 6001–5

Dascal, N., Landau, E. M., Lass, Y. 1984. Xenopus oocyte resting potential, muscarinic responses and the role of calcium and guanosine 3′5′-cyclic monophosphate. *J. Physiol.* 352: 551–74

Dawson, R. M., Jarrott, B. 1981. Response of muscarinic cholinoceptors of guinea pig brain and ileum to chronic administration of carbamate or organophosphate cholinesterase inhibitors. *Biochem. Pharmacol.* 30: 2365–68

Diaz-Arrastia, R., Ashizawa, T., Appel, S. H. 1985. Endogenous inhibitor of ligand binding to the muscarinic acetylcholine receptor. *J. Neurochem.* 44: 622–28

de Belleroche, J., Gardnier, I. M., Hamilton, M. H., Birdsall, N. J. M. 1985. Analysis of muscarinic receptor concentration and subtypes following lesion of rat substantia innominata. *Brain Res.* 340: 201–9

De Lean, A., Stadel, J. M., Lefkowitz, R. J. 1980. A ternary complex model explains the agonist-specific binding properties of adenylate cyclase-coupled *β*-adrenergic receptor. *J. Biol. Chem.* 255: 7108–17

De Miranda, J. F. R., Scheres, H. M. E., Salden, H. J. M., Beld, A. J., Klaassen, A. B. M., Kuijpers, W. 1985. Muscarinic receptors in rat nasal mucosa are predominantly of the low affinity agonist type. *Eur. J. Pharmacol.* 113: 441–45

Dehaye, J.-P., Winand, J., Poloczek, P., Christophe, J. 1984. Characterization of muscarinic cholinergic receptors on rat pancreatic acini by *N*-[^{3}H]methylscopolamine binding. *J. Biol. Chem.* 259: 294–300

Delhaye, M., Desmet, J. M., Taton, G., De Neef, P., Camus, J. C., Fountaine, J., Waelbroeck, M., Robberecht, P., Christophe, J. 1984. A comparison between muscarinic receptor occupancy, adenylate cyclase inhibition, and inotropic response in human heart. *Naunyn-Schmiederbergs Arch. Pharmacol.* 325: 170–75

Dodd, J., Horn, J. P. 1983. Muscarinic inhibition of sympathetic C neurones in the bullfrog. *J. Physiol.* 334: 271–91

Dun, N. J., Kaibara, K., Karczmar, A. G. 1978. Muscarinic and cGMP induced membrane potential changes: Differences in electrogenic mechanisms. *Brain Res.* 150: 658–61

Dunlap, J., Brown, J. H. 1983. Heterogeneity of binding sites on cardiac muscarinic receptors induced by the neuromuscular blocking agents gallamine and pancuronium. *Mol. Pharmacol.* 24: 15–22

Dunlap, J., Brown, J. H. 1984. Differences and similarities in muscarinic receptors of rat heart and retina: Effects of agonists, guanine nucleotides, and *N*-ethylmaleimide. *J. Neurochem.* 43: 214–20

Eckelman, W. C., Reba, R. C., Rzeszotarski, W. J., Gibson, R. E., Hill, T., Holman, B.

L. Budinger, T., Conklin, J. J., Eng, R., Grissom, M. P. 1984. External imaging of cerebral muscarinic acetylcholine receptors. *Science* 223: 291–93

Ehlert, F. J. 1985. The relationship between muscarinic receptor occupancy and adenylate cyclase inhibition in the rabbit myocardium. *Molec. Pharmacol.* 28: 410–21

Ehlert, F. J., Jenden, D. J. 1985. The binding of a 2-chloroethylamine derivative of oxotremorine (BM 123) to muscarinic receptors in the rat cerebral cortex. *Mol. Pharmacol.* 28: 107–19

El-Fakahany, E., Richelson, E. 1983. Effect of some calcium antagonists on muscarinic receptor-mediated cyclic GMP formation. *J. Neurochem.* 40: 705–10

Endoh, M., Maruyama, M., Iijima, T. 1985. Attenuation of muscarinic cholinergic inhibition by islet-activating protein in the heart. *Am. J. Physiol.* (Heart Circ. Physiol. 18) 249: H309–20

Evans, T., Martin, M. W., Hughes, A. R., Harden, T. K. 1984. Guanine nucleotide-sensitive, high affinity binding of carbachol to muscarinic cholinergic receptors of 1321N1 astrocytoma cells is insensitive to pertussis toxin. *Molec. Pharmacol.* 27: 32–37

Feigenbaum, P., El-Fakahany, E. E. 1985. Regulation of muscarinic cholinergic receptor density in neuroblastoma cells by brief exposure to agonist: Possible involvement in desensitization of receptor function. *J. Pharmacol. Exp. Ther.* 233: 134–40

Fisher, S. K., Klinger, P. D., Agranoff, B. W. 1983. Muscarinic agonist binding and phospholipid turnover in brain. *J. Biol. Chem.* 258: 7358–63

Florio, V. A., Sternweis, P. C. 1985. Reconstitution of resolved muscarinic cholinergic receptors with purified GTP-binding proteins. *J. Biol. Chem.* 260: 3477–83

Furchgott, R. F., Zawadzki, J. V. 1980. The obligatory role of endothelial cells in the relaxation of arterial smooth muscle by acetylcholine. *Nature* 288: 373–76

Gainer, M. W., Nathanson, N. M. 1986. Recognition of muscarinic acetylcholine receptor ligands by monoclonal antibodies against propylbenzilylcholine mustard. *Biochem. Pharmacol.* 35: 1209–12

Galper, J. B., Dziekan, L. C., Miura, D. S., Smith, T. W. 1982a. Agonist-induced changes in the modulation of K^+ permeability and beating rate by muscarinic agonists in cultured heart cells. *J. Gen. Physiol.* 80: 231–56

Galper, J. B., Dziekan, L. C., O'Hara, D. S., Smith, T. W. 1982b. The biphasic response of muscarinic cholinergic receptors in cultured heart cells to agonists. *J. Biol. Chem.* 257: 10344–56

Galper, J. B., Klein, W., Catterall, W. A. 1977. Muscarinic acetylcholine receptors in developing chick heart. *J. Biol. Chem.* 252: 8692–99

Galper, J. B., Smith, T. W. 1980. Agonist and guanine nucleotide modulation of muscarinic cholinergic receptors in cultured heart cells. *J. Biol. Chem.* 255: 9571–79

Gazit, H., Silman, I., Dudai, Y. 1979. Administration of an organophosphate causes a decrease in muscarinic receptor levels in rat brain. *Brain Res.* 174: 351–56

Gavish, M., Sokolovsky, M. 1982. Solubilization of muscarinic acetylcholine receptor by zwitterionic detergent from rat brain cortex. *Biochem. Biophys. Res. Commun.* 109: 819–24

Gerthoffer, W. T., Fedan, J. S., Westfall, D. P., Goto, K., Fleming, W. W. 1979. Involvement of the sodium-potassium pump in the mechanism of postjunctional supersensitivity of the vas deferens of the guinea pig. *J. Pharmacol. Exper. Ther.* 210: 27–36

Gil, D. W., Wolfe, G. G. 1984. Pirenzepine distinguishes between muscarinic receptor-mediated phosphoinositide breakdown and inhibition of adenylate cyclase. *J. Pharmacol. Exper. Ther.* 232: 608–16

Giles, W., Noble, J. J. 1976. Changes in membrane currents in bullfrog atrium produced by acetylcholine. *J. Physiol.* 261: 103–23

Gonoi, T. 1980. Changes in cholinergic and adrenergic responses of organ-cultured chick smooth muscle. *Eur. J. Pharmacol.* 68: 287–93

Gorissen, H., Aerts, G., Laduron, P. 1978. Characterization of digitonin-solubilized muscarinic receptor from rat brain. *FEBS Lett.* 96: 64–68

Griffith, T. M., Edwards, D. H., Lewis, M. J., Newby, A. C., Henderson, A. H. 1984. The nature of endothelium-derived vascular relaxant factor. *Nature* 308: 645–47

Gurwitz, D., Kloog, Y., Sokolovsky, M. 1985. High affinity binding of [^{3}H]acetylcholine to muscarinic receptors regional distribution and modulation by guanine nucleotides. *Molec. Pharmacol.* 28: 297–305

Haga, K., Haga, T. 1985. Purification of the muscarinic acetylcholine receptor from porcine brain. *J. Biol. Chem.* 260: 7927–35

Haga, K., Haga, T., Ichiyama, A., Katada, T., Kurose, H., Ui, M. 1985. Functional reconstitution of purified muscarinic receptors and inhibitory guanine nucleotide regulatory protein. *Nature* 316: 731–33

Halvorsen, S. W., Engel, B., Hunter, D. D., Nathanson, N. M. 1983. Development and regulation of cardiac muscarinic acetylcholine receptor number, function, and guanyl nucleotide sensitivity. In *Myocardial Injury*, ed. J. J. Spitzer, pp. 143–58. New York: Plenum

Halvorsen, S. W., Nathanson, N. M. 1981. *In vivo* regulation of muscarinic acetylcholine receptor number and function in embryonic chick heart. *J. Biol. Chem.* 256: 7941–48

Halvorsen, S. W., Nathanson, N. M. 1984. Ontogenesis of physiological responsiveness and guanine nucleotide sensitivity of cardiac muscarinic receptors during chick embryonic development. *Biochemistry* 23: 5813–21

Hammer, R., Berrie, C. P., Birdsall, N. J. M., Burgen, A. S. V., Hulme, E. C. 1980. Pirenzepine distinguishes between different subclasses of muscarinic receptors. *Nature* 283: 90–92

Hammer, R., Giachetti, A. 1982. Muscarinic receptor subtypes: M_1 and M_2 biochemical and functional characterization. *Life Sci.* 31: 2991–98

Harden, T. K., Meeker, R. B., Martin, M. W. 1983. Interaction of a radiolabeled agonist with cardiac muscarinic cholinergic receptors. *J. Pharmacol. Exp. Ther.* 227: 570–77

Harden, T. K., Petch, L. A., Traynelis, S. F., Waldo, G. L. 1985. Agonist-induced alteration in the membrane form of muscarinic cholinergic receptors. *J. Biol. Chem.* 260: 13060–66

Harden, T. K., Scheer, A. G., Smith, M. M. 1982. Differential modification of the interaction of cardiac muscarinic cholinergic amd β-adrenergic receptors with a guanine nucleotide binding component(s). *Mol. Pharmacol.* 21: 570–80

Hartzell, H. C. 1980. Distribution of muscarinic acetylcholine receptors and presynaptic nerve terminals in amphibian heart. *J. Cell Biol.* 86: 6–20

Hartzell, H. C. 1984. Phosphorylation of C-protein in intact amphibian cardiac muscle. *J. Gen. Physiol.* 83: 563–88

Hartzell, H. C., Kueffler, S. W., Stickgold, R., Yoshikama, D. 1977. Synaptic excitation and inhibition resulting from direct action of acetylcholine on two types of chemoreceptors on individual amphibian parasympathetic neurones. *J. Physiol.* 271: 317–46

Hartzell, H. C., Titus, L. 1982. Effects of cholinergic and adrenergic agonists on phosphorylation of a 165,000 dalton myofibrillar protein in intact cardiac muscle. *J. Biol. Chem.* 257: 2111–20

Hata, F., Takeyasu, K., Morikawa, Y., Lai, R. T., Ishida, H., Yoshida, H. 1980. Specific changes in the cholinergic system in guinea-pig vas deferens after denervation. *J. Pharmacol. Exp. Ther.* 215: 716–22

Hazeki, O., Ui, M. 1981. Modification by islet-activating protein of receptor-mediated regulation of cyclic AMP accumulation in isolated rat heart cells. *J. Biol. Chem.* 256: 2856–62

Heisler, S., Desjardins, D., Nguyen, M. H. 1985. Muscarinic receptors in mouse pituitary tumor cells: Prolonged agonist pretreatment decreases receptor content and increases forskolin- and hormone-stimulated cyclic AMP synthesis and adrenocorticotropin secretion. *J. Pharmacol. Exp. Ther.* 232: 232–38

Hill-Smith, I., Purves, R. D. 1978. Synaptic delay in the heart: An ionophoretic study. *J. Physiol.* 279: 31–54

Hokin, M. R., Hokin, L. E. 1954. Effects of acetylcholine on phospholipids in the pancreas. *J. Biol. Chem.* 209: 549–58

Hootman, S. R., Picado-Leonard, T. M. 1985. Effect of proteolytic cleavage on functional properties of muscarinic acetylcholine receptors in rat pancreatic and parotid acinar cells. *Biochem. J.* 231: 617–22

Hootman, S. R., Picado-Leonard, T. M., Burnham, D. B. 1985. Muscarinic acetylcholine receptor structure in acinar cells of mammalian exocrine glands. *J. Biol. Chem.* 260: 4186–94

Hosey, M. M., McMahon, K. K., Danckers, A. M., O'Callahan, C. M., Wong, J., Green, R. D. 1985. Differences in the properties of muscarinic cholinergic receptors in the developing chick myocardium. *J. Pharmacol. Exp. Ther.* 232: 795–801

Hughes, A. R., Martin, M. W., Harden, T. K. 1984. Pertussis toxin differentiates between two mechanisms of attenuation of cyclic AMP accumulation by muscarinic cholinergic receptors. *Proc. Natl. Acad. Sci. USA* 81: 5680–84

Hulme, E. C., Berrie, C. P., Birdsall, N. J. M., Burgen, A. S. V. 1981. Two populations of binding sites for muscarinic antagonists in the rat heart. *Eur. J. Pharmacol.* 73: 137–42

Hulme, E. C., Berrie, C. P., Birdsall, N. J. M., Jameson, M., Stockton, J. M. 1983. Regulation of muscarinic agonist binding by cations and guanine nucleotides. *Eur. J. Pharmacol.* 94: 59–72

Hulme, E. C., Birdsall, N. J. M., Burgen, A. S. V., Mehta, P. 1978. The binding of antagonists to brain muscarinic receptors. *Mol. Pharmacol.* 14: 737–50

Hunter, D. D., Nathanson, N. M. 1984. Decreased physiological sensitivity mediated by newly synthesized muscarinic acetylcholine receptors in the embryonic chick heart. *Proc. Natl. Acad. Sci. USA* 81: 3582–86

Hunter, D. D., Nathanson, N. M. 1985. Assay of muscarinic acetylcholine receptor function in cultured cardiac cells by stimulation of $^{86}Rb^+$ efflux. *Analyt. Biochem.* 149: 392–98

Hurko, O. 1978. Specific [^{3}H]quinuclidinyl benzilate binding activity in digitonin-solubilized preparations from bovine brain. *Arch. Biochem. Biophys.* 190: 434–45

Jagadeesh, G., Sulakhe, P. V. 1985. Gallamine binding to heart M_2 cholinergic receptors does not antagonize cholinergic inhibition of adenylate cyclase in isolated plasma membrane. *Eur. J. Pharmacol.* 109: 311–15

Jakobs, K. H., Aktories, K., Schultz, G. 1979. GTP-dependent inhibition of cardiac adenylate cyclase by muscarinic cholinergic agonists. *Naunyn-Schmiedeberg's Arch. Pharmacol.* 310: 113–19

James, W. L., Klein, W. L. 1982. Autoradiography of dendritic acetylcholine receptors: A method for study of isolated neurons of the adult central nervous system in the turtle. *Neurosci. Lett.* 32: 5–10

Järv, J., Hedlund, B., Bartfai, T. 1979. Isomerization of the muscarinic receptor-antagonist complex. *J. Biol. Chem.* 254: 5595–98

Järv, J., Hedlund, B., Bartfai, T. 1980. Kinetic studies on muscarinic antagonist-agonist competition. *J. Biol. Chem.* 255: 2649–51

Kameyama, M., Hofmann, F., Trautwein, W. 1985. On the mechanism of β-adrenergic regulation of the Ca channel in the guinea-pig heart. *Pflügers Archiv.* 405: 285–93

Kamiya, H., Rotter, A., Jacobowitz, D. M. 1981. Muscarinic receptor binding following cholinergic nerve lesions of the cingulate cortex and hippocampus of the rat. *Brain Res.* 209: 432–39

Kao, L.-S., Schneider, A. S. 1985. Muscarinic receptors on bovine chromaffin cells mediate a rise in cytosolic calcium that is independent of extracellular calcium. *J. Biol. Chem.* 260: 2019–22

Katada, T., Gilman, A. G., Watanabe, Y., Bauer, S., Jakobs, K. H. 1985. Protein kinase C phosphorylates the inhibitory guanine-nucleotide-binding regulatory component and apparently suppresses its function in hormonal inhibition of adenylate cyclase. *Eur. J. Biochem.* 151: 431–37

Kellar, K. J., Martino, A. M., Hall, D. P. Jr., Schwartz, R. D., Taylor, R. L. 1985. High-affinity binding of [^{3}H]acetylcholine to muscarinic cholinergic receptors. *J. Neurosci.* 5: 1577–82

Kennedy, W. R., Sakuta, M., Quick, D. C. 1984. Rodent eccrine sweat glands: A case of multiple efferent innervation. *Neuroscience* 11: 741–49

Kirby, M. L., Aronstam, R. S. 1983. Atropine-induced alterations of normal development of muscarinic receptors in the embryonic chick heart. *J. Molec. Cell Cardiol.* 15: 685–96

Klein, W. L., Nathanson, N. M., Nirenberg, M. 1979. Muscarinic acetylcholine receptor regulation by accelerated rate of receptor loss. *Biochem. Biophys. Res. Commun.* 90: 506–12

Korn, S. J., Martin, M. W., Harden, T. K. 1983. *N*-ethylmaleimide-induced alteration in the interaction of agonists with muscarinic cholinergic receptors of rat brain. *J. Pharmacol. Exp. Ther.* 224: 118–26

Kuno, T. Shirakawa, O., Tanaka, C. 1983. Regulation of the solubilized bovine cerebral cortex muscarinic receptor by GTP and Na^+. *Biochem. Biophys. Res. Commun.* 112: 948–53

Kuromi, H., Hagihara, Y. 1976. Influence of sympathetic nerves on development of responsiveness of the chick smooth muscle to drugs. *Eur. J. Pharmacol.* 36: 55–59

Kurose, H., Katada, T., Amano, T., Ui, M. 1983. Specific uncoupling by islet-activating protein, pertussis toxin, of negative signal transduction via α-adrenergic, cholinergic, and opiate receptors in neuroblastoma X glioma hybrid cells. *J. Biol. Chem.* 258: 4870–75

Labarca, R., Janowsky, A., Patel, J., Paul, S. M. 1984. Phorbol esters inhibit agonist-induced [^{3}H] inositol-1-phosphate accumulation in rat hippocampal slices. *Biochem. Biophys. Res. Commun.* 123: 703–9

Lai, W. S., Ramkumar, V., El-Fakahany, E. E. 1985. Possible allosteric interaction of 4-aminopyridine with rat brain muscarinic acetylcholine receptors. *J. Neurochem.* 44: 1936–42

Large, T. H., Rauh, J. J., DeMello, F. G., Klein, W. L. 1985. Two molecular weight forms of muscarinic acetylcholine receptors in the avian central nervous system: Switch in predominant form during differentiation of synapses. *Proc. Natl. Acad. Sci. USA* 82: 8785–89

Lee, J.-H., El-Fakahany, E. E. 1985a. Heterogeneity of binding of muscarinic receptor antagonists in rat brain homogenates. *J. Pharmacol. Exp. Ther.* 233: 707–14

Lee, J.-H., El-Fakahany, E. E. 1985b. [^{3}H]*N*-

Methylscopolamine binding to muscarinic receptors in intact adult rat brain cell aggregates. *Biochem. Pharmacol.* 34: 4299–4303

Leeb-Lundberg, L. M. F., Cotecchia, S., Lomasney, J. W., DeBernadis, J. F., Lefkowitz, R. J., Caron, M. C. 1985. Phorbol esters promote α_1-adrenergic receptor phosphorylation and receptor uncoupling from inositol phospholipid metabolism. *Proc. Natl. Acad. Sci. USA* 82: 5651–55

Lieber, D., Harbon, S., Guillet, J. G., Andre, C., Strosberg, A. D. 1984. Monoclonal antibodies to purified muscarinic receptor display agonist-like activity. *Proc. Natl. Acad. Sci. USA* 81: 4331–34

Liles, W. C., Hunter, D. D., Meier, K. E., Nathanson, N. M. 1986a. Activation of protein kinase C induces rapid internalization and subsequent degradation of muscarinic acetylcholine receptors in neuroblastoma cells. *J. Biol. Chem.* 261: 5307–13

Liles, W. C., Nathanson, N. M. 1986. Regulation of neuronal muscarinic receptor number by protein glycosylation. *J. Neurochem.* 46: 89–95

Liles, W. C., Taylor, S., Finnell, R., Lai, H., Nathanson, N. M. 1986b. Decreased muscarinic acetylcholine receptor number in the central nervous system of the tottering (*tg/tg*) mouse. *J. Neurochem.* 46: 977–82

Loewi, O. 1921. Über humorale Übertragbarkeit der Herznervenwirkung. *Pflügers Arch. Gesamte Physiol.* 189: 239–42

Lundberg, J. M., Hedlund, B., Bartfai, T. 1982. Vasoactive intestinal polypeptide enhances muscarinic ligand binding in cat submandibular salivary gland. *Nature* 295: 147–49

Luqmani, Y. A., Bradford, H. F., Birdsall, N. J. M., Hulme, E. C. 1979. Depolarization-induced changes in muscarinic cholinergic receptors in synaptosomes. *Nature* 277: 481–83

Luthin, G. R., Wolfe, B. B. 1984a. Comparison of [^{3}H]pirenzepine and [^{3}H]quinuclidinyl-benzilate binding to muscarinic cholinergic receptors in rat brain. *J. Pharmacol. Exp. Ther.* 228: 648–55

Luthin, G. R., Wolfe, B. B. 1984b. [^{3}H]Pirenzepine and [^{3}H]quinuclidinyl binding to brain muscarinic cholinergic receptors—Differences in measured receptor density are not explained by differences in receptor isomerization. *Molec. Pharmacol.* 26: 164–69

Luthin, G. R., Wolfe, B. B. 1985. Characterization of [^{3}H]pirenzepine binding to muscarinic cholinergic receptors solubilized from rat brain. *J. Pharmacol. Exp. Ther.* 234: 37–44

Maloteaux, J. M., Gossuin, A., Pauwels, P. J., Laduron, P. M. 1983. Short-term disappearance of muscarinic cell surface receptors in carbachol-induced desensitization. *FEBS Lett.* 156: 103–7

Marks, M. J., Artman, L. D., Patinkin, D. M., Collins, A. C. 1981. Cholinergic adaptations to chronic oxotremorine infusion. *J. Pharmacol. Exp. Ther.* 218: 337–43

Martin, J. M., Hunter, D. D., Nathanson, N. M. 1985a. Islet activating protein inhibits physiological responses evoked by cardiac muscarinic acetylcholine receptors. Role of GTP-binding proteins in regulation of potassium permeability. *Biochemistry* 24: 7521–25

Martin, M. W., Evans, T., Harden, T. K. 1985b. Further evidence that muscarinic cholinergic receptors of 1321N1 astrocytoma cells couple to a guanine nucleotide regulatory protein that is not N_i. *Biochem. J.* 229: 539–44

Maruyama, Y., Petersen, O. H., Flanagan, P., Pearson, G. T. 1983. Quantification of Ca^{2+}-activated K^+ channels under hormonal control in pig pancreas acinar cells. *Nature* 305: 228–32

Masters, S. B., Harden, T. K., Brown, J. H. 1984. Relationships between phosphoinositide and calcium responses to muscarinic agonists in astrocytoma cells. *Molec. Pharmacol.* 26: 149–55

McKinney, M., Stenstrom, S., Richelson, E. 1984. Muscarinic responses and binding in a murine neuroblastoma clone (N1E-115)-selective loss with subculturing of the low-affinity agonist site mediating cyclic GMP formation. *Molec. Pharmacol.* 26: 156–63

McKinney, M., Stenstrom, S., Richelson, E. 1985. Muscarinic responses and binding in a murine neuroblastoma clone (N1E-115): Mediation of separate responses by high affinity and low affinity agonist-receptor conformations. *Molec. Pharmacol.* 27: 223–35

Meyer, M. R., Gainer, M. W., Nathanson, N. M. 1982. *In vivo* regulation of muscarinic cholinergic receptors in embryonic chick brain. *Molec. Pharmacol.* 21: 280–86

Milligan, G., Strange, P. G. 1985. Muscarinic acetylcholine receptors in neuroblastoma cells: Lack of effect of *Veratrum* alkaloids on receptor number. *J. Neurochem.* 43: 33–41

Misbahuddin, M., Isosaki, M., Houchi, H., Oka, M. 1985. Muscarinic receptor-mediated increase in cytoplasmic free Ca^{2+} in isolated bovine adrenal medullary cells. *FEBS Lett.* 190: 25–28

Morita, K., North, R. A., Tokimasa, T. 1982. Muscarinic agonists inactivate potassium conductances of guinea-pig myenteric neurones. *J. Physiol.* 333: 125–39

Murphy, K. M. M., Sastre, A. 1983. Obligatory role of a tris/choline allosteric site in guanine nucleotide regulation of [^{3}H]-L-QNB binding to muscarinic receptors. *Biochem. Biophys. Res. Commun.* 113: 280–85

Nakamura, T., Ui, M. 1985. Simultaneous inhibitions of inositol phospholipid breakdown, arachidonic acid release, and histamine secretion in mast cells by islet-activating protein, pertussis toxin. *J. Biol. Chem.* 260: 3584–93

Nargeot, J., Lester, H. A., Birdsall, N. J. M., Stockton, J., Wassermann, N. H., Erlanger, B. F. 1982. A photoisomerizable muscarinic antagonist. Studies of binding and of conductance relaxations in frog heart. *J. Gen. Physiol.* 79: 657–78

Nathanson, N. M. 1983. Binding of agonists and antagonists to muscarinic receptors on intact cultured heart cells. *J. Neurochem.* 41: 1545–49

Nathanson, N. M., Holttum, J., Hunter, D. D., Halvorsen, S. W. 1984. Partial agonist activity of oxotremorine at muscarinic acetylcholine receptors in the embryonic chick heart. *J. Pharmacol. Exp. Ther.* 229: 455–58

Nathanson, N. M., Klein, W. L., Nirenberg, M. 1978. Regulation of adenylate cyclase activity mediated by muscarinic acetylcholine receptors. *Proc. Natl. Acad. Sci. USA* 75: 1788–91

Nawrath, H. 1977. Does cyclic GMP mediate the negative inotropic effect of acetylcholine in the heart? *Nature* 267: 72–74

Nishizuka, Y. 1984. The role of protein kinase C in cell surface signal transduction and tumour promotion. *Nature* 308: 693–98

Nordström, Ö, Bartfai, T. 1981. 8-Br-cyclic GMP mimics activation of muscarinic autoreceptor and inhibits acetylcholine release from rat hippocampal slices. *Brain Res.* 213: 467–71

North, R. A., Tokimasa, T. 1983. Depression of calcium-dependent potassium conductance of guinea-pig myenteric neurones by muscarinic agonists. *J. Physiol.* 342: 253–66

Nowak, L. M., MacDonald, R. L. 1983. Muscarine-sensitive voltage-dependent potassium current in cultured murine spinal cord neurons. *Neurosci. Lett.* 35: 85–91

Oettling, G., Schmidt, H., Drews, U. 1985. The muscarinic receptor of chick embryo cells: Correlation between ligand binding and calcium mobilization. *J. Cell Biol.* 100: 1073–81

Ohsako, S., Deguchi, T. 1984. Receptor-mediated regulation of calcium mobilization and cyclic GMP synthesis in neuroblastoma cells. *Biochem. Biophys. Res. Commun.* 122: 333–39

Olinas, M. C., Onali, P., Schwartz, J. P., Neff, N. H., Costa, E. 1984. The muscarinic receptor adenylate cyclase complex of rat striatum: Desensitization following chronic inhibition of acetylcholinesterase activity. *J. Neurochem.* 42: 1439–43

Orellana, S. A., Brown, J. H. 1985. Stimulation of phosphoinositide hydrolysis and inhibition of cyclic AMP formation by muscarinic agonists in developing chick heart. *Biochem. Pharmacol.* 34: 1321–24

Orellana, S. A., Solski, P. A., Brown, J. H. 1985. Phorbol ester inhibits phosphoinositide hydrolysis and calcium mobilization in cultured astrocytoma cells. *J. Biol. Chem.* 260: 5236–39

Oron, Y., Dascal, N., Nadler, E., Lupu, M. 1985. Inositol 1,4,5-triphosphate mimics muscarinic responses in *Xenopus* oocytes. *Nature* 313: 141–43

Pappano, A. J., Hartigan, P. M., Coutu, M. D. 1982. Acetylcholine inhibits the positive inotropic effect of cholera toxin in ventricular muscle. *Am. J. Physiol.* 243: H434–41

Pappano, A. J., Skowronek, C. A. 1974. Reactivity of chick embryo heart to cholinergic agonists during ontogenesis: Decline in desensitization at the onset of cholinergic transmission. *J. Pharmacol. Exp. Ther.* 191: 109–18

Paton, W. O. M., Rang, H. P. 1965. The uptake of atropine and related drugs by intestinal smooth muscle of the guinea-pig in relation to acetylcholine receptors. *Proc. R. Soc. London Ser. B* 163: 1–44

Peterson, G. L., Herron, G. S., Yamaki, M., Fullerton, D. S., Schimerlik, M. I. 1984. Purification of the muscarinic acetylcholine receptor from porcine atria. *Proc. Natl. Acad. Sci. USA* 81: 4993–97

Peterson, G. L., Schimerlik, M. I. 1984. Large scale preparation and characterization of membrane-bound and detergent-solubilized muscarinic acetylcholine receptor from pig atria. *Preparative Biochem.* 14: 33–74

Pfaffinger, P. J., Martin, J. M., Hunter, D. D., Nathanson, N. M., Hille, B. 1985. GTP-binding proteins couple cardiac muscarinic receptors to a K channel. *Nature* 317: 536–38

Rapoport, R. M., Draznin, M. B., Murad, F. 1983. Endothelium-dependent relaxation in rat aorta may be mediated through cyclic GMP-dependent protein phosphorylation. *Nature* 306: 174–76

Reiness, C. G., Hall, Z. W. 1977. Electrical stimulation of denervated muscles reduces

incorporation of methionine into ACh receptor. *Nature* 231 : 296–301
Rodrigues de Miranda, J. F., Scheres, H. M. E., Salden, H. J. M., Beld, A. J., Klaassen, A. B. M., Kuijpers, W. 1985. Muscarinic receptors in rat nasal mucosa are predominately of the low affinity agonist type. *Eur. J. Pharmacol.* 113 : 441–45
Roof, D. J., Applebury, M. L., Sternweis, P. C. 1985. Relationships within the family of GTP-binding proteins isolated from bovine central nervous system. *J. Biol. Chem.* 260 : 16242–49
Rosenberger, L. B., Yamamura, H. I., Roeske, W. R. 1980. The regulation of cardiac muscarinic cholinergic receptors by isoproterenol. *Eur. J. Pharmacol.* 65 : 129–30
Ruess, K.-P., Liefländer, M. 1979. Action of detergents on covalently labelled, membrane bound muscarinic acetylcholine receptor of bovine nucleus caudatus. *Biochem. Biophys. Res. Commun.* 88 : 627–33
Rush, R. A., Crouch, M. F., Morris, C. P., Gannon, B. J. 1982. Neural regulation of muscarinic receptors in chick expansor secondarium muscle. *Nature* 296 : 569–70
Sachs, D. I., Kloog, Y., Korczyn, A. D., Heron, D. S., Sokolovsky, M. 1979. Denervation, supersensitivity, and muscarinic receptors in the cat iris. *Biochem. Pharmacol.* 28 : 1513–18
Sakmann, B., Noma, A., Trautwein, W. 1983. Acetylcholine activation of single muscarinic K^+ channels in isolated pacemaker cells of the mammalian heart. *Nature* 303 : 250–53
Savage, D. D., Rigsbee, L. C., McNamara, J. O. 1985. Knife cuts of entorhinal cortex: Effects on development of amygdaloid kindling and seizure-induced decrease of muscarinic cholinergic receptors. *J. Neurosci.* 5 : 408–13
Schlegel, W., Wuarin, F., Zbaren, C., Wollheim, C. B., Zahnd, G. R. 1985. Pertussis toxin selectively abolishes hormone induced lowering of cytosolic calcium in GH_3 cells. *FEBS Lett.* 189 : 27–32
Schreiber, G., Henis, Y. I., Sokolovsky, M. 1985a. Analysis of ligand binding to receptors by competition kinetics. *J. Biol. Chem.* 260 : 8789–94
Schreiber, G., Henis, Y. I., Sokolovsky, M. 1985b. Rate constants of agonist binding to muscarinic receptors in rat brain medulla. *J. Biol. Chem.* 260 : 8795–8802
Shifrin, G. S., Klein, W. L. 1980. Regulation of muscarinic acetylcholine receptor concentration in cloned neuroblastoma cells. *J. Neurochem.* 34 : 993–99
Shirakawa, O., Tanaka, C. 1985. Molecular characterization of muscarinic receptor subtypes in bovine cerebral cortex by radiation inactivation and molecular exclusion h.p.l.c. *Br. J. Pharmacol.* 86 : 375–83
Shreeve, S. M., Roeske, W. R., Venter, J. C. 1984. Partial functional reconstitution of the cardiac muscarinic cholinergic receptor. *J. Biol. Chem.* 259 : 12398–12402
Siegel, R. E., Fischbach, G. D. 1984. Muscarinic receptors and responses in intact embryonic chick atrial and ventricular heart cells. *Dev. Biol.* 101 : 346–56
Siman, R. G., Klein, W. L. 1979. Cholinergic activity regulates muscarinic receptors in central nervous system cultures. *Proc. Natl. Acad. Sci. USA* 76 : 4141–45
Siman, R. G., Klein, W. L. 1981. Specificity of muscarinic acetylcholine receptor regulation by receptor activity. *J. Neurochem.* 37 : 1099–1108
Snider, R. M., McKinney, M., Forray, C., Richelson, E. 1984. Neurotransmitter receptors mediate cyclic GMP formation by involvement of arachidonic acid and lipoxygenase. *Proc. Natl. Acad. Sci. USA* 81 : 3905–9
Soejima, M., Noma, A. 1984. Mode of regulation of the ACh-sensitive K-channel by the muscarinic receptor in rabbit atrial cells. *Pflügers Arch.* 400 : 424–31
Sokolovsky, M., Egozi, Y., Avissar, S. 1981. Molecular regulation of receptors: Interaction of β-estradiol and progesterone with the muscarinic system. *Proc. Natl. Acad. Sci. USA* 78 : 5554–58
Sorota, S., Tsuji, Y., Tajima, T., Pappano, A. J. 1985. Pertussis toxin treatment blocks hyperpolarization by muscarinic agonists in chick atrium. *Circul. Res.* 57 : 748–58
Sternweis, P. C., Robishaw, J. D. 1984. Isolation of two proteins with high affinity for guanine nucleotides from membranes of bovine brain. *J. Biol. Chem.* 259 : 13806–13
Stockton, J. M., Birdsall, N. J. M., Burgen, A. S. V., Hulme, E. C. 1983. Modification of the binding properties of muscarinic receptors by gallamine. *Molec. Pharmacol.* 23 : 551–57
Strange, P. G., Birdsall, N. J. M., Burgen, A. S. V. 1977. Occupancy of muscarinic acetylcholine receptors stimulates a guanylate cyclase in neuroblastoma cells. *Biochem. Soc. Trans.* 5 : 189–91
Streb, H., Irvine, R. F., Berridge, M. J., Schulz, I. 1983. Release of Ca^{2+} from a non-mitochondrial intracellular store in pancreatic acinar cells by inositol 1,4,5-triphosphate. *Nature* 306 : 67–69
Study, R. E., Breakfield, X. O., Bartfai, T., Greengard, P. 1978. Voltage-sensitive calcium channels regulate guanosine 3′5′-cyclic monophosphate levels in neuro-

blastoma cells. *Proc. Natl. Acad. Sci. USA* 75: 6295–99

Sullivan, J. K., Sorota, S., Zotter, C., Pappano, A. J. 1985. Is muscarinic receptor number in avian ventricle regulated by vagal innervation. In *Cardiac Morphogenesis*, eds. V. J. Ferrans, G. L. Rosenquist, C. Weinstein, pp. 202–7. New York: Elsevier

Swartz, B. E., Woody, C. D. 1979. Correlated effects of acetylcholine and cyclic guanosine monophosphate on membrane properties of mammalian neocortical neurons. *J. Neurobiol.* 10: 465–88

Syrota, A., Comar, D., Paillotin, G., Dary, J. M., Aumont, M. C., Stulzaft, O., Maziere, B. 1985. Muscarinic cholinergic receptor in the human heart evidenced under physiological conditions by position emission tomography. *Proc. Natl. Acad. Sci. USA* 82: 584–88

Takeyasu, K., Uchida, S., Lai, R. T., Higuchi, H., Noguchi, Y., Yoshida, H. 1981. Regulation of muscarinic acetylcholine receptors and contractility of guinea pig vas deferens. *Life Sci.* 28: 527–40

Taniguchi, T., Kurahashi, K., Fujiwara, M. 1983. Alterations in muscarinic cholinergic receptors after preganglionic denervation of the superior cervical ganglion in cats. *J. Pharmacol. Exp. Ther.* 224: 674–78

Taylor, J. E., El-Fakahany, E., Richelson, E. 1979. Long-term regulation of muscarinic acetylcholine receptors on cultured nerve cells. *Life Sci.* 25: 2181–87

Taylor, J. E., Yaksh, T. L., Richelson, E. 1982. Agonist regulation of muscarinic acetylcholine receptors in rat spinalcord. *J. Neurochem.* 39: 521–24

Torphy, T. J., Zheng, C., Peterson, S. M., Fiscus, R. R., Rinard, G. A., Mayer, S. E. 1985. Inhibitory effect of methacholine on drug-induced relaxation, cyclic AMP accumulation, and cyclic AMP-dependent protein kinase activation in canine tracheal smooth muscle. *J. Pharmacol. Exp. Ther.* 233: 409–17

Trautmann, A., Marty, A. 1984. Activation of Ca-dependent K channels by carbamoylcholine in rat lacrimal glands. *Proc. Natl. Acad. Sci. USA* 81: 611–15

Trautwein, W., Taniguchi, J., Noma, A. 1982. The effect of intracellular cyclic nucleotides and calcium on the action potential and acetylcholine response of isolated cardiac cells. *Pflügers Arch.* 392: 307–14

Uchida, S., Matsumoto, K., Takeyasu, K., Higuchi, H., Yoshida, H. 1982. Molecular mechanism of the effects of guanine nucleotide and sulfhydryl reagent on muscarinic receptors in smooth muscles studied by radiation inactivation. *Life Sci.* 31: 210–9

Venter, J. C. 1983. Muscarinic cholinergic receptor structure: Receptor size, membrane orientation, and absence of major phylogenetic structural diversity. *J. Biol. Chem.* 258: 4842–48

Venter, J. C., Eddy, B., Hall, L. M., Fraser, C. M. 1984. Monoclonal antibodies detect the conservation of muscarinic cholinergic receptor structure from *Drosophila* to human brain and detect possible structural homology with α_1-adrenergic receptors. *Proc. Natl. Acad. Sci. USA* 81: 272–76

Vincentini, L. M., Di Virgilio, F., Ambrosini, A., Pozzan, T., Meldolesi, J. 1985. Tumor promoter phorbol 12-myristate, 13-acetate inhibits phosphoinositide hydrolysis and cytosolic Ca^{2+} rise induced by the activation of muscarinic receptors in PC12 cells. *Biochem. Biophys. Res. Commun.* 127: 310–17

Waelbroeck, M., Robberecht, P., Cuatelain, P., Christophe, J. 1982. Rat cardiac muscarinic receptors. 1. Effects of guanine nucleotides on high- and low-affinity binding sites. *Molec. Pharmacol.* 21: 581–88

Wamsley, J. K., Zarbin, M. A., Birdsall, N. J. M., Kuhar, M. J. 1980. Muscarinic cholinergic receptors: Autoradiographic localization of high and low affinity agonist binding sites. *Brain Res.* 200: 1–12

Wastek, G. J., Lopez, J. R., Richelson, E. 1980. Demonstration of a muscarinic receptor-mediated cyclic GMP-dependent hyperpolarization of the membrane potential of mouse neuroblastoma cells using [^{3}H]tetraphenylphosphonium. *Molec. Pharmacol.* 19: 15–20

Watanabe, A. M., McConnaughey, M. M., Strawbridge, R. A., Fleming, J. W., Jones, L. R., Besch, H. R. Jr. 1978. Muscarinic cholinergic receptor modulation of β-adrenoceptor receptor affinity for catecholamines. *J. Biol. Chem.* 253: 4833–36

Watson, M., Yamamura, H. I., Roeske, W. R. 1983. A unique regulatory profile and regional distribution of [^{3}H]pirenzepine binding in the rat provide evidence for distinct M_1 and M_2 muscarinic receptor subtypes. *Life Sci.* 32: 3001–10

Wenger, D. A., Parthasarthy, N., Aronstam, R. S. 1985. Regional heterogeneity of muscarinic acetylcholine receptors from rat brain is retained after detergent solubilization. *Neurosci. Lett.* 54: 65–70

Wess, J., Lambrecht, G., Moser, U., Mutschler, E. 1984. A comparison of the antimuscarinic effects of pirenzepine and *N*-methylatropine of ganglionic and vascular muscarinic receptors in the rat. *Life Sci.* 35: 553–60

Westlind, A., Grynfarh, M., Hedlund, B., Bartfai, T., Fuxe, K. 1981. Muscarinic supersensitivity induced by septal lesions or chronic atropine treatment. *Brain Res.* 225: 131–41

Wojcikiewicz, R. J. H., Dobson, P. R. M., Brown, B. L. 1984. Muscarinic acetylcholine receptor activation causes inhibition of cyclic AMP accumulation, prolactin and growth hormone secretion in GH_3 rat anterior pituitary tumour cells. *Biochim. Biophys. Acta* 805: 25–29

Yamamura, H. I., Snyder, S. H. 1974a. Muscarinic cholinergic binding in rat brain. *Proc. Natl. Acad. Sci. USA* 71: 1725–29

Yamamura, H. I., Snyder, S. H. 1974b. Muscarinic cholinergic receptor binding in the longitudinal muscle of the guinea-pig ileum with [^{3}H]quinuclidinyl benzilate. *Molec. Pharmacol.* 10: 861–67

Yamamura, H. I., Vickroy, T. W., Gehlert, D. R., Wamsley, J. K., Roeske, W. R. 1985. Autoradiographic localization of muscarinic agonist binding sites in the rat central nervous system with (+)-*cis*-[^{3}H]methyldioxolane. *Brain Res.* 325: 340–44

Zarbin, M. A., Wamsley, J. K., Kuhar, M. J. 1982. Axonal transport of muscarinic cholinergic receptors in rat vagus nerve: High and low affinity agonist receptors move in opposite directions and differ in nucleotide sensitivity. *J. Neurosci.* 2: 934–41

Ann. Rev. Neurosci. 1987. 10: 237–67

AN INTEGRATED VIEW OF THE MOLECULAR TOXINOLOGY OF SODIUM CHANNEL GATING IN EXCITABLE CELLS

Gary Strichartz, Thomas Rando, and Ging Kuo Wang

Anesthesia Research Laboratories, Brigham and Women's Hospital, and Department of Pharmacology, Harvard Medical School, Boston, Massachusetts 02115

INTRODUCTION

The subject of this review is the voltage-gated sodium channel, whose function subserves impulse propagation in most excitable cells. Our objective in the review is a concise analysis of the actions of toxins on the gating of sodium channels. From investigations of toxin actions we learn not only how these drugs act but also how channels work, that there are different types of sodium channels, and of the identity of some chemical constituents of Na channels. We concentrate primarily on toxins of natural origin, consider some toxin derivatives and synthetic drugs, and exclude local anesthetics. The subject was reviewed previously in 1980 (Catterall 1980).

Sodium channels catalyze the passage of Na ions through plasma membranes in directions dictated by the electrochemical potential. Channel-mediated permeability can be characterized by two physiological attributes, gating and selectivity, although these need not correspond to separate structural elements. "Gating" phenomena describe the observed kinetics of ion permeability changes, whether measured macroscopically from large ensembles of channels or as the time-averaged behavior of one single channel. Changes in the channel's conformation are implied by kinetic models of gating (Hodgkin & Huxley 1952b), and small, asymmetric capacity currents measured in axons and muscle membranes pro-

0147–006X/87/0301–0237$02.00

vide direct evidence of intra-membranous charge movements accompanying such conformational changes (Bezanilla & Armstrong 1977, Bekkers et al 1984, Neumcke et al 1976, Nonner 1980). The simplest gating scheme posits three conformations (states) of the Na channel: a nonconducting *resting* state (R), a conducting *open* state (O), and a different nonconducting, *inactivated* state (I) (Figure 1). *R* is favored in hyperpolarized membrane, *I* in depolarized membranes (Hodgkin & Huxley 1952a). For most membrane potentials, *O* occurs only transiently. The defining difference between *R* and *I* states is that *R* can be activated and will convert to *O* in response to rapid membrane depolarization, whereas the *I* normally cannot be activated to a conducting channel. Paralleling these constraints in permeability production are the movements of gating charge. Resting channels have sufficiently mobile charges to provide measurable gating current in response to rapid depolarization; inactivated channels have a much reduced gating current, as though most of the charge were immobilized and the conformational changes accompanying activation were restricted, in rate or extent (Armstrong & Bezanilla 1977, Nonner 1980).

True channel kinetics are undoubtedly more complicated. Many intermediate nonconducting states separate *R* and *O* states (Stimers et al 1985); probably more than one conducting state (Sigworth 1981) and almost certainly multiple pathways to plural inactivated states exist (Armstrong & Bezanilla 1977, Vandenberg & Horn 1984). A more essential discrimination considers the choice between different microscopic channel kinetics that underlie macroscopic currents. In the classical analysis of

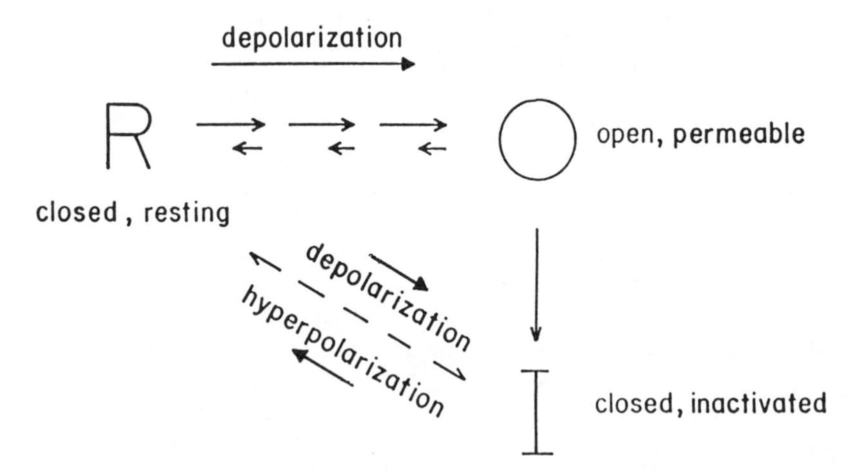

Figure 1 A simple, three-state gating scheme for Na channel kinetics: a nonconducting resting state (R), a conducting open state (O), and a different nonconducting, inactivated state (I).

voltage-clamp records, the rise of currents is taken to represent the synchronous activation of all channels and the decay of currents their synchronous inactivation (Hodgkin & Huxley 1952a, Armstrong & Bezanilla 1977). Modern analyses, based on single channel measurements, indicate that individual channel lifetimes in mammalian neurons are brief compared to the duration of the macroscopic current, and that its shape derives from the dispersion of channel openings following a depolarization (Aldrich et al 1983, Vandenberg & Horn 1984). We return to this question in the section on polypeptide toxins.

"Selectivity" refers to the discrimination among various ions by the channel. Normal channels are highly selective for alkaline metal cations and allow Li^+ through about as easily as Na^+, prefer Na^+ to K^+ by about 10 to 1, and have lower permeability of Rb^+ and almost none to Cs^+ (Chandler & Meves 1965, Hille 1972, Campbell & Hille 1976). A variety of small, monovalent, organic cations also pass through the channels, and their structures provide the basis for a hypothetical "selectivity filter" of narrow dimension and located near the channel's outer opening (Hille 1971). However, the selectivity order can be modified by changes in the internal cation content (Begenisich & Cahalan 1980) and by many of the lipophilic toxins (see below), so it is doubtful that the function of ion discrimination resides in one narrow region of the channel or corresponds uniquely to a single binding site or energy barrier (Hille 1975).

Sodium channels also participate as substrates in biochemical reactions. As expected of proteins in the plasma membrane, they are highly glycosylated. As many as 100 sialic acid residues have been assigned to each channel protein from the electric eel (Miller et al 1983). The physiology of the channel as well as the actions of drugs may be modulated by this dense array of negative charges. Electrical charges on lipids surrounding the channel as well as on the channel itself may also influence drug action. Channels also are acylated (Levinson et al 1986) and are actively phosphorylated by cytoplasmic protein kinase (Costa et al 1982). The physiological consequences of phosphorylation are as yet undemonstrated, but the extensive acylation is probably germane to the binding and actions of lipophilic toxins and anesthetics, as well as the localization and dynamic response of the channel protein within the membrane. For example, esterification of carbonyl residues of the protein provides a means of stabilizing the ester dipoles in the low dielectric core of the membrane. The physiological and pharmacological responses of the Na channel are as intrinsic to the membrane as is the protein itself, and models of structure-function should be cast in the context of the membrane.

Although the sodium channel is generalized here for the sake of pharmacological analysis, it is clear that distinguishably different channels do exist. These "isochannels" are discriminated primarily on pharmacologic grounds; some are insensitive to the guandinium toxins, others are differentially suceptible to alkaloids or peptide toxins.

We have classified the toxins in three categories: (*a*) the peptide toxins, from scorpion venoms for example; (*b*) the broadly inclusive class of lipophilic toxins, including veratrine alkaloids, pyrethroids, and cyclic polyether brevetoxins; and (*c*) the guanidinium toxins, tetrodotoxin and saxitoxin. With few exceptions, this traditional classification respects both chemical properties and pharmacological actions, and in this review we attempt to relate these aspects to each other.

This subject was reviewed most recently from a pharmacological perspective by Catterall (1980) and from a more physiological one by French & Horn (1983).

PEPTIDE NEUROTOXINS: THEIR CHEMISTRY AND PHARMACOLOGY

Sodium channels are strongly affected by various animal venoms, notably those from scorpions, sea anemones, and fish-hunting cone snails. Sodium channel specific toxins isolated from sea anemones and scorpions are small proteins, typically with only about 46–49 and 60–70 amino acid residues, respectively (Rochat et al 1979, 1984, Catterall 1980, Beress 1982, Meves et al 1984b). One smaller sea anemone toxin (ATXIII from *Anemonia sulcata*) with only 27 residues and a few scorpion toxins with 30–40 residues have also been isolated. The μ-conotoxins isolated from *Conus geographus* have only 22 amino acid residues (Olivera et al 1985). No post-translational modifications or unusual amino acids are found in sea anemone or scorpion toxins, whereas in μ-conotoxins unusual amino acids with *trans*-4-hydroxyproline residues and the amination of the C-terminus have been identified. All these neurotoxins contain several disulfide bridges—three for sea anemone, four for scorpion toxins, and three for μ-conotoxins—which render their structure rigid and relatively compact (Fontecilla-Camps et al 1980). Reducing these disulfide bonds appears to destroy their biological activities (Watt & McIntosh 1972, Harbersetzer-Rochat & Sampieri 1976, Rochat et al 1979).

Scorpion Neurotoxins Modulate Na Channel Gating

Two classes of scorpion toxins have been firmly established by toxin binding assays and by their electrophysiological actions. Together, both assays provide convincing evidence that these identified toxins are Na channel specific.

α-SCORPION TOXINS MODULATE Na CHANNEL INACTIVATION GATING The α-scorpion toxins prolong the action potential duration in muscle (Baumgold et al 1983) and in nerve (Koppenhofer & Schmidt 1968a,b). Normal action potentials last only a few milliseconds, whereas in axons treated with α-scorpion toxin, they last as long as several seconds. Included in this class are toxin I (*A. australis* and *B. occitanus*), toxins IV and V (*L. quinquestriatus*), toxins M7 and 2001 (*B. eupeus*), toxin V and var. 1–3 (*C. sculpturatus*), and many others.

Under voltage-clamp, Na inactivation is inhibited by these scorpion toxins, although the total effect can be quite complex (Figure 2A). Firstly, the toxin reverses the apparent voltage dependence of the Na channel

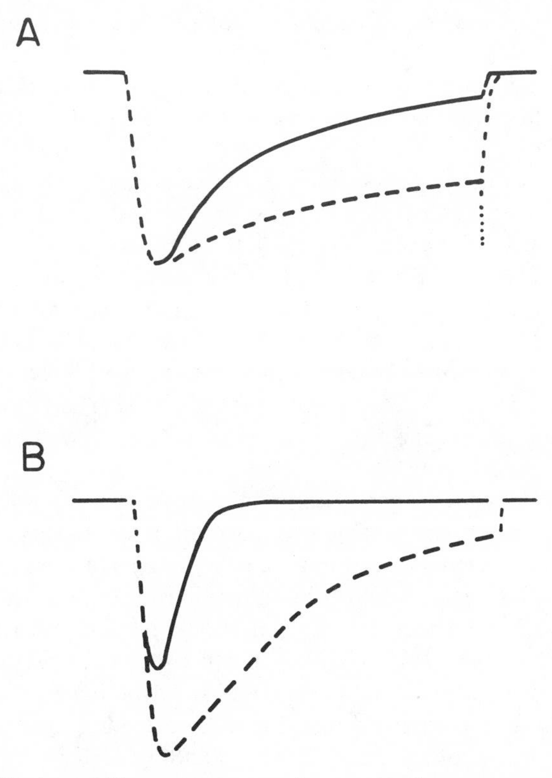

Figure 2 Modifications of Na currents by scorpion α-toxin of *Leiurus quinquestriatus*. *A*. Slowing of inactivation and the elicitation of a non-inactivating conducting state in frog nerve (after Wang & Strichartz 1983). *B*. Increase in peak Na permeability accompanied by slowed inactivation in cultured neuroblastoma cells; steady-state non-inactivating currents are less than 10% of the peak value (Gonoi et al 1984). *Solid lines*, control; *dashed lines*, toxin-treated.

inactivation in myelinated nerve fibers and in invertebrate axons. At more positive potentials, less inactivation than normal occurs, and as a result, larger maintained permeabilities are found as the voltage increases (Koppenhofer & Schmidt 1968b). The steady-state inactivation curve (h_{∞}) has its minimum around -50 mV after toxin treatment. In either direction from this potential, fewer channels are likely to be inactivated. Secondly, α-scorpion toxins also slow both the fast and the slow inactivation components in the node of Ranvier. In general, the fast component is less affected than the slow one (Mozhayeva et al 1980, Wang & Strichartz 1985a). The dose-response curve for the slowing of inactivation is identical to that for the maintained permeability, thus indicating that one bound toxin molecule is responsible for both effects.

Thirdly, the binding of α-scorpion toxin to the Na channel is voltage-dependent. Depolarization of the resting membrane by KCl reduces the toxin-binding affinity of a radiolabelled *Leiurus* toxin (Catterall 1977). Similar voltage-dependent binding was later found for other α-scorpion toxins in various excitable tissues (Catterall 1980, Couraud et al 1982). Originally, it was proposed that the voltage-dependent binding is coupled to the Na channel activation process, since the voltage dependence of the toxin binding closely overlapped that of channel activation (Catterall 1979). However, because the binding experiments required from minutes to hours to reach equilibrium and because the separation of bound from free toxin molecules also requires minutes, Na channels in the depolarized membrane undergo many conformational changes, including activation, fast inactivation, slow and ultraslow inactivation. Therefore, the detailed mechanisms underlying voltage-dependent binding cannot be resolved from such experiments.

Rapid voltage-dependent α-toxin binding has been assayed by voltage-clamp experiments with a rapid time resolution. Mozhayeva and colleagues (1979) first detected a reversal of the effect of an isolated α-scorpion toxin (M7 and M8 of *B. eupeus*) during a depolarization, but this reversal did not correlate with channel activation or inactivation processes. However, since bound toxin slows channel inactivation, they proposed that the voltage-dependent dissociation is coupled to the inactivation transition. Similar effects are observed with the α-toxin from *C. sculpturatus* (toxin V) although the voltage-dependence of binding partially overlaps that of activation (Meves et al 1984a, Strichartz & Wang 1986). The dissociation of toxin from its binding site appears to be relatively fast at large potentials for the above-noted toxins, taking from 50 to 300 msec to complete. In contrast, the rapid dissociation of an isolated *Leiurus* scorpion α-toxin from nerve occurs only at potentials beyond $+20$ mV and is relatively slow for voltages below $+60$ mV (Strichartz & Wang 1986). Application of repetitive depolarizing pulses achieves steady-state reversal (the toxin slowly rebinds between pulses) without decline of Na current due to slow inactivation (Gonoi et al 1984, Strichartz & Wang 1986).

A modulated receptor mechanism accounts for the correlation between the dissociation of different toxins and inactivation of the Na channel (Strichartz & Wang 1986). The scheme proposes that depolarizing conditioning pulses accelerate inactivation and cause subsequent dissociation of the toxin : channel complex, as suggested originally by Mozhayeva et al (1980). Binding of the toxin to an inactivated channel is weak, whereas binding to resting and open channels is strong. The free energy of toxin binding thus increases the energy difference between resting and inactivated channels, stabilizing the non-inactivated states. But a sufficiently large membrane potential can still force inactivation by acting on the relatively weak equivalent electrical charge of this process, which is estimated from these experiments at 0.5 to 0.8 e. Thus, different scorpion α-toxins with varying binding affinities will dissociate from their binding site at different membrane potentials and rates.

The actions of scorpion α-toxins on nerve appear as a slowing of Na channel inactivation and its eventual reversal at large depolarizations (after correction for potential-dependent toxin dissociation). In cultured cells the effects are more complex: The *Leiurus* α-toxin greatly increases the amplitude and slows the inactivation of Na currents in cultured neuroblastoma cells (Figure 2B; Gonoi et al 1984), results that are consistent with brief Na channel openings dispersed throughout the macroscopic current record (see Introduction and Aldrich et al 1983) and prolonged by the bound α-toxin. Assuming an unchanged unit channel conductance, the profound effect of α-toxins on cultured cells must follow from the recruitment of previously non-activatable channels (Renaud et al 1981), perhaps residing in a slowly inactivated population, perhaps existing as a membrane-bound precursor pool and requiring some final "pharmacological annealing" to assume their functional form.

Prolongation of the open state does not account for the total actions of α-toxins, which are both more complex and more exclusive. In cultured cells the addition of simple reagents, such as *N*-bromoacetamide, increases the duration of single-channel openings and also increases the amplitude and delays the time-to-peak of macroscopic Na currents (Patlak & Horn 1982); in contrast, only the amplitude and declining phase kinetics of Na currents are affected by α-toxins. Furthermore, *Leiurus* α-toxin potentiates the steady-state action of lipophilic activators (see section on lipid soluble toxins, below) in cultured cells (Catterall 1977) and in adult nerve (Rando et al 1986), but other agents that also slow channel inactivation (e.g. chloramine-T) do not have this effect. α-Toxins modulate channel gating to produce fast and slow reactions by more complex and, perhaps, tissue-specific mechanisms than the slowing of inactivation per se.

β-SCORPION TOXIN MODULATES THE Na ACTIVATION PROCESS Using crude venom from *Centruroides sculpturatus* scorpion, Cahalan (1975) first demonstrated that an inward Na current appears slowly following transient depolarization of venom-treated axons. At the resting potential (−80 mV), the "venom-induced" Na conductance lasts for fractions of seconds after the "conditioning" depolarization and appears to arise from channels with modified activation transitions that close, from open to resting states, at very slow rates. Channel inactivation recovers almost normally, and thus appears to operate independently of the modified activation process. This conclusion set a limit on the coupling that can exist between the activation and inactivation processes, considered to be independent parallel pathways by Hodgkin & Huxley (1952a).

Somewhat different effects are observed in myelinated nerve fibers with purified β-scorpion toxins from *C. sculpturatus* or *C. suffus suffus*. These toxins are active at concentrations in the nanomolar range, consistent with biochemical binding studies that show a K_D of 0.3 to 3 nM (Couraud et al 1982, Wheeler et al 1983, Meves et al 1982, 1984b). A depolarizing "conditioning" pulse strongly influences the current kinetics during the subsequent test pulse; both activation and the inactivation time-courses are modulated, in contrast to data from whole venom (Hu et al 1983). During a "test" depolarization from a resting membrane, β-toxin-modified sodium currents are slower to activate, require larger depolarizations to activate, and produce a lower maximum macroscopic permeability than control currents. Inactivation is also slowed. Presentation of a conditioning pulse before the test depolarization reverses the effect on activation, which, compared to control currents, is now faster, occurs at a more negative potential, and achieves a greater peak permeability. Inactivation, however, is slowed even further by β-toxin after such conditioning. The forward rate constants for activation appear to be reduced by β-toxin in resting membranes but increased in depolarization-conditioned membranes.

The closing of open channels to the resting state ("deactivation") is markedly slowed by the toxin, and this effect is also increased by depolarization conditioning. Indeed, the number of open channels in the "tail currents" of β-toxin treated nerves increases rapidly after a test depolarization preceded by a conditioning pulse, far faster than the appearance of the "induced current," and is evidence for the existence of multiple open channel states, arrayed in sequence, each with its separate pathway to an inactivated state (Wang & Strichartz 1985b). That multiple open states exist in the node of Ranvier has been previously suggested from fluctuation studies (Sigworth 1981) and in cultured cells by single channel analyses (Nagy et al 1983).

The overall action of β-toxin in depolarization "conditioned" nerves is an acceleration of activation and a slowing of both deactivation and inactivation. The net product is a population of conducting channels that close very slowly at rest and will produce an inward ionic current capable of supporting repetitive impulse firing in response to minimal stimulation.

Slowing of inactivation is achieved by both α- and β-toxins of scorpions. This common effect does not arise from binding to the same site, however, for depolarizing conditioning pulses reverse the effect of α-toxins but enhance that of β-toxins; furthermore, both toxins can bind simultaneously to the same Na channel (Jover et al 1980, Wang & Strichartz 1982, 1983).

γ-TOXIN MODULATES ACTIVATION WITH HIGH AFFINITY Not all scorpion toxins specific for Na channels can be classified as α- or β-scorpion toxins by electrophysiological assays. Even though γ-toxin (isolated from *Tityus serrulatus*) competes with β-scorpion toxin for a common binding site (Wheeler et al 1983), it does not elicit an induced Na current after depolarization, nor does it shift the activation parameter in the depolarized direction (Vijverberg et al 1984). Instead, γ-toxin shifts activation in the hyperpolarized direction, and thereby appears as β-toxin after a conditioning depolarization. Perhaps γ-toxin molecules gain access to sites in resting membranes with which β-toxin normally cannot react. Demonstrably, binding assays cannot distinguish subtle differences in the molecular interactions between various scorpion toxins and their common binding site on the sodium channel.

Sea Anemone Toxins Act Like α-Scorpion Toxins

Sea anemone toxins are among the most studied peptide toxins for the Na channel. With the exception of *Anemonia sulcata* toxin III (ATXIII), the sea anemone toxins are homologous peptides. There is, however, no sequence homology between sea anemone toxins and other isolated scorpion or snake toxins (Beress 1982, Dufton & Rochat 1984). From chemical modifications of these toxins it is concluded that modification of the carboxylate functions of Asp7, 9, and the COOH-terminal Gln47 of ATXII abolishes the toxicity and electrophysiological activity of toxin (Barhanin et al 1981). Despite the loss of toxicity, these derivatives still bind to their receptor, for they are able to compete with the binding of radio-iodinated ATXII. Modification of ε-amino groups and the α-amino function of Gly1 abolishes the toxicity and reduces their binding affinity. ATXII can also be labeled fluorescently and retains its biological activity, although with reduced potency (Barhanin et al 1981, Rack et al 1983).

Like α-scorpion toxin, sea anemone toxin is only active when applied externally (Romey et al 1976). Iontophoretic injection of the positively charged ATXII into the cytoplasm of the crayfish axon elicits no effects on the action potential. The electrophysiological effects of these toxins on macroscopic Na currents in nerve fibers are very similar to those of α-scorpion toxins. They include the inhibition of the Na channel inactivation process (Pelhate et al 1984, Neumcke et al 1985), the presence of Na currents after prolonged depolarization (Warashina & Fujita 1983), and voltage-dependent action (Warashina et al 1981, Strichartz & Wang 1986). In fact, the binding site of sea anemone toxins overlaps with that of some α-scorpion toxins, since both types of toxins compete with each other's binding. This observation might suggest a single binding site, shared by scorpion and anemone α-toxin, that accounts for slowed inactivation. The concept of a single binding site for slowing inactivation is untenable, however, for some anemone α-toxins do not inhibit the binding of the α-scorpion toxin AaHII [e.g. toxin V from *C. sculpturatis* (Wheeler et al 1983)], even though they have α-toxin-like actions on Na currents. This discrepancy between the biochemical binding assay and electrophysiological assay opens the possibility that scorpion toxins may bind to several different sites to produce a common effect and that binding per se does not necessarily alter channel function. A subsequent structural rearrangement of the channel : toxin complex may be required to modify gating, a possibility proposed previously for certain sea anemone toxins (Barhanin et al 1981, Schmidtmayer 1985).

Voltage-dependent binding of anemone toxin may not be detectable directly under voltage-clamp conditions. Prolonged depolarization of crayfish nerve membrane for several minutes reduces the effects of sea anemone toxins in a manner similar to α-scorpion toxins. However, Meves et al (1984a) found only a marginal reversal by voltage of the effects of ATXII in the node of Ranvier. Voltage-dependent binding of ATXII on Na channels in myelinated nerve does occur, but only at very positive potentials ($> +50$ mV) and by repetitive pulses at high frequency (Strichartz & Wang 1986). There are two opinions regarding this discrepancy. The first hypothesizes that at the high concentrations of ATXII required for physiological effect, the rapid on-rate and off-rate combine to produce a rapid equilibration of the toxin binding between depolarizing pulses, so little net reversal of action can be detected. The second opinion is that secondary conformational changes attend the action of anemone toxin after its initial binding, and that voltage-conditioning may reverse these conformational changes but does not produce toxin dissociation. The correct answer may lie between these opinions, depending on the particular toxin and tissue under investigation.

The effects of ATXII on the gating currents in frog myelinated nerve (Neumcke et al 1985) are similar to those of α-scorpion toxins (Nonner 1979). These toxins reduce the "on" charge displacements during depolarization and selectively abolish slower components of the "on" response. In addition, the charge immobilization during depolarization proceeds much more slowly in the toxin-treated fibers. These results can be interpreted as supporting the notion that Na inactivation contributes some charge displacement during depolarization in normal nerve fibers, as proposed for the crayfish axon (Swenson 1983). Since the gating currents were measured in the presence of 300 nM TTX, ATXII and TTX probably bind to different binding sites on the Na channel surface, a conclusion that does not exclude the interaction between these two binding sites (Romey et al 1976, Siemen & Vogel 1983).

μ-Conotoxin Selectively Inhibits Na Currents in Skeletal Muscle

A wide variety of neurotoxins that act specifically on several different kinds of ion channels have been isolated from fish-hunting cones (Olivera et al 1985). μ-Conotoxins selectively inhibit Na channels in skeletal muscle (Cruz et al 1985, Moczydlowski et al 1986b); they compete with ^{3}H-saxitoxin for high affinity receptors in eel electric organ and rat skeletal muscle but have much weaker potencies on neuronal and cardiac Na channels (Moczydlowski et al 1986a, Yanagawa et al 1986).

In addition, μ-conotoxins inhibit macroscopic Na currents in skeletal muscle and single channel currents produced by sodium channels isolated from muscle t-tubule and reconstituted in lipid bilayers. The blocking action of μ-conotoxins on single batrachotoxin-activated Na channels from muscle is quite similar to that of tetrodotoxin (TTX) and saxitoxin (STX) (see below). The inhibition is reversible and voltage-dependent, with a K_D of 100 nM, at 0 mV and 22°C. The forward rate constant is much smaller than that for TTX, probably in part due to the larger mass of the conotoxin molecule (Moczydlowski et al 1986b). As with the guanidinium toxins, no voltage-dependent action of μ-conotoxins is observed directly from the macroscopic current measurements (in BTX-free muscle).

How nerve and muscle Na channels differ at the μ-conotoxin binding site is unknown, but the difference is not detected by STX or TTX. These Na channels derived from different tissues may be coded by two or more different genes (Noda et al 1986), although the primary sequences of the large subunit of the various types of Na channels from rat brain and from eel electric organ are relatively well conserved (Noda et al 1984, Auld et al 1985, Catterall 1986). Alternatively, differences in post-translational modifications of the Na channel may cause structural differences in the toxin binding site. In such cases, tissue-specific Na channels may be created and regulated by excitable cells at different developmental stages.

THE LIPID SOLUBLE TOXINS

The major groups of lipid-soluble toxins affecting Na channels are (*a*) the alkaloid neurotoxins (including batrachotoxin, veratridine, aconitine, and grayanotoxin); (*b*) the pyrethroids (synthetic analogues of the naturally occurring pyrethrins); and (*c*) the brevetoxins. Each group represents a large family of chemically related compounds. Recent work has focused on the most potent members of each family. However, we have much to learn about the chemical nature of the interactions of these toxins with Na channels, and the use of chemically similar compounds that are less potent, have different pharmacological actions, or have been chemically derivatized will certainly contribute to that understanding.

The lipophilic toxins represent the most diverse group of Na channel toxins both in terms of chemical structure and source. From the steroidal alkaloid batrachotoxin and the polycyclic ether brevetoxins, to the cyclopropane-substituted pyrethroids, the one outstanding chemical similarity is their hydrophobicity. It is interesting that with natural origins spanning two phylogenetic kingdoms, these compounds interact so similarly with a single molecular entity, the voltage gated Na channel.

Masutani et al (1981) compared the membrane depolarizing activity of 34 grayanotoxin analogues on frog skeletal muscle. They concluded from the structure-activity relationships that a common feature of active compounds is the presence of three reactive oxygen groups within 5 Å of each other. The authors suggested that two of these would act as proton acceptors and the third as a proton donor. A methyl group in the vicinity of this oxygen triangle also appeared to be essential for activity.

Codding (1982, 1983, 1984) has done structural analyses of veratridine, aconitine, and grayanotoxin. She concluded that the structures of all alkaloids support a model for receptor binding that requires a triangle of reactive oxygens and a proton donor group within 5–6 Å of the triad. The proton donor group is a tertiary nitrogen in batrachotoxin, veratridine, and aconitine, and a hydroxyl group in grayanotoxin.

In support of this model of alkaloid action, reduction of the 3α, 9α hemiketal linkage of batrachotoxin and the 4α, 9α hemiketal linkage of veratridine both drastically reduce the potencies of these compounds in depolarizing membranes (Warnick et al 1975, Ohta et al 1973). In both cases, reduction destroys the triad of reactive oxygens proposed to be essential for activity. Although this model is widely applicable to the alkaloid neurotoxins, there are exceptions. Sabadine and cevacine are veratridine analogues whose chemical structures are identical except that sabadine does not possess the 4α, 9α hemiketal of cevacine; both have veratridine-like activity on frog skeletal muscle (Ohta et al 1973). Grayanotoxin II possesses all the putative necessary elements—a triad of reactive oxygens with a proton donor group (an hydroxyl) in the vicinity—yet it is inactive (Codding 1984).

In this model, a relatively inflexible ring system serves as a framework for specific conformations of reactive oxygens and nitrogens that confer pharmacological activity on the molecules. Such analyses have not been extended to the pyrethroids or brevetoxins. Certainly the brevetoxins are inflexible ring systems with varying patterns of oxygenated substituents and may share certain structural features with the alkaloids. The active pyrethroids do not appear to fit so clearly into this model, but they may also have a less obvious triangle of proton donors and acceptors with the same dimensions specified for the active alkaloids.

Because of their lipophilic nature, these toxins are presumed to bind to the channel at sites buried in the matrix of the lipid bilayer, but the channel protein also may contain hydrophobic pockets in regions close to the membrane surface (Angelides & Brown 1984). Almost all of the lipid-soluble toxins can produce their effects when added to either the extracellular or the cytoplasmic compartment of the nerve, a finding that testifies to the membrane-permeating character of the drugs rather than the locus of their binding sites.

The lipophilic toxins profoundly affect Na channels, modifying virtually every aspect of their physiology: voltage-dependent gating, ion selectivity, and single channel conductance. Likewise, these toxins also modify the interaction of the channel with nearly every other known class of active drug, including polypeptide toxins, local anesthetics, and, recently discovered, the guanidinium toxins. The specific details of this vast array of modifications are summarized below.

The novel characteristics that appear in the presence of lipophilic toxins result from the modification of existing channels and not the creation of new channels. This is shown by the fact that, for every toxin, the modified permeability is noncompetitively inhibited by TTX and that, when studied, a decrease of the unmodified permeability has been shown to parallel an increase of the modified permeability.

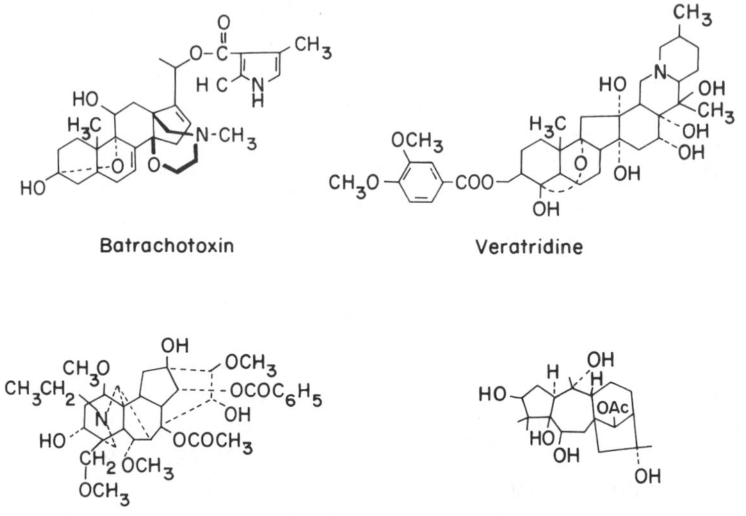

ALKALOIDS

Batrachotoxin

Veratridine

Aconitine

Grayanotoxin I

Figure 3 Structures of the alkaloids.

The Alkaloids

The alkaloid neurotoxins (Figure 3) share the common property of being able to depolarize resting nerve and muscle cells and to increase the resting Na permeability of excitable cells in culture (Ulbricht 1969, Narahashi et al 1971, Catterall 1977). All four classes cause Na channels to open and remain open at rest, but the detailed mechanisms underlying this effect differ.

BATRACHOTOXIN Batrachotoxin (BTX) is isolated from the skin of the Colombian frog, *Phyllobates aurotaenia*. It is the most extensively studied lipid-soluble toxin of Na channels. Batrachotoxin modifies Na channels essentially irreversibly in nerve and muscle preparation but reversibly in cells in culture (Albuquerque & Daly 1976, Khodorov & Revenko 1979, Catterall 1975). Binding to closed channels is extremely slow whereas binding to open channels is much faster, as evidenced by the rapid appearance of modified channels produced by repetitive depolarizations that activate channels (Khodorov & Revenko 1979).

Channels modified by BTX can be distinguished from unmodified channels in the same preparation by their altered gating characteristics. BTX-modified channels activate at potentials 40–50 mV more negative than unmodified channels and do not inactivate (Khodorov 1978, L. Huang et al 1984). In addition to modified gating, these channels have reduced ion selectivity (Khodorov 1978, Khodorov & Revenko 1979, Huang et al 1979, Frelin et al 1981). This change of selectivity must, when more than one permeant ion species is present, lead to a change of the unitary conductance of the channel. A decrease of unitary conductance by BTX has been consistently observed whenever direct comparisons to unmodified channels have been possible (Khodorov et al 1981, Quandt & Narahashi 1982, L. Huang et al 1984).

The pharmacology of Na channels is altered in many ways by BTX. Modified channels have a higher affinity for *Leiurus* scorpion α-toxin than do unmodified channels (Catterall 1977). There seems to be an allosteric coupling between the site(s) of action of alkaloids and those of the polypeptide toxins. BTX-modified channels are less sensitive to inhibition by local anesthetics than are unmodified channels (Khodorov et al 1975, Rando et al 1986). Even the voltage-dependent block of channels by STX is modified (Rando & Strichartz 1986) if not actually induced by BTX (Rando & Strichartz 1985).

Perhaps the most interesting aspect of the pharmacology of BTX comes from studies of the voltage dependence of activation and of intramembrane charge movement, or gating current. According to the simple but elegant model of Hodgkin & Huxley (1952b), Na channels activate by the independent movement of three identical gating particles, with each particle having the equivalent of approximately two electronic charges. In the presence of BTX, the channel activation is better modeled by the movement of a single particle with an equivalent of approximately six electronic charges (Dubois et al 1983). It is as though BTX causes the aggregation of the three particles such that they are no longer independent. Actually, these simple models cannot account for all the data from either BTX-modified or unmodified channels (Neumcke et al 1976, L. Huang et al 1984, Dubois & Schneider 1985), but the data provide a correlation between a gating current change, which presumably represents a change of protein conformation, and a change of channel physiology caused by a neurotoxin.

Batrachotoxin has become an important pharmachological tool for Na channel assays. The use of planar lipid bilayers for the study of single Na channels has developed only with the use of BTX to create non-inactivating channels (Krueger et al 1983, Moczydlowski et al 1984). The synthesis of a radiolabeled BTX derivative, batrachotoxinin-A 20α-benzoate (BTX-B), has made possible binding studies with this lipid-soluble toxin (Brown et al 1981). Catterall and co-workers (1981) have shown that the binding of BTX-B to rat brain synaptosomes (in the presence of *Leiurus* scorpion α-toxin) is antagonized by BTX, veratridine, and aconitine with K_D values of 0.05 μM, 7.0 μM, and 1.2 μM respectively. These K_D values are within a factor of two of the $K_{0.5}$ values for the increase of Na permeability in neuroblastoma cells produced by these alkaloids (Catterall 1975).

VERATRIDINE Veratridine (VTD) is the most potent compound in a mixture of alkaloids, termed veratrum alkaloids, isolated from plants of the family *Lilaceae*. Veratridine modifies Na channels in a manner quite different from BTX. In all preparations studied, the effects of VTD are reversed upon removing the toxin from the bathing medium. Two modes of binding by VTD have been described: a rapid binding to open channels and a slow binding, perhaps to inactivated channels (Ulbricht 1969, Leicht et al 1971, Sutro 1986, Rando 1986a). By whichever pathway modification is achieved, the characteristics of the modified channels appear to be identical (Rando 1986b). The binding of VTD causes the channels to remain open at rest and at more depolarized potentials. The binding of VTD reverses slowly and completely when the membrane is held at potentials somewhat more negative than the resting potential. When the membrane is not under voltage control, depolarization results from the slow modification of channels by VTD and the concomitant increase in resting sodium permeability (Ulbricht 1969, Catterall 1975).

Channels modified by VTD do inactivate, but very slowly and reversibly, so the VTD-modified permeability never decays completely at depolarized potentials (Leicht et al 1971, Rando et al 1986, Sutro 1986). VTD-modified channels may have several different inactivated states, since the just-cited studies identified inactivation with different assays. The decay of VTD-modified currents after a transient depolarization, which has conventionally been considered a deactivation of modified channels, is more accurately described as an inactivation of modified channels (Rando 1986a). Other veratrum alkaloids, some of which reduce the peak Na conductance in voltage-clamped axons but produce no persistent Na conductance, may help to distinguish between various inactivated states (Ohta et al 1973).

Once a channel has been modified by VTD, its properties resemble those of BTX-modified channels, with some quantitative differences. The voltage dependence of activation is shifted by ~ -90 mV in both muscle and nerve (Leibowitz et al 1986, Rando 1986b). The ionic selectivity is less than that of unmodified channels, yet differs from BTX modified channels (Naumov et al 1979, Frelin et al 1981); all alkali metal cations are permeant through VTD-modified channels (Rando 1986b).

ACONITINE Aconitine is purified from a mixture of alkaloids from the plant *Aconitum napellus* (monk's hood). The effects of aconitine have been studied in nerve and muscle preparations, and seem to be less consistent among cell types than those of the other alkaloids. Aconitine-modified channels show shifts in the voltage dependence of activation of ~ -50 mV in nerve and muscle but only ~ -20 mV in neuroblastoma cells (Schmidt & Schmitt 1974, Campbell 1982, Grishchenko et al 1983). In muscle, aconitine-modified channels inactivate completely, but in nerve preparations, inactivation is incomplete (Campbell 1982, Schmidt & Schmitt 1974). In all preparations studied, aconitine reduces the ion selectivity of Na channels (Mozhayeva et al 1977, Campbell 1982, Grishchenko et al 1983).

GRAYANOTOXINS Grayanotoxins (GTX) are the toxic substances found in rhododendron plant and other plants of the family *Ericaceae*. The only thorough voltage-clamp study of the effects of GTX on Na channels was done using the squid giant axon (Seyama & Narahashi 1981). GTX produced a slowly activating current near the resting potential that in some ways resembles the slow VTD modification of the channels. There was also a modification of the rapidly activating Na current that appears more like the modification of channels by BTX. More work needs to be done to clarify the kinetics of GTX action.

GTX-modified channels are less ion selective than unmodified channels and do not inactivate (Hironaka & Narahashi 1977, Frelin et al 1981, Seyama & Narahashi 1981). A novel aspect of the modification by GTX not seen with the other alkaloids is that GTX-modified channels are more resistant to TTX block than are unmodified channels. When 30 nM TTX was applied to an axon treated with GTX, the peak current was "almost completely blocked" whereas the slow current was only "partially decreased"' (Seyama & Narahashi 1981). It remains to be determined whether GTX confers voltage-dependence on the block by TTX, as seems to be the case for batrachotoxin (Rando & Strichartz 1985, Rando & Strichartz 1986).

The Pyrethroids

The pyrethroid insecticides are synthetic analogues of the pyrethrins, neurotoxins isolated from the flowers of the genus *Chrysanthemum*. Although the chemical structures of these compounds vary widely (Figure 4), they all are esters with hydrophobic groups substituted at both the α and the esterified carbons. The substituent on the esterified carbon provides for a division of the pyrethroids into two groups: type II pyrethroids contain a cyano group at this position, type I pyrethroids do not.

Classification of pyrethroids into these two groups was initially based on (*a*) their actions on cercal nerves and (*b*) toxicity symptoms in the cockroach (Gammon et al 1981). This division has held up in studies of pyrethroid action in numerous systems, but recent evidence shows that it is due to a quantitative rather than a qualitative difference between the two groups (Vijverberg et al 1983, Lund & Narahashi 1983).

Both type I and type II pyrethroids modify voltage-dependent Na channels, producing repetitive discharges and depolarizing after-potentials, but not spontaneous depolarizations in various nerve preparations (Narahashi 1985). The interaction of these compounds with the channel is much like that of the alkaloids. Pyrethroids preferentially bind to open channels and produce a steady-state Na current; the pyrethroid-modified currents decay more slowly after a brief, transient depolarization than do unmodified currents; and the time course and amplitude of the peak current is virtually unchanged in the presence of the pyrethroids (Vijverberg et al 1982, 1983, Lund & Narahashi 1983).

Differentiation between the effects of types I and II pyrethroids on Na currents is made by comparing the rates of decay of the modified currents after brief depolarizations. Currents modified by type I pyrethroids decay one to several orders of magnitude faster than currents modified by type II pyrethroids (Vijverberg et al 1983, Lund & Narahashi 1983). However, the range of decay rates within a group is also one to two orders of magnitude, and one study reported that certain pyrethroids did not fit

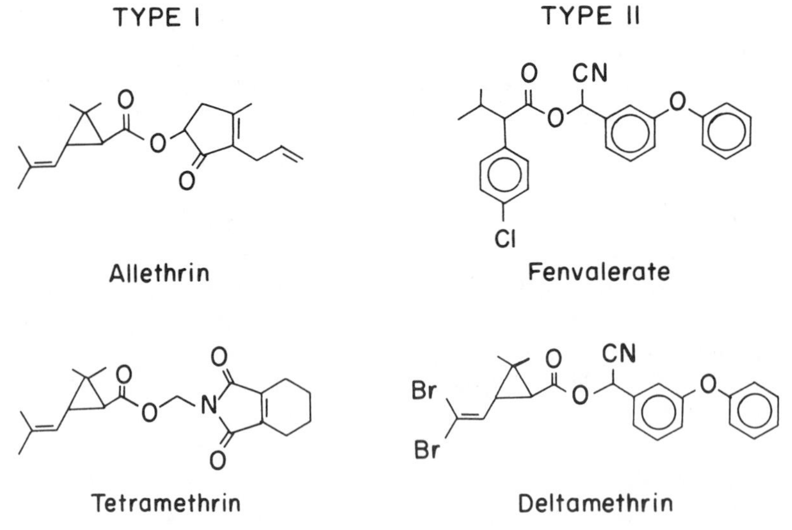

Figure 4 Structures of the pyrethroids.

clearly into one group or the other but were intermediate between the members of the two groups (Lund & Narahashi 1983). Therefore, the two groups are best viewed as representing the polar ends of a continuous spectrum of activity.

Vijverberg et al (1983) applied transition rate theory to describe the decay of the pyrethroid-modified currents. By measuring the rates at different temperatures, they were able to obtain values of the enthalpy and entropy, and thus the free energy, of the first-order process. The decay rates in the presence of the five type I pyrethroids tested had an average time constant of ~ 17 msec (range 5–30 msec), whereas in the presence of the four type II pyrethroids tested, they averaged 1000 msec (463–>1800 msec) at 15°C. Calculated free energies for the underlying processes in the presence of a type I pyrethroid are significantly different from those of a type II. The difference amounted to ~ 9.6 kJ/mole, which is in the range of energy of a single H-bond. It would be interesting to extend this analysis to pyrethroids with intermediate decay rates and to the other lipid soluble toxins.

The pyrethroids will continue to be useful probes of Na channels because most exist as two optical isomers and two geometric isomers. These pairs of isomers can differ greatly in their pharmacologic potency, but inactive isomers are still able to antagonize the actions of their active counterparts. The interactions of the four stereoisomers of tetramethrin have given support to the postulate that there are at least two sites of action of the pyrethroids (Lund & Narahashi 1982).

Soderlund and colleagues (1983), studying binding of radiolabeled stereoisomeric pyrethroids to membrane preparations from mouse brain, were able to demonstrate saturable, stereospecific binding as well as saturable, non-stereospecific binding. Such binding studies are plagued with the tremendous amount of nonsaturable "background" membrane binding exhibited by such a lipophilic molecule. A second approach has been to measure the uptake of ^{22}Na into cells in culture or synaptosomes (Jacques et al 1980, Ghiasuddin & Soderland 1985). Preliminary results indicate that pyrethroids alone do not enhance Na uptake, but that pyrethroids potentiate the uptake induced by alkaloid neurotoxins and polypeptide toxins. The actions of the alkaloids and pyrethroids have many similarities, but, at least in these flux assays, the potentiation of the effects of one by the other suggests separate effector sites.

Brevetoxins

The organism responsible for massive fish kills during "red tides" in the Gulf of Mexico is a dinoflagellate, *Ptychodiscus brevis* (formerly *Gymnodinium breve*), which supplies a host of toxins termed "brevetoxins"

(Figure 5). Brevetoxin B is composed of a single carbon chain that, by virtue of 11 contiguous *trans*-fused ether rings, is locked into a rigid, ladder-like structure (Lin et al 1981). The structures of three other brevetoxins have subsequently been deduced and shown to have very similar chemical structures (Nakanishi 1985, Shimizu et al 1986). Brevetoxin nomenclature has been confused partly because of the taxonomic reclassification and renaming of the organism and partly because of the

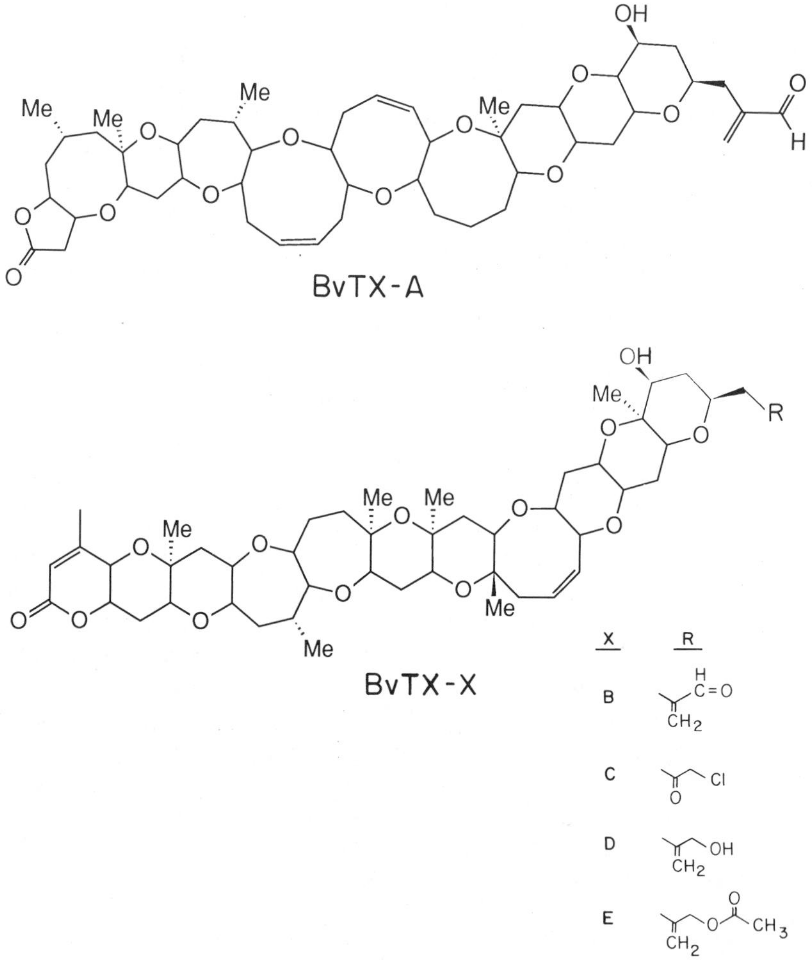

Figure 5 Structures of the brevetoxins (BvTX). The backbone of BvTX-A (Shimizu et al 1986) differs from that of the brevetoxins B through E, which vary only in the substituents (R) on carbon 41 (Nakanishi 1985).

many studies performed before the chemical structures were known. The confusion can now be resolved, but a consistent nomenclature should be adopted. We propose that the old nomenclature be dropped ("BTX" because of the confusion with batrachotoxin, "GB" because it was adopted under the old taxonomic classification, and "T" because it has no identifying significance). Instead, we suggest that brevetoxins be referred to by the abbreviation BvTX. Table 1 gives the old names that would be grouped together in the new classification.

Voltage clamp studies of BvTX-D and BvTX-A showed that the modifications of Na channels by this toxin were similar to those induced by the alkaloids (J. Huang et al 1984; E. Crill, T. Rando and G. Strichartz, paper in preparation). The voltage dependence of Na channel activation was shifted ~ -35 mV and the inactivation process was inhibited. The reversal potential of the BvTX-D modified current was less positive than that of the unmodified current, a phenomenon also seen with alkaloid-modified currents, which indicates an alteration in ion selectivity.

Brevetoxins alone depolarize frog nerve, squid axon, and skeletal muscle in a dose-dependent manner, although with lower efficacy (maximum depolarization) than BTX or VTD (Wu et al 1985; E. Crill, T. Rando, and G. Strichartz, paper in preparation). The potency for this action depends on the toxin structure (see Figure 5); EC50 values follow the order BvTX-A < BvTX-D < BvTX-E < BvTX-B. Channel inhibition as well as activation is produced by brevetoxins, for the efficacy order (BvTX-D > BvTX-B > BvTX-A ~ BvTX-E) does not parallel that of the potency order. Either single brevetoxin molecules have multiple effects when they bind to Na channels or there are multiple binding sites that can accommodate several toxin molecules simultaneously.

The interactions of the brevetoxins with other Na channel toxins indicate some similarities to the alkaloids, the pyrethroids, and the polypeptide α-toxins. The ability of BvTX-E to depolarize crayfish giant axons is, like that of the alkaloids, inhibited by the local anesthetic, procaine, and

Table 1 Designation of brevetoxin nomenclature[a]

New nomenclature	Old nomenclatures
BvTX-A	BTX-A
BvTX-B	BTX-B, GB-2, T_2, T_{34}, T_{47}
BvTX-C	BTX-C
BvTX-D	dihydro-BTX-B, GB-3, T_{17}
BvTX-E	GB-3-acetate

[a] Refer to Figure 5 for structures.

potentiated by the sea anemone toxin, anthopleurin-A (J. Huang et al 1984). However, BvTX-A [the most potent ichthyotoxic brevetoxin (Lin et al 1981)] does not stimulate Na uptake into neuroblastoma cells in the absence of any other toxin (Catterall & Risk 1981). The same study found that BvTX-A, like the pyrethroids and the α-toxins, did potentiate the VTD-induced Na uptake. However, BvTX-A had no effect on ^{125}I-scorpion α-toxin binding (unlike the α-toxins or the alkaloids which inhibit and potentiate the binding, respectively). Interestingly, ^{3}H-STX binding was enhanced in the presence of BvTX-A over the same concentration range at which VTD-induced flux was potentiated (Catterall & Risk 1981).

Other Lipid Soluble Toxins

Two toxins, palytoxin and ciguatoxin, have been studied in preliminary experiments. Palytoxin (PTX), claimed to be one of the most poisonous, nonproteinaceous substances known (Moore & Scheuer 1971), is isolated from marine coelenterates of the zoanthoid species of the genus *Palythoa*. The unusual structure of this toxin has been determined recently (Cha et al 1982). It has a molecular weight of over 2000 daltons, a backbone of 123 carbon centers (including five six-member hemiketals, a seven-member hemiketal, and a seven-member dihemiketal), and possesses 42 hydroxyl moieties.

Palytoxin has been shown to cause a depolarization of a variety of excitable cells by increasing their resting Na permeability (Deguchi et al 1976, Weidmann 1977, Dubois & Cohen 1977). Various investigators have attributed the actions of PTX to an increase of Na permeability via normal voltage-gated Na channels, via TTX-resistant Na channels, and via channels formed *de novo* by the toxin (Dubois & Cohen 1977, Tatsumi et al 1984, Muramatsu et al 1984). Voltage-clamp studies of axonal membrane have shown that in the presence of PTX, Na channels activate at more negative potentials and have a reduced ion selectivity (Dubois & Cohen 1977, Muramatsu et al 1984). The primary argument against the hypothesis that the effects of PTX are mediated by normal voltage-gated Na channels is that the effects of PTX are much less sensitive to block by TTX than are normal channels (Tatsumi et al 1984, Muramatsu et al 1984). However, as described above, the sensitivity of grayanotoxin-modified Na channels to TTX is also less than that of unmodified channels. Perhaps PTX modifies not only the voltage-dependence and ion selectivity of Na channels but the affinity of the channel for TTX as well.

Ciguatoxin, the cause of ciguatera poisoning, is isolated from a dinoflagellate, *Gambierdiscus toxicus*. Effects of ciguatoxin studied under voltage clamp have not been published, but its actions on the steady-state Na permeability of neuroblastoma cells have been partially characterized

(Bidard et al 1984). In this system, ciguatoxin had no effect by itself, but potentiated the Na uptake stimulated by alkaloid toxins, pyrethroids, and polypeptide toxins. Furthermore, ciguatoxin did not inhibit the binding of radiolabeled polypeptide α-toxins to rat brain synaptosomes (Bidard et al 1984).

Finally, a group of lipophilic compounds that have not been reviewed in this article are cocaine (and its many synthetic analogues, the local anesthetics) and other naturally occurring compounds with similar actions on Na channels, including strychnine and oenanthotoxin. These drugs all block Na-dependent action potentials without producing any membrane depolarization. When studied under voltage-clamp conditions, they all inhibit voltage-dependent Na channels by stabilizing them in non-conducting states. Many of these drugs act as competitive inhibitors of the lipophilic "activator" toxins.

The most impressive aspects of the pharmacology of lipid-soluble Na channel toxins are the diversity of structures and the similarities of action. Although the specific interactions of Na channels with VTD, BvTX-A, tetramethrin, etc may differ, the functional modifications of Na channels fall into the same categories: negative shift of the voltage dependence of activation; inhibition of inactivation; slowed decay of the current after brief, transient depolarizations; a decrease of the ion selectivity of the chennel; and a modulation of the binding of other classes of toxins. Thus, perhaps a very small region of the protein to which any individual toxin molecule binds influences the structure of the entire channel. It is noteworthy that these toxins, which presumably associate with a hydrophobic region of the protein, can modify the ion permeation pathway as evidenced by the decrease of ion selectivity.

GUANIDINIUM TOXINS

Traditionally, the guanidinium-containing compounds tetrodotoxin (TTX) and saxitoxin (STX) are regarded as "blocking" agents that reduce the number of conducting Na channels by occupying some site near the outer opening, independently of membrane potential or the presence of other toxins (Ritchie & Rogart 1977). This view requires revision in response to several recent findings:

1. The gating of Na channels is modified by quanidinium toxins; the fraction of channels in a "slow inactivated" state as a result of prolonged depolarization is enhanced and the recovery to an activatable state slowed by TTX and STX (Burnashev et al 1984, Strichartz et al 1986). The recovery rate from this slow inactivation in nerve is orders of magnitude faster than the time-constant for toxin binding to the same tissue, thus indicating

that toxin-bound channels must recover to a conducting state after long depolarizations. Cardiac Na channels show "use-dependent" effects of TTX inhibition (Cohen et al 1981), symptomatic of a reciprocal interaction between channel gating and toxin binding. Gating currents in crayfish axons are also modified by these toxins (Heggeness & Starkus 1986) and their effects on macroscopic Na currents in squid axon are modulated by channel activation reactions (Yeh et al 1986).

2. The actions of certain polypeptide toxins are suppressed by the guanidinium toxins; ATXII's effects do not develop if this toxin is applied to crayfish axons in the presence of TTX (Romey et al 1976), and a scorpion α-toxin also has no effect on frog nerve when co-incubated with TTX (Siemen & Vogel 1983). These actions must be examples of kinetic restriction, for the polypeptide toxins, once applied in the absence of TTX, are not reversed by its subsequent addition.

3. Channels modified by lipophilic toxins have altered reactivity to STX and TTX. BTX-modified channels in planar bilayers manifest a voltage-dependent inhibition by TTX and STX (Krueger et al 1983, Moczydlowski et al 1984); this is also observed in BTX-treated nerve (Rando & Strichartz 1986) but not in normal nerve or in axons where channel open states have been prolonged by chemical treatment (Rando & Strichartz 1985, Strichartz et al 1986). Modest but significant changes in the affinity of STX for lobster axons are wrought by the lipophilic toxins, BTX and aconitine (Strichartz et al 1986). In turn, the guanidinium toxins inhibit, noncompetitively, the binding of BTX derivatives to vesicles from mammalian brain, at their normal blocking concentrations and with a strong temperature dependence that suggests conformational perturbation of the BTX binding site (Brown 1986). Altogether, these observations necessitate a radical change in our model for the actions of guanidinium toxins.

SUMMARY

The neurotoxins that modify Na channels have actions that are characterized by different degrees of specificity (Table 2). These specificities can be correlated with their chemical properties. For example, guanidinium toxins, which are small charged ligands, appear only to "block" Na channels by binding to a site on the external surface. Peptide toxins, which are also positively charged and relatively small, also act from the external solution to modify channel activation and inactivation processes but do not alter ion selectivity. The lipophilic toxins, hydrophobic, neutral drugs, act from either side of the membrane and modify all the functions of Na channels. From such differences, and from the independence of toxin binding as well as toxin action, separate binding sites for these agents have been classified (Catterall 1980).

Recent findings reviewed here suggest that all these toxins share certain features:

1. They differentiate between various states of the channel. Effects of lipophilic activators, polypeptide toxins, and, indeed, even STX and TTX are enhanced or reversed in fractions of seconds under voltage clamp by patterns of membrane potential that selectively populate the channel open state, or the slow or fast inactivated states. Other assays—such as the binding of radiolabeled ligands or the changes of steady-state Na flux that require seconds to minutes of toxin-channel interaction—reveal interactions of the toxins with states of the channel not detected in the usual voltage-clamp analysis. Pharmacological probes may thus reveal channel states or transitions previously unrecognized.

2. The bound toxins appear to interact with one another. The well-documented synergism at equilibrium of α-toxins with lipophilic activators provided a model for allosteric interactions between two separate binding sites (Catterall 1979, 1980). The other toxin interactions are more ephemeral and are characterized by kinetic variations that reflect the availability of reactive channel states. For example, the appearance of β-toxin induced modifications of Na currents is accelerated in the presence of α-toxin (Wang & Strichartz 1983), whereas the modifications of inactivation by α-toxins are prevented by concurrent incubation with tetrodotoxin, although such modifications, once effected, are not reversed by the subsequent addition of TTX. Modifications of gating by lipophilic toxins confer a

Table 2 Summary of sodium channel modifications by toxins

	Na channel property			
Toxin	Activation	Inactivation	Ion selectivity	Unitary conductance
Polypeptide toxins				
α-Toxins	[a]	slow and inhibit	—	—
β-Toxins	Δ V-dependence and rate[b]	slow and inhibit	—	?
γ-Toxins	Δ V-dependence and rate	slow and inhibit	—	?
μ-Conotoxins	—	?	n.m.[c]	blocked
Lipophilic toxins				
Alkaloids	Δ V-dependence and rate[b]	slow and inhibit	decrease	decrease
Pyrethroids	Δ V-dependence and rate	slow and inhibit	?	?
Brevetoxins	Δ V-dependence and rate	slow and inhibit	decrease	?
Guanidinium toxins				
TTX, STX	—	stabilized[d]	n.m.[c]	blocked

[a] No effect.
[b] Modulation depends on membrane potential.
[c] Not measurable (toxin blocks unitary conductance).
[d] Slow inactivation enhanced; fast inactivation unaffected.
? Effect unknown.

selective voltage-dependence on STX and TTX inhibition of open channels that is not observed in drug-free channels. Thus, the allosteric interactions are a common feature of all classes of toxins acting on Na channels, as are the apparent use-dependent actions that arise from state-selective affinities. Since all the binding sites are coupled yet are claimed to be separately arrayed on the channel (Angelides & Nutter 1983), extensive interactions must occur among many parts of this large macromolecule. This may also be true for normal gating phenomena, unaffected by toxins. Channel opening may correspond to conformational changes at the selectivity region, the narrowest part of the channel pore, so that separate structures for gating and ion discriminiation need not be required.

3. Different channel types may selectively interact differently with certain toxins. Neither brevetoxin nor ciguatoxin supports steady-state Na permeability in cultured cells, yet both depolarize axons at rest. Polypeptide α-toxins increase the maximum Na permeability of cultured cells, yet increase only the duration of that permeability in adult nerve and muscle cells. Such differences may arise from intrinsic differences in channel types or from the different expression of channels in culture and in vivo.

Literature Cited

Albuquerque, E. X., Daly, J. W. 1976. Batrachotoxin, a selective probe for channels modulating sodium conductances in electrogenic membranes. In *The Specificity and Action of Animal, Bacterial and Plant Toxins*, ed. P. Cuatrecasas, pp. 297–338. London: Chapman & Hall

Aldrich, R. W., Corey, D. P., Stevens, C. F. 1983. A reinterpretation of mammalian sodium channel gating based on single channel recording. *Nature* 306: 436–41

Angelides, K. J., Brown, G. B. 1984. Fluorescence resonance energy-transfer on the voltage-dependent sodium channel. *J. Biol. Chem.* 259: 6117–26

Angelides, K. J., Nutter, T. J. 1983. Mapping the molecular structure of the voltage-dependent sodium channel. *J. Biol. Chem.* 258: 11958–67

Armstrong, C. M., Bezanilla, F. 1977. Inactivation of the sodium channel. II. Gating current experiments. *J. Gen. Physiol.* 70: 567–90

Auld, V., Marshall, J., Goldin, A., Dowsett, A., Catterall, W., et al. 1985. Cloning and characterization of the gene for alpha subunit of the mammalian voltage-gated sodium channel. *J. Gen. Physiol.* 86: 10a

Barhanin, J., Hugus, M., Schweitz, H., Vincent, J. P., Lazdunski, M. 1981. Structure-function relationships of sea anemone toxin II from *Anemonia sulcata*. *J. Biol. Chem.* 256: 5764–69

Baumgold, J., Parent, J. B., Spector, I. 1983. Development of sodium channels during differentiation of chick skeletal muscle in culture. II. $^{22}Na^+$ uptake and electrophysiological studies. *J. Neurosci.* 3: 1004–13

Begenisich, T. B., Cahalan, M. D. 1980. Sodium channel permeation in squid axons. I. Reversal potential experiments. *J. Physiol.* 307: 217–42

Bekkers, J. M., Greeff, N. G., Keynes, R. D., Neumcke, B. 1984. The effect of local anaesthetics on the components of the asymmetry current in the squid giant axon. *J. Physiol.* 353: 653–68

Beress, L. 1982. Sea anemone toxins: A mini-review. In *Chemistry of Peptides and Proteins*, ed. W. Voelter, E. Wunsch, J. Ovchinnikov, V. Ivanov, 1: 121–26. Berlin/New York: de Gruyter

Bezanilla, F., Armstrong, C. M. 1977. Inactivation of the sodium channel. I. Sodium current experiments. *J. Gen. Physiol.* 70: 567–90

Bidard, J. N., Vijverberg, H. P. M., Frelin,

C., Chungue, E., Legrand, A. M., et al. 1984. Ciguatoxin is a novel type of Na^+ channel toxin. *J. Biol. Chem.* 259: 8353–57

Brown, G. B. 1986. ^{3}H-Batrachotoxinin-A benzoate binding to voltage-sensitive sodium channels: Inhibition by the channel blockers tetrodotoxin and saxitoxin. *J. Neurosci.* 6: 2064–70

Brown, G. B., Tieszen, S. C., Daly, J. W., Warnick, J. E., Albuquerque, E. X. 1981. Batrachotoxinin-A 20-α-benzoate: A new radioactive ligand for voltage sensitive sodium channels. *Cell. Molec. Neurobiol.* 1: 19–40

Burnashev, N., Sokolova, S. N., Khodorov, B. I. 1984. Interaction of potassium ions and tetrodotoxin (TTX) with inactivated sodium channels in isolated rat myocardial cells. *Gen. Physiol. Biophys.* 3: 507–9

Cahalan, M. D. 1975. Modification of sodium channel gating in frog myelinated nerve fibers by *Centruroides sculpturatus* scorpion venom. *J. Physiol.* 244: 511–34

Campbell, D. T. 1982. Modified kinetics and selectivity of sodium channels in frog skeletal muscle fibers treated with aconitine. *J. Gen. Physiol.* 80: 713–31

Campbell, D. T., Hille, B. 1976. Kinetic and pharmacological properties of the sodium channel of frog skeletal muscle. *J. Gen. Physiol.* 67: 309–23

Catterall, W. A. 1975. Activation of the action potential Na^+ ionophore of cultured neuroblastoma cells by veratridine and batrachotoxin. *J. Biol. Chem.* 250: 4053–59

Catterall, W. A. 1977. Membrane potential-dependent binding of scorpion toxin to the action potential Na^+ ionophore. Studies with a toxin derivative prepared by lactoperoxidase-catalyzed iodination. *J. Biol. Chem.* 252: 8660–68

Catterall, W. A. 1977. Activation of the action potential Na^+ ionophore by neurotoxins: An allosteric model. *J. Biol. Chem.* 252: 8669–76

Catterall, W. A. 1979. Binding of scorpion toxin to receptor site associated with sodium channels in frog muscle: Correlation of voltage-dependent binding with activation. *J. Gen. Physiol.* 74: 357–91

Catterall, W. A. 1980. Neurotoxins that act on voltage-sensitive sodium channels in excitable membranes. *Ann. Rev. Pharmacol. Toxicol.* 20: 15–43

Catterall, W. A. 1986. Voltage-dependent gating of sodium channels: Correlating structure and function. *Trends Neurosci.* 9: 7–10

Catterall, W. A., Morrow, C. S., Daly, J. W., Brown, G. B. 1981. Binding of batrachotoxinin A 20-α-benzoate to a receptor site associated with sodium channels in synaptic nerve ending particles. *J. Biol. Chem.* 256: 8922–27

Catterall, W. A., Risk, M. 1981. Toxin $T4_6$ from *Ptychodiscus brevis* (formerly *Gymnodinium breve*) enhances activation of voltage-sensitive sodium channels by veratridine. *Mol. Pharmacol.* 19: 345–48

Cha, J. K., Christ, W. J., Finan, J. M., Fujioka, H., Kishi, Y., et al. 1982. Stereochemistry of palytoxin 4. Complete structure. *J. Am. Chem. Soc.* 104: 7369–71

Chandler, W. K., Meves, H. 1965. Voltage-clamp experiments on internally perfused giant axons. *J. Physiol.* 180: 788–820

Codding, P. W. 1982. Structure and conformation of aconitine. *Acta Crystallogr. B* 38: 2519–22

Codding, P. W. 1983. Structural studies of sodium channel neurotoxins. 2. Crystal structure and absolute centrifugation of veratridine perchlorate. *J. Am. Chem. Soc.* 105: 3172–76

Codding, P. W. 1984. Structural studies of sodium channel neurotoxins. 3. Crystal structures and absolute configurations of grayanotoxin III and α-dihydrograyanotoxin II. *J. Am. Chem. Soc.* 106: 7905–9

Cohen, C. J., Bean, B. P., Colatsky, T. J., Tsien, R. W. 1981. Tetrodotoxin block of sodium channels in rabbit Purkinje fibers. *J. Gen. Physiol.* 78: 383–411

Costa, M. R. C., Casnellie, J. E., Catterall, W. A. 1982. Selective phosphorylation of the α-subunit of the sodium channel by cAMP-dependent protein kinase. *J. Biol. Chem.* 257: 7918–21

Couraud, F., Jover, E., Dubois, J. M., Rochat, H. 1982. Two types of scorpion toxin receptor sites, one related to the activation, the other to the inactivation of the action potential sodium channel. *Toxicon* 20: 9–16

Cruz, L. J., Gray, W. R., Olivera, B. M., Zeikus, R. D., Kerr, L., et al. 1985. *Conus geographus* toxins that discriminate between neuronal and muscle sodium channels. *J. Biol. Chem.* 260: 9280–88

Deguchi, T., Urakawa, N., Takamatsu, S. 1976. Some pharmacological properties of palythoatoxin isolated from the zoanthid, *Palythoa tuberculosa*. In *Animal, Plant, and Microbial Toxins*, ed. A. Ohsaka, K. Hayosh, Y. Sawai, 2: 379–94. New York: Plenum

Dubois, J. M., Cohen, J. B. 1977. Effect of palytoxin on membrane potential and current of frog myelinated fibers. *J. Pharmacol. Exp. Ther.* 201: 148–55

Dubois, J. M., Schneider, M. F. 1985. Kinetics of intramembrane charge movement and conductance activation of batrachotoxin-modified sodium channels in frog node of Ranvier. *J. Gen. Physiol.* 86: 381–94

Dubois, J. M., Schneider, M. F., Khodorov, B. I. 1983. Voltage-dependence of intramembrane charge movement and conductance activation of BTX-modified Na^+ channels in frog node of Ranvier. *J. Gen. Physiol.* 81: 829–44

Dufton, M. J., Rochat, H. 1984. Classification of scorpion toxins according to amino acid composition and sequence. *J. Molec. Evolu.* 20: 120–27

Fontecilla-Camps, J. C., Almassay, R. J., Suddath, F. L., Watt, D. D., Bugg, C. E. 1980. Three-dimensional structure of a protein from scropion venom: A new structural class of neurotoxins. *Proc. Natl. Acad. Sci. USA* 77: 6496–6500

Frelin, C., Vigne, P., Lazdunski, M. 1981. The specificity of the sodium channel for monovalent cations. *Eur. J. Biochem.* 119: 437–42

French, R. J., Horn, R. 1983. Sodium channel gating: Models, mimics and modifiers. *Ann. Rev. Biophys. Bioeng.* 12: 319–56

Gammon, D. W., Brown, M. A., Casida, J. E. 1981. Two classes of pyrethroid action in the cockroach. *Pestic. Biochem. Physiol.* 15: 181–91

Ghiasuddin, S. M., Soderland, D. M. 1985. Pyrethroid insecticides: Potent, stereospecific enhancers of mouse brain sodium channel activation. *Pest. Biochem. Physiol.* 24: 200–6

Gonoi, T., Hille, B., Catterall, W. A. 1984. Voltage-clamp analysis of sodium channels in normal and scorpion toxin-resistance neuroblastoma cells. *J. Neurosci.* 4: 2836–42

Grishchenko, I. I., Naumov, A. P., Zubov, A. N. 1983. Gating and selectivity of aconitine modified sodium channels in neuroblastoma cells. *Neuroscience* 9: 549–54

Habersetzer-Rochat, C., Sampieri, F. 1976. Structure-function relationships of scorpion neurotoxins. *Biochemistry* 15: 2254–61

Heggeness, S. T., Starkus, J. G. 1986. Saxitoxin and tetrodotoxin. Electrostatic effects on gating current in crayfish axons. *Biophys. J.* 49: 629–43

Hille, B. 1971. The permeability of the sodium channel to organic cations in myelinated nerve. *J. Gen. Physiol.* 58: 599–619

Hille, B. 1972. The permeability of sodium channels to metal cations in myelinated nerves. *J. Gen. Physiol.* 59: 637–58

Hille, B. 1975. Ion selectivity, saturation, and block in sodium channels. *J. Gen. Physiol.* 66: 535–60

Hironaka, T., Narahashi, T. 1977. Cation permeability ratios of sodium channels in normal grayanotoxin-treated squid axon membranes. *J. Membr. Biol.* 31: 359–81

Hodgkin, A. L., Huxley, A. F. 1952a. The dual effect of membrane potential on sodium conductance in the giant axon of *Loligo*. *J. Physiol.* 116: 497–506

Hodgkin, A. L., Huxley, A. F. 1952b. A quantitative description of membrane current and its application to conduction and excitation in nerve. *J. Physiol.* 117: 500–44

Hu, S. L., Meves, H., Rubly, N., Watt, D. D. 1983. A quantitative study of the action of *Centruroides sculpturatus* Toxins III and IV on the Na currents on the node of Ranvier. *Pflugers Arch.* 397: 90–99

Huang, J. M. C., Wu, C. H., Baden, D. G. 1984. Depolarizing action of a red-tide dinoflagellate brevetoxin on axonal membranes. *J. Pharmacol. Exp. Ther.* 229: 615–21

Huang, L. Y. M., Catterall, W. A., Ehrenstein, G. 1979. Comparison of ionic selectivity of batrachotoxin-activated channels with different tetrodotoxin dissociation constants. *J. Gen. Physiol.* 73: 839–54

Huang, L. Y. M., Moran, N., Ehrenstein, G. 1984. Gating kinetics of batrachotoxin-modified sodium channels in neuroblastoma cells determined from single-channel measurements. *Biophys. J.* 45: 313–22

Jacques, Y., Romey, G., Cavey, M. T., Kastalovski, B., Lazdunski, M. 1980. Interaction of pyrethroids with the Na^+ channel in mammalian neuronal cells in culture. *Biochim. Biophys. Acta* 600: 882–97

Jover, E., Couraud, F., Rochat, H. 1980. Two types of scorpion neurotoxins characterized by their binding to two separate receptor sites on rat brain synaptosomes. *Biochem. Biophys. Res Commun.* 95: 1607–14

Khodorov, B. I. 1978. Chemicals as tools to study nerve fiber sodium channels: Effects of batrachotoxin and some local anesthetics. In *Membrane Transport Processes*, ed. D. C. Tosteson, Y. A. Orchinnikov, R. Latorre, 2: 153–74. New York: Raven

Khodorov, B. I., Revenko, S. V. 1979. Further analyses of the mechanisms of action of batrachotoxin on the membrane of myelinated nerve. *Neuroscience* 4: 1315–30

Khodorov, B. I., Neumcke, B., Schwarz, W., Stampfli, R. 1981. Fluctuation analysis of Na^+ channels modified by batrachotoxin in myelinated nerve. *Biochim. Biophys. Acta* 648 : 93–99

Khodorov, B. I., Peganov, E. M., Revenko, S. V., Shishkova, L. D. 1975. Sodium currents in voltage clamped nerve fiber of frog under the combined action of batrachotoxin and procaine. *Brain Res* 84 : 541–46

Koppenhoffer, E., Schmidt, H. 1968a. Die wirkung von skorpiongift auf die lonenstrome des Ranvierscher Schnurrings. I. Die Permeabilataten P_{Na} und P_K. *Pflugers Arch.* 303 : 133–49

Koppenhoffer, E., Schmidt, H. 1968b. Die wirkung von skorpiongift auf die ionenstrome des Ranvierschen schnurrings. II. Unvollstandige Natrium-inaktivierung. *Pflugers Arch.* 303 : 150–61

Krueger, B. K., Worley, J. F., French, R. J. 1983. Single sodium channels from rat brain incorporated into planar lipid bilayer membranes. *Nature* 303 : 172–75

Leibowitz, M. D., Sutro, J. B., Hille, B. 1986. Voltage-dependent gating in veratridine-modified Na channels. *J. Gen. Physiol.* 87 : 25–46

Leicht, R., Meves, H., Wellhoner, H. H. 1971. The effect of veratridine on *Helix pomatia* neurones. *Pflugers Arch.* 323 : 50–62

Levinson, S. R., Duch, D. S., Urban, B. W., Recio-Pinto, E. 1986. The sodium channel from *Electrophorus electricus*. In *Tetrodotoxin, Saxitoxin, and the Molecular Biology of the Sodium Channel*, ed. S. R. Levinson, C. Y. Kao. New York: New York Acad. Sci.

Lin, Y. Y., Risk, M., Ray, S. M., VanEngen, D., Clardy, J., et al. 1981. Isolation and structure of brevetoxin B from the "red tide" dinoflagellate *Ptychodiscus brevis* (*Gymnodinium breve*). *J. Am. Chem. Soc.* 103 : 6773–75

Lund, A. F., Narahashi, T. 1982. Dose-dependent interaction of the pyrethroid isomers with sodium channels in squid axon membranes. *Neurotoxicol.* 3 : 11–24

Lund, A. F., Narahashi, T. 1983. Kinetics of sodium channel modification as the basis for the variation in the nerve membrane effects of pyrethroids and DDT analogs. *Pest. Biochem. Physiol.* 20 : 203–16

Masutani, T., Seyama, I., Narahashi, T., Iwasa, J. 1981. Structure-activity relationships for grayanotoxin derivatives in frog skeletal muscle. *J. Pharmacol. Exp. Ther.* 217 : 812–19

Meves, H., Rubly, N., Watt, D. D. 1982. Effect of toxins isolated from the venom of the scorpion *Centruroides sculpturatus* on the Na currents of the node of Ranvier. *Pflugers Arch.* 393 : 56–62

Meves, H., Rubly, N., Watt, D. D. 1984a. Voltage-dependent effect of a scorpion toxin on sodium current inactivation. *Pflugers Arch.* 402 : 24–33

Meves, H., Simard, J. M., Watt, D. D. 1984b. Biochemical and electrophysiological characteristics of toxins isolated from the venom of the scorpion *Centruroides sculpturatus*. *J. Physiol Paris* 79 : 185–91

Miller, J. A., Agnew, W. S., Levinson, S. R. 1983. Principal glycopeptide of the tetrodotoxin/saxitoxin binding protein from *Electrophorus electricus*: Isolation and partial chemical and physical characterization. *Biochemistry* 22 : 462–70

Moczydlowski, E., Hall, S., Garber, S. S., Strichartz, G. R., Miller, C. 1984. Voltage-dependent blockade of muscle Na^+ channels by guanidinium toxins : Effect of toxin charge. *J. Gen. Physiol.* 84 : 687–704

Moczydlowski, E., Olivera, B. M., Gray, W. R., Strichartz, G. R. 1986a. Discrimination of muscle and neuronal Na-channel subtypes by binding competition between ^{3}H-saxitoxin and μ-conotoxins. *Proc. Natl. Acad. Sci. USA* 83 : 5321–25

Moczydlowski, E., Uehara, A., Guo, X., Heiny, J. 1986b. Isochannels and blocking modes of voltage-dependent sodium channels. See Levinson et al 1986

Moore, R. E., Scheuer, P. J. 1971. Palytoxin : A new marine toxin from coelenterate. *Science* 172 : 495–98

Mozhayeva, G. N., Naumov, A. P., Negulyaev, Y. A., Nosyreva, E. D. 1977. The permeability of actonitine-modified Na^+ channels to univalent cations in myelinated nerve. *Biochim. Biophys. Acta* 466 : 461–73

Mozhayeva, G. N., Naumov, A. P., Grishin, E. V., Soldatov, N. M. 1979. Effect of *Buthus eupeus* toxins on sodium channels of Ranvier node membrane. *Biofizika* 24 : 235–41 (in Russian)

Mozhayeva, G. N., Naumov, A. P., Nosyreva, E. D., Grishin, E. V. 1980. Potential-dependent interaction of toxin from venom of the scorpion *Buthus eupeus* with sodium channels in myelinated fibre. *Biochim. Biophys. Acta* 597 : 587–602

Muramatsu, I., Uemura, D., Fujiwara, M., Narahashi, T. 1984. Characteristics of palytoxin-induced depolarization in squid axons. *J. Pharmacol. Exp. Ther.* 231 : 488–94

Nagy, K., Kiss, T., Hof, D. 1983. Single Na channels in mouse neuroblastoma cell membrane. Indications for two open states. *Pflugers Arch.* 399 : 302–8

Nakanishi, K. 1985. The chemistry of brevetoxins : A review. *Toxicon* 23 : 473–79

Narahashi, T. 1985. Nerve membrane ionic

channels as the primary target of pyrethroids. *Neurotoxicology* 6: 3–22

Narahashi, T., Albuquerque, E. X., Deguchi, T. 1971. Effects of BTX on membrane potential and conductance of squid giant axons. *J. Gen. Physiol.* 58: 54–78

Naumov, A. P., Negulayer, Y. A., Nosyreva, E. D. 1979. Changes of selectivity of sodium channels in membrane of nerve treated with veratrine. *Zytologia* 21: 692–96

Neumcke, B., Nonner, W., Stampfli, R. 1976. Asymmetrical displacement current and its relation with the activation of sodium current in the membrane of frog myelinated nerve. *Pflugers Arch.* 363: 193–203

Neumcke, B., Schwarz, W., Stampfli, R. 1985. Comparison of the *Anemonia* toxin II on sodium and gating currents in frog myelinated merve. *Biochim. Biophys. Acta* 814: 111–19

Noda, M., Shimizu, S., Tanabe, T., Takai, T., Kayano, T., et al 1984. Primary structure of *Electrophorus electricus* sodium channel deduced from cDNA sequence. *Nature* 312: 121–27

Noda, M., Ikeda, T., Kayane, T., Suzuki, H., Takeshima, H., et al. 1986. Existence of distinct sodium channel messenger RNAs in rat brain. *Nature* 320: 188–92

Nonner, W. 1979. Effects of *Leiurus* scorpion venom on the "gating" current in myelinated nerve. *Adv. Cytopharmacol.* 3: 345–52

Nonner, W. 1980. Relations between the inactivation of sodium channels and the immobilization of gating charge in frog myelinated nerve. *J. Physiol. London* 299: 573–603

Ohta, M., Narahashi, T., Keeler, R. 1973. Effects of veratrum alkaloids on membrane potential and conductance of squid and crayfish giant axons. *J. Pharmacol. Exp. Ther.* 184: 143–54

Olivera, B. M., Gray, W. R., Zeikus, R., McIntosh, J. M., Varga, J. 1985. Peptide neurotoxins from fish-hunting cone snails. *Science* 230: 1338–43

Patlak, J., Horn, R. 1982. Effect of *N*-bromoacetamide on single sodium channel currents in excised membrane patches. *J. Gen. Physiol.* 79: 333–51

Pelhate, M., Laufer, J., Pichon, Y., Zlotkin, E. 1984. Effects of several sea anemone and scorpion toxins on excitability and ionic currents in the giant axon of the cockroach. *J. Physiol. Paris* 79: 309–17

Quandt, F. N., Narahashi, T. 1982. Modification of single Na^+ channels by BTX. *Proc. Natl. Acad. Sci. USA* 79: 6732–36

Rack, M., Meves, H., Beress, L., Grunhagen, H. H. 1983. Preparation and properties of fluorescence labeled neuro- and cardiotoxin II from the sea anemone (*Anemonia sulcata*). *Toxicon* 21: 231–37

Rando, T. A. 1986a. Rapid and slow interactions between veratridine and sodium channels in frog myelinated nerve. *J. Physiol.* Submitted

Rando, T. A. 1986b. Veratridine-induced modifications of voltage-dependent gating and sodium channels in frog myelinated nerve. *J. Physiol.* Submitted.

Rando, T. A., Strichartz, G. R. 1985. Voltage-dependence of saxitoxin block of Na channels appears to be a property unique to batrachotoxin-modified channels. *J. Gen. Physiol.* 86: 14a

Rando, T. A., Strichartz, G. R. 1986. Saxitoxin blocks batrachotoxin-modified Na channels in the node of Ranvier in a voltage-dependent manner. *Biophys. J.* 49: 7785–94

Rando, T. A., Wang, G. K., Strichartz, G. R. 1986. The interaction between the activator agents batrachotoxin and veratridine and the gating processes of neuronal sodium channels. *Mol. Pharmacol.* 29: 467–77

Renaud, J. F., Romey, G., Lombet, A., Lazdunski, M. 1981. Differentiation of the fast Na^+ channel in embryonic heart cells: Interaction of the channel with neurotoxins. *Proc. Natl. Acad. Sci. USA* 78: 5348–52

Ritchie, J. M., Rogart, R. B. 1977. The binding of saxitoxin and tetrodotoxin to excitable membranes. *Rev. Physiol. Biochem. Pharmacol.* 79: 1–50

Rochat, H., Bernard, P., Couraud, F. 1979. Scorpion toxins: Chemistry and mode of action. *Adv. Cytopharmacol.* 3: 325–34

Rochat, H., Darbon, H., Jover, E., Martin, M. F., Bablito, J., et al. 1984. Interaction of scorpion toxins with the sodium channel. *J. Physiol. Paris* 79: 334–37

Romey, G., Abita, J. P., Schweitz, H., Wunderer, G., Lazdunski, M. 1976. Sea anemone toxin: A tool to study molecular mechanisms of nerve conduction and excitation-secretion coupling. *Proc. Natl. Acad. Sci. USA* 73: 4055–59

Schmidt, H., Schmitt, O. 1974. Effect of aconitine on the sodium permeability at the node of Ranvier. *Pflugers Arch.* 349: 133–48

Schmidtmayer, J. 1985. Behavior of chemically modified sodium channels in frog nerve supports a three-state model of inactivation. *Pflugers Arch.* 404: 21–28

Seyama, I., Narahashi, T. 1981. Modulation of Na^+ channels of squid nerve membranes by grayanotoxin I. *J. Pharmacol. Exp. Ther.* 219: 614–24

Shimizu, Y., Chou, H. N., Bando, H. 1986.

Structure of Brevetoxin-A (GB-1 Toxin), the most potent toxin in the Florida Red Tide organism *Gymnodinium breve* (*Ptychodiscus brevis*). *J. Am. Chem. Soc.* 108: 514–15

Siemen, D., Vogel, W. 1983. Tetrodotoxin interferes with the reaction of scorpion toxin (*Buthus tamulus*) at the sodium channel of the excitable membrane. *Pflugers Arch.* 397: 306–11

Sigworth, F. J. 1981. Covariance of nonstationary sodium current fluctuations at the node of Ranvier. *Biophys. J.* 34: 111–33

Soderland, D. M., Ghiasuddin, S. M., Helmuth, D. W. 1983. Receptor-like stereospecific binding of a pyrethroid insecticide to mouse brain membranes. *Life Sci.* 33: 261–67

Stimers, J. R., Bezanilla, F., Taylor, R. E. 1985. Sodium channel activation in the squid giant axon. Steady-state properties. *J. Gen. Physiol.* 85: 65–82

Strichartz, G. R., Rando, T., Hall, S., Gitschier, J., Hall, L., et al. 1986. On the mechanism by which saxitoxin binds to and blocks sodium channels. See Levinson et al 1986

Strichartz, G. R., Wang, G. K. 1986. Rapid voltage-dependent dissociation of scorpion α-toxins coupled to Na channel inactivation in myelinated nerve. *J. Gen. Physiol.* 88: 413–35

Sutro, J. B. 1986. Kinetics of veratridine action on Na channels of skeletal muscle. *J. Gen. Physiol.* 87: 1–24

Swenson, R. P. 1983. A slow component of gating current in crayfish giant axons resembles inactivation charge movement. *Biophys. J.* 41: 245–49

Tatsumi, M., Takahashi, M., Ohizumi, Y. 1984. Mechanism of palytoxin-induced ^{3}H norepinephrine release from a Ca^+ pheochromocytoma cell line. *Molec. Pharmacol.* 25: 379–83

Ulbricht, W. 1969. The effects of veratridine on excitable membranes in nerve and muscle. *Ergeb. Physiol. Biol. Chem. Exp. Pharmakol.* 61: 17–71

Vandenberg, C. A., Horn, R. 1984. Inactivation viewed through single sodium channels. *J. Gen. Physiol.* 84: 535–64

Vijverberg, H. P. M., Van der Zalm, J. M., Van den Bercken, J. 1982. Similar mode of action of pyrethroids and DDT on sodium channel gating in myelinated nerves. *Nature* 295: 601–3

Vijverberg, H. P. M., Van der Zalm, J. M., Van Kleek, R. G. D. M., Van den Bercken, J. 1983. Temperature- and structure-dependent interaction of pyrethroids with the sodium channels of frog node of Ranvier. *Biochem. Biophys. Acta* 728: 73–82

Vijverberg, H. P. M., Pauron, D., Lazdunski, M. 1984. The effect of *Tityus serrulatus* scorpion toxin γ on Na channels in neuroblastoma cells. *Pflugers Arch.* 401: 297–303

Wang, G. K., Strichartz, G. 1982. Simultaneous modifications of sodium channel gating by two scorpion toxins. *Biophys. J.* 40: 175–79

Wang, G., Strichartz, G. R. 1983. Purification and physiological characterization of neurotoxins from venoms of the scorpions *Centruroides sculpturatus* and *Leiurus quinquestriatus*. *Molec. Pharmacol.* 23: 519–33

Wang, G. K., Strichartz, G. R. 1985a. Kinetic analysis of the action of *Leiurus* scorpion toxin on ionic currents in myelinated nerve. *J. Gen. Physiol.* 86: 739–62

Wang, G. K., Strichartz, G. R. 1985b. The actions of an isolated β-scorpion toxin on Na channel activation. *Biophys. J.* 47: 438a

Warashina, A., Fujita, S., Satake, M. 1981. Potential-dependent effects of sea anemone toxins and scorpion venom of crayfish giant axon. *Pflugers Arch.* 391: 273–76

Warashina, A., Fujita, S. 1983. Effects of sea anemone toxins on the sodium inactivation process in crayfish axons. *J. Gen. Physiol.* 81: 305–23

Warnick, J. E., Albuquerque, E. X., Onur, R., Jansson, S. E., Daly, J., et al. 1975. The pharmacology of batrachotoxin. VII. Structure-activity relationships and the effect of pH. *J. Pharmacol. Exp. Ther.* 193: 232–45

Watt, D. D., McIntosh, M. E. 1972. Effects on lethality of toxins in venom from the scorpion *Centruroides sculpturatus* by group specific reagents. *Toxicon* 10: 173–81

Weidmann, S. 1977. Effects of palytoxin on the electrical activity of dog and rabbit heart. *Experientia* 33: 1487–89

Wheeler, K. P., Watt, D. D., Lazdunski, M. 1983. Classification of Na channel receptors specific for various scorpion toxins. *Pflugers Arch.* 397: 164–65

Wu, C. H., Huang, J. M. C., Vogel, S. M., Luke, V. S., Atchison, W. D., et al. 1985. The actions of *Ptychodiscus brevis* toxins on nerve and muscle membranes. *Toxicon* 23: 481–88

Yanagawa, V., Abe, T., Satake, M. 1986. Blockade of [^{3}H]lysine-tetrodotoxin binding to sodium channel proteins by conotoxin GIII. *Neurosci. Lett.* 64: 7–12

Yeh, J. Z., Salgado, V. L., Narahashi, T. 1986. Use- and voltage-dependent block of sodium channel by saxitoxin. See Levinson et al 1986

Ann. Rev. Neurosci. 1987. 10: 269–95

NEURON SPECIFIC ENOLASE, A CLINICALLY USEFUL MARKER FOR NEURONS AND NEUROENDOCRINE CELLS[1]

Paul J. Marangos

Unit on Neurochemistry, Biological Psychiatry Branch, National Institute of Mental Health, Bethesda, Maryland 20892

Donald E. Schmechel

Division of Neurology and Department of Medicine, Durham Veterans Administration Medical Center, Duke University Medical Center, Durham, North Carolina 27710

Introduction

The rationale for studying cell specific proteins is rooted in the realization that proteins strictly localized or even greatly enriched in a given cell type are likely to be involved in biochemical events that are either specific to that cell type or performed in a different manner by that particular cell. This concept is further strengthened if the cell specific protein has an ontogenetic appearance that parallels that of the differentiation of the cell. The biochemical characterization of the proteins, actin, myosin, and troponin, in muscle as well as hemoglobin in red cells and tubulin in nerve cells constitutes examples illustrating the usefulness of this approach. In all these cases the structural and functional characterization of a given protein provided key insights regarding the function of specific cell types. The elucidation and characterization of nervous system specific proteins is a particularly appropriate research strategy given the enormous cellular diversity and complexity of nervous tissue. The rapid development of techniques in protein purification and in histological procedures such as immunocytochemistry and autoradiography have greatly facilitated progress in the study of brain or nervous system specific proteins and receptors.

In this review we deal with a soluble neuronal protein which we now know to be a cell specific isoenzyme of the glycolytic enzyme enolase (EC 4.2.1.11) and which we have designated neuron specific enolase, or NSE. We attempt to synthesize what has been learned about this protein during the past decade at the structural, functional, and clinical levels. We intend to illustrate the varied types of information to be derived from the study of neuronal or glial proteins, as well as the potential clinically related applications.

Protein maps of brain tissue were first successfully attempted by Moore (Moore & McGregor 1965, Moore 1973). These investigators were the first to show that brain tissue is unique in that it contains a number of highly acidic soluble proteins that are not observed in non-nervous tissue. From these studies two highly acidic brain proteins were purified and characterized, which were designated the 14-3-2 and S-100 proteins (Moore & McGregor 1965, Moore 1973). Early studies utilizing antibodies raised against these two bovine brain proteins suggested that the 14-3-2 protein was neuronal, the S-100 proteins were glial (Moore 1973, Cicero et al 1970, 1972), and that these proteins were present at rather high levels in brain. It has since been established that the 14-3-2 protein is NSE; therefore, its function has been determined and will be the subject of this review.

In order to avoid confusion we mention that the subject of this review has been designated by three names over the past decade. Initially it was called the 14-3-2 protein by Moore (Moore & McGregor 1965). We renamed it neuron specific protein (NSP) upon directly demonstrating its neuronal localization (Pickel et al 1975). When it was shown to have enolase activity we redesignated it neuron specific enolase or NSE (Marangos & Zomzely-Neurath 1976). The designation "enolase" has also been utilized by a number of investigators.

Structural and Immunological Properties of Neuron Specific Enolase

It is a decade since the 14-3-2 protein was shown to be a neuron specific form of the glycolytic enzyme, enolase (Bock & Dissing 1975, Rider & Taylor 1975, Fletcher et al 1976, Marangos et al 1975a, Marangos & Zomzely-Neurath 1976). This realization prompted investigations concerning what other types of enolases were present in vertebrate brain.

Three forms of enolase are present in brain extracts, in marked contrast to the one apparent form present in muscle and liver (Figure 1). The brain is unique in that it contains two acidic forms of enolase. The most acidic of these forms is NSE; the intermediate form is designated as hybrid

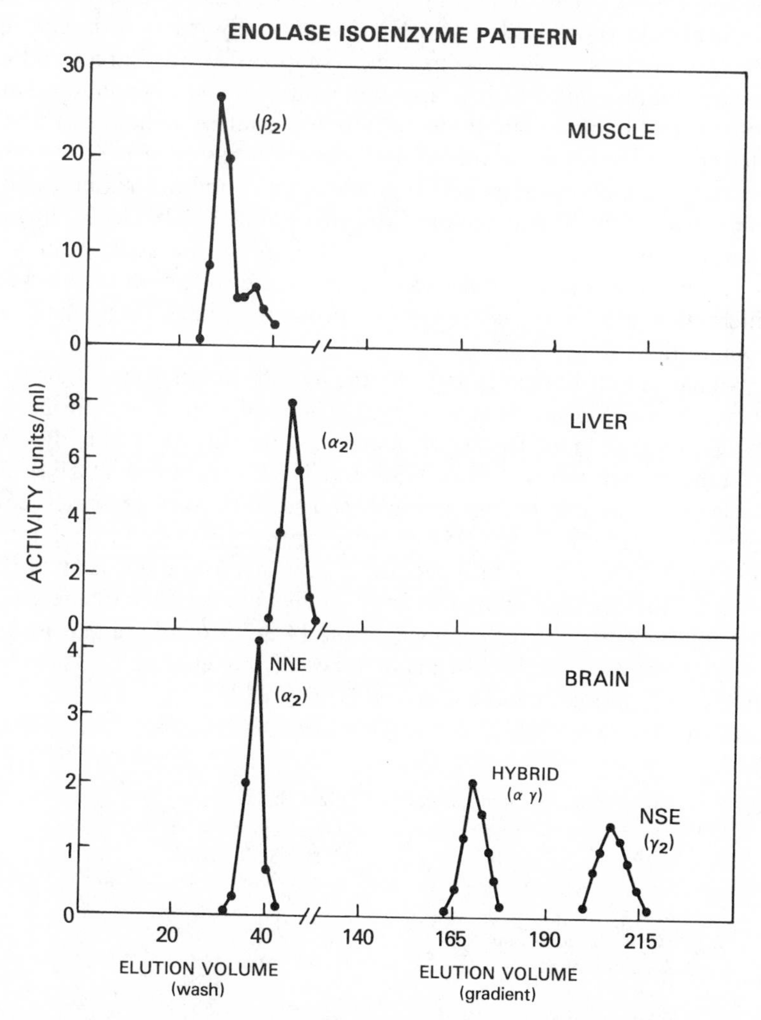

Figure 1 Chromatographic profile of enolase iosoenzymes in various tissues. In each case soluble extracts of the indicated rat tissue were made in trisphosphate buffer (10 mM pH 7.5) and chromatographed on a diethylaminoethyl cellulose column. The wash fraction indicates the buffer elution and the gradient portion is a 0–0.5 M NaCl linear gradient. Similar results are obtained in human tissues.

enolase. The structural properties of these isoenzymes are presented in Table 1, which provides a summary of several previous structural and immunological studies (Marangos et al 1978b, 1977). As shown in Table 1, NSE is a dimer composed of apparently identical subunits, designated δ. The least acidic brain enolase isoenzyme we have called non-neuronal enolase (NNE). This identification was done based on immunocytochemical data showing that anti-NNE serum only reacts with glial cells (see below). This enolase isoenzyme is composed of two larger and less acidic subunits, called α. It is apparent now that the α_2-enolase isolated from rat and human brain is identical to liver enolase, although in studies relating to nervous tissue the NNE designation has persisted. The intermediate brain enolase isoenzyme contains an α and a δ subunit and is referred to as the hybrid enolase (Marangos et al 1978b). As is apparent from the data presented in Table 1, the NSE differs markedly from NNE or liver enolase. The molecular weight and isoelectric points are markedly different, as are the net charge on the proteins as reflected by their electrophoretic mobility.

We have never been able to purify the hybrid enolase from either rat or human brain (Marangos et al 1978b). The final step in our purification procedure is column isoelectric focusing, and this yields highly homogenous preparations of both NSE and NNE (Marangos et al 1975a). Isoelectric focusing of the hybrid enolase, however, generates equal amounts of pure NNE and NSE. Consequently, it has never been purified. Possibly, much of the hybrid enolase in adult nervous tissue is an artifact of tissue homogenization and its levels in vivo are very low (Schmechel & Marangos 1983). The realization that the 14-3-2 protein is a new form of enolase increased the known major types of this enzyme to 3, i.e. liver enolase (α_2), muscle enolase (β_2), and NSE (δ_2).

Table 1 Structural properties of the brain enolases

NNE	Hybrid	NSE	
Mol wt	87,000	82,500	78,000
Subunit composition	$\alpha\alpha$	$\alpha\gamma$	$\gamma\gamma$
Subunit mol wt	43,500	43,500 39,000	39,000
Isoelectric point	7.2	N.D.[a]	4.7
Electrophoretic mobility	0.2	N.D.	0.8
Reactivity with anti-NNE serum	+++	+−	−
Reactivity with anti-NSE serum	−	+−	+++

[a] Not determined.

The most marked difference between NSE and NNE is the apparently complete lack of immunologic cross-reactivity between the two proteins (Marangos et al 1978b, Schmechel et al 1978). Figure 2 shows a standard semi-log plot representing the interaction of human NSE (NSE-H) with anti NSE-H serum. Nanagram amounts of NSE-H inhibit 50% of ^{125}I NSE-H binding to the antiserum, whereas microgram to milligram amounts of pure NNE-H have no effect on binding. This type of result is also obtained when the converse experiment is performed (Marangos et al 1979, Schmechel et al 1978). It is therefore apparent that the α and δ subunit are quite distinct both structurally and immunologically, strongly suggesting that these two proteins are the product of separate genes. Amino acid analysis data further support this as NSE has been shown to have a higher proportion of acidic amino acids than NNE (Marangos et al 1978b).

Functional Properties of Neuron Specific Enolase

As discussed below, NSE is highly localized in neurons and neuroendocrine cells (Pickel et al 1975, Schmechel et al 1978). The obvious question arises, therefore, as to what unique properties this variant of a ubiquitous glycolytic enzyme might have and how these properties serve the physiology of these specialized cell types. The marked structural and immunological difference between NSE and other enolases predicts that the functional properties of NSE would also differ.

Initial studies looking for possible functional differences in NSE at the level of enzyme kinetics were disappointing. When pure NSE and NNE were compared it was found that the K_m for the substrate, 2-phosphoglyceric acid, was very similar, as were the inhibitory potency of fluorophosphate (a relatively specific enolase inhibitor) and the pH optima of the reactions (Marangos & Zomzely-Neurath 1976). A slightly higher affinity for the cofactor, magnesium, was observed for NSE (2.4×10^{-4} vs 6.1×10^{-4} M), but this was judged to be of minor physiologic significance (Marangos & Zomzely-Neurath 1976). Therefore, it appears that from the classical kinetic standpoint NSE is nor markedly distinct from other enolases. This conclusion prompted further studies to look for functionally distinct properties of NSE in areas of the enolase molecule removed from the active site.

Earlier studies of yeast enolase had shown that this enzyme had a rather unique sensitivity to halogens such as chloride and bromide (Gawronski & Westhead 1969). Since chloride is an important ion in relation to neural function, we investigated its effect on pure NSE and NNE. Our results with NNE were similar to those reported earlier for yeast enolase in that millimolar concentrations of chloride and bromide were able to inactivate the enzyme very rapidly via a dissociation of the subunits (Marangos et al

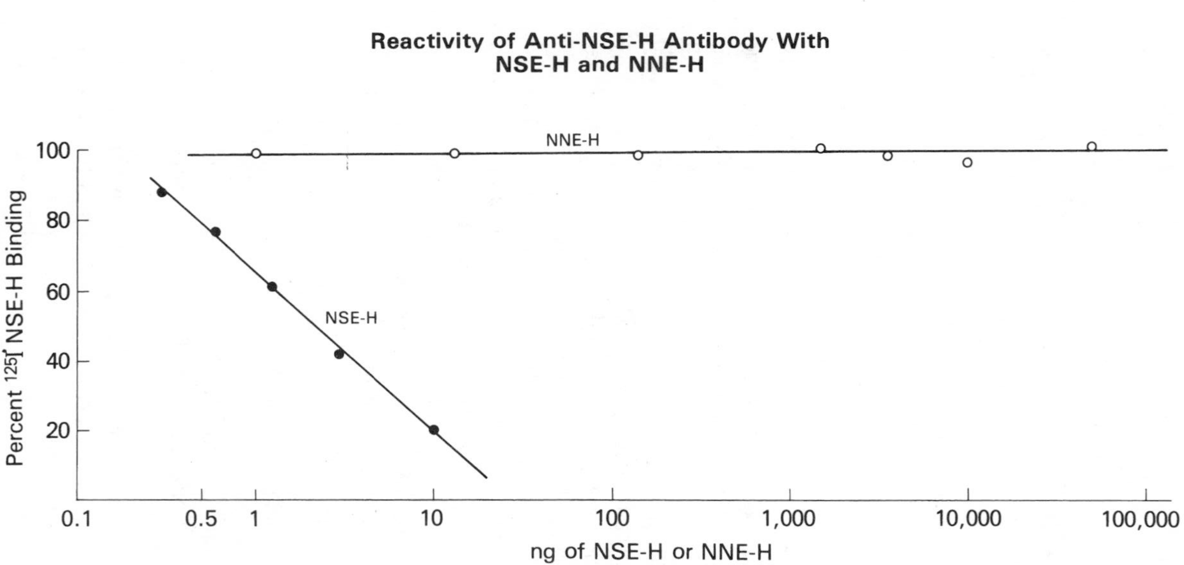

Figure 2 Reactivity of anti-NSE-H antiserum with NSE-H and NNE-H. Pure NNE-H and NSE-H were tested for their ability to inhibit the binding of ^{125}I NSE-H to specific anti-NSE-H antisera. Similar results were obtained by using the rat NSE reagents (rat NSE, rat NNE, and anti-NSE-R antiserum).

1978a). NSE, however, proved to be markedly more stable toward chloride-induced inactivation. The differences observed between NSE and NNE are quite large: 0.5 M KCl has essentially no effect on NSE activity whereas it inactivates virtually 100% of NNE activity. In this same study NSE also proved to be much more resistant to inactivation by high temperature (50°) and much more stable when incubated in the presence of 3 M urea (Marangos et al 1978a). In all these functional comparisons the differences observed between NSE and NNE are very large, i.e. incubation at 50° inactivates virtually all NNE activity within 15 min whereas NSE is unaffected after 1 hr of incubation (Marangos et al 1978a). The functional properties of NSE and NNE are summarized in Table 2.

The picture that emerges from these functional studies indicates that NSE is kinetically quite similar to NNE and other enolases but that it differs markedly from non-nervous tissue enolases in its stability characteristics. This suggests that the active sites of the two proteins are similar; the other functional differences indicate that non-active site regions are very different. The chloride-induced inactivation data are especially interesting since this ion accumulates in nerve cells during periods of repeated depolarizations. It is possible that the marked resistance of NSE towards chloride-induced inactivation may have evolved to accommodate this property of the intracellular milieu in the neuron. If a non-neural enolase were present in the neuron, one would predict that it would be inactivated and glycolysis interrupted precisely when metabolic energy is needed most. NSE may have evolved to function specifically in the neuronal cytoplasmic environment, with its major distinguishing feature being its high degree of stability. Perhaps the switch from NNE to NSE that occurs during neural maturation (see below) involves increases in neuronal chloride levels during

Table 2 Functional properties of NNE and NSE

	NNE	NSE
Substrate K_m	1.3×10^{-4}	1.2×10^{-4}
K_a Mg	6.1×10^{-4}	2.4×10^{-4}
Chloride stability	no	yes
Urea stability (3 M)	no	yes
Temperature stability (50°)	no	yes
Cellular localization	glia and non-neural cells	neurons and neuroendocrine cells
Developmental profile	appearance not correlated to glial maturation	appearance tightly coupled to neural differentiation

the onset of neural activity. Future studies regarding the genetic regulation of NSE expression should provide insights.

Tissue Distribution and Levels of Neuron Specific Enolase

After purifying NSE to homogeneity (Marangos et al 1975a), specific antibodies were raised (Marangos et al 1975b) and radioimmunoassay procedures developed (Marangos et al 1975b, Parma et al 1981). As discussed above, there is essentially no cross-reactivity between anti-NSE serum and NNE and vice versa. This made possible the development of a specific radioimmune assay (RIA) for both NSE and NNE. The most sensitive NSE RIA yet developed utilizes ^{125}I NSE prepared by the Bolton-Hunter method of protein labeling (Parma et al 1981). This double antibody procedure is able to detect picogram amounts of NSE in tissue extracts, serum, or cerebrospinal fluid, as seen in Table 3. The important feature of this assay is its virtually absolute specificity for NSE, as huge excesses of NNE do not react at all (Figure 2). Because anti-NSE serum is unreactive with NNE it is highly unlikely that it will react with other antigens in brain. Immunoelectrophoretic analysis of brain extracts clearly support this since single precipitation arcs comigrating with pure NSE are routinely observed (Marangos et al 1975b, 1978b). Similar results are also obtained when western blots of two-dimensional gels are reacted with anti-NSE serum (Heydorn et al 1985).

The specific assay procedure for NSE has been extensively employed. A summary of some of the results obtained is shown in Table 3. The highest levels of NSE are seen in brain, where they vary from 4 ug/mg

Table 3 NSE levels in various human tissues

	ng NSE/mg soluble protein
Brain (rat and human)	4000–21,000
Peripheral nervous system	200–1200
Adrenal medulla	900
Pineal gland	8500
Pituitary	
anterior	1300
posterior	3400
Liver	<10
Muscle	<10
Serum (human)	4–12
Spinal fluid (human)	1–3

soluble protein to 21 ug/mg soluble protein. Intermediate levels are seen in peripheral nervous tissue and various neuroendocrine glands, while very low levels are seen in non-nervous tissue, serum, and cerebrospinal fluid.

We stress that the quality of any protein RIA is critically dependent on the purity of the antigenused to obtain the antisera. In our laboratory we have extensively documented the high purity of our antigen chromatographically, electrophoretically, by isoelectric focusing, and its behavior during sedimentation equilibrium (Marangos et al 1975a, 1977, 1978b). These precautions coupled with the above data showing no immunologic cross-reactivity between NNE and NSE indicate that each protein can be measured reliably and accurately. Some of the controversy regarding the levels of NSE in various tissues (Marangos et al 1980a, Kato et al 1983, Day & Thompson 1984) may arise from issues such as this.

It is obvious from the results of quantitative studies that NSE is a major brain protein that constitutes between 0.4 to 2.2% of the total soluble protein of brain, depending on the region. Since we know that glial cells do not contain NSE (see below) and that glial cells constitute about half of the cytoplasmic volume in brain, it appears that a very high percentage of the soluble protein in neural cytoplasm is NSE. In some neurons NSE may account for as much as 3–4% of the total soluble protein. This very high level of NSE in the brain allows us to think of it in terms of a very accessible tool for studying neural cells at both the basic and clinical level.

Various endocrine glands such as the adrenal, pituitary, and pineal gland also contain relatively high levels of NSE (Schmechel et al 1979). As we discuss below, the NSE content of these tissues is largely localized in discrete neuroendocrine cell populations. These cells have been termed paraneuronal or APUD (amine precursor uptake and decarboxylation) cells (Pearse 1968, 1976, Polak & Bloom 1979) and are discussed further below. Most non-nervous tissue has very low NSE levels, in the range of 5–10 ng/mg of soluble protein, levels that probably indicate intrinsic innervation. Body fluids such as serum and cerebrospinal fluid also contain very low NSE levels (Table 3). Serum levels generally range from 5–12 ng/ml, while spinal fluid levels are about 2 ng/ml. The neuroendocrine tissues in general contain NSE levels intermediate between those found in brain and non-nervous tissue (Table 3).

Cellular Localization of Neuron Specific Enolase in Neurons and Neuroendocrine Cells

When tissue from the peripheral or central nervous system is properly prepared and reacted with specific antisera to NSE, definite immunoreactivity is seen in neurons of all types (Figure 3). Glial cells, endothelial cells, and other non-neural elements are unstained. These anatomical

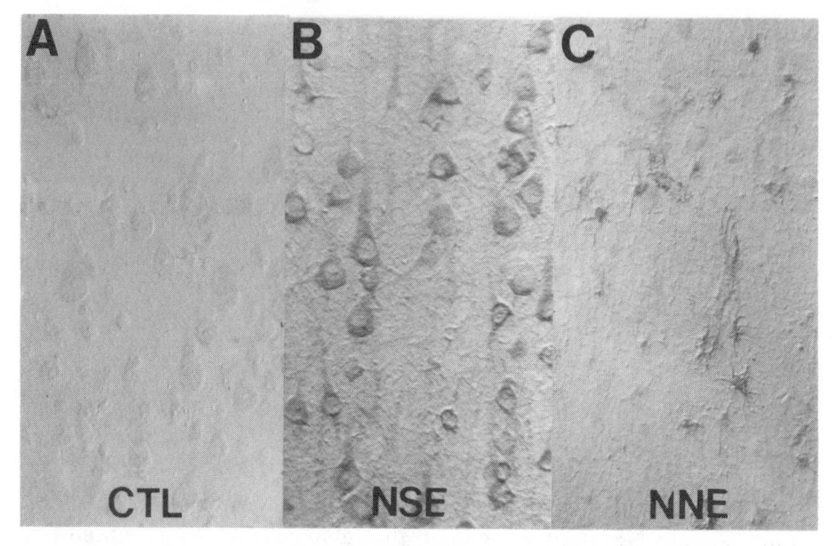

Figure 3 Photomicrographs of immersion-fixed human temporal lobe cortex from fresh surgical specimens. Vibratome sections 25 μ thick were stained with control rabbit serum (A), rabbit anti-NSE-H, and rabbit antihuman NNE (C) by using avidin-biotin peroxidase method. Interference contrast optics show faint cell outlines with control serum (A), but no specific staining. Cortical neurons of all types are strongly NSE-immunoreactive (B), but not NNE immunoreactive (C), glial cells (astrocytes) and their processes, including perivascular and feet, are strongly NNE immunoreactive (C), but not NSE immunoreactive (B). This pattern of enolase localization is typical of nervous tissue and supports relatively strict segregation of NSE (δ_2) in neurons and NNE (α_2) in glial cells.

findings combined with biochemical studies are consistent with a strict localization of the δ subunit, and more particularly the δ_2 isoenzyme, in neurons (Pickel et al 1975, Schmechel et al 1978, Langley et al 1980,

Figure 4 Photomicrographs of perfusion-fixed fetal monkey cerebellum (one hundredth day of gestation—2/3 of gestational period). Vibratome section 25 μ thick were stained with anti-NSE-H (A) and anti-NNE-H (B) by PAP method. The proliferating cells of the external granule layer (EGL) are NNE but not NSE immunoreactive. Purkinje cells and some neurons (Golgi II cells) in the internal granule layer (IGL) are strongly NSE immunoreactive for NSE (A), but not for NNE (example, Pkj label over unstained soma region of unstained Purkinje cell in B). Granule cell neurons in IGL have relatively less NSE immunoreactivity than Purkinje cells and Golgi II cells; this is also typical of adult cerebellum. NNE-stained section (B) shows that Bergmann radial glial cells that occupy the unstained layer in NSE-stained section [denoted by Bgc in (A)], are strongly NNE immunoreactive as in adult cerebellum. There may be a relative decrease in NNE immunoreactivity in the innermost part of the EGL, which is filled with post-mitotic granule cells starting their migration to IGL; this may indicate that neurons not only switch from NNE to NSE during development but may go through a period of lowered total enolase content.

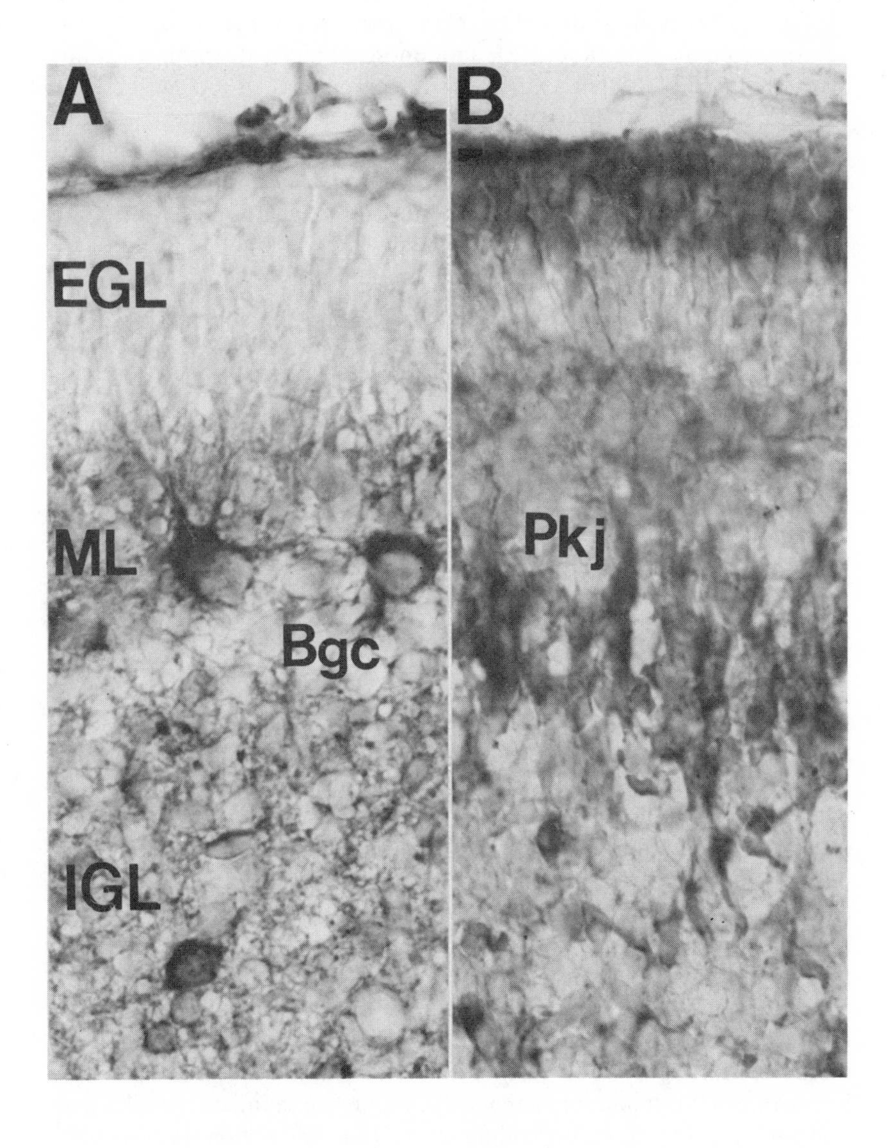
A
B
EGL
ML
Pkj
Bgc
IGL

Ghandour et al 1981; reviewed in Schmechel & Marangos 1983) and hence justify the name "neuron specific enolase" as a convenient common name for this isoenzyme in studies of the nervous system. We emphasize that all neurons contain detectable NSE immunoreactivity with proper fixation and avoidance of protein extraction during processing (Schmechel et al 1980a,b, Schmechel & Marangos 1983). All enolase isoenzymes, including NSE, are soluble cytoplasmic proteins and thus as more difficult than less soluble substances to preserve for anatomical localization studies. The best fixation methods combine properties of coagulation (e.g. acid pH, precipitating agents such as picrate or salicylate) and moderate crosslinking activity (glutaraldehyde in small amounts or formaldehyde) (Schmechel et al 1980a,b, Schmechel & Marangos 1983). The characterization of a cell class as having "low" or "absent" NSE immunoreactivity should only be made after a thorough trial of different fixation and processing protocols.

NSE immunoreactivity fills the cytoplasm, including soma, axon, and dendrites of neurons, but not the nucleus (Figure 3). Ultrastructural studies have confirmed this localization and demonstrated immunoprecipitate dispersed within the cytoplasm as well as next to polyribosomes, microtubules, and external membranes of mitochondria and rough endoplasmic reticulum. NSE immunoreactivity is not seen within the nucleoplasm or Golgi apparatus (Langley et al 1980, Schmechel et al 1980a,b, Trapp et al 1981, Ghandour et al 1981, Vinores et al 1984a). This localization agrees with the biochemical evidence that NSE is a soluble, cytoplasmic enzyme. Nuclear immunoreactivity with antisera to NSE is usually seen in material known to be suboptimal in fixation of processing. Since NSE is very soluble, some aspects of the subcellular localization may merely reflect lability of the antigen during fixation. Moreover, the proposed participation of NSE with other enzymes of intermediate metabolism in a "glycolytic particle" transported by slow axonal transport (Brady & Lasek 1981) has not yet been demonstrated anatomically.

While low dilutions of antisera to NSE (with proper fixation) result in the staining of all neurons, higher dilutions (e.g. 1 : 10,000 or greater in peroxidase anti-peroxidase (PAP) or avidin-biotin protocols) demonstrate that neurons vary widely in intensity of immunoreactivity for NSE (Schmechel & Marangos 1983). Various classes of neurons have a characteristic intensity of NSE immunoreactivity. the difference in intensity varies over a two- to five-fold range, depending on cell class and region. For example, in cerebellum, the relative intensity of staining (from high to low) of neurons is deep cerebellar nucleus neurons, Purkinje cells, Golgi II cells, basket cells, granule cells, and superficial stellate-basket cells (Figure 4). Strong NSE immunoreactivity does not appear to relate to any single variable, such as location, cell size, projection vs local circuit neuron, or transmitter type; rather, these differences appear to denote a very specific metabolic profile in each region that may be related to varying metabolic demands. Table 4 summarizes NSE and NNE cell distribution in nervous tissue.

NSE immunoreactivity is not only seen in neurons but also in (*a*) central neuroendocrine cells, (*b*) peripheral neuroendocrine cells, and to a very limited degree in (*c*) selected cells that are neither nerve cell nor neuroendocrine. The first two categories include not only cells in endocrine glands such as pineal, pituitary, thyroid, pancreas, and adrenal glands

Table 4 Localization of NNE and NSE in nervous tissue

	NNE	NSE
Glial cells		
Astrocytes	+	−
Oligodendrocytes	+	−
Radial glial cells	++	−
Ependymal cells	+	−
Tanocytes	+	−
Neurons		
Cerebellum		
Stellate basket cells	+	++
Purkinje cells	−	+++
Golgi type II	−	++++
Granule cells	−	++
Deep cerebellar neurons	−	++++
Hippocampus		
Non-pyramidal neurons	−	++++
CA 3 pyramids	−	+++
Dentate granule cells	−	++
CA 1 and 2 neurons	+	+
Cortex		
Non-pyramidal neurons	−	++++
Superficial pyramids	+	++
Deep pyramids	−	+++
Thalamus		
Reticular neurons	−	++++
Sensory relay neurons	−	+
Subcortical nuclei		
Compacta neurons (nigra)	−	+
Reticular neurons (nigra)	−	++++
Striatal neurons	−	+
Pallidal neurons	−	++++

(Schmechel 1979), but also cells in the circumventricular organs of the brain and the diffuse neuroendocrine systems of the lung, intestine, thymus gland, and skin (Tapia et al 1981, Wick et al 1983, Schmechel & Marangos 1983; for review see Polak & Marangos 1984). Some neuroendocrine tissue (and presumably cells) may approach neurons in NSE content (e.g. 8 ug/mg protein in pineal) whereas others are probably quite low (e.g. 0.9 ug/mg protein in adrenal medulla). Like neurons, different classes of neuroendocrine cells even in the same gland may vary considerably in apparent NSE content (e.g. pituitary cells) (Van Noordern et al 1984). In addition, NSE immunoreactivity can also be seen in peripheral tissues and cell classes that are clearly neither neuronal nor neuroendocrine. The first example to be discovered were platelets (Marangos et al 1980a). The predominant enolase isoenzyme form in platelets is NNE, with some representation of hybrid enolase ($\alpha\delta$) and $<1\%$ of NSE or δ_2 enolase. This finding demonstrates that immunoreactivity with antisera raised to NSE recognizes not only NSE (δ_2) but hybrid enolase ($\alpha\delta$) and potentially single δ subunits as well. NSE has also been proposed to occur in lymphocytes, smooth muscle cells of visceral organs, juxtaglomercular organ of kidney, and smooth muscle of vessels (Haimoto et al 1985). These examples all contain lower than 1 μg/ml protein of NSE, of which the majority is likely $\alpha\delta$, as for platelets.

Finally, transformed or tumor cells of the above classes of NSE-containing cells (neurons, peripheral, and central neuroendocrine cells) also demonstrate NSE content by RIA and immunocytochemistry (Tapia et al 1981; for review see Polak & Marangos 1984). Significant interest has been generated at the clinical level regarding the rather specific association of NSE with neuroendocrine tumors of the APUDoma designation, as is discussed below.

In summary, NSE can be considered a specific marker for neurons and neuroendocrine cells. The specificity and high titer of many antisera raised to NSE make it possible to demonstrate immunoreactive δ subunit in tissues ranging from 20 μg/mg protein (brain) to 0.1–0.5 μg/mg protein (platelets, other tissues with immunoreactive cells). Liver and skeletal muscle cells show no NSE immunoreactivity, as is consistent with their low NSE content by RIA (<0.01 μg/mg protein). The probable presence of true NSE (i.e. δ_2 isoenzyme) must be substantiated by biochemical separation techniques and the immunocytochemical localization of antisera raised to NNE (α_2). Only rare tissues (e.g. platelets) that are homogeneous or extraordinary techniques (Kato et al 1981) with their own sources of error can directly assist in the cellular localization of enolase isoenzymes. In the future, in situ hybridization with probes for specific mRNA may also help in this regard. We can presently state the following with reasonable confidence:

1. NSE is strictly localized in the central and peripheral nervous system in neurons.
2. The amount of NSE in different classes of neurons may vary by a factor of two- to five-fold.
3. NSE immunoreactivity is also seen in peripheral and central neuroendocrine cells; levels are generally lower than in nerve cells and in many cases this is due to the presence of mainly hybrid enolase. NSE is therefore a very useful probe for virtually all neuroendocrine cells.
4. The sensitivity of immunocytochemical techniques for the localization of antisera directed to NSE is such that tissues with 20–40-fold less NSE content than brain, with some hybrid enolase isoenzyme and $<1\%$ of total enolase present, are also found to be immunoreactive. This sensitivity emphasizes that "NSE" immunoreactivity alone only denotes presence of δ subunit and the presence of the hybrid enolase must be determined with the help of other biochemical information.

Developmental Expression of Neuron Specific Enolase in Neurons: α to δ Switch-over

The striking and specific expression of the NSE (δ_2) isoenzyme in neurons is emphasized by the fact that the latter contains 20 times as much NSE as do some neuroendocrine cells (e.g. adrenal medullary cells), 200 times as much as non-nervous, non-neuroendocrine cells (e.g. platelets), and 1000–2000 times as much as liver or skeletal muscle cells. Two questions then arise: Is the expression of the δ subunit correlated with neuronal differentiation? What is the ontogenic expression of the α subunit in neurons? Early studies performed in neuroblastoma cells showed that NSE levels were low and increased in response to agents that promoted neural differentiation (Marangos & Goodwin 1978). This finding clearly suggested that NSE is correlated with neural differentiation, and it prompted more detailed immunocytochemical studies.

A number of anatomical and biochemical studies have established the timing of NSE expression in neurons. NNE (α_2) is the dominant isoenzymes form in fetal brain (Fletcher et al 1976, Marangos et al 1980b, Jorgensen & Centervall 1982, Yoshida et al 1983, Kato et al 1984). The appearance of NSE occurs with the onset of neurogenesis and continues to increase well after the end of neurogenesis, during the period of early neuronal differentiation (Marangos et al 1980a,b, Jorgensen & Centervall 1982). This data strongly supported a switch-over from α to δ subunit expression in neurons but could not delineate the cellular basis and timing of this switch-over, given the heterogeneous character of neural tissue. Specific immunocytochemistry for NNE and NSE immunoreactivity provides a beginning for fully understanding this switch-over of genetic expression (Schmechel et al 1980a,b, Maxwell et al 1982, Whitehead et al 1982, Chen & Omenn 1984).

Immunocytochemistry has demonstrated the strong and early presence of NNE immunoreactivity in astrocytes and radial glial cells, as well as in the proliferating cell populations (ventricular and subventricular zone and their derivatives) that give rise to neurons (Schmechel et al 1980b). Even in midgestation, some strong NSE immunoreactivity can be seen in early generated neurons whose morphological and functional differentiation is already commenced (Figure 4). These neurons are probably responsible for the "start" of detectable NSE tissue levels, but their signal is swamped by the many less differentiated neurons and precursors. The most important finding is that postmitotic neurons in cerebellum and cerebral cortex can be definitely identified as having low-to-absent NSE immunoreactivity; however, they do have strong and definite NNE immunoreactivity during their migration from the site of final cell division to their eventual adult location (Schmechel et al 1980b). Neurons newly arrived in regions with delayed or prolonged differentiation, such as the superficial layers of cerebral cortex, do not exhibit significant NSE immunoreactivity well after their arrival. Elegant studies in the chick vestibular and auditory system further support the notion that appearance of NSE immunoreactivity (and presumably significant levels of NSE δ_2 isoenzyme) coincide with the onset of functional synaptic activity (Maxwell et al 1982, Whitehead et al 1982). Alternate possibilities such as expression linked to final cell division, cell migration, or even arrival in the adult site are not supported. We emphasize that the close linking of NSE expression to functional activity of the neuron (Figure 5) supports the concept of a close regulation and adaptation of neuronal metabolic enzymes to actual status of the nerve cell (Marangos et al 1978a).

These studies therefore supported a switch-over from α to δ isoenzyme subunit expression in neurons closely linked to the final functional differentiation and activity of the nerve cell. The second question posed above—that of the eventual expression of the α subunit in neurons after differentiation—is still controversial. A considerable population of neurons in the early stages of differentiation probably contain both α and δ subunits (and, therefore, potentially NNE, hybrid, and NSE isoenzymes). Such cells may be the substrate for the early appearance of hybrid enolase ($\alpha\delta$) in biochemical analysis of the NNE to NSE shift during development (Jorgensen & Centervall et al 1982, Kato et al 1984). Our studies in differentiated nervous tissue have not suggested significant NNE immunoreactivity in most mature neurons (compared to astrocytes) (Schmechel et al 1978, 1980b, Schmechel & Marangos 1983). The exceptions to date include stellate-basket cells of cerebellum and some hippocampal pyr-

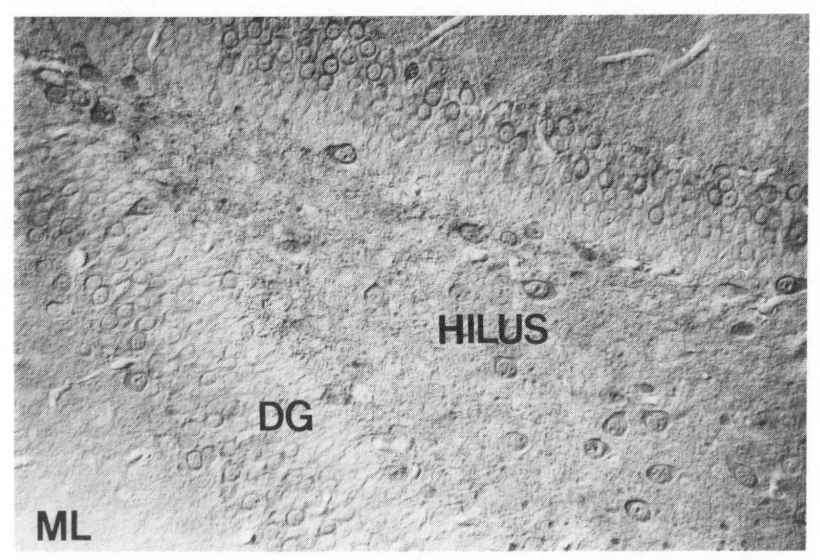

Figure 5 Photomicrograph of 25 μ thick vibratome section of rat hippocampus (2 wk postnatal) stained for localization of rat NSE by using the PAP method. Hilus area of dentate gyrus shows NSE immunoreactive neuropil and scattered cell bodies. Granule cell neurons in the dentate gyrus (DG) granule cell layer show a distinctive and typical pattern of increasing NSE immunoreactivity from inner to outer aspect of the cell layer. This pattern correlates with the developmental gradients of the dentate gyrus granule cell neurons and indicates that NSE content may progressively increase during postnatal differentiation. In the adult rat, the granule cell neurons in this layer show a very uniform degree of NSE immunoreactivity. ML stands for molecular layer.

amidal neurons and cortical pyramidal neurons. NSE is by far the most abundant form of enolase in virtually all adult neurons.

The present picture of enolase isoenzyme expression in neurons offers a striking example of correlating the cell-specific regulation of isoenzymes with specific requirements of the cell. Further work is needed to resolve the issue of α_2 and $\alpha\delta$ levels in adult neurons, the apparent marked variability of δ expression in different classes of neurons, and the status of $\alpha\delta$ expression in various classes of neuroendocrine cells. We predict that the level of relative $\alpha\delta$ expression in neurons will be related to definite metabolic requirements of the neurons and may predict or measure the metabolic status of the nerve cell.

In summary, the following conclusions can be made about the switchover from NNE to (toward) NSE expression in neurons:

1. Neurons develop from proliferating cell populations that are NNE immunoreactive without apparent NSE immunoreactivity.

2. Postmitotic neurons show at least two patterns: (*a*) low-to-absent

immunoreactivity for either NNE or NSE (cerebellum) or (*b*) strong immunoreactivity for NNE and absent immunoreactivity for NSE (cortex). This suggests that the switch from NNE to NSE occurs after final cell division and often after migration, and that over-all enolase levels may be lower. This may be the anatomical basis for the decrease in NNE levels seen during development.

3. Different classes of neurons acquire NSE immunoreactivity in a manner suggesting close correlation of the expression of NSE (δ_2) with functional synaptic activity. The apparent two- to five-fold range in NSE immunoreactivity among different classes of adult neurons suggests further class-specific regulation of NSE levels, perhaps related to differing metabolic requirements.

4. Immunoreactivity for NNE is low in most classes of adult neurons, except for certain definite examples. Since both NNE and NSE antisera detect examples of cells known to contain hybrid enolase ($\alpha\delta$), the anatomical localization is discordant with the significant amounts of hybrid enolase seen in biochemical analysis.

5. Present information suggests that the proportion of $\alpha\delta$ expression, levels of δ expression, and distribution in the three isoenzymes NNE, hybrid, and NSE may be very distinctive for different classes of neurons and neuroendocrine cells. A major or near total switch-over from NNE to NSE occurs in neurons and distinguishes them from neuroendocrine cells, where a less complete switch-over is probable. All three parameters ($\alpha\delta$ proportion, level of δ, and isoenzyme distribution) may allow basic and clinical assessment of the differentiation of neurons and neuroendocrine cells. The α to δ switch can, therefore, be viewed as a marker for neural differentiation.

Clinical Applications

That NSE is so highly localized in neurons and neuroendocrine cells suggests that the monitoring of its levels in biological fluids such as spinal fluid or serum might provide clinically useful information regarding diseases characterized by altered metabolism in or turnover of these cell types. Certain neurologic disorders in which neurons are known to die off, such as Alzheimer's, Huntington's disease and Amyotrophic Laterosclerosis, might be characterized by increases in spinal fluid or serum NSE levels. Convincing evidence of altered spinal fluid or serum NSE levels in patients with diagnosed neurological disorders has been difficult to gather, however. The reason for this difficulty may reside in our inability to determine when cytolytic processes are actually occurring during the often prolonged course of degenerative neurological disorders. Neuronal degeneration and the consequent predicted NSE efflux may occur well before the onset of observable symptoms, perhaps and we are simply taking the biological sample too late to measure altered NSE levels. Elevated spinal fluid NSE levels have been observed in head trauma (Steinberg et al 1984, Dauberschmidt et al 1983, Scarna et al 1982) and following strokes (Hay et al 1984), in which the time course of the process can be more accurately documented. These studies clearly suggest that NSE levels in biological fluids can be a useful parameter for the assessment of neural tissue damage in the brain (Royds et al 1981). Further studies of serum and spinal fluid NSE levels in larger numbers of ischemic patients will provide a clearer picture of the relationship of neurological symptoms and NSE levels and may even reveal that the level of NSE in spinal fluid or serum correlates with the degree of brain damage.

NSE was originally thought to be exclusively localized in neurons (Pickel et al 1975, Schmechel et al 1978). It has recently been shown however, to be present in virtually all of the neuroendocrine, paraneuronal cell types as well (Schmechel et al 1979, Tapia et al 1981, Polak & Marangos 1984). This observation coupled with the fact that transformed neuroendocrine cells found in various neuroendocrine cancers or APUDomas are very rich in NSE (Tapia et al 1981, Odelstad et al 1982, Wick et al 1983, Marangos et al 1982, Dhillon et al 1985) prompted us to investigate the potential clinical utility of serum NSE levels as a diagnostic or prognostic index in these disorders. A large number of neuroendocrine neoplastic tissue types have been found to contain high levels of NSE (Tapia et al 1981, Odelstad et al 1982, Sheppard et al 1984, Beemer et al 1984, Bishop et al 1982, Haimoto et al 1985, Ishiguro et al 1984) by both radioimmunoassay and immunocytochemistry. NSE immunostaining in the periphery generally indicates a neuroendocrine cell (Polak & Marangos 1984), with very few exceptions (Haimoto et al 1985). Therefore, NSE is quite useful in tissue typing.

The additional question arose of whether altered serum NSE levels are characteristic for any of the neuroendocrine neoplasms. Among the patients we have screened, quite an interesting picture has emerged. In most of the neuroendocrine neoplasms, serum NSE levels were apparently normal, but two groups of patients stood out in the first large scale studies: those with small-cell lung cancer and those with pediatric neuroblastoma (Carney et al 1982, Zeltzer et al 1983, 1985).

Several clinical studies have now been performed that show that serum NSE levels are elevated in many small-cell lung cancer patients (Carney et al 1982, Johnson et al 1984, Ariyoshi et al 1983). Although serum NSE levels do not constitute an early detection marker for small-cell lung cancer, this parameter is very useful in following the clinical course of the illness (Carney et al 1982) and in predicting the onset of relapse (Johnson et al

1984). Normal serum NSE levels range from 5–15 ng/ml, whereas levels of as high as 1000 ng/ml have been observed in small-cell patients. More importantly, serum NSE levels correlate well with the clinical course of the illness in that they decline to normal levels during remission and become elevated again during relapse (Johnson et al 1984, Carney et al 1982). Elevated serum NSE levels are observed in greater than 90% of small-cell patients with extensive disease (multiple tumor sites) and in 50–60% of patients with limited disease (tumor localized to lung) (Johnson et al 1984).

Recent studies strongly suggest that serum NSE may be an early detection marker for clinical relapse in small-cell patients (Johnson et al 1984). Johnson et al (1984) reported elevated serum levels 4–12 weeks prior to detection by conventional means (chest X-ray) in patients who were in chemotherapy-induced remission. Serum NSE levels are therefore useful in following the clinical course of small-cell lung cancer patients. This new and convenient methodology may affect the outcome of this neoplasm by enabling more timely reinstitution of chemotherapy. To date no study has been performed using serum NSE levels to determine the timing of chemotherapy treatments.

The results obtained in studies of the second neuroendocrine neoplasm with which significant and reproducible clinical results have been obtained, pediatric neuroblastoma (Zeltzer et al 1983, 1985, Ishiguro et al 1984, Beemer et al 1984, Odelstad et al 1982), are in many ways similar to those reported in small cell lung cancer. Serum NSE levels are elevated in virtually all of the Stage IV (>98%) patients tested and to a lesser degree in the less severe stages. Serum levels can be as high as several thousand nanograms per milliliter, and the levels decrease during remission and increase at relapse (Zeltzer et al 1983, 1985). Serum NSE levels have been particularly useful, in that they appear to be able to distinguish Wilms tumor patients from neuroblastoma patients (Odelstad et al 1982, Zeltzer et al 1985). Studies have also shown that the serum NSE level is highly correlated with survival time (Zeltzer et al 1985) in neuroblastoma patients of all ages and may be able to predict survival in infants less than one year of age. Using 100 ng/ml serum as the cut-off point, it was found that infants having levels below this had a much better prognosis than those with levels above. This finding may prove useful in aiding clinical judgments regarding treatment alternatives for this subpopulation of patients.

Again, as seen with small-cell lung cancer, serum NSE levels are very useful for following the clinical course of pediatric neuroblastoma patients. However, this parameter apparently does not constitute an early detection marker for either disease. Additional diagnostic information is required to discriminate the two neuroendocrine neoplasms. Several small studies have also shown modest elevations of serum NSE levels in pancreatic islet cell carcinoma (Prinz & Marangos 1983). Further larger studies in this are should prove useful. Future studies of serum NSE levels in APUDoma patients should reveal just how generalized the above observations are. NSE immunostaining of neuroendocrine tumor tissue has, however, been shown to constitute a useful pathologic parameter for all APUDomas (Polak & Marangos 1984) and should be useful as a generalized histopathological marker for these tumors. Some of the clinical findings discussed above are summarized in Table 5.

Within the past several years NSE has become a very useful clinical tool for both the identification of neuroendocrine tumor tissue by the pathologist and as a useful index for assessing response to treatment in patients with small-cell lung cancer and neuroblastoma. NSE immunocytochemistry and the serum NSE RIA will probably become routine clinical procedures for neuroendocrine neoplasms.

Molecular Biology of Neuron Specific Enolase

The cell specific nature of NSE and that it is a neuronal differentiation marker make it a very interesting candidate for gene cloning. Characterization of the mechanisms involved in controlling NSE synthesis

Table 5 Current clinical applications of the NSE methodology

Disorder	Clinical finding	Predictive or prognostic value
Neurologic Stroke and head trauma	Elevated NSE level in CSF	NSE level may serve as an index of CNS damage
Neuroendocrine cancers Small cell lung cancer	Elevated serum level which correlates with clinical course	Specific diagnostic factor for small cell lung cancer, monitors drug response and predicts relapse
Pediatric neuroblastoma	Elevated serum level which follows clinical course and predicts survival	Diagnostic and evaluates drug response
Neuroendocrine tumors	Stains all neuroendocrine cells in tumor tissue	Pathological tool for tumor typing

should provide basic insights into the process of neuronal differentiation at the genetic level. The gene for rat NSE has been cloned recently with potential specific DNA probes now available (Sakimura et al 1985a). We have also in our laboratory recently generated cDNA probes specific for human NSE (Ginns et al 1986) and generated sequence data for the human protein (Martin et al 1986). In rat there is a rather high degree of sequence homology between NSE and NNE (Sakimura et al 1985b). This finding is initially somewhat surprising given the apparent total lack of immunologic crossreactivity between NNE and NSE discussed above. The size of the NSE mRNA has been determined to be about 2400 bases, of which about 1200 bases are translated into NSE. The NNE mRNA from rat, although highly homologous, is consistently shorter (1800 bases), having the same number of coded bases as NSE but a shorter 3′ noncoding region (Sakimura et al 1985a,b). The genetic analysis, therefore, indicates that NNE and NSE are the same size, contrary to the direct biochemical analysis of the purified proteins that indicates that NNE is larger than NSE (Marangos et al 1978b). Possibly these differences might be accounted for by some post-translated mechanism.

The recent acquisition of probes for NSE offer the possibility of addressing some important issues concerning neural maturation. It will be important to determine the chromosomal localization of the NSE gene to verify earlier postulates that NNE and NSE are distinct gene products. Characterization at the genetic level of the α to δ subunit switch that occurs during neural differentiation should provide basic insights regarding the molecular basis of neural maturation. The most interesting findings concerning the role of NSE in neural function will probably emerge from future neurogenetic studies.

Discussion

The following summarizes what is currently known about NSE at the biochemical, anatomical, functional, and clinical level:

1. The δ subunit is a marker for all types of neurons.
2. In the periphery the δ subunit serves as a marker for virtually all neuroendocrine or paraneuronal cell types.
3. The α to δ subunit switch is a late event in neural differentiation, making NSE a good index of neural maturation.
4. NSE is unique from other enolases as regards its high degree of stability toward chloride-induced inactivation.
5. Serum and spinal fluid NSE levels are useful clinically in disorders involving neurons or neuroendocrine cells.

All neurons containing NSE at levels that probably are related to their functional activity. This protein is rather unique in that it is a general functional marker for neurons. Although this property has been apparent for a decade, the use of NSE as a general neuronal marker is only now becoming widespread. When we initially designated this protein "NSE" it was not realized that it also exists (at substantially lower levels) in neuroendocrine cells. Although this latter realization has prompted some criticism of the nomenclature, the close structural and functional relationship between neurons and neuroendocrine cells, in our view, warrants retaining this designation for the protein. The low levels of NSE recently observed in certain types of muscle cells (Haimoto et al 1985) also should be viewed in a context that recognizes the relative levels of NSE in these cells. Given the enormously high levels of NSE present in central type neurons (2–3% of soluble neuronal protein) (Marangos et al 1977, 1979) and the strict correlation between its appearance and the acquisition of neuronal function (Schmechel et al 1980b, Whitehead et al 1982), it would appear that the NSE designation is valid.

The presence of NSE in paraneuronal neuroendocrine cells in the periphery has turned out to be highly useful. These cells are generally sparsely distributed in most organ systems, such as the gastrointestinal tract, lung, kidney, and pituitary, thus making the cells difficult to distinguish. NSE immunostaining has helped to visualize these cells and represents a unique way to map the peripheral neuroendocrine cell system (Polak & Marangos 1984; see Marangos et al 1982 for review). It is important to realize that NSE immunostaining and immunoassay are currently the only means of visualizing and quantifying the diffuse neuroendocrine system in its entirety. Virtually all of these cells secrete some type of neuropeptide and they clearly play an important role in homeostatic mechanisms that integrate with the brain (Polak & Bloom 1979). NSE immunostaining has become a criteria for identifying neuroendocrine cells in the periphery and may become important in identifying new putative neuroendocrine cells and their respective peptides. That NSE is present in virtually all neuroendocrine cells adds further support to the view that they are neuronal relatives.

The NSE methodology developed during the past decade provides a good example of the kinds of information and applications that can be derived from the study of cell specific proteins. Not only has NSE become a useful marker for neurons and neuroendocrine cells but it also serves as an index for studying neural differentiation. The availability of the appropriate genetic probes should yield interesting results soon concerning the mechanisms involved in the α to δ subunit switch that occurs during neural development and thereby provide important insights into the process of neuron specific gene expression.

The clinical utility of NSE as an index of the clinical course of various neuroendocrine neoplasms is now obvious. Future studies will reveal whether serum or spinal fluid NSE levels can also be useful in diagnosing the neurological disorders, but current data indicate that this may not be feasible due to the apparently discrepant timing between neural tissue distinction and the appearance of symptoms. NSE levels should be useful in stroke patients, however, since the onset of neural damage can be precisely determined. Preliminary results are encouraging (Steinberg et al 1984, Hay et al 1984, Dauberschmidt et al 1983), and future studies should be designed to determine whether serum or spinal fluid NSE values are correlated with the degree of brain damage.

Classical studies regarding the regulation of glycolysis have not suggested any regulatory role for enolase. It is intriguing to speculate that glycolysis may assume a different role in brain compared to other tissues. Certainly NSE appears to have evolved into a much more stable enzyme than non-neural enolases and it is ideally suited to the neuronal cytoplasmic environment as regards its stability toward chloride-induced inactivation. Since the major source of energy in the vertebrate CNS is glucose and since brain tissue has virtually no glycogen stores, it follows that an uninterrupted glycolytic cycle is necessary to provide precursors for aerobic metabolism. This situation is best illustrated by the absolute dependence of nervous tissue viability on an uninterrupted blood supply to the brain. Central neurons require a continuous supply of both oxygen and glucose and begin to die in a matter of seconds when either is interrupted.

Glycolysis can be visualized as an extremely important gating mechanism of brain energy metabolism. Although the actual energy derived from this pathway is insignificant, it serves to process the only fuel available to the brain (glucose) into a form that can be utilized for oxidative metabolism. Therefore, virtually all of the energy generating capacity of brain is dependent on the functioning of the glycolytic sequence. Certainly the distinctive properties of NSE in relation to non-neural enolases is consistent with the proposal that glycolysis in brain may be a much more stable pathway in that the constituent enzymes may be less labile. The regulation of glycolysis in vertebrate brain is, therefore, of utmost importance in brain physiology. Future studies relating to brain specific forms of other glycolytic enzymes may prove to be important.

ACKNOWLEDGMENTS

Paul J. Marangos acknowledges the excellent and dedicated support of S. Coopersmith for preparing the manuscript and P. Montgomery for technical support.

Literature Cited

Ariyoshi, Y., Kato, K., Ishiguro, Y., Ota, K., Sato, T., Suchi, T. 1983. Evaluation of serum neuron specific enolase as a tumor marker for carcinoma of the lung. *Gann* 74: 219

Beemer, F. A., Vlug, A. M. C., Van Veelen, C. W. M., Rijksen, G., Staal, G. E. J. 1984. Isoenzyme patterns of enolase of childhood tumors. *Cancer* 54

Bishop, A. E., Polak, J. M., Facer, P., Ferri, G.-L., Marangos, P. J., Pearse, A. G. E. 1982. Neuron specific enolase: A common marker for the endocrine cells and innervation of the gut and pancreas. *Gastroenterology* 83: 902–15

Bock, E., Dissing, J. 1975. Demonstration of enolase activity connected to the brain specific protein 14-3-2. *Scand. J. Immunol.* (Suppl. 2)4: 31–36

Brady, S. T., Lasek, R. J. 1981. Nerve specific enolase and creative phosphokinase in axonal transport: Soluble protein and the axoplasmic matrix. *Cell* 23: 515–20

Carney, D. N., Marangos, P. J., Ihde, D. C., Bunn, P. A., Cohen, M. H., Minna, J. D., Gazdar, A. F. 1982. Serum neuron specific enolase: A marker for disease extent and response to therapy in patients with small cell lung cancer. *Lancet* 1: 583–86

Chen, S. H., Omenn, G. S. 1984. Human neuron specific enolase: genetic and developmental studies. *J. Neurogenet.* 1: 159–64

Cicero, T. J., Cowan, W. M., Moore, B. W., Suntzeff, V. 1970. The cellular localization of the two brain specific proteins, S-100 and 14-3-2. *Brain Res.* 18: 25–34

Cicero, T. J., Ferrendelli, J. A., Suntzeff, V., Moore, B. W. 1972. Regional changes in CNS levels of the S-100 and 14-3-2 proteins during development and aging of the mouse. *J. Neurochem.* 19: 2119–25

Dauberschmidt, R., Marangos, P. J., Zinsmeyer, J., Bender, V., Klages, G., Gross, J. 1983. Severe head trauma and the changes of concentration of neuron specific enolase in plasma and in cerebrospinal fluid. *Clin. Chem. Acta* 131: 165–70

Day, I. N. M., Thompson, R. J. 1984. Levels of immunoreactive aldolase C, creative kinase-BB, neuronal and non-neuronal enolase and 14-3-2 protein in circulating human blood cells. *Clin. Chem. Acta* 136: 219–28

Dhillon, A. P., Rode, J., Phillon, D. P., Moss, E., Thompson, R. J., Sprio, S. G., Corrin, B. 1985. Neural markers of carcinoma of the lung. *Br. J. Cancer* 51: 645–52

Fletcher, L., Rider, C. C., Taylor, C. B. 1976. Enolase isoenzymes. III. Chromatographic and immunological characteristics of rat brain enolase. *Biochim. Biophys. Acta* 452: 245–52

Gawronski, H., Westhead, E. W. 1969. Equilibrium and kinetic studies on the reversible dissociation of yeast enolase by neutral salts. *Biochemistry* 8: 4261–70

Ghandour, M. S., Langley, O. K., Keller, A. 1981. A comparative immunohistochemical study of cerebellar enolases: Double labeling techniques and immunoelectromicroscopy. *Exp. Brain Res.* 41: 271–79

Ginns, E. I., Miller, N., Martin, B. M., Fong, K., Abrahamson, L., Winfield, S., Marangos, P. J. 1986. Isolation and sequence analysis of cDNA clones for human nonneuronal and neuron specific enolase. *Fed. Proc.* 45 (Abstr. 826)

Haimoto, H., Takahashi, Y., Koshikawa, T., Nagura, H., Kato, K. 1985. Immunohistochemical localization of γ-enolase in normal human tissues other than neurons and neuroendocrine tissues. *Lab. Invest.* 52: 257–63

Hay, E., Royds, J. A., Aelwyn, G., Davies-Jones, B., Lewtas, N. A., Timperley, W. R., Taylor, C. B. 1984. Cerebrospinal fluid enolase in stroke. *J. Neurol. Neurosurg. Psych.* 47: 724–29

Heydorn, W. E., Creed, J. G., Marangos, P. J., Jacobowitz, D. M. 1985. Identification of NSE in human and rat brain on 2 dimensional polyacrylamide gels. *J. Neurochem.* 44: 201–9

Ishiguro, Y., Kato, K., Takahiro, I., Horisawa, M., Nagaya, M. 1984. Enolase isoenzymes as markers for differential diagnosis of neuroblastoma, khabdomyosarcoma and Wilms' tumor. *Gann* 75: 53–60

Johnson, D. H., Marangos, P. J., Forbes, J. T., Hainsworth, J. D., Welch, R. V., Hande, K. R., Greco, F. A. 1984. Potential utility of serum neuron specific enolase levels in small cell carcinoma of the lung. *Cancer Res.* 44: 5409–11

Jorgensen, O. S., Centervall, G. 1982. $\alpha\gamma$-enolase in the rat: Ontogeny and tissue distribution. *J. Neurochem.* 39: 537–42

Kato, K., Asai, R., Shimizu, A., Suzuki, F., Ariyoshi, Y. 1983. Immunoassay of three enolase isoenzymes in human serum and in blood cells. *Clin. Chem. Acta* 127: 353–58

Kato, K., Suzuki, F., Semba, R. 1981. Determination of brain enolase isoenzymes with an enzyme immunoassay at the level of single neurons. *J. Neurochem.* 37: 998–1005

Kato, K., Suzuki, F., Watanabe, T., Semba,

R., Keino, H. 1984. Developmental profile of three enolase isoenzymes in rat brain: Determination from one cell embryo to adult brain. *Neurochem. Int.* 6: 81–84

Langley, O. K., Ghandour, M. S., Vincendon, G., Gombos, G. 1980. An ultrastructural immunocytochemical study of nerve-specific protein in rat cerabellum. *J. Neurocytol.* 9: 783–98

Marangos, P. J., Campbell, I. C., Schmechel, D. E., Murphy, D. L., Goodwin, F. K. 1980a. Blood platelets contain a neuron specific enolase subunit. *J. Neurochem.* 34: 1254–58

Marangos, P. J., Gazdar, A. F., Carney, D. N. 1982. Neuron specific enolase in human small cell carcinoma cultures. *Cancer Lett.* 15: 67–71

Marangos, P. J., Goodwin, F. K. 1978. Neuron specific protein (NSP) in neuroblastoma cells: Relation to differentiation. *Brain Res.* 145: 49–58

Marangos, P. J., Parma, A. M., Goodwin, F. K. 1978a. Functional properties of neuronal and glial isoenzymes of brain enolase. *J. Neurochem.* 31: 727–32

Marangos, P. J., Polak, J. M., Pearse, A. G. E. 1982. Neuron specific enolase: A probe for neurons and neuroendocrine cells. *Trends Neurosci.* 5: 193–96

Marangos, P. J., Schmechel, D., Parma, A. M., Clark, R. L., Goodwin, F. K. 1979. Measurement of neuronal and non-neuronal enolase of rat, monkey and human tissues. *J. Neurochem.* 33: 319–29

Marangos, P. J., Schmechel, D. E., Parma, A. M., Goodwin, F. K. 1980b. Developmental profile of neuron specific (NSE) and non-neuronal (NNE) enolase in rat and monkey brain. *Brain Res.* 190: 185–93

Marangos, P. J., Zomzely-Neurath, C. 1976. Determination and characterization of neuron specific protein (NSP) associated enolase activity. *Biochem. Biophys. Res. Commun.* 68: 1309–16

Marangos, P. J., Zomzely-Neurath, C., Goodwin, F. K. 1977. Structural and functional properties of neuron specific protein (NSP) from rat, cat and human brain. *J. Neurochem.* 28: 1097–1107

Marangos, P. J., Zomzely-Neurath, C., Luk, C. M., York, C. 1975a. Isolation and characterization of the nervous system specific protein 14-3-2 from rat brain. *J. Biol. Chem.* 250: 1884–1901

Marangos, P. J., Zomzely-Neurath, C., York, C. 1975b. Immunological studies of a nerve specific protein (NSP). *Arch. Biochem. Biophys.* 170: 289–93

Marangos, P. J., Zis, A. P., Clark, R. L., Goodwin, F. K. 1978b. Neuronal non-neuronal and hybrid forms of enolase in brain: Structural, immunological and functional comparisons. *Brain Res.* 150: 117–33

Martin, B. M., Marangos, P. J., Merkle-Lehman, D., Ginns, E. I. 1986. Structural studies of human neuron-specific enolase. *Fed. Proc.* 45 (Abstr. 2146)

Maxwell, G. D., Whitehead, M. C., Connolly, S. M., Marangos, P. J. 1982. Development of neuron-specific enolase immunoreactivity in avian nervous tissue *in vivo* and *in vitro*. *Dev. Brain Res.* 3: 401–19

Moore, B. W. 1973. Brain specific proteins. In *Proteins of the Nervous System*, ed. D. J. Schneider, pp. 1–12. New York: Raven

Moore, B. W., McGregor, D. 1965. Chromatographic and electrophoretic fractionation of soluble proteins of brain and liver. *J. Biol. Chem.* 240: 1647–53

Odelstad, L., Pahlman, S., Lackgren, G., Larsson, E., Grotte, G., Nilsson, K. 1982. Neuron specific enolase: A marker for differential diagnosis of neuroblastoma and Wilms' tumor. *J. Pediatr. Surg.* 17: 381–85

Pearse, A. G. E. 1968. Common cytochemical and ultrastructural characteristics of cells producing polypeptide hormones (the APUD series) and their relevance to thyroid and ultimobranchial C cells and calcitonin. *Proc. R. Soc. London Ser. B* 170: 71–80

Pearse, A. G. E. 1976. Neurotransmission and the APUD concept. In *Chromaffin, Enterochromaffin and Related Cells*, ed. R. E. Coupland, T. Fujita, p. 147. Amsterdam: Elsevier

Parma, A. M., Marangos, P. J., Goodwin, F. K. 1981. A more sensitive radioimmunoassay for neuron specific enolase (NSE) suitable for cerebrospinal fluid determinations. *J. Neurochem.* 36: 1093–96

Pickel, V. M., Reis, D. J., Marangos, P. J., Zomzely-Neurath, C. 1975. Immunocytochemical localization of nervous system specific protein (NSP-R) in rat brain. *Brain Res.* 105: 184–87

Polak, J. M., Bloom, S. R. 1979. The diffuse neuroendocrine system. *J. Histochem. Cytochem.* 27: 1398–1400

Polak, J. M., Marangos, P. J. 1984. Neuron specific enolase, a specific marker for neuroendocrine cells. In *Evolution and Tumour Pathology of the Neuroendocrine System*, ed. S. Falkner, R. Hakanson, F. Sundler, pp. 433–52. Amsterdam: Elsevier

Prinz, R. A., Marangos, P. J. 1983. Serum neuron specific enolase: A serum marker for non-functioning pancreatic islet cell carcinoma. *Am. J. Surg.* 145: 77–81

Rider, C. C., Taylor, C. B. 1975. Evidence

for a new form of enolase in rat brain. *Biochem. Biophys. Res. Commun.* 66: 814–20

Royds, J. A., Timperley, W. R., Taylor, C. B. 1981. Levels of enolase and other enzymes in the cerebrospinal fluid as indices of pathological change. *J. Neurol. Neurosurg. Psychiatr.* 44: 1129–35

Sakimura, K., Kushiya, E., Obinata, M., Odani, S., Takahashi, Y. 1985a. Molecular cloning and the nucleotide sequence of cDNA for neuron-specific enolase messenger RNA of rat brain. *Proc. Natl. Acad. Sci. USA* 82: 7453–57

Sakimura, K., Kushiya, E., Obinata, M., Takahashi, Y. 1985b. Molecular cloning and the nucleotide sequence of cDNA to mRNA for non-neuronal enolase ($\alpha\alpha$ enolase) of rat brain and liver. *Nucl. Acids Res.* 13: 4365–78

Scarna, H., Delafosse, B., Steinberg, R., Debilly, G., Mandrand, B., Keller, A., Pujol, J. F. 1982. Neuron specific enolase as a marker of neuronal lesions during various comas in man. *Neurochem. Int.* 4: 405–11

Schmechel, D. E., Brightman, M. W., Barker, J. L. 1980a. Localization of neuron specific enolase in mouse spinal cord neurons grown in tissue culture. *Brain Res.* 181: 391–400

Schmechel, D. E., Brightman, M. W., Marangos, P. J. 1980b. Neurons switch from non-neuronal (NNE) enolase to neuronal (NSE) enolase during development. *Brain Res.* 190: 195–214

Schmechel, D. E., Marangos, P. J. 1983. Neuron specific enolase as a marker for differentiation in neurons and neuroendocrine cells. In *Current Methods in Cellular Neurobiology*, ed. J. Barker, J. McKelvey, 1: 1–62. New York: Wiley

Schmechel, D. E., Marangos, P. J., Zis, A. P., Brightman, M., Goodwin, F. K. 1978. The brain enolase as specific markers of neuronal and glial cells. *Science* 199: 313–15

Schmechel, D. E., Marangos, P. J., Brightman, M. 1979. Neuron specific enolase is a marker for peripheral and central neuroendocrine cells. *Nature* 276: 834–36

Sheppard, M. N., Corbin, B., Bennett, M. H., Marangos, P. J., Bloom, S. R., Polak, J. M. 1984. Immunocytochemical localization of neuron specific enolase (NSE) in small cell carcinoma and carcinoid tumors of the lung. *Histopathology* 8: 171–81

Steinberg, R., Gueniau, C., Scarna, H., Keller, A., Worcel, M., Pujol, J. F. 1984. Experimental brain ischemia: Neuron-specific enolase level in cerebrospinal fluid as an index of neuronal damage. *J. Neurochem.* 43: 19–24

Tapia, F. J., Polak, J. M., Barbosa, A. J. A., Marangos, P. J., Bloom, S. R., Dermody, C., Pearse, A. G. E. 1981. Neuron specific enolase is produced by neuroendocrine tumors. *Lancet* 1: 808–12

Trapp, B. D., Marangos, P. J., De F. Webster, H. 1981. Immunohistochemical localization and developmental profile of neuron specific enolase (NSE) and non-neuronal enolase (NNE) in aggregating cell cultures of fetal rat brain. *Brain Res.* 220: 121–30

Van Noorden, S., Polak, J. M., Robinson, M., Pearse, A. G. E., Marangos, P. J. 1984. Neuron specific enolase in the pituitary gland. *Neuroendocrinology* 38: 309

Vinores, S. A., Herman, M. M., Rubinstein, L. J., Marangos, P. J. 1984a. Electron microscopic localization of neuron-specific enolase in rat and mouse brain. *J. Histochem. Cytochem.* 32: 1295–1302

Whitehead, M. C., Marangos, P. J., Connolly, S. M., Morest, D. K. 1982. Synapse formation is related to the onset of neuron-specific enolase immunoreactivity in the avian auditory and vestibular system. *Dev. Neurosci.* 5: 298–307

Wick, M. R., Bernd, M. D., Scheithauer, W., Kovacs, K. 1983. Neuron specific enolase in neuroendocrine tumors of the thymus, bronchia and skin. *Am. J. Clin. Pathol.* 79: 703–7

Yoshida, Y., Sakimura, K., Masuda, T., Kushiya, E., Takahashi, Y. 1983. Changes in levels of translatable mRNA for neuron specific enolase and non-neuronal enolase during development of rat brain and liver. *J. Biochem.* 94: 1443–50

Zeltzer, P. M., Marangos, P. J., Parma, A. M., Sather, H., Dalton, A., Siegel, S., Seeger, R. C. 1983. Raised neuron-specific enolase in serum of children with metastatic neuroblastoma. *Lancet* 1(13): 361–63

Zeltzer, P. M., Marangos, P. J., Sather, H., Evans, A., Siegel, S., Wong, K. Y., Dalton, A., Seeger, R., Hammond, D. 1985. Prognostic importance of serum neuron specific enolase in local and widespread neuroblastoma. *Adv. Neuroblast. Res.* 175: 319–29

Ann. Rev. Neurosci. 1987. 10: 297–324

THE NEUROBIOLOGY OF FEVER: Thoughts on Recent Developments

K. E. Cooper

Department of Medical Physiology, Faculty of Medicine,
University of Calgary, Calgary, Alberta, Canada T2N 1N4

INTRODUCTION

The editors of this volume requested a review not all-encompassing but focusing on some of the interesting current issues concerning the role of the nervous system in fever. I have therefore selected a few recent developments in fever research. I quote some more general papers and reviews to which the reader is also referred. The literature cited is only a fraction of the total available, and the subjects covered comprise a small part of the work on fever in the last 20 years. I dwell critically on some methods that have been used in the last two decades because, although they have been extremely useful and have developed in precision and specificity, they are liable to produce artifacts, some of which are often ignored. My hope is to stimulate research into a modification, which may have survival value, of one of the most complex natural homeostatic mechanisms. Our understanding of the neural basis of thermoregulation and its modification in fever is still rudimentary and incomplete.

In preparation for the study of fever, the reader is referred to the following recent reviews for basic information on the neurophysiology of thermosensitive neurons, methods of investigation of thermoregulatory circuits, and thermal signal processing as well as for anatomical and pharmacological investigations: Cabanac (1970), Carpenter (1981), Eisenman (1972), Feldberg (1982), Hensel (1981), Jahns (1977), Palkovits & Zaborszky (1979), Pompeiano (1973), Satinoff (1983), Sutin & McBride (1977), Ruwe et al (1983), Cooper et al (1986), Kluger et al (1985), Dinarello (1984).

0147–006X/87/0301–0297$02.00

Bacterial and Endogenous Pyrogens

Some researchers began to focus on the chemical and physiological mechanisms of fever at the end of the nineteenth century, and such work was sporadically done until the major breakthroughs started to occur in the late 1940s. The peripheral mechanisms of fever include the events that occur between the onset of the infection, or other pathological processes, and the start of the events that cause the reduction of heat loss and the increase of heat production leading to the rise in body temperature. The first important aspect in understanding these mechanisms was the isolation and identification of "bacterial pyrogen," frequently called "endotoxin," the details of which are summarized by Work (1971). These discoveries enabled the study of fever caused by highly purified bacterial products and also enabled the development of assays to detect bacterial pyrogen in solutions and in the circulation.

The bacterial pyrogens are obtained from gram-negative organisms, for example, by extraction in hot phenol water mixtures. Chemically, they are complex lipopolysaccharides having molecular weights close to one million. Their structure is an O-specific chain (polysaccharide), a basal core polysaccharide, and a lipid component (lipid A).

The O-specific side chain confers the immunological characteristics of the O-antigens. The core polysaccharide may contain upwards of five sugars and often includes 2-keto-3-deoxyoctonic acid (KDO) as well as ethanolamine and phosphate groups. The lipid A fraction, possibly together with the KDO fragment, is said to be responsible for the pyrogenic action of the endotoxin.

The next, and most important, discovery concerning the peripheral mechanism of action of bacterial pyrogens was the finding that leucocytes could release a pyrogenic substance on incubation in saline without the presence of bacterial products (Beeson 1948, Bennett & Beeson 1953). Evidence that leucocyte, or endogenous pyrogen, could be released in the circulation of man during fever, induced by endotoxins, was presented by Gerbrandy et al (1954). That endogenous pyrogen is released from leucocytes by de novo synthesis was suggested by Fessler et al (1961), who found that only intact viable leucocytes could make endogenous pyrogen and that preformed endogenous pyrogen could not be released from whole cells by disintegrating them. A great deal of the characterization of leucocyte or endogenous pyrogen was done in the laboratory of the late Professor Barry Wood at Johns Hopkins University, Baltimore, and that work has been further extended there by Murphy & Chesney (1974). The most elegant characterization of the small peptide nature of endogenous pyrogen to date is that by Dinarello & Wolff (1977). This substance, similar or identical to interleukin I, has now been purified, and the amino acid sequence of interleukin I from at least one species (Lomedico et al 1984) is known. Biologically active murine interleukin I is a 156 amino acid peptide of molecular weight ~18,000, but the febrogenic moiety of the molecule still awaits elucidation at the time of writing. The nucleotide sequence of the interleukin I precursor cDNA from human monocytes has also been discovered (Auron et al 1984). Much of our knowledge of the cellular mechanisms of release of endogenous pyrogen has come in the last decade from the work of Atkins et al (1967), Gander & Goodale (1973), and Dinarello et al (1974, 1984), Dinarello (1984).

CNS Involvement in Fever

Direct evidence that intracranial structures are involved in the generation of fever was obtained by Richet (1884), Ott (1884, 1914), and Aronsohn & Sachs (1885), all of whom demonstrated that injury to the "corpus striatum" caused fever. Ott was probably the first to implicate the anterior hypothalamus in the generation of fever. Barbour (1921) demonstrated that altering the temperature of the region of the corpus striatum evoked thermoregulatory responses, which in turn caused opposite changes in the general body temperature. He showed that the "corpus striatum" responded to application of fever-producing or antipyretic substances with the expected changes in body temperature. The localization of the brain areas stimulated in the various ways by the crude techniques of these early experimenters was far from precise. In the late 1930s, better methods localized a very important thermoregulatory function to the anterior hypothalamic area, e.g. Teague & Ranson (1936). The concept, developed from the lesioning work of Ransom (1940), that there is a heat loss center in the anterior hypothalamus and a heat conservation center in the posterior hypothalamus, is no longer accepted. Clark et al (1939) showed that lesions of the posterior hypothalamus caused the animal's temperature to fall uncontrollably in a cold environment, and also that such lesions prevented fevers normally caused by pyrogens.

That the hypothalamic region might be particularly sensitive to the action of endogenous pyrogen was suggested by the experiments of King & Wood (1958), who showed that rabbits got fevers of more rapid onset when endogenous pyrogen was infused slowly into the internal carotid artery than when it was given intravenously. Bennett et al (1957) and Sheth & Borison (1960) demonstrated that intracranial administration of endotoxin causes fever, and Villablanca & Myers (1965) showed that the anterior hypothalamus will respond to local injections of bacterial pyrogen by inducing fever. That endogenous pyrogen can act in the anterior hypothalamus/preoptic region to cause fever was clearly shown by Cooper (1965), Cooper et al (1967), Jackson (1967), and Repin & Kratskin (1967). However, there still has been no convincing demonstration that endogenous pyrogen (interleukin I) enters the brain, and the significance of this is discussed below.

In the last decade, several new concepts have been added to the fever story. For many years, it had been held that fever represents a change in the normal drives that regulate the body temperature. That idea was challenged by Cooper & Veale (1974), who suggested that fever results from an abnormal drive to the heat-conserving and producing mechanisms, possibly arising from substances not usually mediating thermoregulatory processes but acting to modulate the functions of thermoregulatory circuits and their neurotransmitters. That the temperature-regulating mechanism behaves as though some thermostat was reset may result from a regulating mechanism that is functioning normally, but with its synaptic or neuronal functions altered from a nonthermoregulatory source of neuroactive substances to a degree depending on the amount of endogeous pyrogen entering or acting on the brain.

Central Nervous System Chemical Mediators

From the observations of Milton & Wentlandt (1971) and subsequently Feldberg & Gupta (1973) and Vane (1971) came the concept that metabolites of arachidonic acid, particularly the prostaglandins of the E series, might be the final mediators of the CNS fever mechanisms. Milton & Wendlandt (1971) observed fever in cats and rabbits when prostaglandin of the E series (PGE) was injected into the third cerebral ventricle. Feldberg & Gupta found prostaglandin in the cerebrospinal fluid during pyrogen induced fever. Vane demonstrated that an action of aspirin-like antipyretic drugs was the inhibition of prostaglandin synthesis. This notion is still highly controversial, and the evidence related to it is examined below.

The notion of the possible role of shifts in ionic balance, specifically the Ca^{2+}/Na^{+} ratio in the posterior hypothalamus in determining body temperature set point, and its elevation in fever, stems from the work of Feldberg et al (1970), who found that when the ratio of Na^{+}/Ca^{2+} in artificial cerebrospinal fluid perfused through the third ventricle was high in Ca^{2+}, the animal's temperature fell. The converse occurred if the Na^{+} concentration was raised. This notion too remains controversial. Many models have been proposed (e.g. Bligh 1974) to explain the putative role of neurotransmitters in both thermoregulation and fever. Although most of the authors would agree that though these models are oversimplistic, they do serve to focus discussion on possible mechanisms. The rigid proof of the existence of such circuits, however, is still lacking.

Behavioral Fevers

A second important concept relates to other mechanisms of production of fever in mammals and in ectotherms. The work of Hammel et al (1967) demonstrated that an ectotherm, the blue tongued lizard, could maintain body temperatures at preferred levels, or above ambient temperature, by shuttling between warm and cool parts of their environments. Such "behavioral" thermoregulation has been shown to enable ectotherms to get fever, and moreover these fevers have given some clues to the survival value of fever (Kluger 1979).

In addition, the role of behavioral fever in the newborn has proven to be important (Satinoff et al 1976), and its evolutionary significance and mechanism have been well studied recently (Kluger 1979, Laburn et al 1981). The overzealous extrapolation from the discovery that North American desert specimens and some fish (Reynolds et al 1976, 1980) can develop behavioral fevers when infected has received a cautionary reminder by the finding that the African Cordylid lizard does not (Laburn et al 1981). In addition, experiments on the pumpkin seed sunfish (*Lepomis gibbosus*) show that this fish does not get fever in response to *E. coli* injections or intracranial PGE (Marx et al 1984). It seems that the sedative effect of the injected material could cause either hypo- or hyperthermia, according to the side of the shuttle system in which the fish became sedated. This brings into question the interpretation of behavioral thermoregulatory responses to drugs when observations are made using warm/cool shuttle tanks.

Further, it is now realized that "fever," often interpreted just as a rise in body temperature, is only one part of a host defense reaction, now called the "acute phase reaction" (discussed below). More recent work on the suppression of fever in the newborn animal and the possible existence of endogenous antipyresis, together with investigations on the neuroactive substances used in febrogenic pathways, is discussed in detail below.

GENERAL COMMENTS ON METHODS

Numerous putative neurotransmitting substances have been postulated to occur in neuronal pathways involved in thermoregulation and the genesis of fever. Many means of identifying them have been used. While all have been valuable, they are often used without full appreciation of potential errors, and so some brief comments seem in order.

Injection of Substances into the Cerebral Ventricles

Microinjection of substances into a lateral cerebral ventricle often causes some dramatic changes in body temperature, but unless restricted perfusion techniques such as those of Carmichael et al (1964) for perfusion of the ventricular surface of the hypothalamus are followed, anatomical localization of action of such drugs to a periventricular locus is not convincing. Substances vary in their penetration of the ependymal lining of the third ventricle; their diffusion beyond this membrane, if they should penetrate it, varies with the nature of the substance as well as with local mechanisms for its inactivation and excretion. The concentration reaching the third ventricle from a lateral ventricle depends on the rate of secretion of cerebrospinal fluid, the degree of mixing of the substance with the fluid, and the injected volume in relation to the total ventricular volume. The duration of exposure of the hypothalamic ependymal surface to a drug depends on the amount and concentration injected and the rate of flow of cerebrospinal fluid. Once the substance has been washed through the fourth ventricle, it may act on the dorsal surface of the medulla, the ventral surface of the medulla, and anywhere in the circulation of the cerebrospinal fluid along the brain surface or the spinal cord. No safe predictions can be made of neuronal sites of action of a drug administered into a lateral cerebral ventricle.

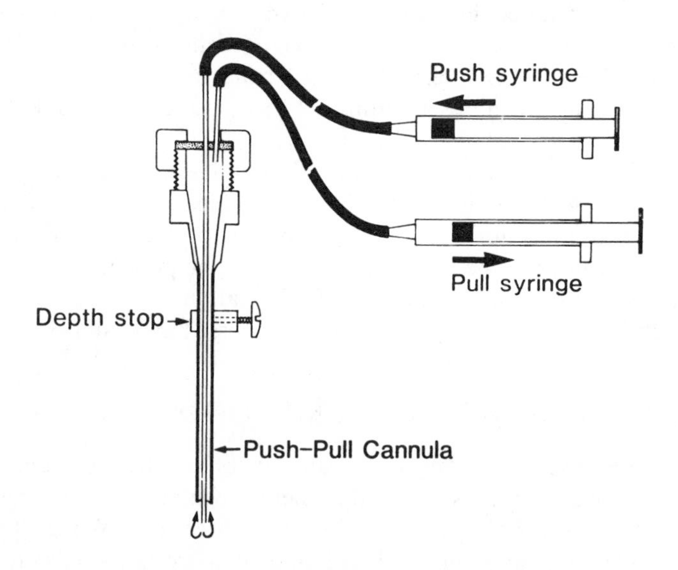

Figure 1 Diagram of a push-pull perfusion device.

Microinjection Studies

Microinjection volumes of 0.5–1.0 μl are the maxima for limiting brain damage, but, even with a fine cannula and very small volumes injected, some significant tissue damage occurs at the locus of the microinjection. Modern stereotaxic techniques can place the tip of the injection cannula with considerable accuracy, but the volume of spread of the injected drug is uncertain. Injection of dyes and measurement of their spread only provide the distance over which a particular colored compound travels in a concentration visible to the human eye, which may not be relevant to the diffusion of a putative transmitter in a concentration at which it has biological action. This is of great importance in studies on small brains. It has been shown (Cooper et al 1967) that under the pressure of microinjection, some fluid may find a plane of cleavage between bands of nerve fibers and travel further than predicted, or leak into the ventricular system. A small injected bolus of a drug may have only a transient action. Material, when injected into or perfused through a locus sensitive to endogenous pyrogen, may cause local release of endogenous pyrogen or PGE from the attracted white cells if the substance is leucotaxic or from glial tissue, and thus it may cause fever only by a secondary action unrelated to properties of the drug other than its leucotaxic, endogenous pyrogen, or PGE-releasing potential. When such substances cause fever, a necessary control is the search for leucocytes near the injection locus at a time immediately preceding the expected fever in a separate group of animals.

"Push-pull" Perfusion

A more useful technique, known as "push-pull" perfusion has been developed (Gaddum 1961, Myers 1970, Myers & Gurley-Orkin 1985). A double-barrelled cannula, with the inner fine tube projecting a fraction of a millimeter beyond the outer tube (Figure 1), is inserted into the brain tissue at the locus to be studied, and the solution of the substance to be administered is injected down the inner tube and withdrawn simultaneously at exactly the same rate up through the space between the inner and the outer tubes. An amount of substance can then be delivered continuously over any desired period at the locus of the tip of the double-barrelled cannula. There is no buildup of pressure, as occurs with microinjection, and the bolus effect of administration is avoided. The device can be used for delivering a substance, or for washing out substances that are released into the tissue. The method may better mimic continuous release from nerve terminals than microinjection, but some tissue damage may occur, and leucotaxic effects are also possible.

SOME CURRENT RESEARCH IN THE NEUROBIOLOGY OF FEVER

What Is the Locus of Action of Endogenous Pyrogen Within the Central Nervous System?

The work of the 1960s (see above) that demonstrated a particular sensitivity of the preoptic area to endogenous pyrogen to cause fever has often been quoted as indicating that endogenous pyrogen acts there during naturally-occurring fevers, or during fever caused by intravenously administered pyrogens. Such a concept assumes that the pyrogen crosses the blood brain barrier to gain access to the preoptic area. Studies with large amounts of radioactively labeled bacterial pyrogens (Rowley et al 1956, Braude et al 1958, Cooper & Cranston 1963) failed to demonstrate entry of the lipopolysaccharide into the brain. So far there has been no evidence that endogenous pyrogen crosses the blood brain barrier. Indeed, endogenous pyrogen (interleukin I) has only recently been available in sufficient quantities to enable labeling studies of its penetration into the brain, and ultrasensitive methods of endogenous pyrogen detection are only just becoming available. In addition, it has only recently been shown that prostaglandins could act in the septal region to cause fever (Williams et al 1977, Ruwe et al 1985) and in the organum vasollosum of the lamina terminalis (Stitt et al 1986), and the possibility that endogenous pyrogen could act in these regions has not yet been clearly established.

If endogenous pyrogen does not cross the blood brain barrier, there are alternative loci of action. It is unlikely that it enters the cerebrospinal fluid and reaches the brain neuropil from the ventricular surface. Rapid perfusion of the ventricular system with artificial cerebrospinal fluid in the rabbit did not alter fever due to intravenous endogenous pyrogen (Feldberg et al 1971); and filling the whole ventricular space of the rabbit brain with inert, pyrogen-free oil exaggerated the fever due to intravenous pyrogens but was without effect on body temperature in the afebrile animal (Cooper & Veale 1972).

Possibly some regions of the blood brain barrier are more permeable than others. Such a tight barrier is said to be absent in certain circumventricular organs, and the barrier might be permeable to small protein molecules (of 15,000 D minimum) (Weindl & Sofroniew 1981) such as endogenous pyrogens. One of these is the organum vasculosum of the lamina terminalis (OVLT), which is in close proximity to the preoptic area and septal regions in small mammals. Blatteis et al (1983) made electrolytic lesions in the OVLT and neighboring tissue in guinea pigs. Following the lesions, intraperitoneal bacterial lipopolysaccharide did not cause fever, as it did in sham-operated animals, and the expected increases of plasma copper and sialic acid that usually accompany fever did not occur. The preoptic area was shown to be intact in the lesioned animals. This work suggested that endogenous pyrogen may pass into the brain through the OVLT.

In contrast, Stitt (1985) made discrete lesions within the OVLT in rabbits and rats and found that the fever responses of the animals to intravenous endogenous pyrogen three days later was greatly enhanced. This enhancement lasted up to 21 days following the lesion. The increased febrile response was in the magnitude, not the duration, of fever. Stitt et al (1983) also found that microinjection of zymosan, a reticuloendothelial cell stimulant, into the OVLT enhanced the fever caused by intravenous endogenous pyrogen. This again suggested a role of the OVLT cells in fever.

Two problems are inherent in the interpretation of the OVLT lesioning: First, how could lesioning exaggerate fever? Second, how can the results of Stitt (1985) and Blatteis et al (1983) be reconciled? Stitt suggests that the lesion within the OVLT might have produced an inflammatory reaction resulting in increased local vascularity and vascular permeability. If the OVLT is a portal of entry of endogenous pyrogen into the brain, then the local inflammatory response would enhance its entry from the systemic circulation, and the observed course of the fevers and of the duration of the period of enhancement could be explained. This result would be consistent with Blatteis's postulate of the OVLT as an entry locus of endogenous pyrogen into the brain. Opposing this hypothesis is the argument that there is a locus at which microinjected endogenous pyrogen acts in the anterior hypothalamus/preoptic area (AH/POA), and bilateral small electrolytic lesions in this region, which also would be expected to increase local vascular permeability, do not enhance fever caused by intravenous endogenous pyrogen. Evidence also suggests that proteins passing through the vascular endothelium of the OVLT do not spread into the brain neural tissue (Weindl 1969). The alternative explanation of Stitt (1985) is that the lesion, by causing an inflammatory response, allows endogenous pyrogen more ready access to endogenous pyrogen receptors within the OVLT itself, and perhaps even sensitizes those receptors. Thus, the OVLT and not the AH/POA could be the primary locus of action of endogenous pyrogen, possibly as a place where a secondary cascade of events such as PGE release could take place. As Stitt points out, this would resolve the problem of the failure so far to demonstrate penetration of endogenous pyrogen into the brain neuropil; and PGE, because it is lipophilic, could be the substance that enters the AH/POA or other location.

With regard to the observations of Blatteis et al (1983), the lesions used in the experiments were more extensive, and they could have destroyed the postulated endogenous pyrogen receptors in the OVLT as well as the route of entry to them. J. Stitt (1986, personal communication), on the basis of the actions of verapamil given into the OVLT and verapamil and nifedipine given intravenously on IL-1 and PGE fevers, has postulated a role for calcium channels in the release of PGE within the OVLT. Further results from use of calcium channel blockers more specific to neural tissue will be exciting.

Are Prostaglandins of the E Series Obligatory Mediators of Fever Within the Brain?

The evidence in favor of PGE as an obligatory mediator of fever in the AH/POA is extensive, but there are also powerful arguments against it. Milton & Wentlandt (1971) found that PGE microinjected into the third ventricle of unanesthetized cats caused fever. Since then it has been shown that microinjection of PGE_1 or PGE_2, particularly PGE_2, into the AH/POA causes a fever of very rapid onset. The doses of PGE required are minute and the effect is dose dependent. The fever-producing activity of PGE in the AH/POA is similar in many species, including cats, rabbits, rats, mice, guinea pigs, chickens, and monkeys. This evidence has been summarized by Milton (1982). A second argument derives from the discovery by Vane (1971) that the aspirin-like antipyretic and anti-inflammatory drugs inhibit prostaglandin synthesis. Later it was shown that the antipyretic drug, indomethacin, does not prevent fever due to PGE microinjection, but does inhibit fever caused by endogenous pyrogen, thus suggesting an action on PGE synthesis.

It is known that PGE synthesis occurs in the central nervous system and increases during fever (Coceani et al 1986). In the rat, a new locus of action of PGE to cause fever, namely the ventro-lateral septum, has recently been reported (Williams et al 1977, Ruwe et al 1985).

Arguments are also drawn from a correlation of PGE-like activity in the cerebrospinal fluid and the body temperature following intravenous pyrogens (Feldberg & Gupta 1973), in which the apparent PGE activity rises following pyrogen administration and falls when antipyretics reduce the body temperature. The arguments against an obligatory role of PGE in fever can be summarized as follows:

1. There is a lack of correlation in some species between the effects of intracerebral PGE and intravenous pyrogens, or intracerebral pyrogens on body temperature. Pittman et al (1974) showed that lambs and sheep that responded to intravenous pyrogens with fever did not respond consistently to hypothalamic administration of PGE. However, large doses of

PGE injected into the lateral cerebral ventricle of the sheep caused fever (Bligh & Milton 1973, Hales et al 1973). Goats in which infusions of PGE were made in large doses into a lateral cerebral ventricle did not get fever but did get antidiuresis and drank copiously (Leskell 1978). Chickens get hypothermia in response to hypothalamic administration of PGE rather than fever; killed bacteria administered intravenously do cause fever (Artunkel et al 1977). However, endotoxin injected into the chicken hypothalamus caused hyperthermia, whereas when injected intravenously it caused hypothermia (Pittman et al 1976); PGE given into the hypothalamus caused a modest fever. The fever due to intrahypothalamic endotoxin was reduced by acetylsalicylic acid, but the hypothermia due to intravenous pyrogen was unaltered. PGE injected into the cerebral ventricles of the echidna lowers its body temperature, but its response to pyrogen is inconsistent (Baird et al 1974).

2. Two antagonists of PGE, which are available under the code names of HR 546 and SC 19220, when administered into the cerebral ventricles abolished the febrile response to PGE injected into the cerebral ventricles, but they did not attenuate fever due to moderate intravenous doses of endogenous pyrogen (Cranston et al 1976).

3. Are the effects of endogenous pyrogen and PGE on hypothalamic thermosensitive single neurons identical? Their effects on thermosensitive hypothalamic single units have been investigated by a number of workers (Wit & Wang 1968, Eisenman 1969, Cabanac et al 1968, Ford 1974, Stitt & Hardy 1975, Jell & Sweatman 1977). The effects caused by intravenous endogenous pyrogen or bacterial pyrogen reported so far have been summarized by Eisenman (1982). In short, 91% of cold sensitive units are facilitated and 94% of warm sensitive units are inhibited (Table 1). A marked disparity between the action of endogenous pyrogen and PGE on thermosensitive neurons was found in hypothalamic slice preparations from guinea pigs by Boulant & Scott (1986). If it can be safely assumed that hypothalamic thermosensitive units play a major role in thermo-

Table 1 Action of bacterial and leucocyte pyrogen on anterior hypothalamic/preoptic area neurons[a]

Neuron type	No. of units	Facilitation (%)	Inhibition (%)	No effect (%)
Warm sensitive	65	3	94	3
Cold sensitive	32	91	0	9
Insensitive	25	0	4	96

[a] From Eisenman (1982).

regulation in the intact conscious animal, these findings provide a reasonable mechanism for inhibition of heat loss and activation of heat conservation pathways. Endogenous pyrogen microinjected into the AH/POA (Schoener & Wang 1975) or the lower brainstem produces more rapid but otherwise similar responses as does intravenous endogenous pyrogen (Sakata et al 1981).

The action of PGE on hypothalamic single units is not so clear. Jell & Sweatman (1977) found patterns of response similar to those obtained with pyrogens, but Stitt & Hardy (1975) were unable to correlate sensitivity to PGE and thermosensitivity. Of course, the unequivocal connection between thermosensitive units in the septum, AH/POA, and lower brain stem, and thermoregulatory behavior is far from established. The action of endogenous pyrogen or PGE on thermosensitive units in these sites does not prove an action on the units normally involved in thermoregulation or fever.

A summary of the available data on the action of PGE on thermosensitive units shows that 73% of warm sensitive neurons were not affected by PGE and almost 42% of cold-sensitive neurons were unaffected. Almost none of the thermally insensitive units in the AH/POA is affected by PGE or endogenous pyrogen, and so an action of PGE on these units is an unlikely mechanism of the raised body temperature (Table 2) (Eisenman 1982).

One new factor could make the interpretation of these data even more difficult: the discovery that AH/POA thermosensitivity is virtually lost in unanesthetized cats during desynchronized sleep (Azzarone et al 1985). If in some of the endogenous pyrogen and PGE experiments on anesthetized animals the thermosensitive unit activity is modified in the same way as in desynchronized sleep, either due to the stage of anesthesia or because of the application of one of the substances, then discrepancies in effects of endogenous pyrogen and PGE could occur. So far, the hypothalamic single unit activity cannot be said to support a role for PGE in fever.

Table 2 Action of PGE_1 and PGE_2 on anterior hypothalamic/preoptic area neurons[a]

Neuron type	No. of units	Facilitation (%)	Inhibition (%)	No effect (%)
Warm sensitive	103	8	19	73
Cold sensitive	41	51	7	42
Insensitive	234	3	2	95

[a] From Eisenman (1982).

4. Dissociation of fever and antipyresis from ventricular cerebrospinal fluid prostaglandin concentrations was found by Cranston et al (1975), who were able to suppress the rise in PGE levels in the cerebrospinal fluid by administration of salicylate before giving intravenous endogenous pyrogen, but the fever was unaltered. The cerebrospinal fluid was obtained from the cisterna magna, so PGE arising from the AH/POA could have been diluted out if cerebrospinal fluid flow was greatly increased in the experiments. PGE from the AH/POA does enter the ventricular cerebrospinal fluid (Veale & Cooper 1974). Filling the ventricular spaces with inert pyrogen free oil increased the febrile response to endogenous pyrogen, thus suggesting that intravenous endogenous pyrogen acts within, or promotes a secondary mediator such as PGE within the hypothalamus rather than entering the tissue via the cerebrospinal fluid (Cooper & Veale 1972). Third ventricular cerebrospinal fluid PGE concentrations were well correlated by Bernheim et al (1980) with body temperatures during endogenous pyrogen induced fevers, but the Crawford et al (1979) data provide no such correlation. The technical problems that have bedevilled accurate PGE assays in cerebrospinal fluid—e.g. failure to allow for protein binding of PGE, and the presence of substances such as hydroperoxides that can sensitize target bioassay tissues to PGE (F. Coceani, personal communication)—make difficult the use of many cerebrospinal fluid prostaglandin studies to test the hypothesis that PGE is an obligatory factor in the brain in fever.

5. Is it possible that the amount of PGE injected into the brain tissue or ventricles is much greater than the endogenous pyrogen releases? Fevers of 1–2°C can result from injections of 20 ng PGE into the AH/POA of the rabbit (Stitt 1973) or of 100 ng bilaterally into the rat ventral septum (Ruwe et al 1985). Even with the lowest amount of PGE injected into the hypothalamus to cause fever, the tissue concentration at its peak, allowing a distribution through a sphere of 1–2 mm diameter, is likely to exceed 600–700 ng/ml. The most accurate resting concentration in cerebrospinal fluid in the afebrile animal is 10–15 pg/ml, and during fever it rises to 100 pg/ml (F. Coceani 1986, personal communication).

The endogenous PGE concentration in the AH/POA required to raise the third ventricular cerebrospinal fluid concentration from 10 to 100 pg/ml will be greater than 100 pg/ml, but it is unlikely to be as high as 700 ng.

6. Ablation of the AH/POA by radio frequency heating prevented fever in the rabbit in response to intraventricular PGE, or to PGE injected into the area of destroyed tissue, but did not prevent fever due to intravenous pyrogens (Veale & Cooper 1975). However, the fever was of slow onset and prolonged, as compared with the rapidly climbing and rapidly defervescing

fever in intact animals. The endogenous pyrogen could have acted in loci that still produced PGE but that are as yet unknown and not reached via the cerebrospinal fluid, or it could have stimulated the activity of a non-PGE dependent pathway. In the squirrel monkey, the febrile response to intracerebroventricular (ICV) bacterial pyrogen and PGE persisted after large AH/POA lesions (Lipton & Trzcinka 1976), though the size of the lesions were probably not of the same relative magnitude as those reported by Veale & Cooper (1975). Andersson et al (1965) reported fever due to peripheral bacterial pyrogen after AH/POA destruction in goats, but the effect of PGE was not investigated.

7. The action of protein synthesis inhibitors on fever was shown by Siegert et al (1976), who showed that systemic administration of cyclohexamide, a protein synthesis inhibitor, reduced or abolished fever; nevertheless this effect (Stitt 1979) could have been due to an action of the drug on peripheral thermoregulation. However, Cranston et al (1981) showed that cyclohexamide given ICV reduced fever due to ICV endogenous pyrogen but did not alter cold defense reactions in rabbits. Cranston et al (1980) showed that intraventricular administration of anisomycin, a protein synthesis inhibitor, suppressed the fever response to endogenous pyrogen given intravenously or into the cerebral ventricles; this finding suggests a process of protein synthesis possibly as an essential step in fever. Ruwe & Myers (1980) also showed that systemic administration of anisomycin suppressed the fever expected in response to intravenous killed *S. typhosa* organisms in the cat. It also prevented fever caused by endotoxin microinjected into the AH/POA, but it did not prevent the hyperthermia caused by similarly injected serotonin or PGE. The anisomycin injection did not modify normal thermoregulation.

Another series of protein synthesis inhibitors, the trichothecene antibiotics, 3,5-diacetoxy-12-hydroxytrichothec-9-ene (DAHT), 3,15-didesacetyl-calonectron (DDAC) and T-2 toxin, were injected into the rabbit cerebral ventricles (Cannon et al 1982). DDAC and T-2 toxin inhibited fever due to ICV-endogenous pyrogen and inhibited leucine incorporation into hypothalamic protein. DAHT did not reduce fever and did not lower leucine incorporation as much. Since four different protein synthesis inhibitors have now been shown to inhibit fever, it seems unlikely that the effect is a pharmacological side effect unrelated to protein synthesis inhibition.

8. Cranston et al (1983) found that mepacrine or parabromophenacylbromide, drugs that inhibit the action of phospholipase A, when given intravenously to rabbits abolished fever for more than an hour after iv injection of endogenous pyrogen. Mitchell et al (1986) suggest that the metabolite, whose synthesis was blocked in Cranston's experiments, might be a nonprostaglandin product of cyclo-oxygenase activity. Because

of its variable effect between species when given ICV, prostacyclin does not seem to be an essential fever mediator. Laburn et al (1977) found that arachidonic acid administered ICV into rabbits caused fever, but when administered with prostaglandin antagonists the fever was only transiently inhibited, whereas the antagonists prevented fever due to ICV PGE. It is possible that endoperoxides, despite conflicting reports of their effects, or thromboxanes might be involved in fever production.

9. Hashimoto et al (1985) compared the action of ICV bacterial pyrogen and PGE on the thresholds for ear vasoconstriction and shivering due to body cooling in rabbits. PGE increased the threshold for ear vasoconstriction, as did bacterial pyrogen. PGE also raised the threshold for shivering, whereas bacterial pyrogen did not. The difference observed suggested a mechanism not involving PGE in the raising of metabolic rate in fever.

10. Mitchell et al (1986) proposed that intracranial PGE may be involved in the initial and rapid increase in temperature at the onset of fever but not necessarily in the late, more sustained, phase of fever. Székely (1978) also suggested this possibility from his finding that indomethacin prevented the first part of a biphasic fever in neonatal guinea pigs but not the second. The slow onset fever, which lacks a fast rising phase, following hypothalamic lesions (Veale & Cooper 1975) is consistent with the hypothesis. Skarnes et al (1981) found PGE in jugular venous blood (plasma) of sheep during the first hour and a half of fever due to intravenous endotoxin but not during the subsequent four and one half hours of fever. Mitchell et al (1986) suggest that PGE might sensitize AH/POA neurons to endogenous pyrogen, nociceptive substances, or other putative mediators of fever.

Objections to these arguments have been made, as recently discussed by Coceani et al (1986). Some problems with the assays for PGE in cerebrospinal fluid call much of the earlier work into question. Also, the interpretation of cerebrospinal PGE levels is difficult if the sample has not been obtained from the region of the anterior recess of the third ventricle. Similarly, it is difficult to equate topically applied PGE with endogenously released PGE, owing to the possible high efficiency of access of endogenous PGE to loci of action, be it at receptors or by cell penetration.

Using more accurate PGE assays, Coceani et al (1986) find close correspondence of fever and cerebrospinal fluid levels of PGE. Their analysis of the evidence does not support a role for a second cyclo-oxygenase product as a fever mediator; for example, intracranially applied thromboxane A analogue produced no consistent change in body temperature, and fever due to ICV endotoxin was not modified by treatment with a thromboxane antagonist.

Intravenous endogenous pyrogen (Coceani et al 1983) or bacterial pyrogen raised thromboxane B_2 levels in the cerebrospinal fluid only after intrathecal administration, but PGE_2 levels rose after intravenous or intrathecal pyrogen. Coceani et al (1986) also point out that if the OVLT were a locus of pyrogen action, particularly to stimulate PGE release, or if for example the brain capillaries were stimulated by endogenous pyrogen to release PGE, the previous work on hypothalamic administration or ICV drugs might become irrelevant. However, endotoxin caused release of PGE from cerebral microvessels and its release was inhibited by IL-1 (endogenous pyrogen). An interesting fever that appears to be mediated by PGE without the involvement of endogenous pyrogen release, namely that due to human interferon, has recently been described (Dinarello et al 1984). Intraventricular interferon caused fever and a rise in cerebrospinal fluid PGE_2 levels in the cat; these effects were blocked by indomethacin. In rats, rabbits and cats, human interferon injected into AH/POA caused fever at ambient temperatures of 3°, 25°, and 30°C, and these fevers were reduced by giving dipyrone, an antipyretic that blocks prostaglandin synthesis peripherally (Ruwe et al 1986). It would be fair to comment now that, concerning the obligatory role of PGE in the brain in fever, the jury remains out.

One additional interesting observation (Alexander et al 1986) is that salicylate infused into the ventral septal area of rats greatly reduces fever due to ICV-PGE. This result could be explained by the notion that salicylate might cause the release of arginine vasopressin within the ventral septal area, which is known to inhibit PGE fever. Thus, in addition to PGE synthesis inhibition, there might be an additional mechanism of salicylate antipyresis. This possibility, if true and if extended to other antipyretics, further confuses the interpretation of experiments that depend on a simple action of antipyretics.

Endogenous Antipyresis and Peptide Hormones

Fever appears to be a drive working on the heat-conserving and heat-producing system of the thermoregulatory system that establishes a new set point for body temperature. This drive may be, possibly through a special chemical pathway, one that is not used in the nonfebrile process of thermoregulation. Further, there is growing evidence that fever, in the sense of a rise in central temperature, is but one part of a complex host response to infection that consists of changes in plasma and intracellular fluid levels of some metal ions and the production of leucotrienes and components of the immune system. Evidence suggests that the whole acute phase reaction is part of a protective defense response necessary to the survival of the infected organism. An excessive rise in body temperature during fever, however, can result in permanent central nervous system damage, if not death, and high fevers in some children may be accompanied by severe convulsions. It should have caused no surprise, then, to the teleologically-minded when evidence began to emerge for the elaboration or secretion of endogenous antipyretic substances within the brain of febrile animals. The initial indication of a potential endogenous antipyretic occurred during studies of fever induced by intravenous endotoxin in newborn lambs and shorn ewes of the Suffolk cross breed at the time of parturition (Pittman et al 1974). The ewes became refractory to intravenous pyrogens from a few days before delivery until a few hours after delivery. The newborn lambs were unable to develop fevers in response to intravenous pyrogens, for several hours after birth and despite the maturity of their thermoregulatory processes. These results led to the search for a possible humoral agent responsible for suppressing fever at these times (Kasting et al 1978, 1979, Cooper et al 1979).

This study led to the observation that arginine vasopressin can be released into the septal region during fever and might function at that locus as a natural endogenous antipyretic substance. [Interestingly, Cushing (1931) describes flushing, sweating, and falling of body temperature in a slightly febrile man following infusion of "pituitrin" into a lateral ventricle.] Heap et al (1981) were unable to find the same degree of suppression of fever at term in a different breed of sheep (Soay and Clun Forest). However, recent experiments on unshorn sheep in Africa (D. Mitchell et al, unpublished results) confirmed Pittman's findings, so perhaps the Soay/Clun sheep display untypical reactions. But C. M. Blatteis and J. R. S. Hales (1986, unpublished results) have found that endotoxin can cause fever in Australian sheep up to parturition. Some differences in subspecies responses, techniques, or, more likely, previously induced immunological responses may account for these variable results.

The evidence for arginine vasopressin as an endogenous antipyretic is weighty but as yet incomplete. It includes (*a*) direct suppression of fever by the application of arginine vasopressin in a sensitive, dose-dependent manner by push-pull perfusion into the region of the septum in animals given intravenous pyrogens; and (*b*) the appearance of arginine vasopressin in the perfusate after perfusion of the septal region with an artificial interstitial fluid, during fever; the vasopressin concentration is inversely related to the magnitude of fever.

Hemorrhage, a well-known stimulus for the release of vasopressin into the circulation, was found to be without effect on normal body temperature but to suppress the fever expected from intravenous pyrogen. The application of an antibody to arginine vasopressin in the septal region led to markedly higher fevers than normally occur from a fixed dose of endotoxin given intravenously (T. Malkinson, personal communication, 1986). Arginine vasopressin itself is without effect on normal body temperature in the sheep when perfused through the ventral septal area. A vasopressin V_1 receptor antagonist, but not a V_2 agonist, perfused through the ventral septal area, enhances fever. The V_2 agonist, 1-desamino-8-D-arginine vasopressin, did not alter the magnitude or the time course of interleukin-I fever (Naylor et al 1986). Some other peptides, such as bombesin, when they are infused into the septum reduce body temperature during fever but often also reduce body temperature when the animal is afebrile.

Zeisberger et al (1980) and Merker et al (1983) have demonstrated that the guinea pig is refractory to endotoxin in that fever is reduced or abolished close to term and for several hours post-partum. They have also demonstrated in guinea pigs, by immuno-cytochemical methods, that arginine vasopressin immunoreactive material increases in the neurons connecting the hypothalamus with the lateral septum and the amygdala during the time when the guinea pigs are refractory to endotoxin and also following the administration of pyrogens. Arginine vasopressin reactivity was increased in fibers projecting from the hypothalamus to the septum and amygdala in the nonpregnant guinea pig during fever (Zeisberger et al 1983).

Arginine vasopressin also suppressed fever caused by intraventricular injection of PGE, when the arginine vasopressin was perfused through the rostral diencephalic loci at which prostaglandins cause fever (Ruwe et al 1985).

Lipton & Glynn (1980) were unable to obtain antipyresis by the application of vasopressin to the septum, and indeed vasopressin increased body temperature when injected into the cerebral ventricles of rabbits. Bernadini et al (1983) also were unable to reduce fever in the rabbit by septal injections of arginine vasopressin. However, comparison between their results and those of Kasting (1979) and his colleagues is difficult because of the different species used. However, recent observations by Naylor et al (1985a) and subsequent work in the same laboratory have demonstrated clearly that arginine vasopressin does suppress fever due to intravenous or ICV pyrogens when it is applied by push-pull perfusion to the septum in the rabbit. This apparent dilemma can be explained by the following three observations:

1. In Lipton's laboratory, early experiments showing the production of a mild fever by intracerebral administration of arginine vasopressin were done by injecting the material into the cerebral ventricles. As discussed above, this method of administration can produce actions at many sites, and actions produced this way are not necessarily part of the normal thermoregulatory or fever process.

2. The administration of arginine vasopressin into the septal area was made by microinjection, and thus it did not provide a prolonged and continuous supply of varopressin such as would be expected from activated nerve terminals in that region.
3. The loci of the injection in the septal area were not the same as those described by Cooper et al (1979) for the sheep and more recently in the rabbit by Naylor et al (1985a). Cooper & Naylor's locus was more posterior and more lateral in the septum.

Further research will lead us to a closer understanding of these differences. Evidence of the endogenous antipyretic action of vasopressin has also been found in rats, in another laboratory (Kovacs & de Wied 1983). Banet & Wieland (1985) found that arginine vasopressin infused into the lateral septum of the rat suppressed the heat production caused by hypothalamic cooling but did not affect vasomotor tone in the skin. The reduction in thermoregulatory increase in heat production that led to a fall of body temperature during cold exposure might explain the antipyretic action (Banet & Wieland 1985). These observations are at variance with more recent work in Veale's laboratory, in which normal thermoregulation was unaffected by arginine vasopressin injected into the ventral septal area. The loci used by Banet & Wieland (1985) are not the same as those reported by Naylor et al (1985a). Wilkinson & Kasting (1986) presented evidence indicating that vasopressin reduces the set point for body temperature in febrile rats.

Recently, Disturnal et al (1985, 1986) have identified neurons in the ventral septal area of the rat, that receive afferent impulses from the paraventricular nucleus and the bed nucleus of the stria terminalis. These neurons are derived from loci known to have arginine-vasopressin-containing neurons. The neurons identified were shown to be connected with thermoresponsive single units that responded to thermal stimulation of scrotal skin. Electrical stimulation of the paraventricular nucleus caused orthodromic inhibition of most septal thermoresponsive units. Neurons projecting from the bed nucleus of the stria terminalis (BST) mostly inhibited warm-responsive and excited cold-responsive ventral septal area units. In addition, afferents from the fornix were mainly inhibitory and from the amygdala were approximately equally excitatory and inhibitory. The demonstration of the afferents from the paraventricular nucleus and BST in this carefully controlled work is consistent with the postulated arginine-vasopressin-containing neurons acting as sources of an endogenous antipyretic. The amygdala receives arginine varopressin immunoreactive projections from the hypothalamus; these can be demonstrated during fever (Zeisberger et al 1980, Merker et al 1983), but their function is still unclear. However, the electrophysiological study of Disturnal et al (1986) adds substance to the proposed connectivity of arginine vasopressin neurons from the hypothalamic area to thermoregulatory circuits.

Only one study has addressed the role of arginine vasopressin in the subhuman primate, namely that of Lee et al (1985), in which macaque monkeys were used. ICV arginine vasopressin had no significant effect on the normal body temperature or on fever, except at a dose of 65 ng, where the mean fall of body temperature of 0.5°C appeared to be due to a larger response in one animal. However, as discussed above, the locus of action of arginine vasopressin as an antipyretic in other species is very precise, and ICV injections may give misleading results. We await similar tissue administration in primates. Arginine vasopressin has also been shown recently (Lipton & Glynn 1980) to cause hyperthermia when injected into the lateral cerebral ventricle of rabbits but is without effect in cats (Rezvani et al 1986), and to cause hypothermia (Naylor et al 1986), or hyperthermia when microinjected into the preoptic area, in the rat. In addition (Naylor et al 1986), microinjection of arginine vasopressin into the AH/POA caused fever, and this hyperthermic action was blocked by an arginine vasopressin V_1 receptor antagonist. Naylor et al (1986) report that arginine vasopressin microinjected into the ventral septal area, substantia innominata, nucleus accumbens, and dorsomedial hypothalamus did not alter body temperature in the afebrile rat.

Arginine vasopressin might be a mediator of fever within the AH/POA and an antipyretic within the ventral septal area. Such a dual function would be consistent with a negative feedback system to modulate the magnitude of fever, a useful function if fever is a beneficial accompaniment of disease. However, a role for arginine vasopressin as a fever mediator as well as an endogenous antipyretic is far from proven.

The hypothesis of arginine vasopressin as an endogenous antipyretic would be further strengthened if severance of the arginine vasopressin pathways from the hypothalamus to the septum, and if possible separately to the amygdala, could be done and were to lead to exaggerated fever. It would be of interest to raise an animal's temperature by physical, nonstressful means and to observe the secretion or nonsecretion of arginine vasopressin into the lateral septum, or to stimulate the hypothalamic-septal arginine vasopressin projections electrically and observe the effect on fever in the conscious animal. It is interesting also that in the push-pull experiments, in which arginine vasopressin was secreted into the perfusate, the perfusion medium was a calcium-free sucrose solution. Perhaps the volume of perfusion fluid was insufficient to wash out much of the interstitial calcium that would be expected to be necessary for release of arginine vasopressin.

If arginine vasopressin is indeed an endogenous antipyretic, is this antipyresis a primary action of arginine vasopressin, or does it act as one step in a peptide cascade? Other peptides may also play a role in the central process of fever and antipyresis, and to imagine a unique role of arginine vasopressin or any other peptide is probably naive.

Another peptide, alpha-melanocyte-stimulating hormone (α-MSH), may play a role as an endogenous antipyretic and may even have a role in normal thermoregulation (Sampson et al 1981). α-MSH was shown by Murphy et al (1983) to have an antipyretic action when given in small doses into the cerebral ventricles. In larger doses it reduced body temperature but without altering body temperature set-point. The level of α-MSH in tissue samples taken at the peak of fever due to intravenous interleukin I was found to be raised in septal tissue but not in material removed from other loci (Lipton 1985). Central administration of an antiserum to α-MSH potentiated fever due to interleukin I. The antipyretic effect of α-MSH is common in all species yet tried (Lipton 1985). Lipton (1985) quotes evidence that α-MSH had no effect on PGE-induced hyperthermia. α-MSH can reduce fever in rabbits when injected into the septal area in doses of 1 μg (Glyn-Ballinger et al 1983) in a locus different from that in which arginine vasopressin causes antipyresis (Ruwe et al 1985). Intraventricular administration of an α-MSH antiserum for three days augments fever caused by intravenous IL-1 (Shih et al 1986). Though important, the evidence implicating α-MSH in endogenous antipyresis still lacks the precise identification of the locus of action, the histochemical demonstration of the projections involved, and the electrophysiological demonstration of potential neuron connectivities—all of which have been more completely demonstrated for arginine vasopressin. It is, however, likely that α-MSH is also an endogenous antipyretic.

Many other peptides have been investigated by intraventricular administration for action on body temperature, e.g. the study by Lipton & Glynn (1980). They proposed that α-MSH, ACTH, oxytocin, vasopressin, and glucagon could have a role in thermoregulation and fever. Intraventricular corticotrophin releasing factor given into the third ventricle reduced fever due to intravenous leucocyte pyrogen (Bernadini et al 1984). However, studies in which naturally occurring substances are given into the cerebral ventricles to cause alterations in body temperature are only useful as initial leads until rigorous proof is obtained of the substances' association with thermoregulatory function. A list of the many peptides that alter body temperature when injected into the brain can be found in Clark (1979).

β-Endorphin injected into the cerebral ventricles of the cat causes a naloxone-sensitive hyperthermia (Clark & Bernadini 1981), but Clark et al (1983) adduce further evidence that it is not involved in normal thermoregulation or fever, since naloxone alone is without effect on temperature regulation or fever.

To assume that arginine vasopressin and α-MSH are necessarily the only brain peptide modulators of body temperature during fever would be naive. Consideration of the published work on these peptides seems to indicate an action at different loci mainly on the duration of fever. The limiting mechanism for the maximum temperature in fever may well involve these and other modulators.

Arginine Vasopressin Kindles Motor Disturbances in the Rat

Intraventricular injection of arginine vasopressin in the rat, undertaken to ascertain whether it might be antipyretic in that species, led to kindling of convulsive activity (Kasting et al 1980). The behavioral motor disturbances were accompanied by electroencephalographic spike discharges. Rats could be primed by endogenous arginine vasopressin release, induced by hemorrhage or hypertonic saline, to respond with seizure-like activity to subsequent small intraventricular doses of arginine vasopressin (Burnard et al 1983a). The seizures so induced indicate one possible mechanism of febrile convulsions, and there is evidence that induced hyperthermic seizures in rats may be arginine vasopressin related (Kasting et al 1980). An arginine vasopressin receptor blocking agent prevented arginine vasopressin induced seizures, but not those caused by pentylenetetrazol (Burnard et al 1983b).

Long Evans rats tended to convulse when made hyperthermic by external heat, whereas Brattleboro rats, which lack arginine vasopressin, did not. Antibody to arginine vasopressin given into the brains of Long Evans rats prevented the convulsions (Veale et al 1984). A locus that is sensitive to topically applied arginine vasopressin, which causes severe motor disturbances, has recently been described in the rat (Naylor et al 1986). This site extends from the diagonal band of Broca to the anterior hypothalamus. There is some speculation regarding a link between the kindling of motor disturbances by intracerebral arginine vasopressin and the occurrence of febrile convulsions in children (Veale et al 1984). To study this possibility, a genetically determined modification of the usual loci of release of arginine vasopressin in fever, or of its receptor behavior, would be required. As yet there is no direct evidence of arginine vasopressin involvement. In rats, the kindling of the motor disturbance is transient, lasting between 7 and 90 days (Burnard & Veale 1986a). The compound $d(CH^2)^5$Tyr(Me) arginine vasopressin blocked the motor disturbances in rats usually kindled by arginine vasopressin, but not those due to pentylenetetrazol or somatostatin (Burnard et al 1986b). A major problem in relating the arginine vasopressin system to human convulsive activity is that so far arginine vasopressin induced convulsions are seen only in the rat.

Space does not permit discussion of the putative central transmitter substances used in the fever neuronal circuits, the reduction of fever responses in older animals, the role of the peripheral nervous system in fever, or the neural mechanisms of behavioral fevers. Less work has been done recently in these areas than in the few topics covered here. What seemed to be a fully understood mechanism in the early 1970s has been opened up in new and exciting ways in the last five years. I look forward to further quantum leaps in our knowledge of the neurobiology of fever in the next few years.

Literature Cited

Alexander, S., Cooper, K. E., Veale, W. L. 1986. *J. Physiol.* In press

Andersson, B., Gale, C. C., Hökfelt, I., Larsson, B. 1965. Acute and chronic effects of pre-optic lesions. *Acta Physiol. Scand.* 65: 45–60

Aronsohn, E., Sachs, J. 1885. Die Beziehungen des Gehirns zur Körperwärme und zum Fieber. *Pflügers Arch.* 37: 232–300

Artunkel, A. A., Marley, E., Stephenson, J. D. 1977. Some effects of prostaglandin E_1 and E_2 injected into the hypothalamus of young chicks: Dissociation between endotoxin fever and the effects of prostaglandins. *Br. J. Pharmacol.* 61: 39–46

Atkins, E., Bodel, P. T., Francis, L. 1967. Release of an endogenous pyrogen in vitro from rabbit mononuclear cells. *J. Exp. Med.* 126: 357–84

Auron, P. E., Webb, A. C., Rosenwasser, L. J., Mucci, S. F., Rich, A., Wolff, S. M., Dinarello, C. A. 1984. Nucleotide sequence of human monocyte interleukin I precursor cDNA. *Proc. Natl. Acad. Sci. USA* 81: 7907–11

Azzaroni, A., Cevolani, D., Ferrari, G., Parmeggiani, P. L. 1985. Thermosensitive neurons during sleep in cats. *J. Physiol.* 369: 60P

Baird, J. A., Hales, J. R. S., Lang, W. J. 1974. Thermoregulatory responses to the injection of monoamines, acetylcholine and prostaglandins into the lateral cerebral ventricle of the echidna. *J. Physiol.* 236: 539–48

Banet, M., Wieland, U. 1985. The effect of intraseptally applied vasopressin on thermoregulation in the rat. *Brain Res. Bull.* 14: 113–16

Barbour, H. G. 1921. The heat-regulating mechanism of the body. *Physiol. Rev.* 1: 295–326

Beeson, P. B. 1948. Temperature elevating effect of a substance obtained from polymorphonuclear leucocytes. *J. Clin. Invest.* 27: 524

Bennett, I. L., Beeson, P. B. 1953. Studies on the pathogenesis of fever. I. The effect of injection of extracts and suspensions of uninfected rabbit tissues upon the body temperature of normal rabbits. *J. Exp. Med.* 98: 477–92

Bennett, I. L., Petersdorf, R. G., Keene, W. R. 1957. Pathogenesis of fever, evidence for a direct cerebral action of bacterial endotoxins. *Trans. Assoc. Am. Physicians* 70: 64–72

Bernadini, G. L., Lipton, J. M., Clark, W. G. 1983. Intracerebroventricular and septal injections of arginine, vasopressin are not antipyretic in the rabbit. *Peptides* 4: 195–98

Bernadini, G. L., Richards, D. B., Lipton, J. M. 1984. Antipyretic effect of centrally administered CRF. *Peptides* 5: 57–59

Bernheim, H. A., Gilbert, T. M., Stitt, J. T. 1980. Prostaglandin E levels in the third ventricular cerebrospinal fluid of rabbits during fever and changes in body temperature. *J. Physiol.* 301: 69–78

Blatteis, C. M., Bealer, S. L., Hunter, W. S., Llanos, Q. J., Ahokas, R. A., Mashburn, T. A. Jr. 1983. Suppression of fever after lesions of the anteroventral third ventricle in guinea pigs. *Brain Res. Bull.* 2: 519–26

Bligh, J. 1974. Neuronal models of hypothalamic temperature regulation. In *Recent Studies of Hypothalamic Function,* ed. K. Lederis, K. E. Cooper, pp. 315–27. Basel: Karger

Bligh, J., Milton, A. S. 1973. The thermoregulatory effects of prostaglandin E_1 when infused into the lateral cerebral ventricle of the Welsh Mountain sheep at different ambient temperatures. *J. Physiol.* 229: 30–31P

Boulant, J. A., Scott, I. M. 1986. Comparison of prostaglandin E_2 and leucocytic pyrogen on hypothalamic neurons in tissue slices. In *Homeostasis and Thermal Stress, Experimental and Therapeutic Advances*, ed. K. Cooper, P. Lomax, E. Schonbaum, W. L. Veale, pp. 78–80. Basel: Karger

Braude, A. I., Zalesky, M., Douglas, N. 1958. The mechanism of tolerance to fever. *J. Clin. Invest.* 37: 880

Burnard, D. M., Pittman, Q. J., Veale, W. L. 1983a. Increased motor disturbances in response to arginine vasopressin following hemorrhage or the administration of hypertonic saline: Evidence for central AVP release in the rat. *Brain Res.* 273: 59–65

Burnard, D. M., Pittman, Q. J., Veale, W. L., 1983b. Prevention of vasopressin-induced convulsions by an antivasopressin antagonist. *Can. J. Physiol. Pharmacol.* 61: A3–A4

Burnard, D. M., Veale, W. L. 1986a. Increased sensitivity to vasopressin-induced motor disturbances: A transient alteration in brain function. *Can. J. Physiol. Pharmacol.* In press

Burnard, D. M., Veale, W. L., Pittman, Q. J. 1986b. Prevention of arginine vasopressin motor disturbances by a potent vasopressor antagonist. *Brain Res.* 362: 40–46

Cabanac, M. 1970. Interaction of cold and warm temperature signals in the brain stem. In *Physiological and Behavioral Temperature Regulation*, ed. J. D. Hardy et al, pp. 549–61. Springfield, Ill: Thomas

Cabanac, M., Stolwijk, J. A. J., Hardy, J. D. 1968. Effect of temperature and pyrogens on single unit activity in the rabbit's brain stem. *J. Appl. Physiol.* 24: 645–48

Cannon, M., Cranston, W. I., Hellon, R. F., Townsend, Y. 1982. Inhibition by tricothecene antibiotics of brain protein synthesis and fever in rabbits. *J. Physiol.* 322: 447–55

Carmichael, E. A., Feldberg, W., Fleischhauer, K. 1964. Methods for perfusing different parts of the cat's cerebral ventricles with drugs. *J. Physiol.* 173: 354–67

Carpenter, D. O. 1981. Ionic and metabolic bases of neuronal thermosensitivity. *Fed. Proc.* 40: 2808–18

Clark, G., Magoun, H. W., Ranson, S. W. 1939. Hypothalamic regulation of body temperature. *J. Neurophysiol.* 2: 61–80

Clark, W. G. 1979. Changes in body temperature after administration of amino acids, peptides, dopamine, neuroleptics and related agents. *Neurosci. Biobehav. Rev.* 3: 179–231

Clark, W. G., Bernadini, G. L. 1981. β-Endorphin induced hyperthermia in the cat. *Peptides* 2: 371–73

Clark, W. G., Pang, I. H., Bernadini. G. L. 1983. Evidence against involvement of β-endorphin in thermoregulation in the cat. *Pharmacol. Biochem. Behav.* 18: 741–45

Coceani, F., Bishai, C. A., Dinarello, C. A., Fitzpatrick, F. A. 1983. Prostaglandin E and thromboxane B in cerebrospinal fluid of afebrile and febrile cat. *Am. J. Physiol.* 244: R785–93

Coceani, F., Bishai, I., Lees, J., Sirko, S. 1986. Prostaglandin E and fever: A continuing debate. *Yale J. Biol. Med.* 59: 169–74

Cooper, K. E. 1965. The role of the hypothalamus in the generation of fever. *Proc. R. Soc. Med.* 58: 740

Cooper, K. E., Cranston, W. I. 1963. Clearance of radioactive bacterial pyrogen from the circulation. *J. Physiol.* 166: 41–42P

Cooper, K. E., Cranston, W. I., Honour, A. J. 1967. Observations on the site and mode of action of pyrogens in the rabbit brain. *J. Physiol.* 191: 325–37

Cooper, K. E., Kasting, N. W., Lederis, K., Veale, W. L. 1979. Evidence supporting a role for endogenous vasopressin in natural suppression of fever in the sheep. *J. Physiol.* 295: 33–45

Cooper, K. E., Veale, W. L. 1972. The effect of an inert oil in the cerebral ventricular system upon fever produced by intravenous pyrogen. *Can. J. Physiol. Pharmacol.* 50: 1066–71

Cooper, K. E., Veale, W. L. 1974. Fever, an abnormal drive to the heat conserving and producing mechanisms. See Bligh 1974, pp. 391–98

Cranston, W. I., Hellon, R. F., Mitchell, D. 1975. A dissociation between fever and prostaglandin concentration in cerebrospinal fluid. *J. Physiol.* 253: 583–92

Cranston, W. I., Duff, G. W., Hellon, R. F., Mitchell, D., Townsend, Y. 1976. Evidence that brain prostaglandin synthesis is not essential in fever. *J. Physiol.* 259: 239–49

Cranston, W. I., Hellon, R. F., Townsend, Y. 1980. Suppression of fever in rabbits by a protein synthesis inhibitor, Anisomycin. *J. Physiol.* 305: 337–44

Cranston, W. I., Gourine, V. N., Townsend, Y. 1981. The effects of intracerebroventricular cycloheximide on protein synthesis and fever in rabbits. *Br. J. Pharmacol.* 73: 6–8

Cranston, W. I., Hellon, R. F., Mitchell, D., Townsend, Y. 1983. Intraventricular injection of drugs which inhibit phospholipase A_2 suppress fever in rabbits. *J. Physiol.* 339: 97–105

Cranston, W. I., Hellon, R. F., Townsend, Y. 1982. Further observations on the suppression of fever by intracerebral action of anisomycin. *J. Physiol.* 322: 441–45

Crawford, J. L., Kennedy, J. I., Lipton, J. M., Ojeda, S. R. 1979. Effects of central administration of probenecid on fever produced by leucocytic pyrogen and PGE_2 in the rabbit. *J. Physiol.* 287: 519–33

Cushing, H. 1931. The reaction to posterior pituitary extract (pituitrin) when introduced into the cerebral ventricles. *Proc. Natl. Acad. Sci. USA* 17: 163–70

Dinarello, C. A. 1984. Interleukin-1. *Rev. Infect. Dis.* 6(1): 51–95

Dinarello, C. A., Goldin, N. P., Wolff, S. M. 1974. Release of an endogenous pyrogen in vitro from rabbit mononuclear cells. *J. Exp. Med.* 126: 357–84

Dinarello, C. A., Bernheim, H. A., Duff, G. W., Le, H. U., Nagabhushan, T. L., Hamilton, N. C., Coceani, F. 1984. Mechanism of fever induced by recombinant human interferon. *J. Clin. Invest.* 74: 906–13

Dinarello, C. A., Wolff, S. M. 1977. Partial purification of human leucocytic pyrogen. *Inflammation* 2: 179–89

Disturnal, J. E., Veale, W. L., Pittman, Q. J. 1985. Electrophysiological analysis of potential arginine vasopressin projections to the ventral septal area of the rat. *Brain Res.* 342: 162–67

Disturnal, J. E., Veale, W. L., Pittman, Q. J. 1986. The ventral septal area: Electrophysiological evidence for a possible role in antipyresis. *Neuroscience.* In press

Eisenman, J. S. 1969. Pyrogen-induced changes in the thermosensitivity of septal and preoptic neurons. *Am. J. Physiol.* 216: 330–34

Eisenman, J. S. 1972. Unit activity studies of thermoresponsive neurons. In *Essays on Temperature Regulation*, ed. J. Bligh, R. E. Moore, pp. 55–69. Amsterdam/London: North-Holland

Eisenman, J. S. 1982. Electrophysiology of the hypothalamus: Thermoregulation and fever. In *Pyretics and Antipyretics*, ed. A. S. Milton, pp. 187–217. Berlin: Springer Verlag

Feldberg, W. 1982. Looking back on some developments in Neurohumoral Physiology. Part II. Approaching the brain from its inner and outer surface. *Fifty Years On. Sherrington Lect.* 16: 27–80

Feldberg, W., Myers, R. D., Veale, W. L. 1970. Perfusion from cerebral ventricle to cisterna magna in the unanaesthetized cat. Effect of calcium on body temperature. *J. Physiol.* 207: 403–16

Feldberg, W., Veale, W. L., Cooper, K. E. 1971. Does leucocyte pyrogen enter the anterior hypothalamus via the cerebrospinal fluid? *Proc. IUPS 25th Int. Congr., Munich*, Vol. 9: 175 (item 511)

Feldberg, W., Gupta, K. P. 1973. Pyrogen fever and prostaglandin like activity in cerebrospinal fluid. *J. Physiol.* 228: 41–53

Fessler, J. H., Cooper, K. E., Cranston, W. I., Vollum, R. L. 1961. Observations on the production of pyrogenic substances by rabbit and human leucocytes. *J. Exp. Med.* 113: 1127–40

Ford, D. M. 1974. A diencephalic island for the study of thermally-responsive neurons in the cat's hypothalamus. *J. Physiol.* 239: 67–68P

Gaddum, J. H. 1961. Push-pull caunulae. *J. Physiol.* 155: 1–2P

Gander, G. W., Goodale, F. 1973. Studies on the endogenous pyrogen released in response to Poly I: Poly C. In *The Pharmacology of Thermoregulation*, ed. E. Schönbaum, P. Lomax. Basel: Karger

Gerbrandy, J., Cranston, W. I., Snell, E. S. 1954. The initial process in the action of bacterial pyrogens in man. *Clin. Sci.* 13: 453–59

Glyn-Ballinger, J. R., Bernadini, G. L., Lipton, J. M. 1983. α-MSH injected into the septal region reduces fever in rabbits. *Peptides* 4: 199–203

Hales, J. R. S., Bennett, J. W., Fawcett, A. A. 1973. Thermoregulatory effects of prostaglandins E_1, E_2, $F_{1\alpha}$, $F_{2\alpha}$ in the sheep. *Pflügers Archiv.* 339: 125–33

Hammel, H. T., Caldwell, F. T., Abrams, R. M. 1967. Regulation of body temperature in the blue tongued lizard. *Science* 156: 1260–63

Hashimoto, M., Nagai, M., Iriki, M. 1985. Comparison of the action of prostaglandin with endotoxin on thermoregulatory response thresholds. *Pflügers Archiv.* 405: 1–4

Heap, R. B., Silver, A., Walters, D. E. 1981. Effects of pregnancy on the febrile responses in sheep. *Q. J. Exp. Physiol.* 66: 129–44

Hensel, H. 1981. *Thermoreception and Temperature Regulation.* London/New York: Academic

Jackson, D. L. 1967. A hypothalamic region responsive to localized injections of pyrogens. *J. Neurophysiol.* 30: 586–602

Jahns, R. 1977. *Leitung und Verarbeitung*

thermischer Information in thermoafferenten System. Bochum: Habil-Schrift
Jell, R. M., Sweatman, P. 1977. Prostaglandin-sensitive neurons in cat hypothalamus: Relation to thermoregulation and to biogenic amines. *Can. J. Physiol. Pharmacol.* 55: 560–67
Kasting, N. W., Veale, W. L., Cooper, K. E. 1978. Evidence for a centrally active endogenous antipyretic near parturition. In *Current Studies of Hypothalamic Function*, ed. W. L. Veale, K. Lederis, pp. 63–71. Basel: Karger
Kasting, N. W., Cooper, K. E., Veale, W. L. 1979. Antipyresis following perfusion of brain sites with vasopressin. *Experentia* 35: 208–9
Kasting, N. W., Veale, W. L., Cooper, K. E. 1980. Convulsive and hypothermic effects of vasopressin in the brain of the rat. *Can. J. Physiol. Pharmacol.* 58: 316–19
King, M. K., Wood, W. B. Jr. 1958. Studies on the pathogenesis of fever. IV. The site of action of leucocyte and circulating endogenous pyrogen. *J. Exp. Med.* 107: 291–303
Kluger, M. J. 1979. *Fever, Its Biology, Evolution and Function*, pp. 129–58. Princeton: Princeton Univ. Press
Kluger, M. J., Oppenheim, J. J., Powanda, M. C., eds. 1985. *The Physiologic, Metabolic and Immunologic Actions of Interleukin-I*. New York: Liss
Kovacs, G. L., De Wied, D. 1983. Hormonally active arginine-vasopressin suppresses endotoxin-induced fever in rats: Lack of effect of oxytocin and a behaviorally active vasopressin fragment. *Neuroendocrinology* 37: 258–61
Laburn, H. P., Mitchell, D., Rosendorf, C. 1977. Effects of prostaglandin antagonism on sodium arachidonate fever in rabbits. *J. Physiol.* 267: 559–70
Laburn, H. P., Mitchell, D., Kenedi, E., Louw, G. N. 1981. Pyrogens fail to produce fever in a cordylid lizard. *Am. J. Physiol.* 241: R198–R202
Lee, T. F., Mora, F., Myers, R. D. 1985. The effect of intracerebroventricular vasopressin on body temperature and endotoxin fever of macaque monkey. *Am. J. Physiol.* 248: R674–R678
Leskell, L. G. 1978. Effects on fluid balance induced by nonfebrile intracerebroventricular infusion of PGE, PGF and arichidonic acid in the goat. *Acta Physiol. Scand.* 104: 225–31
Lin, M. T., Wang, T. I., Chan, H. K. 1983. A prostaglandin-adrenergic link occurs in the hypothalamic pathways which mediate the fever induced by vasopressin in the rat. *J. Neural. Trans.* 56: 21–31
Lipton, J. M. 1985. Antagonism of IL-I fever by the neuropeptide α-MSH. See Kluger et al 1985, pp. 121–32
Lipton, J. M., Glynn, J. R. 1980. Central administration of peptides alters thermoregulation in the rabbit. *Peptides* 1: 15–18
Lipton, J. M., Trzcinka, G. P. 1976. Persistence of febrile response to pyrogens after PO/AH lesions in squirrel monkeys. *Am. J. Physiol.* 231: 1638–48
Lomedico, P. T., Gubler, U., Hellman, C. P., Dukovich, M., Giri, J. G., et al. 1984. Cloning and expression of murine interleukin-I, cDNA in Escherechia coli. *Nature* 312: 458–62
Marx, J., Hilbig, R., Rahman, H. 1984. Endotoxin and Prostaglandin E fail to induce fever in a teleost fish. *Comp. Biochem. Physiol.* 77A: 483–87
Merker, G., Zeisberger, E., Blahser, S., Kraunig, M. 1983. Immunocytochemical reaction of vasopressin containing neurons during development of fever in the guinea pig. *Naunyn-Schmiedeberg's Arch. Pharmacol.* 322(Suppl.): R81
Milton, A. S. 1982. Prostaglandins in fever and the mode of action of antipyretic drugs. See Eisenman 1982, pp. 257–303
Milton, A. S., Wendlandt, S. 1971. Effects on body temperature of prostaglandins of the A, E and F series on injection into the third ventricle of unanaesthetized cats and rabbits. *J. Physiol.* 218: 325–36
Mitchell, D., Laburn, H. P., Cooper, K. E., Hellon, R. F., Cranston, W. I., Townsend, Y. 1986. Is prostaglandin E the neural mediation of the febrile response? The case against a proven obligatory role. *Yale J. Biol. Med.* 59: 159–68
Murphy, M. T., Richards, D. B., Lipton, J. M. 1983. Antipyretic potency of centrally administered-melanocyte stimulating hormone. *Science* 221: 192–93
Murphy, P. A., Chesney, P. J. 1974. Further purification of rabbit leucocyte pyrogen. *J. Lab. Clin. Med.* 83: 310–22
Myers, R. D. 1970. An improved push-pull cannula system for perfusing an isolated region of brain. *Physiol. Behav.* 5: 243–46
Myers, R. D., Gurley-Orkin, L. 1985. New "micro push-pull" catheter system for localized perfusion of diminutive structures. *Brain Res. Bull.* 14: 477–83
Naylor, A. M., Gubitz, G. J., Dinarello, C. A., Veale, W. L. 1986. Central effects of vasopressin and 1-Desamino-8-D-arginine vasopressin (DDAUP) on interleukin-I fever in the rat. *Brain Res. Bull.* In press
Naylor, A. M., Ruwe, W. D., Kohut, A. F., Veale, W. L. 1985a. Perfusion of vasopressin within the ventral septum of the rabbit suppresses endotoxin fever. *Brain Res. Bull.* 15: 209–13

Naylor, A. M., Ruwe, W. D., Burnard, D. M., McNeely, P. D., Turner, S. L., Pittman, Q. J., Veale, W. L. 1985b. Vasopressin-induced motor disturbances: Localization of a sensitive forebrain site in the rat. *Brain Res.* 361: 242–46

Ott, I. 1884. The relation of the nervous system to the temperature of the body. *J. Nerv. Ment. Dis.* 2: 141–52

Ott, I. 1914. *Fever, Its Thermotaxis and Metabolism*. New York: Hoeber

Palkovits, M., Zaborszky, L. 1979. Neural connections of the hypothalamus. *Handb. Hypothalamus* 1: 379–509

Pittman, Q. J., Cooper, K. E., Veale, W. L., Van Petten, G. R. 1974. Observations on the development of the febrile reponse to pyrogens in the sheep. *Clin. Sci. Molec. Med.* 46: 591–602

Pittman, Q. J., Veale, W. L., Cockeram, A. W., Cooper, K. E. 1976. Changes in body temperature produced by prostaglandins and pyrogens in the chicken. *Am. J. Physiol.* 230: 1284–87

Pompeiano, O. 1973. Reticular formation. *Handb. Sensory Physiol.* 2: 382–488

Ranson, S. W. 1940. Regulation of body temperature. *Assoc. Res. Nerv. Ment. Dis. Proc.* 20: 342–99

Repin, I. S., Kratskin, I. L. 1967. An analysis of the hyperthermic mechanism of fever. *Fiziol. Zh. SSSR* 53: 1206–11

Reynolds, W. W., Casterlin, M. E., Covert, J. B. 1976. Behavioral fever in teleost fishes. *Nature* 259: 41–42

Reynolds, W. W., Casterlin, M. E., Covert, J. B. 1980. Behaviorally mediated fever in aquatic ectotherms. In *Fever*, ed. J. M. Lipton, pp. 207–12. New York: Raven

Rezvani, A. H., Denbow, D. M., Myers, R. D. 1986. α-Melanocyte-stimulating hormone infused ICV fails to affect body temperature or endotoxin fever in the cat. Submitted for publication

Richet, C. 1884. De l'influence des lésions du cerveau sur la température. *CR Acad. Sci.* 98: 827–29

Rowley, D., Howard, J. G., Jenkin, C. R. 1956. The fate of 32p labelled bacterial lipopolysaccharide in laboratory animals. *Lancet* 1: 366–67

Ruwe, W. D., Myers, R. D. 1980. The role of protein synthesis in the hypothalamic mechanism mediating pyrogen fever. *Brain Res. Bull.* 5: 735–43

Ruwe, W. D., Naylor, A. M., Veale, W. L. 1985. Perfusion of vasopressin within the rat brain suppresses prostaglandin E-hyperthermia. *Brain Res.* 338: 219–24

Ruwe, W. D., Naylor, A. M., Veale, W. L., Dinarello, C. A. 1986. Site and mechanism of action of interferon-induced fever. *Can. J. Physiol. Pharmacol.* In press

Ruwe, W. D., Veale, W. L., Cooper, K. E. 1983. Peptide neurohormones: Their role in thermoregulation and fever. *Can. J. Biochem. Cell. Biol.* 61: 579–93

Sakata, Y., Morimoto, A., Takase, Y., Murakami, N. 1981. Direct effects of endogenous pyrogen on medullary temperature sensitive neurons in rabbits. *Jpn. J. Physiol.* 31: 247–57

Sampson, W. K., Lipton, J. M., Zimmer, J. A., Glyn, J. R. 1981. The effect of fever on central α-MSH concentrations in the rabbit. *Peptides* 2: 419–23

Satinoff, E., McEwen, G. N. Jr., Williams, B. A. 1976. Behavioral fever in newborn rabbits. *Science* 193: 1139–40

Satinoff, E. 1983. A re-evaluation of the concept of the homeostatic organization of temperature regulation. *Handb. Behav. Neurobiol.* 6: 443–71

Schoener, E. P., Wang, S. C. 1975. Leucocytic pyrogen and sodium acetylsalicylate on hypothalamic neurons in the cat. *Am. J. Physiol.* 229: 185–90

Sheth, U. K., Borison, H. L. 1960. Central pyrogenic action of *Salmonella typhosa* lipopolysaccharide injected into the lateral cerebral ventricle in cats. *J. Pharmacol. Exp. Therap.* 130: 411–17

Shih, S. T., Khorram, O., Lipton, J. M., McCann, S. 1986. Central administration of α-MSH antiserum augments fever in the rabbit. *Am. J. Physiol.* 250: R803–R806.5

Siegert, R., Phillip-Dormston, W. K., Radsak, K., Menzel, H. 1976. Mechanism of fever induction in rabbits. *Infect. Immun.* 14: 1130–37

Skarnes, R. C., Brown, S. K., Hull, S. S., McCracken, J. A. 1981. Role of prostaglandin E in the biphasic fever response to endotoxin. *J. Exp. Med.* 154: 1212–24

Stitt, J. T. 1973. Prostaglandin E fever induced in rabbits. *J. Physiol.* 232: 163–79

Stitt, J. T. 1979. The effect of cyclohexamide on temperature regulation and fever production in the rabbit. In *Thermoregulatory Mechanisms and Their Therapeutic Implications*, ed. B. Cox, P. Lomox, S. Milton, E. Schonbaum. Basel: Karger

Stitt, J. T. 1985. Evidence for the involvement of the organism vasculosum laminae terminalis in the febrile response of rabbits and rats. *J. Physiol.* 368: 501–11

Stitt, J. T. 1986. Prostaglandin E as the neural mediator of the febrile response. *Yale J. Biol. Med.* 59: 137–49

Stitt, J. T., Hardy, J. D. 1975. Microelectrophoresis of PGE on single units in the rabbit hypothalamus. *Am. J. Physiol.* 229: 240–45

Stitt, J. T., Shimada, S. G., Bernheim, H. A. 1983. Microinjection of zymosan and lipopolysaccharide into the organum vasculosum laminae terminalis of rats enhances their febrile responsiveness to endogenous pyrogen. *Proc. Int. Physiol. Congr. Satellite Symp. Thermoregulation, Brisbane, Australia*, p. 113

Sutin, J., McBride, R. L. 1977. Anatomical analysis of neuronal connectivity. *Meth. Psychobiol.*, pp. 1–26

Székely, M. 1978. Endotoxin fever in the new-born guinea pig and the modulating effects of indomethacin and p-chlorophenylalanine. *J. Physiol.* 281: 467–76

Teague, R. S., Ranson, S. W. 1936. The role of the anterior hypothalamus in temperature regulation. *Am. J. Physiol.* 117: 562–70

Vane, J. R. 1971. Inhibition of prostaglandin synthesis as a mechanism of action for aspirin-like drugs. *Nature New Biol.* 231: 232–35

Veale, W. L., Cooper, K. E. 1974. Prostaglandin in cerebrospinal fluid following perfusion of hypothalamic tissue. *J. Appl. Physiol.* 37: 942–45

Veale, W. L., Cooper, K. E. 1975. Comparison of sites of action of prostaglandin and leucocyte pyrogen in brain. In *Temperature Regulation and Drug Action*, ed. P. Lomax, E. Schönbaum, J. Jacob, pp. 218–26. Basel: Karger

Veale, W. L., Cooper, K. E., Ruwe, W. D. 1984. Vasopressin: Its role in antipyresis and febrile convulsion. *Brain Res. Bull.* 12: 161–65

Villablanca, J., Myers, R. D. 1965. Fever produced by microinjection of typhoid vaccine into hypothalamus of cats. *Am. J. Physiol.* 208: 703–7

Weindl, A. 1969. Electron microscopic observations on the organum vasculosum of the lamina terminalis after intravenous injection of horseradish peroxidase. *Neurology* 19: 295

Weindl, A., Sofroniew, M. V. 1981. Relation of neuropeptides to mammalian circumventricular organs. In *Neurosecretion and Brain Peptides*, ed. J. B. Martin, S. Reichlin, K. L. Bick, pp. 303–20. New York: Raven

Wilkinson, M. F., Kasting, N. W. 1986. Centrally applied vasopressin is antipyretic due to its effects on febrile set point. *Proc. IUPS 30th Congr.* 16(119.07): 60

Williams, J. W., Rudy, T. A., Yaksh, T. I., Viswanathan, C. T. 1977. An extensive exploration of the rat brain for sites mediating prostaglandin induced hypothermia. *Brain Res.* 120: 251–62

Wit, A., Wang, S. C. 1968. Temperature-sensitive neurons in preoptic/anterior hypothalamic region: Actions of pyrogens and acetylsalicylate. *Am. J. Physiol.* 215: 1160–69

Work, E. 1971. Production, chemistry and properties of bacterial pyrogens and endotoxins. In *Pyrogens and Fever*, ed. G. E. W. Wolstenholme, J. Burch, pp. 23–46. Edinburgh/London: Churchill Livingstone

Zeisberger, E., Merker, G., Blahser, S. 1980. Fever response in the guinea pig before and after parturition and its relationship to the antipyretic reaction of the pregnant sheep. *Brain Res.* 212: 379–92

Zeisberger, E., Merker, G., Blahser, S., Krannig, M. 1983. Changes in activity of vasopressin neurons during fever in the guinea pig. *Neurosci. Lett.* (Suppl.) 14: S414

Ann. Rev. Neurosci. 1987. 10 : 325–62

THE ORGANIZATION AND FUNCTION OF THE VOMERONASAL SYSTEM

M. Halpern

Department of Anatomy and Cell Biology, State University of New York, Health Sciences Center at Brooklyn, Brooklyn, New York 11203

INTRODUCTION

The vomeronasal (VN) organ is a chemoreceptive structure situated at the base of the nasal septum of most terrestrial vertebrates. Structurally, the VN system is very similar to the main olfactory system. Functionally, the VN and main olfactory systems have different roles in the execution of several species-typical behaviors that depend on reception of chemosignals emitted by conspecifics or prey.

Although the VN organ (Jacobson's organ) was discovered and described in the early part of the nineteenth century and occasional studies of its morphology, functional significance, and physiology appeared in the scientific literature during the first half of this century, the current explosion of interest can be attributed to the simultaneous emergence of two ideas in the early 1970s. During the 1950s and 1960s, behavioral endocrinologists had discovered and attempted to explain several pheromonal effects in mice. These pheromonal effects were found to be under the control of chemical stimuli in the urine of conspecifics and were absent in mice with their olfactory bulbs removed. By 1970 there was a general consensus that the pheromonal effects depended on a functional olfactory system. At the time it was known that the bipolar neurons of the olfactory epithelium terminated in the glomerular layer of the main olfactory bulb and that the bipolar neurons of the VN organ terminated in the accessory olfactory bulb; however, scant attention was given to the fact that the olfactory bulbs are part of the CNS and therefore not simply a relay in the olfactory pathway and that olfactory bulbectomy denervated both main and accessory olfactory systems.

0147–006X/87/0301–0325$02.00

In 1970 Winans & Scalia documented the segregation in the projections of the main and accessory olfactory bulbs in a mammal, the rabbit, and suggested that the pheromonal effects that had previously been ascribed to olfaction might depend on a functional VN system. This idea was further elaborated by Raisman (1972) and Scalia & Winans (1975, 1976) into an explicit statement of the dual olfactory hypothesis, which proposed that two parallel pathways could be traced from the VN organ and the main olfactory epithelium into the telencephalon and diencephalon of all vertebrates possessing these two nasal chemoreceptive systems. A corollary of this hypothesis is that each system should be involved in distinct behavioral domains. Anatomical, behavioral, and physiological research on the VN system since 1970 has been concentrated on testing the dual olfactory hypothesis and its corollary.

Several general approaches have been particularly fruitful in elucidating the differences between the VN and main olfactory systems. These include consideration of the similarities and differences in the anatomy of the two systems, particularly their gross morphology, ultrastructure, central nervous system connections, and transmitter/peptide content. A second approach has attempted to identify behaviors or components of behavioral sequences that rely primarily or exclusively on one or the other system. A third approach has been concerned with chemical characterization of the substances that stimulate the VN system but not the main olfactory system. Finally, an approach, still very much in its infancy, consists of attempts to physiologically identify activity in the VN system that correlates with peripheral activation. The present review attempts to update our knowledge based on the results of these four approaches.

The role of the VN system in vertebrate behavior and/or reproductive physiology has been the subject of several reviews since 1970. Burghardt (1970) provided an excellent comprehensive review of reptilian vomeronasal systems and functional correlates. Madison (1977) reviewed the literature on chemical communication in amphibians and reptiles, Halpern (1980a) reviewed the use of allelochemics (substances communicated between members of different species) by nonaquatic vertebrates, and Halpern (1983) and Halpern & Kubie (1984) reviewed the nasal chemical senses in snakes. The most comprehensive review of the literature on the mammalian VN system and its role in reproductive behavior was published by Wysocki (1979). In the same year, Keverne (1979) reviewed the mammalian reproductive literature as it relates to the VN system and proposed that the main olfactory pathways are involved in processing chemical cues of a complex nature. These responses to main olfactory stimulation are modifiable, for example, by past experience. Furthermore, Keverne suggested that the VN system is primarily involved in processing chemical signals influencing "emotive" behaviors that occur without conscious perception and that can be secondary to neuroendocrine changes. Additional reviews of the literature on the role of the mammalian VN system have been published by Johns (1980), Meredith (1980, 1983), and Johnston (1983, 1985). Because of the wealth of previous reviews, the present one emphasizes studies not previously covered. No attempt is made here to include the early history of research on the VN system, as that has been very adequately handled (Burghardt 1970, Wysocki, 1979). To the best of my knowledge this is the first review that combines information about the VN system in mammals and nonmammals.

ANATOMY OF THE VOMERONASAL ORGAN

Whereas the olfactory epithelium lines the dorsal posterior aspect of the nasal cavity and is easily accessible to airborne odorant molecules, in most vertebrates the VN organ is sequestered in a nasal diverticulum that in embryogenesis may have become totally separated from the nasal cavity. In most mammals the VN organ is a cigar-shaped tube contained in a bony capsule located along the ventral edge of the nasal septum. Posteriorly the tube ends blindly and anteriorly it frequently communicates via its own duct with the nasopalatine duct, thus permitting communication with the nasal cavity, oral cavity or both. The VN organ possesses a lumen that is typically crescentric in cross-section with a convex lateral wall lined with ciliated pseudostratified nonsensory epithelium and a concave medial wall lined with a specialized neurosensory epithelium (Barber & Raisman 1974, Bertmar 1981a,b, Cooper & Bhatnagar 1976, Hunter et al 1984, Loo 1977, Maier 1980). In reptiles that possess a VN organ, it is most commonly described as hemispheric in shape with a duct leading directly into the oral cavity. The reptilian VN organ contains a crescentic shaped lumen whose dorso-caudal aspect is concave and contains the neurosensory epithelium. The convex, rostro-ventral aspect of the VN organ, the mushroom body, is lined by a ciliated stratified nonsensory epithelium (Gabe & Saint Girons 1976, Wang & Halpern 1980a).

The systematic presence of vomeronasal organs and their mode of communication with the oral and nasal cavities was recently reviewed by Bertmar (1981a). Generally, VN organs appear in terrestrial vertebrates, are absent in some aquatic vertebrates (e.g. fish), and are absent or vestigial in some adult forms of flying and arboreal species (e.g. birds). Within a systematic group, variation in the development of the VN organ may correlate with ecological niche. For example, among lizards (Gabe & Saint Girons 1976, Saint Girons 1975) a wide variation in development of the VN organ is observed, with arboreal species demonstrating a regressed or absent VN organ whereas fossorial forms possess exceptionally well developed VN organs. Among prosimians and primates, variation also has been noted (Hunter et al 1984, Maier 1980). The prosimian *Tupaia glis* and lower primates, e.g. *Microcebus*, possess highly developed VN organs. The platyrrhini have, by contrast, a reduced but functional VN organ. In adult catarrhini no VN organ is present. The human fetus possesses a VN organ (Bossy 1980, Kreutzer & Jafek 1980, Nakashima et al 1985), and in many human adults a patent VN pit and duct system can be observed (Johnson et al 1985). However, there is no evidence at present that the human VN epithelium is functional or maintains connections with the central nervous system.

The structure of the VN neuroepithelium has been described recently in representatives of virtually every major vertebrate group at the light microscopic and ultrastructural level (Adams & Weikamp 1984, Bhatnagar et al 1982, Ciges et al 1977, Kolnberger 1971, Kolnberger & Altner 1971, Kratzing 1971a,b, Loo & Kanagasuntheram 1972, Miragall et al 1979, Naguro & Breipohl 1982, Taniguchi & Mikami 1985, Taniguchi & Mochizuki 1982, Vaccarezza et al 1981, Wang & Halpern 1980a,b) There is a remarkable conservation of structure through widely separated groups. The VN neuroepithelium of all vertebrates studied to date lines a lumen and consists of three types of cells: supporting, sensory, and basal (undifferentiated). Cells closest to the lumen are considered to be more apically situated than cells closer to the basal lamina. The organization of these cells into recognizable layers varies, but two general schemata prevail. In one, supporting cells and sensory cells traverse the depth of the epithelium and the basal cells are relatively few and located primarily at the margins of the neuroepithelium where it joins the nonsensory epithelium (e.g. in mice). In the second, a highly segregated lamination exists in which supporting cells are separated from the deeper layers; the receptor cells traverse the entire epithelium; and the basal cells are numerous and situated deep in the epithelium (e.g. in snakes). In this latter type of epithelium the receptor cells and basal cells are sequestered into basket-like columns by connective tissue septae, and each column is further subdivided by longitudinally running axons and dendrites into subcolumnar arrays (Wang & Halpern, 1980a,b).

The vast majority of the ultrastructural studies on the VN epithelium describe microvilli, with some modifications noted on the surface of the receptor cells and an absence of cilia on these same cells (Bannister 1968, Miragall et al 1979, Taniguchi & Mikami 1985). Exceptions are the presence of cilia on receptor cells in immature animals (Kratzing 1971a), an occasional cilia reported on isolated cells (Bhatnagar et al 1982), and a single study in which ciliated receptor cells are described in the dog VN organ (Adams & Weikamp 1984). Since the cilia on main olfactory receptor cells are believed to be the site of olfactory receptors, the almost uniform finding of an absence of cilia on VN receptor cells represents a major difference between the two systems and implies that chemoreception transduction does not require cilia.

Receptor cell turnover in the main olfactory system has been well documented (e.g. Graziadei 1973). Evidence for turnover of VN receptor cells derives from several observations that have been made in a number of species, but is best described in snakes (Wang & Halpern 1980a). The basal (undifferentiated) cells at the base of the epithelium have all the cytological features of stem cells, i.e. large nucleus, scant perikaryal cytoplasm, and a paucity of cytoplasmic organelles. These cells have been observed in various stages of mitosis. Receptor cells differ in cytological characteristics, depending on their relative apical position in the epithelium. More deeply situated receptor cells appear less mature than the receptor cells lying above them. This immaturity is marked by scanter rough and smooth endoplasmic reticulum, few inclusion bodies, a less well-developed Golgi apparatus, and absent or short apical and basal processes. The most superficially located bipolar neurons have characteristics suggestive of aging or degenerating neurons. They frequently display crenated nuclei, dispersed rough endoplasmic reticulum, and large quantities of lysosomes and inclusion bodies, and appear swollen with smooth endoplasmic reticulum. On rare occasions it is possible to observe a receptor cell lying between adjacent supporting cells, apparently in the process of being extruded into the lumen of the VN organ (Wang & Halpern 1980a,b). The basal cells of the VN organ of snakes (Wang et al 1979) and mice (Barber & Raisman 1978a, Wilson & Raisman 1980) incorporate ^{3}H-thymidine. Labeled cells are first seen in the basal cell compartment, but with protracted intervals between injection of the tracer and sacrifice of the animal, labeled cells may be observed in the receptor cell compartment. In snakes, labeled cells are found closer to the surface, with long (21 days) survival times (Wang et al 1979), and in mice, labeled columns of receptor cells are found further away from the neurosensory-nonsensory epithelial boundary region with increasing survival time (Barber & Raisman 1978a). Wilson & Raisman (1980) estimate, based on counts of numbers of labeled cells, that the entire VN epithelium of mice turns over approximately every two to three months. Since the axons of bipolar neurons make synaptic connections in the accessory olfactory bulb (AOB), these observations raise the interesting question of how the specificity of connections are maintained during reinnervation.

The VN organ has an organizational substructure orthogonal to its lamination. Columns of cells in the mouse VN organ are generated more or less synchronously. In contrast, in the VN organ of snakes, where a columnar organization is created by connective tissue septae that invaginate the epithelium, cells within a structural column are generated at different times. The structural columns of snakes appear to be genetic columns, since bipolar neurons in a column are all generated from basal cells in that same column. It is not yet known whether there are functional correlates to these different types of columnar organization.

A consistent pattern of first order VN afferents is found across species. The unmyelinated axons of the bipolar receptor neurons of the epithelium pierce its basal lamina, form the compact, relatively long, vomeronasal nerves that penetrate the cribriform plate, and terminate in the glomerular layer of the accessory olfactory bulb (Barber & Raisman 1974, Barber & Field 1975, Barber et al 1978, Wang & Halpern 1982b, Meredith 1982). A similar pattern exists in the main olfactory system; the unmyelinated axons of bipolar neurons form the short olfactory fila that terminate in the glomerular layer of the main olfactory bulb. The rabbit vomeronasal nerve contains two types of axons that can be differentiated based on their antigenic properties. Two monoclonal antibodies raised originally in mouse to rabbit olfactory bulb homogenates form complexes with antigens on the surface of axons terminating in either the lateral or medial accessory olfactory bulb (Imamura et al 1985). Previously, there was no evidence that a distinction could be made among groups of VN nerve fibers. In addition to its innervation by the VN nerve, the VN organ (and main olfactory mucosa) appears to be innervated by the terminal nerve (Bojsen-Moller 1975), which travels in close proximity to the VN nerve.

A technique used to study the turn-over process in the VN organ involves section of the VN nerve. This procedure results in retrogade degeneration of the bipolar neurons of the VN epithelium (Barber 1981, Barber & Raisman 1978b, Kubie et al 1978, Wang & Halpern 1982a). In mice (Barber 1981, Barber & Raisman 1978b), neurosensory cells disappear and numerous mitoses are observed throughout the depth of the epithelium eight days following nerve cut. Supporting cells remain, and sensory cells at the margins of the neurosensory epithelium appear normal. Epithelial repopulation occurs first in the basal portion of the receptor cell layer, and later, by 32 days, the entire depth of the epithelium becomes repopulated. Double labeling studies with ^{3}H-thymidine and horseradish peroxidase demonstrate that cells labeled with ^{3}H-thymidine, and therefore generated during the regeneration process, develop axons that terminate in the AOB.

A similar process is seen in garter snakes, where all bipolar neurons in a column degenerate following section of their axons (Wang & Halpern 1982a,b). The necrotic cells are extruded from the epithelium or phagocytosed within two weeks. Supporting cells undergo minor morphological changes. Basal cells immediately begin to increase their mitotic activity and fill the vacated columns. When the columns are approximately two thirds filled (four weeks following axotomy), the most apical cells begin to differentiate into neurons. By the eighth postoperative week the columns are filled and the most apical six to ten cells are morphologically bipolar neurons. At this time, ^{3}H proline injected into the VN organ results in incorporation of the amino acid into protein and transport of the labeled protein to the AOB.

ACCESSORY OLFACTORY BULBS

Accessory olfactory bulbs are very similar among vertebrates and possess essentially the same laminar pattern as that of simplified main olfactory bulbs. The most superficial layer is composed of VN nerve fibers that pass into the glomerular layer. The external plexiform layer is reduced when compared to the main olfactory bulb, and the "mitral" cell layer is thicker than the monolayer present in the main olfactory bulb (Switzer & Johnson 1977). Tufted cells appear to be absent in the AOB. An internal plexiform layer and granule cell layer are both present. The size of the AOB has been compared among species of bats (Frahm 1981, Frahm & Bhatnagar 1980) and found to correlate with preferred diet. The *Phyllostomatidae*, which comprise frugivorous, nectarivorous, and sanguivorous bats, have well-developed VN systems whereas the insectivorous, piscivorous, and carnivorous bats have reduced or nonfunctional VN systems.

The chemical cytoarchitecture of the AOB was recently reviewed by Macrides & Davis (1983), and studies cited by them are not repeated here. Interestingly, several hypothalamic releasing hormones have been found in the AOB. Luteinizing hormone releasing hormone (LHRH) immunoreactivity has been found in rat (Witkin & Silverman 1983), hamster (Jennes & Stumpf 1980, Phillips et al 1980, 1982), and new world primates (Witkin 1985). Thyrotropin releasing hormone (TRH) is found in the AOB of rats in the highest concentrations of any structure in the neuroaxis (Manaker et al 1985, Mantyh & Hunt 1985), and somatostatin-positive mitral cells and somatostatin-positive fibers have been observed in the rat AOB (Vincent et al 1985). How to interpret these findings is not obvious, but a study by Dluzen & Ramirez (1983) is suggestive. As discussed below, that the VN system mediates endocrinological and behavioral responses to sex-related odors is now well established. Dluzen & Ramirez (1983) exposed male mice to other male or to female mice and determined the LHRH, TRH, norepinephrine, and dopamine content of the anterior dorsal olfactory bulbs, which include only the main olfactory bulb, and the posterior dorsal olfactory bulb, which primarily includes the AOB.

Under control conditions, concentrations of LHRH and norepinephrine were higher in the region including the AOB whereas concentrations of TRH and DA did not differ in the two regions. LHRH increased selectively in the region including the AOB after exposure to male or female mice. Dopamine levels also increased selectively in this region when males were exposed to females but not to other males. Norepinephrine levels increased in both regions when the males were exposed to females but not to males. TRH changes were not significant regardless of stimulus conditions. These results demonstrate that chemical changes occur in the region of the AOB in parallel with behavioral changes that are discussed below.

DEVELOPMENT

Cuschieri & Bannister (1975) studied the development of the VN organ in the mouse. The organ is first recognizable as a thickening in the epithelium of the medial wall of the olfactory pit at 11 days of gestation. This epithelium invaginates, forming first a groove and then, by fusion of the lips, a tube closed posteriorly and opening anteriorly into the main olfactory pit. The hamster VN organ develops at a similar time in gestation (Taniguchi et al 1982a). Early in fetal life, the sensory epithelium is made up entirely of undifferentiated cells. From eight to ten days after birth most neurosensory cells acquire microvilli, suggesting that full development of the sensory epithelium is achieved during the postnatal period. In contrast, the hamster olfactory epithelium has an "adult-like" appearance one day after birth (Taniguichi et al 1982b). Olfactory marker protein, a protein found only in olfactory and vomeronasal neurosensory cells, first appears in the VN organ of rats four days after birth. Interestingly, this is much later than its appearance in the main olfactory epithelium, where it is found at 18 days of gestation (Farbman & Margolis 1980). In human embryos, the VN groove is first observed at about 37 days of gestation. The VN organ and nerve are observed later, but no AOB is evident (Bossy 1980).

Wilson & Raisman (1980) observed an increase of 43% in the total number of neurosensory cells in the VN organ of the mouse between one and four months of age, followed by a fall of 21% between four and eight months of age. These changes were accompanied by an increase in the volume of the glomerular layer of the AOB (Wilson & Raisman 1981).

Although the VN organ appears to develop later than the main olfactory epithelium, neurogenesis of the AOB has been reported to precede main olfactory bulb neurogenesis in rat (Bayer 1983) and mouse (Hinds 1967). Output neurons precede intrinsic neurons, and granular cell neurogenesis continues into postnatal periods. In the tertiary regions of the main and accessory olfactory systems, e.g. pyriform cortex and amygdala, neurogenetic gradients do not systematically vary with respect to the system, but demonstrate a rostral-caudal gradient, with more rostral structures developing and maturing earlier than more caudal structures (Bayer 1980, Ten Donkelaar et al 1979, Leonard 1975).

Analysis of the developmental pattern of structures comprising the VN and main olfactory systems does not lead to a clear-cut answer to the questions of which system develops first or when each system becomes functional. An experiment by Pedersen et al (1984) demonstrated that the AOB of rat fetuses incorporate 2-deoxyglucose in amounts greater than other regions of the forebrain such as the main olfactory bulb. However, since this study did not employ odor delivery as part of the experimental design, no evidence was obtained that the increased uptake in 2-deoxyglucose was related to the sensory function of the VN system. As suggested by Shepherd (1985), 2-deoxyglucose uptake in the one to two days prior to birth may reflect increased activity in centrifugal fibers. Interestingly, differential 2-deoxyglucose uptake does not occur in the AOB of postnatal or adult rats during odor exposure, suckling, or electrical stimulation of the VN nerves or in hamster AOB during exposure to females (Shepherd 1985). Since some of these experimental manipulations clearly stimulate the VN system, one must approach both positive and negative results of glucose uptake studies with caution.

The VN system is sexually dimorphic in rats, and this dimorphism can be altered with perinatal manipulations of the endocrine system. The VN organ and neurosensory epithelial volume is greater in male rats at six months of age than in females. These differences can be eliminated by removing the testicles of the males and androgenizing the females on the day of birth (Segovia & Guillamon 1982). In postpubertal rats, gonadectomy results in reduction in the height of the VN epithelium and a decrease in the nuclear size of bipolar neurons in males and females (Segovia et al 1984b). The volume of the AOB and the number of mitral cells in the AOB of three-month-old rats shows a sexual difference, with males having larger AOBs and more mitral cells than females. These sex differences can be eliminated by orchidectomizing males and androgenizing the females on the day of birth (Segovia et al 1984a, Valencia et al 1986).

SECONDARY PROJECTIONS OF THE VN SYSTEM

In mammals, the AOB projects to the bed nucleus of the accessory olfactory tract, the medial nucleus, and the posteromedial cortical nucleus of the amygdala, and the bed nucleus of the stria terminalis (e.g. Broadwell 1975, Davis et al 1978, deOlmos et al 1978, Devor 1976, Price 1973, Scalia

& Winans 1975, Shammah-Lagnado & Negrao 1981, Skeen & Hall 1977, Winans & Scalia 1970). In reptiles, the AOB projects via the accessory olfactory tract to the nucleus sphericus, a presumed homologue of the "vomeronasal amygdala" (Halpern 1976, Heimer 1969, Ulinski & Peterson 1981). In amphibians, the AOB also projects, via the ventrolateral olfactory tract, to a presumptive homologue of the amygdala (Northcutt & Royce 1975, Scalia 1972). Centrifugal fibers to the AOB exist, and these arise primarily from AOB target areas (Barber & Field 1975, Broadwell & Jacobowitz 1976, Conrad & Pfaff 1976, Davis et al 1978, Halpern 1980, Kevetter & Winans 1981, Kretteck & Price 1977, 1978a, Raisman 1972), the locus coeruleus (adrenergic), and perhaps the raphe nuclei (serotinergic) (Broadwell & Jacobowitz 1976).

The major targets of the mammalian main olfactory bulb projections are the anterior olfactory nucleus, olfactory cortex, olfactory tubercle, antero-lateral cortical amygdaloid nucleus, and lateral entorhinal cortex (e.g. Scalia & Winans 1975). The main olfactory pathway may interact with the accessory olfactory pathway at the level of the amygdala. The pyriform cortex projects to the endopiriform nucleus, which in turn projects to the medial and posteromedial cortical amygdaloid nuclei (Kretteck & Price 1978a,b).

The tertiary projections of the VN system are primarily to the preoptic-anterior hypothalamic junction, ventromedial hypothalamic nucleus, and ventral premammillary nucleus (deOlmos & Ingram 1972, Halpern 1980b, Kevetter & Winans 1981, Kretteck & Price 1977, 1978a). These projections provide an anatomical substrate for the endocrinological and behavioral effects to be described below. The principal tertiary projections of the main olfactory system are to the hippocampal complex, frontal cortex, dorsomedial thalamus, and hypothalamus (e.g. Kretteck & Price 1977, 1978a).

VOMERONASAL MEDIATION OF PRIMER PHEROMONE EFFECTS

The functional significance of the vomeronasal system has been particularly well elucidated in analysis of behavioral changes that depend on chemosignals to modify activity of the hypothalamic-pituitary axis. Primer pheromones are chemical substances secreted or excreted by an individual that have an endocrinological effect on other members of the same species. In the 1950s and 1960s, a number of such effects were described, primarily in mice, and related to reproductive functions. These included suppression of estrus in group-housed females (Lee-Boot effect), induction of estrus and estrus synchrony produced by male odors (Whitten effect), acceleration of puberty in female mice produced by male odors (Vandenbergh effect), and pregnancy block caused by odor of a "strange" male (Bruce effect). As discussed below, evidence indicates that all of these effects are mediated by the VN system. The connections thought to be essential are those from the VN organ to the AOB to the corticomedial amygdala and into the neuroendocrine hypothalamus. Keverne (1982) suggested that all of the primer pheromones induce a change in hypothalamic dopamine turnover, which in turn affects the release of luteinizing hormone releasing hormone (LHRH) and prolactin inhibiting factor. Each primer pheromone effect appears to result from changes in secretion of luteinizing hormone (LH) or prolactin. In general, male pheromones are thought to reduce prolactin levels and increase LH, while female pheromones are thought to increase prolactin secretion. Some of the details of the studies that led Keverne to this generalization are detailed below; it appears to be a good generalization that has the quality of parsimony in explaining several diverse phenomena.

Lee-Boot Effect

Group-housed female mice (8–12 to a cage) exhibit a suppression of estrus. This effect can be eliminated or reversed by removal of the VN organ (Reynolds & Keverne 1979, Ingersoll 1981). The reversing effect of VN deafferentation is mimicked by injection of the dopamine agonist, Bromocriptine. Since Bromocriptine lowers serum prolactin levels, this is probably the mechanism of action. The effect of VN organ removal can be overcome or reversed by administration of haloperidol, a dopamine antagonist that elevates plasma prolactin levels. Cessation of haloperidol administration results in return to cycling (Reynolds & Keverne 1979). VN organ removal has no direct effect on estrus behavior of female mice, since group-housed mice (six to a cage) with and without VN organs cycle similarly (Lepri et al 1985).

Whitten Effect

The Whitten effect is the induction of estrus in female rodents made anestrus by group housing or exposure to continuous light. This response can be caused by males or their urine alone and is dependent on a functional VN system (Johns et al 1978). Similarly, group-housed female rats exhibit a shorter estrous cycle when exposed to male odors; this response is also dependent on a functional VN system (Gallego & Sanchez Criado 1979, Sanchez-Criado 1982, Mora et al 1985). Ovariectomized estrogen-primed female rats display an LH surge following exposure to male rat odors. Accessory olfactory bulb lesions or VN organ removal prevent this odor-mediated LH surge (Beltramino & Taleisnik 1983).

The neural mechanisms underlying the VN mediated LH surge and subsequent reflex ovulation have been well studied. An LH surge can be produced in ovariectomized estrogen-primed female rats by stimulation of the AOB (Beltramino & Taleisnik 1979). Stimulation of the medial and cortical amygdaloid nuclei facilitate a LH release (Velasco & Taleisnik 1969, Beltramino & Taleisnik 1978), as does stimulation of the medial portion of the bed nucleus of the stria terminalis (Beltramino & Taleisnik 1980, Velasco & Taleisnik 1969). The LH surge in response to exposure to male bedding or to stimulation of the medial amygdala can be prevented by section of the stria terminalis, lesion of the ventral premammillary nucleus bilaterally, or lesion of the ventral premammillary nucleus on the side ipsilateral to the medial amygdaloid nucleus stimulus or ipsilateral to the functional VN system in an animal with contralateral VN organ removal (Beltramino & Taleisnik 1985). These data suggest that the LH surge in response to male odors and the consequent ovulation depend on a pathway that originates in the VN organ and includes the ipsilateral AOB, cortico-medial amygdala, stria terminalis, and ventral premammillary nucleus of the hypothalamus.

Intact male mice also exhibit an LH surge following exposure to female mice or their urine odors. VN organ removal blocks the reflex release of LH in response to female urine, but not to the female animal, suggesting that other cues emanating from the female are sufficient to stimulate the hormonal response in sexually experienced males (Coquelin et al 1984).

Vandenbergh Effect

Exposure of immature female mice to the odor of male mice at some critical period in development results in a rapid increase in their uterine weights and accelerates the onset of puberty (Vandenbergh effect). Males do not require a functional VN system to produce the puberty-accelerating pheromone (Lepri et al 1985). Bilateral olfactory bulbectomy, unilateral bulbectomy with contralateral VN nerve section (Keneko et al 1980), and bilateral VNO removal (Lomas & Keverne 1982) in female mice result in failure of male urine odors to accelerate puberty. Similar findings have been reported in rats (Sanchez-Criado & Gallego 1979, Sanchez-Criado 1982). Lowering prolactin levels by injection of Bromocriptine advances onset of puberty in both intact and VN lesioned female mice (Lomas & Keverne 1982), suggesting that this pheromonally mediated endocrine effect is caused by VN-stimulated perterbations in hypothalamic dopamine activity.

Group-housed females produce a puberty-delaying pheromone (Drickamer 1986) that is not present in the urine of group-housed female mice whose VN organs have been removed (Lepri et al 1985).

Bruce Effect

Fertilized ova of female mice fail to implant in the uterus if, following copulation and some critical period, the female is exposed to a "strange" male that differs significantly from the "stud" male (Bruce effect). The Bruce effect is dependent on a functional VN system (Bellringer et al 1980, Ingersoll 1981, Rajendren & Dominic 1984) but not a functional olfactory system (Lloyd-Thomas & Keverne 1982, Rajendren & Dominic 1986). Bromocriptine can mimic the Bruce effect in females not exposed to a "strange" male (Bellringer et al 1980). A female can be tricked into responding to the "stud" as if it were "strange" if the stud is immediately removed after copulation and replaced by the "strange" male. When the female is subsequently reexposed to the "stud" she responds with implantation failure (Keverne & de la Riva 1982). This implantation failure does not occur if the "strange male" remains with the female. The anatomical substrate mediating the failure of the familiar male to cause implantation failure under normal conditions may be a noradrenergic input to the AOB via the medial olfactory stria (Keverne & de la Riva 1982, Keverne 1983). These noradrenergic projections appear to be important for imprinting of chemosensory information. If this noradrenergic projection is severed, e.g. by a small infusion of the neurotoxin 6-hydroxydopamine into the medial olfactory stria, the AOB is depleted of more than 70% of its noradrenalin. Under these conditions the female mouse continues to respond to the pheromone producing the "olfactory" block to pregnancy, but she fails to recognize or become imprinted to the odors of the stud male, and consequently his pheromone also blocks pregnancy (Keverne & de la Riva 1982, Keverne 1983).

These studies strongly support the idea that the olfactory block to pregnancy is based on chemical cues perceived by the VN organ of a recently impregnated female. These cues signal a mismatch between the odors of a stud male and a novel male. Imprinting of the odor of stud male depends on intact noradrenergic input to the AOB during the period immediately following copulation. Perception of the mismatch probably affects dopamine synthesis and release in the hypothalamus, which in turn decreases prolactin secretion and results in implantation failure.

Female mice spend more time investigating the bedding of intact males compared to that of castrated males. Zinc sulfate irrigation of the main olfactory epithelium, with its consequent coagulation necrosis and anosmia, eliminates this preference. However, these anosmic animals still demonstrate a Bruce effect. In contrast, VN system lesions, nerve cut, or organ removal have no effect on investigation time or preference for intact male bedding over castrate bedding but do prevent occurrence of the

Bruce effect (Lloyd-Thomas & Keverne 1982). This dissociation of effects suggests that the lesions of the VN system that block the effect of male pheromones on female reproduction do not do so by preventing the female from discriminating or responding differentially to the odor cues in male urine. Conversely, the absence of such discriminative abilities, consequent to main olfactory deafferentation, does not prevent occurrence of the chemosensory block to pregnancy.

The cumulative evidence from studies on the mouse primer pheromone effects and their comparable effects in rats strongly supports the position of Keverne (1983) that the VN system provides the critical neural pathway from the periphery to the hypothalamus. The connections within this system provide for modulation of LH and prolactin release that is the endocrinological basis for the pheromone effects.

VN nerve transection does not invariably lead to endocrinological deficits. Such lesions in male rats (Sanchez-Barcelo et al 1985) and mice (Wysocki et al 1982, Wysocki 1982) do not have an effect on body weight, testes, accessory sex glands, prostrate weights, or serum testosterone levels.

ROLE OF VOMERONASAL ORGAN IN RELEASER OR SIGNALING PHEROMONE EFFECTS

Intraspecific chemical communication occurs in a variety of situations that do not involve the response delay that accompanies the endocrinological changes described above. The chemosignals responsible for such responses are typically called releaser or signaling pheromones and responses to them occur relatively rapidly. Mating sequences and behaviors associated with aggregation, aggression, maternal care, and infant-mother recognition occur in response to these pheromones in many vertebrate groups.

Hamster Sexual Behavior

The first demonstration of differential involvement of the VN system in sexual behavior was a report by Powers & Winans (1975) that severe mating deficits developed in sexually experienced male hamsters following VN nerve cuts. These deficits, only developed in approximately one third of the hamsters, however. Although zinc sulfate lavage of the nasal cavity alone had no effect on mating behavior, combined VN nerve cut and zinc sulfate lavage produced mating deficits in 100% of the hamsters. The variability in the effect of VN nerve cuts was not found to result from differential involvement of the main olfactory bulb in the lesions (Winans & Powers 1977). The authors of these studies concluded that both the VN system and main olfactory system are sensitive to female hamster pheromones, and that VN nerve section irrepairably reduces the arousal necessary for mating in some male hamsters; whereas for others, the main olfactory system is capable of mediating sufficient arousal from pheromonal stimuli. Johnston (1985) reported that in comparison to intact male hamsters, adult male hamsters with their VN organs removed demonstrated an increased latency in the time taken to mount females and an increase in time spent sniffing the vaginal region of females.

There have been some conflicting findings concerning the effects of VN system lesions on copulatory behavior in hamsters. Johnston & Rasmussen (1984), Meredith (1980), and Murphy (1980) have reported, using different testing conditions, that males without VN organs continue to mate. Some of the differences in results may be a reflection of different levels of preoperative sexual experience in the male hamsters (Meredith 1983, Keverne et al 1986), a suggestion strongly supported by a recent study in which Meredith (1986) demonstrated that VN organ removal prior to sexual experience has severe effects on male hamster mating behavior, whereas such lesions in previously experienced males have minimal effects.

Male hamsters prefer females of their own species and exhibit this preference to the vaginal secretions of the females as well. VN nerve cuts have no effect on this preference nor is there a deficit in responding to the secretions in general (Murphy 1980). Female hamster vaginal secretions have been found to have an attractant pheromone, containing dimethyl disulfide, as well as an investigation-maintaining (or mounting) pheromone. Male hamsters tested with female hamster vaginal secretions placed on environmental surfaces are attracted to the treated surfaces and spend time investigating them. Zinc sulfate lavage of the main olfactory epithelium eliminates the response to the attraction components of the secretion when they are placed on environmental surfaces but not when they are placed onto anesthetized, castrated males. VN deafferentation produces a deficit in investigation of secretions only in animals exhibiting a mating deficit (Powers et al 1979).

Female hamster vaginal secretions have been fractionated into volatile components and a high molecular weight "nonvolatile" component. These components stimulate different aspects of sexual behavior in the male hamster, and their reception may be mediated by two distinct chemosensory systems, the main olfactory and VN systems. The volatile components appear to be associated with the attraction pheromone, whereas the high molecular weight component is associated with the "investigation-maintaining" or mounting pheromone. Zinc sulfate treatment of the olfactory epithelium affects male hamster response to volatiles in female hamster vaginal secretion, but has no effect on mounting behavior. VN nerve cuts, however, have no effect on attraction (O'Connell & Meredith 1984). Vomeronasal organ removal causes a loss of differential responses to the high molecular weight component of female hamster vaginal discharge, but does not critically affect response to unfractionated discharge (Clancy et al 1984a). These studies suggest that female hamsters attract and maintain the interest of male hamsters with two distinct components of their vaginal discharge. However, it may be simplistic to think purely in terms of volatile vs nonvolatile components, as VN nerve cuts or organ removal significantly reduces sexual behavior in male hamsters even when volatiles alone are guiding the behavior (O'Connell & Meredith 1984, M. Meredith personal communication).

Structures of the VN system located centrally are involved in transmission of information from the VN organ in the periphery to the preoptic area and hypothalamus, and damage to these structures adversely affects behaviors dependent on a functional VN system. Male hamster mating behavior can be totally disrupted by placing electrolytic lesions in the medial nucleus of the amygdala (Lehman & Winans 1982, Lehman et al 1980), a lesion that also diminishes male sniffing and licking during investigation of the female hamster's anogenital region. The medial amygdaloid nucleus projects to the hypothalamus via both the stria terminalis and the ventral amygdalo-fugal pathway. Bilateral destruction of the stria terminalis does not eliminate copulatory behavior (Lehman & Winans 1983). The VN and olfactory information required for male hamster copulatory behavior transmitted via connections in the medial amygdaloid nucleus appear to be relayed to the preoptic portion of the bed nucleus of the stria terminalis by way of the ventral amygdalo-fugal pathway (Winans et al 1982).

Two studies concerning the role of the hamster VN system in the recognition of individuals used different experimental paradigms and have come to contradictory conclusions. Sexually satiated male hamsters show a preference for novel females (Coolidge effect). VN organ removal does not interfere with expression of this preference, but zinc sulfate irrigation of the main olfactory epithelium does. These results led Johnston & Rasmussen (1984) to conclude that a functional VN system was not necessary for individual recognition of female hamsters. Steel & Keverne (1985) made use of the observation that a male hamster exposed to a female hamster's vaginal secretion subsequently shows a greater interest (sniffs and investigates) in the donor of the secretion compared to a hamster that was not the donor (Steel 1984). Male hamsters have to be able to lick the secretions in order to make the subsequent differential response to the donor animal, suggesting that a nonvolatile component of the secretion is the critical discriminative stimulus. Male hamsters with their VN organs removed fail to respond selectively to the donors of the vaginal secretions they had previously sampled, whereas zinc sulfate treated and sham control animals did make differential responses. Thus, for this behavioral task, it appears that a functional VN system is critical for recognition of individuals. There is no obvious explanation for the apparent discrepancy in the results of these two studies. However, one of the differences in the two procedures is that Steel & Keverne monitored individual recognition by presentation of the female hamster vaginal secretions alone during the "exposure" period. If, as discussed below, VN stimulation is a critical unconditioned stimulus when other cues are absent, the deficit in the Steel & Keverne study can be explained. Conversely, in the Johnston & Rasmussen study, all testing is done with intact females that provide, in addition to the VN-mediated cues, other cues that may function as unconditioned stimuli.

Mouse Sex Behavior

Several aspects of male mouse sexual behavior appear to depend on information obtained through the VN system. Male mice with VN organs removed have a severe copulatory deficity compared to controls (Clancy et al 1984b) and do not exhibit the testosterone surge normally observed in male mice following exposure to the odor of a female mouse (Wysocki et al 1983). Additionally, male mice emit ultrasounds in the presence of female mice or their odors. Disruption of the VN system, but not of the olfactory system, leads to a consistent reduction of ultrasound emissions in the presence of female mice (Bean 1982a). If the males are sexually experienced prior to VN organ removal, they still demonstrate that they can discriminate females from males by emitting more ultrasounds to female stimuli. However, if the males have had no social experience prior to VN organ removal they do not respond to male or female stimulus animals with ultrasounds, and these lesioned animals never learn to emit ultrasounds to females even after lengthy periods of cohabitation. The ultrasound production deficit is not secondary to a hormonal deficit produced by VN organ ablation, since endocrine organ weights and circulating androgen levels are not significantly different for VN lesioned mice as compared to sham controls, and testosterone treatment does not increase vocalization to female stimuli (Wysocki 1982, Wysocki et al 1982).

Sexual Behavior in Other Mammalian Species

Mating deficits or depressed responsiveness to reproduction-related odors have been reported in a number of other species following interference with the VN system. Male guinea pigs with VN organs removed exhibit normal sex behavior but spend less time investigating urine, and demonstrate a depression in their differential response to female urine as compared to male urine. Responsiveness to urine declines over time, an effect that is reminiscent of extinction (Beauchamp et al 1982, 1985).

Bilateral ablation of the AOB target areas, the medial and posterior cortical amygdaloid nuclei, fails to affect sexual behavior of male rats when they are tested with ovariectomized, estradiol benzoate and progesterone treated females but does result in lengthened ejaculatory latencies when they are tested with ovariectomized females treated only with estradiol benzoate (Perkins et al 1980). These results were interpreted by the authors as suggesting that the females treated with only estradiol benzoate were less attractive and less solicitous than the females receiving progesterone treatment in addition, and therefore the absence of VN mediated information had a more pronounced effect. Accessory olfactory bulb lesions in female rats appear to have no adverse effect on sexual receptivity (Kelche & Aron 1984).

Some feminized, estrogen and progesterone treated male rats display lordosis in response to mounts by intact male rats. This response is increased in frequency of occurrence following AOB ablation, suggesting that the accessory olfactory system may normally inhibit feminine behavior in male rats (Schaeffer et al 1986).

Sexual Behavior in Reptiles

The effects of VN deafferentation are most dramatically observed in snakes. Bilateral VN nerve cuts but not bilateral main olfactory nerve cuts completely abolish male sexual behavior in garter snakes (Kubie et al 1978, Halpern & Kubie 1983). Blocking odorant access to the VN organ in garter snakes (Halpern & Kubie 1983) or in the adder, *Vipera berus* (Andren 1982), also causes complete cessation of courtship behaviors. Male snakes are known to respond to a pheromone on the surface of females with tongue flicking and courtship displays. This pheromone does not appear to have an attractant component that is perceived at a distance, since male snakes do not selectively approach females based on distal cues and the male snake must make direct contact with the female before it displays courtship behavior. Thus snake reproductive behavior appears to be uniquely dependent on nonvolatile pheromonal cues that depend entirely on the VN system for their detection.

Nonsexual Behaviors

A number of nonsexual social behaviors depend on the appropriate response of an individual to conspecific odors. The VN system has been implicated in some of these and not others.

AGGREGATION Garter snakes respond to conspecific odors in nonreproductive setting as well. They use these odors to aggregate and to locate preferred shelter sites. The VN system was deafferented by several techniques to study its role in these behaviors (Burghardt 1980, Heller & Halpern 1982). Methods included removing the tongue, which disables the mechanism of odor delivery to the VN organ, suturing closed the VN ducts, which prevents access to the VN organ, or severing the VN nerves. All deafferentation procedures caused severe deficits in garter snake aggregation and shelter selection behavior.

AGONISTIC BEHAVIOR A functional VN system appears to be important for expression of aggressive behavior in snakes (Andren 1982) and mice (Ingersoll 1981, Clancy et al 1984b, Bean 1982b, Wysocki et al 1986). Lesions of the medial amygdaloid nucleus but not the cortical amygdaloid nucleus of rats produces a deficit in learned avoidance of a dominant opponent (Luiten et al 1985), suggesting that this nucleus may transmit information about conspecifics that mediates appropriate dominant or submissive behavior. Klemm et al (1984a) have also implicated the VN system in aggressive behavior of steers.

MATERNAL BEHAVIOR Chemical communication is an important part of mother-infant interactions. At least two aspects are involved: maternal recognition of the young and infant recognition of the mother or a mother substitute. Normal maternal behavior appears to be unaffected by severe disruption of the VN system of mice (Lepri et al 1985) or rats (Jirik-Babb et al 1984). However, virgin rats (Fleming et al 1979), who would normally require a week or more of exposure to rat pups to display maternal-like behavior, decrease the latency to exhibit maternal behavior when their VN systems are deafferented. Combined VN and olfactory bulb lesions result in an even greater reduction in latency to show this maternal behavior. These results were interpreted by the authors as suggesting that both VN and main olfactory systems were involved in tonic inhibition of maternal behavior in virgin rats. Supportive evidence for this interpretation has been obtained in hamsters (Marques 1979). Virgin female hamsters, when presented with pups, will either kill them or carry them. VN nerve cuts convert approximately 68% of the killers to carriers, and zinc sulfate irrigation of the main olfactory epithelium in addition to the VN nerve cuts converts all the remaining females that formerly killed pups into females that will carry them. Zinc sulfate treatment alone converts 28% of the killers into carriers. Again, these authors conclude that both VN and main olfactory system play a role in the behavior of virgin females to pups.

NEWBORN BEHAVIOR Newborn rabbits locate and attach to their mother's nipples by using chemical cues from a "nipple search pheromone." A neutral stimulus, citral, can be paired with suckling behavior and become a conditioned nipple attachment stimulus. VN organ removal in newborns does not interfere with nipple attachment in the presence of the natural nipple attachment pheromone or in the presence of conditioned citral odors (Hudson & Distel 1986). This study demonstrates that in newborn rabbits the VN system is not critical for nipple attachment. The VN system in rats does not appear to be sufficient to maintain normal pup behavior. Zinc sulfate irrigation of the main olfactory epithelium without disruption of the VN system results in weight loss and deficiencies in attachement and suckling behavior (Singh et al 1976). These studies suggest that infant appetitive behavior does not depend on nor can it be maintained by the VN system. In constrast, Teicher et al (1984) report that VN deafferentation causes disorientation of newborn rat pups; instead of orienting to the ventral surface as controls do, experimental pups orient to the mother's dorsal surface. However, mice with VN organs removed neonatally (and surviving beyond 24 hours) have growth rates comparable to controls (Wysocki et al 1986), suggesting that any deficits in orientation are minimal.

VOMERONASAL ROLE IN PREY RECOGNITION AND RESPONSE TO PREY ODORS

Snakes depend on chemical cues for detecting and searching for appropriate prey (e.g. Burghardt 1970). In vipers and rattlesnakes, where prey detection is mediated via vision and infrared receptors, post-strike searching behavior and prey identification appears to be mediated by the chemical senses. Cutting the tongue tips (and thereby impeding access of chemicals to the vomeronasal organ) significantly reduces trailing efficiency (Dullemeijer 1961). Graves & Duvall (1985a) recently reported that rattlesnakes with their VN ducts sutured closed fail to strike prey, suggesting that VN stimulation may be important in the pre-strike phase of prey acquisition as well as in the post-strike searching phase.

Several studies by Gordon Burghardt and his colleagues (reviewed in Burghardt 1970) have demonstrated that naive newborn members of several snake species preferentially respond to aqueous washes of prey that form the diet of adult members of their species. Olfaction and vision are not essential for differential responses to these prey washes (Burghardt & Hess 1968), but tongueless snakes suppress both feeding and response to prey washes (Burghardt & Pruitt 1975). VN nerve cuts, but not olfactory nerve cuts, in adult garter snakes also result in a failure of discrimination of prey washes and a prey attack deficit (Halpern & Frumin 1979).

Garter snakes will follow prey extract trails in a four- or two-choice maze for food rewards. They follow intense trails more accurately than weaker trails, and when following intense trails tongue flick more rapidly than when following weaker trails (Kubie & Halpern 1978, Halpern & Kubie 1983). Snakes can follow dry or wet trails, but require tongue contact with the trail to follow it accurately, i.e. if the trail is placed below a perforated false floor, the volatiles from the extract, if they exist, are not sufficient to maintain accurate trail following (Kubie & Halpern 1978). VN nerve cuts or VN duct sutures, but not olfactory nerve section, result in failure to follow prey trails accurately or to demonstrate an elevated tongue flick rate to trails of high concentration extracts (Kubie & Halpern 1979). These studies demonstrate that the snake VN system is essential for differential responding to certain prey chemicals.

In the studies described above direct lingual contact with the chemical stimulus was required for accurate trail following; garter snakes do, however, respond to airborne odorants with increased tongue flicking (Burghardt 1977, Halpern & Kubie 1983, J. Halpern et al 1985) and, on occasion, attack (Burghardt 1977). Both VN nerve and main olfactory nerve section result in diminished tongue flicks to airborne odorants. However snakes with VN nerve cuts still increase their tongue flick rates in the presence of the odorants, demonstrating that they can discriminate the presence or absence of the airborne stimuli. Snakes without a functional main olfactory system do not demonstrate an elevated tongue flick rate in the presence of airborne odorants (J. Halpern et al 1985). These results suggest that the increased tongue flick rate in response to airborne odorants observed previously (Burghardt 1977, 1980, Halpern & Kubie 1983) may be mediated by both olfactory and VN systems in snakes. Since olfactory nerve lesions appear to have a more devastating effect on the response to airborne odorants, it is possible that the initial increase in tongue flick rate following odor stimulation depends on the olfactory system, whereas the maintained increase in tongue flick rate depends on a functional VN system.

IS VOMERONASAL STIMULATION INTRINSICALLY REWARDING?

A set of observations reported for several species tested in a variety of behaviors following VN nerve lesions or VN organ removal has strongly implicated the VN system in the transmission of information that is reinforcing. The first finding, reported by Kubie & Halpern (1979), was that following VN nerve lesions, trailing behavior of garter snakes trained to follow prey trails for food rewards is immediately reduced to chance levels, but prey attack and reward ingestion extinguish over time. Typically, following a VN nerve lesion, a snake will ingest worm bits for several trials; on subsequent trials the snake may strike the bits and spit them out. Eventually the snake will ignore worm bits altogether. These findings suggest that the vomeronasally mediated stimuli arising from the prey trail or the prey bits are intrinsically reinforcing. If VN stimulation has intrinsic rewarding properties, then one should be able to train snakes to make a discriminitive response in order to obtain access to a VN stimulus even when no consumatory behavior follows. In such a study garter snakes were trained to discriminate between two visual patterns (dots vs stripes) to gain 30 seconds of access to a dish coated with dry prey extract, a substance known to stimulate the VN system. Snakes learned this task in fewer than 70 trials and learned the reversed task as well (Halpern et al 1985).

Recent findings in mammals are similar to those first found in snakes and support the idea that the VN system is involved in mediating reception of reinforcing stimuli. Male guinea pig head-bobbing in response to female urine, a vomeronasally mediated behavior, is resistant to extinction in intact animals but shows an extinction pattern similar to the one seen in garter snakes following VN deafferentation (Beauchamp et al 1985). Similarly, male mice with VN organs removed show a decrement in mounting behavior as a function of postoperative experience with receptive females (Wysocki et al 1986) and the mating behavior of male hamsters decreases with experience after VN nerve lesions (Winans & Powers 1977).

Interference with reception of reinforcing stimuli should cause not only extinction but also should interfere with new learning. For example, in the behaviors discussed below, deafferentation of the VN system consistently results in deficits in behavior that utilizes chemosignals to establish new response patterns. Aggression in male mice is modified by experience and the absence of a functional VN system prevents subsequent utilization of information gained during agonistic encounters. In male mice VN organ removal results in severe deficits in agonistic behavior, and the severity of the deficit is inversely related to the extent of agonistic experience prior to surgery. Intact mice become increasingly aggressive with training, but few mice with their VN organs removed learn to attack or fight. Similarly, lactating female mice exhibit a high incidence of aggression toward male intruders. Experienced females who have been aggressive prior to VN organ removal remain aggressive. If, however, VN organ removal precedes experiences that would normally elicit aggression, they will not subsequently exhibit aggression (Wysocki et al 1986). As mentioned above, male mice produce ultrasounds to female mice and, with experience, produce ultrasounds, in response to the urine of female mice. The effects of VN organ removal is related to the amount of presurgical experience the male has with females, with naive males exhibiting the greatest deficit (Bean

1982a, Wysocki et al 1982, Wysocki 1982, Keverne et al 1986). Intact male mice paired with perfumed mice can be conditioned to emit ultrasounds to the perfume stimulus. Such conditioning is not possible in mice with their VN organs removed. Taken together these studies strongly support the idea that chemical stimuli sensed by the vomeronasal system are capable of acting as unconditioned stimuli that reinforce behavior. In the absence of VN mediation, behaviors previously conditioned to VN stimuli extinguish if no other reinforcing stimuli are present. No conditioning is possible in naive animals if vomeronasally mediated stimuli are the sole source of reinforcement.

ODORANT ACCESS TO THE VOMERONASAL ORGAN

With few exceptions the VN organs of vertebrates are sequestered from the nasal and oral cavities and the ducts leading to them are exceedingly narrow. A number of early authors suggested that animals must actively enlarge the openings to the organs, activate a pumping or suction mechanism, or deliver odorants directly into the organ. Recent studies support these ideas.

Snakes sample environmental substrates with their tongues and as the tongue is retracted into its lingual sheath, the tongue tips travel by the openings of the VN ducts. Tongue flicks are capable of delivering high concentrations of chemicals to the VN organs, including nonvolatiles such as proline. Delivery of odorants is dependent on lingual contact with the source of the nonvolatile and is prevented if the VN ducts are sutured closed (Halpern & Kubie 1980). Cutting the tongue at its base severely attenuates the amount of material entering the VN organ, but does not prevent all access when the material is tapped onto the snout of the animal (Halpern & Kubie 1980). Gillingham & Clarke (1981) used cinematographic recording of rat snakes during tongue flicking to demonstrate that the anterior processes of the mouth, which are contacted by the ventral surface of the tongue, are elevated following each retraction of the tongue into the lingual sheath. The anterior processes are situated just below the orifice of the VN duct, and it is possible that the anterior processes provide a mechanism for transfer of substances from the tongue into the VN duct. Support for this possibility comes from the observation that removal of the anterior processes in garter snakes results in a deficit in prey detection in an "open field" apparatus (Gillingham & Clarke 1981). Graves & Duvall (1983) describe mouth gaping and head shaking as two behaviors that increase in prairie rattlesnakes upon exposure to conspecific odors and prey. Following VN duct suture these behaviors increase in frequency, a finding that led Graves & Duvall (1985b) to hypothesize that mouth gaping and head shaking aid in the delivery of odorants to the VN organ. Although tongue flicking is associated with VN stimulation, it is also associated with movement (Kubie & Halpern 1975, 1978) and caution should be exercised when interpreting experimental results that show a change in tongue flick rate or latency. A demonstration of an increase in tongue flick rate does not by itself indicate that the stimulus that led to the increase in tongue flick rate was sensed by the vomeronasal system.

The tongue is also reported to be used by bulls to effect transfer of vaginal odors to the VN organ. Bulls produce tongue compression strokes of the hard palate while investigating the vulvar region of cows. Contact between the bull's nose and/or upper lip and the cow's vulva is maintained throughout this behavior (Jacobs et al 1981).

Nose rubbing and licking may also be used to bring substances in close proximity to the VN duct or associated structures. Male guinea pigs investigate conspecific secretions or excretions by placing their snouts in direct contact with the material under investigation. The guinea pig moves its head forward and back across the sample at a rate of one to four cycles per second, a behavior that has been called head-bobbing. Direct contact with the stimulus is critical since placing the sample below a screen results in cessation of head-bobbing (Beauchamp et al 1980). When rhodamine, a nonvolatile fluorescent dye, is added to female urine and males are permitted to headbob to the urine, the VN organ of the male is filled with fluorescent rhodamine, demonstrating that a nonvolatile substance is capable of entering the VN organ of this mammal (Wysocki et al 1980). Rhodamine has also been observed in the VN organ of male mice and voles exposed to rhodamine-adulterated urine and in the VN organ of pine and meadow voles following self-grooming or social grooming of rhodamine-smeared fur (Wysocki et al 1985). Rhodamine also enters the VN organ during eating and drinking, and may enter the organ by passive diffusion after death when painted on the external nares (Wysocki et al 1985). The apparent facility by which this nonvolatile dye enters the VN organ does not appear to be related to its relatively small molecular weight, since rhodamine conjugated to bovine serum albumin readily enters the VN organ (Wysocki et al 1985).

Facial grimacing that acts to enlarge the openings to the VN organ will facilitate odorant access if those odorants are already in the nose or the mouth. Flehmen is a behavior, characteristic of ungulates exposed to conspecific urine, that is comprised of head extension, upper lip retraction, baring of the gum and frequently opening of the mouth. Flehmen, which occurs in a variety of mammals other than ungulates, has been proposed as a mechanism for delivery of odorants to the VN organ (Estes 1972). The history of this idea and a review of the pertinent literature appeared recently (Ladewig & Hart 1982) and is not repeated here. Flehmen has been observed to failitate entry of substances into the VN organ (Melese-d'Hospital & Hart 1985). Tracer dye placed in the mouth is found in the anterior (nonsensory) and posterior (neurosensory) portions of the VN organ of goats if flehmen occurs following dye placement. In the absence of flehmen only the anterior portion of the VN organ contains dye (Ladewig & Hart 1980). Occlusion of the VN ducts in male cats leads to a reduction of flehmen bouts following urine sampling (Verberne 1976), and flehmen in a number of mammalian species has recently been associated with sampling of sex-related odors (Crowell-Davis & Houpt 1985, Gaughwin 1979, Ladewig & Hart 1980, Reinhardt 1983, Verberne & De Boer 1976).

Access and egress of substances to and from the VN organ of some mammals appears to be under the control of the autonomic nervous system. Electrical stimulation of cervical sympathetic nerve in cats causes suction of fluid into the lumen of the VN organ, and stimulation of parasympathetic fibers causes secretion from the VN duct (Eccles 1982). In hamsters, the VN organ is incased in bone, and the large blood vessels located within the bony capsule are capable of dilation and constriction. Autonomic afferents controlling vasomotor activity run in the nasopalatine nerve, a branch of the maxillary trigeminal nerve. Stimulation of the nasopalatine nerve causes constriction of the blood vessels in the VN organ, reducing blood volume in the bony capsule, enlarging the VN lumen, and effecting aspiration of fluid into the VN organ via the VN duct. Relaxation of blood vessels or active dilation has the reverse effect, forcing fluids out through the VN pore (Meredith & O'Connell 1979, Meredith 1980, 1982, Meredith et al 1980). The VN pump, as the above mechanism is now called, can also be activated by injection of adrenalin into the carotid artery (Meredith 1982) and is adequate to transport stimuli to the VN receptor cells. In anesthetized preparations, odor-related activation of the AOB occurs when the VN pump is active, but not when it has been inactivated (Meredith 1982). Further evidence that odorant access to the VN organ of hamsters is dependent on a functioning pump mechanism is obtained from studies demonstrating that horseradish peroxidase infused into the nasal cavity gets into the VN organ, and is transported to the AOB, only when the VN pump is activated (Meredith 1982). Furthermore, cutting the nasopalatine nerve, and thereby inactivating the pump, results in behavioral deficits similar to those produced by severing the male hamster's VN nerve (Meredith et al 1980, Meredith 1982). It is thus quite clear that the hamster VN pump is essential for normal VN stimulation and that passive diffusion of chemicals from the nose does not normally occur at levels sufficient to support VN-dependent behaviors.

CHEMICAL CHARACTERIZATION OF SUBSTANCES WHOSE DETECTION IS VOMERONASALLY MEDIATED

One of the fundamental differences between the VN and main olfactory system may be the chemical nature of substances that stimulate them. The *sine qua non* of olfactory stimuli is that they are volatile. At present, only a few substances have been identified that naturally stimulate the VN system and whose detection lead to behavioral or endocrinological change. In almost all such cases the substances have been found to be of relatively high molecular weight, nonvolatile, and to contain proteins.

Female hamster vaginal secretions contain, in addition to the attractant pheromone dimethyl disulfide (Singer et al 1976), a high molecular weight fraction (15,000–60,000 daltons) with low volatility that acts as a mounting pheromone and whose detection requires a functional VN system (Clancy et al 1984a, Singer et al 1980, 1984b). This portion of female hamster vaginal discharge contains proteins with molecular weights in excess of 10,000 daltons whose presence varies with phase of the estrous cycle and are reduced in concentration following ovariectomy or hypophysectomy. Following digestion with the proteolytic enzyme pronase, the aphrodisiac activity of the high molecular weight fraction is significantly reduced (Singer et al 1984a,b).

The androgen-dependent puberty-accelerating pheromone in male mouse urine is found in a high molecular weight fraction (Novotny et al 1980). This substance is heat labile, nondialyzable, precipitated by ammonium sulfate, and not extractable in ether. These characteristics suggest that the active component is either a protein or a small molecule associated with a protein. Following digestion with pronase, the pheromone initially retains its activity, but loses its activity following dialysis, suggesting that the active site is a portion of a protein or a substance associated with a protein (Vandenbergh et al 1975).

Preliminary attempts to characterize vomeronasally mediated urinary pheromones have been reported in elephants (Rasmussen et al 1982), guinea pigs (Beauchamp et al 1980), deer (Crump et al 1984), and mice (Marchlewska-Koj 1980, 1983). In most of these reports large, nonvolatile, protein-containing substances have been identified in the active fractions of the pheromone-containing urine.

Warm water washings of the outer surfaces of earthworm contain a chemoattractant that garter snakes respond to with increased tongue flicking and attack. A functional VN system is critical for response to this chemoattractant (Reformato et al 1983). Separated on Sephadex G-75 or AcA 44 gel exclusion columns, the active component of the chemoattractant is found in a high molecular weight fraction, is relatively heat-insensitive, contains carbohydrates and protein, is precipitated by ammonium sulfate, and is not dialyzable through membranes with cut-offs as high as 50,000 mol weight (Halpern et al 1984, Reformato et al 1983, Sheffield et al 1968). The chemoattractant is very sensitive to alkaline hydrolysis and less sensitive to acid hydrolysis (Kirschenbaum et al 1985). Amino acid and carbohydrate analyses of the active fraction of the chemoattractant demonstrates a close resemblance between it and earthworm cuticle collagen. Purified earthworm cuticle collagen is a snake chemoattractant and its chemical and physical properties are indistinguishable from the chemoattractant in earthworm wash (Kirschenbaum et al 1986). Furthermore, warm water washes of decuticlized earthworms do not contain a chemoattractant for snakes. These results offer strong support for the proposition that a component of the earthworm chemoattractant for snakes is structurally related to the glycoprotein earthworm cuticle collagen (Kirschenbaum et al 1986). Sodium periodate treatment of earthworm wash destroys its chemoattractant properties, thus suggesting that the carbohydrates on the collagen molecule may be critical for recognition by the VN system (Halpern et al 1986).

ELECTROPHYSIOLOGICAL CHARACTERIZATION OF THE VOMERONASAL SYSTEM

Studies using electrophysiological techniques to characterize receptor properties, synaptic interactions, or mechanisms of sensory coding in the VN system have been sparse (Adrian 1954, 1955, Altner & Müller 1968, Tucker 1960, 1963, 1971). This approach to studying the VN system is presently in its infancy, and relatively little is yet understood.

Müller (1971) recorded electroolfactograms (EOG) from the VN organ of lizards (*Lacerta*) and found that lower members of homologous series of alcohols, aldehydes, and fatty acids were particularly effective in producing higher EOG amplitudes. The EOGs recorded from the VN organs of mice and amphibians did not differ from those recorded from the main olfactory epithelium. Beurman (1977) was unsuccessful in his attempts to record an odor evoked potential from the AOB of gopher and box turtles to a geraniol stimulus. Single unit recording from the AOB of partially anesthetized snakes failed to show changes in firing rate following tongue flicking, but did show an increased firing rate when swabs coated with prey extracts were pressed to the roof of the mouth (Meredith & Burghardt 1978). Multiunit activity from the AOB of freely moving snakes increased following tongue flicking (Meredith & Burghardt 1978). Unfortunately, none of these studies have had sufficient success to aid in the understanding of how the sensory message is coded and transmitted to the CNS. Experiments in which an odor is artificially directed onto the epithelium may answer the question of whether the odor is capable of activating the epithelium, but has little relevance in a normal, awake animal if that odor never has access to the VN organ.

Large, sustained electrographic activity was recorded from the VN capsule of bulls in response to cow urine (Klemm et al 1984b). However, these results are difficult to interpret since unilateral perfusion of one VN organ with the stimulus produced a concurrent response in the contralateral organ. Physiological saline produced a response, although smaller and less sustained, and control testing of electrical activity recorded from a rubber tube also produced a response.

Meredith & O'Connell (1979) reported activation of AOB units by simultaneous stimulation of the nasopalatine nerve (to activate the VN pump) and continuous flow of dimethyl disulfide vapor over the VN pore. The significance of this finding has yet to be determined, since the VN system does not appear to be critical to the detection of or response to dimethyl disulfide.

The most successful electrophysiological studies of the VN system have not employed natural stimulation, but have attempted to characterize the interactions between neuronal elements in the AOB of the rabbit by using electrical stimulation of the VN nerve, amygdala, or olfactory tract (MacLeod & Reinhardt 1983, Reinhardt et al 1983). Such stimulation evokes characteristic and reproducible field potentials in the AOB. VN nerve stimulation reliably evokes a complex field potential in the AOB consisting of a compount action potential followed by four negative waves. Amygdaloid stimulation evokes a long latency (20 msec), long duration, biphasic potential that is initially negative (about 40 ms in duration), followed by a long positive wave (about 70 ms in duration) (MacLeod & Reinhardt 1983). The primary synapse in the glomerular layer is invariably excitatory. Intracellular unit recording of projection cells identified by antidromic invasion from amygdaloid stimulation was used to characterize the nature of centripetal and centrifugal influences on the "mitral" cells. VN nerve stimulation produced an initial EPSP and associated spike discharges, followed by an IPSP and a consequent suppression of spontaneous firing for up to 100 msec. IPSPs were always evoked by amygdaloid stimulation. Stimulation of the lateral olfactory tract rarely produced an antidromic spike. These studies indicate that information processing in the AOB is very similar to that described for the main olfactory bulb.

Clearly, a considerable gap exists in our understanding of stimulus transduction, coding, and transmission in the VN system. The next major step in this area of research will be combining electrophysiological recording techniques with delivery of odorants that have biological significance to the organism. Such studies will permit a neuron-to-neuron tracking of the effects of these stimuli on the activity of the VN system.

The exponential increase of interest in the structure and function of the VN system that began in the early 1970s has yielded an impressive literature. Real progress has been made in elucidating the neuronal pathway of the VN system and the role of the system in behavioral responses to chemical signals. What is still less clear is precisely how the main olfactory and VN systems interact to yield appropriate responses of organisms to biologically significant chemical signals. Recent research advances have been characterized by an interdisciplinary approach to problems requiring sophisticated behavioral, endocrinological, anatomical, electrophysiological, and biochemical analyses. Continuation of this interdisciplinary approach should yield new insights into the organization and functional significance of the VN system.

ACKNOWLEDGMENTS

The author's research is supported by the National Institutes of Health (NS11713). Thomas Cox, David Holtzman, John Kubie, and Charles J. Wysocki read an earlier version of this manuscript and made valuable comments and suggestions. Rose Kraus typed the manuscript.

Literature Cited

Adams, D. R., Weikamp, M. D. 1984. The canine vomeronasal organ. *J. Anat.* 138: 771–87

Adrian, E. D. 1954. Synchronized discharges from the organ of Jacobsen. *J. Physiol.* 126: 28–29

Adrian, E. D. 1955. Synchronized activity in the vomero-nasal nerves with a note on the function of the organ of Jacobsen. *Pflugers Arch.* 260: 188–92

Altner, H., Müller, W. 1968. Elektrophysiologische und electronenmikroskopische untersuichugen an der Reichochleimhaut des Jacobsonchen Organs von Eideschsen (*Lacerta*). *Z. Vergl. Physiol.* 60: 151–55

Andren, C. 1982. The role of the vomeronasal organs in the reproductive behavior of the adder *Vipera berus*. *Copeia*, pp. 148–57

Bannister, L. H. 1968. Fine structure of the sensory endings in the vomeronasal organ of the slow-worm *Anguis fragilis*. *Nature* 217: 275–76

Barber, P. C. 1981. Axonal growth by newly-formed vomeronasal neurosensory cells in the normal adult mouse. *Brain Res.* 216: 229–37

Barber, P. C., Field, P. M. 1975. Autoradiographic demonstration of afferent connections of the accessory olfactory bulb in the mouse. *Brain Res.* 85: 201–3

Barber, P. C., Parry, D. M., Field, P. M., Raisman, G. 1978. Electron microscope autoradiographic evidence for specific transneuronal transport in the mouse accessory olfactory bulb. *Brain Res.* 152: 283–302

Barber, P. C., Raisman, G. 1974. An autoradiographic investigation of the projection of the vomeronasal organ to the accessory olfactory bulb in the mouse. *Brain Res.* 81: 21–30

Barber, P. C., Raisman, G. 1978a. Cell division in the vomeronasal organ of the adult mouse. *Brain Res.* 141: 57–66

Barber, P. C., Raisman, G. 1978b. Replacement of receptor neurons after section of the vomeronasal nerves in the adult mouse. *Brain Res.* 147: 297–313

Bayer, S. A. 1980. Quantitative 3H-thymidine radiographic analyses of neurogenesis in the rat amygdala. *J. Comp. Neurol.* 194: 845–75

Bayer, S. A. 1983. 3H-Thymidine-radiographic studies of neurogenesis in the rat olfactory bulb. *Exp. Brain Res.* 50: 329–40

Bean, N. J. 1982a. Olfactory and vomeronasal mediation of ultrasonic vocalizations in male mice. *Physiol. Behav.* 28: 31–37

Bean, N. J. 1982b. Modulation of agonistic behavior by the dual olfactory system in male mice. *Physiol. Behav.* 29: 433–37

Beauchamp, G. K., Martin, I. G., Wysocki, C. J., Wellington, J. L. 1982. Chemoinvestigatory and sexual behavior of male guinea pigs following vomeronasal organ removal. *Physiol. Behav.* 29: 329–36

Beauchamp, G. K., Wellington, J. L., Wysocki, C. J., Brand, J. G., Kubie, J. L., Smith, A. B. III. 1980. Chemical communication in the guinea pig: Urinary components of low volatility and their access to the vomeronasal organ. In *Chemical Signals Vertebrates and Aquatic Invertebrates*, ed. D. Muller-Schwarze, R. M. Silverstein, pp. 327-39. New York: Plenum

Beauchamp, G. K., Wysocki, C. J., Wellington, J. L. 1985. Extinction of response to urine odor as a consequence of vomeronasal organ removal in male guinea pigs. *Behav. Neurosc.* 99: 950–55

Bellringer, J. F., Pratt, H. P. M., Keverne, E. B. 1980. Involvement of the vomeronasal organ and prolactin in pheromonal induction of delayed implantation in mice. *J. Reprod. Fertil.* 59: 223–28

Beltramino, C., Taleisnik, S. 1978. Facilitatory and inhibitory effects of electrochemical stimulation of the amygdala on the release of luteinizing hormone. *Brain Res.* 144: 95–107

Beltramino, C., Taleisnik, S. 1979. Effect of electrochemical stimulation in the olfactory bulbs on the release of gonadotropin hormones in rats. *Neuroendocrinology* 28: 320–28

Beltramino, C., Taleisnik, S. 1980. Dual action of electrochemical stimulation of the bed nucleus of the stria terminalis on the release of LH. *Neuroendocrinology* 30: 238–42

Beltramino, C., Taleisnik, S. 1983. Release of LH in the female rat by olfactory stimuli effect of the removal of the vomeronasal organs or lesioning of the accessory olfactory bulbs. *Neuroendocrinology* 36: 53–58

Beltramino, C., Taleisnik, S. 1985. Ventral premammillary nuclei mediate pheromonal-induced LH release stimuli in the rat. *Neuroendocrinology* 41: 119–24

Bertmar, G. 1981a. Evolution of vomeronasal organs in vertebrates. *Evolution* 35: 359–66

Bertmar, G. 1981b. Variations in size and structure of vomeronasal organs in reindeer *Rangifer tarandus tarandus L. Arch. Biol. Bruxelles* 92: 343–66

Beuerman, R. W. 1977. Slow potentials in the turtle olfactory bulb in response to odor stimulation of the nose and electrical stimulation of the olfactory nerve. *Brain Res.* 128: 429–45

Bhatnager, K. P., Matulionis, D. H., Breipohl, W. 1982. Fine structure of the vomeronasal neuroepithelium of bats: A comparative study. *Acta Anatomica* 112: 158–77

Bojsen-Moller, F. 1975. Demonstration of terminalis, olfactory, trigeminal and perivascular nerves in the rat nasal septum. *J. Comp. Neurol.* 159: 245–56

Bossy, J. 1980. Development of olfactory and related structures in staged human embryos. *Anat. Embryol.* 161: 225–36

Breipohl, W., Mendoza, A. S., Miragall, F. 1982. Freeze-fracturing studies on the main and vomeronasal olfactory sensory epithelia in NMRI-mice. In *Olfaction and Endocrine Regulation*, ed. W. Breipohl, pp. 309–22. London: IRL Press

Broadwell, R. D. 1975. Olfactory relationships of the telencephalon and diencephalon in the rabbit. I. An autoradiographic study of the efferent connections of the main and accessory olfactory bulbs. *J. Comp. Neurol.* 163: 329–46

Broadwell, R. D., Jacobowitz, D. M. 1976. Olfactory relationships of the telencephalon and diencephalon in the rabbit. III. The ipsilateral centrifugal fibers to the olfactory bulbar and retrobulbar formations. *J. Comp. Neurol.* 170: 321–46

Burghardt, G. M. 1970. Chemical reception

in reptiles. In *Advances in Chemoreception, 1: Communication by Chemical Signals*, ed. J. W. Johnson, D. R. Moulton, A. Turk, pp. 241–308. New York: Appleton-Century-Crofts

Burghardt, G. M. 1977. The ontogeny, evolution, and stimulus control of feeding in humans and reptiles. In *The Chemical Senses and Nutrition*, ed. M. R. Kare, O. Maller, pp. 253–75. New York: Academic

Burghardt, G. M. 1980. Behavioral and stimulus correlates of vomeronasal functioning in reptiles: Feeding, grouping, sex, and tongue use. See Beauchamp et al 1980, pp. 275–301

Burghardt, G. M., Hess, E. H. 1968. Factors influencing the chemical release of prey attack in newborn snakes. *J. Comp. Physiol. Psychol.* 66: 289–95

Burghardt, G. M., Pruitt, C. H. 1975. Role of the tongue and senses in feeding of naive and experienced garter snakes. *Physiol. Behav.* 14: 185–94

Ciges, M., Labella, T., Gayoso, M., Sanchez, G. 1977. Ultrastructure of the organ of Jacobson and comparative study with olfactory mucosa. *Acta Otolaryngol.* 83: 47–58

Clancy, A., N., Macrides, F., Singer, A. G., Agosta, W. C. 1984a. Male hamster copulatory responses to a high molecular weight fraction of vaginal discharge: Effects of vomeronasal organ removal. *Physiol. Behav.* 33: 653–60

Clancy, A. N., Coquelin, A., Macrides, F., Gorski, R. A., Noble, E. P. 1984b. Sexual behavior and aggression in male mice: Involvement of the vomeronasal system. *J. Neurosci.* 4: 2222–29

Conrad, L. C. A., Pfaff, D. W. 1976. Efferents from medial basal forebrain and hypothalamus in the rat. An autoradiographic study of the medial preoptic area. *J. Comp. Neurol.* 169: 185–220

Cooper, J. G., Bhatnagar, K. P. 1976. Comparative anatomy of the vomeronasal organ complex in bats. *J. Anat.* 122: 571–601

Coquelin, A., Clancy, A. N., Macrides, F., Noble, E. P., Gorski, R. A. 1984. Pheromonally induced release of luteinizing hormone in male mice: Involvement of the vomeronasal system. *J. Neurosci.* 4: 2230–36

Crowell-Davis, S., Houpt, K. A. 1985. The ontogeny of flehmen in horses. *Anim. Behav.* 33: 739–45

Crump, D., Swigar, A. A., West, J. R., Silverstein, R. M., Muller-Schwarze, D., Altieri, R. 1984. Urine fractions that release flehmen in black-tailed deer. *J. Chem. Ecol.* 10: 203–15

Cuschieri, A., Bannister, L. H. 1975. The development of the olfactory mucosa in the mouse: Light microscopy. *J. Anat.* 119: 277–86

Davis, B. J., Macrides, F., Youngs, W. M., Schneider, S. P., Rosene, D. L. 1978. Efferents and centrifugal afferents to the main and accessory olfactory bulbs in hamster. *Brain Res. Bull.* 3: 59–72

deOlmos, J. S., Ingram, W. R. 1972. The projection field of the stria terminalis in the rat brain. An experimental study. *J. Comp. Neurol.* 146: 303–34

deOlmos, J., Hardy, H., Heimer, L. 1978. The afferent connections of the main and accessory olfactory bulb formations in the rat: An experimental HRP-study. *J. Comp. Neurol.* 181: 213–44

Devor, M. 1976. Fiber trajectories of olfactory bulb efferents in the hamster. *J. Comp. Neurol.* 166: 31–48

Dluzen, D. E., Ramirez, V. D. 1983. Localized and discrete changes in neuropeptide (LHRH and TRH) and neurotransmitter (NE and DA) concentrations within the olfactory bulbs as a function of social interaction. *Horm. Behav.* 17: 139–45

Drickamer, L. C. 1986. Puberty-influencing chemosignals in house mice: Ecological and evolutionary considerations. In *Chemical Signals in Vertebrates, Vol. 4, Ecology, Evolution, and Comparative Biology*, ed. D. Duvall, D. Muller-Schwarze, R. M. Silverstein, pp. 1–15 New York: Plenum

Dullemeijer, P. 1961. Some remarks on the feeding behavior of rattlesnakes. *Konikl. Nederl. Acad. Van Wetenschappen Proc. Ser. C* 64: 383–96

Eccles, R. 1982. Autonomic innervation of the vomeronasal organ of the cat. *Physiol. Behav.* 28: 1011–15

Estes, R. D. 1972. The role of the vomeronasal organ in mammalian reproduction. *Extrait Mammal.* 36: 317–30

Farbman, A. I., Margolis, F. L. 1980. Olfactory marker protein during ontogeny: Immunohistochemical localization *Devel. Biol.* 74: 205–15

Fleming, A., Vaccarino, F., Tambosso, L., Chee, P. 1979. Vomeronasal and olfactory system modulation of maternal behavior in the rat. *Science* 203: 372–74

Frahm, H. D. 1981. Volumetric comparison of the accessory olfactory bulb in bats. *Acta Anat.* 109: 173–83

Frahm, H. D., Bhatnagar, K. P. 1980. Comparative morphology of the accessory olfactory bulb in bats. *J. Anat.* 130: 349–65

Gabe, M., Saint Girons, H. 1976. Contribution a la morphologie comparee des fosses nasales et de leurs annexes chez les lepidosauriens. *Mem. Mus. Nat. D'Hist. Nat. Ser. A* 98: 1–87

Gallego, A., Sanchez Criado, J. E. 1979. Pheromonal regulation of the estrous cycle in the female rat by the accessory olfactory system. *Acta Endocrinol. Copenhagen Suppl.* 225: 256

Gaughwin, M. D. 1979. The occurrence of flehmen in a marsupial—the hairy-nosed wombat (*Lasiorhinus latifrons*). *Anim. Behav.* 27: 1063–65

Gillingham, J. C., Clark, D. L. 1981. Snake tongue-flicking: Transfer mechanics to Jacobson's organ. *Can. J. Zool.* 59: 1651–57

Graves, B., Duvall, D. 1983. Occurrence and function of prairie rattlesnake mouth gaping in a non-feeding context. *J. Exp. Zool.* 227: 471–74

Graves, B. M., Duvall, D. 1985a. Mouth gaping and head shaking by prairie rattlesnakes are associated with vomeronasal organ olfaction. *Copeia* 1985: 496–97

Graves, B. M., Duvall, D. 1985b. Avomic prairie rattlesnakes (*Crotalus viridis*) fail to attack rodent prey. *Z. Tierpsychol.* 67: 161–66

Graziadei, P. P. C. 1973. Cell dynamics in the olfactory mucosa. *Tiss. Cell* 5: 113–31

Halpern, J., Erichsen, E., Halpern, M. 1985. Role of olfactory and vomeronasal senses on garter snake response to airborne odorants. *Soc. Neurosci. Abstr.* 11: 1221

Halpern, M. 1976. The efferent connections of the olfactory bulb and accessory olfactory bulb in the snakes, *Thamnophis sirtalis* and *Thamnophis parietalis*. *J. Morphol.* 150: 553–78

Halpern, M. 1980a. Chemical ecology in terrestrial vertebrates. In *Animals and Environmental Fitness*, ed. R. Gilles, pp. 263–82. New York: Pergamon

Halpern, M. 1980b. The telencephalon of snakes. In *Comparative Neurology of the Telencephalon*, ed. S. O. E. Ebbesson, pp. 257–95. New York: Plenum

Halpern, M. 1983. Nasal chemical senses in snakes. In *Advances in Vertebrate Neuroethology*, ed. J.-P. Ewert, R. R. Capranica, D. J. Ingle, pp. 141–76. New York: Plenum

Halpern, M., Frumin, N. 1979. Roles of the vomeronasal and olfactory systems in prey attack and feeding in adult garter snakes. *Physiol. Behav.* 22: 1183–89

Halpern, M., Kubie, J. L. 1980. Chemical access to the vomeronasal organs of garter snakes. *Physiol. Behav.* 24: 367–71

Halpern, M., Kubie, J. L. 1983. Snake tongue flicking behavior: Clues to vomeronasal system functions. In *Chemical Signals in Vertebrates*, ed. D. Muller-Schwarze, R. M. Silverstein, 3: 45–72. New York: Plenum.

Halpern, M., Kubie, J. L. 1984. The role of the ophidian vomeronasal system in species-typical behavior. *Trends Neurosci.* 7: 472–77

Halpern, M., Schulman, N., Scribani, L., Kirschenbaum, D. M. 1984. Characterization of vomeronasally-mediated response eliciting components of earthworm wash II. *Pharmacol. Biochem. Behav.* 21: 655–62

Halpern, M., Schulman, N., Kirschenbaum, D. M. 1986. Characteristics of earthworm washings detected by the vomeronasal system of snakes. In *Chemical Signals in Vertebrates: Ecology, Evolution, and Comparative Biology*, Vol. 4, ed. D. Duvall, D. Muller-Schwarze, R. M. Silverstein. New York: Plenum. In press

Halpern, M., Scribani, L., Kubie, J. L. 1985. Vomeronasal stimuli can be reinforcing. *Abstr. Assoc. Chemorecept. Sci., Sarasota, Fla.* April, 1985

Heimer, L. 1969. The secondary olfactory connections in mammals, reptiles and sharks. *Ann. N.Y. Acad. Sci.* 167: 129–46

Heller, S. B., Halpern, M. 1982. Laboratory observations of aggregative behavior of garter snakes, *Thamnophis sirtalis*: Roles of the visual, olfactory, and vomeronasal senses. *J. Comp. Physiol. Psychol.* 96: 967–83

Hinds, J. W. 1967. Autoradiographic study of histogenesis in the mouse olfactory bulb. I. Time of origin of neurons and neuroglia. *J. Comp. Neurol.* 134: 287–304.

Hudson, R., Distel, H. 1986. The pheromonal release of suckling in rabbits does not depend on the vomeronasal organ. *Physiol. Behav.* 37: 123–28

Hunter, A. J., Fleming, D., Dixson, A. F. 1984. The structure of the vomeronasal organ and nasopalatine ducts in *Aotus trivirgatus* and some other primate species. *J. Anat.* 138: 217–25

Imamura, K., Mori, K., Fujita, S. C., Obata, K. 1985. Immunochemical identification of subgroups of vomeronasal nerve fibers and their segregated terminations in the accessory olfactory bulb. *Brain Res.* 328: 362–66

Ingersoll, D. W. 1981. Role of the vomeronasal organ in murine priming and signalling chemocommunication systems. *Dissert. Abstr.* 41B: 3215

Jacobs, V. L., Sis, R. F., Chenoweth, P. J., Klemm, W. R., Sherry, C. J. 1981. Structures of the bovine vomeronasal complex and its relationship to the palate: Tongue manipulation. *Acta Anat.* 110: 48–58

Jennes, L., Stumpf, W. E. 1980. LHRH-systems in the brain of the golden hamster. *Cell Tissue Res.* 209: 239–56

Jirik-Babb, P., Manaker, S., Tucker, A. M., Hofer, M. A. 1984. The role of the accessory and main olfactory systems in maternal behavior of the primiparous rat. *Behav. Neural Biol.* 40: 170–78

Johns, M. A. 1980. The role of the vomeronasal system in mammalian reproductive physiology. See Beauchamp et al 1980, pp. 341–64

Johns, M. A., Feder, H. H., Komisaruk, B. R., Mayer, A. D. 1978. Urine-induced reflex ovulation in anovulatory rats may be a vomeronasal effect. *Nature* 272: 446–48

Johnson, A., Josephson, R., Hawke, M. 1985. Clinical and histological evidence for the presence of the vomeronasal (Jacobson's) organ in adult humans. *J. Otolaryng.* 14: 71–79

Johnston, R. E. 1983. Chemical signals and reproductive behavior. In *Pheromones and Reproduction in Mammals*, ed. J. G. Vandenbergh, pp. 95–112. New York: Academic

Johnston, R. E. 1985. Olfactory and vomeronasal mechanisms of communication. In *Taste, Olfaction and the Central Nervous System*, ed. D. W. Pfaff, pp. 322–46. New York: Rockefeller Univ. Press

Johnston, R. E., Rasmussen, K. 1984. Individual recognition of female hamsters by males: Role of chemical cues of the olfactory and vomeronasal systems. *Physiol. Behav.* 33: 95–104

Kaneko, N., Debski, E. A., Wilson, M. C., Whitten, W. K. 1980. Puberty acceleration in mice. II. Evidence that the vomeronasal organ is a receptor for the primer pheromone in male mouse urine. *Biol. Reprod.* 22: 873–78

Kelche, C., Aron, C. 1984. Olfactory cues and accessory olfactory bulb lesion: Effects on sexual behavior in the cyclic female rat. *Physiol. Behav.* 33: 45–48

Keverne, E. B. 1979. The dual olfactory projections and their significance for behaviour. In *Chemical Ecology: Odour Communication in Animals*, ed. F. J. Ritter, pp. 75–83. Amsterdam: Elsevier/North-Holland Biomed. Press

Keverne, E. B. 1982. The accessory olfactory system and its role in pheromonally mediated changes in prolactin. In *Olfaction and Endocrine Regulation*, ed. W. Breipohl, pp. 127–40. London: IRL Press

Keverne, E. B. 1983. Pheromonal influences on the endocrine regulation of reproduction. *Trends Neurosci.* 6: 381–84

Keverne, E. B., de la Riva, C. 1982. Pheromones in mice: Reciprocal interactions between the nose and brain. *Nature* 296: 148–50

Keverne, E. B., Murphy, C. L., Silver, W. L., Wysocki, C. J., Meredith, M. 1986. Non-olfactory chemoreceptors of the nose: Recent advances in understanding the vomeronasal and trigeminal systems. *Chem. Senses* 11: 119–33

Kevetter, G. A., Winans, S. S. 1981. Connections of the corticomedial amygdala in the golden hamster. I. Efferents of the "Vomeronasal Amygdala." *J. Comp. Neurol.* 197: 81–98

Kirschenbaum, D. M., Schulman, N., Yao, P., Halpern, M. 1985. Chemo-attractant for the garter snake: Characterization of vomeronasally mediated response-eliciting components of earthworm wash-III. *Comp. Biochem. Physiol.* 82B: 447–53

Kirschenbaum, D. M., Schulman, N., Halpern, M. 1986. Earthworms produce a collagen-like substance detected by the garter snake vomeronasal system. *Proc. Natl. Acad. Sci. USA* 83: 1213–16

Klemm, W. R., Sherry, C. J., Sis, R. F., Morris, D. L. 1984b. Electrophysiologic recording from bovine vomeronasal capsule under spontaneous and stimulated conditions. *Brain Res. Bull.* 12: 275–82

Klemm, W. R., Sherry, C. J., Sis, R. F., Schake, L. M., Waxman, A. B. 1984a. Evidence of a role for the vomeronasal organ in social hierarchy in feedlot cattle. *Appl. Anim. Behav. Sci.* 12: 53–62

Kolnberger, I. 1971. Vergleichende untersuchungen am riechepithel, insbesondere des Jacobsonchen organs von amphibien, reptilien und saugetieren. *Z. Zellforsch.* 122: 53–67

Kolnberger, I., Altner, H. 1971. Ciliary-structure precursor bodies as stable constituents in the sensory cells of the vomeronasal organ of reptiles and mammals. *Z. Zellforsch.* 118: 254–62

Kratzing, J. E. 1971a. The fine structure of the sensory epithelium of the vomeronasal organ in suckling rats. *Aust. J. Biol. Sci.* 24: 787–96

Kratzing, J. E. 1971b. The structure of the vomeronasal organ in the sheep. *J. Anat.* 108: 247–60

Kratzing, J. E. 1975. The fine structure of the olfactory and vomeronasal organs of a lizard (*Tiliqua scincoides scincoides*). *Cell Tissue Res.* 156: 239–52

Krettek, J. E., Price, J. L. 1977. Projections from the amygdaloid complex to the cerebral cortex and thalamus in the rat and cat. *J. Comp. Neurol.* 172: 687–722

Krettek, J. E., Price, J. L. 1978a. Amygdaloid projections to subcortical structures within the basal forebrain and brainstem in the rat and cat. *J. Comp. Neurol.* 178: 225–54

Krettek, J. E., Price, J. L. 1978b. A description of the amygdaloid complex in the rat and cat with observations on intra-amygdaloid connections. *J. Comp. Neurol.* 178: 255–80

Kreutzer, E. W., Jafek, B. W. 1980. The vomeronasal organ of Jacobson in the human embryo and fetus. *Otolaryngol. Head Neck Surg.* 88: 119–23

Kubie, J. L., Halpern, M. 1975. Laboratory observations of trailing behavior in garter snakes. *J. Comp. Physiol. Psych.* 89: 667–74

Kubie, J. L., Halpern, M. 1978. Garter snake trailing behavior: Effects of varying prey extract concentration and mode of prey extract presentation. *J. Comp. Physiol. Psychol.* 92: 362–73

Kubie, J. L., Halpern, M. 1979. The chemical senses involved in garter snake prey trailing. *J. Comp. Physiol. Psychol.* 93: 648–67

Kubie, J. L., Vagvolgyi, A., Halpern, M. 1978. The roles of the vomeronasal and olfactory systems in the courtship behavior of male garter snakes. *J. Comp. Physiol. Psychol.* 92: 627–41

Ladewig, J., Hart, B. L. 1980. Flehmen and vomeronasal organ function in male goats. *Physiol. Behav.* 24: 1067–71

Ladewig, J., Hart, B. L. 1982. Flehmen and vomeronasal function. In *Olfaction and Endocrine Regulation*, ed. W. Breipohl, pp. 237–47. London: IRL Press

Lehman, M. N., Winans, S. 1982. Vomeronasal and olfactory pathways to the amygdala controlling male hamster sexual behavior: Autoradiographic and behavioral analyses. *Brain Res.* 240: 27–41

Lehman, M., N., Winans, S. S. 1983. Evidence for a ventral non-strial pathway from the amygdala to the bed nucleus of the stria terminalis in the male golden hamster. *Brain Res.* 268: 139–46

Lehman, M. N., Winans, S. S., Powers, J. B. 1980. Medial nucleus of the amygdala mediates chemosensory control of male hamster sexual behavior. *Science.* 210: 557–60

Leonard, C. M. 1975. Developmental changes in olfactory bulb projections revealed by degeneration argyrophilia. *J. Comp. Neurol.* 162: 467–86

Lepri, J. J., Wysocki, C. J., Vandenbergh, J. G. 1985. Mouse vomeronasal organ: Effects on chemosignal production and maternal behavior. *Physiol. Behav.* 35: 809–14

Lloyd-Thomas, A., Keverne, E. B. 1982. Role of the brain and accessory olfactory system in the block to pregnancy in mice. *Neuroscience* 7: 907–13

Lomas, D. E., Keverne, E. B. 1982. Role of the vomeronasal organ and prolactin in the acceleration of puberty in female mice. *J. Reprod. Fert.* 66: 101–7

Loo, S. K. 1977. Electron-dense microvilli in the vomeronasal organ of the tree shrew. *Acta Anat.* 98: 221–23

Loo, S. K., Kanagasuntheram, R. 1972. The vomeronasal organ in tree shrew and slow loris. *J. Anat.* 112: 165–72

Luiten, P. G. M., Koolhass, J. M., de Boer, S., Koopmans, S. J. 1985. The corticomedial amygdala in the central nervous system organization of agonistic behavior. *Brain Res.* 332: 283–97

MacLeod, N. K., Reinhardt, W. 1983. An electrophysiological study of the accessory olfactory bulb in the rabbit-I. Analysis of electrically evoked potential fields. *Neuroscience* 10: 119–29

Macrides, R., Davis, B. J. 1983. The olfactory bulb. In *Chemical Neuroanatomy*, ed. P. C. Emson, pp. 391–426. New York: Raven

Madison, D. M. 1977. Chemical communication in amphibians and reptiles. In *Chemical Signals in Vertebrates*, ed. D. Muller-Schwarze, M. M. Mozell, pp. 135–68. New York: Plenum

Maier, W. 1980. Nasal structures in old and new world primates. In *Evolutionary Biology of the New World Monkeys and Continental Drift*, ed. R. L. Ciochon, A. B. Chiarelli, pp. 219–41. New York: Plenum

Manaker, S., Winokur, A., Rostene, W. H., Rainbow, T. C. 1985. Autoradiographic localization of thryotropin-releasing hormone receptors in the rat central nervous system. *J. Neuroscl.* 5: 167–74

Mantyh, P. W., Hunt, S. P. 1985. Thyrotropin-releasing hormone (TRH) receptors: Localization by light microscopic autoradiography in rat brain using (3H) (3-Me-His2) TRH as the radioligand. *J. Neurosci.* 5: 551–61

Marchlewska-Koj, A. 1980. Partial isolation of pregnancy block pheromone in mice. See Beauchamp et al 1980, pp. 413–14

Marchlewska-Koj, A. 1983. Pregnancy blocking by pheromones. In *Pheromones and Reproduction in Mammals*, ed. J. G. Vandenbergh, pp. 151–74. New York: Academic

Marques, D. M. 1979. Roles of the main olfactory and vomeronasal systems in the response of the female hamster to young. *Behav. Neural Biol.* 26: 311–29

Melese-d'Hospital, P. Y., Hart, B. L. 1985. Vomeronasal organ cannulation in male goats: Evidence for transport of fluid from oral cavity to vomeronasal organ during flehmen. *Physiol. Behav.* 35: 941–44

Meredith, M. 1980. The vomeronasal organ and accessory olfactory system in the hamster. See Beauchamp et al 1980, pp. 303–26

Meredith, M. 1982. Stimulus access and other processes involved in nasal chemosensory function: Potential substrates for neural and hormonal influence. in *Olfaction and Endocrine Regulation*, ed. W. Breipohl, pp. 223–36. London: IRL Press

Meredith, M. 1983. Sensory physiology of pheromone communication. In *Pheromones and Reproduction in Mammals*, ed. J. G. Vandenbergh, pp. 199–252. New York: Academic

Meredith, M. 1986. Vomeronasal organ removal before sexual experience impairs male hamster mating behavior. *Physiol. Behav.* 36: 737–43

Meredith, M., Burghardt, G. M. 1978. Electrophysiological studies of the tongue and accessory olfactory bulb in garter snakes. *Physiol. Behav.* 21: 1001–8

Meredith, M., Marques, D. M., O'Connell, R. J., Stern, F. L. 1980. Vomeronasal pump: Significance for male hamster sexual behavior. *Science* 207: 1224–26

Meredith, M., O'Connell, R. J. 1979. Efferent control of stimulus access to the hamster vomeronasal organ. *J. Physiol.* 286: 301–16

Miragall, F., Breipohl, W., Bhatnager, K. P. 1979. Ultrastructural investigation on the cell membranes of the vomeronasal organ in the rat. *Cell Tissue Res.* 200: 397–408

Mora, O. A., Sanchez-Criado, J. E., Guisado, S. 1985. Role of the vomeronasal organ on the estral cycle reduction by pheromones in the rat. *Rev. Espan. Fisiol.* 41: 305–10

Müller, W. 1971. Vergleichende Electrophysioligische Untersuchungen an den Sinnessepithelien des Jacobsonschen Organs und der Nase von Amphibien (*Rana*), Reptilien (*Lacerta*) und Saugetieren (*Mus*). *Z. Vergl. Physiologie.* 72: 370–85

Murphy, M. R. 1980. Sexual preferences of male hamsters: Importance of preweaning and adult experience, vaginal secretion, and olfactory or vomeronasal sensation. *Behav. Neural Biol.* 30: 323–40

Naguro, T., Breipohl, W. 1982. The vomeronasal epithelia of NMRI mouse. A scanning electronmicroscopic study. *Cell Tissue Res.* 227: 519–34

Nakashima, T., Kimmelman, C. P., Snow, J. lB. 1985. Vomeronasal organs and nerves of Jacobson in the human fetus. *Acta Otolaryngol.* 99: 226–71

Northcutt, R. G., Royce, G. J. 1975. Olfactory bulb projections in the bullfrog *Rana catesbiana. J. Morphol.* 145: 251–68

Novotny, M., Jorgenson, J. W., Carmack, M., Wilson, S. R., Boyse, E. A., Yamazaki, K., Wilson, M., Beamer, W., Whitten, W. D. 1980. Chemical studies of the primer mouse pheromones. See Beauchamp et al 1980, pp. 377–90

O'Connell, R. J., Meredith, M. 1984. Effects of volatile and non volatile chemical signals on male sex behaviors mediated by the main and accessory olfactory system. *Behav. Neurosci.* 98: 1083–93

Pedersen, P. E., Stewart, W. B., Greer, C. A., Shepherd, G. M. 1984. Evidence for olfactory function in utero. *Science* 221: 478–80

Perkins, M. S., Perkins, M. N., Hitt, J. C. 1980. Effects of stimulus female on sexual behavior of male rats given olfactory tubercle and corticomedial amygdaloid lesions. *Physiol. Behav.* 25: 495–500

Phillips, H. S., Ho, B. T., Linner, J. G. 1982. Ultrastructural localization of LH-RH immunoreactive synapses in the hamster accessory olfactory bulb. *Brain Res.* 246: 193–204

Phillips, H. S., Hostetter, G., Kerdelhue, B., Kozlowski, G. P. 1980. Immunocytochemical localization of LHRH in central olfactory pathways of hamster. *Brain Res.* 193: 574–79

Powers, J. B., Fields, R. B., Winans, S. S. 1979. Olfactory and vomeronasal system participation in male hamsters' attraction to female vaginal secretions. *Physiol. Behav.* 22: 77–84

Powers, J. B., Winans, S. S. 1975. Vomeronasal organ: Critical role in mediating sexual behavior in the male hamster. *Science* 187: 961–63

Price, J. L. 1973. An autoradiographic study of complementary laminar patterns of termination of afferent fibers to the olfactory cortex. *J. Comp. Neurol.* 150: 87–108

Raisman, G. 1972. An experimental study of the projection of the amygdala to the accessory olfactory bulb and its relationship to the concept of a dual olfactory system. *Exper. Brain Res.* 14: 395–408

Rajendren, G., Dominic, C. J. 1984. Role of the vomeronasal organ in the male-induced implantation failure (The Bruce Effect) in mice. *Arch. Biol. Bruxelles* 95: 1–9

Rajendren, G., Dominic, C. J. 1986. Effect of bilateral transection of the lateral olfactory tract on the male-induced implantation failure (the Bruce effect) in mice. *Physiol. Behav.* 36: 587–90

Rasmussen, L. E., Schmidt, M., Henneous, R., Groves, D. 1982. Asian bull elephants: Flehmen-like responses to extractable components in female elephant estrous urine. *Science* 217: 159–62

Reformato, L. S., Kirschenbaum, D. M., Halpern, M. 1983. Preliminary characterization of response-eliciting components of earthworm extract. *Pharmacol. Biochem. Behav.* 18: 247–54

Reinhardt, V. 1983. Flehmen, mounting and copulation among members of a semi-wild cattle herd. *Anim. Behav.* 31: 641–50

Reinhardt, W., MacLeod, N. K., Ladewig, J. Ellendorff, F. 1983. An electrophysiological study of the accessory olfactory bulb in the rabbit. II. Input-output relations as assessed from analysis of intra- and extracellular unit recordings. *Neuroscience* 10: 131–39

Reynolds, J., Keverne, E. B. 1979. The accessory olfactory system and its role in the pheromonally mediated suppression of oestrus in grouped mice. *J. Reprod. Fert.* 57: 31–35

Saint Girons, M. H. 1975. Histologie Comparee-Developpement respectif de l'epithelium sensoriel du cavum et de l'organe de Jacobson chez les Lepidosauriens. *CR Acad. Sci. Paris* 280: 721–24

Sanchez-Barcelo, E. J., Mediavilla, M. D., Sanchez-Criado, J. E., Cos, S., Cortines, M. D. G. 1985. Antigonadal actions of olfactory and light deprivation. 1. Effects of blindness combined with olfactory bulb deafferentation, transection of vomeronasal nerves, or bulbectomy. *J. Pineal Res.* 2: 177–90

Sanchez-Criado, J. E. 1982. Involvement of the vomeronasal system in the reproductive physiology of the rat. In *Olfaction and Endocrine Regulation*, ed. W. Breipohl, pp. 209–21. London: IRL Press

Sanchez-Criado, J. E., Gallego, A. 1979. Male induced precocious-puberty in the female rat. Role of the vomeronasal system. *Acta Endocrinol. Copenhagen Suppl.* 225: 255

Scalia, F. 1972. The projection of the accessory olfactory bulb in the frog. *Brain Res.* 36: 409–11

Scalia, F., Winans, S. S. 1975. The differential projections of the olfactory bulb and accessory olfactory bulb in mammals. *J. Comp. Neurol.* 161: 31–56

Scalia, R., Winans, S. S. 1976. New perspectives on the morphology of the olfactory system: Olfactory and vomeronasal pathways in mammals. In *Mammalian Olfaction, Reproductive Processes, and Behavior*, ed. R. L. Doty, pp. 7–28. New York: Academic

Schaeffer, C., Roos, J., Aron, C. 1986. Accessory olfactory bulb lesions and lordosis behavior in the male rat feminized with ovarian hormones. *Horm. Behav.* 20: 118–27

Segovia, S., Guillamon, A. 1982. Effects of sex steroids on the development of the vomeronasal organ in the rat. *Devel. Brain Res.* 5: 209–12

Segovia, S., Orensanz, L. M., Valencia, A., Guillamon, A. 1984a. Effects of sex steroids on the development of the accessory olfactory bulb in the rat. A volumetric study. *Devel. Brain Res.* 16: 312–14

Segovia, S., Paniagua, R., Nistal, M., Guillamon, A. 1984b. Effects of postpuberal gonadectomy on the neurosensorial epithelium of the vomeronasal organ in the rat. *Devel. Brain Res.* 14: 289–91

Shammah-Lagnado, S. J., Negrao, N. 1981. Efferent connections of the olfactory bulb in the oppossum (*Didelphis marsupialis aurita*): A Fink-Heimer study. *J. Comp. Neurol.* 201: 51–63

Sheffield, L. P., Law, J. M., Burghardt, G. M. 1968. On the nature of chemical food sign stimuli for newborn garter snakes. *Commun. Behav. Biol.* 2: 7–12

Shepherd, G. M. 1985. Are there labeled lines in the olfactory pathway? In *Taste, Olfaction and the Central Nervous System*, ed. D. W. Pfaff, pp. 307–21. New York: Rockefeller Univ. Press

Singer, A. G., Agosta, W. C., O'Connell, R. J., Pfaffman, C., Bowen, D. V., Field, F. H. 1976. Dimethyl disulfide: An attractant pheromone in hamster vaginal secretion. *Science* 191: 948–49

Singer, A. G., Clancy, A. N., Macrides, F., Agosta, W. C. 1984a. Chemical studies of hamster vaginal discharge: Effects of endocrine ablation and protein digestion on behaviorally active macromolecular fractions. *Physiol. Behav.* 33: 639–43

Singer, A. G., Clancy, A. N., Macrides, F., Agosta, W. C. 1984b. Chemical studies of hamster vaginal discharge: Male behavioral responses to a high molecular weight fraction require physical contact. *Physiol. Behav.* 33: 645–51

Singer, A. G., Macrides, F., Agosta, W. C. 1980. Chemical studies of hamster reproductive pheromones. See Beauchamp et al 1980, pp. 365–75

Singh, P. J., Tucker, A. M., Hofer, M. A. 1976. Effects of nasal ZnSO4 irrigation and olfactory bulbectomy on rat pups. *Physiol. Behav.* 17: 373–82

Skeen, L. C., Hall, W. C. 1977. Efferent projections of the main and the accessory olfactory bulb in the tree shrew (*Tupaia glis*). *J. Comp. Neurol.* 172: 1–36

Steel, E. 1984. Effect of the odour of vaginal secretion on non-copulatory behaviour of male hamsters (*Mesocricetus auratus*). *Anim. Behav.* 32: 597–608

Steel, E., Keverne, E. B. 1985. Effect of

female odour on male hamsters mediated by the vomeronasal organ. *Physiol. Behav.* 35: 195–200
Switzer, R. C. III, Johnson, J. I. Jr. 1977. Absence of mitral cells in monolayer in monotremes. Varitions in vertebrate olfactory bulbs. *Acta Anat.* 99: 36–42
Taniguchi, K., Mikami, S. 1985. Structure of the epithelia of the vomeronasal organ of horse and cattle. *Cell Tissue Res.* 240: 41–48
Taniguchi, K., Mochizuki, K. 1983. Comparative morphological studies on the vomeronasal organ in rats, mice and rabbits. *Jpn. J. Vet. Sci.* 45: 67–76
Taniguchi, K., Mochizuki, K. 1982. Morphological studies on the vomeronasal organ in the golden hamster. *Jpn. J. Vet. Sci.* 44: 419–26
Taniguchi, K., Taniguchi, K., Mochizuki, K. 1982a. Developmental studies on the vomeronasal organ in the golden hamster. *Jpn. J. Vet. Sci.* 44: 709–16
Taniguchi, K., Taniguchi, K., Mochizuki, K. 1982b. Comparative developmental studies on the fine structure of the vomeronasal sensory and olfactory epithelia in the golden hamster. *Jpn. J. Vet. Sci.* 44: 881–90
Teicher, M. H., Shaywitz, B. A., Lumia, A. R. 1984. Olfactory and vomeronasal system mediation of maternal recognition in the developing rat. *Devel. Brain Res.* 12: 97–110
Ten Donkelaar, H. J., Lammers, G. J., Gribnau, A. A. M. 1979. Neurogenesis in the amygdaloid nuclear complex in a rodent (the Chinese hamster). *Brain Res.* 165: 348–53
Tucker, D. 1960. Simultaneous recordings from vomeronasal and olfactory nerves. *Physiologist* 3: 167
Tucker, D. 1963. Olfactory vomeronasal and trigeminal receptor responses to odorants. In *Olfaction and Taste*, ed. Y. Zotterman, pp. 45–69. New York: Macmillan
Tucker, D. 1971. Nonolfactory responses from the nasal cavity: Jacobson's organ and the trigeminal system. In *Olfaction*, ed. L. M. Beidler, pp. 151–81. New York: Springer-Verlag
Ulinski, P. S., Peterson, E. H. 1981. Patterns of olfactory projections in the desert iguana, *Dipsosaurus dorsalis*. *J. Morphol.* 168: 189–227
Vaccarezza, O. L., Sepich, L. N., Tramezzani, J. H. 1981. The vomeronasal organ of the rat. *J. Anat.* 132: 167–85
Valencia, A., Segovia, S., Guillamon, A. 1986. Effects of sex steroids on the development of the accessory olfactory bulb mitral cells in the rat. *Devel. Brain Res.* 24: 287–90
Vandenbergh, J. G., Whitsett, J. M., Lombardi, J. R. 1975. Partial isolation of a pheromone accelerating puberty in female mice. *J. Reprod. Fert.* 43: 515–23
Velasco, M. E., Taleisnik, S. 1969. Release of gonadotropins induced by amygdaloid stimulation in the rat. *Endocrinology* 84: 132–39
Verberne, G. 1976. Chemocommunication among domestic cats, mediated by the olfactory and vomeronasal senses. II. The relation between the function of Jacobson's organ (vomeronasal organ) and Flehmen behavior. *Z. Tierpsychol.* 42: 113–28
Verberne, G., de Boer, J. 1976. Chemocommunication among domestic cats, mediated by the olfactory and vomeronasal senses. I. Chemocommunication. *Z. Tierpsychol.* 42: 86–109
Vincent, S. R., McIntosh, C. H. S., Buchan, A. M. J., Brown, J. C. 1985. Central somatostatin systems revealed with monoclonal antibodies. *J. Comp. Neurol.* 238: 169–86
Wang, R. T., Halpern, M. 1980a. Light and electron microscopic observations on the normal structure of the vomeronasal organ of garter snakes. *J. Morphol.* 164: 47–67
Wang, R. T., Halpern, M. 1980b. Scanning electron microscopic studies of the surface morphology of the vomeronasal epithelium and olfactory epithelium of garter snakes. *Am. J. Anat.* 157: 339–428
Wang, R. T., Halpern, M. 1982a. Neurogenesis in the vomeronasal epithelium of adult garter snakes. 1. Degeneration of bipolar neurons and proliferation of undifferentiated cells following experimental vomeronasal axotomy. *Brain Res.* 237: 23–39
Wang, R. T., Halpern, M. 1982b. Neurogenesis in the vomeronasal epithelium of adult garter snakes. 2. Reconstitution of the bipolar neuron layer following experimental vomeronasal axotomy. *Brain Res.* 237: 41–59
Wang, R. T., Vagvolgyi, A., Mendelsohn, B., Halpern, M. 1979. Further observations on postnatal neurogenesis in the snake's vomeronasal epithelium: Use of ^{3}H-thymidine autoradiography to trace the differentiation, maturation and movement of bipolar neurons. *Neurosci. Abstr.* 5: 103
Wilson, K. C. P., Raisman, G. 1980. Age-related changes in the neurosensory epithelium of the mouse vomeronasal organ: Extended period of postnatal growth in size and evidence for rapid cell turnover in the adult. *Brain Res.* 185: 103–13

Wilson, K. C. P., Raisman, G. 1981. Estimation of numbers of vomeronasal synapses in the glomerular layer of the accessory olfactory bulb of the mouse at different ages. *Brain Res.* 205: 245–53

Winans, S. S., Lehman, M. N., Powers, J. B. 1982. Vomeronasal and olfactory CNS pathways which control male hamster mating behavior. In *Olfaction and Endocrine Regulation*, ed. W. Breipohl, pp. 23–34. London: IRL Press

Winans, S. S., Powers, J. B. 1977. Olfactory and vomeronasal deafferentation of male hamsters: Histological and behavioral analyses. *Brain Res.* 126: 325–44

Winans, S. S. Scalia, F. 1970. Amygaloid nucleus: New afferent input from the vomeronasal organ. *Science* 170: 330–32

Witkin, J. W. 1985. Luteinizing hormone-releasing hormone in olfactory bulbs of primates. *Am. J. Primat.* 8: 309–15

Witkin, J. W., Silverman, A.-J. 1983. Luteinizing hormone-releasing hormone (LHRH) in rat olfactory systems. *J. Comp. Neurol.* 218: 426–32

Wysocki, C. J. 1979. Neurobehavioral evidence for the involvement of the vomeronasal system in mammalian reproduction. *Neurosci. Biobehav. Rev.* 3: 301–41

Wysocki, C. J. 1982. The vomeronasal organ: Its influence upon reproductive behavior and underlying endocrine systems. In *Olfaction and Endocrine Regulation*, ed. W. Breipohl, pp. 195–208. London: IRL Press

Wysocki, C. J., Beauchamp, G. K., Reidinger, R. R., Wellington, J. L. 1985. Access of large and nonvolatile molecules to the vomeronasal organ of mammals during social and feeding behaviors. *J. Chem. Ecol.* 11: 1147–59

Wysocki, C. J., Bean, N. J., Beauchamp, G. K. 1986. The mammalian vomeronasal system: Its role in learning and social behaviors. In *Chemical Signals in Vertebrates: Ecology, Evolution, and Comparative Biology*, ed. D. Duvall, D. Muller-Schwarze, R. M. Silverstein, Vol. 4. New York: Plenum. In press

Wysocki, C. J., Katz, Y., Bernhard, R. 1983. Male vomeronasal organ mediates female-induced testosterone surges in mice. *Biol. Reprod.* 28: 917–22

Wysocki, C. J., Nyby, J., Whitney, G., Beauchamp, G. K., Katz, Y. 1982. The vomeronasal organ: Primary role in mouse chemosensory gender recognition. *Physiol. Behav.* 29: 315–27

Wysocki, C. J., Wellington, J. L., Beauchamp, G. K. 1980. Access of urinary nonvaolatiles to the mammalian vomeronasal organ. *Science* 207: 781–84

Ann. Rev. Neurosci. 1987. 10: 363–401

VISUAL PROCESSING IN MONKEY EXTRASTRIATE CORTEX

John H. R. Maunsell

Department of Physiology and Center for Visual Science, University of Rochester, Rochester, New York 14642

William T. Newsome

Department of Neurobiology and Behavior, State University of New York, Stony Brook, New York 11794

INTRODUCTION

The neuronal processes that lead to visual perception have attracted intense interest since Kuffler's studies of receptive field organization in cat retinal ganglion cells over three decades ago (Kuffler 1953). A variety of anatomical and physiological approaches have been employed to analyze the organization of the visual pathway between the retina and striate cortex (V1) and the transformations of visual information that occur at each stage (see Hubel & Wiesel 1977, Stone 1983, Shapley & Lennie 1985). The growth in understanding of the retinostriate pathway has been accompanied by increasing interest in visual processing in the expanse of extrastriate cortex beyond V1. Studies of extrastriate cortex in many species showed that it comprises a mosaic of visual areas that can be distinguished by several anatomical and physiological criteria (reviewed by Kaas 1978, Zeki 1978, Cowey 1979, Van Essen 1979, 1985, Wagor et al 1980, Tusa et al 1981).

The literature in this field is large, and we do not attempt to review all relevant studies. Rather, we concern ourselves with three recent developments that have yielded insight into information processing and flow within extrastriate cortex. The first of these is the convergence of anatom-

0147–006X/87/0301–0363$02.00

ical, physiological, and behavioral results to suggest the existence of distinct streams of processing in extrastriate cortex. This idea has provided a useful conceptual framework for organizing an increasing body of observations and for guiding new experiments. The second development concerns physiological properties that reveal transformations in the information encoded at different stages in visual cortex. Such observations provide clues about the operations performed on the visual image by cortex beyond V1. Experiments with alert, trained animals have led to the third development: the discovery of major and widespread extraretinal influences on neuronal responses even at early stages of the extrastriate pathways. These influences are likely to constitute a major aspect of the neuronal processing in extrastriate cortex and have been neglected in previous studies with anesthetized preparations.

We restrict our discussion to primates because most of the work relevant to these topics has been performed in monkeys. Our discussion focuses on results from the macaque monkey because of the extensive physiological data available for this species. Because of space limitations, we are unable to consider some important topics of active investigation such as stereoscopic mechanisms (reviewed by Poggio & Poggio 1984) and interactions between extrastriate cortex and other brain regions (e.g. Tigges & Tigges 1985).

THE ORGANIZATION OF PRIMATE VISUAL CORTEX

Early evidence that primate extrastriate cortex contains many identifiable visual areas came largely from experiments by Allman & Kaas in the owl monkey and by Zeki in the macaque. In a series of studies Allman & Kaas (1971, 1974a,b, 1975, 1976) showed that owl monkey visual cortex is composed of at least eight visual areas, most of which were found to contain a topographic mapping of the contralateral visual hemifield. In

Figure 1 The location of macaque cortical visual areas. The positions of several visual areas are illustrated on lateral and ventromedial views of the cerebral hemispheres. Outline drawings at the top indicate the major sulci. At least some portion of every visual area is buried within the cortical sulci, and some areas are entirely hidden from view. In this diagram and those that follow, *thin lines* indicate the borders of areas that are reasonably well established, while *dashed lines* mark borders whose position is less well defined. Abbreviations for visual areas: AIT, anterior inferotemporal; DP, dorsal prelunate; PIT, posterior inferotemporal; PO parieto-occipital; PS, prostriata; VA, ventral anterior; VP, ventral posterior. Abbreviations for sulci: A, arcuate; Ca, calcarine; Ce, central; Ci, cingulate; Co, collateral; IO, inferior occipital; IP, intraparietal; Lu, lunate; La, lateral; OT, occipitotemporal; P, principal; PO, parieto-occipital; ST, superior temporal.

the macaque monkey, Zeki and others demonstrated that visual cortex is subdivided in a related, but not identical manner (Zeki 1969, Cragg 1969, Dubner & Zeki 1971, Zeki 1971, 1975). It is now clear that visually responsive cortex in the macaque occupies a large, contiguous region that includes the entire occipital lobe and large portions of the temporal and

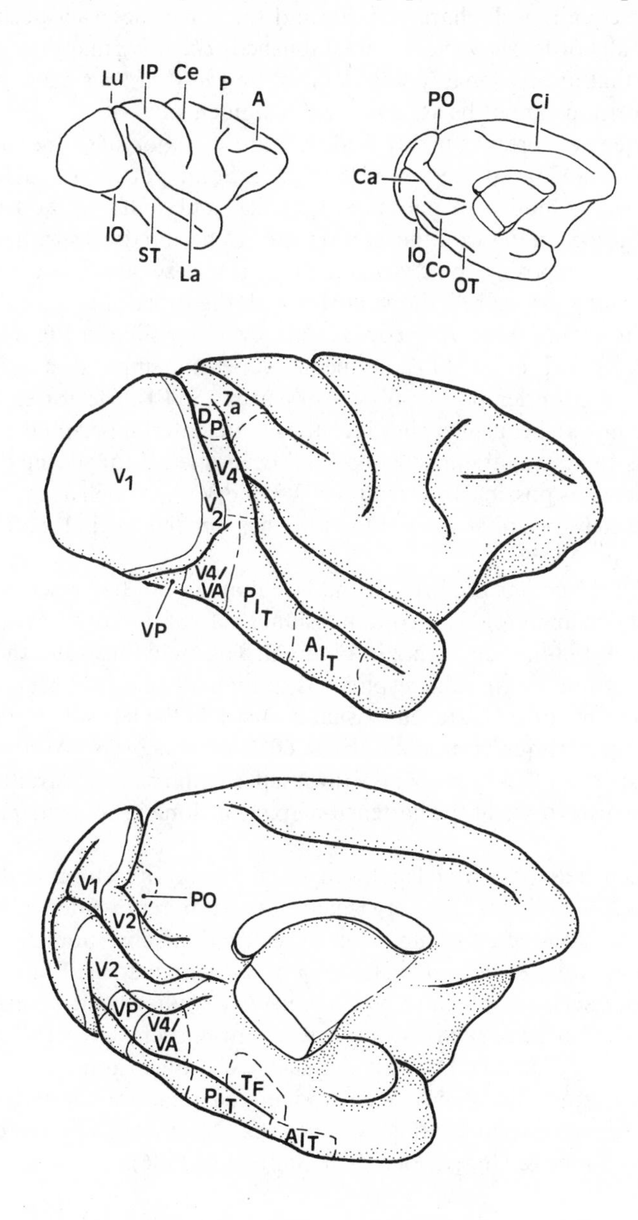

parietal lobes as well (Macko et al 1982). Although some parts of visual cortex remain to be explored, almost 20 areas have already been identified. The location of some of these areas can be seen in Figure 1. The upper of the two drawings is a lateral view of the right cerebral hemisphere, while the lower drawing is a ventromedial view of the left hemisphere. Not all areas are equally well characterized, and for some the exact position or extent of all borders have not been established. *Thin lines* mark the borders of areas that are reasonably well defined, and *dashed lines* mark borders whose position cannot be assigned with confidence.

Macaque neocortex is highly folded, and the major sulci are identified in smaller drawings at the top of the figure. Some portion of each of the visual areas is hidden in one or more sulci, and some visual areas are entirely buried. In Figure 2 the lateral view has been redrawn with selected sulci opened, exposing some visual areas that otherwise are not seen. The upper drawing in Figure 2 shows cortex with the superior temporal sulcus opened, revealing three visual areas that are not visible in Figure 1: the middle temporal area (MT), the medial superior temporal area (MST), and the superior temporal polysensory area (STP). The lower half of Figure 2 shows a hemisphere in which the lunate, inferior occipital, parieto-occipital, and intraparietal sulci have all been opened to varying degrees. In this view it is possible to see parts of V3 and V3A, as well as the portion of the parieto-occipital area (PO) that lies on the medial wall of the intraparietal sulcus.

The shape and surface area of the visual areas can be presented more accurately on an unfolded, two-dimensional map of the cortex (Van Essen & Maunsell 1980), such as that in Figure 3. This map illustrates the entire neocortex from the right hemisphere as though all sulci had been opened and the entire cortex flattened. Visual cortex fills the left side of the map and in the macaque comprises about 60% of neocortex. Although V1 abuts V2 in the intact brain, V1 is shown separated from extrastriate cortex to reduce distortions in the flattened map, as is done in some maps of the earth.

Nineteen areas have been labeled in visual cortex. In sections that follow we discuss the different types of visual information represented in some of these areas. However, it is important to realize that substantial differences exist in the extent of our understanding of individual areas. For example, some areas, such as MT, have been intensively studied using several techniques, while others, such as the ventral intraparietal area (VIP), have been tentatively identified by only one such technique. Future experiments may demonstrate that some currently identified areas are not unitary. For instance, there is accumulating evidence that MST contains two distinct areas (Desimone & Ungerleider 1986, Saito et al 1986).

Van Essen (1985) has recently discussed in detail the extent to which each of these visual areas has been characterized; here we mainly wish to emphasize the types of uncertainty that exist for many areas. Demonstrating that a region of cortex contains a single, identifiable visual area is rarely straightforward. In principle an area can be identified by any of several criteria, including topographic organization, anatomical connections, neuronal response properties, architectonics, and behavioral deficits resulting from ablation. In practice these techniques are often difficult to

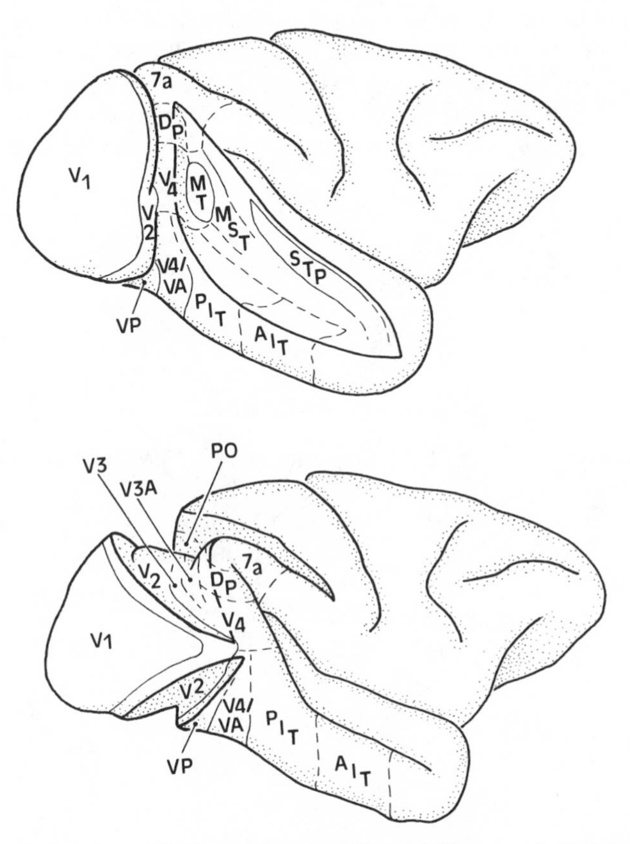

Figure 2 The location of visual areas buried within sulci. The upper half of the figure is a lateral view in which the superior temporal sulcus has been opened to show the positions of visual areas within it. Areas of the superior temporal sulcus that are not visible in the intact view are the middle temporal area (MT), the medial superior temporal area (MST), and the superior temporal polysensory area (STP). The lower half of the figure is a view in which the lunate, intraparietal, parieto-occipital, and inferior occipital sulci have been opened, exposing parts of V3, V3A, and the ventral posterior area (VP). A portion of the parieto-occipital area (PO) can be seen on the medial wall of the intraparietal sulcus.

apply, and their results can be contradictory. For example, V3 and the ventral posterior area (VP) together contain one complete representation of the visual hemifield, but striking differences in their connections, response properties, and architectonics suggest that it may be more appropriate to treat them as different areas (Burkhalter et al 1986). The use of topographic criteria is further complicated in that the representation of the visual field may be either too disorderly to be dependable or simply nonexistent (e.g. Robinson et al 1978, Desimone & Gross 1979), and similar problems occur with other criteria. As a result, identification and localization of specific areas is a complex process and for many areas is not yet complete. The tentative state of our knowledge concerning several visual areas suggests that our subdivision of extrastriate cortex will change with time. However, identification of 19 areas reflects substantial progress in our understanding of the organization of visual cortex beyond V1.

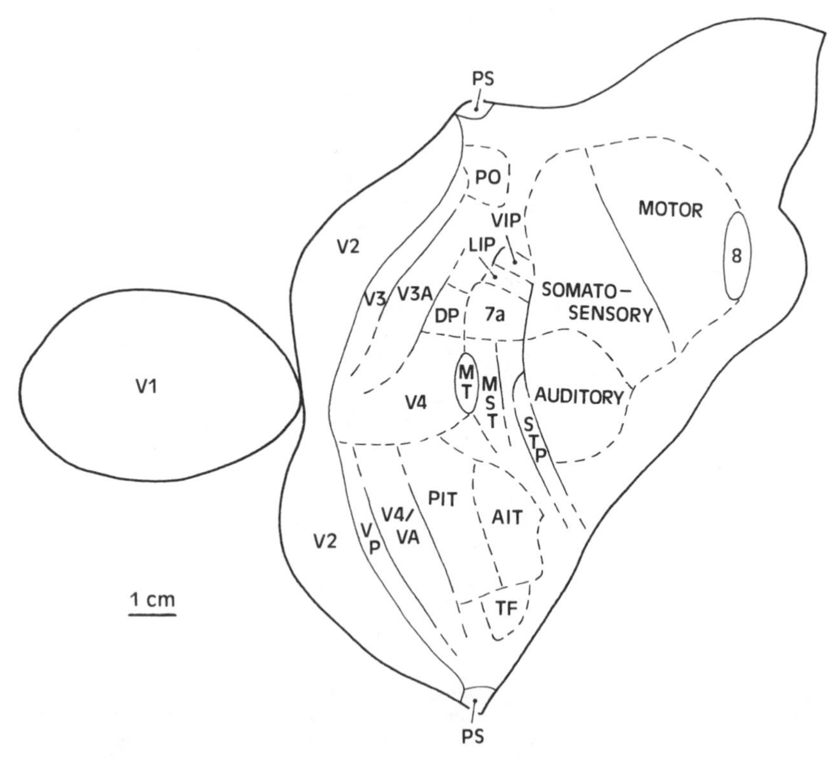

Figure 3 A map of the entire cortex from a right cerebral hemisphere. Visual cortex occupies the left (posterior) side of the map and comprises more than half of cortex. In this flattened representation it is possible to see the relative dimensions of the different cortical areas. Individual areas have not been drawn for other modalities, with the exception of area 8 (the frontal eye fields).

Hierarchical Organization

The identification of this large number of cortical visual areas has led to the discovery of a larger number of connections that transmit information between them. The connections of macaque visual cortex have been recently reviewed by Van Essen (1985), who compiled 84 identified or suspected pathways connecting the 19 visual areas shown in Figure 3. While this number is small compared to the number of possible connections, it is nonetheless sufficient to make the visual cortex a dauntingly complex system to study. Fortunately, macaque visual cortex appears to adhere to organizational principles that simplify its examination. In particular, it is possible to assign cortical areas to different levels of processing based on the laminar distribution of the neurons that connect them.

Several investigators have observed that ascending projections can be distinguished from descending projections by the cortical layers in which they originate and terminate (Jones & Wise 1977, Tigges et al 1977, 1981, Wong-Riley 1978, Rockland & Pandya 1979, Jones et al 1978, Weller & Kaas 1981, Friedman 1983). Ascending projections, which transmit information away from primary sensory areas, arise primarily in the superficial layers of cortex and terminate primarily in layer 4 and the lower part of layer 3. Descending projections, which carry signals back toward primary sensory areas, originate largely from neurons in the deep layers and end primarily in the superficial and deep layers. Most connections among the extrastriate cortical areas have one of these patterns and can be assigned as forward (ascending) or feedback (descending) by these anatomical criteria alone.

These assignments can be used to construct a hierarchy of visual areas by placing each area on a level that puts it above all areas from which it receives a forward projection, and below all those from which it receives a feedback projection (Maunsell & Van Essen 1983c). The hierarchy of macaque cortical areas in Figure 4 does not include some of the areas from Figure 3 because the laminar distributions of their connections are not known. It is important to note that for such a hierarchy to exist, internally consistent rules must be embodied within the total pattern of connections. For example, among reciprocally interconnected pairs of areas there are no cases where both projections are the forward type or both are feedback. It is also necessary that transitivity apply: If area A sends a forward projection to area B, which in turn sends a forward projection to area C, then a projection from A to C must have the forward pattern of laminar termination. All adequately studied connections among visual cortical areas in the macaque conform to this scheme. Although a hierarchical

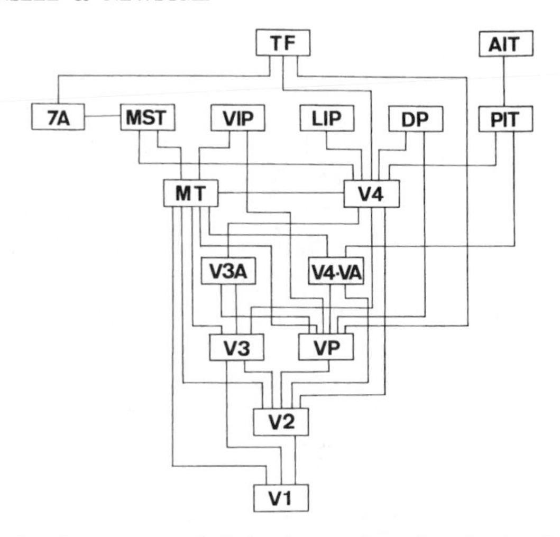

Figure 4 Hierarchy of macaque cortical visual areas. Areas have been assigned to different levels based on anatomical criteria by which connections can be assigned as being forward or feedback. Each area is one level above the highest level from which it receives forward input, and below all levels from which it receives feedback. Concomitantly, each area is above all areas to which it sends a feedback projection, and below those to which it sends a forward projection. The hierarchy is based on connections from numerous studies that were tabulated by Van Essen (1985) and includes only major connections. For clarity, the connections that have been demonstrated in one direction only are not distinguished from others. The connections that exist between V4 and MT do not fit either the forward or feedback category, and have been assigned as "lateral," leaving these two areas on the same level (see Maunsell & Van Essen 1983c).

organization of this nature may not be general for cortex of all mammalian species (see Symonds & Rosenquist 1984), it provides a valuable simplifying framework for considering the flow of information within primate extrastriate cortex.

Two Streams of Processing

Extrastriate areas differ in the types of visual information that they process. The initial evidence for selective processing of visual information in primate extrastriate cortex was obtained in the macaque by Zeki and his collaborators. Dubner & Zeki (1971) found that a region of V1-recipient cortex in the superior temporal sulcus contained a preponderance of neurons that were highly selective for the direction of stimulus motion yet relatively unselective for stimulus color or form (Dubner & Zeki 1971, Zeki 1974b). We refer to this area as MT, the middle temporal visual area, because it is homologous with MT in the owl monkey (Allman & Kaas 1971 ; see Baker et al 1981, Maunsell & Van Essen 1983a, Albright 1984), but it has also been called the motion area of the superior temporal sulcus (Zeki 1974b) and V5 (Zeki 1983c). In other studies, Zeki (1973, 1977) identified another area, V4, which was described as containing many neurons selective for color or orientation but few selective for direction of motion. These studies formed the basis for a proposal that each extrastriate visual area was responsible for processing a different type of visual information, such as motion or color (Zeki 1975, 1978).

Recent studies in several laboratories have confirmed early findings of substantial differences in physiological properties between certain visual areas, although the segregation frequently is not as complete as originally envisioned (e.g. Poggio & Fischer 1977, Petersen et al 1980, Schein et al 1982). At the same time, however, some visual areas do process very similar types of visual information (see below). These two observations have been integrated in a notion of functional streams of processing in visual cortex. In this view, visual cortex contains parallel streams of processing that analyze different types of visual information. Each stream involves several areas, with each area representing a different level of processing for a particular type of information.

The idea that primate visual cortex contains distinct streams of processing was formulated by Ungerleider & Mishkin (1982), who described two streams that diverge in the early levels of extrastriate cortex, one leading to visual cortex of the parietal lobe and the other to visual cortex of the temporal lobe. They suggested that these streams are associated with different visual capabilities—the parietal stream is involved in visual assessment of spatial relationships, and the temporal stream is concerned with visual recognition of objects. Support for this hypothesis comes from several lines of research, including anatomical, physiological, and behavioral experiments. In humans, clinical observations indicate that damage to the parietal cortex can affect visual perception of position or movement, yet leave object recognition unimpaired (Ratcliff & Davies-Jones 1972, Damasio & Benton 1979, Zihl et al 1983). In contrast, temporal lobe lesions can produce specific deficits related to object recognition, such as an inability to recognize faces (Meadows 1974a,b, Pearlman et al 1979, Damasio et al 1982, Joynt et al 1985). A corresponding separation of visual functions between the temporal and parietal cortices exists in the macaque monkey. Many studies have shown that lesions of temporal cortex impair visual discrimination of objects, whether the discriminanda differ in color, orientation, brightness, pattern, or shape (see Gross 1973a,b, Dean 1976, 1982, Wilson 1978, Gross et al 1981, Ungerleider & Mishkin 1982, Mishkin et al 1983). Lesions of the parietal cortex, however, leave object discrimination capacities largely intact and instead specifically affect the ability to do tasks related to visual assessment of the location of objects, such as the visual guidance of hand movements (Pohl 1973, Buchbinder et al 1980, Ungerleider & Mishkin 1982, Weiskrantz & Saunders 1984). Because the parietal and temporal cortices receive inputs from different parts of extrastriate cortex, Ungerleider & Mishkin suggested that these differences in lesion effects reflect the fact that parietal and temporal regions are associated with cortical pathways that process different classes of visual information (see Mishkin et al 1983).

Anatomical and physiological studies now indicate that functional streams of processing exist even at the earliest levels of visual cortex. These streams are present within distinct subdivisions of V1 and V2 and are largely segregated into separate sets of visual areas in later stages of extrastriate cortex. In this review we emphasize two streams for which current evidence is most compelling. One of these streams contains a high proportion of neurons that are selective for properties related to motion, such as direction and speed. The other exhibits more selectivity for color and orientation. We use the terms *motion pathway* and *color and form pathway* (Van Essen & Maunsell 1983) to describe these streams of processing. While these labels are useful for the purposes of discussion, they are unlikely to be completely descriptive. For example, the motion pathway may not be strictly limited to analyzing motion and may not incorporate every aspect of motion. These streams of processing are related to, but are probably not identical with, those described by Ungerleider & Mishkin (1982). The motion pathway appears to contribute visual inputs to parietal cortex, though the relevant connections are not yet fully established. Similarly, the color and form pathway provides inputs to the temporal cortex. However, the motion pathway and the color and form pathways as currently defined probably represent subsets of the information that reaches parietal or temporal cortex and are unlikely to mediate all visual behaviors associated with these brain regions.

Figure 5 shows the areas that we include in the motion and the color and form pathways and their relationships. The motion pathway begins in layer 4b in V1, which contains a high proportion of direction-selective neurons and little or no color sensitivity (Dow 1974, Blasdel & Fitzpatrick 1984, Livingstone & Hubel 1984, Movshon & Newsome 1984, Michael 1985). This layer gives rise to most of the direct projection from V1 to MT. The color and form pathway is found in the superficial layers of V1. These layers contain an array of dot-like patches of cytochrome oxidase rich tissue that have come to be called "blobs" (Horton & Hubel 1980, Humphrey & Hendrickson 1980). Livingstone & Hubel (1984) found that the blobs contain a higher proportion of neurons with color selectivity than the interblob regions. Orientation selectivity, which is taken to convey information about form, was far more common in the interblob regions.

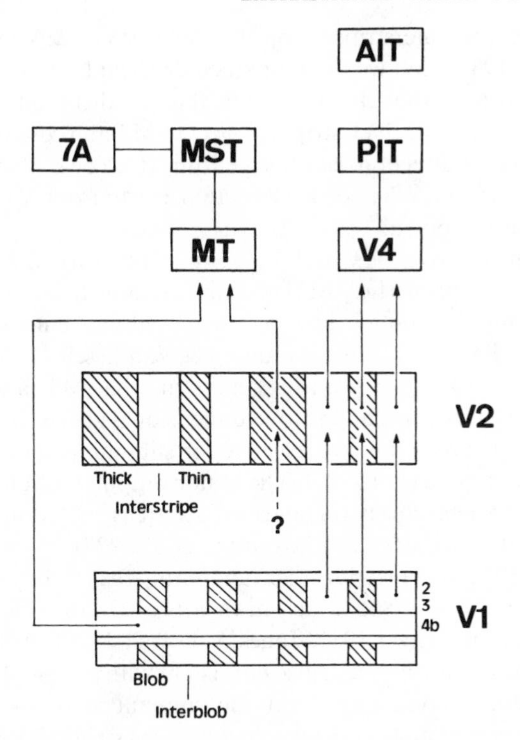

Figure 5 Major components of the motion pathway and the color and form pathway. Segregation of the two pathways is evident in V1, where the color and form pathway arises from the blobs and inter-blob regions in layers 2 and 3. These regions project to the thin stripes and interstripes in V2, which in turn project to V4. The outputs of V4 lead to PIT and AIT. The motion pathway is found in layer 4b in V1, which projects directly to MT. MT also receives a projection from the thick stripes in V2, whose inputs are not yet established. The outputs of MT lead to areas including MST and area 7a.

They suggested that the blob/interblob segregation may represent separate streams for processing color and form information.

Corresponding subdivisions associated with each stream can be found in V2 as well. V2 has regions of high and low cytochrome oxidase activity that are arranged in alternating thin and thick stripes of cytochrome oxidase rich tissue and are separated by interstripe regions with little activity (Tootell et al 1983). The thin stripes receive input from V1 cytochrome oxidase blobs, while the V1 interblob regions project to the V2 interstripe regions (Livingstone & Hubel 1984). Mirroring the functional segregation observed in V1, the V2 thin stripes are enriched in color selectivity while the V2 interstripes contain a relatively high proportion of orientation selective neurons (Hubel & Livingstone 1985). However, both the thin stripes and interstripes project to V4 (DeYoe & Van Essen 1985, Shipp & Zeki 1985), and we therefore include them both in a single color and form pathway, although it is possible that parallel subdivisions of this pathway persist. The V2 thick stripes project to MT and contain a relatively high proportion of direction-selective neurons (DeYoe & Van Essen 1985, Shipp & Zeki 1985). Whether a projection exists from V1 to the thick stripes remains unknown.

As mentioned above, MT and V4 contain neurons with significantly different response properties. MT neurons are selective not only for the direction of stimulus motion but also for speed and binocular disparity (Zeki 1974a,b, Baker et al 1981, Maunsell & Van Essen 1983a,b, Albright 1984, Felleman & Kaas 1984), thus suggesting that MT is well suited to the analysis of visual motion in three dimensional space. In contrast, V4 contains many neurons that are selective for stimulus color or orientation (Zeki 1978, Desimone et al 1985). The major outputs of MT appear to be relayed to the parietal cortex via intermediate areas, including MST (Jones & Powell 1970, Leichnitz 1980, Mesulam et al 1977). V4 sends a major projection to PIT, the posterior inferotemporal area (Rockland & Pandya 1979, Desimone et al 1980), which in turn projects to AIT, the anterior inferotemporal area (Iwai 1981, Jones & Powell 1970). A separation of motion and color information is also consistent with psychophysical observations that stimuli consisting of only isoluminant colors cannot generate a strong sensation of motion (Ramachandran & Gregory 1978, Cavanagh et al 1984).

It is possible that these pathways originate at earlier stages than V1. The retinostriate pathway contains two markedly different streams that are segregated into the parvocellular and magnocellular subdivisions of the lateral geniculate nucleus (see Stone 1983, Derrington & Lennie 1984, Derrington et al 1984). These subdivisions project to different sublaminae in V1, with layer 4cα receiving magnocellular input and layer 4cβ receiving parvocellular input (Hubel & Wiesel 1972). Some evidence suggests that the motion and the color and form pathways arise more or less directly from these subcortical streams (see Maunsell 1986). For example, layer 4cα is strongly connected with layer 4b (Lund & Boothe 1975, Fitzpatrick et al 1985), which projects directly to MT. Similarities between some response properties of neurons in the motion pathway and those in the magnocellular layers of the lateral geniculate nucleus have been noted (Motter & Mountcastle 1981, Maunsell & Van Essen 1983a). In contrast, layer 4cβ, which receives parvocellular input, has stronger connections with superficial layers (Fitzpatrick et al 1985), which give rise to the color and form pathway. It is thus possible that the primate visual system contains two subsystems that operate largely in parallel from early subcortical stages to the highest levels of processing in the visual cortex. This striking possibility might be tested by examining responses in the motion or color and form pathway while blocking transmission through one subdivision of the lateral geniculate nucleus.

Although evidence from several different approaches suggests that separate streams of processing exist for motion and for color and form, it seems unlikely that these streams are completely independent. The separation of physiological response properties between the streams, while pronounced, is not complete. For example, direction-selective cells are found in the color and form pathway (Zeki 1978, DeYoe & Van Essen 1985), and neurons in the motion pathway are not devoid of orientation selectivity (Albright 1984). Neither is the anatomical segregation complete, since some connections exist between MT and V4 (Rockland & Pandya 1979, Maunsell & Van Essen 1983c, Ungerleider & Desimone 1986) and between parietal and temporal cortex (Desimone et al 1980, Seltzer & Pandya 1984). In addition, psychophysical (Mayhew & Anstis 1972) and behavioral (Iwai 1985) studies also suggest that some functional overlap exists between the two streams.

While the separation between these streams is not absolute, the evidence discussed above strongly suggests that distinct pathways play a prominent role in processing different types of visual information. Consideration of how physiological properties differ between areas within these pathways adds further support for this scheme of organization. In the sections that follow we consider in detail the transformations of visual signals that occur in these pathways, concerning ourselves first with motion and then with color and form.

THE MOTION PATHWAY

The transformations of the visual image that occur along the motion pathway do not appear to result in increased selectivity for basic parameters such as direction or speed. To the contrary, a recent quantitative study by Albright (1984) has shown that direction tuning in MT is somewhat broader than that in V1. Similarly, retinotopic specificity decreases progressively in successive levels of the pathway. The average receptive field area in MT is 100 times that in V1 (Gattass & Gross 1981), and this trend continues in MST, where receptive fields covering a full quadrant of the visual field can be found (Van Essen et al 1981, Tanaka et al 1986, Desimone & Ungerleider 1986). Rather than sharpening basic tuning curves, the transformation of information between areas in the motion pathway appears to elaborate new, more complex response properties. In the next sections we consider several transformations that occur on the motion pathway, beginning with those that distinguish MT from V1 and subsequently proceeding to MST and area 7a. Although the motion pathway includes V2 (and probably other areas as well), physiological data are not yet available concerning the specific contributions of other areas to motion processing.

Motion of Complex Patterns

Visually guided behavior frequently depends on accurate information about the motion of complex patterns and objects. However, extraction of such information by the visual system is not as straightforward as it might seem. Consider for example the motion of the complex plaid pattern in Figure 6C. This plaid was created by superimposing the sinusoidal gratings in Figure 6A and 6B. The motion of each component grating is orthogonal to its orientation as indicated by the *arrows*. However, human observers viewing the plaid perceive rightward motion that is different from that of either oriented component (Adelson & Movshon 1982). The neuronal representation of the motion of the plaid is particularly interesting in light of the strongly orientation-selective responses of most V1 neurons. If motion sensitive neurons in V1 respond only to the oriented components of the plaid stimulus, their discharges will encode the motion of the component gratings (A and B) rather than the motion of the complex pattern (C). Because the motion of the pattern is the behaviorally relevant datum, the visual system must extract this information at higher levels of processing.

Movshon and colleagues (1986) have used the stimuli illustrated in Figure 6A–C to analyze the responses of direction-selective neurons in cat and monkey cortex. They found that while neurons in V1 encode the motion of the oriented components, a population of neurons in MT encodes the unitary motion of the entire pattern. The responses of each type

→

Figure 6 Direction selectivity for components and patterns in visual cortical neurons. A–C: Stimuli used to distinguish component direction selective responses from pattern direction selective responses. A sinusoidal grating moving upward and to the right (A) superimposed on a grating moving downward and to the right (B) results in a perception of a plaid pattern moving directly to the right (C). D–E: Direction tuning curves for a component direction selective neuron in cat striate cortex in response to a single sinusoidal grating (D) and a sinusoidal plaid (E). This neuron responded to the motion of the components of the plaid rather than to the unitary motion of the plaid itself (see text). This response pattern is typical of V1 neurons in cat and monkey. F–G: Direction tuning curves for a pattern direction selective neuron in macaque MT in response to a single grating (F) and plaid (G). This neuron responded to the unitary motion of the plaid pattern. The *dashed lines* in E and G indicate predicted tuning curves for a component direction selective response, assuming that the neurons respond to each component of the plaid in the same way that they respond to the single grating (D and F). Reprinted with permission from Movshon et al (1986).

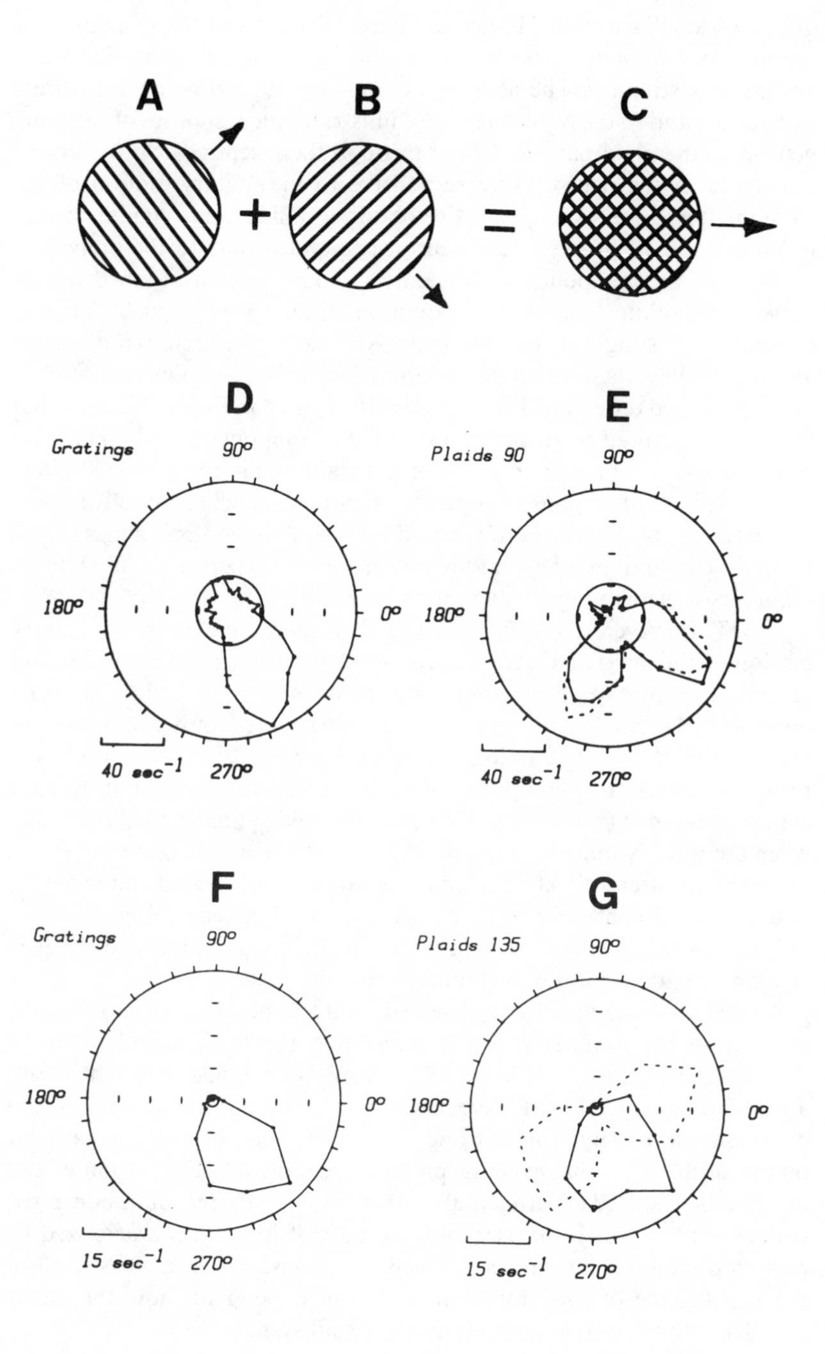
A
B
C
D
E
Gratings
90°
180°
0°
270°
40 sec⁻¹
Plaids 90
90°
180°
0°
270°
40 sec⁻¹
F
G
Gratings
90°
180°
0°
270°
15 sec⁻¹
Plaids 135
90°
180°
0°
270°
15 sec⁻¹

of neuron are illustrated in Figure 6. Figure 6D is a polar plot of a direction tuning curve obtained from a V1 neuron by using a drifting sine wave grating as a stimulus. The neuron responded to a narrow range of directions down and to the right. Figure 6E illustrates the responses of the same neuron to the plaid pattern drifted through the receptive field in various directions. If the neuron were responsive to the unitary motion of the whole pattern, one would expect to obtain a tuning curve similar to that in Figure 6D. If, however, the neuron responded only to the individual motions of the components, the neuron should yield a bi-lobed tuning curve, responding first when the motion of the plaid is such that one component grating traverses the receptive field in the preferred direction and responding again when the motion of the other component grating is in the preferred direction. The responses illustrated in Figure 6E show that the neuron responded to the motion of the components and yielded no response when the unitary motion of the plaid was in the preferred direction. This type of response, termed "component direction selective," was characteristic of the V1 neurons tested by Movshon and colleagues (1986) in both cat and monkey, although the motion perceived by human observers was unambiguously that of the plaid.

In MT, however, about 20% of the neurons responded to the unitary motion of the plaid and were accordingly described as "pattern direction selective." Figure 6F illustrates the direction tuning curve obtained from one such MT neuron by using a single sine wave grating stimulus. In Figure 6G, the dotted bi-lobed tuning curve indicates the predicted response pattern for a component direction-selective neuron, but the data actually obtained (*solid curve*) show that this neuron responded optimally when the unitary motion of the plaid was in the neuron's preferred direction. Thus, pattern direction-selective neurons in MT encode information about the motion of complex patterns that is not signaled by single neurons in V1. The responses of these neurons are strikingly parallel to the perception of pattern motion by human observers.

Several investigators have observed neurons in MT whose preferred orientation for stationary slits is parallel to the preferred direction of motion (Baker et al 1981, Maunsell & Van Essen 1983a, Albright 1984), and Albright argued that these neurons are those identified as pattern direction-selective by Movshon and colleagues. This observation may bear on the mechanisms that generate pattern direction-selective responses and needs to be resolved experimentally. Another issue that must be addressed is the extent to which the ascending outputs of MT are characterized by pattern or component direction-selective responses. A firm assessment of the significance of this transformation will depend on how the newly encoded information is used within the visual system.

Speed Selectivity

The human visual system excels at extracting information about the speed of moving stimuli. Human observers reliably discriminate stimuli differing in speed by as little as 5% and maintain this capability across a broad range of speeds in all parts of the visual field (McKee 1981, McKee & Nakayama 1984). A simple hypothesis is that this perceptual capability is dependent on speed-tuned cortical neurons. However, several studies have shown that the speed tuning of single neurons in cat and monkey V1 varies substantially, depending on the spatial structure of the stimulus; their responses are not, therefore, invariant with respect to speed. Rather, V1 neurons are invariant in their responses to spatial and temporal frequency (Tolhurst & Movshon 1975, Holub & Morton-Gibson 1981, Bisti et al 1985, Foster et al 1985). For example, a neuron that responds preferentially to a sinusoidal grating of 4 cycles/deg will exhibit the same preferred spatial frequency regardless of the temporal drift rate at which the measurements are made. When tested with a temporal drift rate of 2 cycles/sec, therefore, this neuron will respond best to a speed of 0.5 deg/sec (speed = temporal frequency/spatial frequency), but it would respond best to a speed of 4 deg/sec were a temporal frequency of 16 cycles/sec used. Such neurons whose "preferred" speed changes with spatial and temporal frequency contrast strikingly to the performance of psychophysical observers who accurately judge the speed of sinusoidal gratings despite random variations in spatial and temporal frequency (McKee et al 1986).

Preliminary evidence indicates that in the monkey a small class of MT neurons responds in a manner that may account for the spatio-temporal independence of speed perception. Spatial and temporal tuning are not independent in these neurons, but covary so as to maintain a constant preferred speed over a broad range of spatial and temporal frequencies (Newsome et al 1983). Such neurons appear to represent a higher level of processing, where afferent neurons tuned to particular combinations of spatial and temporal frequency converge to build invariant responses to speed. Speed invariance may not be characteristic of all biological visual systems, however, because behavioral studies of the fly suggest that spatial and temporal tuning are entirely separable in its visual system (see Hildreth & Koch 1987 this volume).

Spatial Scale of Directional Interactions

A recent analysis of motion selectivity in monkey cortex by using stroboscopic stimuli revealed substantial differences in the spatial extent of directional interactions in V1 and MT (Mikami et al 1986a,b). In this study, a slit of light was stroboscopically stepped across the receptive field of direction-selective neurons in both the preferred and null directions. The distance (spatial interval) and time (temporal interval) between successive flashes of the slit were systematically varied, and the responses were analyzed to determine the maximum spatial and temporal intervals for which direction selectivity could be obtained in each neuron.

The major findings were (*a*) that MT neurons maintained direction selectivity over spatial intervals that were, on the average, three times as large as those for V1 neurons at corresponding eccentricities, and (*b*) that MT and V1 neurons maintained direction selectivity over a similar range of temporal intervals. Consistent with these observations, MT neurons are direction selective for smooth motions of higher speeds ($\Delta x/\Delta t$) than V1 neurons. The maximum speed for direction selectivity to smooth motion is positively correlated with the maximum spatial interval in both MT and V1, and is twice as large (on the average) in MT as in V1 (Mikami et al 1986b). Preferred speeds are also higher in MT than in V1 (Figure 9 of Van Essen 1985). Thus, it appears that a major role of MT is to extend direction selectivity to a higher range of speeds by increasing the spatial scale of motion-sensing mechanisms. This trend may continue in cortical areas beyond MT. Preliminary observations on visually responsive neurons in Area 7a suggest that directional interactions can occur for spatial separations of 10–20 deg (Mountcastle et al 1984, Motter et al 1985). These values represent a substantial expansion of the spatial interactions underlying direction selectivity over those observed in MT.

The responses of cortical neurons to stroboscopic motion also suggest possible neural substrates for the phenomenon of apparent motion. Human observers perceive motion for successively flashed visual stimuli separated by appropriate spatial and temporal intervals (see review by Nakayama 1985). The spatio-temporal dependence of motion selectivity in MT neurons corresponded closely to that for perception of motion by human observers at high (8–32 deg/sec) apparent speeds, while both MT and V1 neurons were congruent with human performance at low (1–4 deg/sec) apparent speeds (Newsome et al 1986). It will be of interest to determine whether lesion studies confirm separable roles for V1 and MT over different ranges of speed in the manner suggested by the physiological experiments.

Surround Antagonism

It has recently become clear that the responses of many neurons in visual cortex can be greatly influenced by stimuli outside the classically defined receptive field. The classical receptive field is the region in which a single stimulus, such as a spot or bar, can evoke a response. Stimuli in the surrounding regions, by definition, cannot generate a response alone, but can greatly modulate the response to a stimulus in the classical receptive field (reviewed by Allman et al 1985b). Several investigations have reported the existence of neurons whose responses to stimulus motion are modulated in a directionally antagonistic manner by motion in the surround (Sterling & Wickelgren 1969, Frost et al 1981, Frost & Nakayama 1983, von Grünau & Frost 1983). In monkeys, investigation has been directed toward MT and to a lesser extent, MST and V2. In owl monkey MT, Allman and colleagues (1985a) stimulated the classical receptive field with motion of a random dot pattern in the preferred direction while simultaneously stimulating a large surround region with a second random dot pattern. They found that for most MT neurons, the effects of surround stimulation on the responses to stimulation of the classical receptive field were dependent on the direction and speed of the surround pattern. The most common effect was the presence of an antagonistic, direction-selective surround that suppressed the response of a neuron to optimal stimulation of the classical receptive field when surround motion was in the same direction as motion in the classical receptive field. For many of these neurons, surround motion in the direction opposite that of classical receptive field motion facilitated the response to stimulation of the classical receptive field. Tanaka and colleagues (1986) observed similar effects in macaque MT. As yet, neurons in monkey V1 have not been examined for direction specific antagonistic surrounds. Whether such effects arise strictly from neuronal interactions in extrastriate cortex or whether such processing exists in V1 as well is not yet clear.

The antagonistic effects reported in these studies may play a role in a variety of visual functions requiring relative motion cues (see review by Allman et al 1985b). For example, MT neurons with antagonistic surrounds respond well to local motion within the classical receptive field, but not to global motion of extended textures or of the entire field. As such they are well suited for distinguishing a moving figure from background and for distinguishing object motion in the world from the global motion induced by a viewer's movement through the world.

Rotation

Tanaka, Saito and their colleagues recently found that the dorsal half of MST contains neurons selective for higher order motion phenomena such as rotation and changing size. These properties were not observed in MT (Tanaka et al 1986, Saito et al 1986). Neurons selective for the direction of rotatory motion have been reported previously in parietal cortex (Leinonen 1980, Sakata et al 1985, 1986), frontal cortex (Rizzolatti et al 1981), and the superior temporal polysensory area (STP) (Bruce et al 1981, Jeeves et al 1983). Fourteen percent of the MST neurons studied by Saito and colleagues responded well to rotatory motion but not to linear movements of the same visual stimuli. Most responded selectively either to clockwise or to counterclockwise motion. Other tests showed that selectivity for rotation did not result from a preference for different directions of motion in different subregions of the receptive field, since small rotating texture patterns elicited the same response when placed in any portion of the receptive field. Neurons in both MST and area 7a are also reported to respond to rotations in depth, that is, in planes that are not perpendicular to the line of sight (Saito et al 1986, Sakata et al 1985, 1986). Rotation-selective neurons were not observed in the ventral half of MST, and this fact supports the argument that MST comprises two distinct visual areas (Desimone & Ungerleider 1986).

Rotation-selective responses constitute a property not evident at earlier stages in the motion pathway. However, because rotating patterns contain different linear velocities in different regions of the stimulus, it is possible that these neurons are simply responding to velocity shear. This issue needs to be addressed before selectivity for rotation per se is firmly established. Selectivity for rotation in depth is more problematic. The difficulty is that rotation of the visual displays in depth is accompanied by several cues such as changes in shape of the entire display. It is possible that an apparent selectivity for rotation in depth arises from selectivity for other cues that are not necessarily associated with rotation in depth. The use of random dot displays in which the degree of rotation may be smoothly varied while holding other cues invariant holds promise for addressing these issues.

Changing Size

Zeki (1974b) reported that neurons in MT of the macaque encode a monocular cue for motion in depth. These neurons responded preferentially to two edges moving in opposite directions within the receptive field. Because changing size is a potent cue for motion in depth, these neurons can provide a signal of motion toward (separating edges) or away from (approaching edges) the viewer. The cue is monocular because selective responses are obtained with stimulation of either eye alone. Similar response properties have been found in the extrastriate cortex of the cat (Regan & Cynader 1979).

Selectivity for changing size was found by Saito and colleagues (1986) in about 15% of MST neurons. They found that this response to changing size occurs within small subregions of the receptive field, thus showing that the mechanism for detecting changing size is reproduced continuously throughout the receptive field, as was the sensitivity for rotation described above. Control experiments indicated that the neurons respond poorly, if at all, to unidirectional motion of the stimulus in the receptive field and that they did not respond to the changing light flux that accompanies changing size. These neurons seem to require simultaneous stimulation by motion in at least two different directions before emitting a substantial response. Thus the responses to changing size appear to represent another convergence of low-level motion cues to create higher level representations of important environmental events. Adaptation of neurons sensitive to changing size could account for the psychophysical observation that perception of changing size can be adapted independently of perception of simple linear stimulus motion (Regan & Beverly 1978a,b).

Opponent Vector Organization

Another major transformation in neuronal responses to visual motion is represented by "opponent vector" organization of receptive fields in area 7a (Motter & Mountcastle 1981) and in STP (Bruce et al 1981). Although the precise connections of 7a and STP with motion-related areas such as MT and MST are not yet established, some anatomical evidence suggests that both 7a and STP receive inputs from MST (see Desimone & Ungerleider 1986). Opponent vector neurons have large, frequently bilateral receptive fields in which the preferred direction of motion varies systematically from subregion to subregion within the receptive field. In one population of neurons the preferred directions are oriented relative to the fixation point so that each neuron responds best to motion toward the center of gaze regardless of the position of the stimulus in the receptive field. Another population responds best to motion away from the center of gaze. This type of response has not been observed in earlier levels of the motion pathway. Neurons with opponent vector responses (Motter & Mountcastle 1981, Mountcastle et al 1984) are likely to respond maximally to radial flow of the entire visual field either away from or toward the fixation point. Radial flow of this nature generally occurs as the animal moves through the environment, and such optical flow signals are known to be useful for perception of depth and as a cue for postural adjustment (see review by Nakayama 1985).

Motion-Related Behavioral Deficits

Recent studies employing small chemical lesions restricted to MT provided the first behavioral evidence for localization of motion analysis within extrastriate cortex. Newsome and colleagues (1985a) made small lesions in an identified portion of the topographic representation in MT, and observed selective (though transient) deficits in the monkey's use of visual motion to guide eye movements. When a moving target appeared in the affected portion of the visual field, both smooth pursuit eye movements and saccadic eye movements were impaired in a manner suggesting underestimation of the speed of the moving target. The affected portion of the visual field corresponded to the topographic location at which the MT lesion was made; eye movements to targets in other regions were normal. In contrast, saccadic eye movements to stationary targets were normal at all points in the visual field including the region in which responses to moving targets were impaired. These effects of MT lesions were clearly different from those of restricted V1 lesions, which impaired eye movements to both moving and stationary targets (e.g. Mohler & Wurtz 1977, Goldberg et al 1982, Segraves et al 1983, Newsome et al 1985b).

These results were extended in a subsequent series of experiments in which Dürsteler and colleagues (1986) tested the effects of MT lesions on the pursuit of visual targets that were stabilized on the retina by using the monkey's eye position to move the target. MT lesions had no effect when the monkey pursued a target with a constant position error of one degree from the fovea. If, however, the experimenters added a constant velocity error to the stabilized image, MT lesions impaired the increase in eye speed with which the monkey normally responded to such velocity errors. Again, MT lesions had selective effects on the monkey's response to motion while leaving its response to static position unimpaired. (The implications of these results for visual control of the pursuit system are considered by Lisberger et al 1987, this volume.)

Wilson and colleagues (1977, 1979) examined the effects of suction ablations of MT in the bushbaby and the macaque. They concluded that animals with MT lesions were deficient in their ability to search visual space for behaviorally significant visual stimuli. While several aspects of these experiments pose difficulties for interpretation of the results (incomplete lesions, involvement of the optic radiation, lack of eye movement control), deficits in visuo-spatial search are broadly consistent with the known behavioral functions of parietal areas that are likely to receive inputs from the motion pathway. In another lesion study of cortex in the superior temporal sulcus, Collin & Cowey (1980) failed to find post-lesion deficits in movement detection thresholds. However, the lesions did not include all of MT, and the detection task may not have specifically tested the contributions of MT because it required no judgment about direction or speed of motion, parameters for which MT neurons are notably selective.

While lesion experiments are subject to serious pitfalls, the success of some of the experiments described above suggests that a more thorough behavioral analysis of the motion pathway may now be feasible. The discovery of several physiological transformations at higher levels of the pathway, coupled with apparent psychophysical correlates, has led to specific hypotheses about the functional significance of these transformations. It is reasonable to anticipate that lesions of motion-related cortical areas in animals trained on appropriate psychophysical tasks will greatly clarify how physiological processing in these areas contributes to visual perception.

THE COLOR AND FORM PATHWAY

The color and form pathway has been less intensely studied than the motion pathway, but significant transformations of visual information have nonetheless been found within it. Several aspects of these transformations are similar to those seen in the motion pathway. For example, selectivities for basic stimulus parameters such as wavelength (deMonasterio & Schien 1982) or orientation (Desimone et al 1985) do not appear to increase along the color and form pathway. Also, retinotopic specificity decreases in successive levels of the color and form pathway: Compared to receptive fields in V1, fields in V4 are about 30 times larger (Van Essen & Zeki 1978, Maguire & Baizer 1984, Desimone & Gross 1985), and fields in AIT are well over 100 times larger (Desimone & Gross 1979). The similarities between the pathways suggest that the pathways use common computational strategies for processing information. In the sections that follow we consider evidence that the transformations along this pathway, like those in the motion pathway, are directed toward generating properties different and more complex than those that exist in earlier levels.

Although the transformations performed by the color and form pathway and the motion pathway may prove to be similar, the types of information that they process plainly differ. While color sensitivity is not readily observed in the motion pathway, high proportions of color selective neurons have been described in the blobs in V1, in the thin stripes in V2, and in V4. The percentage of color selective neurons reported for V4 has varied greatly between laboratories, due largely to differing criteria (see Desimone et al 1985, Van Essen 1985) and perhaps partly to the existence of distinct subdivisions within V4 (Zeki 1971, 1977, 1983c, Van Essen & Zeki 1978, Schein et al 1982, Maguire & Baizer 1984). While it was once thought that this area is primarily devoted to color analysis, it is now apparent that many V4 neurons lack color selectivity and are instead more sensitive to orientation or spatial frequency (Schein et al 1982, Desimone et al 1985). Little is known about the color specificity of neurons in the later stages of the color and form pathway to which V4 projects. Although neurons in AIT have been reported to distinguish different colors (Mikami & Kubota 1980, Fuster & Jervey 1982, Desimone et al 1984), these observations have generally been incidental, and no attempts have been made to control carefully for luminance or spectral content.

This pathway has been associated with form processing based in part on the orientation and spatial frequency selectivities that are prevalent in V1, V2, and V4. A continuation of form analysis in the inferotemporal areas is suggested by both neurophysiological and behavioral studies. Orientation selective neurons are found in PIT and AIT (Gross et al 1972, Desimone et al 1984), although their proportions are not yet established. These areas also contain neurons that are selective for other parameters relevant to the analysis of form, such as length, width, size, shape, and texture. As discussed above, lesions of the inferotemporal cortex cause deficits in discriminations of shape, size, or pattern.

Surround Mechanisms

Several distinctive physiological properties that exist in the color and form pathway appear to result from antagonistic surrounds. Receptive fields of V4 neurons have large surrounds that can greatly influence the responses to stimuli within the classically defined receptive field (Desimone & Schein 1983, Moran et al 1983, Schein et al 1983). Desimone and collaborators (1985) found that surround effects are often antagonistic and that they can show specificity for either the color or the spatial frequency of the surround stimulus. DeYoe and colleagues (1986) recently found a similar effect in V2, where neuronal responses are inhibited by patterns of oriented line segments in their surrounds only when the orientation of the segments matches the neuron's preferred orientation. Neurons with antagonistic, stimulus specific surrounds may contribute to discriminating figure from ground by responding to patterned objects on different backgrounds but not to uniformly patterned fields.

Von der Heydt et al (1984) tested the responses of neurons in macaque V1 and V2 to figures in which humans perceive illusory contours. Such figures are constructed so that an illusory edge or bar appears to span a region that is in fact blank. These investigators reported that when figures were positioned so that only the illusory contour entered a neuron's receptive field, some V2 neurons responded as though the contour were a real edge. V1 neurons did not respond to this stimulation. Although responses to an illusory contour depend on inputs from regions surrounding the classical receptive field, these inputs must differ from those mediating the surround effects described above because they are capable of driving neurons without direct stimulation of the classical receptive field. Responses to illusory contours therefore suggest that surround mechanisms may have greater scope than previously recognized.

Surround antagonism in the color and form pathway corresponds closely to that described in the motion pathway and probably reflects a common computational strategy. Because the magnitude of surround effects has not been extensively examined in V1, it is not yet clear whether they play a major role in striate cortex as well. Although most research has focused on responses to stimulation of the classical receptive field, understanding the role of surrounds in shaping the visual response will undoubtedly represent an important area of study in the future.

Color Constancy

A perceptual capability that may depend on surround mechanisms is color constancy. In natural viewing conditions objects appear to have much the same color even when changes in illumination cause them to reflect light of different wavelength compositions (Jameson & Hurvich 1959, 1977, Land & McCann 1971, McCann et al 1976, Land 1977). The power of this compensating mechanism has been vividly demonstrated by Land using a montage of differently colored patches that he calls a "Mondrian" (see McCann et al 1976). The Mondrian is first illuminated with three projectors that emit narrow bandwidths of short, medium, and long wavelength light. Under these conditions the radiance of two of the patches, say a yellow and a green one, is measured to determine how much of each of the three illuminating wavelengths they reflect. The intensities of the three projectors are then individually adjusted so that the mixture of wavelengths reflected by the yellow patch precisely matches the mixture initially reflected by the green patch. If perceived color depended strictly on wavelength composition, the original yellow patch would now appear green. Instead, it still appears yellowish. This remarkable consistency in the perceived color does not occur if the surrounding colors are masked off and the patch is viewed in isolation. Thus, the perceived color of a surface is not a simple function of the wavelengths it reflects, but depends greatly on the wavelengths reflected by surrounding surfaces. Although this phenomenon has been referred to as color constancy, it should be noted that the compensating mechanism is not perfect. The perceived color may change somewhat under different lighting conditions, although it retains its basic hue.

Neurophysiological studies of visual cortex have largely neglected the distinction between wavelength composition and perceived color, leaving the neuronal basis of color constancy obscure. The mechanisms that generate this color compensation could in principle exist within the retina itself (Land 1977), but behavioral observations suggest that color constancy probably depends on the cerebral cortex (Land et al 1983). Recently Zeki (1980, 1983a,b) used a Mondrian stimulus to search for neuronal correlates of color constancy in macaque V1 and V4, and reported that neurons in V1 responded to the simple wavelength composition of stimuli, while neurons in V4 appeared to signal the color perceived by a human observer. These intriguing observations are consistent with the notion that color constancy depends in part on transformations of signals that occur along the color and form pathway. However, there are alternative explanations for these results. The effects described in V4 depended on stimulating regions surrounding the classical receptive field (Zeki 1983a), but the surrounds of receptive fields in V1 were apparently not stimulated in an equivalent manner. Mondrian stimuli of fixed size were used to test neurons in both V1 and V4. Because receptive fields in V1 cover only about 1/30 the area of those in V4, a single patch of the Mondrian may have covered a critical region of the surround as well as the classical receptive field of the V1 neurons. The observed differences between V1 and V4 may therefore simply reflect differences in the spatial extent of receptive fields and their surrounds, rather than the appearance of a new property at the level of V4. Although the results of these experiments are of great interest, they must be interpreted cautiously until such issues are resolved.

Wild and colleagues (1985) made lesions in macaque monkeys that involved a restricted region of cortex including V4. The animals were impaired at discriminating colors under different illumination conditions, and these workers suggested that V4 is necessary for color constancy. While this is an attractive hypothesis, the data are not yet conclusive. V4 lesions also affect simple hue discrimination (Heywood & Cowey 1985), and the extent to which this impairment may affect performance on color constancy tasks is not known. Although much remains to be learned about the cortical contributions to color constancy, the recognition of a distinction between the wavelength composition and perceived color is likely to prove important in interpreting the neural representation of color in the later stages of visual cortex.

Face Selective Neurons

The later stages of the color and form pathway in PIT and AIT include many neurons whose stimulus preference is not immediately obvious: 40% of neurons in AIT have been reported to respond in a nonspecific manner to all visual stimuli (Desimone et al 1984). However, a comparable fraction appears to respond preferentially to particular complex objects, such as patterns, hands, or faces (Gross et al 1972, 1979, Rolls et al 1977, Ridley et al 1977, Bruce et al 1981, Rolls 1984, Perrett et al 1982, 1984, 1985, Baylis et al 1985). Neurons responding preferentially to complex visual objects have not been reported in the early stages of the color and form pathway. This implies a substantial transformation of the signals present at early levels and suggests that the inferotemporal cortex contributes to later stages of form analysis.

Among neurons that appear specific for complex objects, those selective for faces have attracted considerable attention. These neurons represent a small fraction of cells in the inferotemporal cortex (Desimone et al 1984), but they are apparently concentrated in a region in AIT near the fundus of the superior temporal sulcus, where as many as a quarter to a third of responsive neurons have been categorized as face selective (Desimone et al 1984, Rolls 1984, Perrett et al 1985). Neurons have been identified as face selective on the basis of a number of distinctive properties. Face selective neurons respond more strongly to faces than to other complex stimuli and generally do not respond to simple stimuli such as bars or gratings (Bruce et al 1981, Desimone et al 1984). They are sensitive to manipulations that affect the ability to identify a face, such as removing different facial features (Perrett et al 1982, Desimone et al 1984), yet most show no pronounced sensitivity to changes that have little effect on recognition (e.g. color, size, orientation, distance, spatial frequency content, direction of motion, or position within their receptive fields) (see Perrett et al 1982, Desimone et al 1984, Rolls 1984, Rolls et al 1985). Although face selective neurons generally respond to a variety of faces, the majority respond more or less strongly to particular faces (Perrett et al 1984, 1985, Rolls 1984, Baylis et al 1985).

Because faces are such complex patterns, it is difficult to prove conclusively that a neuron encodes information that is specifically related to faces. It is possible that these neurons are signaling information about either a broader class of complex objects or some simple pattern that is common to faces. However, demonstration that these neurons are sensitive to a variety of changes that adversely affect the identification of a face while being relatively insensitive to as many as seven other manipulations that do not alter recognition, presents a strong case that their activity conveys face related, and perhaps face specific information. Further characterization of the response properties of these neurons using systematic and repeatable stimulus manipulations will strengthen these findings. There is currently no physiological understanding of intermediate transformations by which face selective responses might be generated from the far simpler response properties described in early stages. Investigation of the processes that occur between these levels may improve our understanding of the functional role of these neurons.

Face selective cells are obvious candidates for contributing to the discrimination and identification of faces or individuals. The existence of these neurons does not, however, necessitate a theory of perception based on "grandmother" cells that are specific to one particular face (see Barlow 1972). There are simply too few neurons in the brain to represent the entire visual world in this way (Ballard et al 1983). In addition, computational studies have shown that cortex-like mechanisms can identify complex patterns without using individual cells to signal particular objects (see Ballard 1986). In systems of this sort, a neuron might contribute to object identification by signaling a class of frequently encountered complex objects as opposed to specific instances of those objects. How the nervous system represents and recognizes particular visual objects remains one of the most formidable questions facing integrative neurophysiology.

EXTRARETINAL INPUTS

To date most investigations of information processing in visual cortex have been dominated by a strictly sensory perspective. Neurons have been described primarily as extracting and filtering different types of information from the visual scene. While this approach has contributed most of our current understanding of cortical function, neuronal activity in visual cortex is also influenced by other inputs that are not of direct retinal origin. Early studies of extraretinal inputs to visual neurons involved analysis of an "enhancement effect" in which visual responses were modulated by the animal's use of the stimulus: A neuron's response to a particular stimulus was stronger when a monkey attended to that stimulus than when the animal was attending elsewhere. Pronounced enhancement effects were observed in parietal cortex (Goldberg & Robinson 1977, Yin & Mountcastle 1977, 1978, Robinson et al 1978, Bushnell et al 1981), the frontal eye fields (Wurtz & Mohler 1976a, Goldberg & Bushnell 1981), and the superior colliculus (Goldberg & Wurtz 1972, Wurtz & Mohler 1976b).

Initial experiments in the early stages of macaque visual cortex, however, revealed no specific enhancement effects. Those neurons that did show an effect were modulated when the animal attended to any visual stimulus, thus suggesting that the change might simply reflect a level of arousal rather than a process specifically related to visual attention (see Wurtz et al 1980). Several studies support the idea that neurons in V1 are not strongly influenced by extraretinal inputs (Wurtz & Mohler 1976a, Hänny & Schiller 1986), but it has recently become clear that intermediate levels of visual cortex, like later stages, receive widespread and robust extraretinal inputs. For example, V4, only two stages removed from V1, has been shown to contain neurons with spatially selective enhancement effects similar to those previously demonstrated in the parietal cortex (Fischer & Boch 1981, 1985). It has also become obvious that extraretinal inputs are not restricted to modulating visual responses. Although a thorough survey of these inputs is beyond the scope of this review, the following examples illustrate the prevalence and variety of extraretinal effects that can be found in the extrastriate cortex.

Moran & Desimone (1985) showed that neurons in V4 are influenced in a highly specific manner by attention. They found that if a red bar and a green bar were simultaneously flashed within the receptive field of a neuron that preferred red stimuli, the neuron responded well when the animal attended to the red bar but not when it attended to the non-preferred green bar. These results suggest that attention can, in effect, select a particular stimulus to be encoded by a neuron's response even though competing stimuli are simultaneously present in its receptive field. Extraretinal effects of this sort are common in V4: More than half the neurons tested showed attentional specificity. Other studies of this region of cortex have also found effects of attention that are specific to particular aspects of the stimuli (Braitman 1984, Hochstein & Maunsell 1985, Hänny & Schiller 1986).

A different type of extraretinal input was demonstrated by Fuster & Jervey (1982), who trained an animal to do a delayed match-to-sample task. At the start of each trial the animal was briefly cued with a color that he was required to remember during a delay period of more than 15 sec. At the end of the delay, several colors were presented and the animal had to select the one that had been cued. Among the neurons recorded in AIT were some whose rate of discharge during the delay depended on which color was cued. This difference in activity was not immediately related to a sensory input because no stimulus was present during the delay period. The activity during the delay appeared to encode task-specific information about the stimulus to which the animal should respond rather than signaling the presence of a visual stimulus.

Further indication that neurons in the visual cortex can encode task-specific information was found by Hänny et al (1986), who tested neurons in V4 by using a variation of the match-to-sample task in which an animal was required to match orientations. In these tests the animal was cued non-visually by having him feel the orientation of a grooved plate that he could not see. More than half the neurons in V4 responded in an orientation-selective fashion to the cue stimulus, although this information was not provided visually. Other experiments showed that these signals were also present if the orientation was cued visually before the start of each trial, and that they were not present when the animal felt the grooved plate while performing a task in which its orientation had no relevance. Thus it appears that many neurons in V4, like those in AIT, can convey signals that are not visual in any direct sense, but instead are related to a particular task.

In the motion pathway, preliminary investigations have found little or no evidence for extraretinal inputs to MT (Newsome & Wurtz 1981,

Wurtz et al 1984), but several such inputs exist in MST. Among the MST neurons exhibiting extraretinal effects, the neurons with signals related to pursuit eye movements are the best characterized. Visual tracking neurons were first reported in the parietal lobe by Hyvärinen & Poranen (1974) and by Mountcastle and colleagues (Mountcastle et al 1975; Lynch et al 1977). Robinson and colleagues (1978) found that many tracking neurons received strong visual inputs and suggested that the tracking responses could be explained solely on the basis of retinal signals. This was shown to be the case for many tracking neurons in area 7a (Sakata et al 1983), but a population of neurons located in and around MST in the depths of the superior temporal sulcus receive an extraretinal input related to pursuit eye movements in addition to their visual inputs (Sakata et al 1983, 1985, Kawano et al 1984, Kawano & Sasaki 1984, Wurtz et al 1984). This pursuit signal cannot be explained by passive visual properties because it exists during short intervals of pursuit in total darkness (Sakata et al 1983) and during pursuit of stabilized retinal images (Wurtz & Newsome 1985). Rather, the pursuit signal results from extraretinal inputs and may provide an internal representation of eye velocity (Kawano et al 1984).

A final example of extraretinal influences on visual responses are recent findings in area 7a that suggest a mechanism for representing the visual world in stable, head-centered coordinates. In earlier stages of the visual pathway, visual space is represented in retinotopic coordinates so that the cortical representation of a stationary object changes with each eye movement. Most motor acts, however, are directed toward locations in space. Andersen & Mountcastle (1983) have examined neurons in area 7a that are visually responsive, with discrete (though large) receptive fields in which the strength of the visual response depends on the direction in which the monkey's eyes point. Because a neuron will not respond unless the eyes point in a particular direction *and* a stimulus falls on a particular retinal region, a neuron will be strongly driven only by visual stimuli within a restricted range of locations in head-centered space. By combining the activity of neurons that respond to a common location in space but for a wide range of eye positions, an eye-position independent representation of visual space can be achieved (Andersen et al 1985). As yet there is no evidence that neuronal activity is combined in this manner in single neurons, but the aggregate activity of neurons in area 7a encodes the necessary information.

Collectively, these examples show that extraretinal signals exist in many extrastriate visual areas and appear to play a role that goes beyond simple modulation of visual signals. It is likely that extraretinal signals represent a substantial component of the information present in extrastriate cortex of the behaving animal. While the anesthetized preparation has certain advantages, many cortical signals are observable only in alert animals. Further insight about this largely unexplored dimension of cortical activity is likely to be essential for understanding the function of extrastriate visual cortex.

CONCLUDING COMMENTS

Several new insights about information flow and processing in the extrastriate cortex have emerged in the last few years. The current notion of cortical streams of visual processing is a modification of the proposal by Zeki (1975, 1978) that extrastriate areas operate in parallel, with each area processing a different type of information. Evidence now suggests that the organization of cortical visual areas encompasses both parallel and serial relationships. Visual cortex appears to contain parallel streams of processing that consist of serially connected areas, with each area representing a different level of processing. The case for serial processing in extrastriate cortex is supported both by the pattern of ascending and descending anatomical connections in the cortical hierarchy (Figure 4) and the increasingly complex properties of neurons at successive stages of each stream of processing.

While the concept of streams introduces a serial aspect to the organization of extrastriate visual areas, it would be incorrect to view processing within a stream as a strictly sequential set of operations. Anatomical experiments indicate that processing within each stream is not completely serial. More elaborate forms of processing are suggested both by the projections that skip levels within a stream and by the ubiquitous feedback connections. Anatomical and physiological evidence for interconnections between the streams also points to more complex interactions. While much remains to be learned about the functional implications of streams of processing, the identification of a level of organization between that of individual cortical areas and the entire visual cortex should help considerably in evaluating the cortical processes underlying vision.

A major need for future research is to identify additional transformations of visual information in the extrastriate cortex. The transformations that have been observed to date probably represent only a small fraction of the total. Impetus for discovery of new physiological transformations is likely to come from visual psychophysics and computational neuroscience. Studies of properties such as pattern direction selectivity in MT illustrate the potential for contributions from psychophysical and computational approaches. Psychophysics yields important information concerning the response of the entire visual system to complex stimuli, and can provide valuable hints about the types of transformations and signals that occur in visual cortex. Computational studies attempt to specify rigorously the computational steps required to implement a particular capability, and can suggest neuronal operations that might be necessary. Simulations of such operations may help us deduce the significance of signals encoded in cortical neurons. We expect that interactions with these disciplines will become increasingly valuable as physiologists pursue higher-order aspects of integrative neural function.

Finally, investigation of extraretinal signals in the visual cortex is likely to become an area of particular interest in the future. Robust extraretinal signals are apparently common in most extrastriate visual areas, suggesting that they represent a fundamental aspect of cortical processing. We currently have only the most basic understanding of these extraretinal inputs. Many questions remain unanswered about the origin of extraretinal signals and their flow within visual cortex. More intriguing still are questions about how extraretinal signals interact with visual signals in cortex. Although systematic investigation of these interactions poses many challenges, both in the execution of experiments and in the interpretation of results, information about this aspect of cortical function may ultimately prove essential to understanding the neural processes leading to visual perception.

Acknowledgments

We are indebted to Drs. R. A. Andersen, R. Desimone, R. A. Eatock, S. G. Lisberger, W. H. Merigan, T. Pasternak, G. Sclar, D. C. Van Essen, and R. H. Wurtz for helpful comments and suggestions. Preparation of this review was supported by Center Grant EY-01319 to the Center for Visual Science at the University of Rochester and EY-05603 and a Sloan Research Fellowship to W. T. N.

Literature Cited

Adelson, M., Movshon, J. A. 1982. Phenomenal coherence of moving visual patterns. *Nature* 300: 523–25

Albright, T. D. 1984. Direction and orientation selectivity of neurons in visual area MT of the macaque. *J. Neurophysiol.* 52: 1106–30

Allman, J. M., Kaas, J. H. 1971. A representation of the visual field in the caudal third of the middle temporal gyrus of the owl monkey (*Aotus trivirgatus*). *Brain Res.* 31: 85–105

Allman, J. M., Kaas, J. H. 1974a. The organization of the second visual area (V II) in the owl monkey: A second order transformation of the visual hemifield. *Brain Res.* 76: 247–65

Allman, J. M., Kaas, J. H. 1974b. A crescent-shaped cortical visual area surrounding the middle temporal area (MT) in the owl monkey (*Aotus trivirgatus*). *Brain Res.* 81: 199–213

Allman, J. M., Kaas, J. H. 1975. The dorsomedial cortical visual area: A third tier area in the occipital lobe of the owl monkey (*Aotus trivirgatus*). *Brain Res.* 100: 473–87

Allman, J. M., Kaas, J. H. 1976. Representation of the visual field on the medial wall of occipital-parietal cortex in the owl monkey. *Science* 191: 572–75

Allman, J. M., Meizin, F., McGuinness, E.

1985a. Direction and velocity-specific responses from beyond the classical receptive field in the middle temporal visual area (MT). *Perception* 14: 105–26

Allman, J. M., Meizin, F., McGuinness, E. 1985b. Stimulus specific responses from beyond the classical receptive field: Neurophysiological mechanisms for local-global comparisons in visual neurons. *Ann. Rev. Neurosci.* 8: 407–30

Andersen, R. A., Essick, G. K., Siegel, R. M. 1985. The encoding of spatial location by posterior parietal neurons. *Science* 230: 456–58

Andersen, R. A., Mountcastle, V. B. 1983. The influence of the angle of gaze upon the excitability of the light-sensitive neurons of the posterior parietal cortex. *J. Neurosci.* 3: 532–48

Baker, J. F., Petersen, S. E., Newsome, W. T., Allman, J. M. 1981. Visual response properties of neurons in four extrastriate visual areas of the owl monkey (*Aotus trivirgatus*). *J. Neurophysiol.* 45: 397–416

Barlow, H. 1972. Single units and sensation: A neuron doctrine for perceptual psychology? *Perception* 1: 371–94

Ballard, D. H. 1986. Cortical connections and parallel processing: Structure and function. *Behav. Brain Sci.* 9: 67–120

Ballard, D. H., Hinton, G. E., Sejnowski, T. J. 1983. Parallel visual computation. *Nature* 306: 21–26

Baylis, G. C., Rolls, E. T., Leonard, C. M. 1985. Selectivity between faces in the responses of a population of neurons in the cortex in the superior temporal sulcus of the monkey. *Brain Res.* 342: 91–102

Bisti, S., Carmignoto, G., Galli, L., Maffei, L. 1985. Spatial-frequency characteristics of neurons of area 18 in the cat: Dependence on the velocity of the visual stimulus. *J. Physiol.* 359: 259–68

Blasdel, G. G., Fitzpatrick, D. 1984. Physiological organization of layer 4 in macaque striate cortex. *J. Neurosci.* 4: 880–95

Braitman, D. J. 1984. Activity of neurons in monkey posterior temporal cortex during multidimensional visual discrimination tasks. *Brain Res.* 307: 17–28

Bruce, C. J., Desimone, R., Gross, C. G. 1981. Visual properties of neurons in a polysensory area in superior temporal sulcus of the macaque. *J. Neurophysiol.* 46: 369–84

Buchbinder, S., Dixon, B., Hwang, Y. W., May, J. G., Glickstein, M. 1980. The effects of cortical lesions on visual guidance of the hand. *Soc. Neurosci. Abstr.* 6: 675

Burkhalter, A., Felleman, D. J., Newsome, W. T., Van Essen, D. C. 1986. Anatomical and physiological asymmetries related to visual areas V3 and VP in macaque extrastriate cortex. *Vision Res.* 26: 63–80

Bushnell, M. C., Goldberg, M. E., Robinson, D. L. 1981. Behavioral enhancement of visual responses in monkey cerebral cortex. I. Modulation in posterior parietal cortex related to selective visual attention. *J. Neurophysiol.* 46: 755–72

Cavanagh, P., Tyler, C. W., Favreau, O. E. 1984. Perceived velocity of moving chromatic gratings. *J. Opt. Soc. Am.* 1: 893–99

Collin, N. G., Cowey, A. 1980. The effect of ablation of frontal eye-fields and superior colliculi on visual stability and movement discrimination in rhesus monkeys. *Exp. Brain Res.* 40: 251–60

Cowey, A. 1979. Cortical maps and visual perception. *Q. J. Exp. Psychol.* 31: 1–17

Cragg, B. G. 1969. The topography of the afferent projections in circumstriate visual cortex of the monkey studied by the Nauta method. *Vision Res.* 9: 733–47

Damasio, A. R., Benton, A. L. 1979. Impairments of hand movements under visual guidance. *Neurology* 29: 170–8

Damasio, A. R., Damasio, H., Van Hoesen, G. W. 1982. Prosopagnosia: Anatomical basis and behavioral mechanisms. *Neurology* 32: 331–41

Dean, P. 1976. Effects of inferotemporal lesions on the behavior of monkeys. *Psychol. Bull.* 83: 41–71

Dean, P. 1982. Visual behavior in monkeys with inferotemporal lesions. See Ingle et al 1982, pp. 587–628

deMonasterio, F. M., Schein, S. J. 1982. Spectral bandwidths of color-opponent cells of geniculocortical pathway of macaque monkeys. *J. Neurophysiol.* 47: 214–24

Derrington, A. M., Lennie, P. 1984. Spatial and temporal contrast sensitivities of neurons in lateral geniculate nucleus of macaque. *J. Physiol.* 357: 219–40

Derrington, A. M., Krauskopf, J., Lennie, P. 1984. Chromatic mechanisms in lateral geniculate nucleus of macaque. *J. Physiol.* 357: 241–65

Desimone, R., Albright, T. D., Gross, C. G., Bruce, C. 1984. Stimulus-selective properties of inferior temporal neurons in the macaque. *J. Neurophysiol.* 4: 2051–62

Desimone, R., Fleming, J., Gross, C. G. 1980. Prestriate afferents to inferotemporal cortex: An HRP study. *Brain Res.* 184: 41–55

Desimone, R., Gross, C. G. 1979. Visual areas in the temporal cortex of the macaque. *Brain Res.* 178: 363–80

Desimone, R., Schein, S. J. 1983. Receptive field properties of neurons in visual area

V4 of the macaque. *Soc. Neurosci. Abstr.* 9: 153

Desimone, R., Schein, S. J., Moran, J., Ungerleider, L. G. 1985. Contour, color and shape analysis beyond the striate cortex. *Vision Res.* 25: 441–52

Desimone, R., Ungerleider L. G. 1986. Multiple visual areas in the caudal superior temporal sulcus of the macaque. *J. Comp. Neurol.* 248: 164–89

DeYoe, E. A., Knierim, J., Sagi, D., Julesz, B., Van Essen, D. C. 1986. Single unit responses to static and dynamic texture patterns in macaque V2 and V1 cortex. *Invest. Ophthalmol. Vis. Sci.* 27: 18

DeYoe, E. A., Van Essen, D. C. 1985. Segregation of efferent connections and receptive field properties in visual area V2 of the macaque. *Nature* 317: 58–61

Dow, B. M. 1974. Functional classes of cells and their laminar distribution in monkey visual cortex. *J. Neurophysiol.* 37: 927–46

Dubner, R., Zeki, S. M. 1971. Response properties and receptive fields of cells in an anatomically defined region of the superior temporal sulcus. *Brain Res.* 35: 528–32

Dürsteler, M., Wurtz, R. H., Newsome, W. T. 1986. Directional and retinotopic pursuit deficits following lesions of the foveal representation within the superior temporal sulcus of the macaque monkey. *J. Neurophysiol.* In press

Edelman, G. M., Gall, W. E., Cowan, W. M., eds. 1984. *Dynamic Aspects of Neocortical Function.* New York: Wiley

Felleman, D. J., Kaas, J. H. 1984. Receptive-field properties of neurons in middle temporal visual area (MT) of owl monkeys. *J. Neurophysiol* 52: 488–513

Fischer, B., Boch, R. 1981. Selection of visual targets activates prelunate cortical cells in rhesus monkey. *Exp. Brain Res.* 41: 431–33

Fischer, B., Boch, R. 1985. Peripheral attention versus central fixation: Modulation of the visual activity of prelunate cortical cells of the rhesus monkey. *Brain Res.* 345: 111–23

Fitzpatrick, D., Lund, J. S., Blasdel, G. G. 1985. Intrinsic connections of macaque striate cortex: Afferent and efferent connections of lamina 4C. *J. Neurosci.* 5: 3329–49

Foster, K. H., Gaska, J. P., Nagler, M., Pollen, D. A. 1985. Spatial and temporal frequency selectivity of neurones in visual cortical areas V1 and V2 of the macaque monkey. *J. Physiol.* 365: 331–63

Friedman, D. P. 1983. Laminar patterns of termination of cortico-cortical afferents in the somatosensory system. *Brain Res.* 273: 147–51

Frost, B. J., Nakayama, K. 1983. Single visual neurons code opposing motion independent of direction. *Science* 220: 774–75

Frost, B. J., Scilley, P. L., Wong, S. C. P. 1981. Moving background patterns reveal double-opponency of directionally specific pigeon tectal neurons. *Exp. Brain Res.* 43: 173–85

Fuster, J. M., Jervey, J. P. 1982. Neuronal firing in the inferotemporal cortex of the monkey in a visual memory task. *J. Neurophysiol.* 2: 361–75

Gattass, R., Gross, C. G. 1981. Visual topography of striate projection zone (MT) in posterior superior temporal sulcus of the macaque. *J. Neurophysiol.* 46: 621–38

Goldberg, M. E., Bruce, C. J., Ungerleider, L., Mishkin, M. 1982. Role of the striate cortex in generation of smooth pursuit eye movements. *Neurology* 12: 113

Goldberg, M. E., Bushnell, M. C. 1981. Behavioral enhancement of visual responses in monkey cerebral cortex. II. Modulation in frontal eye field specifically related to saccades. *J. Neurophysiol.* 46: 773–87

Goldberg, M. E., Robinson, D. L. 1977. Visual responses of neurons in monkey inferior parietal lobule: The physiological substrate of attention and neglect. *Neurology* 27: 350

Goldberg, M. E., Wurtz, R. H. 1972. Activity of superior colliculus in behaving monkey. II. Effect of attention on neuronal responses. *J. Neurophysiol.* 35: 560–74

Gross, C. G. 1973a. Visual functions of inferotemporal cortex. In *Handb. Sensory Physiol.* 7(3): 451–82

Gross, C. G. 1973b. Inferotemporal cortex and vision. *Prog. Physiol. Psychol.* 5: 77–123

Gross, C. G., Bender, D. B., Gerstein, G. L. 1979. Activity of inferior temporal neurons in behaving monkeys. *Neuropsychology* 17: 215–29

Gross, C. G., Bruce, C. J., Desimone, R., Fleming, J., Gattass, R. 1981. Cortical visual areas of the temporal lobe. See Woolsey 1981, 2: 187–216

Gross, C. G., Rocha-Miranda, C. E., Bender, D. B. 1972. Visual properties of neurons in inferotemporal cortex of the macaque. *J. Neurophysiol.* 35: 96–111

Hänny, P. E., Maunsell, J. H. R., Schiller, P. H. 1986. State dependent activity in monkey visual cortex: II. Visual and non-visual factors in V4. Submitted

Hänny, P. E., Schiller, P. H. 1986. State dependent activity in monkey visual cortex: I. Effects in V1 and V4. Submitted

Heywood, C. A., Cowey, A. 1985. Disturbances of pattern and hue discrimination following removal of the "colour" area in primates. *Neurosci. Lett.* 21: S11

Hildreth, E. C., Koch, C. 1987. The analysis of visual motion: From computational theory to neuronal mechanisms. *Ann. Rev. Neurosci.* 10: 477–533

Hochstein, S., Maunsell, J. H. R. 1985. Dimensional attention effects in the responses of V4 neurons of the macaque monkeys. *Soc. Neurosci. Abstr.* 11: 1244

Holub, R. A., Martin-Gibson, M. 1981. Response of visual cortical neurons of the cat to moving sinusoidal gratings: Response-contrast functions and spatiotemporal interactions. *J. Neurophysiol.* 46: 1244–59

Horton, J. C., Hubel, D. H. 1980. Cytochrome oxidase stain preferentially labels intersections of ocular dominance and vertical orientation columns in macaque striate cortex. *Soc. Neurosci. Abstr.* 6: 315

Hubel, D. H., Livingstone, M. S. 1985. Complex-unoriented cells in a subregion of primate area 18. *Nature* 315: 325–27

Hubel, D. H., Wiesel, T. N. 1972. Laminar and columnar distribution of geniculocortical fibers in macaque monkey. *J. Comp. Neurol.* 146: 421–50

Hubel, D. H., Wiesel, T. N. 1977. Functional architecture of macaque monkey visual cortex. *Proc. R. Soc. London Ser. B* 198: 1–59

Humphrey, A. L., Hendrickson, A. E. 1980. Radial zones of high metabolic activity in squirrel monkey striate cortex. *Soc. Neurosci. Abstr.* 6: 315

Hyvärinen, J., Poranen, A. 1974. Function of parietal associative area 7 as revealed from cellular discharges in alert monkeys. *Brain* 97: 673–92

Ingle, D. J., Goodale, M. A., Mansfield, R. J. W. 1982. *Analysis of Visual Behavior*. Cambridge, Mass.: MIT Press

Iwai, E. 1981. Visual mechanisms in the temporal and prestriate association cortices of the monkey. *Adv. Physiol. Sci.* 17: 279–86

Iwai, E. 1985. Neuropsychological basis of pattern vision in macaque monkeys. *Vision Res.* 25: 425–39

Jameson, D., Hurvich, L. M. 1959. Perceived color and its dependence on focal, surrounding, and preceding stimulus variables. *J. Opt. Soc. Am.* 49: 890–98

Jameson, D., Hurvich, L. M. 1977. Color adaptation: Sensitivity, contrast, afterimages. In *Handb. Sensory Physiol.* 7(4): 568–81

Jeeves, M. A., Milner, A. D., Perrett, D. I., Smith, P. A. J. 1983. Visual cells responsive to direction of motion and stimulus form in the anterior superior temporal sulcus of the macaque monkey. *Proc. Roy. Soc. London Ser. B* 341: 80P

Jones, E. G., Coulter, J. D., Hendry, S. H. C. 1978. Intracortical connectivity of architectonic fields in the somatic sensory, motor and parietal cortex of monkeys. *J. Comp. Neurol.* 181: 291–348

Jones, E. G., Powell, T. P. S. 1970. An anatomical study of converging sensory pathways within the cerebral cortex of the monkey. *Brain* 93: 793–802

Jones, E. G., Wise, S. P. 1977. Size, laminar and columnar distribution of efferent cells in the sensory motor cortex of monkeys. *J. Comp. Neurol.* 175: 391–438

Joynt, R. J., Honch, G. W., Rubin, A. J., Trudell, R. G. 1985. Occipital lobe syndromes. *Handb. Clinical Neurol.* 1: 45–62

Kaas, J. H. 1978. The organization of visual cortex in primates. In *Sensory Systems of Primates*, ed. C. R. Noback, pp. 151–79. New York: Plenum

Kawano, K., Sasaki, M., Yamashita, M. 1984. Response properties of neurons in posterior parietal cortex of monkey during visual-vestibular stimulation. I. Visual tracking neurons. *J. Neurophysiol.* 51: 340–51

Kawano, K., Sasaki, M. 1984. Response properties of neurons in posterior parietal cortex of monkey during visual-vestibular stimulation. II. Optokinetic neurons. *J. Neurophysiol.* 51: 352–60

Kuffler, S. W. 1953. Discharge patterns and functional organization of mammalian retina. *J. Neurophysiol.* 16: 37–68

Land, E. H. 1977. The retinex theory of color vision. *Sci. Am.* 237(6): 108–28

Land, E. H., McCann, J. J. 1971. Lightness and retinex theory. *J. Opt. Soc. Am.* 61: 1–11

Land, E. H., Hubel, D. H., Livingstone, M. S., Perry, S. H., Burns, M. M. 1983. Colour-generating interactions across the corpus callosum. *Nature* 303: 616–18

Leichnitz, G. R. 1980. An interhemispheric columnar projection between two cortical multi-sensory convergence areas (inferior parietal lobule and prefrontal cortex): An anterograde study in macaque using HRP gel. *Neurosci. Lett.* 18: 119–24

Leinonen, L. 1980. Functional properties of neurons in the posterior part of area 7 in awake monkeys. *Acta Physiol. Scand.* 108: 301–8

Lisberger, S. G., Morris, E. J., Tychsen, L. 1987. Visual motion processing and sensory-motor integration for smooth pursuit eye movements. *Ann. Rev. Neurosci.* 10: 97–129

Livingstone, M. S., Hubel, D. H. 1984. Anatomy and physiology of a color system in the primate visual cortex. *J. Neurosci.* 4: 309–56

Lund, J. S., Boothe, R. G. 1975. Interlaminar connections and pyramidal neuron organisation in the visual cortex, area 17, of the macaque monkey. *J. Comp. Neurol.* 159: 305–34

Lynch, J. C., Mountcastle, V. B., Talbot, W. H., Yin, T. C. T. 1977. Parietal lobe mechanisms for directed visual attention. *J. Neurophysiol.* 40: 362–89

Macko, K. A., Jarvis, C. D., Kennedy, C., Miyaoka, M., Shinohara, M., Sokoloff, L., Mishkin, M. 1982. Mapping the primate visual system with [2-14C] deoxyglucose. *Science* 218: 394–97

Maguire, W. M., Baizer, J. S. 1984. Visuotopic organization of the prelunate gyrus in rhesus monkey. *J. Neurosci.* 4: 1690–1704

Maunsell, J. H. R. 1986. Physiological evidence for two visual systems. In *Matters of Intelligence*, ed. L. Vaina. Dordrecht: Reidel. In press

Maunsell, J. H. R., Van Essen, D. C. 1983a. Functional properties of neurons in the middle temporal visual area (MT) of the macaque monkey: I. Selectivity for stimulus direction, speed and orientation. *J. Neurophysiol.* 49: 1127–47

Maunsell, J. H. R., Van Essen, D. C. 1983b. Functional properties of neurons in the middle temporal visual area (MT) of the macaque monkey: II. Binocular interactions and the sensitivity to binocular disparity. *J. Neurophysiol.* 49: 1148–67

Maunsell, J. H. R., Van Essen, D. C. 1983c. The connections of the middle temporal visual area (MT) and their relationship to a cortical hierarchy in the macaque monkey. *J. Neurophysiol.* 3: 2563–86

Mayhew, J. E. W., Anstis, S. M. 1972. Movement after-effects contingent on color, intensity, and pattern. *Percept. Psychobiol.* 12: 77–85

McCann, J. J., McKee, S. P., Taylor, R. H. 1976. Quantitative studies in retinex theory. *Vision Res.* 16: 445–58

McKee, S. P. 1981. A local mechanism for differential velocity detection. *Vision Res.* 21: 491–500

McKee, S. P., Nakayama, K. 1984. The detection of motion in the peripheral visual field. *Vision Res.* 24: 25–32

McKee, S. P., Silverman, G. H., Nakayama, K. 1986. Precise velocity discrimination despite random variations in temporal frequency and contrast. *Vision Res.* In press

Meadows, J. C. 1974a. Disturbed perception of colours associated with localized cerebral lesions. *Brain* 97: 615–32

Meadows, J. C. 1974b. The anatomical basis of prosopagnosia. *J. Neurol. Neurosurg. Psychiatr.* 37: 489–501

Mesulam, M. M., Van Hoesen, G. W., Pandya, D. N., Geschwind, N. 1977. Limbic and sensory connections of the inferior parietal lobule (area PG) in the rhesus monkey: A study with a new method for horseradish peroxidase histochemistry. *Brain Res.* 136: 393–414

Michael, C. R. 1985. Laminar segregation of color cells in the monkey striate cortex. *Vision Res.* 25: 415–23

Mikami, A., Kubota, K. 1980. Inferotemporal neuron activities during color discrimination with delay. *Brain Res.* 182: 65–78

Mikami, A., Newsome, W. T., Wurtz, R. H. 1986. Motion selectivity in macaque visual cortex: I. Mechanisms of direction and speed selectivity in extrastriate area MT. *J. Neurophysiol.* 55: 1308–27

Mikami, A., Newsome, W. T., Wurtz, R. H. 1986b. Motion selectivity in macaque visual cortex: II. Spatio-temporal range of directional interactions in MT and V1. *J. Neurophysiol.* 55: 1328–39

Mishkin, M., Ungerleider, L. G., Macko, K. A. 1983. Object vision and spatial vision: Two cortical pathways. *Trends Neurosci.* 6: 414–17

Mohler, C. W., Wurtz, R. H. 1977. Role of striate cortex and superior colliculus in visual guidance of saccadic eye movements in monkeys. *J. Neurophysiol.* 40: 74–94

Moran, J., Desimone, R. 1985. Selective attention gates visual processing in the extrastriate cortex. *Science* 229: 782–84

Moran, J., Desimone, R., Schein, S. J., Mishkin, M. 1983. Suppression from ipsilateral visual field in area V4 of the macaque. *Soc. Neurosci. Abstr.* 9: 957

Motter, B. C., Steinmetz, M. A., Mountcastle, V. B. 1985. Directional sensitivity of parietal visual neurons to moving stimuli depends upon the extent of the field traversed by the moving stimuli. *Soc. Neurosci. Abstr.* 11: 1011

Motter, B. C., Mountcastle, V. B. 1981. The functional properties of the light-sensitive neurons of the posterior parietal cortex studied in waking monkeys: Foveal sparing and opponent vector organization. *J. Neurosci.* 1: 3–26

Mountcastle, V. B., Lynch, J. C., Georgopoulos, A., Sakata, H., Acuna, C. 1975. Posterior parietal association cortex of the monkey: Command function for operations within extrapersonal space. *J. Neurophysiol.* 38: 871–908

Mountcastle, V. B., Motter, B. C., Steinmetz, M. A., Duffy, J. 1984. Looking and seeing: The visual functions of the parietal lobe. See Edelman et al 1984, pp. 159–93

Movshon, J. A., Adelson, E. H., Gizzi, M.

S., Newsome, W. T. 1986. The analysis of moving visual patterns. In *Pattern Recognition Mechanisms*, ed. C. Chagas, R. Gattass, C. Gross, pp. 117–51. New York. Springer-Verlag

Movshon, J. A., Newsome, W. T. 1984. Functional characteristics of striate cortical neurons projecting to MT in the macaque. *Soc. Neurosci. Abstr.* 10: 933

Nakayama, K. 1985. Biological image motion processing: A review. *Vision Res.* 25: 625–60

Newsome, W. T., Gizzi, M. S., Movshon, J. A. 1983. Spatial and temporal properties of neurons in macaque MT. *Invest. Ophthalmol. Vis. Sci.* 24: 106

Newsome, W. T., Mikami, A., Wurtz, R. H. 1986. Motion selectivity in macaque visual cortex. III. Psychophysics and physiology of apparent motion. *J. Neurophysiol.* 55: 1340–51

Newsome, W. T., Wurtz, R. H. 1981. Response properties of single neurons in the middle temporal visual area (MT) of alert macaque monkeys. *Soc. Neurosci. Abstr.* 7: 832

Newsome, W. T., Wurtz, R. H., Dürsteler, M. R., Mikami, A. 1985a. Deficits in visual motion perception following ibotenic acid lesions of the middle temporal visual area of the macaque monkey. *J. Neurosci.* 5: 825–40

Newsome, W. T., Wurtz, R. H., Dürsteler, M. R., Mikami, A. 1985b. Punctate chemical lesions of striate cortex in the macaque monkey: Effect on visually guided saccades. *Exp. Brain Res.* 58: 392–99

Pearlman, A. L., Birch, J., Meadows, J. C. 1979. Cerebral color blindness: An acquired defect in hue discrimination. *Ann. Neurol.* 5: 253–61

Perrett, D. I., Rolls, E. T., Caan, W. 1982. Visual neurons responsive to faces in the monkey temporal cortex. *Exp. Brain Res.* 47: 329–42

Perrett, D. I., Smith, P. A. J., Potter, D. D., Mistlin, A. J., Head, A. S., Milner, A. D., Jeeves, M. A. 1984. Neurons responsive to faces in the temporal cortex: Studies of functional organization, sensitivity to identity and relation to perception. *Human Neurobiol.* 3: 197–208

Perrett, D. I., Smith, P. A. J., Potter, D. D., Mistlin, A. J., Head, A. S., Milner, A. D., Jeeves, M. A. 1985. Visual cells in the temporal cortex sensitive to face view and gaze direction. *Proc. R. Soc. London Ser. B* 223: 293–317

Peters, A., Jones, E. G., eds. 1985. *Cerebral Cortex*. New York: Plenum

Petersen, S. E., Baker, J. F., Allman, J. M. 1980. Dimensional selectivity of neurons in the dorsolateral visual area of the owl monkey. *Brain Res.* 197: 507–11

Poggio, G. F., Fischer, B. 1977. Binocular interactions and depth sensitivity in striate and prestriate cortex of behaving rhesus monkey. *J. Neurophysiol.* 40: 1392–1405

Poggio, G. F., Poggio, T. 1984. The analysis of stereopsis. *Ann. Rev. Neurosci.* 7: 379–412

Pohl, W. 1973. Dissociation of spatial discrimination deficits following frontal and parietal lesions in monkeys. *J. Comp. Physiol. Psychol.* 82: 227–39

Ramachandran, V. S., Gregory, R. L. 1978. Does colour provide an input to human motion perception? *Nature* 275: 55–56

Ratcliff, G., Davies-Jones, G. A. B. 1972. Defective visual localization in focal brain wounds. *Brain* 95: 49–60

Regan, D., Beverly, K. 1978a. Illusory motion in depth: After-effect of adaptation to changing size. *Vision Res.* 18: 209–12

Regan, D., Beverly K. 1978b. Looming detectors in the human visual pathway. *Vision Res.* 18: 415–21

Regan, D., Cynader, M. 1979. Neurons in area 18 of cat visual cortex selectively sensitive to changing size: Nonlinear interactions between responses to two edges. *Vision Res.* 19: 699–711

Ridley, R. M., Hester, N. S., Ettlinger, G. 1977. Stimulus- and response-dependent units from the occipital and temporal lobes of the unanaesthetized monkey performing learnt visual tasks. *Exp. Brain Res.* 27: 539–52

Rizzolatti, G., Scandolara, C., Matelli, M., Gentilucci, M. 1981. Afferent properties of periarcuate neurons in macaque monkeys. II. Visual responses. *Behav. Brain Res.* 2: 147–63

Robinson, D. L., Goldberg, M. E., Stanton, G. B. 1978. Parietal association cortex in the primate: Sensory mechanisms and behavioral modulations. *J. Neurophysiol.* 41: 910–32

Rockland, K. S., Pandya, D. N. 1979. Laminar origins and terminations of cortical connections of the occipital lobe in the rhesus monkey. *Brain Res.* 179: 3–20

Rolls, E. T., Baylis, G. C., Leonard, C. M. 1985. Role of low and high spatial frequencies in the face-selective responses of neurons in the cortex in the superior temporal sulcus in the monkey. *Vision Res.* 25: 1021–35

Rolls, E. T., Judge, S. J., Sanghera, M. K. 1977. Activity of neurons in the inferotemporal cortex of the alert monkey. *Brain Res.* 130: 229–38

Rolls, E. T. 1984. Neurons in the cortex of the temporal lobe and in the amygdala of the monkey with responses selective for faces. *Human Neurobiol.* 3: 209–22

Saito, H., Yukie, M., Tanaka, K., Hikosaka, K., Fukada, Y., Iwai, E. 1986. Integration of direction signals of image motion in the superior temporal sulcus of the macaque monkey. *J. Neurosci.* 6: 145–57

Sakata, H., Shibutani, H., Ito, Y., Tsurugai, K. 1986. Parietal cortical neurons responding to rotary movement of visual stimulus in space. *Exp. Brain Res.* 61: 658–63

Sakata, H., Shibutani, H., Kawano, K. 1983. Functional properties of visual tracking neurons in posterior parietal association cortex of the monkey. *J. Neurophysiol.* 49: 1364–80

Sakata, H., Shibutani, H., Kawano, K., Harrington, T. L. 1985. Neural mechanisms of space vision in the parietal association cortex of the monkey. *Vision Res.* 25: 453–63

Schein, S. J., Desimone, R., deMonasterio, F. M. 1983. Spectral properties of V4 cells in macaque monkey. *Invest. Ophthalmol. Vis. Sci.* 24: 107

Schein, S. J., Marrocco, R. T., deMonasterio, F. M. 1982. Is there a high concentration of color-selective cells in area V4 of monkey visual cortex? *J. Neurophysiol.* 47: 193–213

Segraves, M. A., Deng, S.-Y., Bruce, C. J. Ungerleider, L. G., Mishkin, M., Goldberg, M. E. 1983. Monkeys with striate lesions have saccadic dysmetria to moving targets. *Invest. Ophthalmol. Vis. Sci.* 24: 25

Seltzer, B., Pandya, D. N. 1984. Further observations on parieto-temporal connections in the rhesus monkey. *Exp. Brain Res.* 55: 301–12

Shapley, R., Lennie, P. 1985. Spatial frequency analysis in the visual system. *Ann. Rev. Neurosci.* 8: 547–83

Shipp, S., Zeki, S. 1985. Segregation of pathways leading from area V2 to areas V4 and V5 of macaque monkey visual cortex. *Nature* 315: 322–25

Sterling, P., Wickelgren, B. G. 1969. Visual receptive fields in the superior colliculus of the cat. *J. Neurophysiol.* 32: 1–15

Stone, J. 1983. *Parallel Processing in the Visual System.* New York: Plenum

Symonds, L. L., Rosenquist, A. C. 1984. Laminar origins of visual corticocortical connections in the cat. *J. Comp. Neurol.* 229: 39–47

Tanaka, K., Hikosaka, H., Saito, H., Yukie, Y., Fukada, Y., Iwai, E. 1986. Analysis of local and wide-field movements in the superior temporal visual areas of the macaque monkey. *J. Neurosci.* 6: 134–44

Tigges, J., Tigges, M. 1985. Subcortical sources of direct projections to visual cortex. See Peters & Jones 1985, 3: 351–78

Tigges, J., Tigges, M., Perachio, A. A. 1977. Complementary laminar terminations of afferents to area 17 originating in area 18 and the lateral geniculate nucleus in squirrel monkey. *J. Comp. Neurol.* 176: 87–100

Tigges, J., Tigges, M., Anschel, S., Cross, N. A., Letbetter, W. D., McBride, R. L. 1981. Areal and laminar distributions of neurons interconnecting the central visual cortical areas 17, 18, 19 and MT in the squirrel monkey (*Saimiri*). *J. Comp. Neurol.* 202: 539–60

Tolhurst, D. J., Movshon, J. A. 1975. Spatial and temporal contrast sensitivity of striate cortex neurons. *Nature* 176: 87–100

Tootell, R. B. H., Silverman, M. S., DeValois, R. L., Jacobs, G. H. 1983. Functional organization of the second cortical visual area in primates. *Science* 220: 737–39

Tusa, R. J., Palmer, L. A., Rosenquist, A. C. 1981. Multiple cortical visual areas. See Woolsey 1981, 2: 1–32

Ungerleider, L. G., Desimone, R. 1986. Cortical connections of visual area MT in the macaque. *J. Comp. Neurol.* 248: 190–222

Ungerleider, L. G., Mishkin, M. 1982. Two cortical visual systems. See Ingle et al 1982, pp. 549–80

Van Essen, D. C. 1979. Visual areas of the mammalian cerebral cortex. *Ann. Rev. Neurosci.* 2: 227–63

Van Essen, D. C. 1985. Functional organization of primate visual cortex. See Peters & Jones 1985, 3: 259–329

Van Essen, D. C., Maunsell, J. H. R. 1980. Two-dimensional maps of the cerebral cortex. *J. Comp. Neurol.* 191: 255–81

Van Essen, D. C., Maunsell, J. H. R. 1983. Hierarchical organization and functional streams in the visual cortex. *Trends Neurosci.* 6: 370–75

Van Essen, D. C., Maunsell, J. H. R., Bixby, J. L. 1981. The middle temporal visual area in the macaque: Myeloarchitecture, connections, functional properties and topographic representation. *J. Comp. Neurol.* 199: 293–326

Van Essen, D. C., Zeki, S. M. 1978. The topographic organization of rhesus monkey prestriate cortex. *J. Physiol.* 277: 193–226

von der Heydt, R., Peterhand, E., Baumgartner, G. 1984. Illusory contours and cortical neuron responses. *Science* 224: 1260–62

von Grünau, M., Frost, B. J. 1983. Double-opponent-process mechanism underlying RF-structure of directionally specific cells of cat lateral suprasylvian visual area. *Exp. Brain Res.* 49: 84–92

Wagor, E., Mangini, N. S., Pearlman, A. L. 1980. Retinal organization of striate and extrastriate visual cortex in the mouse. *J. Comp. Neurol.* 193: 187–202

Weiskrantz, L., Saunders, R. C. 1984. Impairments of visual object transforms in monkeys. *Brain* 107: 1033–72

Weller, R. E., Kaas, J. H. 1981. Cortical and subcortical connections of the visual cortex in primates. See Woolsey 1981, 2: 121–55

Wild, H. M., Butler, S. R., Carden, D., Kulikowski, J. J. 1985. Primate cortical area V4 important for colour constancy but not wavelength discrimination. *Nature* 313: 133–35

Wilson, M. 1978. Visual systems: Pulvinar-extrastriate cortex. *Handb. Behav. Neurobiol.* 1: 209–47

Wilson, M., Wilson, W. A., Remez, R. 1977. Effects of prestriate, inferotemporal and superior temporal sulcus lesions on attention and gaze shifts in rhesus monkeys. *J. Comp. Physiol. Psychol.* 91: 1261–71

Wilson, M., Keys, W., Johnston, T. D. 1979. Middle temporal cortical visual area and visuospatial function in *Galago senegalensis*. *J. Comp. Physiol. Psychol.* 93: 247–59

Woolsey, C. N., ed. 1981. *Cortical Sensory Organization*. Clifton, NJ: Humana

Wong-Riley, M. 1978. Reciprocal connections between striate and prestriate cortex in squirrel monkey as demonstrated by combined peroxidase histochemistry and autoradiography. *Brain Res.* 147: 159–64

Wurtz, R. H., Goldberg, M. E., Robinson, D. L. 1980. Behavioral modulation of visual responses in the monkey: Stimulus selection for attention and movement. *Prog. Psychobiol. Physiol. Psychol.* 9: 43–83

Wurtz, R. H., Mohler, C. W. 1976a. Enhancement of visual responses in monkey striate cortex and frontal eye fields. *J. Neurophysiol.* 39: 766–72

Wurtz, R. H., Mohler, C. W. 1976b. Organization of monkey superior colliculus: Enhanced visual response of superficial layer cells. *J. Neurophysiol.* 39: 745–62

Wurtz, R. H., Newsome, W. T. 1985. Divergent signals encoded by neurons in extrastriate areas MT and MST during smooth pursuit eye movements. *Soc. Neurosci. Abstr.* 11: 1246

Wurtz, R. H., Richmond, B. J., Newsome, W. T. 1984. Modulation of cortical visual processing by attention, perception, and movement. See Edelman et al 1984, pp. 195–217

Yin, T. C. T., Mountcastle, V. B. 1977. Visual input to the visuomotor mechanisms of the monkey's parietal lobe. *Science* 197: 1381–83

Yin, T. C. T., Mountcastle, V. B. 1978. Mechanisms of neuronal integration in the parietal lobe for visual attention. *Fed. Proc.* 37: 2251–57

Zeki, S. M. 1969. Representation of central visual fields in prestriate cortex of monkey. *Brain Res.* 14: 271–91

Zeki, S. M. 1971. Cortical projections from two prestriate areas in the monkey. *Brain Res.* 34: 19–35

Zeki, S. M. 1973. Colour coding in rhesus monkey prestriate cortex. *Brain Res.* 53: 422–27

Zeki, S. M. 1974a. Cells responding to changing image size and disparity in the cortex of the rhesus monkey. *J. Physiol.* 242: 827–41

Zeki, S. M. 1974b. Functional organization of a visual area in the posterior bank of the superior temporal sulcus of the rhesus monkey. *J. Physiol.* 236: 549–73

Zeki, S. M. 1975. The functional organization of projections from striate to prestriate visual cortex in the rhesus monkey. *Cold Spring Harbor Symp. Quant. Biol.* 40: 591–600

Zeki, S. M. 1977. Colour coding in the superior temporal sulcus of rhesus monkey visual cortex. *Proc. R. Soc. London Ser. B* 197: 195–223

Zeki, S. M. 1978. Uniformity and diversity of structure and function in rhesus monkey prestriate visual cortex. *J. Physiol.* 277: 273–90

Zeki, S. M. 1980. The representation of colours in the cerebral cortex. *Nature* 284: 412–18

Zeki, S. M. 1983a. Colour coding in the cerebral cortex: The reaction of cells in monkey visual cortex to wavelengths and colours. *Neurosci.* 9: 741–65

Zeki, S. M. 1983b. Colour coding in the cerebral cortex: The responses of wavelength-selective and colour-coded cells in monkey visual cortex to changes in wavelength composition. *Neurosci.* 9: 767–81

Zeki, S. M. 1983c. The distribution of wavelength and orientation selective cells in different areas of monkey visual cortex. *Proc. R. Soc. London Ser. B* 217: 449–70

Zihl, J., von Cramon, D., Mai, N. 1983. Selective disturbance of movement vision after bilateral brain damage. *Brain* 106: 313–40

Ann. Rev. Neurosci. 1987. 10: 403–57

DEVELOPMENTAL REGULATION OF NICOTINIC ACETYLCHOLINE RECEPTORS

Stephen M. Schuetze

Department of Biological Sciences, Columbia University, New York, New York 10027

Lorna W. Role

Department of Anatomy and Cell Biology, and Center for Neurobiology and Behavior, College of Physicians and Surgeons of Columbia University, New York, New York 10032

INTRODUCTION

The last ten years have brought many important advances in our understanding of nicotinic acetylcholine receptors (AChR). Early progress was catalyzed by the discovery of a specific, high-affinity ligand, α-bungarotoxin (α-BGT), and tissue sources rich in receptor, the electric organs of *Torpedo* and *Electrophorus*. The more recent addition of specific antibodies to the receptor, cDNA probes for the receptor subunit messages, and patch clamp technology have revolutionized study of the AChR.

These tools have revealed significant developmental changes in the distribution and properties of AChRs, which can be summarized as follows. During myogenesis, AChRs are found over the entire surface of uninnervated myotubes. Soon after the nerve contacts a muscle fiber, a cluster of AChRs forms beneath the nerve terminals. These embryonic synaptic AChRs are metabolically unstable and, when activated by ACh, their ion

0147–006X/87/0301–0403$02.00

channels stay open longer and conduct less current than adult endplate receptors. Within days to weeks, depending on the species, extrajunctional AChRs disappear, while synaptic AChRs become metabolically stable and their ion channels express adult properties. Recent studies have also shown that AChR function at mature synapses can be modulated by other transmitters and by some drugs.

In this paper we review developmental regulation of the number, distribution, turnover, and channel properties of the AChR at the nerve-muscle synapse. Where information is available, regulation of the neuronal nicotinic receptor is considered also. We conclude with a brief discussion on the modulation of AChR function at mature synapses.

SYNTHESIS, INSERTION, AND DEGRADATION OF ACETYLCHOLINE RECEPTORS

The best-characterized AChR is that found in *Torpedo* electric organ. It is an $\sim$275 kD pentamer comprised of four distinct subunits, α, β, γ, δ, in the stoichiometry $\alpha_2\beta\gamma\delta$ (Weill et al 1974, Reynolds & Karlin 1978, Raftery et al 1980, McCarthy et al 1986). Skeletal muscle AChRs are similar, except that in at least one species the γ subunit is replaced by a fifth type of subunit, called ε (Takai et al 1985, Mishina et al 1986). Each subunit is believed to cross the membrane four or five times (Anderson & Blobel 1981, Claudio et al 1983, Finer-Moore & Stroud 1984, Lindstrom et al 1984, McCarthy et al 1986). The receptor complex has two ACh binding sites, one on each of the two α-subunits (Karlin et al 1975, Neubig & Cohen 1979, Sine & Taylor 1980, Hall et al 1983). A cytoplasmic protein, called 43 kD protein, is commonly associated with the AChRs (Froehner et al 1981, Neubig & Cohen 1979, St. John et al 1982).

Early studies have shown that AChRs appear on the muscle surface about 1.5–3 hr after their synthesis is completed. AChR accumulation is blocked by inhibitors of either protein synthesis or glycosylation (see Fambrough 1979, 1983 for reviews). More recent work has revealed important details of the post-translational modifications and assembly of the receptor. These post-translational modifications may be important regulatory steps in the neural induction of AChR synthesis (Merlie 1984, Anderson 1986, Merlie & Smith 1986).

Cell-free translation systems show that each of the four distinct subunits of the AChR is translated from a separate mRNA, inserted into the endoplasmic reticulum, and glycosylated (reviewed in Anderson 1986, Merlie & Smith 1986, Salpeter & Loring 1986). Some processing in the Golgi is inferred from the presence of complex oligosaccharides on the mature receptor. Receptor subunits are subject to additional modifications, including phosphorylation, disulfide bond formation, methylation, and fatty acid acylation. Some of these post-translational modifications are implicated in the regulation of receptor function. Newly synthesized α subunits do not bind α-BGT. Apparently only 30% of the α chains acquire this ability and the rest are degraded rapidly. Finally, the AChR complex is assembled, but the site of assembly is not yet known. Only complete receptor molecules are transported to the cell surface.

Clathrin-coated vesicles have been implicated as a possible route for shuttling and inserting assembled AChRs into the plasma membrane (Bursztajn & Fischbach 1984, Bursztajn 1984). AChR-loaded vesicles are visualized with horseradish peroxidase (HRP) conjugated to α-BGT. HRP-stained coated vesicles are seen in the vicinity of AChR clusters, and their number is reduced by protein synthesis inhibitors.

AChRs incorporated into the surface membrane are later internalized and degraded, probably by lysosomal enzymes (reviewed in Fambrough 1979, Salpeter & Loring 1986; see also Libby et al 1980, Hyman & Froehner 1983, Clementi et al 1983). The degradation products are released into the extracellular fluid (Berg & Hall 1975, Devreotes & Fambrough 1975). Extrajunctional AChRs and AChRs at immature endplates have a metabolic half life of about one day. In contrast, AChRs at mature endplates are roughly ten-fold more stable. This is considered in more detail below.

The above description only outlines some of the key features of AChR biosynthesis and degradation. The reader is referred to several reviews that treat the subject more completely (Fambrough 1979, Karlin 1980, 1983, Changeux 1981, Conti-Tronconi & Raftery 1982, Fairclough et al 1983, Changeux et al 1984, Anderson 1986, Salpeter & Loring 1986, Merlie et al 1987, Merlie & Sanes 1986, Merlie & Smith 1986).

CONTROL OF THE NUMBER AND DISTRIBUTION OF ACETYLCHOLINE RECEPTORS

The small area (less than 0.1% of the total) of a mature skeletal muscle fiber that is contacted by motor nerve terminals is highly specialized. The hallmark of the synaptic region is the cluster of AChRs organized along the tops of the junctional folds at the impressive packing density of about 10,000 receptors per square micrometer of membrane (Peper & McMahan 1972, Anderson & Cohen 1974, Kuffler & Yoshikami 1975, Fertuck & Salpeter 1976, Matthews-Bellinger & Salpeter 1978). There is a steep gradient of AChR distribution: AChRs within adjacent extrasynaptic regions are present at levels 1/1000 to 1/10,000 that of the synaptic density (Fertuck & Salpeter 1976). Similarly, early experiments by Kuffler and his colleagues suggest that ACh sensitivity is localized to the subsynaptic region on autonomic neurons (Harris et al 1971). The following sections discuss the mechanisms that underlie the regulation of the number and distribution of AChRs on muscle and neurons.

Formation of Receptor Clusters at Developing Nerve-Muscle Synapses

During the development of both chick and rat muscle, motoneurons arrive when the muscle mass is still relatively undifferentiated (Landmesser & Morris 1975, Dennis et al 1981). Therefore, inductive interactions between nerve and muscle may start even before synaptogenesis, suggesting that the mechanisms involved in the initial formation of AChR clusters might differ from those controlling AChR distribution in adults (Dennis 1981). The following discussion is presented with this distinction in mind. We consider in turn the organization of AChR clusters during development, their induction by aneural stimuli, the possible roles of muscle activity and soluble neurotrophic factors in regulating AChRs, and the stabilization of AChR clusters.

METHODS FOR STUDYING THE DISTRIBUTION OF ACETYLCHOLINE RECEPTORS Iontophoretic mapping of ACh sensitivity reveals the distribution of functional AChRs. In this technique an intracellular microelectrode is used to measure membrane potential. The local sensitivity to ACh is assayed as the membrane depolarization (or membrane current) induced by ACh applied to a discrete region on the muscle surface from a second pipette. Under optimal conditions the resolution of this technique is <10 μm (Kuffler & Yoshikami 1975).

Alternatively, the AChR distribution can be mapped morphologically using specific AChR ligands, such as α-BGT, coupled to a fluorophore, an electron dense marker, or ^{125}I. This provides a quantitative, high resolution assessment of AChR distribution at the light or electron microscopic level. Other probes include specific antibodies against the main immunogenic region (MIR) of the α subunit of the *Electrophorus* receptor (e.g. mAb 35; Lindstrom 1983). Furthermore, regions synthesizing AChRs can be localized by probing for subunit message with cDNA clones (e.g. Ballivet et al 1982, Numa et al 1983; reviewed in Merlie et al 1987, Anderson 1986).

INDUCTION OF RECEPTOR CLUSTERS BY INNERVATION AChR clusters form rapidly after the arrival of the motor nerve (Blackshaw & Warner 1976, Kullberg et al 1977, Braithwaite & Harris 1979, Chow & Cohen 1983, Kidokoro et al 1980, Creazzo & Sohal 1983, Frank & Fischbach 1979,

Role et al 1982, 1987). Although some areas of high AChR density are seen before innervation (Fischbach & Cohen 1973, Sytkowski et al 1973, Anderson & Cohen 1977, Jacob & Lentz 1979), the ingrowing motoneuron does not seek out pre-existing clusters but rather induces a new aggregate (Anderson & Cohen 1977, Frank & Fischbach 1979). Following the onset of transmission, the number of extrajunctional AChRs declines steadily (Diamond & Miledi 1962, Bevan & Steinbach 1977).

The growing nerve terminal is capable of release before the formation of AChR clusters (Blackshaw & Warner 1976, Kullberg et al 1977, Bevan & Steinbach 1977, Anderson et al 1979, Frank & Fischbach 1979, Cohen 1980, Role et al 1982, 1987, Kidokoro 1980, Kidokoro & Yeh 1982, Hume et al 1983, Young & Poo 1983a). This suggests a potentially important inductive influence of neurally-released substances on cluster formation (reviewed in Dennis 1981). Several studies suggest that this neural influence may initially extend beyond the region of nerve terminal contact. Induced receptor clusters at newly formed synapses extend past the boundaries of the terminal arbor, whereas those at the mature endplate colocalize precisely with the region of presynaptic contact (Bevan & Steinbach 1977, Steinbach 1981a, Matthews-Bellinger & Salpeter 1983, Ishikawa et al 1983). New synaptic clusters are initially comprised of mini-aggregates of receptor that subsequently increase in area and then coalesce to form the more compact clusters seen at mature contacts (Steinbach 1981a, Kuromi et al 1985, Role et al 1985, 1987, Kidokoro & Brass 1985).

Apparently, only cholinergic neurons have the capacity to induce AChR clusters. Cholinergic motoneurons from the spinal cord and parasympathetic neurons induce the same number of neurite-associated receptor clusters along contacted myotubes in vitro (Role et al 1985, 1987). Other cholinergic neurons also cluster AChRs (Nelson et al 1976, Schubert et al 1977, Nurse & O'Lague 1975), whereas non-motoneuron spinal cord cells or sensory neurons do not even when their processes overlie myotubes for hundreds of micrometers (Cohen & Weldon 1980, Kidokoro et al 1980, Role et al 1985). Furthermore, extracts of spinal cord and brain increase the number and clustering of AChRs on cultured myotubes, but extracts of sensory ganglia or non-neural tissues have no effect (see below and Christian et al 1978, Podleski et al 1978, Jessel et al 1979, Axelrod et al 1981, Connolly et al 1982, Salpeter et al 1982a,b, Wallace et al 1985).

The capacity of the nerve to induce AChR clusters may be restricted to particular regions along the length of the neurite (Anderson & Cohen 1977, Anderson et al 1977, 1979, Cohen & Fischbach 1977, Frank & Fischbach 1979). In vitro, growth cones extending along myotubes are associated with a disproportionately high number of receptor clusters (Role et al 1982, 1987). The growth cone might release factors that cluster

AChRs locally or it might provide some local contact or activity that mobilizes receptors into the forming cluster (see below).

MECHANISM OF ACETYLCHOLINE RECEPTOR CLUSTER FORMATION AT DEVELOPING CONTACTS The mechanisms involved in the induction of AChR clusters are complex. As outlined above, some AChRs are diffusely distributed and others are clustered at relatively high density on uninnervated muscle both in vivo and in vitro (Fischbach & Cohen 1973, Sytkowski et al 1973, Anderson & Cohen 1977, Jacob & Lentz 1979). In addition, AChRs are continuously synthesized and inserted into the muscle plasma membrane (Devreotes et al 1977; reviewed in Fambrough 1983). Thus, the induction of an AChR cluster by the incoming nerve may involve lateral diffusion of pre-existing surface receptors to the contact site, changes in receptor turnover and insertion, or both.

Lateral diffusion of acetylcholine receptors in the membrane AChR mobility in the lipid bilayer has been measured with two techniques. Fluorescence photobleaching recovery (Axelrod et al 1976, Kidokoro & Brass 1985) measures the rate at which AChRs tagged with fluorescently labeled α-BGT move into an area of membrane that has been bleached with a microlaser beam. AChR mobility may also be estimated physiologically by measuring the rate at which ACh sensitivity recovers following a partial block with α-BGT (Poo 1982, Young & Poo 1983b; reviewed in Poo 1985). Both techniques show that the diffusely distributed AChRs are mobile in the lipid bilayer, but the estimated rates in the same preparation differ by a factor of 10 (3×10^{-9} cm^2/sec, Poo 1982; $\sim 3 \times 10^{-10}$ cm^2/sec, Kidokoro & Brass 1985). Nevertheless, even the lower mobility values are consistent with receptor diffusion playing some role in the formation of AChR aggregates (Kidokoro & Brass 1985).

Direct evidence for a contribution of pre-existing surface receptors comes from experiments in which AChRs are labeled with fluorescent α-BGT before the nerve arrives. New neurally-induced clusters form, at least in part, from the redistribution of prelabeled receptors (Anderson et al 1977, Kuromi & Kidokoro 1984a, Ziskind-Conhaim et al 1984, Role et al 1985). Both diffusely distributed AChRs (Anderson et al 1977, Ziskind-Conhaim et al 1984) and AChRs from adjacent noninnervated clusters (Kuromi & Kidokoro 1984a, Kidokoro & Brass 1985) may contribute to the subneural cluster. Nearby noninnervated clusters disperse after nerve-muscle contact but before the new cluster appears (Moody-Corbett & Cohen 1982, Kuromi & Kidokoro 1984a). Recent experiments have demonstrated that cholinergic neurons are more effective than noncholinergic neurons in accelerating the break-up of extrajunctional AChR clusters (Kidokoro & Brass 1985). However, it has not yet been shown directly that AChRs from these pre-existing clusters move to the newly formed contacts.

A possible mechanism for cluster formation is that the contact site traps freely diffusing AChRs (Edwards & Frisch 1976). This "diffusion trap" model requires high levels of AChR mobility in the surface membrane, but with some revision it apparently can account for the observed rates of cluster formation in the frog (Poo 1982, Young & Poo 1983b, Kuromi et al 1985, Kidokoro & Brass 1985; reviewed in Fraser & Poo 1982). Although rat muscle AChRs are not sufficiently mobile to form clusters by passive diffusion-trapping alone, this mechanism does contribute to cluster formation (Stya & Axelrod 1983). In sum, passive diffusion of AChRs contributes to cluster formation, but other mechanisms are probably involved also.

Another possible mechanism for mobilizing AChRs to synaptic sites is electromigration (Fraser & Poo 1982, Poo 1985). This model proposes that the electric field associated with currents at newly formed synapses electrophoreses AChRs to the subsynaptic region. In fact, a steady electric field does cause AChR accumulation at the cathodal side of the muscle cells (Orida & Poo 1978, Fraser & Poo 1982). However, since neurite associated AChR clusters still form in preparations bathed in blocking concentrations of α-BGT or curare (Cohen 1972, Anderson & Cohen 1977), it is unlikely that the ACh-induced current accounts for cluster formation during synaptogenesis.

Control of acetylcholine receptor synthesis and insertion In principle, nerve terminals might control the number and distribution of AChRs by altering any of the steps in AChR synthesis, assembly, insertion, or degradation (reviewed in Salpeter & Loring 1986, Steinbach & Bloch 1986). Several studies suggest that newly inserted AChRs contribute to the formation of the subneural receptor cluster (Salpeter & Harris 1983, Ziskind-Conhaim et al 1984, Role et al 1985, Kidokoro & Brass 1985). However, there is some disagreement on the relative contribution of pre-existing surface receptors and new receptors to synaptic clusters.

In the chick, approximately 60% of the AChRs at neurite-muscle contacts are newly inserted within the first 8 hr (Role et al 1985). Likewise, both pre-existing and newly inserted receptors contribute to clusters at developing rat and frog synapses (Ziskind-Conhaim et al 1984, Kidokoro & Brass 1985). At uninnervated clusters, only 20% of the AChRs are new in chick (Role et al 1985), whereas in frog new and old AChRs contribute about equally in synaptic and extrasynaptic areas (Kidokoro & Brass 1985).

The mechanism of the neural induction of new AChRs is not known.

Localized synthesis and insertion of receptors at the synaptic site has been proposed as an attractive mechanism because of its simplicity (Fischbach et al 1984, Rubin et al 1987, Anderson 1986, Merlie & Sanes 1986). Several observations are consistent with this notion. Intracellular AChRs and AChR message are localized at high density in the synaptic region (Merlie & Sanes 1985, Pestronk 1985). Furthermore, Rubin and his colleagues have found that AChR clusters frequently colocalize with myoplasmic nuclei (Englander & Rubin 1985, 1986, Rubin et al 1987). Merlie & Sanes and their collaborators have suggested that local release of a neurotrophic substance may selectively activate transcription of AChR message by nuclei near the synapse (Merlie et al 1987, Merlie & Sanes 1986).

Local AChR insertion is also suggested by experiments showing AChR-loaded coated vesicles at neurally induced receptor clusters (Bursztajn 1984) and, in preliminary experiments, at identified synapses in vitro (Bursztajn & Fischbach 1984). Although periods of rapid increase in AChR number at developing mouse nerve-muscle junctions were not correlated with increased incidence of coated vesicles (Matthews-Bellinger & Salpeter 1983), they still might be a major route of AChR insertion because very few vesicles would be needed to maintain the observed receptor density (Salpeter & Loring 1986). Salpeter & Harris (1983) have suggested that the large number of vesicle profiles seen at the junction between dense and nondense membrane within each junctional fold and at the bottom of the folds is consistent with AChRs being inserted in these regions and then diffusing to the tops of the folds.

Initial Formation of Acetylcholine Receptor Clusters in the Absence of Nerve

A variety of non-neural stimuli induce the formation of AChR clusters. Clusters form along silk thread (Jones & Vrbova 1974), beneath polycationic latex beads (Peng et al 1981, Peng & Cheng 1982, Peng & Phelan 1984), and at attachment plaques between muscle and substratum in vitro (Moody-Corbett & Cohen 1981, Bloch & Geiger 1980). Furthermore, clusters form spontaneously on the surface of uninnervated myotubes (see above). Contacts between latex beads and muscle, like nerve-muscle contacts, are associated with basal lamina, membrane infoldings, 43 kD protein, coated vesicles, and membrane-associated cytoplasmic density (Peng & Cheng 1982).

Aneural AChR clusters form from both pre-existing and newly inserted AChRs (Bloch 1979, Bursztajn et al 1985). Aneural clusters will not form in the presence of concanavalin A, protein synthesis inhibitors, or agents that disrupt micro-tubules, such as colchicine (Bloch 1979, 1983, Burstein & Shainberg 1979, Connolly 1984). The formation of bead-induced clusters apparently requires Ca^{2+} entry (Peng 1984) and the formation of AChR clusters in aneural cultures requires both Ca^{2+} and ATP synthesis (Bloch 1979, 1983, Bursztajn et al 1985). High external Ca^{2+} increases the size of AChR clusters and the rate at which they form (Bursztajn et al 1985). Nerve-induced AChR clusters are similarly dependent on extracellular Ca^{2+} and energy metabolism, at least early in development (Bloch & Steinbach 1981). These studies indicate that energy metabolism and external Ca^{2+} are important for receptor aggregation.

The interpretation of the experiments is complicated, however. First, the results to date do not indicate whether these drug treatments affect AChR synthesis, insertion, or mobilization to the cluster. More importantly, the specific property of the aneural stimuli that is responsible for receptor clustering must be identified before the model systems provide insight into the physiological mechanisms for cluster formation.

Effect of Muscle Activity, Neurotrophic Factors, and Basal Lamina on the Number and Distribution of Acetylcholine Receptors

The nerve plays an important role in the control of junctional and extrajunctional AChR density. Although the relative importance of muscle activity and neural "trophic" factors has been much debated, it now appears that both regulate AChR number and distribution, but probably through very different mechanisms. The cumulative data suggest that muscle activity plays a critical role in regulating extrajunctional AChRs, whereas neurotrophic influences are more important in establishing and maintaining early synaptic clusters. More recently, elements associated with the extracellular matrix have been implicated in organizing AChRs. These topics are discussed below.

MUSCLE ACTIVITY The most dramatic example of the role of activity in regulating AChR distribution is denervation supersensitivity (Axelsson & Thesleff 1959, reviewed in Fambrough 1979). Agents that block action potentials (e.g. tetrodotoxin) or neuromuscular transmission (e.g. α-BGT) increase the rate of AChR synthesis and the number of surface AChRs (reviewed in Fambrough 1979). This suggests that activity may be responsible for the developmental loss of extrajunctional AChRs (Diamond & Miledi 1962, Bevan & Steinbach 1977). In fact, Burden (1977a) has shown that the loss of extrajunctional AChRs in developing chick muscle is inhibited by treatment with curare.

Direct electrical stimulation of the muscle prevents or reverses the increase in extrajunctional AChR in denervated muscle (Lomo & Westgaard 1975). Furthermore, conditioned medium from electrically stimulated muscle cells may specifically inhibit AChR synthesis (Shainberg & Isac 1984). Some recent data suggest that activity may regulate receptor synthesis via transcriptional control. Denervation increases total AChR message levels in rat muscle (Merlie et al 1984), and tetrodotoxin increases message levels in chick muscle (Klarsfeld & Changeux 1985). However, an alternative interpretation of these results is that activity affects message stability (Anderson 1986). It is interesting to consider these results together with those of Merlie & Sanes (1985) that synaptic regions are enriched in AChR message. Perhaps neurotrophic factors promote transcription by synaptic nuclei, and activity depresses transcription by extrasynaptic nuclei (Merlie et al 1987, Anderson 1986).

NEUROTROPHIC FACTORS The possible importance of soluble neurotrophic factors comes from early observations that both the total number of AChRs and the number of AChR clusters are increased on myotubes close to spinal cord explants in vitro (Cohen & Fischbach 1977, Podleski et al 1978). Several "factors" have since been identified that differ in their molecular weight, their biochemistry, and their effects on AChR number and clustering (Christian et al 1978, Podleski et al 1978, Jessel et al 1979, Salpeter et al 1982b, Connolly et al 1982, Buc Caron et al 1983, Neugebauer et al 1985, Usdin & Fischbach 1986).

In their crudest form (brain extract), neurally derived factors mimic several aspects of AChR regulation during synaptogenesis. Within a few hours after their addition to myotubes in vitro, soluble brain factors mobilize pre-existing AChRs, stimulate AChR synthesis, and generate AChR clusters with near-junctional site densities (Connolly et al 1982, Buc Caron et al 1983, Olek et al 1983, Salpeter et al 1982a, Usdin & Fischbach 1986). Just as in synaptogenesis, factor-induced clusters form from mini-aggregates that increase in size and coalesce to form a more compact structure (Olek et al 1983). At least in some cases, these clusters are associated with other synaptic specializations, such as basal lamina antigens (Sanes et al 1984).

One of the factors derived from brain extract appears to be a trypsin-sensitive polypeptide of about 42 kD that increases AChR synthesis and clustering. This factor has been purified over 60,000-fold by using high pressure liquid chromatography and runs as a single band on SDS-polyacrylamide gels (Usdin & Fischbach 1986). The material eluted from this gel retains activity and induces an increase in AChR synthesis with the same time course observed at developing synapses (Fischbach et al 1984, Usdin & Fischbach 1986).

Two low-molecular weight factors that increase the number of AChRs on myotubes have been described. One is a trypsin-sensitive peptide whose molecular weight is estimated at 1000 D (Jessel et al 1979, Buc Caron et al 1983). Like the 42 kD factor, this material increases both AChR synthesis and clustering on chick myotubes. Ascorbic acid is another component of brain extract that increases AChR number on rat myotubes (Knaack & Podleski 1985, Knaack et al 1986). Neither of these low molecular weight factors affects receptor degradation (Jessel et al 1979, Knaack et al 1986).

Other factors have also been reported (Christian et al 1978, Podleski et al 1978, Jessel et al 1979, Axelrod et al 1981, Bauer et al 1981, 1985, Salpeter et al 1982b, Neugebauer et al 1985). Apparently the major effect of these materials is to increase the number of AChR aggregates. The active factors appear to be of relatively high molecular weight, but further chemical characterization has not yet been reported. Finally, a factor previously reported to increase AChR production has since been shown to increase protein synthesis in general. This material, called sciatin, appears to be a transferrin-like polypeptide (Markelonis et al 1982a,b).

BASAL LAMINA If the muscle cell is damaged at the time of denervation, it will regenerate within its basal lamina scaffolding. Even in the absence of the nerve and Schwann cells, regenerating muscles elaborate postsynaptic specializations, including the AChR cluster at the original synaptic site (Marshall et al 1977, Sanes et al 1978, Burden et al 1979, McMahan et al 1980). This observation demonstrates that some synapse-specific components of basal lamina are involved in reforming the AChR cluster at its original site (Burden et al 1979, Nitkin et al 1983, McMahan & Slater 1984).

Although the basal lamina is present early in the development of the synapse, some evidence suggests that AChR clusters appear prior to it (Weinberg et al 1981, Chiu & Sanes 1984; reviewed in Sanes & Chiu 1983, Sanes 1983). Morphologically identifiable basal lamina and antigens specific to it appear after AChRs cluster (Chiu & Sanes 1984). However, these assays may not be sufficiently sensitive to detect early events in the formation of the basal lamina. Indeed, recent work suggests that morphologically identifiable basal lamina may appear within hours after cluster formation (Olek et al 1986). Like the formation of AChR clusters, the initial appearance of synaptic basal lamina may be controlled by the nerve (Sanes et al 1984, Anderson et al 1984). In addition, recent studies of the heparin sulfate proteoglycan component of basal lamina show that AChR clusters and adjacent patches of proteoglycan appear at about the same time and grow at nerve contacts together (Anderson et al 1984).

In an effort to identify the specific basal lamina molecules that regulate AChR aggregation, McMahan and his colleagues have characterized a factor derived from basement-membrane-enriched fractions of *Torpedo* electric organ (Wallace et al 1982, 1985, Rubin & McMahan 1982, Nitkin et al 1983, Godfrey et al 1984, Fallon et al 1985). The activity has been purified about 10,000-fold and has an apparent molecular weight of 50–100 kD. Beginning about 3 hr after its addition, the factor increases the number of AChR clusters on myotubes in vitro 3–20-fold. Clusters are formed by the redistribution of pre-existing surface receptors. Antibodies against this factor block its AChR clustering activity and cross-react with a component of basal lamina in frog muscle and with molecules concentrated at the neuromuscular junction of *Torpedo*. This factor is probably responsible for basal lamina induction of AChR clusters in regenerating muscle and may play a role in AChR organization at developing synapses.

Maintenance and Stabilization of Acetylcholine Receptor Clusters on Muscle

STABILITY OF ACETYLCHOLINE RECEPTOR AGGREGATES AChR clusters at newly formed synapses are disrupted by removal of extracellular Ca^{2+}, chronic carbachol exposure, and elevated external K^+ (Bloch & Steinbach 1981). In contrast, mature synaptic contacts are resistant to these treatments. Similarly, AChR clusters at developing synapses disperse within hours or days after denervation (Slater 1982, Kuromi & Kidokoro 1984), but clusters at mature endplates remain intact for 2 wk or more after removal of the nerve (Hartzell & Fambrough 1972, Frank et al 1975, Ko et al 1977, Braithwaite & Harris 1979, Steinbach 1981b, Slater 1982, Labovitz & Robbins 1983). Apparently, some continuing action of the nerve is required for cluster maintenance at newly-formed endplates, but not at mature contacts. Presumably, at the adult junction other factors maintain the AChR distribution in the absence of the nerve (Slater 1982).

Maturation may represent the acquisition of critical extracellular matrix material such as synapse-specific basal lamina (reviewed in Sanes & Chiu 1983, Nitkin et al 1983). Alternatively, it may reflect the final "anchoring" of AChRs by cytoskeletal elements (reviewed in Froehner 1986). Morphological studies reveal a cytoplasmic fibrous material associated with synaptic areas in vivo and in vitro (Fertuck & Salpeter 1976, Weldon & Cohen 1979, Cartaud et al 1981, Hirokawa & Heuser 1982, Peng 1983). AChRs have been postulated to be immobilized via attachment to these fibers (e.g. Prives et al 1982). Specific antibodies directed against the cytoskeletal elements, vinculin, α-actinin, and filamin, stain the neuromuscular junction within a few days after the onset of synaptogenesis in the rat (Bloch & Hall 1983). Disruption of certain cytoskeletal elements increases receptor mobility and breaks up clusters (Stya & Axelrod 1983,

Connolly 1984). Agents that disrupt actin filaments, but not those that interfere with microtubules, appear to disperse newly formed receptor clusters (Connolly 1984, Connolly & Graham 1985).

Several investigators have proposed that the "43 kD protein" anchors AChRs by its interaction with cytoskeletal components. The 43 kD protein is a peripheral membrane protein associated with the AChR (Neubig & Cohen 1979, Froehner et al 1981). It is located on the inner aspect of the membrane (St. John et al 1982, Porter & Froehner 1983) and is sufficiently close to the AChR β subunit to allow chemical cross-linking (Burden et al 1983). Morphological studies show that 43 kD protein precisely colocalizes with the receptor (Nghiem et al 1983, Sealock et al 1984, Peng & Froehner 1985). The 43 kD protein also binds actin (Walker et al 1984) and may be responsible for limiting AChR mobility (Lo et al 1980, Cartaud et al 1981, 1983, Rousselet et al 1982) by anchoring AChRs to cytoskeletal components. It will be important to define the molecular relationship among the AChR, 43 kD protein, and cytoskeletal elements in cluster stabilization.

Cytoskeletal elements probably play an important role in maintaining and re-establishing AChR clusters, but their function in regulating AChRs during synaptogenesis is less well established. Some data suggest that cytoskeletal elements might not appear until after AChRs cluster, but the techniques might not be sufficiently sensitive to detect their initial appearance (Hall et al 1981, Matthews-Bellinger & Salpeter 1983, Connolly 1984; reviewed in Froehner 1986, Salpeter & Loring 1986).

METABOLIC STABILIZATION OF ACETYLCHOLINE RECEPTORS At about the same time that the AChR aggregate becomes resistant to disruption, the metabolic stability of synaptic AChRs also increases. The degradation rate of AChRs is usually studied by labeling receptors with ^{125}I-α-BGT and following the fate of the radioactive iodine. A typical experimental paradigm is to pulse-label AChRs on muscles of several animals with ^{125}I-α-BGT, and to measure the amount of radioactivity remaining on the muscle at various times thereafter. The validity of these procedures has been tested by measuring the degradation rate of AChRs labeled with amino acid isotopes, and the results indicate that AChRs with bound α-BGT are degraded at about the same rate as AChRs without toxin (Merlie et al 1976, Gardner & Fambrough 1979).

These studies have shown that AChRs in extrajunctional regions of rat and chick muscle fibers are degraded rapidly with a metabolic half-life of about one day (reviewed in Fambrough 1979, Salpeter & Loring 1986). This has been demonstrated for extrajunctional AChRs on embryonic myotubes, whether clustered or diffuse (Devreotes & Fambrough 1975,

Burden 1977a,b, Steinbach et al 1979, Reiness & Weinberg 1981, Salpeter et al 1982b), and for AChRs that appear in adult muscle following denervation (Berg & Hall 1975, Chang & Huang 1975). In short, extrajunctional AChRs are degraded rapidly at all ages.

In contrast, AChRs within synaptic clusters become stabilized during endplate maturation. The first systematic study of AChR turnover at developing endplates was done in chicks (Burden 1977a,b). From the time of synapse formation up to about 1 wk after hatching, chick endplate AChRs have the same brief half-life ($\sim$30 hr) as extrajunctional receptors. However, by 3 wk after hatching, endplate receptors (but not extrajunctional AChRs) become stable ($t_{1/2} \geq 5$ days).

Later work showed that stabilization also occurs in rats, but with a different time course. AChRs in rat diaphragms cluster beneath nerve terminals during embryonic days 15–16, shortly after the myotubes are formed. Like extrajunctional receptors, AChRs at these newly formed endplates have a half-life of about 32–35 hr. By embryonic day 21 (about 1 day before birth) the degradation rate of endplate AChRs decreases about sevenfold to the slow rate characteristic of adults ($t_{1/2}$ = 6–11 days, Reiness & Weinberg 1981; see also Steinbach et al 1979, Berg & Hall 1975, Michler & Sakmann 1980, Chang & Huang 1975, Loring & Salpeter 1980, Salpeter & Harris 1983). Again, extrajunctional receptors are not stabilized. AChR stabilization also occurs during synapse formation in *Xenopus* myotomal muscle in vivo (Brehm 1986).

Under normal conditions, endplate AChRs continue to be degraded slowly throughout adulthood. EM autoradiography studies indicate that mouse endplate AChRs are degraded slowly whether they are at the top of the folds, where they are tightly packed, or at the bottoms of the folds, where their density is much lower (Salpeter & Harris 1983). However, some evidence indicates that even mature endplates contain a subpopulation of AChRs that are degraded rapidly (Stanley & Drachman 1983). In addition, denervation leads to an acceleration of endplate AChR degradation (reviewed in Fambrough 1979, Salpeter & Loring 1986).

Mechanism of metabolic stabilization of acetylcholine receptors It is not known how AChRs become stable, but several obvious possibilities can be ruled out. First, the AChR degradation rate is not a simple consequence of AChR clustering beneath nerve terminals. The decrease in degradation rate occurs days (rats and *Xenopus*) or even weeks (chicks) after the formation of synaptic AChR clusters (Burden 1977a,b, Bevan & Steinbach 1977, Reiness & Weinberg 1981, Matthews-Bellinger & Salpeter 1983, Brehm 1986). Second, stable and unstable AChRs are unlikely to represent different gene products. As discussed elsewhere (Salpeter & Loring 1986), studies of adult muscles indicate that denervation can destabilize pre-existing endplate AChRs, and reinnervation can confer stability to previously unstable receptors. Third, the degradation rate appears to be independent of the mechanism governing AChR channel properties. Metabolic stabilization occurs distinctly before (rats) or in the absence of (chicks) developmental changes in AChR channel open time (see below). Moreover, changes in AChR channel properties occur in *Xenopus* myocytes in the absence of any change in degradation rate (Brehm et al 1983).

Salpeter & Loring (1986) postulate that the endplate contains anchoring sites that bind AChRs reversibly and with high affinity. According to their model, the receptor can be degraded only when it is not attached to the anchor. If, for example, AChRs were bound to such sites 90% of the time, their degradation rate would decrease by a similar amount. The potential role of cytoskeletal elements and AChR-associated proteins as anchors is discussed in the preceding section.

Summary of acetylcholine receptor stabilization Early in muscle development, both junctional and extrajunctional AChRs are degraded rapidly. Days to weeks after receptors cluster beneath the nerve terminal, some unknown mechanism selectively stabilizes endplate AChRs. In rats, metabolic stabilization precedes the formation of postsynaptic folds and developmental changes in AChR channels, indicating that AChR degradation is independent of these other features of endplate maturation. However, decreased degradation does begin at roughly the same time at which AChR clusters become resistant to disruption (Bloch & Steinbach 1981). Thus, the same mechanism that makes receptor clusters resistant to disruption may also promote metabolic stabilization.

Summary of Skeletal Muscle Acetylcholine Receptor Clustering Studies

In sum, these studies indicate that the regulation of the number and distribution of AChRs at the nerve-muscle synapse is complex. Although the nerve certainly promotes the development of AChR clusters, clusters can also form in the absence of neural contact. Mature endplate clusters can be maintained without the nerve, and the dramatic effects of denervation on extrajunctional AChRs are regulated by muscle activity. However, AChR cluster formation during synaptogenesis is critically influenced by the nerve, and young clusters will disperse if denervated. Initially, the influence of the nerve apparently extends beyond the limits of nerve terminal contact. Highly localized effects on AChR synthesis and insertion, including possible selective activation of synaptic nuclei, may also be under neural control, with specific regions along the neurite specialized for AChR cluster induction.

Control of the Number and Distribution of Acetylcholine Receptors at Synapses Between Neurons

Compared to muscle AChRs, little is known about neuronal nicotinic AChRs. Studies of neuronal AChRs have been severely hampered by the lack of suitable probes. The recent development of specific antibodies and the isolation of a cDNA clone for a putative neuronal AChR suggest promising new directions (see below). The following discussion outlines techniques that have been employed to study neuronal AChR distribution, and describes what is known about the regulation of these AChRs during synaptogenesis and following denervation. Many neurons have both muscarinic and nicotinic AChRs. We consider here only the regulation of nicotinic AChRs.

FUNCTIONAL ASSAYS OF ACETYLCHOLINE RECEPTORS Most studies of the distribution of neuronal AChRs have employed intracellular recording to measure the depolarization evoked by iontophoretic ACh pulses (Kuffler et al 1971, Harris et al 1971, Roper 1976, Brenner & Martin 1976, O'Lague et al 1978, Dennis & Sargent 1979, Crean et al 1982, Pelligrino & Simonneau 1984). As the resolution of this technique is $\sim 10\ \mu m$, it provides only a crude map of ACh sensitivity because neuronal somas are only $\sim 30\ \mu m$ in diameter.

In other studies, the total ACh sensitivity of neurons is estimated by measuring changes in membrane potential (Smith et al 1983; P. M. Dunn and L. M. Marshall, personal communication) or inward current (Role 1984a,b, 1985) evoked by diffusely applied ACh. Diffusely applied ACh exposes all the AChRs on a neuron ($\sim 200\ \mu m$ radius) to the drug and assays overall ACh sensitivity. One advantage of this approach is that it permits quantitative pharmacology of the receptor, because drugs can be presented at known concentrations by several different techniques (Choi & Fischbach 1981, Clapham & Neher 1984).

α-BUNGAROTOXIN AS A NEURONAL ACETYLCHOLINE RECEPTOR PROBE The utility of α-BGT as a probe for the neuronal nicotinic receptor has been a topic of considerable debate. Many studies have demonstrated that α-BGT binds to both central and peripheral cholinoceptive neurons with high affinity (reviewed in Morley et al 1979, Morley & Kemp 1981, Oswald & Freeman 1981). Most functional studies, however, show that the toxin does not block AChRs at central (Duggan et al 1976, Goodman & Spitzer 1979, Brown et al 1983) or peripheral (Chou & Lee 1969, unpublished results of Z. Hall and G. Pilar cited in Landmesser & Pilar 1974, Ko et al 1976, Brown & Fumagalli 1977, Bursztajn & Gershon 1977, Patrick & Stallcup 1977a,b, Carbonetto et al 1978, Kouvelas et al 1978, Ascher et al

Loring & Zigmond 1985, Halvorsen & Berg 1986). The 66 amino acid peptide has been sequenced and found to have greatest homology with the "long" class of postsynaptic neurotoxins. Some unusual sequences have also been identified that may underlie its blocking activity (Grant & Chiapinelli 1985, Loring & Zigmond 1985). Since the material can be iodinated and yet retain binding activity, it is potentially a powerful tool for studying neuronal AChR distribution. Recent work by Loring & Zigmond (1985) shows that this toxin is synaptically localized in the chick ciliary ganglion.

RADIOLABELED AGONISTS AS PROBES FOR NEURONAL NICOTINIC ACETYLCHOLINE RECEPTORS High affinity binding of nicotinic agonists to putative neuronal AChRs has been demonstrated in rodent brain membrane fractions and in brain sections (Yoshida & Imura 1979, Marks & Collins 1982, Schwartz et al 1982, Costa & Murphy 1983, Rainbow et al 1984, Clarke et al 1984, 1985, Schwartz & Kellar 1983). The binding is high affinity, stereoselective, and has the appropriate pharmacology. In general, the autoradiographic data on AChR distribution revealed by ^{3}H-ACh in the presence of atropine and by ^{3}H-nicotine match well (Clarke et al 1985), and these binding sites are found in systems believed to be cholinergic (compare Armstrong et al 1983 and Mesulam et al 1983 with Clarke et al 1985 and Schwartz et al 1982). ^{3}H-ACh and ^{3}H-nicotine labeling patterns generally differ from that of ^{125}I-BGT (Schwartz et al 1982, Clarke et al 1985). Thus, the radiolabeled agonists may label a distinct set of nicotinic binding sites that may be the functional AChRs of neurons.

ANTI-ACETYLCHOLINE RECEPTOR ANTIBODIES Recent work of Berg and Lindstrom and their collaborators suggests that a monoclonal antibody directed against the α subunit of the *Electrophorus* AChR (mAb 35) is a valuable probe for neuronal AChRs (Jacob et al 1984, Smith et al 1985, 1986; D. K. Berg, personal communication). The antigen is found on ciliary and sympathetic neurons but not in heart, liver, spinal cord, or dorsal root ganglia (Smith et al 1985). The antigen does not coprecipitate with the α-BGT binding component (Smith et al 1985). Detergent extracts reveal that the mAb 35 antigen sediments at about 10 S (Smith et al 1985), a value close to that reported for AChR monomers (Changeux et al 1984). In addition, the cross-reacting component reaches peak levels during ganglionic synaptogenesis (between embryonic day 8–12 in the ciliary ganglion) and, in contrast to α-BGT, is predominantly localized at synapses (Jacob et al 1984). The mAb 35 component also binds concanavalin A, which blocks neuronal AChR function (Messing et al 1984, Smith et al 1985). Polyclonal antibodies raised against the mAb 35 antigen from brain block ACh-induced depolarization in ciliary ganglion neurons (J.

Stollberg, P. J. Whiting, J. M. Lindstrom, and D. K. Berg, personal communication). Use of these probes to examine the regulation of AChR number on ciliary ganglion neurons is discussed below.

cDNA PROBES TO THE NEURONAL ACETYLCHOLINE RECEPTOR Patrick and Heinemann and their colleagues (Boulter et al 1986) have recently cloned a cDNA derived from PC12 cells that appears to code for the α subunit of a neuronal nicotinic receptor. The clone contains extensive regions of sequence homology with the mouse muscle AChR α subunit, including the two cysteines believed to be part of the ACh binding site (Karlin & Cowburn 1973, Kao et al 1984). However, the amino acid sequence near these residues is not highly conserved and may underlie the pharmacological differences between neuronal and muscle AChRs (Boulter et al 1986). This probe is promising for studying the regulation of AChR expression in neurons.

REGULATION OF THE NUMBER AND DISTRIBUTION OF NEURONAL ACETYLCHOLINE RECEPTORS WITH INNERVATION Iontophoretic maps of ACh sensitivity on sympathetic and parasympathetic neurons of several species and on some cholinoceptive leech neurons reveal nonuniform ACh sensitivity over the cell surface (Harris et al 1971, Kuffler et al 1971, Roper 1976, Roper et al 1976, O'Lague et al 1978, Dennis & Sargent 1979, Crean et al 1982, Pellegrino & Simonneau 1984, Dunn & Marshall 1985). Peaks of sensitivity correlate with the locations of presynaptic boutons (Harris et al 1971, Kuffler et al 1971, Roper 1976, Roper et al 1976). AChRs visualized with antireceptor antibody and with HRP-α-BGT colocalize with morphologically identified active zones in frog sympathetic ganglion at the EM level (Marshall 1981). The mAb 35 component (Jacob et al 1984) and the binding sites for the blocking toxin (Loring & Zigmond 1985) are also localized at synapses in chick ciliary ganglia. In sum, neuronal AChRs, like muscle AChRs, are probably concentrated at the synapse. This raises the question of how synaptic AChR clusters form during development.

Preliminary studies suggest that the ACh sensitivity of chick sympathetic neurons in vitro increases strikingly following innervation (Role 1985, and unpublished observations). The maximum ACh-induced currents of innervated neurons is 10-fold higher than that of neurons grown in the absence of preganglionic input. Increased sensitivity is detected in cells even when innervation occurs in the continued presence of the AChR antagonist, *d*-tubocurarine, so it is unlikely that ACh itself induces this change (L. W. Role, unpublished observations). The ACh sensitivity of frog sympathetic neurons in vivo increases steadily following the development of synaptic input. The onset of ACh sensitivity may be triggered by synaptogenesis, but it is not yet possible to determine whether innervation precedes all detectable ACh sensitivity (P. M. Dunn and L. M. Marshall, personal communication). Overall, these preliminary studies are consistent with an inductive effect of the nerve on postsynaptic ACh sensitivity. This is similar to what is seen at the nerve muscle synapse, although it is not yet clear whether neuronal AChRs are at first diffusely distributed and then clustered following innervation.

Several other factors may regulate the ACh sensitivity of ciliary ganglion neurons grown in vitro (Nishi & Berg 1981a,b, Smith et al 1983, 1986). Cells grown in high K^+ are 13 times less sensitive to ACh than cells grown in medium with eye extract and normal concentrations of K^+ (Smith et al 1983, 1986). A similar, but less dramatic, decrease in the levels of ^{125}I-mAb 35 binding occurs with K^+ (Smith et al 1986). Furthermore, the percentage of neurons that receive spontaneous synaptic input from other ciliary ganglion cells is only 4% with high K^+ as opposed to $>70\%$ with eye extract (Smith et al 1983). Therefore, the observed effects on ACh sensitivity and AChR antibody binding sites might be due to differences in innervation of the neurons. In any case, the comodulation of ACh sensitivity and the levels of mAb 35 binding activity observed with most (though not all) of the conditions tested supports the notion that the antigen is the functional neuronal AChR (Smith et al 1986).

CHANGES IN THE NUMBER AND DISTRIBUTION OF NEURONAL ACETYLCHOLINE RECEPTORS FOLLOWING DENERVATION Denervation of mammalian (see Cannon & Rosenblueth 1949 for review of early studies) and amphibian autonomic ganglia has been reported to increase ACh sensitivity (Kuffler et al 1971, Roper 1976, Dennis & Sargent 1979). Using focal ACh application, Kuffler et al (1971) found peaks of ACh sensitivity at synaptic sites on normal parasympathetic ganglia but found uniformly high sensitivity following denervation.

These results suggest that denervation increases the number of AChRs or changes their distribution, but other experiments suggest another interpretation. Although denervation causes an 18-fold increase in ACh sensitivity of the neurons, it does not change their sensitivity to carbachol, a nonhydrolyzable cholinergic agonist. This suggests a decrease in acetylcholinesterase (AChE) activity (Brown 1969, Dunn & Marshall 1985). Consistent with this, Koelle & Ruch (1983) have shown that AChE activity decreases in denervated cat superior cervical ganglia. Furthermore, in contrast to Kuffler et al (1971), Dunn & Marshall (1985) found no change in the distribution of ACh sensitivity with denervation in sympathetic ganglia. Although different autonomic ganglia may respond differently to denervation, previous data on denervation supersensitivity of autonomic neurons should be reconsidered.

CHANGES IN ACETYLCHOLINE SENSITIVITY OF NEURONS FOLLOWING REMOVAL OF THE TARGET Two lines of evidence suggest that the target also regulates neuronal ACh sensitivity. First, axotomy of autonomic neurons decreases their ACh sensitivity. This is evidenced both by a decrease in the mean amplitude minature synaptic potential and by a decrease in the amplitude of responses to iontophoresed ACh (Purves 1975, Brenner & Johnson 1976, Brenner & Martin 1976). Second, studies of ciliary ganglion neurons in vitro indicate that five times as many neurons respond to iontophoretically applied ACh in cultures grown for 2 wk with target compared to neurons grown alone (Crean et al 1982). The mechanism underlying the apparent target dependence of ACh sensitivity in autonomic neurons is not known. Whether the target affects AChR number or AChE activity and whether functional contact with target is required are still undetermined.

Summary of Neuronal Acetylcholine Receptor Distribution

Like muscle AChRs, neuronal nicotinic AChRs apparently are localized to the synapse and may be induced by innervation. In other respects, such as the lack of neuronal denervation supersensitivity, neuronal AChRs appear to be regulated very differently from skeletal muscle AChRs. Although the utility of α-BGT as a probe for neuronal AChRs is questionable, new probes seem much more promising.

CHANNEL PROPERTIES OF ACETYLCHOLINE RECEPTORS IN SKELETAL MUSCLE AND NEURONS

As discussed above, both the distribution and the metabolic stability of muscle AChRs change during development. Other studies have shown that their functional properties also change. This was first suggested by the pioneering "noise analysis" studies of AChR function in adult grass frogs by Katz & Miledi (1972). Their results, which have been confirmed and extended by others (Neher & Sakmann 1976, Dreyer et al 1976b, Sakmann 1978), show that endplate channels differ from those that appear extrajunctionally after denervation. When activated by ACh, endplate channels conduct larger currents but close more quickly than extrajunctional channels. The similarity between embryonic AChRs and extrajunctional AChRs of denervated muscles in their degradation rate and distribution suggested that they might also have similar channel properties.

In fact, AChR channel properties were subsequently shown to change during endplate maturation from "extrajunctional-like" to "endplate-like." That is, AChR channels in endplates show a developmental decrease in their mean open time (τ) and an increase in their single-channel conductance (γ). Surprisingly, these studies also revealed that AChRs in *extra-junctional* regions of developing muscle fibers change in parallel with those at endplates. Therefore, the common practice of describing AChR channels as either "extrajunctional type" or "junctional type" is misleading.

The following discussion focuses on AChR channels in rat and *Xenopus* muscle, the two best-characterized systems, but similar changes probably occur in other species, including human (Bevan et al 1978, Cull-Candy et al 1978, 1982, Adams & Bevan 1985), mouse (Dreyer et al 1976a, Steele & Steinbach 1986), and bovine (Mishina et al 1986) muscle. An exception is chick muscle, in which AChR channels express a long τ throughout endplate development (Schuetze 1980).

At Least Two Kinds of Acetylcholine-Sensitive Channels Exist in Developing Muscle

Like grass frogs, rat muscles can express at least two types of AChR channels (Sakmann 1978, Sakmann & Brenner 1978, Schuetze & Fischbach 1978, Fischbach & Schuetze 1980, Michler & Sakmann 1981, Hamill & Sakmann 1981, Siegelbaum et al 1984, Vicini & Schuetze 1985; also see below). Embryonic muscle expresses AChRs with a single-channel conductance (γ) of $\sim$35 pS and a mean channel open time (τ) of $\sim$6 msec at room temperature. In contrast, adult endplates express AChRs with a γ of $\sim$50 pS and a τ of $\sim$1 msec. Both types of AChR channels can adopt a subconductance state of $\sim$10 pS, but apparently this happens only rarely under physiological conditions (Hamill & Sakmann 1981). For economy, embryonic-like and adult-like AChRs are referred to here as *slow channels* and *fast channels*, respectively. Earlier studies (Sakmann 1978, Sakmann & Brenner 1978, Fischbach & Schuetze 1980) estimated the τ of slow channels at $\sim$3–4 msec rather than $\sim$6–7 msec, probably as a result of unresolved fast channel activity at endplates that appeared to have only slow channels.

AChR channels in developing *Xenopus* myotomal muscle also fall into two categories that differ in γ and τ (Kullberg et al 1981, Brehm et al 1982, 1984a,b, Leonard et al 1984, Kullberg & Kasprzak 1985, Greenberg et al 1985). At the resting membrane potential, one type has a γ of $\sim$33 pS and a τ of 2–3 msec, and the other has a γ of 48 pS and a τ of <1 msec. At hyperpolarized potentials, the conductances increase to 48 pS (slow) and 60 pS (fast) (Brehm et al 1984a). At least two channel populations also are expressed by human (Bevan et al 1978, Cull-Candy et al 1978, 1982, Adams & Bevan 1985), mouse (Dreyer et al 1976a, Steele & Steinbach 1986), and fetal calf (Mishina et al 1986) muscle.

Single-channel recordings have revealed more detailed biophysical differences between fast and slow channels. First, the distribution of open times (more properly, burst durations) appears to be singly exponential for adult channels, and doubly exponential for embryonic channels (Siegelbaum et al 1984, Brehm et al 1984a,b). However, recent data suggest that fast channels also may have a second class of very brief openings that are often missed because of the limited frequency response of the recording equipment (F. Jaramillo and S. M. Schuetze, unpublished observations). Second, openings of both types of channels are interrupted by one or more very brief closings (cf Colquhoun & Sakmann 1981, 1985), and these "gaps" appear to be even briefer in fast channels than in slow ones (F. Jaramillo and S. M. Schuetze, unpublished observations).

Measuring the Relative Numbers of Fast and Slow Channels

Three approaches have been used to measure the relative proportion of fast and slow channels in muscle cells. Each technique estimates the relative number of fast and slow channel *openings* induced by ACh, not the actual numbers of fast and slow channels in the membrane. The three methods give comparable results, but each has its own advantages and disadvantages.

SINGLE-CHANNEL RECORDING The most direct, unambiguous technique is counting the number of high-γ, fast events and low-γ, slow events in patch clamp recordings. This technique also provides a detailed picture of the gating kinetics of fast and slow channels. Although it is the most powerful approach, some disadvantages of patch clamping are that it is comparatively tedious, it assays AChRs in only a small patch of membrane, and many preparations, such as intact muscles, require enzymatic cleaning of the surface membrane. Furthermore, in most preparations it is difficult, if not impossible, to record from single channels within the endplate itself.

ANALYSIS OF MINIATURE ENDPLATE CURRENTS A fast, relatively simple approach is the analysis of miniature endplate currents (MEPCs). MECPs can be recorded either with a conventional two-electrode voltage clamp (e.g. Michler & Sakmann 1980), or with a focal extracellular electrode, similar to a patch electrode, placed at the edge of the endplate (e.g. Vicini & Schuetze 1985). At endplates that have a single channel population and enough AChE to remove ACh promptly, the mean AChR channel open time (τ) is equal to the time constant of MEPC decay (Magleby & Stevens 1972, Anderson & Stevens 1973). At endplates with two channel populations, MEPC decays have two exponential components whose time constants reflect the two τ's and whose amplitudes reflect the relative number of fast and slow channel openings (Fischbach & Schuetze 1980, Michler

& Sakmann 1980, Kullberg & Kasprzak 1985, Vicini & Schuetze 1985). With care, estimates obtained with this technique are reproducible to within 10%, but they tend to overestimate the fraction of slow channels (see Vicini & Schuetze 1985). Disadvantages of MEPC analysis are that it is less direct than single-channel recording and that it can assay only endplate AChRs.

ACETYLCHOLINE NOISE ANALYSIS A third approach is fluctuation analysis (noise analysis), that is, spectral analysis of random variations in macroscopic membrane currents induced by long ($\sim$30 sec) ACh pulses (Katz & Miledi 1972, Anderson & Stevens 1973). Like MEPC analysis, noise analysis also indicates the relative numbers of fast and slow channels and their respective τ's (Sakmann & Brenner 1978, Fischbach & Schuetze 1980, Brehm et al 1982, Brenner & Sakmann 1983, Brenner et al 1983, Kullberg & Kasprzak 1985, Vicini & Schuetze 1985). It offers the advantage of being able to assay channels in both endplate and non-endplate regions (provided the receptor density is high enough), but it is somewhat less precise than MEPC analysis (S. M. Schuetze and S. Vicini, unpublished observations). The details of noise analysis are complex and are not discussed here (see Anderson & Stevens 1973, Kullberg & Kasprzak 1985, Vicini & Schuetze 1985 for details).

ACh noise signals most often are recorded by using a two-electrode voltage clamp to measure the membrane current evoked by iontophoretically applied ACh (e.g. Sakmann & Brenner 1978). This method yields information about both τ and γ. One limitation of this approach, which is a particular problem in studies of denervated muscle, is that it is difficult to activate endplate AChRs without also activating nearby extrajunctional AChRs. An alternative approach is focal extracellular recording. In this method, an electrode filled with dilute ACh is lowered onto the membrane being assayed, and the same electrode both applies ACh and records the resulting inward currents (Schuetze et al 1978, Schuetze 1980). ACh may be applied by passive diffusion or by pressure ejection (Kullberg & Kasprzak 1985). This technique assays only those AChRs beneath the electrode tip, thereby permitting independent comparisons of AChRs in neighboring membrane regions (Schuetze 1980). Extracellular recording sacrifices information about γ and absolute current amplitudes, but in return if offers simplicity, excellent frequency response, and less cellular damage.

In several studies (e.g. Sakmann & Brenner 1978, Kullberg & Kasprzak 1985, Vicini & Schuetze 1985), two or more of these techniques have been applied to the same preparation, and in each case the different approaches give comparable results.

Regulation of Acetylcholine Receptor Channels in Developing Rat Muscle

ACETYLCHOLINE RECEPTOR CHANNELS AT RAT ENDPLATES The fraction of slow channels at rat endplates decreases during early postnatal life (Sakmann & Brenner 1978, Fischbach & Schuetze 1980, Michler & Sakmann 1980, Vicini & Schuetze 1985). In the most extensive study, MEPCs were recorded extracellularly from 282 developing soleus endplates (Vicini & Schuetze 1985). The fraction of slow channels decreases steadily from virtually 100% immediately after birth to <20% three weeks later. Similar results were seen with noise analysis. The more sensitive techniques used in this study reveal that these changes in AChR channel properties begin earlier and end later than reported in earlier studies (Sakmann & Brenner 1978, Fischbach & Schuetze 1980, Michler & Sakmann 1980).

The change in channel properties at individual endplates cannot be determined from population studies because during the transition period, some endplates contain all fast channels, others within the same muscle have all slow channels, and still others have various proportions of both (Vicini & Schuetze 1985). Because of this asynchrony, individual endplates must be observed over time, a virtually impossible task for endplates developing in situ.

However, long-term recordings are possible at endplates in vitro. In a recent study (Schuetze & Vicini 1986), MEPCs were recorded repeatedly from individual endplates in isolated neonatal soleus muscles for up to 24 hr. At 57 out of 97 developing endplates, the relative amplitude of the fast MEPC decay component increased by between 10% (the minimum amount judged significant) and 53% over several hours. MEPCs changed significantly at 2/3 of the endplates that had both channel types when first tested, but not at all at endplates with virtually all slow or all fast channels. That is, MEPCs changed only at endplates that apparently had begun to acquire fast channels in vivo. The results are consistent with a change in channel properties in vitro (half-time of about 7 hr) at about half of the endplates assayed. Noise analysis experiments also indicated rapid changes in channel properties.

Changes in AChR channel properties probably occur more slowly in vivo. During the postnatal transition period, most endplates have significant levels of both fast and slow channels; few appear either fully mature or fully immature. If the changes at individual endplates in vivo were as rapid as the in vitro data suggest, endplates with high levels of both fast and slow channels should have been much rarer than was actually observed. One possibility consistent with all the data is that the shift in channel properties is not a smooth, continuous process but rather an intermittent one in which periods of change are interrupted by periods of stability. In this view, the 40 out of 97 endplates that did not change in the in vitro studies were in such a stable period.

ACETYLCHOLINE RECEPTOR CHANNELS IN NON-ENDPLATE REGIONS OF RATS The developmental changes in AChR channel properties in rat muscle (as in *Xenopus*—see below) are not limited to the endplate region. Unlike adult muscle fibers, innervated neonatal muscle fibers frequently contain many AChRs in extrajunctional regions (Bevan & Steinbach 1977). These usually disappear a few days after birth, but their loss is slowed when the contralateral muscle is denervated (cf Steinbach 1981b). Focal extracellular noise analysis experiments have shown that these extrajunctional AChRs have the same mix of fast and slow channels as endplate AChRs on the same innervated fibers (Schuetze & Vicini 1984). In a few fibers the vast majority of both endplate AChRs and extrajunctional AChRs were clearly fast.

P. Brehm and R. Kullberg (personal communication) recently confirmed and extended these results. They patch-clamped adult mouse flexor digitorum brevis muscle fibers at various points along their lengths. Although AChRs were present in extrajunctional regions only at a very low density, the majority of those that were detected were high-γ, fast channels regardless of their distance from the endplate.

REGULATION OF RAT ACETYLCHOLINE RECEPTOR CHANNELS BY NERVE AND MUSCLE ACTIVITY Denervating rat soleus muscles shortly after birth, when endplates still have mostly slow channels, blocks the normal developmental appearance of fast channels. Schuetze & Vicini (1984) found that 2–14 days after neonatal denervation, the former endplate regions, identified by a histochemical stain for AChE, had primarily slow channels. It could not be determined whether denervation prevented or merely delayed the appearance of fast channels. Within 2–3 wk after neonatal denervation, muscles became too atrophied and covered with connective tissue to study.

Studies of endplate formation in adults (Brenner et al 1983) suggest that denervation exerts these effects via the loss of muscle activity rather than the absence of a neural factor. Foreign nerves can innervate denervated adult muscles fibers at ectopic sites (away from the "old" endplate). AChRs at new ectopic endplates also undergo changes in their channel properties during the 2–3 wk after innervation (Brenner & Sakmann 1978, 1983). To test the role of the nerve in this process, Brenner et al (1983) cut the foreign nerve before fast channels appeared and stimulated the denervated muscles electrically with chronically implanted electrodes. Noise recordings showed clearly that the ectopic endplate regions acquired fast channels normally. Unfortunately, an absolute requirement for muscle activity was not demonstrated, because for technical reasons unstimulated muscles could not be tested.

Further evidence that the acquisition of fast channels requires innervation or activity comes from studies of embryonic rat myotubes in vitro. Siegelbaum et al (1984) reported that no more than 2–3% of the channel openings were fast, even in myotubes studied 2 wk after plating. Similar results were found by Fischbach & Schuetze (1980) and F. Jaramillo, L. L. Rubin, and S. M. Schuetze (unpublished observations). Hamill & Sakmann (1981), however, reported up to 40% fast channels on embryonic rat muscle in vitro. The reason for this discrepancy is not known, but perhaps there was a high level of spontaneous muscle activity in some cultures that promoted the expression of fast channels.

Apparently neither the nerve nor muscle activity is required for the continued expression of fast channels in adult endplates. When adult rat muscles are denervated, at least half of the channels at former endplates are fast even 18–20 days after denervation (Brenner & Sakmann 1983), when most of these receptors presumably were replaced at least once (cf Levitt & Salpeter 1981). In contrast, the newly-synthesized AChRs that appear in extrajunctional regions have channel properties similar to those of embryonic AChRs (Sakmann 1978, Fischbach & Schuetze 1980, but see *Additional Types of Acetylcholine Receptor Channels*, below). Brenner & Sakmann (1983) propose that some feature unique to mature endplates promotes fast gating.

In short, the combined results of the denervation studies on neonatal and adult ectopic endplates suggest that the appearance of fast AChR channels requires muscle activity, but not the continued presence of the nerve. However, activity has not been shown directly to be necessary for adults to acquire fast channels, nor has activity been demonstrated to promote the appearance of fast channels at endplates denervated neonatally. This is important, because AChR channels at neonatal endplates might be regulated differently than at ectopic endplates developing in adulthood. Finally, the continued presence of fast channels at adult endplates even several weeks after denervation suggests that innervation or activity may be more important for acquiring fast channels than for maintaining their expression.

Regulation of Acetylcholine Receptor Channels in Developing Xenopus *Myotomal Muscle*

ACETYLCHOLINE RECEPTOR CHANNELS IN *XENOPUS* MYOTOMAL MUSCLE IN VIVO The proportion of fast channels in *Xenopus* myotomal muscle increases during development, much as in rat, although on a somewhat faster time scale (Kullberg et al 1981, Brehm et al 1984a, Kullberg &

Kasprzak 1985). Noise analysis and MEPC studies have shown a decrease in τ from ~ 3 msec (room temperature) at new endplates to ~ 1 msec at stage 42 (a little over two days later; Kullberg et al 1981, Kullberg & Kasprzak 1985). MEPC recordings indicate a further decrease in τ to ~ 0.7 msec by stage 50 (about 2 wk later).

As in rat, extrajunctional regions of *Xenopus* myotomal muscle also express high levels of fast channels (Kullberg et al 1981, Kullberg & Kasprzak 1985). Indeed, developmental changes in the channel properties of nonendplate AChRs were first described in this system. This has been shown with both ACh noise (Kullberg et al 1981, Kullberg & Kasprzak 1985) and single-channel recordings (Brehm et al 1984a). Prior to stage 33, virtually all channels are the slow, low-γ type. The fraction of fast channels increases to $\geq 65\%$ of the total by about stage 48 (5–6 days later; Kullberg & Kasprzak 1985). The γ and τ of both fast and slow channels remain the same throughout this period, but Leonard et al (1984) have found that the very first low-γ AChRs that are expressed by newly formed myocytes have an unusually long τ (see below).

REGULATION OF *XENOPUS* ACETYLCHOLINE RECEPTOR CHANNELS BY NERVE AND MUSCLE ACTIVITY Recently, Kullberg et al (1985) found that raising embryos in the presence of the Na^+ channel blocker, tetrodotoxin, which completely immobilizes the developing frogs, does not block the appearance of fast channels. Their results, based on MEPC analysis, indicate that muscle activity is not a requirement for expressing fast channels in *Xenopus*. However, Cohen et al (1984) found that raising *Xenopus* embryos in the local anesthetic, tricaine, which also immobilizes embryos, does block the normal decrease in MEPC decay time. The reason for this discrepancy is not clear. In both studies, endplates accumulated AChE normally, suggesting that differences in ACh hydrolysis were not responsible for these contrasting results.

Other evidence that innervation is not required for expressing fast channels in *Xenopus* comes from in vitro studies. ACh noise analysis experiments show that, in contrast to rat muscle, *Xenopus* myotomal muscle in vitro expresses high levels of fast channels (Brehm et al 1982). This has been confirmed in more recent patch-clamp studies (Brehm et al 1984b, Leonard et al 1984, Greenberg et al 1985). When cultures are prepared from stage 17–20 embryos, in which myotomes have not yet been innervated, about 85–95% of the channel openings recorded from newly plated myocytes are slow. One day later 40% are fast, and five days after plating $>56\%$ are fast. The γ and τ of each channel population remain unchanged. In short, changes in AChR channel properties in uninnervated myocytes in vitro are similar to those of innervated myocytes in vivo.

The *Xenopus* culture system is unusual, however, in that innervation appears to inhibit and even reverse the appearance of fast channels in *Xenopus* myocytes in vitro. Brehm et al (1982) studied myocytes cocultured with neurons and found that about 10% of the channels were fast versus ~70% in sister cultures that had no neurons. Slow channels were equally prevalent in junctional and extrajunctional regions. Apparently this effect is mediated by some factor in the medium, because noninnervated myotubes in the cocultures also had few fast channels, and in one experiment this was mimicked by nerve-conditioned medium.

In summary, these studies indicate that neither innervation nor muscle activity is required for *Xenopus* myocytes to express fast channels. Possibly, expression of fast channels is part of the inherent developmental program of *Xenopus* myocytes. However, the loss of fast channels in nerve-muscle cocultures suggests that if such a program exists, it may be subject to revision by unidentified factors.

Possible Mechanisms for Regulating Acetylcholine Receptor Channel Properties

In principle, fast and slow channels could be identical molecules in different membrane environments (one-channel hypothesis); they could be different gene products (two-channel hypothesis); or fast channels could be derived from slow ones via a post-translational modification (conversion hypothesis). These various possibilities are not necessarily mutually exclusive, though most authors tend to favor only a single alternative. At present, the most direct experiments favor the two-channel hypothesis.

ONE-CHANNEL HYPOTHESIS According to the one-channel hypothesis, an individual AChR can express either adult-like or embryonic-like channel properties depending upon its microenvironment. Some developmental change in the structure of the endplate region is supposed to promote fast, adult-like gating. It has been proposed that the formation of postsynaptic membrane folds might change AChR channel properties, perhaps as a consequence of the anchoring of receptors to the tops of the folds (Fischbach & Schuetze 1980, Brenner & Sakmann 1983, Brenner et al 1983). A fairly large but indirect body of evidence supports this idea.

1. Population studies indicate that postsynaptic folds form in parallel with the appearance of fast channels both at neonatal rat endplates (Kelly & Zachs 1969, Korneliussen & Jansen 1976) and at adult rat ectopic endplates (Brenner et al 1983).
2. When adult rat muscles are denervated, endplate regions continue to have both fast channels and membrane folds for weeks, whereas the

newly-appearing extrajunctional channels are slow (Brenner & Sakmann 1983).

3. Adult chick endplates express neither folds (Atsumi 1971) nor fast channels (Schuetze 1980).

However, four lines of evidence argue against the one-channel hypothesis. First, Vicini & Schuetze (1985) found no correlation between the extent of postsynaptic membrane folding and the percentage of adult-type channels in individual rat endplates. Second, approximately equal numbers of fast channels can be found in junctional and extrajunctional regions of developing *Xenopus* (Kullberg et al 1981, Brehm et al 1984a,b) and rat (Schuetze & Vicini 1984) muscle fibers, indicating that fast channels do not require any endplate-specific specialization. Third, both channel types can be observed in individual patch clamp records, thus demonstrating that fast and slow channels can exist side by side (Hamill & Sakmann 1981, Brehm et al 1984a,b, Vicini & Schuetze 1985). Finally, in bovine muscle, differences in AChR subunit composition can account for the developmental changes in channel properties (see below).

TWO-CHANNEL HYPOTHESIS Evidence indicating that fast and slow channels in rats differ in their structure came from studies using antibodies from myasthenic patients. Myasthenic sera typically have some antibodies that bind embryonic rat AChRs selectively (Dwyer et al 1981, Reiness & Hall 1981, Hall et al 1983), but in one exceptional serum virtually all the antibodies are of this type (Hall et al 1985). Immunocytochemical studies showed that these antibodies bind to endplate receptors in neonatal rats but not in adults, and the developmental loss of binding activity parallels the loss of slow channels.

These antibodies block channel function when they bind to AChRs (Maricq et al 1985), a trait that made possible a critical experiment. When rat soleus endplates that contained both types of channels were treated with these antibodies, the contribution of slow channels to endplate currents and to noise records decreased strikingly and selectively (Schuetze et al 1985). This result strongly suggests that some determinant on the extracellular portion of the molecule distinguishes rat AChRs with fast and slow channels.

Experiments performed on *Xenopus* myocytes indicate that the two types of AChRs are different proteins that are synthesized independently. Both Brehm et al (1985) and Carlson et al (1985) have reported that the appearance of fast channels in *Xenopus* cultures, which normally happens spontaneously, is blocked by adding the protein synthesis inhibitor, cycloheximide, to the culture medium. For example, Brehm et al (1985) found that the level of fast channels increased from 12% to 40% over 24 hr in control cultures but remained unchanged in sister cultures treated with inhibitor. Carlson et al (1985) reported similar results in cultures treated with either cycloheximide or tunicamycin, an inhibitor of glycosylation.

More recently, Mishina et al (1986) have shown directly that bovine AChRs with fast and slow channels are distinct proteins with different subunit compositions. In addition to the α, β, γ, and δ subunits, bovine muscle expresses a fifth type of AChR subunit called ε that is homologous to γ (Takai et al 1985). Two sets of experiments strongly suggest that a switch from an $\alpha_2\beta\gamma\delta$-subunit composition to $\alpha_2\beta\varepsilon\delta$-subunit composition underlies the developmental changes in channel properties.

First, single-channel recordings show that a substitution of the γ subunit by ε mimics all the known developmental changes in AChR channel properties. These experiments were performed by transcribing cDNAs encoding the various calf AChR subunits in vitro and injecting a mixture of the subunit-specific mRNAs into *Xenopus* oocytes. When given a complete set of AChR mRNAs, oocytes translate the messages, assemble the subunits, and insert functional AChRs into their surface membranes (Mishina et al 1984). The channel properties of the "foreign" AChRs can be studied by single-channel recording (Sakmann et al 1985).

When oocytes were injected with a mixture of mRNAs encoding the bovine α, β, γ, and δ subunits, they expressed AChRs that were physiologically identical to those of fetal bovine muscle fibers in vitro (Mishina et al 1986). In both cases the AChR channels displayed the same τ (13 msec), the same γ (40 pS), the same voltage dependence, and the same ionic selectivity. In addition, oocytes injected with a mixture of bovine α, β, ε, and δ mRNAs expressed AChRs that were physiologically indistinguishable from those expressed at adult bovine endplates ($\tau = 6$ msec, $\gamma = 60$ pS; note that the bovine adult τ is long compared to that of rats).

Other experiments showed that γ- and ε-subunit mRNAs are expressed at the appropriate times during development (Mishina et al 1986). Blot hybridization analysis of developing bovine diaphragm mRNA indicates that γ-mRNA is abundant at early fetal stages (3–5 mo of gestation) but is not expressed after birth. The pattern of ε-mRNA expression is just the opposite: It cannot be detected at 3–4 mo of gestation, but is relatively abundant after birth. At intermediate times (7–8 mo of gestation) both γ-mRNA and ε-mRNA are expressed in quantity. (This is in contrast to an earlier report (Takai et al 1985) that suggested that ε-mRNA was found only early in development and not in adulthood.)

In short, these experiments provide strong evidence that the developmental changes in AChR channel properties in calf muscle are due to the loss of γ-subunit mRNA coincident with an increase in the amount of ε-subunit mRNA. This change in gene expression is probably regulated at the transcriptional level, but developmental changes in message stability have not been ruled out.

CHANNEL CONVERSION The evidence for channel conversion is indirect. Michler & Sakmann (1980) measured both the AChR degradation rate and the time course of changes in AChR channel properties in developing rat muscles and concluded that the rate of AChR turnover was too slow to account for the appearance of fast channels. However, the more extensive data of Vicini & Schuetze (1985) on the same preparation indicate that the functional changes are slower than reported by Michler & Sakmann and occur with a time course consistent with the replacement of slow channels with fast ones.

However, the rapid changes in MEPC decays at endplates in excised rat soleus muscles maintained in vitro appear to occur faster than AChR replacement (Schuetze & Vicini 1986). As discussed above, in this system the relative fraction of fast channels often increases strikingly within a few hours. Although the rate of receptor turnover was not measured directly in this study, endplate AChRs have a half-life of ~7–11 days and remain stable when muscles are placed in vitro (Steinbach et al 1979). Thus, AChR turnover is slow compared to the changes in MEPC decays. This suggests that in addition to replacing slow channels with fast ones, there may be a mechanism for modifying the channel properties of embryonic-type AChRs.

CONCLUSIONS FROM DEVELOPMENTAL STUDIES OF ACETYLCHOLINE RECEPTOR CHANNEL PROPERTIES The recent data of Mishina et al (1986) provide convincing evidence that AChR channel properties in bovine muscle can be regulated by differential expression of γ and ε subunits. Other species may employ a similar mechanism. Indirect evidence suggests that other means of modifying channel properties may also exist.

During development, junctional and extrajunctional membrane regions contain equal numbers of fast and slow channels. This suggests that all the nuclei within each muscle fiber begin to express ε-subunit mRNA in parallel. After denervation, however, fast channels are found only in endplate regions and slow channels are expressed elsewhere. Therefore, perhaps endplate nuclei in mature muscle fibers express ε-type AChRs constitutively, whereas extrajunctional nuclei express γ-type AChRs in inactive fibers and few or no AChRs in active ones.

Additional Types of Acetylcholine Receptor Channels

Most of the studies reviewed above are consistent with the idea that the skeletal muscle of any given species expresses two types of AChR channels, fast and slow. In this view, the proportion of fast and slow AChR channels may vary, but not the channel properties of each type. Several studies indicate that this idea is too simplistic.

First, the properties of endplate channels apparently vary among muscles with different muscle properties. For example, Miledi & Uchitel (1981) used noise analysis and endplate current decays to study gating in different muscles in frogs (*Rana temporaria*). They found a briefer τ and higher γ at fast twitch fiber endplates (5 msec, 16 pS at 5°C) than at nontwitch, slow fiber endplates (15 msec, 8 pS). Fibers with structural and functional properties intermediate between these fiber types had intermediate AChR channel properties (7 msec, 13 pS). A correlation between contraction velocity and MEPC decay rates has also been suggested in garter snake (Dionne & Parsons 1978, 1981) and *Xenopus* (Kullberg & Owens 1986) muscles. Whether these differences in channel properties are due to different AChR structures or to differences in the AChR's membrane environment remains unknown.

Second, increasing evidence suggests that muscles can express more than one type of low-conductance, slow AChR channel. Leonard et al (1984) found that when young *Xenopus* myocytes that have not yet expressed AChRs are placed into cell culture, the first AChRs that appear have low-γ channels with an unusually long τ. Two to three days later, the τ of low-γ channels is about threefold shorter, as reported by many others (see above). Steele & Steinbach (1986) have found that the τ of extrajunctional AChRs in denervated adult mouse muscle is shorter than the τ of embryonic AChRs, even though both have the same low γ. Finally, Auerbach & Lingle (1986) have found three different "burst modes" of *Xenopus* AChR gating (this was also true for AChRs with high-γ channels). In all these cases, it is not clear whether there is one type of low-γ AChR channel that can operate differently in different environments, or whether there are two or more types of low-γ AChR channels that differ in their structure.

The results of most studies of AChR gating can be explained on the basis of just two types of AChR channels, whose properties remain the same throughout development. However, accumulating evidence supports additional types of AChR channels, and it will be important to determine how they are regulated.

Functional Role of Developmental Changes in Channel Properties

The physiological importance of the changes in AChR channel properties has not been established, but some possibilities can be offered. Slow channels may promote synapse maturation by increasing muscle activity. Spontaneous miniature endplate potentials are larger and slower in neonatal rat soleus muscle than in adult muscle, and occasionally they are superthreshold and initiate fiber contraction (Diamond & Miledi 1962, Dennis et al 1981; S. M. Schuetze and S. Vicini, unpublished data). Computer simulations indicate that the high resistance of the small neonatal fibers and the long open times of embryonic AChRs contribute roughly equally to the large size of miniature endplate potentials in young rats (S. M. Schuetze and S. Vicini, unpublished data). Several events in synaptic development have been shown to depend on muscle activity (reviewed by Dennis 1981), and slow channels may help promote synapse maturation by generating spontaneous contractions. Slow channels may also increase the safety factor for evoked neuromuscular transmission at newly formed endplates.

Then why shift to fast channels? One possibility is that fast channels may be required to limit calcium entry at the endplate region to a nontoxic level. Salpeter et al (1982a) have found that poisoning synaptic AChE, which prolongs endplate currents, leads to myopathy at the endplate region within hours. Their data suggest that slowing the synaptic currents increases calcium influx at the endplate region, and this in turn activates calcium-sensitive intracellular proteases. The myopathy can be prevented by removing extracellular calcium or by blocking AChRs with α-BGT before poisoning AChE.

Failing to express fast channels might also lead to excessive calcium influx. Engel et al (1982) have reported a disease in which patients have many of the same symptoms as myasthenia gravis, such as muscle fatigue, but no circulating anti-AChR antibodies. These patients have abnormally slow synaptic currents, though their AChE levels appear to be normal. Engel et al suggest that the primary cause of the disease is the failure to express fast channels and have termed the disease "slow channel syndrome."

Regulation of Acetylcholine Receptor Channel Properties in Neurons

At least some types of ACh-sensitive neurons also have two types of channels that differ in open time. Both noise analysis and analysis of evoked synaptic currents indicate that individual rat submandibular ganglion neurons have both fast ($\tau \sim 5$–9 msec at 20°C) and slow ($\tau \sim 27$–45 msec) AChR channels (Rang 1981). The fast channels contribute slightly over half of the total peak amplitude of evoked synaptic currents. Curiously, spontaneous miniature synaptic currents have only a fast decay phase. Rang (1981) suggests that fast and slow channels are spatially separate, and that spontaneous transmitter release occurs only at regions with fast channels. Subsequently, Gray & Magnus (1983) found that both types of channels are found in newborn rat submandibular ganglion cells grown in vitro. In contrast to newborn skeletal muscle, the neonatal neurons did not appear to be enriched for either channel type compared to adult neurons.

Using ACh noise analysis, Derkach et al (1983) also detected two AChR channel populations, similar to those described by Rang, in rabbit superior cervical ganglion neurons. However, in this system both evoked and spontaneous synaptic currents had singly exponential, slow decays. This finding is consistent with the idea that the slow channels are synaptic and the fast ones are extrasynaptic, a striking contrast to the results reported by Rang.

More recently, Marshall (1986) has studied AChR channels in bullfrog lumbar sympathetic ganglia. These ganglia contain two kinds of principal neurons (B and C) that differ in size and in the velocity at which they conduct action potentials. They also are innervated by distinct preganglionic axons (B and C fibers). Noise analysis and synaptic current analysis indicate that B cells contain a homogeneous population of fast AChR channels ($\tau \sim 5$ msec at 22°C), whereas C cells have a single population of slow channels ($\tau \sim 10$ msec). When B cells are denervated, their AChR channels remain fast. This suggests that B cells (and perhaps also C cells) express a single type of channel whether innervated or denervated.

Marshall (1985) went on to study denervated B cells that were reinnervated by C fibers. The results of both noise analysis and synaptic current analysis experiments showed clearly that these cross-innervated B cells lost their fast channels and acquired the slow channels characteristic of C cells. This study provides the first evidence that the gating kinetics of AChR channels can be regulated by their presynaptic input, and is reminiscent of older studies showing presynaptic control of the contractile properties of skeletal muscle fibers. It will be important to find the mechanism behind this switch.

MODULATION OF ACETYLCHOLINE RECEPTOR FUNCTION

Modulation of Acetylcholine Receptors by Peptides and Amines

Substance P has been shown to alter the activation of the neuronal AChR at both central and peripheral synapses. The cholinergic synapse of motoneuron axon-collatorals on the Renshaw cell is mediated by both nicotinic and muscarinic AChRs; the former are selectively inhibited by substance P (Belcher & Ryall 1977, reviewed in Ryall 1982). Nicotinic activation of chromaffin cells (Mizobe et al 1979, Role et al 1981, Clapham & Neher 1984, Boksa & Livett 1984, 1985), ciliary ganglion neurons (Role 1984a, Margiotta & Berg 1986), PC 12 cells (Stallcup & Patrick 1980, Boyd et al 1983, Simasko et al 1985), and sympathetic neurons (Akasu et al 1983b, Role 1984a) is inhibited by substance P. The peptide has also been shown to decrease miniature EPSP amplitude at the Mauthner fiber-giant synapse of the hatchetfish (Steinacker & Highstein 1976) and at the nerve-muscle synapse of the frog (see below and Akasu et al 1984).

Examination of the rate of decay of macroscopic inward currents in voltage clamped chick neurons and chromaffin cells suggests that one effect of the peptide may be an increase in the rate of inactivation of the neuronal AChRs (Clapham & Neher 1984, Role 1984a). Neuronal AChR channels appear in bursts and clusters with agonist induced desensitization (Clapham & Neher 1984), as is seen with muscle AChRs (Sakmann et al 1980). Substance P decreases ACh current in part by increasing the interburst interval and decreasing burst duration of the AChR channels. In addition, substance P causes a decrease in the mean AChR channel open time (Clapham & Neher 1984). Consistent with this, substance P induces a decrease in the EPSC half duration in innervated sympathetic neurons (Role 1984b), although the peptide had no effect on time course of miniature endplate potentials at the frog nerve-muscle synapse (Akasu et al 1983b).

The effects of substance P on cholinergic activation are dose dependent, reversible, and specific for the nicotinic receptor in chromaffin cells, sympathetic neurons, and ciliary ganglion neurons (Mizobe et al 1979, Role et al 1981, Akasu et al 1983b, Clapham & Neher 1984, Role 1984a,b, Margiotta & Berg 1986). Half maximal effects of the peptide are seen at about 0.5–1 μM (Mizobe et al 1979, Role et al 1981, Akasu et al 1983b, Role 1984b). Since immunoreactive substance P is present in some autonomic ganglia in which the effects of the peptide on the AChRs are seen (Erichsen et al 1982, Hayashi et al 1983), substance P may function physiologically to modulate synaptic activation of the neurons. Preliminary studies in innervated sympathetic neurons in vitro and in cardiac ganglion neurons in vivo suggest that this might be the case (Role 1984b, Bowers et al 1984).

Koketsu and his colleagues observe AChR modulation by substance P in the frog that may be quite different and is apparently a very general phenomenon. ATP, ADP, luteinizing hormone releasing hormone (LHRH), epinephrine, serotonin, and substance P all modulate peak ACh responses in frog sympathetic ganglia but have no effect on the rate of ACh current decay (Koketsu et al 1982, Akasu et al 1983a,b, 1984, Akasu & Koketsu 1985). These investigators also find that substance P (as well as LHRH, catecholamines, and serotonin) decreases peak ACh responses at the nerve-muscle synapse of the frog, without affecting either ACh or endplate current decay rate (Akasu et al 1983b).

Both somatostatin and enkephalin have also been reported to decrease the nicotinic activation of catecholamine release from chromaffin cells

(Mizobe et al 1979, Kumakura et al 1980, Role et al 1981, Saiani & Guidotti 1982). Although the concentrations required for inhibition by opiate peptides are high, the effect is blocked by specific antagonists, and the pharmacology of AChR inhibition agrees well with the kinetics of opiate ligand binding. The presence of opiate peptides in both presynaptic and chromaffin cell elements in the adrenal gland (Schultzberg et al 1978, Viveros et al 1979) are consistent with high local concentrations of the peptide, suggesting a possible physiological role for opiates in chromaffin cell receptor regulation despite previous observations to the contrary (e.g. Dean et al 1982).

Modulation of ACh responses by small amines such as serotonin (5HT) has recently been reported in *Aplysia* (Simmons & Koester 1986). 5HT increases the ACh response of the RB cells of the abdominal ganglia in a dose dependent manner. The effect is specific in that other amines, such as histamine and dopamine, do not affect the ACh response of the neurons. Since this effect of 5HT is specific to the class of cholinergic neurons that receives 5HT input in vivo, it may represent a physiological modulation of nicotinic transmission.

Modulation of Skeletal Muscle Acetylcholine Receptor

Studies of AChRs from *Torpedo* electric organ have demonstrated that all four subunits are phosphoproteins (Gordon et al 1977, Teichberg et al 1977, Huganir & Greengard 1983, Huganir et al 1984). A tyrosine-specific protein kinase phosphorylates the β, γ, and δ subunits, protein kinase C phosphorylates serine residues on the α and δ subunits, and cAMP-dependent protein kinase phosphorylates the γ and δ subunits (reviewed in Huganir et al 1984). Recently, several groups have begun to study the effects of AChR phosphorylation on channel gating. In other systems, protein kinases have clear effects on ionic channel gating (reviewed in Browning et al 1985 and Siegelbaum & Tsien 1983), though in virtually all cases the critical phosphoprotein has not been identified and may not be the channel protein itself.

Eusebi et al (1985) have studied the effects of agents that stimulate protein kinase C, such as the tumor promoter 12-O-tetradecanoyl-phorbol-13-acetate (TPA) and glyceryl dioleate, on the ACh sensitivity of chick and mouse myotubes in vitro. They found that the ACh sensitivity began to fall within minutes after adding these agents. In some cases ACh sensitivity fell to 10–50% of that of controls, and recovered again after washout. In mouse, but not chick, myotubes there was recovery even in the continued presence of the drugs, indicating complex effects of kinase activation. Eusebi et al (1985) also noted an increase in the rate of desensitization after stimulating protein kinase C. However, they did not demonstrate that these results were due to phosphorylation of the AChR itself.

More recently, Middleton et al (1986) and Albuquerque et al (1986) studied the effects of forskolin, a potent activator of adenylate cyclase, on AChR function at rat soleus endplates. Stimulating adenylate cyclase raises the intracellular level of cAMP, thereby activating cAMP-dependent protein kinase. Both groups found that forskolin enhances AChR desensitization. Middleton et al (1986) found that 10–100 μM forskolin increased the rate of desensitization evoked by iontophoretic ACh pulses by two orders of magnitude with no immediate decrease in ACh sensitivity. Albuquerque et al (1986) found similar but smaller effects with lower doses of forskolin. In all cases forskolin did not densensitize AChRs directly, but rather increased the rate of AChR desensitization induced by agonist.

These results probably were due to AChR phosphorylation by cAMP-dependent kinase. Chemical analogs of forskolin that are less potent in activating adenylate cyclase were also less effective in promoting desensitization (Albuquerque et al 1986). In addition, the effects of forskolin were potentiated by inhibitors of cAMP phosphodiesterase, and these inhibitors had small, but qualitatively similar, effects when used alone (Middleton et al 1986). Finally, forskolin treatment stimulates phosphorylation of AChRs in rat myotubes in vitro with a time course and concentration dependence appropriate to the physiological effects of the drug (unpublished data of D. Anthony, R. Huganir, P. Middleton, K. Miles, L. Rubin, and S. Schuetze).

Much stronger evidence that AChR phosphorylation by cAMP-dependent kinase speeds desensitization comes from recent work of Huganir et al (1986) on *Torpedo* AChRs. These workers used rapid quench-flow and stop-flow kinetic techniques to study ion transport by purified AChRs that were reconstituted into phospholipid vesicles. They compared the properties of nonphosphorylated AChRs with those of AChRs phosphorylated by cAMP-dependent protein kinase. The results showed clearly that for any given concentration of agonist, the fast phase of desensitization was much more pronounced for the phosphorylated AChRs.

FUTURE DIRECTIONS

During synaptic development, skeletal muscle AChRs change their distribution, metabolic stability, and functional properties. Neural nicotinic AChRs undergo developmental changes as well. Even though the AChR is by far the best-characterized chemically gated channel, most of these changes in AChRs are not yet understood at the molecular level.

Developmental changes in AChR structure can be studied by isolating cDNA clones encoding the various AChR subunits. Use of these clones in an appropriate expression system should elucidate the functional consequences of structural changes. Indeed, recent experiments using this approach indicate that the developmental changes in AChR channel properties are due to a switch in the expression of the ε- and γ-subunits. Similar experiments should prove useful for comparing embryonic AChRs with those that appear extrasynaptically in adults following denervation, for comparing AChRs expressed by different muscles, and for studying the structural and functional properties of neuronal AChRs. Site-directed mutagenesis of the cDNA clones should provide additional insights into AChR channel function.

In situ hybridization with probes complementary to AChR message should prove useful in determining the role of receptor gene expression in cluster formation during synaptogenesis. Adequate resolution and sensitivity will enable us to determine whether synaptic nuclei are selectively activated by innervation and whether transcription by extrasynaptic nuclei is regulated by muscle activity. Similarly, use of probes specific for particular subunits should make possible the determination of whether different nuclei express AChRs of different subunit composition. This will pave the way for exploring the mechanisms that might regulate such differential gene expression.

However, some of the developmental regulation and modulation of the AChRs might be due to post-translational modifications, to changes in the receptor's membrane environment, or to interactions between the AChR and other cellular elements. Changes in the distribution and stability of AChRs are clearly influenced by other molecules. Likely candidates include neurally derived factors, 43 kD protein, cytoskeletal proteins, and basal lamina. Determining the precise structure, developmental expression, and distribution of these molecules is critical to understanding their role in regulating AChRs. Finally, it will be important to determine the molecular interactions between the AChR and these other proteins, and to explore possible mechanisms by which AChRs can be modified while in the surface membrane.

ACKNOWLEDGMENTS

We wish to thank D. Berg, P. Brehm, J. Dodd, K. Dunlap, G. Fischbach, T. Jessel, L. Marshall, L. Rubin, and A. Silverman for their critical readings of the manuscript. The preparation of this review and some of the work described was supported by the National Institutes of Health (L. W. R. and S. M. S.), the Alfred P. Sloan Foundation (L. W. R. and S. M. S.), the Klingenstein Foundation (L. W. R.), the Muscular Dystrophy

Association (S. M. S.), R. Pollock (S. M. S.), and E. B. Rothenberg (L. W. R.).

Literature Cited

Adams, D. J., Bevan, S. 1985. Some properties of acetylcholine receptors in human cultured myotubes. *Proc. R. Soc. London Ser. B* 224: 183–96

Akasu, T., Koketsu, K. 1985. Effect of adenosine triphosphate on the sensitivity of the nicotinic acetylcholine-receptor in the bullfrog sympathetic ganglion cell. *Br. J. Pharmacol.* 84: 525–31

Akasu, T., Kojima, M., Koketsu, K. 1983a. Luteinizing hormone-releasing hormone modulates nicotinic ACh-receptor sensitivity in amphibian cholinergic transmission. *Brain Res.* 279: 347–51

Akasu, T., Kojima, M., Koketsu, K. 1983b. Substance P modulates the sensitivity of the nicotinic receptor in amphibian cholinergic transmission. *Br. J. Pharmacol.* 80: 123–31

Akasu, T., Ohta, Y., Koketsu, K. 1984. Neuropeptides facilitate the desensitization of nicotinic acetylcholine-receptor in frog skeletal muscle endplate. *Brain Res.* 290: 342–47

Albuquerque, E. X., Deshpande, S. S., Aracava, Y., Alkondon, M., Daly, J. W. 1986. A possible involvement of cyclic AMP in the expression of desensitization of the nicotinic acetylcholine receptor. A study with forskolin and its analogs. *FEBS Lett.* 199: 113–20

Anderson, C. R., Stevens, C. F. 1973. Voltage clamp analysis of acetylcholine produced end-plate current fluctuations at frog neuromuscular junction. *J. Physiol. London* 235: 655–91

Anderson, D. J. 1986. Molecular biology of the acetylcholine receptor: Structure and regulation of biogenesis. In *The Vertebrate Neuromuscular Junction*, ed. M. Salpeter. New York/Liss. In press

Anderson, D. J., Blobel, G. 1981. In vitro synthesis, glycosylation and membrane insertion of the four subunits of *Torpedo* acetylcholine receptor. *Proc. Natl. Acad. Sci. USA* 78: 5598–5602

Anderson, M. J., Cohen, M. W. 1974. Fluorescent staining of acetylcholine receptors in vertebrate skeletal muscle. *J. Physiol. London* 237: 385–400

Anderson, M. J., Cohen, M. W. 1977. Nerve-induced and spontaneous redistribution of acetylcholine receptors on cultured muscle cells. *J. Physiol. London* 268: 757–73

Anderson, M. J., Cohen, M. W., Zorychta, E. 1977. Effects of innervation on the distribution of acetylcholine receptors on cultured muscle cells. *J. Physiol. London* 268: 731–56

Anderson, M. J., Kidokoro, Y., Gruener, R. 1979. Correlation between acetylcholine receptor localization and spontaneous synaptic potentials in cultures of nerve and muscle. *Brain Res.* 166: 185–90

Anderson, M. J., Klier, F. G., Tanguay, K. E. 1984. Acetylcholine receptor aggregation parallels the deposition of a basal lamina proteoglycan during development of the neuromuscular junctions. *J. Cell Biol.* 99: 1769–84

Armstrong, D. M., Saper, C. B., Levey, A. I., Wainer, B. H., Terry, R. D. 1983. Distribution of cholinergic neurons in rat brain: Demonstrated by the immunocytochemical localization of choline acetyltransferase. *J. Comp. Neurol.* 216: 53–68

Ascher, P. W., Large, W. A., Rang, H. P. 1979. Studies on the mechanism of action of acetylcholine antagonists on rat parasympathetic ganglion cells. *J. Physiol. London* 295: 139–70

Atsumi, S. 1971. The histogenesis of motor neurons with special reference to the correlation of their endplate formation. I. The development of endplates in the intercostal muscle in the chick embryo. *Acta Anat.* 80: 161–82

Auerbach, A., Lingle, C. J. 1986. Heterogeneous kinetic properties of acetylcholine receptor channels in *Xenopus* myocytes. *J. Physiol. London* 378: 119–40

Axelrod, D., Bauer, H., Stya, M., Christian, C. N. 1981. A factor from neurons induces partial immobilization of nonclustered acetylcholine receptors on cultured muscle cells. *J. Cell Biol.* 88: 459–62

Axelrod, D., Ravdin, P., Koppel, D. E., Schles-singer, J., Webb, W. W., et al. 1976. Lateral motion of fluorescently labeled acetylcholine receptors in membranes of developing muscle fibers. *Proc. Natl. Acad. Sci. USA* 73: 4594–98

Axelsson, J., Thesleff, F. 1959. A study of supersensitivity in denervated mammalian skeletal muscle. *J. Physiol. London* 147: 178–93

Ballivet, M., Patrick, J., Lee, J., Heinemann, S. 1982. Molecular cloning of a cDNA coding for the gamma subunit of *Torpedo* acetylcholine receptor. *Proc. Natl. Acad. Sci. USA* 79: 4466–70

Bauer, H. C., Daniels, M. P., Pudimat, P. A., Jacques, L., Sugiyama, H., Christian, C. N. 1981. Characterization and partial purification of a neuronal factor which increases acetylcholine receptor aggregation on cultured muscle cells. *Brain Res.* 209: 395–404

Bauer, H. C., Hasegawa, S., Sonderegger, P., Daniels, M. P., Pudimat, P. 1985. Specificity of neuronal factors which aggregate acetylcholine receptors on cultured myotubes. *Exp. Cell Res.* 175: 288–92

Belcher, G., Ryall, R. W. 1977. Substance P and Renshaw cells: A new concept of inhibitory synaptic interactions. *J. Physiol. London* 272: 105–19

Berg, D. K., Hall, Z. W. 1975. Loss of α-bungarotoxin from junctional and extrajunctional acetylcholine receptors in rat diaphragm muscle in vivo and in organ culture. *J. Physiol. London* 252: 771–89

Betz, H. 1983. Regulation of α-bungarotoxin receptor accumulation in chick retina cultures: Effects of membrane depolarization, cyclic nucleotide derivatives and Ca^{2+}. *J. Neurosci.* 3: 1333–34

Bevan, S., Kullberg, R. W., Rice, J. 1978. Acetylcholine-induced conductance fluctuations in cultured human myotubes. *Nature* 273: 469–71

Bevan, S., Steinbach, J. H. 1977. The distribution of α-bungarotoxin binding sites on mammalian skeletal muscle developing in vivo. *J. Physiol. London* 267: 195–213

Blackshaw, S., Warner, A. 1976. Onset of acetylcholine sensitivity and endplate activity in developing myotome muscles of *Xenopus*. *Nature* 262: 217–18

Bloch, R. J. 1979. Dispersal and reformation of acetylcholine receptor clusters of cultured rat myotubes treated with inhibitors of energy metabolism. *J. Cell Biol.* 82: 626–43

Bloch, R. J. 1983. Acetylcholine receptor clustering in rat myotubes: Requirement for Ca^{2+} and effects of drugs which depolymerize microtubules. *J. Neurosci.* 3: 2670–80

Bloch, R. J., Geiger, G. 1980. The localization of acetylcholine receptor clusters in areas of cell-substrate contact in cultures of rat myotubes. *Cell* 21: 25–35

Bloch, R. J., Hall, Z. W. 1983. Cytoskeletal components of the vertebrate neuromuscular junction: Vinculin, α-actinin and filamin. *J. Cell Biol.* 97: 217–23

Bloch, R. J., Steinbach, J. H. 1981. Reversible loss of acetylcholine receptor clusters at the developing rat neuromuscular junction. *Devel. Biol.* 81: 386–91

Boksa, P., Livett, B. G. 1984. Substance P protects against desensitization of the nicotinic response in isolated adrenal chromaffin cells. *J. Neurochem.* 42: 618–27

Boksa, P., Livett, B. G. 1985. The substance P receptor subtype modulating catecholamine release from adrenal chromaffin cells. *Brain Res.* 332: 29–38

Boulter, J., Evans, K., Goldman, D., Martin, G., Treco, D., et al. 1986. Isolation of a cDNA clone coding for a possible neural nicotinic acetylcholine receptor α-subunit. *Nature* 319: 368–74

Bowers, C. W., Jan, L. Y., Jan, Y. N. 1984. Substance P-like peptide in preganglionic sympathetic nerve terminal of the bullfrog. *Soc. Neurosci. Abstr.* 14: 1120

Boyd, N. D., Anthony, M. P., Leeman, S. E. 1983. Structure-activity relationships of the inhibitory actions of substance P on nicotinic cholinergic receptors. *Soc. Neurosci. Abstr.* 13: 142

Braithwaite, A. W., Harris, A. J. 1979. Neural influence on acetylcholine receptor clusters in embryonic development of skeletal muscles. *Nature* 279: 549–51

Breer, H., Kleene, R., Hinz, G. 1985. Molecular forms and subunit of the acetylcholine receptor in the central nervous system of insects. *J. Neurosci.* 5: 3386–92

Brehm, P. 1986. Alteration in the rate of receptor degradation during development of *Xenopus* myotomal muscle. *Biophysical J.* 49: 362a (Abstr.)

Brehm, P., Bates, L. Kream, R., Moody-Corbett, F. 1985. Inhibition of acetylcholine receptor incorporation blocks developmental changes in channel gating in *Xenopus* muscle. *Soc. Neurosci. Abstr.* 11: 849

Brehm, P., Kidokoro, Y., Moody-Corbett, F. 1984a. Acetylcholine receptor channel properties during development of *Xenopus* muscle cells in culture. *J. Physiol. London* 357: 203–17

Brehm, P., Kullberg, R., Moody-Corbett, F. 1984b. Properties of non-junctional acetylcholine receptor channels on innervated muscle of *Xenopus laevis*. *J. Physiol. London* 350: 631–48

Brehm, P., Steinbach, J. H., Kidokoro, Y. 1982. Channel open time of acetylcholine receptors on *Xenopus* muscle cells in dissociated cell culture. *Devel. Biol.* 91: 93–102

Brehm, P., Yeh, E., Patrick, J., Kidokoro, Y. 1983. Metabolism of acetylcholine receptors on embryonic amphibian muscle. *J. Neurosci.* 3: 101–7

Brenner, H. R., Johnson, E. W. 1976.

Physiological and morphological effects of post-ganglionic axotomy on presynaptic nerve terminals. *J. Physiol. London* 260: 143–58

Brenner, H. R., Martin, A. R. 1976. Reduction in acetylcholine sensitivity of axotomized ciliary ganglion cells. *J. Physiol. London* 260: 159–75

Brenner, H. R., Meier, Th., Widmer, B. 1983. Early action of nerve determines motor endplate differentiation in rat muscle. *Nature* 305: 536–37

Brenner, H. R., Sakmann, B. 1978. Gating properties of acetylcholine receptor in newly formed neuromuscular synapses. *Nature* 271: 366–68

Brenner, H. R., Sakmann, B. 1983. Neurotrophic control of channel properties at neuromuscular synapses of rat muscle. *J. Physiol. London* 337: 159–71

Brown, D. A. 1969. Responses of normal and denervated cat superior cervical ganglia to some stimulant compounds. *J. Physiol. London* 201: 225–36

Brown, D. A., Docherty, R. J., Halliwell, J. V. 1983. Chemical transmission in the rat interpeduncular nucleus in vitro. *J. Physiol. London* 341: 655–70

Brown, D. A., Fumagalli, L. 1977. Dissociation of α-bungarotoxin binding and receptor block in the rat superior cervical ganglion. *Brain Res.* 129: 165–68

Browning, M. D., Huganir, R., Greengard, P. 1985. Protein phosphorylation and neuronal function. *J. Neurochem.* 45: 11–23

Buc Caron, M. H., Nystrom, P., Fischbach, G. D. 1983. Induction of acetylcholine receptor synthesis and aggregation: Partial purification of low molecular weight activity. *Devel. Biol.* 95: 378–86

Burden, S. 1977a. Acetylcholine receptors at the neuromuscular junction: Developmental change in receptor turnover. *Devel. Biol.* 61: 79–85

Burden, S. 1977b. Development of the neuromuscular junction in the chick embryo: The number, distribution, and stability of acetylcholine receptors. *Devel. Biol.* 57: 317–29

Burden, S. J., De Palma, R. L., Gottesman, G. S. 1983. Crosslinking of proteins in acetylcholine receptor-rich membranes: Association between the β-subunit and the 43 Kd subsynaptic protein. *Cell* 35: 687–92

Burden, S. J., Sargent, P. B., McMahan, U. J. 1979. Acetylcholine receptors in regenerating muscle accumulate at original synaptic sites in the absence of nerve. *J. Cell Biol.* 82: 412–25

Burstein, M., Shainberg, A. 1979. Concanavalin A inhibits fusion of myoblasts and appearance of acetylcholine receptors in muscle cultures. *FEBS Lett.* 103: 33–37

Bursztajn, S. 1984. Coated vesicles are associated with acetylcholine receptors at nerve-muscle contacts. *J. Neurocytol.* 13: 503–18

Bursztajn, S., Berman, S. A., McManaman, J. L., Watson, M. L. 1985. Insertion and internalization of acetylcholine receptors at clustered and diffuse domains on cultured myotubes. *J. Cell Biol.* 101: 104–11

Bursztajn, S., Fischbach, G. D. 1984. Evidence that coated vesicles transport acetylcholine receptors to the surface membrane of chick myotubes. *J. Cell Biol.* 98: 498–506

Bursztajn, S., Gershon, M. D. 1977. Discrimination between nicotinic receptors in vertebrate ganglia and skeletal muscle by α-bungarotoxin and cobra venoms. *J. Physiol. London* 269: 17–31

Cannon, W. B., Rosenblueth, A. 1949. *The Supersensitivity of Denervated Structures. A Law of Denervation.* New York/MacMillan

Carbonetto, S. T., Fambrough, D. M., Muller, K. J. 1978. Nonequivalence of α-bungarotoxin receptors and acetylcholine receptors in chick sympathetic neurons. *Proc. Natl. Acad. Sci. USA* 75: 1016–20

Carlson, C. G., Leonard, R. J., Nakajima, S. 1985. The aneural development of the acetylcholine receptor in the presence of agents which block protein synthesis and glycosylation. *Soc. Neurosci. Abstr.* 11: 156

Cartaud, J., Kordeli, C., Nghiem, H. O., Changeux, J. P. 1983. La proteine 43000 daltons: Piece intermediaire assurant l'ancarage du recepteur cholinergique au cytosquelette sous-neural? *CR Acad. Sci. Paris* 297: 285–89

Cartaud, J., Sobel, A., Rousselet, A., Devaux, P. F., Changeux, J. P. 1981. Consequences of alkaline treatment for the ultrastructure of the acetylcholine receptor-rich membranes from *Torpedo marmorata* electric organ. *J. Cell Biol.* 90: 418–26

Chang, C. C., Huang, M. C. 1975. Turnover of junctional and extrajunctional acetylcholine receptors of the rat diaphragm. *Nature* 253: 643–44

Changeux, J. P. 1981. The acetylcholine receptor: An "allosteric" membrane protein. *Harvey Lect.* 75: 85–254

Changeux, J. P., Devillers-Thiery, A., Chemouilli, P. 1984. The acetylcholine receptor: An allosteric protein engaged in intercellular communication. *Science* 225: 1335–45

Chiappinelli, V. A. 1983. κ-Bungarotoxin:

A probe for the neuronal nicotinic receptor in the avian ciliary ganglion. *Brain Res.* 277: 9–21

Chiappinelli, V. A., Cohen, J. B., Zigmond, R. E. 1981. The effects of α- and β-neurotoxins from the venom of various snakes on transmission in autonomic ganglia. *Brain Res.* 211: 107–26

Chiappinelli, V. A., Zigmond, R. E. 1978. α-Bungarotoxin blocks nicotinic transmission in the avian ciliary ganglion. *Proc. Natl. Acad. Sci. USA* 75: 2999–3003

Chiu, A. Y., Sanes, J. R. 1984. Development of basal lamina in synaptic and extrasynaptic portions of embryonic rat muscle. *Dev. Biol.* 103: 456–67

Choi, D. W., Fischbach, G. D. 1981. GABA conductance of chick spinal cord and dorsal root ganglion neurons in cell culture. *J. Neurophysiol.* 45: 605–20

Chou, T. C., Lee, C. Y. 1969. Effect of whole and fractioned cobra venom of sympathetic ganglionic transmission. *Eur. J. Pharmacol.* 8: 326–30

Chow, I., Cohen, M. W. 1983. Developmental changes in the distribution of acetylcholine receptors on the myotomes of *Xenopus laevis*. *J. Physiol. London* 339: 553–71

Christian, C. N., Daniels, M. P., Sugiyama, H., Vogel, Z., Jacques, L., Nelson, P. G. 1978. A factor from neurons increases the number of acetylcholine receptor aggregates on cultured muscle cells. *Proc. Natl. Acad. Sci. USA* 75: 4011–15

Clapham, D. E., Neher, E. 1984. Substance P reduces acetylcholine induced currents in isolated bovine chromaffin cells. *J. Physiol. London* 347: 255–77

Clarke, P. B. S., Pert, C. B., Pert, A. 1984. Autoradiographic distribution of nicotine receptors in rat brain. *Brain Res.* 323: 390–95

Clarke, P. B. S., Schwartz, R. D., Paul, S. M., Pert, C. B., Pert, A. 1985. Nicotinic binding in rat brain: Autoradiographic comparison of ^{125}-α-bungarotoxin. *J. Neurosci.* 5: 1307–15

Claudio, T., Ballivet, M., Patrick, J., Heinemann, S. 1983. Nucleotide and deduced amino acid sequences of *Torpedo californica* acetylcholine receptor γ subunit. *Proc. Natl. Acad. Sci. USA* 80: 1111–15

Clementi, F., Sher, E., Erroi, A. 1983. Acetylcholine receptor degradation: Study of mechanism of action of inhibitory drugs. *Eur. J. Cell Biol.* 29: 274–80

Cohen, M. W. 1972. The development of neuromuscular connexions in the presence of D-tubocurarine. *Brain Res.* 41: 457–63

Cohen, M. W., Greschner, M., Tucci, M. 1984. In vivo development of cholinesterase at a neuromuscular junction in the absence of motor activity in *Xenopus laevis*. *J. Physiol. London* 348: 57–66

Cohen, M. W., Weldon, P. R. 1980. Localization of acetylcholine receptors and synaptic ultrastructure at nerve muscle contacts in culture: Dependence on nerve type. *J. Cell Biol.* 86: 388–401

Cohen, S. A. 1980. Early nerve-muscle synapses in vitro release transmitter over postsynaptic membrane having low acetylcholine sensitivity. *Proc. Natl. Acad. Sci. USA* 77: 644–48

Cohen, S. A., Fischbach, G. D. 1977. Cluster of acetylcholine receptors located at identified nerve-muscle synapses in vitro. *Devel. Biol.* 59: 24–38

Colquhoun, D., Sakmann, B. 1981. Fluctuations in the microsecond time range of the current through single acetylcholine receptor ion channels. *Nature* 294: 464–66

Colquhoun, D., Sakmann, B. 1985. Fast events in single-channel currents activated by acetylcholine and its analogues at the frog muscle end-plate. *J. Physiol. London* 369: 501–57

Connolly, J. A., St. John, P. A., Fischbach, G. D. 1982. Extracts of electric lobe and electric organ from *Torpedo californica* increase the total number as well as the number of aggregates of chick myotube acetylcholine receptors. *J. Neurosci.* 2: 1207–13

Connolly, J. A. 1984. Is there an endogenous bungarotoxin-like molecule in the vertebrate central nervous system? *Brain Res.* 323: 307–10

Connolly, J. A., Graham, A. J. 1985. Actin filaments and acetylcholine receptor clusters in embryonic chick myotubes. *Eur. J. Cell Biol.* 37: 191–95

Conti-Tronconi, B. M., Raftery, M. A. 1982. The nicotinic cholinergic receptor: Correlation of molecular structure with functional properties. *Ann. Rev. Biochem.* 51: 491–530

Costa, E., Guidotti, A., Hanbauer, I., Saiani, L. 1983. Modulation of nicotinic receptor function by opiate recognition sites highly selective for Met 5-enkephalin [ARG-^{6}PHE^7]. *Fed. Proc.* 42: 2946–62

Costa, L. G., Murphy, S. D. 1983. ^{3}H-nicotinic binding in rat brain: Alteration after chronic acetylcholinesterase inhibition. *J. Pharmacol. Exp. Ther.* 226: 392–97

Crean, G., Pilar, G., Tuttle, J. B., Vaca, K. 1982. Enhanced chemosensitivity of chick parasympathetic neurones in co-culture with myotubes. *J. Physiol. London* 331: 87–104

Creazzo, T. L., Sohal, G. S. 1983. Neural control of embryonic acetylcholine receptor and skeletal muscle. *Cell Tissue Res.* 228: 1–12

Cull-Candy, S. G., Miledi, R., Trautmann, A. 1978. Acetylcholine-induced channels and transmitter release at human endplates. *Nature* 271: 74–75

Cull-Candy, S. G., Miledi, R., Uchitel, O. D. 1982. Properties of junctional and extrajunctional acetylcholine-receptor channels in organ cultured human muscle fibres. *J. Physiol. London* 333: 251–67

Dean, D. M., Lemaire, S., Livett, B. G. 1982. Evidence that inhibition of nicotine mediated catecholamine secretion from adrenal chromaffin cells by enkephalin beta endorphin, dynorphin (1–13) and opiates is not mediated via specific opiate receptors. *J. Neurochem.* 38: 606–13

Dennis, M. J. 1981. Development of the neuromuscular junction: Inductive interactions between cells. *Ann. Rev. Neurosci.* 4: 43–68

Dennis, M. J., Sargent, P. B. 1979. Loss of extrasynaptic acetylcholine sensitivity upon reinnervation of parasympathetic ganglion cells. *J. Physiol. London* 289: 263–75

Dennis, M. J., Ziskind-Conhaim, L., Harris, A. J. 1981. Development of neuromuscular junctions in rat embryos. *Devel. Biol.* 81: 266–79

Derkach, V. A., Selyanko, A. A., Skok, V. I. 1983. Acetylcholine-induced current fluctuations and fast excitatory post-synaptic currents in rabbit sympathetic neurons. *J. Physiol. London* 336: 511–26

Devreotes, P. N., Fambrough, D. M. 1975. Acetylcholine receptor turnover in membranes of developing muscle fibers. *J. Cell Biol.* 65: 335–58

Devreotes, P. N., Gardner, J. M., Fambrough, D. M. 1977. Kinetics of biosynthesis of acetylcholine receptor and subsequent incorporation into plasma membrane of cultured chick skeletal muscle. *Cell* 10: 365–73

Diamond, J., Miledi, R. 1962. A study of foetal and new-born rat muscle fibers. *J. Physiol. London* 162: 393–408

Dionne, V. E., Parsons, R. L. 1978. Synaptic channel gating differences at snake twitch and slow neuromuscular junctions. *Nature* 274: 902–4

Dionne, V. E., Parsons, R. L. 1981. Characteristics of the acetylcholine-operated channel at twitch and slow fibre neuromuscular junctions of the garter snake. *J. Physiol. London* 310: 145–58

Dreyer, F., Mueller, K.-D., Peper, K., Sterz, R. 1976a. The *M. omohyoideus* of the mouse as a convenient mammalian muscle preparation. *Pfluegers Arch.* 367: 115–22

Dreyer, F., Walther, Chr., Peper, K. 1976b. Junctional and extrajunctional acetylcholine receptors in normal and denervated frog muscle fibers. Noise analysis experiments with different agonists. *Pfluegers Arch.* 366: 1–9

Duggan, A. W., Hall, J. G., Lee, C. Y. 1976. Alpha-bungarotoxin, cobra neurotoxin and excitation of Renshaw cells by acetylcholine. *Brain Res.* 107: 166–70

Dun, N. J., Kraczmar, A. G. 1980. Blockade of ACh potentials by α-bungarotoxin in rat superior cervical ganglion cells. *Brain Res.* 196: 536–40

Dunn, P. M., Marshall, L. M. 1985. Lack of nicotinic supersensitivity in frog sympathetic neurones following denervation. *J. Physiol. London* 363: 211–25

Dwyer, D. S., Bradley, R. J., Furner, R. L., Kemp, G. E. 1981. Immunochemical properties of junctional and extrajunctional acetylcholine receptor. *Brain Res.* 217: 23–40

Edwards, C., Frisch, H. L. 1976. A model for the localization of acetylcholine receptors at the muscle endplate. *J. Neurobiol.* 7: 377–81

Engel, A. G., Lambert, E. H., Mulder, D. M., Torres, C. F., Sahashi, K., et al. 1982. A newly recognized congenital myasthenic syndrome attributed to a prolonged open time of the acetylcholine-induced ion channel. *Ann. Neurol.* 11: 553–69

Englander, L. L., Rubin, L. L. 1985. Acetylcholine receptor clustering and nuclear movement in muscle fibers in culture. *J. Cell Biol.* 101: 130a (Abstr.)

Englander, L. L., Rubin, L. L. 1986. Acetylcholine receptor clustering and nuclear movement in muscle fibers in culture. *J. Cell Biol.* In press

Erichsen, J. T., Karten, H. J., Eldred, W. D., Brecha, N. C. 1982. Localization of substance P-like and enkephalin-like immunoreactivity within the preganglionic terminals of the avian ciliary ganglion: Light and electron microscopy. *J. Neurosci.* 2: 994–1003

Eusebi, F., Molinaro, M., Zani, B. M. 1985. Agents that activate protein kinase C reduce acetylcholine sensitivity in cultured myotubes. *J. Cell Biol.* 100: 1339–42

Fairclough, R. H., Finer-Moore, J., Love, R. A., Kristofferson, D., Desmeules, P. J., Stroud, R. M. 1983. Subunit organization and structure of an acetylcholine receptor. *Cold Spring Harbor Symp. Quant. Biol.* 48: 9–20

Fallon, J. R., Nitkin, R. M., Reist, N. E., Wallace, B. G., McMahan, U. J. 1985. Acetylcholine receptor-aggregating factor is similar to molecules concentrated at neuromuscular junctions. *Nature* 315: 571–74

Fambrough, D. M. 1979. Control of acetylcholine receptors in skeletal muscle. *Physiol. Rev.* 59: 165–227

Fambrough, D. M. 1983. Biosynthesis and intracellular transport of acetylcholine receptors. *Met. Enzym.* 96: 331–52

Farley, G. R., Morley, B. J., Javel, E., Gorga, M. P. 1983. Single-unit responses to cholinergic agents in the rat inferior colliculus. *Hear. Res.* 11: 73–91

Fertuck, H. C., Salpeter, M. M. 1976. Quantitation of junctional and extrajunctional acetylcholine receptors by electron microscope autoradiography after ^{125}I-α-bungarotoxin binding at mouse neuromuscular junctions. *J. Cell Biol.* 69: 144–58

Finer-Moore, J., Stroud, R. M. 1984. Amphipathic analysis and possible formation of the ion channel in an acetylcholine receptor. *Proc. Natl. Acad. Sci. USA* 81: 155–59

Fischbach, G. D., Cohen, S. A. 1973. The distribution of acetylcholine sensitivity over uninnervated and innervated muscle fibres grown in cell culture. *Devel. Biol.* 31: 147–62

Fischbach, G. D., Role, L. R., Hume, R. I. 1984. The accumulation of ACh receptors at nerve-muscle synapses in culture. In *Cellular and Molecular Biology of Neuronal Development*, ed. I. Black, pp. 107–15. New York: Plenum

Fischbach, G. D., Schuetze, S. M. 1980. A post-natal decrease in acetylcholine channel open time at rat end-plates. *J. Physiol. London* 303: 125–37

Frank, E., Fischbach, G. D. 1979. Early events in neuromuscular junction formation in vitro. Induction of acetylcholine receptor clusters in the postsynaptic membrane and morphology of newly formed nerve-muscle synapses. *J. Cell Biol.* 83: 143–58

Frank, E., Gautvik, K., Sommerschild, H. 1975. Persistence of junctional acetylcholine receptors following denervation. *Cold Spring Harbor Symp. Quant. Biol.* 40: 275–81

Fraser, S. E., Poo, M.-M. 1982. Development, maintenance and modulation of patterned membrane topography: Models based on the acetylcholine receptors. *Curr. Top. Devel. Biol.* 17: 77–100

Froehner, S. C. 1986. The role of the postsynaptic cytoskeleton in AChR organization. *Trends Neurosci.* 9: 37–41

Froehner, S. C., Goulbrandsen, V., Hyman, C., Jeng, A. Y., Neubig, R. R., Cohen, J. B. 1981. Immunofluorescence localization at the mammalian neuromuscular junction of the Mr 43,000 protein of *Torpedo* postsynaptic membrane. *Proc. Natl. Acad. Sci. USA* 78: 5230–34

Fumagalli, L., De Renzis, G. 1984. Extrasynaptic localization of α-bungarotoxin receptors in the rat superior cervical ganglion. *Neurochem. Int.* 6: 355–64

Gardner, J. M., Fambrough, D. M. 1979. Acetylcholine receptor degradation measured by density labeling: Effects of cholinergic ligands and evidence against recycling. *Cell* 16: 661–74

Godfrey, E. W., Nitkin, R. M., Wallace, B. G., Rubin, L. L., McMahan, U. J. 1984. Components of *Torpedo* electric organ and muscle that cause aggregation of acetylcholine receptors on cultured muscle cells. *J. Cell Biol.* 99: 615–27

Goodman, C. S., Spitzer, N. C. 1979. Embryonic development of identified neurones: differentiation from neuroblast to neurone. *Nature* 280: 208–14

Gordon, A. S., Davis, C. G., Milfay, D., Diamond, I. 1977. Phosphorylation of acetylcholine receptor by endogenous membrane protein kinase in receptor-enriched membranes of *Torpedo californica*. *Nature* 267: 539–40

Grant, G. A., Chiappinelli, V. A. 1985. κ-Bungarotoxin: Complete amino acid sequence of a neuronal nicotinic receptor probe. *Biochemistry* 24: 1532–37

Gray, P. T. A., Magnus, C. J. 1983. The culture of rat submandibular ganglion neurons and their macroscopic responses to acetylcholine. *Proc. R. Soc. London Ser. B* 220: 265–71

Greenberg, A. S., Nakajima, S., Nakajima, Y. 1985. Functional properties of newly inserted acetylcholine receptors in embryonic *Xenopus* muscle cells. *Devel. Brain Res.* 19: 289–96

Hall, Z. W., Gorin, P. D., Silberstein, L., Bennett, C. 1985. A postnatal change in the immunological properties of the acetylcholine receptor at rat muscle endplates. *J. Neurosci.* 5: 730–34

Hall, Z. W., Lubit, B. W., Schwartz, J. H. 1981. Cytoplasmic actin in postsynaptic structures at the neuromuscular junction. *J. Cell Biol.* 90: 789–92

Hall, Z. W., Roisin, M. P., Gu, Y., Gorin, P. D. 1983. A developmental change in the immunological properties of acetylcholine receptors at the rat neuromuscular junction. *Cold Spring Harbor Symp. Quant. Biol.* 48: 101–8

Halvorsen, S. W., Berg, D. K. 1986. Identification of a nicotinic acetylcholine receptor on neurons using an α-neurotoxin that blocks receptor function. *J. Neurosci.* 6: 3405–12

Hamill, O. P., Marty, A., Neher, E., Sakmann, B., Sigworth, F. J. 1981. Improved patch-clamp techniques for high-resolution current recording from cells and cell-free membrane patches. *Pfluegers Arch.* 391 : 85–100

Hamill, O. P., Sakmann, B. 1981. Multiple conductance states of single acetylcholine receptor channels in embryonic muscle cells. *Nature* 294 : 462–64

Harris, A. J., Kuffler, S. W., Dennis, M. J. 1971. Differential chemosensitivity of synaptic and extrasynaptic areas on the neuronal surface membrane in parasympathetic neurons of the frog, tested by microapplication of acetylcholine. *Proc. R. Soc. London Ser. B* 177 : 541–53

Harrow, I. D., Hue, B., Gepner, J. I., Hall, L. M., Sattelle, D. B. 1980. An α-bungarotoxin-sensitive acetylcholine receptor in the CNS of the cockroach, *Periplaneta americana*. In *Insect Neurobiology and Pesticide Action*, pp. 137–44. London: Soc. Chem. Indust.

Hartzell, H. C., Fambrough, D. M. 1972. Acetylcholine receptors: Distribution and extrajunctional density in rat diaphragm after denervation correlated with acetylcholine sensitivity. *J. Gen. Physiol.* 60 : 248–62

Hayashi, M., Edgar, D., Thoenen, H. 1983. The development of substance P, somatostatin and vasoactive intestinal polypeptide in sympathetic and spinal sensory ganglia of the chick embryo. *Neuroscience* 10 : 31–39

Hirokawa, N., Heuser, J. E. 1982. Internal and external differentiations of the postsynaptic membrane at the neuromuscular junction. *J. Neurocytol.* 11 : 487–510

Huganir, R. L., Delcour, A. H., Greengard, P., Hess, G. P. 1986. Phosphorylation of the nicotinic acetylcholine receptor regulates its rate of desensitization. *Nature* 321 : 774–76

Huganir, R. L., Greengard, P. 1983. cAMP-dependent protein kinase phosphorylates the nicotinic acetylcholine receptor. *Proc. Natl. Acad. Sci. USA* 80 : 1130–34

Huganir, R. L., Miles, K., Greengard, P. 1984. Phosphorylation of the nicotinic acetylcholine receptor by an endogenous tyrosine-specific protein kinase. *Proc. Natl. Acad. Sci. USA* 81 : 6968–72

Hume, R. I., Role, L. W., Fischbach, G. D. 1983. Acetylcholine release from growth cones detected with patches of acetylcholine receptor-rich membranes. *Nature* 305 : 632–34

Hyman, C., Froehner, S. C. 1983. Degradation of acetylcholine receptor in muscle cells: Effect of leupeptin on turnover rate, intracellular pool sizes, and receptor properties. *J. Cell Biol.* 96 : 1316–24

Ishikawa, Y., Masuko, S., Shimada, Y. 1983. Acetylcholine receptors and motor nerve terminals in developing chick skeletal muscles as revealed by fluorescence microscopy. *Devel. Brain Res.* 8 : 111–18

Jacob, M. H., Berg, D. K. 1983. The ultrastructural localization α-bungarotoxin binding sites in relation to synapses on chick ciliary ganglion neurons. *J. Neurosci.* 3 : 260–71

Jacob, M. H., Berg, D. K., Lindstrom, J. M. 1984. Shared antigenic determinant between the *Electrophorus* acetylcholine receptor and a synaptic component on chick ciliary ganglion neurons. *Proc. Natl. Acad. Sci. USA* 81 : 3223–27

Jacob, M., Lentz, T. L. 1979. Localization of acetylcholine receptors by means of horseradish peroxidase-α-bungarotoxin during formation and development of the neuromuscular junction in the chick embryo. *J. Cell Biol.* 82 : 195–211

Jessel, T. M., Siegel, R. E., Fischbach, G. D. 1979. Induction of acetylcholine receptors on cultured skeletal muscle by a factor extracted from brain and spinal cord. *Proc. Natl. Acad. Sci. USA* 76 : 5397–5401

Jones, R., Vrbova, G. 1974. Two factors responsible for the development of denervation hypersensitivity. *J. Physiol. London* 236 : 517–38

Kao, P. N., Dwork, A. J., Kaldany, R. R. J., Silver, M., Wideman, J., et al. 1984. Identification of the α-subunit half cysteine specifically labeled by an affinity reagent for the acetylcholine receptor binding site. *J. Biol. Chem.* 259 : 11662–65

Karlin, A. 1980. Molecular properties of nicotinic acetylcholine receptors. In *The Cell Surface and Neuronal Function*, ed. C. W. Cotman, G. Poste, G. L. Nicholson, pp. 191–260. Amsterdam: Elsevier/North Holland

Karlin, A. 1983. The anatomy of a receptor. *Neurosci. Comment.* 1 : 111–23

Karlin, A., Cowburn, D. 1973. The affinity labeling of partially purified acetylcholine receptor from electric tissue of *Electrophorus*. *Proc. Natl. Acad. Sci. USA* 70 : 3636–40

Karlin, A., Holtzman, E., Yodh, N., Lobel, P., Wall, J., Hainfeld, J. 1983. The arrangement of the subunits of the acetylcholine receptor of *Torpedo californica*. *J. Biol. Chem.* 258 : 6678–81

Karlin, A., Weill, C. L., McNamee, M. G., Valderrama, R. 1975. Facets of the structures of acetylcholine receptors from *Electrophorus* and *Torpedo*. *Cold Spring Harbor Symp. Quant. Biol.* 40 : 203–10

Kato, E., Narahashi, T. 1982. Low sensitivity of the neuroblastoma cell cholinergic receptors to erabutoxins and α-bungarotoxin. *Brain Res.* 245 : 159–62

Katz, B., Miledi, R. 1972. The statistical nature of the acetylcholine potential and its molecular components. *J. Physiol. London* 224: 665–99

Kehoe, J., Sealock, R., Bon, C. 1976. Effects of α-toxins from *Bungarus multicinctus* and *Bungarus caeruleus* on cholinergic responses in *Aplysia* neurons. *Brain Res.* 107: 527–40

Kelly, A. M., Zachs, S. I. 1969. The fine structure of motor endplate morphogenesis. *J. Cell Biol.* 42: 154–69

Kidokoro, Y. 1980. Developmental changes of spontaneous synaptic potential properties in the rat neuromuscular contact formed in culture. *Devel. Biol.* 78: 231–41

Kidokoro, Y., Anderson, M. J., Gruener, R. 1980. Changes in synaptic potential properties during acetylcholine receptor accumulation and neuronspecific interactions in *Xenopus* nerve-muscle cell culture. *Devel. Biol.* 78: 464–83

Kidokoro, Y., Yeh, E. 1982. Initial synaptic transmission at the growth cone in *Xenopus* nerve-muscle cultures. *Proc. Natl. Acad. Sci. USA* 79: 6727–31

Kidokoro, Y., Brass, B. 1985. Redistribution of acetylcholine receptors during neuromuscular junction formation in *Xenopus* cultures. *J. Physiol. Paris* 80: 212–20

Klarsfeld, A., Changeux, J. P. 1985. Activity regulates the levels of acetylcholine receptor α-subunit mRNA in cultured chicken myotubes. *Proc. Natl. Acad. Sci. USA* 82: 4558–62

Knaack, D., Podleski, T. Ascorbic acid mediates acetylcholine receptor increase induced by brain extract on myogenic cells. *Proc. Natl. Acad. Sci. USA* 82: 575–79

Knaack, D., Shen, I., Salpeter, M. M., Podleski, T. R. 1986. Selective effects of ascorbic acid on acetylcholine receptor number and distribution. *J. Cell Biol.* 102: 795–802

Ko, C. P., Burton, H. S., Bunge, R. L. P. 1976. Synaptic transmission between rat spinal cord explants and dissociated superior cervical ganglion neurons in tissue culture. *Brain Res.* 117: 437–60

Ko, P. K., Anderson, M. J., Cohen, M. W. 1977. Denervated skeletal muscle fibers develop discrete patches of high acetylcholine receptor density. *Science* 196: 540–42

Koelle, G. B., Ruch, G. A. 1983. Demonstration of a neurotrophic factor for the maintenance of acetylcholinesterase and butyrylcholinesterase in the preganglionically denervated superior cervical ganglion of the cat. *Proc. Natl. Acad. Sci. USA* 80: 3106–10

Koketsu, K., Akasu, T., Miyagawa, M., Hirai, K. 1982. Modulation of nicotinic transmission by biogenic amines in bullfrog sympathetic ganglia. *J. Auton. Nerv. Sys.* 6: 47–53

Korneliussen, H., Jansen, J. K. S. 1976. Morphological aspects of the elimination of polyneuronal innervation of skeletal muscle fibres in newborn rats. *J. Neurocytol.* 5: 591–604

Kouvelas, E. D., Dichter, M. A., Greene, L. A. 1978. Chick sympathetic neurons develop receptors for α-bungarotoxin in vitro, but the toxin does not block nicotinic receptors. *Brain Res.* 154: 83–93

Kuffler, S. W., Dennis, M. J., Harris, A. J. 1971. The development of chemosensitivity in extrasynaptic areas of the neuronal surface after denervation of parasympathetic ganglion cells in the heart of the frog. *Proc. R. Soc. London Ser. B* 177: 555–63

Kuffler, S. W., Yoshikami, D. 1975. The distribution of acetylcholine sensitivity at the post-synaptic membrane of vertebrate skeletal twitch muscles: Iontophoretic mapping in the micron range. *J. Physiol. London* 244: 703–30

Kullberg, R. W., Brehm, P., Steinbach, J. H. 1981. Nonjunctional acetylcholine receptor channel open time decreases during development of *Xenopus* muscle. *Nature* 289: 411–13

Kullberg, R., Kasprzak, H. 1985. Gating kinetics of nonjunctional acetylcholine receptor channels in developing *Xenopus* muscle. *J. Neurosci.* 5: 970–76

Kullberg, R., Lentz, T., Cohen, M. 1977. Development of mytomal neuromuscular junction in *Xenopus laevis*: An electrophysiological and fine structural study. *Devel. Biol.* 60: 101–29

Kullberg, R., Owens, J. L. 1986. Comparative development of endplate currents in two muscles of *Xenopus laevis*. *J. Physiol. London* 374: 413–27

Kullberg, R., Owens, J. L., Vickers, J. 1985. Development of synaptic currents in immobilized muscle of *Xenopus laevis*. *J. Physiol. London* 364: 57–68

Kumakura, K., Karoum, F., Guidotti, A., Costa, E. 1980. Modulation of nicotinic receptors by opiate receptor agonists in cultured adrenal chromaffin cells. *Nature* 283: 489–92

Kuromi, H., Brass, B., Kidokoro, Y. 1985. Formation of acetylcholine receptor clusters at neuromuscular junction in *Xenopus* cultures. *Devel. Biol.* 109: 165–76

Kuromi, H., Kidokoro, Y. 1984a. Nerve disperses preexisting acetylcholine receptor clusters prior to induction of receptor accumulation in *Xenopus* muscle cultures. *Devel. Biol.* 103: 53–61

Kuromi, H., Kidokoro, Y. 1984b. Denervation disperses acetylcholine receptor clusters at the neuromuscular junction in *Xenopus* cultures. *Devel. Biol.* 104: 421–27

Labovitz, S. S., Robbins, N. 1983. A maturational increase in rat neuromuscular junctional acetylcholine receptors despite disuse or denervation. *Brain Res.* 266: 155–58

Landmesser, L. T., Morris, D. G. 1975. The development of functional innervation in the hind limb of the chick embryo. *J. Physiol. London* 249: 301–26

Landmesser, L., Pilar, G. 1974. Synaptic transmission and cell death during normal ganglonic development. *J. Physiol. London* 241: 737–49

Leah, J., Dvorak, D., Kidson, C. 1980. Development of sensitivity to acetylcholine in cultured chick embryo sympathetic ganglion neurons. *Neurosci. Lett.* 19: 73–77

Lees, G., Beadle, D. J., Botham, R. P. 1983. Cholinergic receptors on cultured neurones from the central nervous system of embryonic cockroaches. *Brain Res.* 288: 49–59

Lentz, T. L., Chester, J. 1977. Localization of acetylcholine receptors in central synapses. *J. Cell Biol.* 75: 258–67

Leonard, R. J., Nakajima, S., Nakajima, Y., Takahashi, T. 1984. Differential development of two classes of acetylcholine receptors in *Xenopus* muscle in culture. *Science* 226: 55–57

Levitt, T. A., Salpeter, M. M. 1981. Denervated endplates have a dual population of junctional acetylcholine receptors. *Nature* 291: 239–41

Libby, P., Bursztajn, S., Goldberg, A. L. 1980. Degradation of the acetylcholine receptor in cultured muscle cells: Selective inhibitors and the fate of undegraded receptors. *Cell* 19: 481–91

Lindstrom, J. 1983. Using monoclonal antibodies to study acetylcholine receptors and myasthenia gravis. *Neurosci. Comment.* 1: 139–56

Lindstrom, J., Criado, M., Hochschwender, S., Fox, J. L., Sarin, V. 1984. Immunochemical tests of acetylcholine receptor subunit models. *Nature* 311: 573–75

Lo, M. M. S., Garland, B., Lamprecht, J., Barnard, E. A. 1980. Rotational mobility of the membrane-bound acetylcholine receptor of *Torpedo* electric organ measured by phosphorescence depolarization. *FEBS Lett.* 111: 407–12

Lømo, T., Westgaard, R. H. 1975. Control of ACh sensitivity in rat muscle fibers. *Cold Spring Harbor Symp. Quant. Biol.* 40: 263–74

Loring, R. H., Salpeter, M. M. 1980. Denervation increases turnover rate of junctional acetylcholine receptors. *Proc. Natl. Acad. Sci. USA* 77: 2293–97

Loring, R. H., Dahm, L. M., Zigmond, R. E. 1985. Localization of α-bungarotoxin binding sites in the ciliary ganglion of the embryonic chick: An autoradiographic study at the light and electron microscopic level. *Neuroscience* 14: 645–60

Loring, R. H., Zigmond, R. E. 1985. Amino acid sequence of a neurotoxin that blocks neuronal nicotinic receptors and the localization of its binding sites in chick ciliary ganglion. *Soc. Neurosci. Abstr.* 11: 92

Magleby, K. L., Stevens, C. F. 1972. A quantitative description of endplate currents. *J. Physiol. London* 223: 173–97

Margiotta, J. F., Berg, D. K. 1982. Functional synapses are established between ciliary ganglion neurons in dissociated cell culture. *Nature* 296: 152–54

Margiotta, J. F., Berg, D. K. 1986. Enkephalin and substance P modulate synaptic properties of chick ciliary ganglion neurons in cell culture. *Neuroscience* 18: 175–82

Maricq, A. V., Gu, Y., Hestrin, S., Hall, Z. 1985. The effects of a myasthenic serum on the acetylcholine receptors of C2 myotubes. II. Functional inactivation of the receptor. *J. Neurosci.* 5: 1917–24

Markelonis, G. J., Bradshaw, R. A., Oh, T. H., Johnson, J. L., Bates, O. J. 1982a. Sciatin is a transferrin-like polypeptide. *J. Neurochem.* 39: 315–20

Markelonis, G. J., Oh, T. H., Eldefrawi, M. E., Guth, L. 1982b. Sciatin: A myotrophic protein increases the number of acetylcholine receptors and receptor clusters in cultured skeletal muscle. *Devel. Biol.* 89: 353–61

Marks, M. J., Collins, A. C. 1982. Characterization of nicotine binding in mouse brain and comparison with the binding of alpha-bungarotoxin and quinuclidinyl benzilate. *Molec. Pharmacol.* 22: 554–64

Marshall, L. M. 1981. Synaptic localization of L-bungarotoxin binding which blocks nicotinic transmission at frog sympathetic neurons. *Proc. Natl. Acad. Sci. USA* 78: 1948–52

Marshall, L. M. 1985. Presynaptic control of synaptic channel kinetics in sympathetic neurones. *Nature* 317: 621–23

Marshall, L. M. 1986. Different synaptic channel kinetics in sympathetic B and C neurons of the bullfrog. *J. Neurosci.* 6: 590–93

Marshall, L. M., Sanes, J. R., McMahan, U. J. 1977. Reinnervation of original synaptic sites on muscle fiber basement membrane after disruption of muscle cells. *Proc. Natl. Acad. Sci. USA* 74: 3073–77

Matthews-Bellinger, J., Salpeter, M. M. 1978. Distribution of acetylcholine receptors at frog neuromuscular junctions with a discussion of some physiological implications. *J. Physiol. London* 279: 197–213

Matthews-Bellinger, J. A., Salpeter, M. M. 1983. Fine structural distribution of acetylcholine receptors at developing mouse neuromuscular junctions. *J. Neurosci.* 3: 644–57

McCarthy, M. P., Earnest, J. P., Young, E. F., Choe, S., Stroud, R. M. 1986. The molecular neurobiology of the acetylcholine receptor. *Ann. Rev. Neurosci.* 9: 383–413

McGeer, P. L., Eccles, J. C., McGeer, E. G. 1978. *Molecular Neurobiology of the Mammalian Brain*, pp. 141–82. New York: Plenum Press

McMahan, U. J., Slater, C. R. 1984. The influence of basal lamina on the accumulation of acetylcholine receptors at synaptic sites in regenerating muscle. *J. Cell Biol.* 98: 1453–73

McMahan, U. J., Sargent, P. B., Rubin, L. L., Burden, S. J. 1980. Factors that influence the organization of acetylcholine receptors in regenerating muscle are associated with the basal lamina at the neuromuscular junction. In *Ontogenesis and Functional Mechanisms of Peripheral Synapses*, ed. J. Taxi, pp. 345–54. Amsterdam: Elsevier/North Holland

Merlie, J. P. 1984. Biogenesis of the acetylcholine receptor, a multisubunit integral membrane protein. *Cell* 36: 573–75

Merlie, J. P., Buonanno, A., Carlin, B., Covault, J., Crowder, C. M., et al. 1986. Coordinate regulation of expression of synaptic proteins in skeletal muscle: Studies with ACh receptor. In *Proteins in Excitable Membranes*, ed. B. Hille, D. Fambrough. New York: Wiley. In press

Merlie, J. P., Changeux, J.-P., Gros, F. 1976. Acetylcholine receptor degradation measured by pulse chase labeling. *Nature* 264: 74–76

Merlie, J. P., Isenberg, I. E., Russell, S. D., Sanes, J. R. 1984. Denervation supersensitivity in skeletal muscle: Analysis with a cloned cDNA probe. *J. Cell Biol.* 99: 332–35

Merlie, J. P., Sanes, J. R. 1985. Concentration of acetylcholine receptor mRNA in synaptic regions of adult muscle fibres. *Nature* 317: 66–68

Merlie, J. P., Sanes, J. R. 1986. Regulation of synapse specific genes. In *Molecular Aspects of Neurobiology*, ed. E. Kandel, R. Levi-Montalcini, pp. 75–80. Berlin/Heidelberg: Springer-Verlag

Merlie, J. P., Smith, M. M. 1986. Synthesis and assembly of acetylcholine receptor, a multisubunit membrane glycoprotein. *J. Memb. Biol.* 91: 1–10

Messing, A., Bizzini, B., Gonatas, N. K. 1984. Concanavalin A inhibits nicotinic acetylcholine receptor function in cultured chick ciliary ganglion neurons. *Brain Res.* 303: 241–49

Messing, A., Gonatas, N. K. 1983. Extra synaptic localization of alpha-bungarotoxin receptors in cultured chick ciliary ganglion neurons. *Brain Res.* 269: 172–76

Mesulam, M. M., Mufson, E. J., Wainer, B. H., Levey, A. I. 1983. Central cholinergic pathways in the rat: An overview based on an alternative nomenclature. *Neuroscience* 10: 1185–1201

Michler, A., Sakmann, B. 1980. Receptor stability and channel conversion in the subsynaptic membrane of the developing mammalian neuromuscular junction. *Devel. Biol.* 80: 1–17

Middleton, P., Jaramillo, F., Schuetze, S. M. 1986. Forskolin increases the rate of acetylcholine receptor desensitization at rat soleus endplates. *Proc. Natl. Acad. Sci. USA* 83: 4967–71

Miledi, R., Uchitel, O. D. 1981. Properties of postsynaptic channels induced by acetylcholine in different frog muscle fibres. *Nature* 291: 162–65

Mills, A., Wonnacott, S. 1984. Antibodies to nicotinic acetylcholine receptors used to probe the structural and functional relationships between brain alpha-bungarotoxin binding sites and nicotinic receptors. *Neurochem. Int.* 6: 249–57

Mishina, M., Kurosaki, T., Tobimatsu, T., Morimoto, Y., Noda, M., et al. 1984. Expression of functional acetylcholine receptor from cloned cDNAs. *Nature* 307: 604–8

Mishina, M., Takai, T., Imoto, K., Noda, M., Takahashi, T., et al. 1986. Molecular distinction between fetal and adult forms of muscle acetylcholine receptor. *Nature* 321: 406–11

Mizobe, F., Kozousek, V., Dean, D. M., Livett, B. G. 1979. Pharmacological characterization of adrenal paraneurons: Substance P and somatostatin as inhibitory modulators of the nicotinic response. *Brain Res.* 178: 555–66

Moody-Corbett, F., Cohen, M. W. 1982. Influence of nerve on the formation and survival of acetylcholine receptor and acetylcholinesterase patches on embryonic *Xenopus* muscle cells in culture. *J. Neurosci.* 2: 633–46

Morley, B. J., Kemp, G. E. 1981. Characterization of a putative nicotinic acetylcholine

receptor in mammalian brain. *Brain Res. Revs.* 3: 81–104
Morley, B. J., Kemp, G. E., Salvaterra, P. M. 1979. α-Bungarotoxin binding sites in the CNS. *Life Sci.* 24: 859–62
Neher, E., Sakmann, B. 1976. Noise analysis of drug induced voltage clamp currents in denervated frog muscle fibres. *J. Physiol. London* 258: 705–29
Nelson, P., Christian, C., Nirenberg, M. 1976. Synapse formation between clonal neuroblastoma X glioma-hybrid cells and striated muscle cells. *Proc. Natl. Acad. Sci. USA* 73: 123–27
Neubig, R. R., Cohen, J. B. 1979. Equilibrium binding of ^{3}H-acetylcholine by *Torpedo* postsynaptic membranes: Stoichiometry and ligand interaction. *Biochemistry* 18: 5464–75
Neugebauer, K., Salpeter, M. M., Podleski, T. R. 1985. Differential responses of L5 and rat primary muscle cells to factors in rat brain extract. *Brain Res.* 346: 58–69
Nghiem, H. O., Cartaud, J., Dubreuil, C., Kordeli, C., Buttin, G., Changeux, J. P. 1983. Production and characterization of a monoclonal antibody directed against the 43,000-dalton V(1) polypeptide from *Torpedo marmorata* electric organ. *Proc. Natl. Acad. Sci. USA* 80: 6403–7
Nishi, R., Berg, D. K. 1981a. Two components from eye tissue that differentially stimulate the growth and development of ciliary ganglion neurons in cell culture. *J. Neurosci.* 1: 505–13
Nishi, R., Berg, D. K. 1981b. Effects of high K^{+} concentrations on the growth and development of ciliary ganglion neurons in cell culture. *Devel. Biol.* 87: 301–7
Nitkin, R. M., Wallace, B. G., Spira, M. E., Godfrey, E. W., McMahan, U. J. 1983. Molecular components of the synaptic basal lamina that direct differentiation of regenerating neuromuscular junctions. *Cold Spring Harbor Symp. Quant. Biol.* 48: 653–65
Numa, S., Noda, M., Takahashi, H., Tanabe, Y., Toyosato, M., et al. 1983. Molecular structure of the nicotinic acetylcholine receptor. *Cold Spring Harbor Symp. Quant. Biol.* 48: 57–69
Nurse, C., O'Lague, P. 1975. Formation of cholinergic synapses between dissociated sympathetic neurons and skeletal myotubes of rat in cell culture. *Proc. Natl. Acad. Sci. USA* 72: 1955–59
O'Lague, P. H., Potter, D. D., Furshpan, E. J. 1978. Studies on rat sympathetic neurons developing in cell culture. *Devel. Biol.* 67: 384–443
Olek, A. J., Ling, A., Daniels, M. 1986. Development of ultrastructural specialization during the formation of ACh receptor aggregation on cultured myotubes. *J. Neurosci.* 6: 487–97
Olek, A. J., Pudimat, P. A., Daniels, M. P. 1983. Direct observation of the rapid aggregation of acetylcholine receptors on identified cultured myotubes after exposure to embryonic brain extract. *Cell* 34: 255–64
Orida, N., Poo, M. M. 1978. Electrophoretic movement and localisation of acetylcholine receptors in the embryonic muscle cell membrane. *Nature* 275: 31–35
Oswald, R. E., Freeman, J. A. 1981. Alpha bungarotoxin binding and central nervous system nicotinic acetylcholine receptors. *Neuroscience* 6: 1–14
Patrick, J., Stallcup, W. B. 1977a. Immunological distinction between acetylcholine receptor and the α-bungarotoxin binding component on sympathetic neurons. *Proc. Natl. Acad. Sci. USA* 74: 4689–92
Patrick, J., Stallcup, W. B. 1977b. α-Bungarotoxin binding and cholinergic receptor function on a rat sympathetic nerve line. *J. Biol. Chem.* 252: 8629–33
Pellegrino, M., Simonneau, M. 1984. Distribution of receptors for acetylcholine and 5-hydroxytryptamine on identified leech neurones growing in culture. *J. Physiol. London* 352: 669–84
Peng, H. B. 1983. Cytoskeletal organization of the presynaptic nerve terminal and the acetylcholine receptor cluster in cell cultures. *J. Cell Biol.* 97: 489–98
Peng, H. B. 1984. Participation of calcium and calmodulin in the formation of acetylcholine receptor clusters. *J. Cell Biol.* 98: 550–57
Peng, H. B., Cheng, P. C. 1982. Formation of postsynaptic specializations induced by latex beads in cultured muscle cells. *J. Neurosci.* 2: 1760–74
Peng, H. B., Cheng, P. C., Luther, P. W. 1981. Formation of ACh receptor clusters induced by positively charged latex beads. *Nature* 292: 831–34
Peng, H. B., Froehner, S. C. 1985. Association of the postsynaptic 43K protein with newly formed acetylcholine receptor clusters in cultured muscle cells. *J. Cell Biol.* 100: 1698–1705
Peng, H. B., Phelan, K. A. 1984. Early cytoplasmic specialization at the presumptive acetylcholine receptor cluster: a meshwork of thin filaments. *J. Cell Biol.* 99: 344–49
Peper, K., McMahan, U. J. 1972. Distribution of acetylcholine receptors in the vicinity of nerve terminals on skeletal muscle of the frog. *Proc. R. Soc. London Ser. B* 181: 431–40
Pestronk, A. 1985. Intracellular acetyl-

choline receptors in skeletal muscles of the adult rat. *J. Neurosci.* 5: 1111–17
Podleski, T. R., Axelrod, D., Ravdin, P., Greenberg, I., Johnson, M. M., Salpeter, M. M. 1978. Nerve extract induces increase and redistribution of acetylcholine receptors on cloned muscle cells. *Proc. Natl. Acad. Sci. USA* 75: 2035–39
Poo, M. M. 1982. Rapid lateral diffusion of functional ACh receptors in embryonic muscle cell membrane. *Nature* 295: 332–34
Poo, M. M. 1985. Mobility and localization of proteins in excitable membranes. *Ann. Rev. Neurosci.* 8: 369–406
Porter, S., Froehner, S. C. 1983. Characterization of the Mr 43,000 proteins associated with acetylcholine receptor rich membranes. *J. Biol. Chem.* 258: 10034–40
Prives, J., Fulton, A. B., Penman, S., Daniels, M. P., Christian, C. N. 1982. Interaction of the cytoskeletal framework with acetylcholine receptors on the surface of embryonic muscle cells in culture. *J. Cell Biol.* 92: 231–36
Purves, D. 1975. Functional and structural changes in mammalian sympathetic neurones following interruption of their axons. *J. Physiol. London* 252: 429–63
Quik, M., Lamarca, M. V. 1982. Blockade of transmission in rat sympathetic ganglia by a toxin which co-purifies with L-bungarotoxin. *Brain Res.* 238: 385–99
Raftery, M. A., Hunkapiller, M. W., Strader, C. D., Hood, L. E. 1980. Acetylcholine receptor: Complex of homologous subunits. *Science* 208: 1454–57
Rainbow, T. C., Schwartz, R. D., Parsons, B., Kellar, K. J. 1984. Quantitative autoradiography of [3-H]-acetylcholine binding sites in rat brain. *Neurosci. Lett.* 50: 193–96
Rang, H. P. 1981. The characteristics of synaptic currents and responses to acetylcholine of rat submandibular ganglion cells. *J. Physiol. London* 311: 23–55
Ravdin, P. M., Berg, D. K. 1979. Inhibition of neuronal acetylcholine sensitivity by α-toxins from *Bungarus multicinctus* venom. *Proc. Natl. Acad. Sci. USA* 76: 2072–76
Ravdin, P. M., Nitkin, R. M., Berg, D. K. 1981. Internalization of α-bungarotoxin on neurons induced by a neurotoxin that blocks neuronal acetylcholine sensitivity. *J. Neurosci.* 1: 849–61
Reiness, C. G., Hall, Z. W. 1981. The developmental change in immunological properties of the acetylcholine receptor in rat muscle. *Devel. Biol.* 81: 324–31
Reiness, C. G., Weinberg, C. B. 1981. Metabolic stabilization of acetylcholine receptors at newly formed neuromuscular junctions in rat. *Devel. Biol.* 84: 247–54
Reynolds, J. A., Karlin, A. 1978. Molecular weight in detergent solution of acetylcholine receptor from *Torpedo californica.* *Biochemistry* 17: 2035–38
Role, L. W. 1984a. Substance P modulation of acetylcholine-induced currents in embryonic chicken sympathetic and ciliary ganglion neurons. *Proc. Natl. Acad. Sci. USA* 81: 2924–28
Role, L. W. 1984b. Peptide modulation of ACh currents in autonomic neurons. *Soc. Neurosci. Abstr.* 10: 1117 (Abstr.)
Role, L. W. 1985. Synaptic input increases acetylcholine sensitivity of embryonic chick sympathetic neurons in vitro. *Soc. Neurosci. Abstr.* 11: 774
Role, L. W., Hume, R. I., Fischbach, G. D. 1982. Transmitter release and receptor aggregation at ciliary neuron-muscle synapses. *Soc. Neurosci. Abstr.* 8: 129
Role, L. W., Leeman, S. E., Perlman, R. L. 1981. Somatostatin and substance P inhibit catecholamine secretion from isolated cells of guinea-pig adrenal medulla. *Neuroscience* 6: 1813–21
Role, L. W., Matossian, V. R., O'Brien, R. J., Fischbach, G. D. 1985. On the mechanism of acetylcholine receptor accumulation at newly formed synapses on chick myotubes. *J. Neurosci.* 5: 2197–2204
Role, L. W., Roufa, D., Fischbach, G. D. 1987. The distribution of acetylcholine receptor clusters and sites of transmitter release along chick ciliary ganglion neurite-myotube contacts in culture. *J. Cell Biol.* In press
Roper, S. 1976. The acetylcholine sensitivity of the surface membrane of multiply-innervated parasympathetic ganglion cells in the mudpuppy before and after partial denervation. *J. Physiol. London* 254: 455–73
Roper, S., Purves, McMahan, U. J. 1976. Synaptic organization and acetylcholine sensitivity of multiply innervated autonomic ganglion cells. *Cold Spring Harbor Symp. Quant. Biol.* 40: 283–95
Rousselet, A., Cartaud, J., Devaux, P. F., Changeux, J. P. 1982. The rotational diffusion of the acetylcholine receptor in *Torpedo marmorata* membrane fragments studied with a spin-labelled alpha-toxin: Importance of the 43000 protein(s). *EMBO J.* 1: 439–45
Rubin, L. L., Anthony, D. T., Englander, L. L., Lappin, R. L., Lieberburg, I. M. 1986. Molecular modifications during nerve-muscle synapse formation. *Prog. Brain Res.* 71: 383–89
Rubin, L. L., McMahan, U. J. 1982. Regeneration of the neuromuscular junction: Steps toward defining the molecular

basis of the interaction between nerve and muscle. In *Disorders of the Motor Unit*, ed. D. L. Schotland, pp. 187–96. New York: Wiley

Ryall, R. W. 1982. Modulation of cholinergic transmission by substance P. In *Substance P in the Nervous System. CIBA Found. Symp.* 91: 267–80

Saiani, L., Guidotti, A. 1982. Opiate receptor-mediated inhibition of catecholamine release in primary cultures of bovine adrenal chromaffin cells. *J. Neurochem.* 39: 1669–76

Saiani, L., Kageyama, H., Conti-Tronconi, B. M., Guidotti, A. 1984. Purification and characterization of a bungarotoxin polypeptide which blocks nicotinic receptor function in primary culture of adrenal chromaffin cells. *Molec. Pharmacol.* 25: 327–34

Sakmann, B. 1978. Acetylcholine-induced ionic channels in rat skeletal muscle. *Fed. Proc.* 37: 2654–59

Sakmann, B., Brenner, H. R. 1978. Change in synaptic channel gating during neuromuscular development. *Nature* 276: 401–2

Sakmann, B., Methfessel, C., Mishina, M., Takahashi, T., Takai, T., et al. 1985. Role of acetylcholine receptor subunits in gating of the channel. *Nature* 318: 538–43

Sakmann, B., Patlak, J., Neher, E. 1980. Single acetylcholine-activated channels show burst kinetics in presence of desensitizing concentrations of agonist. *Nature* 286: 71–73

Salpeter, M. M., Harris, R. 1983. Distribution and turnover rate of acetylcholine receptors throughout the junction folds at a vertebrate neuromuscular junction. *J. Cell Biol.* 96: 1781–85

Salpeter, M. M., Leonard, J. P., Kasprzak, H. 1982a. Agonist-induced postsynaptic myopathy. *Neurosci. Commun.* 1: 73–83

Salpeter, M. M., Loring, R. H. 1986. Nicotinic acetylcholine receptor in vertebrate muscle: Properties, distribution and neural control. *Prog. Neurobiol.* 25: 297–325

Salpeter, M. M., Spanton, S., Holley, K., Podleski, T. R. 1982b. Brain extract causes acetylcholine receptor redistribution which mimics some early events at developing neuromuscular junctions. *J. Cell Biol.* 93: 417–25

Sanes, J. R. 1983. Roles of extracellular matrix in neural development. *Ann. Rev. Physiol.* 45: 581–600

Sanes, J. R., Chiu, A. Y. 1983. The basal lamina of the neuromuscular junction. *Cold Spring Harbor Symp. Quant. Biol.* 48: 667–78

Sanes, J. R., Feldman, D. H., Cheney, J. M., Lawrence, J. C. 1984. Brain extract induces synaptic characteristics in the basal lamina of cultured myotubes. *J. Neurosci.* 4: 464–73

Sanes, J. R., Marshall, L. M., McMahan, U. J. 1978. Reinnervation of muscle fiber basal lamina after removal of myofibers. Differentiation of regenerating axons at original synaptic sites. *J. Cell Biol.* 78: 176–98

Sattelle, D. B., David, J. A., Harrow, I. D., Hue, B. 1980. Actions of α-bungarotoxin on identified insect central neurons. In *Receptors for Neurotransmitters, Hormones and Pheromones in Insects*, ed. D. B. Sattelle, L. M. Hall, J. G. Hildebrand, pp. 125–39. Amsterdam: Elsevier/North Holland

Schubert, D., Heinemann, S., Kidokoro, Y. 1977. Cholinergic metabolism and synapse formation by a rat nerve cell line. *Proc. Natl. Acad. Sci. USA* 74: 2579–83

Schuetze, S. M. 1980. The acetylcholine channel open time in chick muscle is not decreased following innervation. *J. Physiol. London* 303: 111–24

Schuetze, S. M., Fischbach, G. D. 1978. Channel open time decreases postnatally in rat synaptic acetylcholine receptors. *Neurosci. Abstr.* 4: 374

Schuetze, S. M., Frank, E. F., Fischbach, G. D. 1978. Channel open time and metabolic stability of synaptic and extrasynaptic acetylcholine receptors on cultured chick myotubes. *Proc. Natl. Acad. Sci. USA* 75: 520–23

Schuetze, S. M., Vicini, S. 1984. Neonatal denervation inhibits the normal postnatal decrease in endplate channel open time. *J. Neurosci.* 4: 2297–2302

Schuetze, S. M., Vicini, S. 1986. Apparent acetylcholine receptor channel conversion at individual rat soleus endplates in vitro. *J. Physiol. London* 375: 153–67

Schuetze, S. M., Vicini, S., Hall, Z. W. 1985. Myasthenic serum selectively blocks acetylcholine receptors with long channel open times at developing rat endplates. *Proc. Natl. Acad. Sci. USA* 82: 2533–37

Schultzberg, M., Lundberg, J. M., Hökfelt, T., Terenius, L., Brandt, J., Elde, R. P., Goldstein, M. 1978. Enkephalin like immunoreactivity in gland cells and nerve terminals of adrenal medulla. *Neuroscience* 3: 1169–86

Schwartz, I. R., Bok, P. 1979. Electron microscopic localization of ^{125}I-α-bungarotoxin binding sites in the outer plexiform layer of the goldfish retina. *J. Neurocytol.* 8: 53–66

Schwartz, R. D., Kellar, K. J. 1983. Nicotinic cholinergic receptor binding sites in the brain: Regulation in vivo. *Science* 220: 214–16

Schwartz, R. D., McGee, R., Kellar, K. J. 1982. Nicotinic cholinergic receptors labeled by ^{3}H-acetylcholine in rat brain. *Molec. Pharmacol.* 22: 56–62

Sealock, R., Wray, B. E., Froehner, S. C. 1984. Ultrastructural localization of the Mr 43,000 protein and the acetylcholine receptor in *Torpedo* postsynaptic membranes using monoclonal antibodies. *J. Cell Biol.* 98: 2239–44

Shainberg, A., Isac, A. 1984. Inhibition of acetylcholine receptor synthesis by conditioned medium of electrically stimulated muscle cultures. *Brain Res.* 308: 373–76

Siegelbaum, S. A., Trautmann, A., Koenig, J. 1984. Single acetylcholine-activated channel currents in developing muscle cells. *Devel. Biol.* 104: 366–79

Siegelbaum, S. A., Tsien, R. W. 1983. Modulation of gated ion channels as a mode of transmitter action. *Trends Neurosci.* 6: 307–13

Simasko, S. M., Soares, J. R., Weiland, G. A. 1985. Structure-activity relationship for substance P inhibition of carbamylcholine-stimulated $^{22}Na^+$ flux in neuronal PC12 and non-neuronal BC3H1 cell lines. *J. Pharmacol. Exp. Therap.* 235: 601–5

Simmons, L. K., Koester, J. 1986. Serotonin enhances the excitatory acetylcholine responses in the RB cell cluster of *Aplysia californica. J. Neurosci.* 6: 774–81

Sine, S. M., Taylor, P. 1980. Relationship between agonist occupation and the permeability response of the cholinergic receptor revealed by bound cobra α-toxin. *J. Cell Biol.* 255: 10144–56

Slater, C. R. 1982. Neural influence on the postnatal changes in acetylcholine receptor distribution at nerve-muscle junctions in the mouse. *Devel. Biol.* 94: 23–30

Smith, M. A., Margiotta, J. F., Berg, D. K. 1983. Differential regulation of acetylcholine sensitivity and α-bungarotoxin-binding sites on ciliary ganglion neurons in cell culture. *J. Neurosci.* 3: 2395–2402

Smith, M. A., Margiotta, J. F., Franco, A., Lindstrom, J. M., Berg, D. K. 1986. Cholinergic modulation of an acetylcholine receptor-like antigen on the surface of chick ciliary ganglion neurons in cell culture. *J. Neurosci.* 6: 946–53

Smith, M. A., Stollberg, J., Lindstrom, J. M., Berg, D. K. 1985. Characterization of a component in chick ciliary ganglia that cross-reacts with monoclonal antibodies to muscle and electric organ acetylcholine receptor. *J. Neurosci.* 5: 2726–31

Smollen, A. J. 1983. Specific binding of α-bungarotoxin to synaptic membranes in rat sympathetic ganglion: Computer best-fit analysis of electron microscope radioautographs. *Brain Res.* 289: 177–88

St. John, P. A., Froehner, S. C., Goodenough, P. A., Cohen, J. P. 1982. Nicotinic postsynaptic membranes from *Torpedo*: Sidedness, permeability to macromolecules and topography of major polypeptides. *J. Cell Biol.* 92: 333–42

Stallcup, W. B., Patrick, J. 1980. Substance P enhances cholinergic receptor desensitization in a clonal nerve cell line. *Proc. Natl. Acad. Sci. USA* 77: 634–38

Stanley, E. F., Drachman, D. B. 1983. Rapid degradation of "new" acetylcholine receptors at neuromuscular junctions. *Science* 222: 67–69

Steele, J. A., Steinbach, J. H. 1986. Single channel studies reveal three classes of acetylcholine-activated channels in mouse skeletal muscle. *Biophys. J.* 49: 361a (Abstr.)

Steinacker, A., Highstein, S. M. 1976. Pre- and postsynaptic action of substance P at the Mauthner fiber-giant fiber synapse in the hatchetfish. *Brain Res.* 114: 128–33

Steinbach, J. H. 1981a. Developmental changes in acetylcholine receptor aggregates at rat skeletal neuromuscular junctions. *Devel. Biol.* 84: 267–76

Steinbach, J. H. 1981b. Neuromuscular junctions and α-bungarotoxin-binding sites in denervated and contralateral cat skeletal muscles. *J. Physiol. London* 313: 513–28

Steinbach, J. H., Bloch, R. J. 1986. The distribution of acetylcholine receptors on vertebrate skeletal muscle cells. In *Receptors in Cellular Recognition and Developmental Processes*, ed. R. M. Gorczynski, pp. 183–213. New York: Academic

Steinbach, J. H., Merlie, J., Heinemann, S., Bloch, R. 1979. Degradation of junctional and extrajunctional acetylcholine receptors by developing rat skeletal muscle. *Proc. Natl. Acad. Sci. USA* 76: 3547–51

Stya, M., Axelrod, D. 1983. Diffusely distributed acetylcholine receptors can participate in cluster formation on cultured rat myotubes. *Proc. Natl. Acad. Sci. USA* 80: 449–53

Swanson, L. W., Lindstrom, J., Tzartos, S., Schmued, L. C., O'Leary, D. D. M., Cowan, W. M. 1983. Immunohistochemical localization of monoclonal antibodies to the nicotinic acetylcholine receptor in chick midbrain. *Proc. Natl. Acad. Sci. USA* 80: 4532–36

Sytkowski, A. J., Vogel, Z., Nirenberg, M. W. 1973. Development of acetylcholine receptor clusters on cultured muscle cells. *Proc. Natl. Acad. Sci. USA* 70: 270–74

Takai, T., Noda, M., Mishina, M., Shimizu, S., Furutani, Y., et al. 1985. Cloning, sequencing and expression of cDNA for a novel subunit of acetylcholine receptor from calf muscle. *Nature* 315: 761–64

Teichberg, V. I., Sobel, A., Changeux, J.-P. 1977. In vitro phosphorylation of the acetylcholine receptor. *Nature* 267: 540–42

Usdin, T. B., Fischbach, G. D. 1986. Purification and characterization of a polypeptide from chick brain that promotes the accumulation of acetylcholine receptors in chick myotubes. *J. Cell Biol.* 103: 493–507

Vicini, S., Schuetze, S. M. 1985. Gating properties of acetylcholine properties at developing rat endplates. *J. Neurosci.* 5: 2212–24

Viveros, O. H., Diliberto, E. J., Hazum, E. J., Hazum, E., Chang, K. J. 1979. Opiate like materials in adrenal medulla: Evidence for storage and secretion with catecholamines. *Molec. Pharmacol.* 16: 1101–8

Vogel, Z., Maloney, G. J., Ling, A., Daniels, M. P. 1977. Identification of synaptic acetylcholine receptor sites in retina with peroxidase-labeled α-bungarotoxin. *Proc. Natl. Acad. Sci. USA* 74: 3268–72

Walker, J. H., Boustead, C. M., Witzemann, V. 1984. The 43-K protein, VI, associated with acetylcholine receptor containing membrane fragments is an actin-binding protein. *EMBO J.* 3: 2287–90

Wallace, B. G., Godfrey, E. W. Nitkin, R. M., Rubin, L. L., McMahan, U. J. 1982. An extract of extracellular matrix fraction that organizes acetylcholine receptors. In *Muscle Development. Molecular and Cellular Control*, ed. M. L. Pearson, H. F. Epstein, pp. 469–95. New York: Cold Spring Harbor Lab. Press

Wallace, B. G., Nitkin, R. M., Reist, N. E., Fallon, J. R., Moayeri, N. N., McMahan, U. J. 1985. Aggregates of acetylcholinesterase induced by acetylcholine receptor-aggregating factor. *Nature* 315: 574–77

Weill, C. L., McNamee, M. G., Karlin, A. 1974. Affinity labeling of purified acetylcholine receptor from *Torpedo californica*. *Biophys. Res. Commun.* 61: 997–1003

Weinberg, C. B., Reiness, C. G., Hall, Z. W. 1981. Topographical segregation of old and new acetylcholine receptors at developing ectopic endplates in adult rat muscle. *J. Cell Biol.* 88: 215–20

Weldon, P. R., Cohen, M. W. 1979. Development of synaptic ultrastructure at neuromuscular contacts in an amphibian cell culture system. *J. Neurocytol.* 8: 239–59

Yoshida, K., Imura, H. 1979. Nicotinic cholinergic receptors in brain synaptosomes. *Brain Res.* 172: 453–59

Young, S. H., Poo, M.-M. 1983a. Topographical rearrangement of ACh receptors alters channel kinetics. *Nature* 305: 634–37

Young, S. H., Poo, M.-M. 1983b. Rapid lateral diffusion of extrajunctional acetylcholine receptors in *Xenopus* tadpole myotomes. *J. Neurosci.* 3: 225–31

Zatz, M., Brownstein, M. J. 1981. Injection of α-bungarotoxin near the superchiasriatic nucleus blocks the effects of light on nocturnal pineal enzyme activity. *Brain Res.* 213: 438–42

Ziskind-Conhaim, L., Geffen, I., Hall, Z. W. 1984. Redistribution of acetylcholine receptors on developing rat myotubes. *J. Neurosci.* 4: 2346–49

Ann. Rev. Neurosci. 1987. 10: 459–76

MOLECULAR MECHANISMS FOR MEMORY: Second-Messenger Induced Modifications of Protein Kinases in Nerve Cells

James H. Schwartz and Steven M. Greenberg

Howard Hughes Medical Institute, Columbia University College of Physicians and Surgeons, New York, New York 10032

Dedication: In memory of Fritz Lipmann, who first isolated and identified serine phosphate from protein (Lipmann & Levine 1932).

INTRODUCTION

The protein kinases that are activated by Ca^{2+}/calmodulin, Ca^{2+}/phospholipid, and cAMP are now thought to govern many types of slow (or modulatory) synaptic mechanisms and to mediate many forms of short-term synaptic plasticity (Reichardt & Kelly 1983, Nairn et al 1985, Nishizuka 1986). These processes, which do not depend on the synthesis of new proteins and can endure from minutes to a substantial part of an hour, are the neurophysiological correlates of short-term memory (see Kandel & Schwartz 1982, Schwartz et al 1983, Goelet et al 1986). Unlike the rapid conformational changes in the polypeptides that constitute chemically gated ion channels and that last only fractions of seconds after the stimulating ligand is removed, the slow onset of short-term synaptic mechanisms on the one hand and their persistence for periods of time exceeding even the longest allosteric rearrangements of protein molecules on the other (Frieden 1970) reflect the receptor stimulated appearance, accumulation, and relatively slow dissipation of free Ca^{2+}, diacylglycerol, or cAMP within the stimulated nerve cell. As second messengers, these mol-

0147–006X/87/0301–0459$02.00

ecules operate by maintaining one or more of the three protein kinases in an active state. The kinetic features of the synaptic mechanisms, in turn, parallel the kinetic properties of the governing protein kinase because the phosphorylation step appears to determine the time course of the regulatory cascade (Schwartz et al 1983). For this reason, any molecular mechanism that prolongs the time that the kinase is active would be expected to prolong the duration of the synaptic mechanism, and, as a consequence, to prolong the span of the behavioral modification.

We review three types of post-translational modifications of these kinase molecules, brought about by receptor-mediated second messenger mobilization or synthesis, that might prolong their action: autophosphorylation, alteration of subcellular distribution, and specific proteolysis. In most instances, the post-translational modifications thus far described also tend to render the kinase reaction less dependent upon the initial stimulating second messenger; in the extreme, the kinase can become fully autonomous. As a consequence, after appropriate previous exposure to the stimulus, the enzyme can continue to phosphorylate protein substrates even when the second messenger falls to the concentration that had initially been present in the naive neuron. These modifications can also produce a primed kinase: prior stimulation can enhance the sensitivity of the enzyme, teaching it to respond to lower concentrations of the second messenger.

Persistence and priming both might be considered forms of molecular memory. Crick (1984) and Lisman (1985) have put forward theories involving stimulus-induced post-translational modifications of specific regulatory enzymes that could result in essentially permanent memory without changes in gene expression. None of the demonstrated modifications of the three kinases, however, precisely fits the requirements postulated for producing enduring changes; in addition, considerable evidence suggests that new protein synthesis is needed for acquisition of long-term memory (see Goelet et al 1986). The three kinds of protein modifications reviewed below might instead participate in consolidation, bridging between the relatively brief accumulation and turnover of small second-messenger molecules and the induction and axonal transport of new wisdom macromolecules to the learned synapse.

DESCRIPTION OF THE PROTEIN KINASES

Common Features

Categorized as possessing broad substrate specificities (Nairn et al 1985), the Ca^{2+}/calmodulin-, Ca^{2+}/phospholipid-, and cAMP-dependent protein kinases transfer the terminal phosphoryl group of ATP to target proteins at selected serine and threonine residues. Substrate specificity is produced in part by amino acid sequences in the neighborhood of the acceptor hydroxy amino acids (see Nestler & Greengard 1984). It is clear that other factors determine specificity as well, since the three kinases (which, for example, share synapsin I, tyrosine hydroxylase, and microtubule-associated protein 2 as substrates) phosphorylate some sites in common and some that are singular (Nairn et al 1985, Tsuyama et al 1986). All of these kinases are autophosphorylated, but none of the enzymes has been reported to be a substrate for either of the other two.

Although there are similar enzyme molecules in other tissues, specific forms of each of the three kinases exist in brain (Nairn et al 1985, Nishizuka 1986). All are fairly abundant in brain, constituting at a minimum 0.01–0.1% of total protein (Ca^{2+}/phospholipid- and cAMP-dependent; Hartl & Roskoski 1982, Kikkawa et al 1982) and as much as 1–4% in some brain regions (Ca^{2+}/calmodulin; Erondu & Kennedy 1985). (Many other types of protein kinases are present in nervous tissue, but their distribution and substrates are more restricted, and less information is available about any specific role they might play in synaptic function.) Some portion of the total of each type of kinase has been found to be associated with particulate fractions by assay of enzyme activity or with membranous or cytoskeletal subcellular organelles by histochemical or immunocytochemical procedures (De Camilli et al 1986, Lohmann & Walter 1984, Kikkawa et al 1982, Kennedy et al 1983a, Sahyoun et al 1985, Ouimet et al 1984, Girard et al 1985, 1986, Saitoh & Schwartz 1983, 1985). Unlike some varieties of protein-tyrosine kinases (Hunter & Cooper 1985), however, none of the second messenger-stimulated kinase molecules are intrinsic membrane proteins. One of our aims is to suggest that extent of autophosphorylation, cellular localization, and state of proteolysis alter the activities of these kinases. Consequently, the determinations of subcellular distribution, substrate specificity, and even regional distribution, reviewed briefly below for each kinase, may have to be reinterpreted.

Ca^{2+}/Calmodulin-dependent Protein Kinase

In rat brain, the type II kinase holoenzyme (M_r about 650,000) consists of a variable combination of α (M_r 50,000–55,000) and β (M_r 60,000–65,000) subunits (Bennett et al 1983, Yamauchi & Fugisawa 1983, Goldenring et al 1983, Schulman 1984, McGuinness et al 1985). In *Aplysia* (Saitoh & Schwartz 1985, DeRiemer et al 1984) and *Drosophila* (Willmund et al 1986), only one type of subunit has been identified (M_r 50,000–55,000). Highest concentrations have been found in parts of the nervous system thought to mediate learning: the hippocampus in vertebrates (Ouimet et al 1984, Erondu & Kennedy 1985) and sensory neurons in *Aplysia* (Saitoh

& Schwartz 1983). An immunocytochemical survey reveals that the kinase is associated with vesicles and plasma membrane in presynaptic terminals, while the most intense staining was observed in cell bodies and dendritic spines (Ouimet et al 1984). Biochemical and immunocytochemical studies (Grab et al 1981, Kennedy et al 1983b, Ouimet et al 1984) have localized the kinase to postsynaptic densities. In both vertebrates and in *Aplysia*, at least half of the enzyme activity in the tissue is recovered in cytosolic fractions (Kennedy et al 1983a, Saitoh & Schwartz 1985) with great regional variability in the partitioning between soluble and particulate fractions (McGuinness et al 1985). Activation by Ca^{2+}/calmodulin is a complex process (reviewed below) that depends on the enzyme's association with the membrane-cytoskeleton (Saitoh & Schwartz 1983, 1985, Le Vine et al 1986), on conditions of assay, and on prior exposure to Ca^{2+}/calmodulin and ATP (Shields et al 1984, Miller & Kennedy 1986, Lai et al 1986, Lou et al 1986, Schworer et al 1986). Several enzyme intermediates with different states of activation can be inferred, and, because several sites on both the α and β subunits (Bennett et al 1983) and on the *Aplysia* M_r 55,000 subunit (T. Saitoh and J. H. Schwartz, unpublished observation) can be phosphorylated, these states of activation are likely to correspond to the extents to which the different sites on the enzyme are autophosphorylated.

Ca^{2+}/Phospholipid-dependent Protein Kinase

In most tissues, Ca^{2+}/phospholipid-dependent protein kinase appears as an M_r 78,000–80,000 monomer (Kikkawa et al 1982, Girard et al 1986). At least three closely related forms are expressed in brain (Coussens et al 1986, Knopf et al 1986). The kinase is present in highest concentrations in brain (hippocampus, neocortex, and cerebellum) in both pre- and postsynaptic components (Girard et al 1986, Worley et al 1986). Subcellular fractionation and immunocytochemical studies at the ultrastructural level indicate that the kinase is enriched in presynaptic terminals; it is found both in soluble and particulate fractions and is associated with both the plasma and nuclear membranes (Kikkawa et al 1982, Girard et al 1985, 1986, Wood et al 1986). Activation requires phospholipid, diacylglycerol, and Ca^{2+} (Hannun et al 1985). The requirement for phospholipid is satisfied by association of the kinase with membrane; this association is promoted by Ca^{2+}. Diacylglycerol (1–2 molecules per enzyme), produced through receptor-mediated phospholipase C-catalyzed turnover of inositol phospholipids in the membrane, activates the kinase by greatly lowering the requirement for Ca^{2+} (Kishimoto et al 1980). Tumor-promoting phorbol esters can also efficiently activate the kinase. The coordinate production of inositol-1,4,5-trisphosphate (IP_3) during phospholipid breakdown can lead to the mobilization of Ca^{2+} from intracellular stores (Berridge & Irvine 1984).

cAMP-dependent Protein Kinase

The holoenzyme, which is inactive, consists of two identical M_r 41,000 catalytic (C) subunits and two regulatory (R) subunits. The two C subunits become active when released from the inhibition exerted by the R subunit pair. Dissociation occurs by the cooperative binding of two molecules of cAMP to each of the two R subunits (Nairn et al 1985, Nestler & Greengard 1984). Although there is some evidence for two closely related forms (Uhler et al 1986), the C subunits from various tissues are quite similar. Like some protein-tyrosine kinases, to which they share a high degree of homology (Hunter & Cooper 1985), their *N*-terminus is blocked by a myristic acid residue (Carr et al 1982). Individual holoenzyme molecules differ from tissue to tissue in type of R subunits: while each holoenzyme tetramer (R_2C_2) contains two identical R subunits, cells can contain more than one version of R subunit. In addition to the ubiquitous R_I (M_r 45,000–49,000) and R_{II} (M_r 52,000–57,000) subunits, vertebrate and invertebrate brain and germ cells contain specialized forms (Eppler et al 1982, Palazzolo 1985, Jahnsen et al 1986). Some forms of R subunit are autophosphorylated, and these recombine less readily with C subunits than do their dephosphorylated counterparts (Rangel-Aldao & Rosen 1976a).

In nervous tissues, about half of the holoenzyme is soluble, the rest particulate (Rubin et al 1979, Eppler et al 1986). This distribution is determined by the particular type of R subunit, since free C subunit is soluble (Corbin et al 1977). Specific subcellular localization of particular R subunit types has been described both in vertebrates (Theuerkauf & Vallee 1982, De Camilli et al 1986, Nigg et al 1985) and in invertebrates (Eppler et al 1982, Greenberg et al 1987).

MODIFICATIONS THAT PROLONG ACTIVATION

Autophosphorylation

GENERAL FEATURES Phosphorylation is a common post-translational mechanism for altering the activity of proteins. Among the earliest characterized examples of enzyme regulation by this mechanism is the cAMP-dependent phosphorylation of phosphorylase kinase and glycogen synthase, which act in concert to promote the formation of glucose in skeletal muscle (see Cohen 1983). Autophosphorylation denotes the self-catalyzed acceptance of the phosphoryl group by the kinase molecule itself. The phosphorylation in principle can either be intramolecular (catalyzed by the same protein molecule that acts as the acceptor) or intermolecular (catalyzed by another kinase molecule of the same type). Intramolecular autophosphorylation can occur as a catalytic intermediate of certain enzymes (Schwartz 1963), or it can be regulatory, as in phosphorylase kinase (King et al 1983) and the insulin receptor (Rosen et al 1983), where it stimulates enzyme activity.

Many proteins contain more than one acceptor site for phosphorylation. In general, phosphorylation at different sites within a protein produces different effects on that protein's function (see Cohen 1983, Nestler & Greengard 1984). Ca^{2+}/calmodulin-dependent protein kinase II has several sites on each subunit, the phosphorylation of which appears to produce opposing effects on kinase activity, as reviewed below. Untangling the causes and effects of phosphorylation at each of the sites is necessary for understanding how the protein is regulated.

A kinase activated by intermolecular autophosphorylation can act as a molecular switch, flipping, when the rate of autophosphorylation exceeds the rate of dephosphorylation, from an inactive state to a stable activated enzyme (Lisman 1985). Thus, much attention has been paid to phosphorylation reactions that activate kinases and their potential for stability.

Ca^{2+}/CALMODULIN-DEPENDENT PROTEIN KINASE Autophosphorylation of the Ca^{2+}/calmodulin-dependent kinase can eliminate the enzyme's requirements for Ca^{2+} and calmodulin, thereby producing an autonomous enzyme that is active in the absence of its second-messenger ligands (Saitoh & Schwartz 1985, Miller & Kennedy 1986, Lai et al 1986, Schworer et al 1986, Lou et al 1986). Both the α and β subunits are autophosphorylated, but only a few of the approximately 30 available sites in the holoenzyme need to be phosphorylated to eliminate the requirement for Ca^{2+} and calmodulin (Miller & Kennedy 1986, Lai et al 1986). When studied with the purified enzyme, the autophosphorylation appears to be intramolecular (Kuret & Schulman 1985, Miller & Kennedy 1986, Lai et al 1986, Colbran et al 1986). The mechanism may, however, be affected by the conditions in the cell. For example, activation of the enzyme attached to the cytoskeleton differs from that of the purified, soluble kinase in that calmodulin acts cooperatively to activate the attached form (Le Vine et al 1986). In the cell, kinase holoenzymes could exchange subunits with each other; exchange would convert the intramolecular phosphorylation into a two-step intermolecular activation process, as suggested by Miller & Kennedy (1986).

The autonomy of the kinase is abolished by dephosphorylation, which in vitro can be catalyzed by purified phosphoprotein phosphatase 1 or 2A (Shields et al 1984, Schworer et al 1986, Lai et al 1986). Another possible mechanism of inactivation may be autophosphorylation at additional sites.

Phosphorylation at other sites has been found to occur in vitro preferentially in incubations with concentrations of ATP less than 100 μM (about an order of magnitude lower than that in the cell (Lou et al 1986), and is favored at higher temperatures (30°C rather than 0°C) (H. Schulman, personal communication). Autophosphorylation occurring at inactivating sites probably accounts for the 65% loss of enzyme activity reported by Miller & Kennedy (1986) and for the apparent thermolability of the phosphorylated enzyme reported by Lai et al (1986). Attempts to identify these presumptively different sites by tryptic fingerprinting of ^{32}P-labeled phosphopeptides from the kinase phosphorylated at both low and high concentrations of ATP resulted in some labeled peptides that are similar and some that differ (Lou et al 1986). These results suggest that ATP, in addition to being a substrate, acts as an allosteric modulator of the kinase, determining the site of autophosphorylation.

Autophosphorylation of the kinase has not yet been directly related to any learning process in vertebrates. The subunit of the kinase in *Drosophila*, however, appears to undergo stable changes in phosphorylation and subcellular localization in response to long-term visual adaptation (Willmund et al 1986). Long-lasting changes in the kinase's autophosphorylation also result from septal kindling of rat brain (Goldenring et al 1986).

Ca^{2+}/PHOSPHOLIPID-DEPENDENT PROTEIN KINASE The Ca^{2+}/phospholipid-dependent kinase, when activated, is autophosphorylated at approximately two sites per monomer (Kikkawa et al 1982, Huang et al 1986). Autophosphorylation has been reported to increase the enzyme's affinity for both Ca^{2+} and phorbol ester (Huang et al 1986), to decrease its affinity for histone substrates (D. Mochly-Rosen and D. Koshland, personal communication), and to reverse its binding to the membrane (Wolf et al 1985a).

cAMP-DEPENDENT PROTEIN KINASE The regulatory subunit R_{II} of the vertebrate cAMP-dependent kinase (Robinson-Steiner et al 1984) as well as several forms of the regulatory subunits in *Aplysia* (Eppler et al 1982) and *Drosophila* (Foster et al 1984) can be phosphorylated by the catalytic subunit. With the bovine heart R_{II} subunit, phosphorylation can occur either in the absence or in the presence of cAMP; in its absence the reaction is intramolecular, in its presence, intermolecular (Rangel-Aldao & Rosen 1976b). In vitro dephosphorylation is greatly accelerated by cAMP, suggesting that the R_{II} subunit in the holoenzyme is phosphorylated to a greater extent than is the free regulatory subunit.

Autophosphorylation decreases the rate at which the R_{II} subunit recombines with the C subunit (Rangel-Aldao & Rosen 1976a), and increases the affinity of the kinase for cAMP (Hofmann et al 1975). Both of these effects would enhance and prolong the phosphorylation of substrates by the C subunit within the cell. The C subunit phosphorylates itself as well, with no known functional consequences (Shoji et al 1979).

The R_I and R_{II} subunits are also phosphorylated at positions not phosphorylated by the C subunit (Geahlen & Krebs 1980, Hemmings et al 1982). The phosphorylation of the R_I subunit, however, is greatly diminished in mutant cells lacking C subunit and decreased in wild-type cells by treatments that dissociate the holoenzyme (Steinberg & Agard 1981a).

Changes in Subcellular Localization

GENERAL FEATURES Upon activation, all three kinases change their distribution between soluble and particulate cellular fractions. It is clear that kinases do not have access in the cell to all of their potential substrates (see Nestler & Greengard 1984). Although it is venturesome to be certain about the precise physiologic function of translocations between these imprecisely defined biochemical compartments, changes in intracellular localization are likely to affect the availability of particular protein substrates to the kinases.

Ca^{2+}/PHOSPHOLIPID-DEPENDENT PROTEIN KINASE The Ca^{2+}/phospholipid-dependent protein kinase moves to cell membranes upon activation (Kraft & Anderson 1983). Translocation requires Ca^{2+} and is enhanced by diacylglycerol or phorbol esters but can occur in their absence (Wolf et al 1985b, Hannun et al 1985). Binding of the kinase to membrane is proposed to involve the coordinate interaction of the carboxyl groups from four phosphatidylserine molecules with a Ca^{2+}-enzyme complex (Ganong et al 1986).

The transient elevation of free Ca^{2+} induced by IP_3 may prime the kinase by effecting its translocation, thereby allowing it to be activated by concomitant or subsequent appearance of diacylglycerol in the membrane. Association with the membrane is much faster at high concentrations of Ca^{2+} (full translocation in approximately 30 sec) than with phorbol esters (translocation in 10 min) (Wolf et al 1985b). When the IP_3-induced transient increase in Ca^{2+} in adrenal glomerulosa cells is blocked, receptor-stimulated activation of the kinase does not reach a maximum, presumably because less kinase moves to the membrane in the absence of the priming step (Kojima et al 1985). Priming of the kinase can outlast the initial stimulus, creating a type of short-term memory: adrenal cells that were stimulated and then have returned to the basal state show a heightened response when exposed to the stimulating agonist or to a Ca^{2+} ionophore (Barrett et al 1986).

Ca^{2+}/CALMODULIN-DEPENDENT PROTEIN KINASE Partitioning of Ca^{2+}/calmodulin-dependent protein kinase between cytosol and particulate fractions is variable in different brain regions, but in vertebrates differences in this distribution have not yet been found to correspond to any physiologic treatment. The kinase in *Aplysia* nervous tissue, however, appears in the cytosol fraction prepared from neurons that have been exposed to treatments that elevate intracellular cAMP (Saitoh & Schwartz 1983). The kinase is released from the isolated membrane-cytoskeletal fraction upon exposure (under phosphorylating conditions) to either cAMP or Ca^{2+}/calmodulin (Saitoh & Schwartz 1983, 1985). Translocation may involve phosphorylation of proteins that anchor kinase to the cytoskeleton with subsequent release of the enzyme. Sahyoun et al (1986a) identified binding proteins in rat brain postsynaptic densities that, when phosphorylated, lose their ability to bind kinase. The enzyme is also released under conditions of extraction that do not preserve the cytoskeleton (Saitoh & Schwartz 1985). Translocation in itself is not sufficient to activate the enzyme: release of the kinase can occur without activation, as after treatments that elevate cAMP, and activation can occur after release, as has been demonstrated with the soluble purified enzyme.

cAMP-DEPENDENT PROTEIN KINASE The regulatory subunits of the cAMP-dependent kinase are thought to determine the subcellular localization of the kinase holoenzyme. They associate in a characteristic manner with soluble and particulate fractions and with specific cellular organelles (Corbin et al 1977, Rubin et al 1979, Eppler et al 1982, Theuerkauf & Vallee 1982, De Camilli et al 1986). The free C subunit is soluble; thus activation of a membrane-associated kinase holoenzyme shifts the C subunit from the particulate to the soluble fraction. In several non-neural cell types, cAMP-dependent kinase has been reported to move to the nucleus after prolonged stimulation with analogues of cAMP (Lohmann & Walter 1984). Although the mechanism of this translocation is unknown, it might mediate cAMP-dependent changes in gene expression in neurons.

Proteolytic Cleavage

GENERAL FEATURES Specific proteolytic cleavage is the familiar mechanism for processing membrane and secretory proteins: it produces trypsin from trypsinogen and neuroactive peptides from polyproteins (see Cohen 1983). Cleavage of the cytoskeletal protein, fodrin, has also been suggested as the molecular basis of the increase in glutamate binding to hippocampal synaptic membranes during long-term potentiation (see Lynch & Baudry 1984). Unlike the modifications reviewed above, proteolysis is not reversible and thus would necessarily result in relatively long-lasting synaptic changes.

cAMP-DEPENDENT PROTEIN KINASE Degradation of R subunits can be induced by elevating intracellular cAMP, by the application of certain cAMP analogues, or by the genetic removal of functional C subunits (Steinberg & Agard 1981b, Greenberg et al 1986). The absence of functioning C subunits in mutant S49 lymphoma cells results in a decrease in the half-life of the R_I subunit from 8.4 hr to less than 1 hr. In *Aplysia* synaptosomes, treatment with cAMP analogues for 2 hr reduces by half all types of R subunit; the amount of C subunits is not significantly reduced by this treatment. Long-term sensitization training of *Aplysia* results in a loss of R subunits in the neurons that mediate sensitization, again with no significant change in the C subunit.

The increased lability of R subunits is likely to be the result of increased susceptibility to proteolytic cleavage when not protected by a bound C subunit (Potter & Taylor 1980). Breakdown of the R subunits is also influenced by other factors: dibutyryl cAMP diminishes rather than enhances the rate of turnover (Steinberg & Agard 1981b, M. M. Gottesman, personal communication). All R subunits studied contain a region highly sensitive to proteolysis (Potter & Taylor 1979a, Corbin et al 1978). This region lies near the domain in the R molecule that interacts with the C subunit (Potter & Taylor 1979b, Steinberg 1983), and treatment of porcine R_I subunit with thermolysin can produce an M_r 35,000 fragment that binds cAMP but not C subunit (Potter & Taylor 1979a). Prolonged elevation of cAMP might thus result in selective loss of R subunits and increased catalytic activity that outlasts the cAMP signal.

Ca^{2+}/PHOSPHOLIPID-DEPENDENT PROTEIN KINASE Specific proteolysis results in an altered Ca^{2+}/lipid-dependent kinase. Treatment of the kinase with the Ca^{2+}-activated thiol protease, calpain, produces a fragment (designated protein kinase M) (M_r 51,000 by SDS gel electrophoresis or 60,000–65,000 by gel filtration) that is active in the absence of Ca^{2+} and phospholipid (Kishimoto et al 1983, Melloni et al 1985). Cleavage and activation are stimulated by Ca^{2+} and by the presence of phospholipid or membranes, and occur under conditions in which both the kinase and calpain move to the membrane. The fragment generated no longer remains bound to membrane, either in the presence or absence of Ca^{2+} and diacylglycerol; some activity may also be lost (Melloni et al 1985).

There is some indication that an autonomous kinase molecule is produced in vivo as well as in vitro. Application of phorbol esters to neutrophils results in the appearance, essentially complete after 10 min, of kinase activity that is independent of Ca^{2+} and phospholipid, and which is associated with a soluble polypeptide similar in size to the proteolytic fragment (Melloni et al 1986). Calpain inhibitors block this effect of phorbol esters but do not affect movement of the kinase to the membrane. The autonomous kinase activity in neutrophils is stable, indicating that the kinase can remain in this active state for some time after the stimulating second messengers are dissipated. Treatment of a rat pituitary cell line with phorbol ester also reduces the amount of intact kinase (Ballester & Rosen 1985).

Ca^{2+}/CALMODULIN-DEPENDENT PROTEIN KINASE There is no evidence that the Ca^{2+}/calmodulin-dependent kinase is subjected to specific proteolysis. The size of the phosphorylated, autonomous enzyme is not reduced, and treatment of purified kinase with proteases does not produce autonomy (Lou et al 1986). In contrast, Edelman et al (1986) have found that treatment of the Ca^{2+}/calmodulin-dependent myosin light chain kinase from skeletal muscle with proteases generates a catalytically active fragment that no longer requires calmodulin for activity.

PERSPECTIVES: PHYSIOLOGIC RELEVANCE

General Considerations

The hypothesis guiding much current research on the mechanisms of learning and memory is that behavioral modifications of both short and long duration result from changes in synaptic transmission caused by previous experience. Progress in the understanding of many aspects of synaptic transmission has been enormous in the past few decades, making it possible to forge plausible links between molecular events at synapses and their regulation on the one hand and physiologic and behavioral phenomena on the other. Mobilization of synaptic vesicles and alteration in the performance of presynaptic ion channels or postsynaptic receptors, brought about by receptor-mediated second messenger pathways, are the favored explanations for short-term processes; rendering these modifications durable by changes in gene expression is favored for the long-term processes (see Goelet et al 1986).

Besides causing phosphorylation of specific synaptic components (for example, channel proteins), activation of second messenger pathways also results in modification of a wide range of proteins that do not directly produce changes in membrane potential, transmitter release, or receptor activity. Indeed, many of the neuron's proteins are altered by post-translational mechanisms, and it might be questioned whether the modifications of the three kinases that we have considered have any specific or profound physiologic significance or are in any way related to psychological memory. If significant, these molecular mechanisms should parallel the memory, first, in being initiated by the same stimulating conditions and, second, in working toward the same behavioral goal, altering the neuron's capacity to respond for a period longer than the duration of the stimulus. The modifications reviewed here are initiated by stimulation of receptors, and have the potential of prolonging the actions of the kinases and of making them less dependent upon the continued presence of second messengers. Persistence of action for relatively long periods of time raises the question of whether any of these post-translational mechanisms are sufficient (in the absence of specific changes in gene expression) to account for long-term, permanent memory. Although the requirements of the theories thus far advanced are not satisfied, the definitive answer to this question must be postponed until the performance of the kinases within cells is better understood.

Ca^{2+}/Calmodulin-dependent Protein Kinase

Autophosphorylation of Ca^{2+}/calmodulin-dependent protein kinase has many of the properties required of a stable molecular switch: elevation of intracellular Ca^{2+} would cause a burst of autophosphorylation, activate the kinase, and produce long-lasting facilitation of neurotransmitter release. One feature of this mechanism for memory is that it can be confined to a single synapse of a neuron. A change in the neuron's gene expression cannot by itself account for branch-specific modulation, if it actually occurs.

There is some indication that influx of Ca^{2+} triggers long-term synaptic changes. Glutamate receptors sensitive to *N*-methyl-D-aspartate (NMDA), which in rat hippocampus appear to be required for long-term potentiation (LTP) (Collingridge et al 1983) and place learning (Morris et al 1986), mediate entry of Ca^{2+} (MacDermott et al 1986). Also, injection of the kinase into the squid giant synapse enhances neurotransmitter release whereas injection of dephosphorylated synapsin depresses release (Llinas et al 1985). These experiments suggest that the dephosphorylated form of the cytoskeletal kinase substrate, synapsin I (Baines & Bennett 1985), blocks the mobilization of synaptic vesicles; upon phosphorylation by the kinase, the block would disintegrate, making the vesicles available for exocytosis. It is tempting to suggest because of their great abundance that kinase molecules themselves might also form a structural impediment to vesicle release. By this reasoning, autophosphorylation at activating sites on the kinase subunits would cause them to dissociate, thereby promoting mobilization of vesicles. Continued phosphorylation at inactivating sites would safeguard susceptible protein substrates in the cytosol, but permit the anemnestic program to continue by preventing reformation of the putative blocking cytoskeletal structure.

cAMP-dependent Protein Kinase

The demonstration that long-term training in *Aplysia* reduces the number of R subunits relative to C indicates that a covalent modification of a protein kinase—in this instance, proteolytic degradation of the regulatory component—can participate in long-term memory. Loss of R subunits should result in an activation of the kinase that outlasts the rise in cAMP that initiated it. As yet, no evidence has been found for permanent activation of the kinase; autonomy produced by proteolysis would last only until R subunit molecules are replenished by protein synthesis (see Lohmann & Walter 1984) and reach the synapse or until the free C subunit is degraded.

Ca^{2+}/Phospholipid-dependent Protein Kinase

We reviewed two mechanisms that prolong the activity of the Ca^{2+}/phospholipid-dependent kinase: attachment to membrane (Barrett et al 1986) and proteolytic cleavage (Melloni et al 1986). Translocation could be expected to promote phosphorylation of membrane-associated proteins, and there is evidence that the activated, translocated kinase affects membrane channel activity (Akers et al 1986, Malenka et al 1986, Baraban et al 1985, Alkon et al 1986, Farley & Auerbach 1986). Released from the membrane by proteolytic cleavage, the kinase should instead phosphorylate soluble or even nuclear substrates preferentially (Butler et al 1986, Sahyoun et al 1986b). It would be interesting to know whether particular temporal sequences in the elevation of Ca^{2+} and diacylglycerol favor one or the other of these qualitatively different outcomes.

ACKNOWLEDGMENTS

We thank Daniel Koshland, Howard Rasmussen, Howard Schulman, Thomas Soderling, and Rolf Willmund for sharing their recent experiments with us before publication, and Hagan Bayley, Eric Kandel, Michael Gottesman, Robert Steinberg, and Tsunao Saitoh for their comments on the manuscript. Steven Greenberg is supported by NIH Training Grant GM07367-11.

Literature Cited

Akers, R. F., Lovinger, D. M., Colley, P. A., Linden, D. J., Routtenberg, A. 1986. Translocation of protein kinase C activity may mediate hippocampal long-term potentiation. *Science* 231: 587–89

Alkon, D. L., Kubota, M., Neary, J. T., Naito, S., Coulter, D., Rasmussen, H. 1986. C-kinase activation prolongs Ca^{2+}-dependent inactivation of K^+ currents. *Biochem. Biophys. Res. Commun.* 134: 1245–53

Baines, A. J., Bennett, V. 1985. Synapsin I is

a spectrin-binding protein immunologically related to erythrocyte protein 4.1. *Nature* 315: 410–11

Ballester, R., Rosen, O. M. 1985. Fate of immunoprecipitable protein kinase C in GH_3 cells treated with phorbol 12-myristate 13-acetate. *J. Biol. Chem.* 260: 15194–99

Baraban, J. M., Snyder, S. H., Alger, B. E. 1985. Protein kinase C regulates ionic conductance in hippocampal pyramidal neurons: Electrophysiological effects of phorbol esters. *Proc. Natl. Acad. Sci. USA* 82: 2538–42

Barrett, P., Kojima, I., Kojima, K., Zawalich, K., Isales, C. M., Rasmussen, H. 1986. Short term memory in the calcium messenger system: Evidence for a sustained activation of C-kinase in adrenal glomerulosa cells. *Biochem. J.* 238: 905–12

Bennett, M. K., Erondu, N. E., Kennedy, M. B. 1983. Purification and characterization of a calmodulin-dependent protein kinase that is highly concentrated in brain. *J. Biol. Chem.* 258: 12735–44

Berridge, M. J., Irvine, R. F. 1984. Inositol triphosphate, a novel second messenger in cellular signal transduction. *Nature* 312: 315–21

Butler, A. P., Byus, C. V., Slaga, T. J. 1986. Phosphorylation of histones is stimulated by phorbol esters in quiescent Reuber H35 hepatoma cells. *J. Biol. Chem.* 261: 9421–25

Carr, S. A., Biemann, K., Shoji, S., Parmelee, D. C., Titani, K. 1982. *n*-Tetradecanoyl is the NH_2-terminal blocking group of the catalytic subunit of cyclic AMP-dependent protein kinase from bovine cardiac muscle. *Proc. Natl. Acad. Sci. USA* 79: 6128–31

Cohen, P. 1983. *Control of Enzyme Activity*. London: Chapman & Hall. 96 pp. 2nd ed.

Colbran, R. J., Schworer, C. M., Soderling, T. R. 1986. Autophosphorylation sites and the Ca^{2+}-independent activity of Ca^{2+} (calmodulin)-dependent protein kinase II. Abst. No. 159, 6th Intl. Conf. Cyclic Nucleotides, Calcium Protein Phosphoryl., Bethesda, MD, Sept. 2–7

Collingridge, G. L., Kehl, S. J., McLennan, H. 1983. The antagonism of amino acid-induced excitations of rat hippocampal CA1 neurones *in vitro*. *J. Physiol. London* 334: 19–31

Corbin, J. D., Sugden, P. H., Lincoln, T. M., Keely, S. L. 1977. Compartmentalization of adenosine 3′:5′-monophosphate and adenosine 3′:5′-monophosphate-dependent protein kinase in heart tissue. *J. Biol. Chem.* 252: 3854–61

Corbin, J. D., Sugden, P. H., West, L., Flockhart, D. A., Lincoln, T. M., McCarthy, D. 1978. Studies on the properties and mode of action of the purified regulatory subunit of bovine heart adenosine 3′:5′-monophosphate-dependent protein kinase. *J. Biol. Chem.* 253: 3997–4003

Coussens, L., Parker, P. J., Rhee, L., Yang-feng, T. L., Chen, E., et al. 1986. Multiple, distinct forms of bovine and human protein kinase C suggest diversity in cellular signaling pathways. *Science* 233: 859–66

Crick, F. 1984. Memory and molecular turnover. *Nature* 312: 101

De Camilli, P., Moretti, M., Denis Dorini, S., Walter, U., Lohmann, S. M. 1986. Heterogenous distribution of the cAMP receptor protein R_{II} in the nervous system. Evidence for its intracellular accumulation on microtubules, microtubule-organizing centers and in the area of the Golgi complex. *J. Cell Biol.* 103: 189–203

DeRiemer, S. A., Kaczmarek, L. K., Lai, Y., McGuinness, T. L., Greengard, P. 1984. Calcium/calmodulin-dependent protein phosphorylation in the nervous system of *Aplysia*. *J. Neurosci.* 4: 1618–25

Edelman, A. M., Takio, K., Blumenthal, D. K., Hansen, R. S., Walsh, K. A., et al. 1986. Characterization of the calmodulin-binding and catalytic domains in skeletal muscle myosin light chain kinase. *J. Biol. Chem.* 260: 11275–85

Eppler, C. M., Palazzolo, M. J., Schwartz, J. H. 1982. Characterization and localization of adenosine 3′:5′ monophosphate-binding proteins in the nervous system of *Aplysia*. *J. Neurosci.* 2: 1692–1704

Eppler, C. M., Bayley, H., Greenberg, S. M., Schwartz, J. H. 1986. Structural studies on a family of cAMP-binding proteins in the nervous system of *Aplysia*. *J. Cell Biol.* 102: 320–31

Erondu, N. E., Kennedy, M. B. 1985. Regional distribution of type II Ca^{2+}/calmodulin-dependent protein kinase in rat brain. *J. Neurosci.* 5: 3270–77

Farley, J., Auerbach, S. 1986. Protein kinase C activation induces conductance changes in *Hermissenda* photoreceptors like those seen in associative learning. *Nature* 319: 220–23

Foster, J. L., Guttman, J. J., Hall, L. M., Rosen, O. M. 1984. *Drosophila* cAMP-dependent protein kinase. *J. Biol. Chem.* 259: 13049–55

Frieden, C. 1970. Kinetic aspects of regulation of metabolic processes. The hysteretic enzyme concept. *J. Biol. Chem.* 245: 5788–99

Ganong, B. R., Loomis, C. R., Hannun, Y. A., Bell, R. M. 1986. Specificity and mechanism of protein kinase C activation by *sn*-1,2-diacylglycerols. *Proc. Natl. Acad. Sci. USA* 83: 1184–88

Geahlen, R. L., Krebs, E. G. 1980. Studies on the phosphorylation of the type I cAMP-dependent protein kinase. *J. Biol. Chem.* 255: 9375–79

Geahlen, R. L., Allen, S. M., Krebs, E. G. 1981. Effect of phosphorylation on the regulatory subunit of the type I cAMP-dependent protein kinase. *J. Biol. Chem.* 256: 4536–40

Girard, P. R., Mazzei, G. J., Wood, J. G., Kuo, J. F. 1985. Polyclonal antibodies to phospholipid/Ca^{2+}-dependent protein kinase and immunocytochemical localization of the enzyme in rat brain. *Proc. Natl. Acad. Sci. USA* 82: 3030–34

Girard, P. R., Mazzei, G. J., Kuo, J. F. 1986. Immunological quantitation of phospholipid/Ca^{2+}-dependent protein kinase and its fragments. *J. Biol. Chem.* 261: 370–75

Goelet, P., Castellucci, V. F., Schacher, S., Kandel, E. R. 1986. The long and the short of long-term memory—a molecular framework. *Nature* 322: 419–22

Goldenring, J. R., Gonzalez, F., McGuire, J. S. Jr., De Lorenzo, R. J. 1983. Purification and characterization of a calmodulin-dependent kinase from rat brain cytosol able to phosphorylate tubulin and microtubule-associated proteins. *J. Biol. Chem.* 258: 12632–40

Goldenring, J. R., Wasterlain, C. G., Oestreicher, A. B., de Graan, P. N. E., Farber, D. B., Glaser, G., De Lorenzo, R. J. 1986. Kindling induces a long-lasting change in the activity of a hippocampal membrane calmodulin-dependent protein kinase system. *Brain Res.* 377: 47–53

Grab, D. J., Carlin, R. K., Siekevitz, P. 1981. Function of calmodulin in postsynaptic densities II. Presence of a calmodulin-activatable protein kinase activity. *J. Cell Biol.* 89: 440–48

Greenberg, S. M., Bayley, H., Castellucci, V. F., Schwartz, J. H. 1986. Loss of the regulatory subunit at the cyclic AMP-dependent protein kinase in *Aplysia* neurons after sensitizing treatments. *Soc. Neurosci. Abstr.* 12: 1339

Greenberg, S. M., Bernier, L., Schwartz, J. H. 1987. Distribution of cAMP and cAMP-dependent protein kinases in *Aplysia* sensory neurons. *J. Neurosci.* 7: 291–301

Hannun, Y. A., Loomis, C. R., Bell, R. M. 1985. Activation of protein kinase C by Triton X-100 mixed micelles containing diacylglycerol and phosphatidylserine. *J. Biol. Chem.* 260: 10039–43

Hartl, F. T., Roskoski, R. Jr. 1982. Adenosine 3′,5′-cyclic monophosphate protein kinase from bovine brain: Inactivation of the catalytic subunit and holoenzyme by 7-chloro-4-nitro-2,1,3-benzoxadiazole. *Biochemistry* 21: 5175–83

Hemmings, B. A., Aitken, A., Cohen, P., Rymond, M., Hofmann, F. 1982. Phosphorylation of the type II regulatory subunit of cyclic-AMP-dependent protein kinase by glycogen synthase kinase 3 and glycogen synthase kinase 5. *Eur. J. Biochem.* 127: 473–81

Hofmann, F., Beavo, J. A., Bechtel, P. J., Krebs, E. G. 1975. Comparison of adenosine 3′: 5′-cyclic monophosphate-dependent protein kinases from rabbit skeletal and bovine heart muscle. *J. Biol. Chem.* 250: 7795–7801

Huang, K.-P., Chan, K.-F. J., Singh, T. J., Nakabayashi, H., Huang, F. L. 1986. Autophosphorylation of rat brain Ca^{2+}-activated and phospholipid-dependent protein kinase. *J. Biol. Chem.* 261: 12134–40

Hunter, T., Cooper, J. A. 1985. Protein-tyrosine kinases. *Ann. Rev. Biochem.* 54: 897–930

Jahnsen, T., Hedin, L., Lohmann, S. M., Walter, U., Richards, J. S. 1986. The neural type II regulatory subunit of cAMP-dependent protein kinase is present and regulated by hormones in the rat ovary. *J. Biol. Chem.* 261: 6637–39

Kandel, E. R., Schwartz, J. H. 1982. Molecular biology of learning: Modulation of transmitter release. *Science* 218: 433–43

Kennedy, M. B., McGuinness, T., Greengard, P. 1983a. A calcium/calmodulin-dependent protein kinase from mammalian brain that phosphorylates synapsin I: Partial purification and characterization. *J. Neurosci.* 3: 818–31

Kennedy, M. B., Bennett, M. K., Erondu, N. E. 1983b. Biochemical and immunochemical evidence that the "major postsynaptic density protein" is a subunit of a calmodulin-dependent protein kinase. *Proc. Natl. Acad. Sci. USA* 80: 7357–61

Kikkawa, U., Takai, Y., Minakuchi, R., Inohara, S., Nishizuka, Y. 1982. Calcium-activated phospholipid-dependent protein kinase from rat brain: Subcellular distribution, purification and properties. *J. Biol. Chem.* 257: 13341–48

King, M. M., Fitzgerald, T. J., Carlson, G. M. 1983. Characterization of initial autophosphorylation events in rabbit skeletal muscle phosphorylase kinase. *J. Biol. Chem.* 258: 9925–30

Kishimoto, A., Takai, Y., Mori, T., Kikkawa, U., Nishizuka, Y. 1980. Activation of calcium and phospholipid-dependent protein kinase by diacylglycerol, its possible relation to phosphatidylinositol turnover. *J. Biol. Chem.* 255: 2273–76

Kishimoto, A., Kajikawa, N., Shiotu, M.,

Nishizuka, Y. 1983. Proteolytic activation of calcium-activated, phospholipid-dependent protein kinase by calcium-dependent neutral protease. *J. Biol. Chem.* 258: 1156–64

Knopf, J. L., Lee, M.-H., Sultzman, L. A., Kriz, R. W., Loomis, C. R., Hewick, R. M., Bell, R. M. 1986. Cloning and expression of multiple protein kinase C cDNAs. *Cell* 46: 491–502

Kojima, I., Kojima, K., Rasmussen, H. 1985. Role of calcium fluxes in the sustained phase of angiotensin II-mediated aldosterone secretion from adrenal glomerulosa cells. *J. Biol. Chem.* 260: 9177–84

Kraft, A. S., Anderson, W. B. 1983. Phorbol esters increase the amount of Ca^{2+}, phospholipid-dependent protein kinase associated with plasma membrane. *Nature* 301: 621–23

Kuret, J., Schulman, H. 1985. Mechanism of autophosphorylation of the multifunctional Ca^{++}/calmodulin-dependent protein kinase. *J. Biol. Chem.* 260: 6427–33

Lai, Y., Nairn, A. C., Greengard, P. 1986. Autophosphorylation reversibly regulates the Ca^{2+}/calmodulin-dependence of Ca^{2+}/calmodulin-dependent protein kinase II. *Proc. Natl. Acad. Sci. USA* 83: 4253–57

Le Vine, H. III, Sahyoun, N. E., Cuatrecasas, P. 1986. Binding of calmodulin to the neuronal cytoskeletal protein kinase type II cooperatively stimulates autophosphorylation. *Proc. Natl. Acad. Sci. USA* 83: 2253–57

Lipmann, F., Levine, P. A. 1932. Serine-phosphoric acid obtained on hydrolysis of vitellinic acid. *J. Biol. Chem.* 98: 109–14

Lisman, J. E. 1985. A mechanism for memory storage insensitive to molecular turnover: A bistable autophosphorylating kinase. *Proc. Natl. Acad. Sci. USA* 82: 3055–57

Llinas, R., McGuinness, T. L., Leonard, C. S., Sugimori, M., Greengard, P. 1985. Intraterminal injection of synapsin I or calcium/calmodulin-dependent protein kinase II alters neurotransmitter release at the squid giant synapse. *Proc. Natl. Acad. Sci. USA* 82: 3035–39

Lohmann, S. M., Walter, U. 1984. Regulation of the cellular and subcellular concentrations and distribution of cyclic nucleotide-dependent protein kinases. *Adv. Cyc. Nuc. Res.* 18: 63–117

Lou, L. L., Lloyd, S. J., Schulman, H. 1986. Activation of the multifunctional Ca^{2+}/calmodulin-dependent protein kinase by autophosphorylation. ATP modulates production of an autonomous kinase. *Proc. Natl. Acad. Sci. USA* 83: 9497–9501

Lynch, G., Baudry, M. 1984. The biochemistry of memory: A new and specific hypothesis. *Science* 212: 1057–63

Malenka, R. C., Madison, D. V., Nicoll, R. A. 1986. Potentiation of synaptic transmission in the hippocampus by phorbol esters. *Nature* 321: 175–77

MacDermott, A. B., Mayer, M. L., Westbrook, G. L., Smith, S. J., Barker, J. L. 1986. NMDA-receptor activation increases cytoplasmic calcium concentration in cultured spinal cord neurones. *Nature* 321: 519–22

McGuinness, T. L., Lai, Y., Greengard, P. 1985. Ca^{2+}/calmodulin-dependent protein kinase II: Isozymic forms from rat forebrain and cerebellum. *J. Biol. Chem.* 260: 1696–1704

Melloni, E., Pontremoli, S., Michetti, M., Sacco, O., Sparatore, B., et al. 1985. Binding of protein kinase C to neutrophil membranes in the presence of Ca^{2+} and its activation by a Ca^{2+}-requiring proteinase. *Proc. Natl. Acad. Sci. USA* 82: 6435–39

Melloni, E., Pontremoli, S., Michetti, M., Sacco, O., Sparatore, B., et al. 1986. The involvement of calpain in the activation of protein kinase C in neutrophils stimulated by phorbol myristic acid. *J. Biol. Chem.* 261: 4101–5

Miller, S. G., Kennedy, M. B. 1986. Regulation of brain type II Ca^{2+}/calmodulin-dependent protein kinase by autophosphorylation: A Ca^{2+}-triggered molecular switch. *Cell* 44: 861–70

Morris, R. G. M., Anderson, E., Lynch, G. S., Baudry, M. 1986. Selective impairment of learning and blockade of long-term potentiation by an *N*-methyl-D-aspartate receptor antagonist, AP5. *Nature* 319: 774–76

Nairn, A. C., Hemmings, H. C. Jr., Greengard, P. 1985. Protein kinases in the brain. *Ann. Rev. Biochem.* 54: 931–76

Nestler, E. J., Greengard, P. 1984. *Protein Phosphorylation in the Nervous System.* New York: Wiley. 398 pp.

Nigg, E. A., Schafer, G., Hilz, H., Eppenberger, H. M. 1985. Cyclic AMP-dependent protein kinase type II is associated with the Golgi complex and with centrosomes. *Cell* 41: 1039–51

Nishizuka, Y. 1986. Studies and perspectives of protein kinase C. *Science* 233: 305–12

Ouimet, C. C., McGuinness, T. L., Greengard, P. 1984. Immunocytochemical localization of calcium/calmodulin-dependent protein kinase II in rat brain. *Proc. Natl. Acad. Sci. USA* 81: 5604–8

Palazzolo, M. J. 1985. *Third Messenger Molecules in* Aplysia. PhD thesis. Columbia Univ., New York. 194 pp.

Potter, R. L., Taylor, S. S. 1979a. Relationships between structural domains and function in the regulatory subunits of cAMP-dependent protein kinases I and II from brain skeletal muscle. *J. Biol. Chem.* 254: 2413–18

Potter, R. L., Taylor, S. S. 1979b. Correlation of the cAMP binding domain with a site of autophosphorylation on the regulatory subunit of cAMP-dependent protein kinase II from porcine skeletal muscle. *J. Biol. Chem.* 254: 9000–5

Potter, R. L., Taylor, S. S. 1980. The structural domains of cAMP-dependent protein kinase I: Characterization of two sites of proteolytic cleavage and homologies to cAMP-dependent kinase II. *J. Biol. Chem.* 255: 9706–12

Rangel-Aldao, R., Rosen, O. M. 1976a. Dissociation and reassociation of the phosphorylated and nonphosphorylated form of adenosine 3′:5′-monophosphate-dependent protein kinase from bovine cardiac muscle. *J. Biol. Chem.* 251: 3375–80

Rangel-Aldao, R., Rosen, O. M. 1976b. Mechanism of self-phosphorylation of adenosine 3′:5′-monophosphate-dependent protein kinase from bovine cardiac muscle. *J. Biol. Chem.* 251: 7526–29

Reichardt, L. F., Kelly, R. B. 1983. A molecular description of nerve terminal function. *Ann. Rev. Biochem.* 52: 871–926

Robinson-Steiner, A. M., Beebe, S. J., Rannels, S. R., Corbin, J. D. 1984. Microheterogeneity of type II cAMP-dependent protein kinase in various mammalian species and tissues. *J. Biol. Chem.* 259: 10596–10605

Rosen, O. M., Herrera, R., Olowe, Y., Petruzzelli, L. M., Cobb, M. H. 1983. Phosphorylation activates the insulin receptor tyrosine kinase. *Proc. Natl. Acad. Sci. USA* 80: 3237–40

Rubin, C. S., Rangel-Aldao, R., Sarkar, D., Erlichman, J., Fleischer, N. 1979. Characterization and comparison of membrane-associated and cytosolic cAMP-dependent protein kinases. *J. Biol. Chem.* 254: 3797–3805

Sahyoun, N., Le Vine, H. III, Bronson, D., Siegel-Greenstein, F., Cuatrecasas, P. 1985. Cytoskeletal calmodulin-dependent protein kinase. Characterization, solubilization, and purification from rat brain. *J. Biol. Chem.* 260: 1230–37

Sahyoun, N., Le Vine, H. III, McDonald, B., Cuatrecasas, P. 1986a. Specific postsynaptic density proteins bind tubulin and calmodulin-dependent protein kinase II. *J. Biol. Chem.* 261: 12339–44

Sahyoun, N., Wolf, M., Besterman, J., Hsieh, T.-S., Sander, M., et al. 1986b. Protein kinase C phosphorylates topoisomerase II: Topoisomerase activation and its possible role in phorbol ester-induced differentiation of HL-60 cells. *Proc. Natl. Acad. Sci. USA* 83: 1603–7

Saitoh, T., Schwartz, J. H. 1983. Serotonin alters the subcellular distribution of a Ca^{2+}/calmodulin-binding protein in neurons of *Aplysia*. *Proc. Natl. Acad. Sci. USA* 80: 6708–12

Saitoh, T., Schwartz, J. H. 1985. Phosphorylation-dependent subcellular translocation of a Ca^{2+}/calmodulin-dependent protein kinase produces an autonomous enzyme in *Aplysia* neurons. *J. Cell Biol.* 100: 835–42

Schulman, H. 1984. Phosphorylation of microtubule-associated proteins by a Ca^{2+}/calmodulin-dependent protein kinase. *J. Cell Biol.* 99: 11–19

Schwartz, J. H. 1963. The phosphorylation of alkaline phosphatase. *Proc. Natl. Acad. Sci. USA* 49: 871–78

Schwartz, J. H., Bernier, L., Castellucci, V. F., Palazzolo, M., Saitoh, T., et al. 1983. What molecular steps determine the time course of the memory for short-term sensitization of *Aplysia*? *Cold Spring Harbor Symp. Quant. Biol.* 48: 811–19

Schworer, C. M., Colbran, R. J., Soderling, T. R. 1986. Reversible generation of a Ca^{2+}-independent form of Ca^{2+} (calmodulin)-dependent protein kinase II by an autophosphorylation mechanism. *J. Biol. Chem.* 261: 8581–84

Shields, S. M., Vernon, P. J., Kelly, P. T. 1984. Autophosphorylation of calmodulin-kinase II in synaptic junctions modulates endogenous kinase activity. *J. Neurochem.* 43: 1599–1609

Shoji, S., Titani, K., Demaille, J. G., Fischer, E. H. 1979. Sequence of two phosphorylated sites in the catalytic subunit of bovine cardiac muscle adenosine 3′:5′-monophosphate-dependent protein kinase. *J. Biol. Chem.* 254: 6211–14

Steinberg, R. A. 1983. Sites of phosphorylation and mutation in regulatory subunit of cyclic AMP-dependent protein kinase from S49 mouse lymphoma cells: Mapping to structural domains. *J. Cell Biol.* 97: 1072–80

Steinberg, R. A., Agard, D. A. 1981a. Studies on the phosphorylation and synthesis of type I regulatory subunit of cyclic AMP-dependent protein kinase in intact S49 mouse lymphoma cells. *J. Biol. Chem.* 256: 11356–64

Steinberg, R. A., Agard, D. A. 1981b. Turnover of regulatory subunit of cyclic AMP-dependent protein kinase in S49 mouse lymphoma cells: Regulation by catalytic subunit and analogs of cyclic AMP. *J. Biol. Chem.* 256: 10731–34

Theuerkauf, W. E., Vallee, R. B. 1982. Molecular characterization of the cAMP-dependent protein kinase bound to microtubule-associated protein 2. *J. Biol. Chem.* 257: 3284–90

Tsuyama, S., Bramblett, G. J., Huang, K.-P., Flavin, M. 1986. Calcium/phospholipid-dependent kinase recognizes sites in microtubule-associated protein 2 which are phosphorylated in living brain and are not accessible to other kinases. *J. Biol. Chem.* 261: 4110–16

Uhler, M. D., Carmichael, D. F., Lee, D. C., Chrivia, J. C., Krebs, E. G., McKnight, G. S. 1986. Isolation of cDNA clones coding for the catalytic subunit of mouse cAMP-dependent protein kinase. *Proc. Natl. Acad. Sci. USA* 83: 1300–4

Walters, E. T., Byrne, J. H., Carew, T. J., Kandel, E. R. 1983. Mechanoafferent neurons innervating tail of *Aplysia*. II. Modulation by sensitizing stimuli. *J. Neurophysiol.* 50: 1543–59

Willmund, R., Mitschulat, H., Schneider, K. 1986. Longlasting modulation of Ca^{2+}-stimulated autophosphorylation and subcellular distribution of the Ca^{2+}/calmodulin-dependent protein kinase in the brain of *Drosophila*. *Proc. Natl. Acad. Sci. USA* 83: 9789–93

Wolf, M., Cuatrecasas, P., Sahyoun, N. 1985a. Interaction of protein kinase C with membranes is regulated by Ca^{++}, phorbol esters, and ATP. *J. Biol. Chem.* 260: 15718–22

Wolf, M., Le Vine, H. III, May, W. S. Jr., Cuatrecasas, P., Sahyoun, N. 1985b. A model for intracellular translocation of protein kinase C involving synergism between Ca^{+2} and phorbol esters. *Nature* 317: 546–49

Wood, J. G., Girard, P. K., Mazzei, G. J., Kuo, J. F. 1986 Localization of protein kinase C in identified neuronal components of rat brain. *J. Neurosci.* 6: 2571–77

Worley, P. F., Baraban, J. M., Snyder, S. H. 1986. Heterogeneous localization of protein kinase C in rat brain: Autoradiographic analysis of phorbol ester binding. *J. Neurosci.* 6: 199–207

Yamauchi, T., Fujisawa, H. 1983. Purification and characterization of the brain calmodulin-dependent protein kinase (Kinase II) which is involved in the activation of tryptophan 5-monooxygenase. *Eur. J. Biochem.* 123: 15–21

Ann. Rev. Neurosci. 1987. 10: 477–533

THE ANALYSIS OF VISUAL MOTION: From Computational Theory to Neuronal Mechanisms

Ellen C. Hildreth and Christof Koch[1]

Center for Biological Information Processing and Artificial Intelligence Laboratory, Massachusetts Institute of Technology, Cambridge, Massachusetts 02139

INTRODUCTION

The measurement and use of visual motion is one of the most fundamental abilities of biological vision systems, serving many essential functions. For example, a sudden movement in the scene might indicate an approaching predator or a desirable prey. The rapid expansion of features in the visual field can signal an object about to collide with the observer. Discontinuities in motion often occur at the locations of object boundaries and can be used to carve up the scene into distinct objects. Motion signals provide input to centers controlling eye movements, allowing objects of interest to be tracked through the scene. Relative movement can be used to infer the three-dimensional (3-D) structure and motion of object surfaces, and the movement of the observer relative to the scene, allowing biological systems to navigate quickly and efficiently through the environment. More generally, the analysis of visual motion helps us to maintain continuity of our perception of the constantly changing environment around us.

This article reviews our current understanding of a number of aspects of visual motion analysis in biological systems, from a computational perspective. We illustrate the kinds of insights that have been gained through computational studies and how they can be integrated with experimental studies from psychology and the neurosciences to understand the

[1] Present address: Division of Biology, 216–76, California Institute of Technology, Pasadena, California 91125.

0147–006X/87/0301–0477$02.00

particular computations used by biological systems to analyze motion. In the remainder of this introduction, we briefly describe the computational approach to the study of vision and discuss the areas of motion analysis that are addressed in this review.

The Computational Study of Vision

One of the most important tenets underlying a computational approach to the study of biological vision is the belief that the brain, like a computer, can be thought of as a machine that processes information extracted from the environment that results in some sort of action. Like Aristotle, Galen, and Descartes before us, we often think of the brain in terms of our most successful machines, which today happen to be digital computers. We must be careful in making such an analogy, however. The electrochemical environment of neurons, their means of transmitting information, and their overall architecture is very different from that of the wires and etched crystals of semiconducting material that comprise computers. The Turing machine, a core concept of computer science, works in a discrete mode in a world determined by classical physics. Such a machine can only approximate the truly analog operations of biological hardware in a world governed by the laws of quantum physics.

Although their hardware differs greatly, both biological systems and machines can perform similar functions that rely on the same mathematical and physical principles. Thus, there exists a level of description of the tasks performed by these two systems that is independent of the underlying hardware. In order to understand how natural or artificial systems can solve problems like sensing motion or depth or manipulating the environment, we must understand the nature of the problem—for example, whether it can be solved at all and what constraints the physical world imposes on the solution—before we can fully understand the detailed procedures used to find a solution.

A computational approach to the study of biological systems, based on the founding principles of the field of Artificial Intelligence, was elucidated by Marr & Poggio (1977, Marr 1982). Marr was attracted to the field of Artificial Intelligence after experiencing certain limitations of other theoretical approaches to brain research in his early work on the cerebellum (Marr 1969). Although his model for learning in the cerebellum has led to important experimental work (for example, Ito 1984), Marr abandoned this line of research after realizing that it did not shed light on how complex motor behavior can actually be achieved.

In his later work in computational vision, Marr elucidated three distinct levels of analysis that are necessary for understanding an information processing task:

1. A *computational theory* analyzes what problem is being solved and why, and investigates the natural constraints that the physical world imposes on the solution to the problem.
2. An *algorithm* is a detailed step-by-step procedure that represents one method for yielding the solution indicated by the theory.
3. An *implementation* is a physical realization of the algorithm by some mechanism or hardware.

These levels could suggest a prescription for conducting research on complex problems; that is, one first formulates a theory, then derives an algorithm, and lastly designs a mechanism that implements the algorithm:

theory $\Rightarrow$ algorithm $\Rightarrow$ mechanism.

Despite the initial success of this approach, research over the past few years has shown that computational theories, even if complemented by psychophysical experiments revealing how humans perform visual tasks, have inherent limitations in understanding the brain. In particular, the nature of the hardware can profoundly influence the type of algorithm needed to solve a particular problem. Thus, while the computational theory and properties of the hardware can often be studied independently, the algorithmic level is influenced by both. A given computation, such as the computation of stereo depth or motion, usually can be performed by several different algorithms. These algorithms depend not only on the nature of the computation itself, but also on the properties and limitations of the hardware in which the algorithm is implemented. Thus, in order to explain the functions of a visual system at its different levels, not only must the abstract, computational nature of a task be understood but also the properties of the underlying hardware. The flow of information is therefore in both directions:

theory $\Rightarrow$ algorithm $\Leftarrow$ mechanism.

These observations stress the importance of integrating the results of computational studies with those of experimental studies of biological vision systems.

Other introductions to the computational approach described here can be found, for example, in Poggio (1984), Morgan (1985), Ullman (1986), and Hildreth & Hollerbach (1985). The latter review also addresses the limitations and successes of the computational approach in the area of motor control.

Other "Computational" Approaches to the Study of Biological Systems

The term *computational* is often used within the neurosciences to denote very different concepts. For example, certain neural modeling approaches that study how neuronal networks can operate and how these operations can be extrapolated to explain higher brain functions frequently are termed "computational." Examples of this include the seminal work by McCulloch & Pitts (1943) on neuronal networks, the work on perceptrons (Minsky & Papert 1969) and parallel "connectionist" networks (Ballard 1986), as well as Marr's original work on the cerebellum. The word "computational" in this case refers to the detailed working of specialized hardware, such as linear threshold automata, rather than to an analysis of information processing at a level independent of the underlying hardware. Similarly, connectionist theories refer directly to neuronal hardware and therefore lack the characteristics of Marr's notion of a computational theory (Koch 1986). Although they have made important contributions to automata theory and theoretical cybernetics, we want to emphasize a distinction between these approaches and that described by Marr & Poggio (1977, Marr 1982). It is of course essential to understand the properties of the biological hardware—neurons, dendrites, synapses, channels, etc—in order to understand what algorithms the brain uses to analyze its environment, and a substantial fraction of this article is devoted to aspects of neuronal hardware. We believe, however, that to fully understand a complex information processing system, it is necessary first to understand the nature of the tasks the system is required to perform.

Finally, "computational" is used in yet another sense, as in *computational chemistry* or *computational biophysics*. This term generally refers to the existensive use of computers to simulate a given chemical or biophysical system, such as the reconstruction of the tertiary structure of simple proteins by using the principles of quantum physics and chemistry (Clementi 1985) or the simulation of the electrical properties of an array of pyramidal cells in the hippocampus (Traub et al 1984). In the following pages we refer frequently to such simulations of biophysical circuits.

Overview of Visual Motion Analysis

The pattern of movement in a changing image is not given to the visual system directly, but must be inferred from the changing intensities that reach the eye. The 3-D shape of object surfaces, the locations of object boundaries, and the movement of the observer relative to the scene can in turn be inferred from the pattern of image motion. Typically, the overall analysis of motion is divided into two stages: first, the measurement of movement in the changing two-dimensional (2-D) image, and second, the use of motion measurements, for example to recover the 3-D layout of the environment. It is not clear whether motion analysis in biological systems is necessarily performed in two distinct stages, but this division has served to facilitate theoretical studies of motion analysis and to focus empirical questions for perceptual and physiological studies.

The measurement of movement can itself be divided into multiple stages and may be performed in different ways in biological systems. In the human visual system alone, motion may be measured by at least two processes, termed *short-range* and *long-range processes* (for example, Braddick 1974, 1980). The short-range process analyzes continuous motion, or motion presented discretely but with small spatial and temporal displacements from one moment to the next. The long-range process may then analyze motion over larger spatial and temporal displacements, as in apparent motion. Evidence indicates that these two processes interact at some stage (Clatworthy & Frisby 1973, Green & von Grünau 1983), but initially they may be somewhat independent.

The subsequent uses of motion measurements impose different requirements on the precision and completeness with which image motion must be represented. The localization of object boundaries requires the detection of sharp changes in direction or speed of movement, but may not need a precise representation of absolute velocities everywhere. Object tracking requires knowledge of the gross translation of an object, but not information about the detailed relative movements that take place within the object. The recovery of the accurate 3-D shape of a moving object, on the other hand, appears to require a more precise and complete estimate of the local variations of motion across object surfaces. Motion analysis in the human visual system may ultimately involve the interaction of many processes, some fast but rough, others slow but more accurate, and still others that are specialized for specific tasks such as detecting object boundaries or looming motion. These processes must work together in a way that provides a versatile and robust motion analysis system.

In this review, we first discuss the earliest stage of motion measurement. We discuss two important theoretical models of motion detection, correlation and gradient models, and present relevant psychophysical and physiological data regarding biological motion detectors. We then discuss at length possible biophysical mechanisms that implement the computations underlying motion discrimination in retinal and cortical neurons. Later stages of motion measurement are then discussed in a subsequent section, which addresses the computation of an instantaneous 2-D velocity field, long-range motion correspondence, and the detection of motion discontinuities. Finally, we discuss the recovery of 3-D structure from relative motion. This article is not intended as an exhaustive overview of work on motion analysis. Rather, we highlight some of the areas that exhibit fruitful interactions between computational and experimental studies. Two recent reviews of motion analysis include the surveys by Barron

(1984), focusing on computational methods for deriving and interpreting optical flow, and by Nakayama (1985) focusing primarily on the psychophysics and physiology of motion.

EARLY MOTION DETECTION AND MEASUREMENT

Detecting Motion: Theory

Before motion can be used to reconstruct the 3-D structure of objects, the vision system must first reliably detect and measure relative motion in the 2-D image. What types of schemes have been proposed for this initial detection? How are these schemes related? What are their computational properties? The most general property of any motion discrimination system is that the underlying operation must be nonlinear. As first noted by Poggio & Reichardt (1973), no linear operation can extract the direction of motion of a moving stimulus. The schemes proposed for motion detection fall broadly into two classes: (*a*) correlation-like schemes (Hassenstein & Reichardt 1956, Poggio & Reichardt 1973, van Santen & Sperling 1984) and (*b*) gradient schemes (Fennema & Thompson 1979, Horn & Schunck 1981, Marr & Ullman 1981). As we shall see, most biological motion detection schemes cannot reliably measure velocity even for one-dimensional motions, since their output typically depends on contrast and on a mixture of velocity and spatial structure of the moving pattern (Reichardt et al 1983).

CORRELATION MODELS The best known motion detection scheme is based on research done over the last 30 years on movement perception in insects. On the basis of open- and closed-loop experiments performed first on the beetle, *Chlorophanus*, and later on the fruitfly, *Drosophila*, and the housefly, *Musca Domestica*, a number of researchers, most notably W. Reichardt, were led to the following conclusions regarding motion discrimination in insects (Hassenstein & Reichardt 1956, Varju & Reichardt 1967, Götz 1968, 1972, Reichardt 1969, Poggio & Reichardt 1976, Reichardt & Guo 1986):

1. A sequence of two light stimuli impinging on adjacent receptors is the elementary event that evokes an optomotor response.
2. The relation between the stimulus input to these two receptors and the optomotor output follows the rule of algebraic sign multiplication. For instance, stimulating receptor 1 with alternating dark to light changes

and receptor 2 with light to dark transitions leads to a turning response of the insect opposite to the direction of stimulus successions, while dark to light transitions presented to both receptors elicits a turning response in the direction of the stimulus succession.

3. The strength of the optomotor response is proportional to the product of the two stimuli.

On the basis of these experimental conclusions, a minimum mathematical model of motion perception in insects was formulated. Figure 1a shows a modified version of this correlation model. The image is sampled by a receptor with a point-like receptive field. The input to the receptor can thus be described by $I(t)$. The output of the receptor is subsequently passed through a linear high-pass filter, removing steady-state components of the output of the receptor, before being multiplied with a low- or band-pass filtered signal from a neighboring receptor. Thus, at this stage the signal strength is given by

$$R(t) = \int_{-\infty}^{+\infty} \int_{-\infty}^{+\infty} W(t_1, t_2) I(t-t_1) I(t-t_2) \, dt_1 \, dt_2$$

where $W(t_1, t_2)$ represents the lumped transfer-function for the different filters. Subsequently, the output of the multiplication operation is integrated over time. A little analysis will show that the output of this stage is equivalent to the autocorrelation of the input function $I(t)$. Let us assume that the low-pass filter actually corresponds to a fixed delay $\delta t > 0$. We are then essentially multiplying a linearly transformed version of $I(t)$ with itself, but shifted by the total amount $\Delta t = \delta t + \Delta x / v$ (where $\Delta x > 0$ is the spacing between the receptors and v the velocity of the stimulus), and integrating the resulting function over time. For a range of negative velocities, i.e. movement from the right to the left, Δt will be very small and the final output of this subunit will be large. For positive velocities, that is for movements in the opposite direction, the two functions $I(t)$ and $I(t + \Delta t)$ are out of synchrony and their product, integrated over time, will be small. The output of this subunit is then subtracted from the output of the complementary subunit to yield the total detector response. It follows that if the output of the right subunit exceeds the output of the left subunit, the detector response is positive, indicating rightward motion; likewise, if the output of the left subunit exceeds the output of the right subunit, detector response is negative, indicating leftward motion.

This theoretical model has a number of properties that can be tested experimentally. Two of the most interesting are *phase invariance* and *spatial aliasing* (for an overview see Reichardt 1969).

Imagine a light pattern consisting of a number of superimposed sinu-

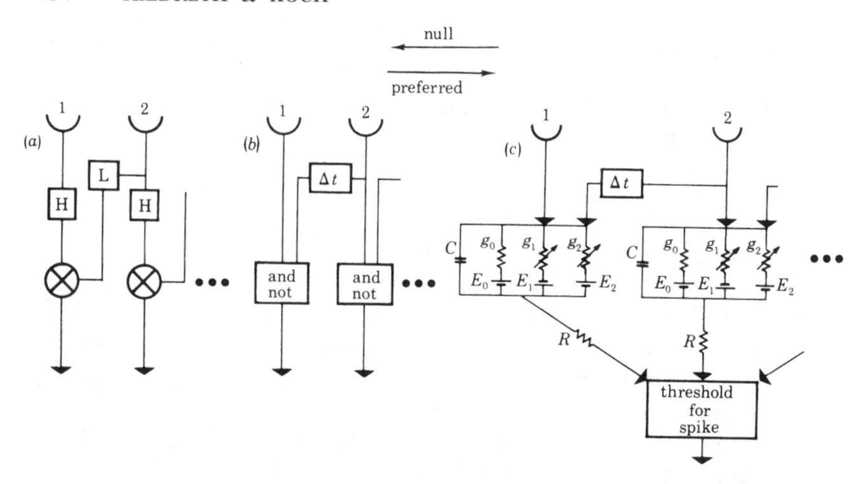

Figure 1 (*a*) A direction-selective subunit of the correlation model of Hassenstein & Reichardt (1956) as modified by Kirschfeld (1972). The two inputs are multiplied after low pass filtering with different time constants. If an average operation is made on the output, the overall operation is equivalent to cross-correlation of the two inputs. Subsequently, the time-averaged response of this subunit is subtracted from the response of a similar but mirror-symmetric subunit to yield the final movement-sensitive response. (*b*) The functional scheme proposed by Barlow & Levick (1965) to account for direction selectivity in the rabbit retina. A pure delay Δt is not necessary: A low pass filtering operation is sufficient. (*c*) The equivalent electrical circuit of the synaptic interaction assumed to underly direction selectivity as proposed by Torre & Poggio (1978). The interaction implemented by the circuit is of the type $g_1 - \alpha g_1 g_2$, where g_1 and g_2 represent the excitatory and inhibitory synaptic inputs. From Torre & Poggio (1978).

soidal gratings of different spatial frequencies. Because the process of autocorrelation, i.e. multiplication and subsequent integration, destroys all of the information that is inherent to the specification of the phases of the gratings, the output of the motion detectors is invariant to any changes in the phase relations of the sinusoidal gratings. Since any pattern $I(t)$ can be decomposed into its Fourier components, it follows that this class of motion detectors does not sense the relative position of the Fourier components. This important result has been tested and confirmed in experiments with the beetle, *Chlorophanus*, the fruitfly, *Drosophila*, and with *Musca* by evaluation of the time-averaged optomotor reactions to the angular motion of a fixed pattern painted on the inside of a drum. Moreover, the total time-averaged response is simply given by the sum of the time-averaged response to the individual Fourier components (Poggio & Reichardt 1973). Figure 2a shows the angular distribution of the brightness of two distinct patterns, obtained by superposition of the different Fourier components. These patterns only differ with respect to their phase

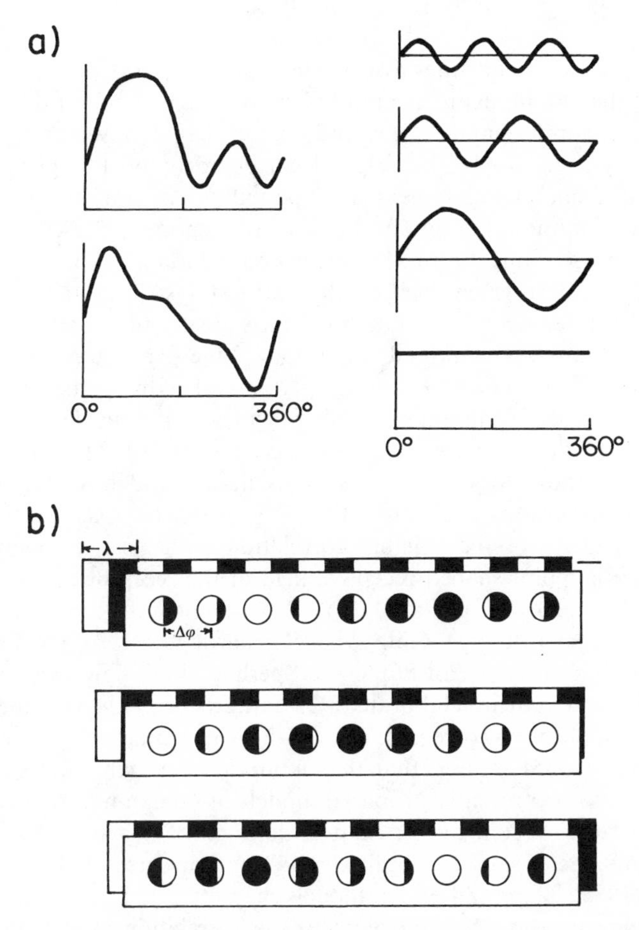

Figure 2 Two experimental predictions of the correlation model. (*a*) Phase invariance: The *left* part of the figure shows two different light patterns received by a photoreceptor at different angular positions of the environment. Both distributions contain the same set of Fourier components shown in the *right* part of the figure, but with different phases. However, insects like the housefly, the fruitfly or the beetle respond with the same optomotor reaction to both patterns. At the moment, it is not known whether direction-selective cells in the mammalian visual system show phase invariance. (*b*) Inverse motion perception: interference phenomena in the insect eye elicited by a moving pattern with a comparatively small spatial wavelength. When the distance $\Delta\varphi$ between input channels in the insect's eye is between one half and one spatial period λ of the pattern of excitation, the correlation model signals the incorrect direction of motion. The insect is compelled to follow this apparent motion in the direction opposite to the "true" direction of motion. Redrawn from Götz (1972).

relations. Yet the fruitfly reacts equally to motions of the two patterns (Götz 1972).

For any particular sinewave grating, the temporal phase difference between the two inputs to the multiplication will depend on the distance between its input channels, Δx, and on the spatial wavelength λ of the sinewave grating used. The original correlation model displays spatial aliasing: If one changes the spatial period of the grating, but not its direction of motion, the sign of the detector response reverses, indicating an incorrect motion. Within the wavelength region $\lambda > 2\Delta x$, the moving sinusoidal pattern is resolved by the receptor system as the number of samples received per period λ at any time is greater than or equal to two. If, however $\lambda < 2\Delta x$, optimal resolution of the periodic pattern breaks down, since less than two samples per wave length of the pattern are observed (see also Shannon's sampling theorem), and the detector signals the incorrect direction for $\Delta x < \lambda < 2\Delta x$ (Figure 2b). This inversion of apparent motion does occur in various insects and has been used to determine the grating constant of the receptor spacing (Reichardt 1969).

This property of the original correlation model can be avoided by replacing the point-shaped receptive field of the receptor in the original Reichardt model with a spatial-dependent receptive field of finite extent (Fermi & Reichardt 1963, Götz 1965, Reichardt et al 1983, van Santen & Sperling 1984, 1985). Van Santen & Sperling show how to choose the receptive field in their "elaborated Reichardt detector" so that the sign of the detector output is correct for any drifting sinewave grating. Van Santen & Sperling (1985) showed that the elaborated Reichardt model is fully equivalent to two recently proposed models of human motion detection: an elaborated version of the motion detector of Watson & Ahumada (1985) and the "spatiotemporal energy" motion detector of Adelson & Bergen (1985). These and similar models characterized by a multiplication-like nonlinearity are all equivalent to the correlation model (Poggio & Reichardt 1973).

GRADIENT MODELS Gradient schemes rely on the relationship between the spatial and temporal gradients of image intensity. In the case of the one-dimensional movement of an intensity profile $I(x, t)$ over a small displacement $\mathrm{d}x$ in time $\mathrm{d}t$, the temporal derivative of image intensity $I_t \approx (I(x, t+\mathrm{d}t)-I(x, t))/\mathrm{d}t$ and the spatial derivative of the intensity $I_x \approx (I(x+\mathrm{d}x, t)-I(x, t))\mathrm{d}x$ are related by

$$v = \frac{\mathrm{d}x}{\mathrm{d}t} = -\frac{I_t}{I_x}$$

where v is the velocity of the pattern. This method was originally proposed by Limb & Murphy (1975) and later extended by Fennema & Thompson (1979). The approach carries over to the 2-D case (Horn & Schunk 1981). Here, however, due to a fundamental limitation in the measurement process, termed the *aperture problem* (discussed below), only the component of the velocity in the direction of the brightness gradient can be measured. If we assume that the motion measurement process occurs along an edge, only the velocity component at right angles to the edge can be recovered. It is given by

$$v = -\frac{I_t}{\sqrt{I_x^2 + I_y^2}}$$

where I_x, I_y are the spatial derivatives in the x and y directions. This equation is strictly only correct for rigid, translating patterns, with no rotation, seen under parallel projection (Schunck 1984). For sufficiently small temporal and spatial displacements dx, dy, and dt, however, the equation approximates the correct one. Gradient schemes suffer from the disadvantage that they require computation of the derivatives of the intensity values, an operation that is sensitive to noise.

A quantized version of the gradient scheme was proposed by Marr & Ullman (1981). This model operates on locations in the images where the light intensity changes significantly. Marr & Hildreth's analysis (1980) showed that zero-crossings, that is locations where the Laplacian of the image is zero, correspond closely to intensity edges in the original image. Marr & Ullman track the motion of zero-crossing in the following way. An edge detector S of the Marr & Hildreth type signals the absence or presence of a zero-crossing at location x. This detector has two variants, one for transitions from dark to light (termed a *light-on edge*) and one for light to dark transitions (*light-off edge*). A second type of detector, termed a T *unit*, samples the temporal derivative of the intensity in approximately the same patch of the visual field as the edge-detecting unit. One version of this unit, T^+, only signals when the temporal derivative is positive, that is when a light-on edge has moved to the left or a light-off edge moves to the right, whereas T^- only responds to a reduction in light intensity. Combining the output of an S and a T unit conjunctively yields a set of detectors signaling the left (or rightward) motion of light-on or light-off edges. Marr & Ullman tentatively identify the edge detecting S units with sustained X-like on- or off-center cells and the T^+ and T^- units with transient Y-like on- or off-center cells. Computer experiments on some images have shown that this gradient scheme can recover motion information from image sequences. Note that their model, different from other gradient schemes, does not provide an estimate of the local velocity, but only its sign, that is the direction of motion, although some measure of velocity could be extracted.[2]

MOTION PRIMITIVES What are the primitives used to detect and measure motion, and at what stage in the analysis of the image does the detection of motion take place? For instance, are the initial measurements of the light intensity in the photoreceptors taken as primitives, or are the measurements extracted after the filtering and smoothing of the visual input at the stage of the retinal ganglion cells or even cortical cells? Finally, more symbolic primitives such as zero-crossings, edges, and line segments or even endpoints, corners, breaks, local deformities of objects, or discontinuities in line orientation could also be used. The advantage of matching more symbolic tokens, such as zero-crossings, across the image is that these tokens mark interesting points in an image, for instance locations where the image intensity changes most. Tokens are generally far more stable to changes and noise in the illumination than the original intensities or some filtered version of them. Moreover, since tokens presumably are sparsely distributed in the image, far fewer points must be matched and ambiguities can be avoided. If, however, large areas of the image contain no tokens, for instance if the light intensity changes little, these areas will not have any motion measurements assigned initially (these areas could be filled in later on). A further disadvantage of symbolic primitives is that they must be unambiguously identified before they can be matched, thus preventing an early computation of motion.

For the visual system of the fly, the experimental evidence suggests that the primitive is simply some measure of local intensity flux (Reichardt et al 1983). For the short-range motion system, Hildreth (1984) discusses the evidence that motion measurement relies on the detection of the movement of zero-crossings, or some similar measure operating on the smoothed intensity values, and that the limits on spatial and temporal displacements observed empirically in the short-range motion system are the consequence of the limited spatial and temporal extent of the initial filtering (see also Marr & Ullman 1981). Much more work needs to be done, however, before the question of the primitives used by the motion system can be answered.

Detecting Motion: Psychophysics

Both gradient and correlation schemes are local, involving only limited parts of the visual scene, and are therefore likely to provide a dominant input to the short-range process, which appears to operate on motion

[2] It can be shown formally that for small contrast amplitudes, the correlation-model and the gradient scheme are equivalent (T. Poggio, personal communication).

restricted to a spatial range of up to 10–15 minutes of visual arc and an interstimulus interval less than 80–100 msec (Braddick 1974, 1980). Since these separations in space and time are small, establishing correspondence between items in consecutive images is considerably easier than in the long-range process (see next section). Finally, the short-range process is assumed to operate directly on the light intensities, filtered intensities, or on edges or zero-crossings. Interestingly, color seems to provide little if any input to the short-range process (Ramachandran & Gregory 1978). In the following, we discuss the (limited) human perception evidence that has been used to discriminate between the various models of motion computation discussed above.

One of the main properties of the Reichardt correlation model is that its output responds not only to pattern velocity but also to structural properties of the pattern contrast. This property allows the motion detector to be used as pattern discriminator, at least in flies (Reichardt et al 1983, Reichardt & Guo 1986). Specifically, it can be shown (for instance, in Poggio & Reichardt 1973, 1976) that the time-averaged response of the correlation subunit depends on the ratio, for each spatial Fourier component, of the pattern velocity v and the spatial wavelength λ of the stimulus used. Thus wavelength and velocity trade off against each other and, as a consequence, the correlation model cannot reliably measure the speed of movement. This property, first confirmed with behavioral experiments for the fly, *Musca Domestica* (Eckert 1973), also seems to extend to the human visual system. If subjects fixate a point while square or sinusoidal gratings of variable spatial wavelength are moved past the fixation point at various speeds, their perception of velocity depends linearly on both the speed and spatial frequency of the gratings (Diener et al 1976, Burr & Ross 1982). These experiments seem consistent with a multiplicative-like second-order correlation model.

A striking prediction of the original Reichardt model is motion inversion: If the wavelength of the stimulus pattern is less than twice the separation between input channels, the insect will perceive motion in the direction opposite to the true direction of motion (Reichardt 1969, Götz 1972). Since humans, in contrast to insects, generally do not seem to show spatial aliasing, the point-like receptive field assumption of the original correlation model must be abandoned in favor of extended receptive fields (Fermi & Reichardt 1963). It can then be shown that motion reversal can be prevented (see, for instance, van Santen & Sperling 1984). Van Santen & Sperling (1984, 1985) test this "elaborated Reichardt" model with a number of psychophysical experiments. In particular, by varying the contrast of neighboring vertically oriented bars moving in a horizontal direction, they show that the total response of the subject depends on the product of the amplitudes of the two bars, a finding that offers support for the multiplication principle.

Psychophysical evidence in favor of the gradient scheme is presented by Moulden & Begg (1986). In one particularly ingenious experiment, they show polarity and direction-specific effects on motion discrimination in response to adaptation to a nonmoving, spatially homogeneous stimulus, and provide evidence for channels tuned to detect an increase or decrease in the light intensity [Marr & Ullman's (1981) T^+ and T^- units]. Thus, the current psychophysical evidence does not decisively favor a particular theory.

Detecting Motion: Circuitry and Biophysics

Having discussed some of the algorithms proposed to underlie motion detection, we discuss in more detail the biophysical mechanisms that may be used for motion detection. Numerous nerve cells in the visual system of both invertebrates and vertebrates respond differentially to motion. Moving a visual stimulus, say a dark bar on a light background, in the *preferred direction* elicits a vigorous response from the cell whereas movement in the opposite direction, termed the *null direction*, yields no significant response. *Directional-selective cells*, first described in the frog's retina in a classical paper by Maturana et al (1960), have subsequently been identified in the third optic ganglion of the house fly (for a review of the extensive literature see Hausen 1982a,b), in the retina of pigeons (Maturana & Frenk 1963, Holden 1977), rabbits (Barlow et al 1964, Barlow & Levick 1965), ground squirrels (Michael 1966), and cats (Stone & Fabian 1966, Cleland & Levick 1974), and in the visual cortex of both cats and monkeys (Hubel & Wiesel 1959, 1962, Schiller et al 1976, Orban et al 1981). Analyzing these cells afford us the opportunity to study the elementary biophysical events underlying a well characterized but nonlinear (that is, nontrivial) operation in single nerve cells.

In most mammals, except cats and primates, the first cells that seem to discriminate the direction of motion are the retinal ganglion cells. Thus, in the rabbit's retina approximately one quarter of the ganglion cells can be described as direction selective. In the cat retina, however, less than 1% of the physiological identified ganglion cells are direction selective (Rodieck 1979), while no such cells have been reported in the monkey's retina.[3] Since neither cells in the A and A1 layers of the lateral geniculate nucleus (LGN) of the cat nor cells in the magno- and parvo-cellular layers in the monkey are strongly direction selective, the appearance of

[3] Due to the inevitable electrode bias, this does not necessarily imply that such cells do not exist in the primate retina.

substantial numbers of direction-selective neurons in the primary cortex of both animals strongly suggests that this property arises first in the cortex.

COMPUTING THE DIRECTION OF MOTION IN THE RETINA

Early experiments Barlow & Levick (1965) systematically explored directional selectivity in the retina of the rabbit by using extracellular recordings. About 20% of the ganglion cells in the visual streak give both on and off responses to stationary, flashed stimuli and are direction selective for moving stimuli. These cells therefore compute the direction of motion independent of the contrast of the stimulus (i.e. dark stimulus on a light background or vice versa). A smaller proportion of ganglion cells ($\approx 7\%$) are direction selective and of the on-type, that is, they respond only to light-on edges. These cells project to the accessory optic system in the midbrain and are believed to be crucial for the control of the optokinetic nystagmus (Oyster et al 1972) and image stabilization (Simpson 1984). Off-type direction-selective cells have been reported in neither rabbit nor cat, although they are found in the turtle. Two important conclusions can be drawn from Barlow & Levick's (1965) report. First, inhibition is crucial for direction selectivity. On the basis of this evidence Barlow & Levick proposed that sequence discrimination is based upon a scheme whereby the response to the null direction is vetoed by appropriate neighboring inputs (the and-not gate in Figure 1b). Directionality is achieved by an asymmetric delay—or by a low pass filter—between excitatory and inhibitory channels from the photoreceptors to the ganglion cell. This model can be considered as an instance of the Reichardt correlation model. Second, this veto operation must occur within small independent subunits distributed throughout the receptive field of the cell, since movement of a bar over 0.25° to 0.5° elicits a direction-selective response (whereas the whole receptive field subtends 4.5°; Barlow & Levick 1965). Thus, the site of the veto operation is extensively replicated throughout the receptive field of the direction-selective cell. Confirming evidence for the critical role of inhibition comes from experiments in which inhibition is blocked with pharmacological agents (Caldwell et al 1978, Ariel & Daw 1982, Ariel & Adolph 1985), a situation resulting in an equal response for both preferred and null directions (see below).

A biophysical model We can now ask how this operation is implemented at the level of the hardware, i.e. at the level of retinal cells. Torre & Poggio (1978) proposed a specific biophysical mechanism implementing the neural equivalent of a veto operation.

When two neighboring regions of a dendritic tree experience simultaneous conductance changes, induced by synaptic inputs, the resulting postsynaptic potential is generally not the sum of the potentials generated by each synapse alone; that is, synaptic inputs may interact in a highly nonlinear fashion. This is particularly true for an inhibitory synaptic input that increases the membrane conductance with an associated ionic battery that reverses at, or very near, the resting potential E_{rest} of the cell. Activating this type of inhibition, called *silent* or *shunting* inhibition, is similar to opening a hole in the membrane: Its effect is only noticed if the intracellular potential is substantially different from E_{rest}. Torre & Poggio (1978) showed in a lumped electrical model of the membrane of the cell that silent inhibition can cancel effectively the excitatory postsynaptic potential (EPSP) induced by an excitatory synapse without hyperpolarizing the membrane. Moreover, for small synaptic conductance inputs the interaction between excitation and silent inhibition is multiplication-like, thereby approximating the nonlinear operation underlying the correlation scheme (see legend to Figure 1c). Pairs of excitatory and inhibitory synapses distributed throughout the dendritic tree may compute the direction of motion at many independent sites throughout the receptive field of the cell, in agreement with the physiological data. Since nonlinearity of the interaction is an essential requirement of this scheme, Torre & Poggio suggest that the optimal location for excitation and inhibition are fine distal dendrites or spines of the direction-selective ganglion cell.

Because this analysis left out the precise conditions required to produce effective and specific nonlinear interactions in a dendritic tree, Koch et al (1982, 1983) used one-dimensional cable theory to analyze the interaction between time-varying excitatory and inhibitory synaptic inputs in a morphologically characterized cat retinal ganglion cell (of the δ type; see Boycott & Wässle 1974). They were able to prove rigorously in the case of steady state synaptic conductance inputs that in a passive and branched dendritic tree the most effective location for silent inhibition (most effective in terms of reducing an EPSP) must always be on the direct path between the location of the excitatory synapse and the soma.

Detailed biophysical simulations of highly branched and passive neurons show that this *on-the-path* condition can be quite specific. If the amplitude of the inhibitory conductance change is above a critical value, inhibition can reduce excitation by as much as a factor of 10, as long as inhibition is located between the excitatory synapse and the soma. Inhibition more than about 10 μm behind excitation or on a neighboring branch 10 or 20 μm off the direct path is ineffective in reducing excitation significantly. This specificity in terms of spatial positioning of excitatory and inhibitory synapses carries over into the temporal domain. For maximal effect, inhibition must last at least as long as excitation and the inhibitory and excitatory conductance changes must occur nearly synchronously (Koch et al 1983, Segev & Parnas 1983). Finally, the *on-the-path* condition is also valid in the presence of action potentials: In order for silent inhibition to block the propagation of a spike past a branching point, it must be located at most 5 μm from the branch point (O'Donnell et al 1985). Since such a precise mapping imposes stringent conditions on the specificity of the positioning of synapses during development of the retinal circuitry, one simple developmental rule would be that a pair of excitatory and inhibitory inputs originating from interacting photoreceptors should contact the ganglion cell dendrite close to one another.

The specificity of silent inhibition contrasts with the action of a hyperpolarizing synaptic input (i.e. a conductance change with an associated battery below E_{rest}). In this case, the interaction between excitation and inhibition will be much more linear, that is, the inhibitory synapse will reduce the EPSP generated by the excitatory synapse by an amount roughly proportional to the inhibitory conductance change with less regard to the relative spatial positioning of excitatory and inhibitory synapses (Koch et al 1982, O'Donnell et al 1985, Koch & Poggio 1986).

Critical predictions of the model How does the model fare against experimental evidence? The following lists some of the most important predictions:

1. On-Off direction-selective cells receive distinct excitatory and inhibitory synaptic inputs. The reversal potential of the inhibitory input is close to the resting potential of the cell (probably acting via a $GABA_A$ receptor).
2. Bicucculin should abolish direction selectivity.
3. Inhibitory synapses are not more distal to the soma than excitatory synapses.
4. Direction selectivity is computed at many independent sites in the dendritic tree before spike initiation at the axonal hillock.
5. The direction-selective cell should show a δ-like morphology, with a highly branched, bistratified dendritic tree with small diameter dendrites or possibly spines.
6. On-Off direction-selective cells are expected to show little interaction between a dark bar/spot and light bar/spot moving in opposite directions within the receptive field.

Currently, the main support for this hypothesis derives from intracellular recordings in retinal ganglion cells from the turtle (Marchiafava 1979) and the bullfrog (Watanabe & Murakami 1984). Moving a spot or bar in the preferred direction gives rise to a somatic EPSP with superimposed action potentials whereas null direction stimulation results in a smaller EPSP without a hyperpolarization. The reduced somatic EPSP in the null direction appears to be caused by an inhibitory process that increases the membrane conductance with an associated reversal potential at or very near the resting potential of the cell. This silent inhibition is revealed by injecting a steady-state depolarizing current into the soma, giving rise to a hyperpolarization (see Figure 3). Preliminary evidence from rabbit ganglion cells indicates the presence of a similar inhibitory input (F. Amthor, personal communication).

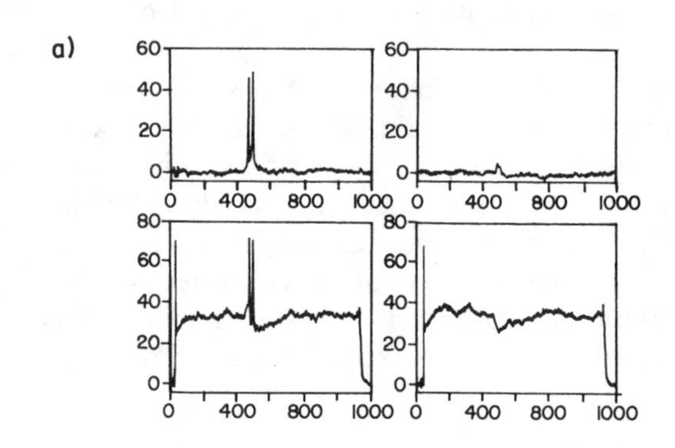

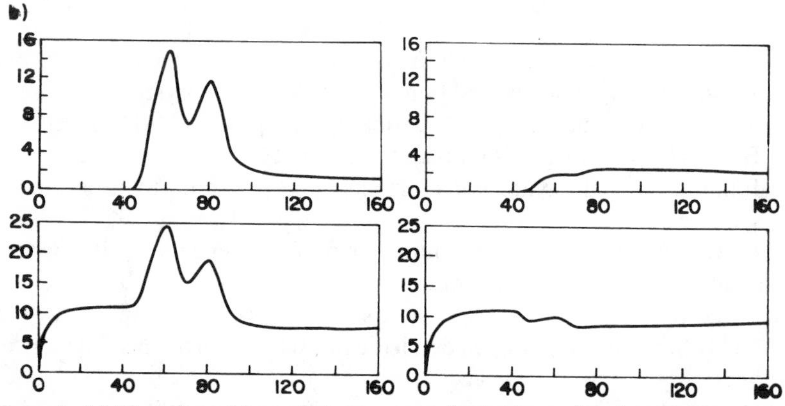

Figure 3 (*a*) The effect of intracellular current injection upon the photoresponse in an intracellular recorded direction-selective turtle ganglion cell. The response in the preferred and null directions are shown in the *left* and *right* part of (*a*). The *lower* record shows the photoresponse while 0.23 nÅ current was being injected into the soma. Adapted from Marchifava (1979). (*b*) Simulated intracellular potential at the soma of the reconstructed rabbit on-off direction-selective ganglion cell shown in Figure 4, assuming a purely passive membrane. The two distinct peaks correspond to the leading edge, receiving on input, and the trailing edge, receiving off input. In the *bottom* half, a step current of 0.091 nÅ was injected into the soma. Preferred direction is left and null direction right. From Koch et al (1986).

Within the last few years, two groups have determined the structure of on-off direction-selective ganglion cells. Using a fluorescent stain, Jensen & DeVoe (1983) visualized these cells in the turtle retina. Amthor et al (1984) used horseradish peroxidase (HRP) in the rabbit. The overall morphology of these cells is similar in the two species. Rabbit direction-selective ganglion cells have several distinct features that allow visual identification on purely morphological grounds (Figure 4a).

1. These cells have two levels of dendritic ramification. This observation is consistent with studies that have divided the inner plexiform layer into on and off laminae (Famiglietti & Kolb 1976).
2. The dendritic branches of the direction-selective cells are of very small diameter relative to other rabbit ganglion cells. Moreover, the dendrites carry spines or spine-like structures.
3. The dendritic branching pattern is quite complex, with dendrites forming apparent loops.

Note that although the cell drawn in Figure 4a has an asymmetric placement of the soma with respect to the dendritic tree, preferred and null directions do not appear to be predictable from the gross dendritic morphology of these cells. Thus, the morphology of direction-selective cells agrees well with previous predictions (Koch et al 1982).

In order to model massive synaptic input to a direction-selective ganglion cell, the passive electrical properties of the anatomically reconstructed cell shown in Figure 4a was simulated on the basis of one-dimensional cable theory (O'Donnell et al 1985, Koch et al 1986b). The computation of the voltages is carried out by a circuit simulation program, SPICE, first applied to biophysical circuit modeling by Segev et al (1985). Figure 3 shows the resulting somatic depolarization in the absence and in the presence of a depolarizing current step injected at the soma, in comparison with experimental records obtained from turtle ganglion cells (Marchiafava 1979). The intracellular potential can also be displayed in color throughout the entire cell (O'Donnell et al 1985, Koch et al 1986b).

Presynaptic circuitry How much do we know about the origin and properties of the excitatory and inhibitory inputs to direction-selective cells? Considerable evidence implicates acetylcholine (ACh) as the excitatory neurotransmitter underlying direction selectivity in the rabbit retina (Ariel & Adolph 1985). If all synaptic transmission in the perfused retina is blocked by pharmacological manipulation of the bathing medium, on-off direction-selective cells can be driven by direct application of ACh, thus implying that these cells are the postsynaptic target for cholinergic synapses. Ariel & Daw (1982) found that upon application of physostigmine, a drug that inhibits the hydrolysis of ACh after it has bound to the postsynaptic membrane, ganglion cells lose their ability to discriminate motion. Other properties like speed and size specificity and radial grating inhibition do not seem to be affected. This result may at first seem paradoxical, since physostigmine increases the effectiveness of ACh. One simple explanation is that this increased effectiveness during null direction serves to overcome the inhibition and to initiate action potentials at the soma. In turtle retina, similar experiments yield similar results (Ariel & Adolph 1985).

Recently, Masland and colleagues (Masland et al 1984, Tauchi & Masland 1984) identified two unique populations of cholinergic amacrine cells. In the rabbit retina, the only cells synthesizing and releasing ACh are two groups of amacrine cells distributed in the on and off layers. Using radioactive labeled ACh, Masland et al demonstrated that these two subtypes of amacrine cells release ACh transiently either at the onset (cells in the on layer) or at the offset of light (cells in the off layer). Because the cells have a unique morphology reminiscent of fireworks, they are called *starbust amacrine cells*. These cells appear to be presynaptic to bistratified ganglion cells, with the morphological attributes of the direction-selective cells of Amthor et al (1984).

The inhibitory input for motion discrimination is believed to be mediated by the neurotransmitter, γ-aminobutyric acid (GABA). Caldwell et al (1978) and Ariel & Daw (1982) infused picrotoxin, a potent antagonist of GABA, into the rabbit retina. Within minutes after the start of drug infusion, the response of direction-selective cells in the null direction increased dramatically, so that the cell became equally responsive to move-

Figure 4 (*a*) Camera lucida drawing of an HRP-injected on-off direction-selective cell in the visual streak of the rabbit retina. The dendritic fields have been drawn in two parts: "Outer" refers to the part of the inner plexiform layer (IPL) closest to the inner nuclear layer, where the cells of the off pathway make synaptic connections, while "inner" is the layer closest to the ganglion cell layer where the on pathway is connected. There are no obvious asymmetries in the cell that are correlated with the preferred direction. Adapted from Amthor et al (1984). (*b*) A simplified schematic of the excitatory pathway from the outer plexiform layer (OPL) to the on-off direction-selective ganglion cell in the rabbit. Depolarizing (on) and Hyperpolarizing (off) bipolar cells convey the visual information from the OPL to the on or off part of the IPL. Here they most likely synapse either directly, possibly using glutamate or aspartate as excitatory neurotransmitter, or indirectly, via other amacrine cells, onto the cholinergic starburst amacrine cells. These amacrine cells feed in turn directly onto the bistratified on-off ganglion cells. (*c*) Possible sites for the computations underlying motion discrimination. GABAergic amacrine cells can veto the excitatory pathway at the level of the ganglion cell (1), the starburst amacrine cells (2), or bipolar cells (3). Current evidence seems to favor site (1). The on and off pathways are segregated up to the cell body of the on-off direction-selective cell. From Koch et al (1986b).

ment in both directions. A few minutes after drug infusion was discontinued, the cell again became direction selective. In the turtle retina, direct application of ACh leads to spontaneous firing in direction-selective cells during blockage of synaptic transmission via a low calcium concentration and EGTA (Ariel & Adolph 1985). This Ach-induced spike activity can be suppressed by GABA, thus indicating that both ACh and GABA receptors must coexist on the membrane of turtle direction-selective ganglion cells. In the rat retina, the only cells staining for glutamic acid decarboxylase (GAD; the rate-limiting enzyme for the synthesis of GABA) are amacrine cells (Vaughn et al 1981). These cells make synapses onto

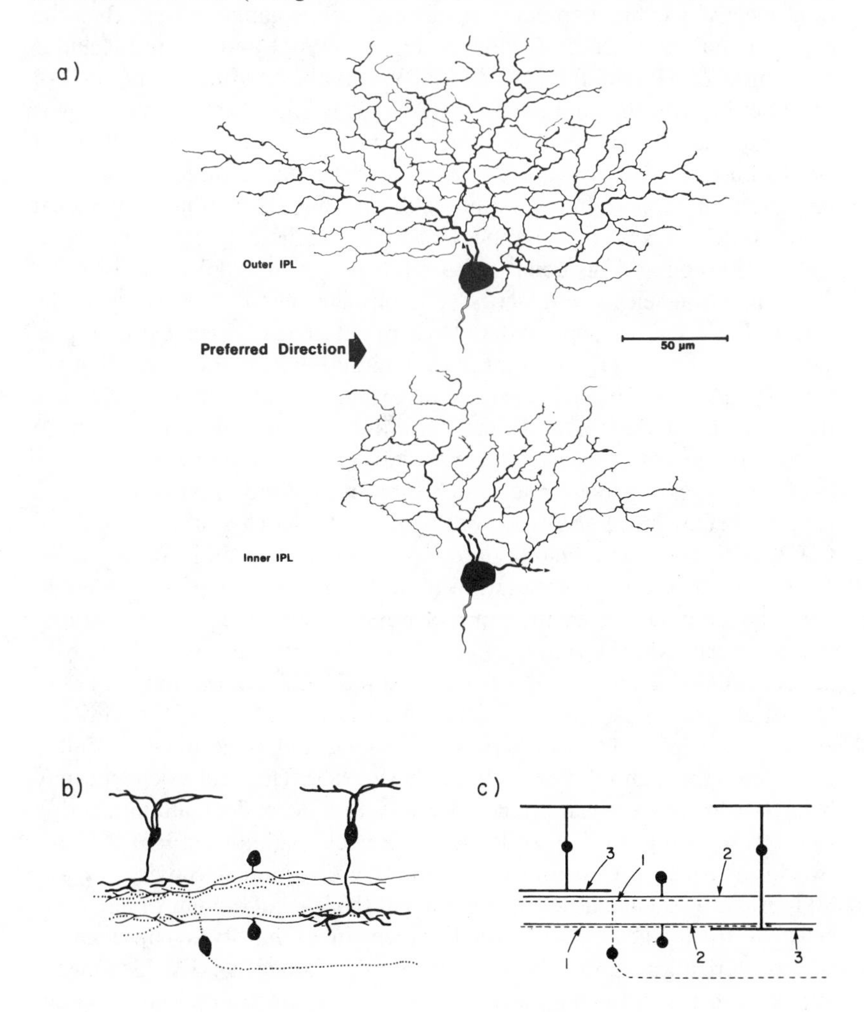

processes of bipolar, amacrine and ganglion cells in descending order of frequency.

Thus, at least in the turtle and rabbit retina, the excitatory and the inhibitory inputs to direction-selective ganglion cells appear to derive from cholinergic and GABAergic amacrine cells. This finding does not exclude, however, direct input from bipolar cells that may be responsible, for instance, for the center-surround organization of direction selective cells.

Alternative models What are the alternative models for the neuronal operations underlying motion discrimination? If one assumes that direction selectivity is first expressed at the level of the ganglion cells, then the experimental evidence of Barlow & Levick (1965) and the intracellular recordings of Marchiafava (1979) and Watanabe & Murakami (1984) in conjunction with the pharmacology (Ariel & Adolph 1985) argue in favor of our postsynaptic, silent inhibition scheme. Although both Werblin (1970) and Marchiafava (1979) have failed to record direction-selective responses in bipolar or amacrine cells, the possibility that the critical computations occur presynaptic to ganglion cells cannot be excluded. Indeed, DeVoe and his collaborators (DeVoe et al 1985) have recorded from direction-selective amacrine and bipolar cells in the retina of the turtle. Their evidence points toward an alternative or coexistent presynaptic site for the critical computation underlying direction selectivity in the turtle. A second piece of evidence favoring a presynaptic arrangement is the influence of GABA on ACh. GABA inhibits the light evoked release of ACh in the rabbit retina (Massey & Neal 1979; see Figure 4).

Other classes of presynaptic models for motion discrimination have been proposed (Dowling 1979, Koch & Poggio 1986, Koch et al 1986b). Since GABAergic processes synapse onto bipolar, amacrine, and ganglion cells, the site of the critical computation underlying direction selectivity could either be a bipolar cell exciting the starburst amacrine cell or the starburst amacrine cell itself. Starburst amacrine cells have dendrites that are probably decoupled from each other and the soma (Miller & Bloomfield 1983). Only the distal-most portion of the dendrites give rise to conventional chemical synaptic output, whereas the bipolar and amacrine cell input is distributed throughout the cell (Famiglietti 1983). Thus, each dendrite may behave from an electrical point of view as an independent subunit, acting as the morphological basis of Barlow & Levick's subunits (1965). At least two biophysical mechanisms could underlie direction selectivity: (*a*) the and-not veto scheme, now implemented at the level of bipolar or amacrine cells, or (*b*) a linear interaction between an excitatory synapse and a hyperpolarizing synapse followed by synaptic rectification (Koch & Poggio 1986). In this case, the nonlinearity essential for direction selectivity (Poggio & Reichardt 1973) would be implemented by a synaptic transduction mechanism that only allows transmission of depolarizing events. For these presynaptic models, the release of neurotransmitter, whether from the bipolar onto the amacrine cell or from the amacrine onto the ganglion cell, would in itself be direction selective.

We would like to point out that both pre- and postsynaptic models may turn out to be correct. For instance, the direction-selective bipolar and amacrine cells recorded by DeVoe et al (1985) have a smaller velocity range than direction-selective ganglion cells. Thus, a rough estimate of the direction of a moving stimuli could be computed at the level of bipolar/amacrine cells, while ganglion cells would perform similar but finer measurements.

COMPUTING MOTION IN THE VISUAL CORTEX Much more work has been done on the biophysical mechanisms underlying direction selectivity in the retina than in the cortex. Therefore, our discussion of cortical mechanisms will necessarily be brief. As mentioned above, cells in the primary visual cortex of cats and primates are likely to compute the direction of motion *de nouveau*, since the geniculate input shows no evidence of direction selectivity. Moreover, if the inhibition mediated by local interneurons is removed by application of bicuculline, an antagonist of GABA (Sillito 1977, Sillito et al 1980), direction selectivity of cortical cells is severely reduced or abolished.[4] This experiment, similar to Ariel & Daw's experiment in the retina (1982), underscores the importance of inhibition for direction discrimination.

An extension of the veto mechanism outlined above has been proposed to underlie direction selectivity in the visual cortex (Poggio 1982, Koch & Poggio 1985). The basic idea is as follows: A single LGN on-center neuron (or a row of such cells) excites a cell in area V1 whenever a light-on stimulus falls within its receptive field center. A neighboring on-center LGN cell reduces the activity of the cortical neuron by a delayed silent inhibition. Since it is unlikely that LGN cells have an inhibitory effect on their postsynaptic targets, the second geniculate cell excites an interneuron, possibly in layer 4c, which in turn inhibits the direction-selective cell. This seems plausible in light of the fact that direction-selective cells in the primate cortex first occur one synapse beyond layer 4c, i.e. in layer 4b (Dow 1974). If the silent inhibition is located either on the direct path between excitation and the soma or very near the excitatory synapse, it will effectively veto excitation in the null direction. Adding a similar but

[4] The crucial nature of inhibition for motion discrimination seems to be well preserved across species. Injecting picrotoxin, a GABA antagonist, into the third optic ganglion of the blowfly, *Calliphora Erythrocephala*, abolishes motion discrimination at both the cellular and the behavioral levels (Bülthoff & Bülthoff 1986).

inverted circuit constructed of geniculate off-center neurons endows our cortical neuron with direction selectivity for both light-on and light-off edges moving in the same direction—the most common type of direction-selective cell (the S2 cell of Schiller et al 1976). These off-center neurons, whose receptive fields overlap with the fields of their on-center counterparts, map onto a different part of the dendritic tree of the direction-selective cortical cell. This prediction, i.e. that direction selectivity for light-on and light-off edges results from the independent convergence from the geniculate, is supported by experiments done by Schiller (1982) in the monkey and by Sherk & Horton (1984) in the cat, using the pharmacological agent APB. APB infusion into the retina reversibly blocks the on pathway at the level of the retinal outer plexiform layer and eliminates the response of the cortical direction-selective cell to light edges while leaving the response to dark edges intact.

One intriguing possibility is that dendritic spines might be the specialized sites for the synaptic veto operation to take place. 5–20% of spines on cortical cells have been reported to carry symmetrical and asymmetrical synaptic profiles on the same spine (see, for instance, Jones & Powell 1969, Sloper & Powell 1979). Such an arrangement can be used to perform a highly tuned temporal discrimination operation, essentially without influencing the rest of the neuron (Koch & Poggio 1983). With a fast excitatory and a much slower inhibitory conductance change simultaneously occurring on the same spine, inhibition will effectively veto excitation if it sets in before the start of excitation (null direction). Activating the inhibition some fraction of a millisecond after the start of excitation will not influence excitation to any significant degree (preferred direction).

Very recently Saito et al (1986) have proposed that a more complex type of motion discrimination, namely cells in the superior temporal sulcus of the macaque monkey that respond only to either expanding or contracting size change of patterns or to rotation of patterns in one direction, is based on local synaptic veto operations occurring at numerous independent sites in the dendritic tree of these cells. Finally, Warren et al (1986) recently proposed that the synaptic veto mechanism underlies the direction selective response to cells in the somatosensory cortex of awake monkeys when wheels with surface grating are rolled over their skin.

Open Questions

The evidence still seems insufficient to press a clear-cut case for either the correlation or the gradient scheme for human motion discrimination. In fact, both schemes may be used by the human visual system. Since the physiological and behavioral data seem to indicate the validity of the correlation model for invertebrates and a large class of vertebrates, it may be hypothesized that the Reichardt correlation model, possibly implemented via the synaptic veto mechanisms of Torre & Poggio (1978), is used in the primate retina to endow some cells with direction selectivity. These cells, which cannot exist in very large numbers, project to the superior colliculus and from there possibly to the cortex. Motion discrimination in the cortex could be computed *de nouveau* within simple cells in the striate cortex by use of a different scheme, for instance the gradient scheme of Marr & Ullman (1981) or the implementation of the correlation model based on and-not type of synaptic logic (Poggio 1982, Koch & Poggio 1985). Psychophysical experiments may thus be unable to separate these two models. Clearly, what is needed are physiological experiments, e.g. single cell recordings using some of the psychophysical paradigms, to identify unambiguously the algorithm used to detect motion.

In the section on the biophysical mechanisms possibly underlying direction selectivity we have discussed the strengths and limitations of simulating biophysical hardware, that is neurons. Modeling the events underlying a particular computation at the cellular level can give us valuable insights into the elementary operations underlying information processing at the single cell level, operations that cannot be resolved by present experimental techniques because of the small distances and the brief times involved. Thus, the major justification of this approach is its predictive power. Computer simulations should provide a number of detailed predictions that can be evaluated experimentally. Ideally, these predictions should be nontrivial and should rule out alternative explanations.

The major drawback of this approach is that any model is only as good as its fundamental assumptions. For instance, most of the studies addressing properties of the synaptic veto operation assume the absence of any significant electrical nonlinearity, such as dendritic spikes. This proviso must be taken into account when comparing experiments with the theoretical predictions, and the effect of this simplifying assumption on the mechanism in question must be carefully assessed (see O'Donnell et al 1985). Biophysical models of the electrical properties of neurons depend on a host of parameters and assumptions, most of which are poorly characterized. Thus, the foremost requirement of any detailed model of cellular properties must be robustness: Varying some parameter, such as the membrane resistance, by a given amount should not lead to drastically changed properties in the circuit except if some critical, and specified, value has been crossed. Ideally, one would like to show that some particular behavior occurs for a broad range of parameters and is not overly sensitive to any one of them. If the model's behavior varies dramatically by changing a parameter, for instance the location of inhibition with respect to excitation, this dependency should be studied carefully, since it may lead to interesting predictions. Any model that overly constrains a parameter seems biologically unreasonable.

THE INTEGRATION OF EARLY MOTION MEASUREMENTS

Solving the Aperture Problem

The motion detection mechanisms described in the preceding section provide only partial information about the 2-D pattern of movement in the changing image, due to a problem often referred to as the *aperture problem* (Wallach 1976, Fennema & Thompson 1979, Burt & Sperling 1981, Horn & Schunck 1981, Marr & Ullman 1981, Adelson & Movshon 1982). Consider the computation of the projected 2-D velocity field for the rotating wireframe object illustrated in Figure 5a. Suppose that the movement of features on the object were first detected by using operations that examine only a limited area of the image, such as those performed by neural mechanisms with spatially limited receptive fields. The information provided by such mechanisms is illustrated in Figure 5b. The extended edge E moves across the image, and its movement is observed through a window defined by the circular aperture A. Through this window, it is only possible to observe the movement of the edge in the direction perpendicular to its orientation. The component of motion along the orientation of the edge is invisible through this limited aperture. Thus it is not possible to distinguish between motions in the directions b, c, and d. This failure to distinguish between motions when the object is viewed through a small window has been referred to as the *aperture problem* and is inherent in any motion detection operation that examines only a limited area of the image.

As a consequence of the aperture problem, the measurement of motion

---

Figure 5 The aperture problem in motion measurement. (*a*) On the *left* are three views of a wire-frame object undergoing rotation around a central vertical axis. On the *right*, the *arrows* along the contours of the object represent the instantaneous velocity field at one position in the object's trajectory. For simplicity, an orthographic projection is used. (*b*) An operation that views the moving edge E through the local aperture A can compute only the component of motion c in the direction perpendicular to the orientation of the edge. The true motion of the edge is ambiguous. (*c*) The circle undergoes pure translation to the right; the *arrows* represent the perpendicular components of velocity that can be measured from the changing image. (*d*) The vector **v** represents the perpendicular component of velocity at some location in the image. The true velocity at that location must project to the line l perpendicular to **v**; examples are shown with *dotted arrows*. (*e*) The curve C rotates, translates, and deforms over time to yield the curve C′. The velocity of the point p is ambiguous.

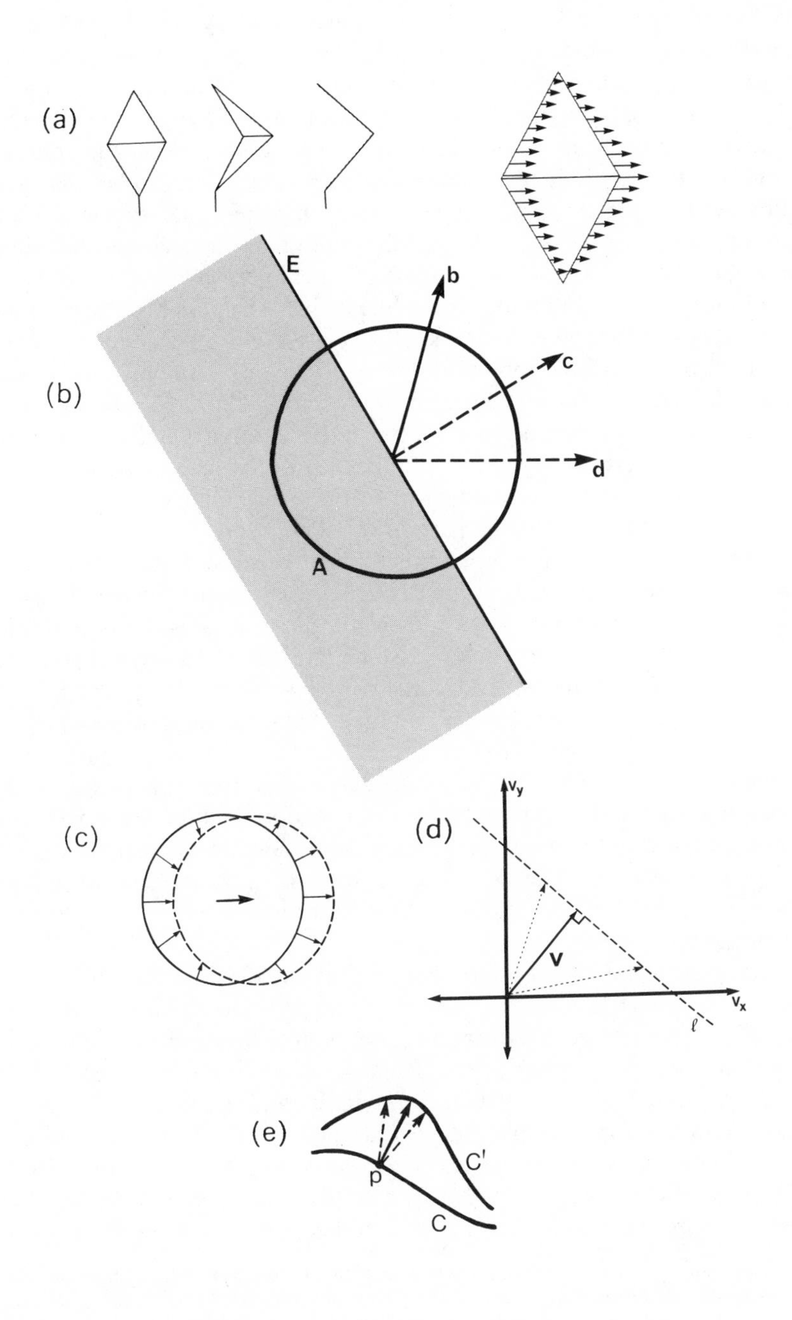
(a)
(b)
E
b
c
d
A
(c)
(d)
v_y
V
v_x
ℓ
(e)
p
C′
C

in the changing image requires two stages of analysis: The first stage measures components of motion in the direction perpendicular to image features; the second combines these components of motion to compute the full 2-D pattern of movement in the image. In Figure 5c, a circle undergoes pure translation to the right. The arrows along the contour represent the perpendicular components of velocity that can be measured directly from the changing image. These component measurements each provide some constraint on the possible motion of the circle, as illustrated in Figure 5d. The vector **v** represents the local perpendicular component of motion at a particular location in the image. The possible true motions at that location are given by the set of velocity vectors whose endpoint lies along the line *l* oriented perpendicular to the vector **v**. Examples of possible true velocities are indicated by the dotted vectors. The movement of image features such as corners or small spots can be measured directly. In general, however, the first measurements of movement provide only partial information about the true movement of features in the image, and must be combined to compute the full pattern of 2-D motion.

The measurement of movement is difficult because in theory there are infinitely many patterns of motion that are consistent with a given changing image. For example, in Figure 5e, the contour C rotates, translates, and deforms to yield the contour C' at some later time. The true motion of the point p is ambiguous. Additional constraint is required to identify a unique solution.[4] It should also be noted that in general, it may not be possible to recover the 2-D projection of the true 3-D field of motions of points in space, from the changing image intensities. Factors such as changing illumination, specularities, and shadows can generate patterns of optical flow in the image that do not correspond to the real movement of surface features. The additional constraint used to measure image motion can yield at best a solution that is most plausible from a physical standpoint.

Many physical assumptions could provide the addition constraint needed to compute a unique pattern of image motion. One possibility is the assumption of pure translation. That is, it is assumed that velocity is constant over small areas of the image. This assumption has been used both in computer vision studies and in biological models of motion measurement (for example, Lappin & Bell 1976, Pantle & Picciano 1976, Fennema & Thompson 1979, Anstis 1980, Marr & Ullman 1981, Thompson & Barnard 1981, Adelson & Movshon 1982). Methods that assume pure translation may be used to detect sudden movements or to

[5] Like many early vision problems, the measurement of motion is an ill-posed problem, as formalized by Hadamard (Poggio et al 1985). A body of mathematics known as *regularization theory* may serve to unify the solution to many ill-posed problems in vision.

track objects across the visual field. These tasks may require only a rough estimate of the overall translation of objects across the image. Tasks such as the recovery of 3-D structure from motion require a more detailed measurement of relative motion in the image. The analysis of variations in motion such as those illustrated in Figure 5a requires the use of a more general physical assumption.

Davis et al (1983) proposed a computational method for solving the aperture problem that assumes that the pattern of image motion can be approximated locally by rigid motion in the image plane. In more recent studies, the local image motions have been modeled by second-order polynomials in the image coordinates (Wohn 1984, Waxman & Wohn 1985, Wohn & Waxman 1985, Waxman 1986). This approach implicitly assumes that the image locally represents the projection of a quadric surface patch in motion.

Other computational studies have assumed that velocity varies smoothly across the image (Horn & Schunck 1981, Hildreth 1984, Nagel 1984, Nagel & Enkelmann 1984, 1986, Anandan & Weiss 1985, Scott 1986). The assumption rests on the principle that physical surfaces are generally smooth; that is, variations in the structure of a surface are usually small, compared with the distance of the surface from the viewer. When surfaces move, nearby points tend to move with similar velocities. There exist discontinuities in movement at object boundaries, but most of the image is the projection of relatively smooth surfaces. Thus, it is natural to assume that image velocities vary smoothly over most of the visual field. A unique pattern of movement can be obtained by computing a velocity field that is consistent with the changing image and has the least amount of variation possible. In other words, a pattern of movement is derived for which nearby points in the image move with velocities that are as similar as possible.

The use of the smoothness assumption for motion measurement has several important attributes from a computational perspective. First, it allows general motion to be analyzed. Surfaces can be rigid or nonrigid, undergoing any movement in space. It is always possible to compute a projected velocity field that preserves the variation in the local pattern of movement. Second, the smoothness assumption can be embodied in the motion measurement computation in a way that guarantees a unique solution (Hildreth 1984). Third, the velocity field of least variation can be computed straightforwardly, using standard computer algorithms (Horn & Schunck 1981, Hildreth 1984, Nagel & Enkelmann 1984, Anandan & Weiss 1985), as well as simple analog resistive networks (Poggio et al 1985, Poggio & Koch 1985).

From the perspective of perceptual psychology, one can ask whether the human visual system derives patterns of movement that are consistent with those predicted by a computation that uses the smoothness assumption. In particular, one can ask whether an incorrect pattern of motion is perceived in situations in which a computer algorithm also fails. The method for computing the velocity field suggested by Hildreth (1984) is guaranteed to yield the correct solution for at least two classes of motion: (*a*) pure translation and (*b*) general motion (translation and rotation) of rigid 3-D objects whose edges are essentially straight. For example, the computation yields the correct velocity field for the moving object of Figure 5a. For smooth curves undergoing rotation, this computation sometimes yields a solution that differs from the correct projected velocity field. The human visual system also appears to derive an incorrect perception of motion in these situations (Hildreth 1984). Comparisons between the results of computational models and perceptual behavior have so far been only qualitative, however. Open questions remain regarding whether the human visual system maintains a local representation of the pattern of image motions, and whether perceived motion is quantitatively consistent with that expected from a computation that uses the smoothness constraint. Perceptual studies indicate that when visual patterns undergo uniform translation, human observers can match velocity directions to a resolution of about 1° (Levinson & Sekuler 1976, Nakayama & Silverman 1983). It is not yet known, however, whether such precision of velocity direction is also obtained when the velocity field varies continuously across the visual field.

A second issue that arises regarding the solution to the aperture problem is the question of whether the early motion measurements are integrated over 2-D areas of the image or along connected contours such as edges. Models such as that suggested by Horn & Schunck (1981) integrate these measurements over areas, while the model proposed by Hildreth (1984) integrates motion measurements along connected contours. This issue was addressed in a recent perceptual study by Nakayama & Silverman (1984a). Their study used a simple distorted line, oscillating up and down. When viewed alone, a central diagonal section of the line appeared to move in an oblique direction, so that the entire figure appeared nonrigid. The figure could be made to appear to move rigidly up and down by the introduction of additional features that were unambiguously moving up and down. Nakayama & Silverman introduced both breaks on the contour and short segments off the contour. They found that both the breaks on the line and the segments off the line could cause the central part of the line to appear to move up and down, but the features on the contour had a much stronger effect, in that their distance from the center could be very large. The segments had to be very close to the line in order to exert any influence on the perception of its motion. These phenomena suggest that the integration of motion constraints along contours may play a stronger role in the human visual system, an observation that is also supported by perceptual demonstrations presented by Hildreth (1984).

The local perpendicular components of motion are not always combined by the human visual system. The conditions governing whether or not these measurements are combined were studied by Adelson & Movshon (1982) and by Nakayama & Silverman (1983). In the Adelson & Movshon study, the stimulus patterns consisted by two superimposed sinewave gratings at different orientations, moving in the direction perpendicular to their orientations. Together, the two gratings formed a single rigid pattern, moving in a direction consistent with the constraints imposed by the two components. Under some conditions, the gratings did not form a single coherent pattern perceptually; rather, the two components appeared to split and move independently of one another. The coherence of the combined pattern was found to decrease with an increase in any of the following factors: (*a*) the difference in contrast between the two gratings, (*b*) the angle between the primary directions of the gratings, (*c*) the difference between the two spatial frequencies, and (*d*) the speed of movement of the overall pattern. In a later study by Adelson (1984), it was shown that the two components of motion would also appear to split if they were presented on different depth planes. This observation suggests that stereo disparity enters into the solution to the aperture problem in motion. Nakayama & Silverman (1983), by using stimuli consisting of sinewave lines, demonstrated that two components of motion tend not to be combined if their orientations are very similar (i.e. they differ by at most about 30°). These perceptual studies suggest that early measurements of the perpendicular components of motion are not always combined by the human visual system. Under some conditions, they will remain separate, resulting in a perception of motion that corresponds directly to the pattern of components. More generally, these studies provide implicit support for the notion that motion measurement takes place in two stages, with the first stage providing the perpendicular components of motion and the second stage combining these components into a single coherent pattern of motion. More explicit psychophysical support for a two-stage motion measurement computation is presented in Movshon et al (1985).

The motion measurement problem can also be examined from a physiological perspective. Early movement detectors in biological systems have spatially limited receptive fields and therefore face the aperture problem. Stimulated by a theoretical analysis of the aperture problem, Movshon et al (1985) sought and found direct physiological evidence for a two-stage motion measurement computation in the primate visual system. Two visual areas that include an abundance of motion-sensitive neurons are cortical areas V1 and the middle temporal area of extrastriate cortex (MT), located in the posterior bank of the superior temporal sulcus (for example, see Maunsell & Van Essen 1983, Van Essen & Maunsell 1983, Allman et al 1985, Saito et al 1986). The explicit role of area MT in the cortical analysis of visual motion was confirmed recently by Newsome et al (1985), who showed that small restricted chemical lesions in area MT of the macaque monkey led to a behavioral deficit in the monkey's ability to match the velocity of smooth pursuit eye movements with the velocity of visual targets. Moreover, lesions in the cat's Claire-Bishop area, which is assumed to correspond to area MT in the macaque anatomically, led to a much reduced ability of behaving cats to distinguish small moving figures from both moving and stationary surround (Strauss & van Seelen 1986). Movshon et al (1985) explored the type of motion analysis taking place in the primate's MT by using the same stimulus with superimposed sinewave gratings used by Adelson & Movshon (1982). The results of these experiments indicate that the selectivity of neurons in area V1 for direction of movement is such that they could provide only the component of motion in the direction perpendicular to the orientation of image features. These neurons essentially only respond to a single component of the combined grating pattern and their response is uninfluenced by the presence of the second grating. Area MT, however, contains a subpopulation of cells, referred to as *pattern cells*, that appear to respond to the 2-D direction of motion of the combined grating pattern. For example, imagine a sinewave grating moving diagonally up this page (bottom left to top right) and a second pattern, superimposed on the first, moving diagonally down the page (top left to bottom right). A neuron in V1 whose best direction is diagonally upward would respond to the superimposed patterns as though the downward moving diagonal was not even present. A pattern cell in MT, however, would respond to the superimposed patterns as though they were moving directly across the page from left to right. Thus, these pattern cells serve to combine motion components to compute the real 2-D direction of velocity of a moving pattern. These experiments do not yet distinguish between the use of the simple assumption of pure translation, as suggested in the study by Movshon et al (1985), and a more general assumption such as smoothness. Stimulus patterns undergoing more complicated motions are required to make such a distinction. If the pattern cells in area MT employ the assumption of smoothness in their computation of motion, one would expect to find direct interaction between pattern cells that analyze nearby areas of the visual field.

Poggio & Koch (1985, Poggio et al 1985) presented hypothetical neural implementations of regularization algorithms in terms of very simple linear, electrical or chemical, analog networks. In particular, they proposed an implementation for the computation of the smoothest velocity field as suggested by Hildreth (1984). From these networks, a neural circuit is then designed that behaves in a similar way. Examples of the electrical and neural networks are shown in Figure 6. In the network of Figure 6a, the currents I_i and conductances g and g_i represent measurements of the perpendicular components of velocity and other properties of a moving contour obtained directly from the image. The voltages V_i represent the tangential component of velocity (i.e. the component of velocity in the direction parallel to the orientation of features in the image) that is recovered by the computation of the full 2-D velocity field. These analog resistive networks allow a fast computation of the smoothest velocity field and are guaranteed to converge to the correct solution (Poggio & Koch 1985). In the corresponding neural implementation of Figure 6b, the tangential component of the velocity field is represented by the voltages

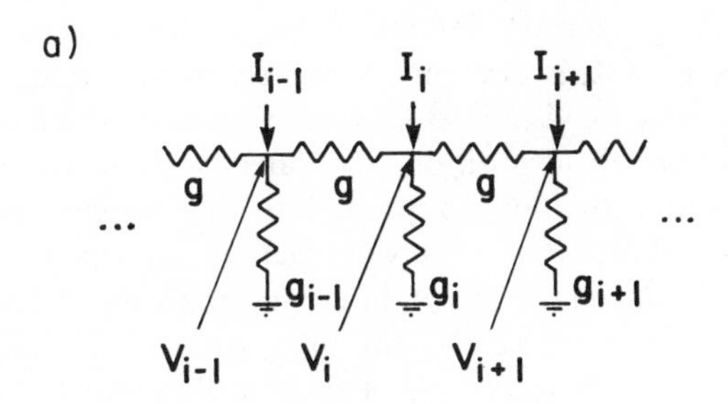

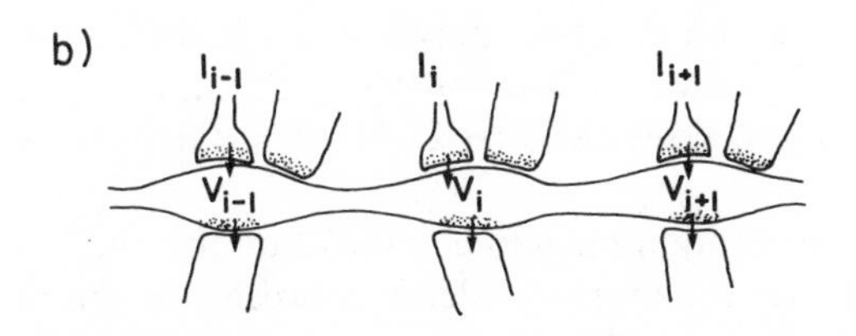

Figure 6 Analog models of the velocity field computation. (*a*) A simple resistive network that computes the smoothest velocity field. The conductances g and g_i and the currents I_i represent properties of a moving contour that are measured directly from the image. In particular, g_i is proportional to the square of the contrast of the contour at location *i*. The 2-D velocity field along the contour is represented implicitly by the combination of these inputs and the resulting voltages V_i. (*b*) A hypothetical neural implementation of the circuit shown in (*a*). Synaptic mediated currents I_i, and additional inputs R_i (possibly a $GABA_A$ type of synapse) represent properties of a moving contour. The resulting voltages V_i, sampled by dendro-dendritic synapses, together with the input currents, represent local velocities along the contour. Redrawn from Poggio & Koch (1985).

V_i along a dendrite, which are sampled by dendro-dendritic synapses. Measurements from the image are represented by synaptically mediated current injections I_i and other synaptic inputs R_i (for instance, a silent $GABA_A$ type inhibitory synapse) that control the membrane resistance. The full 2-D velocity field is represented implicitly by the combination of the currents I_i and the voltages V_i. This hypothetical neural implementation was not intended as a specific model for the measurement of motion in area MT. Rather, its intent was to show that it is possible for neural hardware to exploit a model of this computation that incorporates a general assumption such as smoothness of the velocity field. Models such as this can help to focus experimental questions regarding the actual neural circuitry in areas such as MT.

Long-Range Motion Correspondence

The preceding section addressed computational models that might underlie the short-range process. The computation of a velocity field requires that motion in the image be roughly continuous. The perception of motion by the human visual system does not however, require that objects move continuously across the visual field. Motion can be inferred when features are presented discretely at positions separated by up to several degrees of visual angle[6] and with long temporal intervals between presentations. There are many visual patterns that yield qualitatively different perceptions of motion, depending on the size of the spatial and temporal displacements between frames (for example, Ternus 1926, Anstis 1970, 1980, Braddick 1974, 1980, Anstis & Rogers 1975, Pantle & Picciano 1976, Petersik & Pantle 1979, Shepard & Judd 1976, Burt & Sperling 1981, Green & von Grünau 1983, Hildreth 1984, Anstis & Mather 1985). Although the short- and long-range motion processes may interact at some stage (Clatworthy & Frisby 1973, Green & von Grünau 1983), there is evidence that they are initially distinct processes (Mather et al 1985, Gregory 1985, Anstis & Mather 1985).

The long-range motion phenomena illustrate the ability of the human visual system to derive a correspondence between elements in the changing image, over considerable distances and temporal intervals. Under these conditions, there is no continuous motion of elements across the image to be measured directly. A correspondence computation is therefore likely to underlie the long-range motion process. Two issues arise regarding this computation: First, what features in the image are matched from one moment to the next, and second, how is a unique correspondence of

[6] In fact, commissurotomy patients perceive apparent motion even when the two locations are lying on each side of the vertical meridian at a distance of up to 9° (Ramachandran et al 1986).

features established? Just as with the velocity field computation, very many possible matchings between features in two images can exist, and additional constraints must be imposed to compute a single correspondence that is most plausible from a physical standpoint.

The possible image features that could form the matching elements span a wide range, from simple edge and line segments, points, and blobs, to texture boundaries, subjective contours, and groups of primitive features, and even to structured forms or entire objects. Motion measurement schemes used in computer vision, reviewed for example in Thompson & Barnard (1981), Ullman (1981a), and Barron (1984), have considered most of these possible matching elements. In general, the earlier tokens such as edge and line segments are easier to compute, but entail greater ambiguity in the matching of these tokens from one moment to the next. The use of primitive tokens also allows the correspondence process to operate on arbitrary objects undergoing complex shape changes. More complex tokens such as structured forms can simplify the correspondence process, but more computation is required to extract these features from the image, and there is less flexibility in the types of motion that can be analyzed.

Perceptual studies suggest that many long-range motion phenomena can be explained in terms of a correspondence of elements such as edges, bars, line terminations, and points (Ullman 1979). The human visual system can also establish a correspondence between groups of primitive elements even when the constituents of the groups are not the same (Riley 1981), subjective contours and texture boundaries (Ramachandran et al 1973, Riley 1981) and subjective surfaces (Ramachandran 1985). Properties of primitive elements such as orientation, contrast, and size can influence the correspondence computation (for example, Frisby 1972, Kolers 1972, Ullman 1979, 1981b), although a correspondence can be established between objects that differ significantly in their components (Navon 1976, Anstis & Mather 1985). Chen (1985) has suggested that topological features such as connectivity, closure, and the presence of holes can play a role in motion correspondence, but it is not clear whether these properties are made explicit in the description of the matching elements, or whether they are reflected in the constraints that are used to establish a unique correspondence of elements between frames.

The rules or constraints that are used by the human visual system to establish a correspondence of elements between frames have also been explored in many studies. Early perceptual studies focused on the role of the time and distance between elements in successive frames (for example, Ternus 1926, Kolers 1972, Burt & Sperling 1981). When the elements in motion are isolated dots, each dot in general "prefers" to match its nearest neighbor in the subsequent frame, although this constraint sometimes can be violated locally when a field of dots in motion interacts (Ullman 1979, Burt & Sperling 1981). The distance metric that is used in the correspondence process appears to be based on 2-D distances between elements rather than 3-D distances (Ullman 1979, Mutch et al 1983, Tarr & Pinker 1985). Ramachandran & Anstis (1983, 1985) showed that "inertia" can influence correspondence; that is, in ambiguous situations, moving elements will tend to maintain the same direction of motion over time.

A computational model of correspondence presented by Ullman (1979) assumes independence of the matching elements. Subsequent studies have revealed situations in which the independence assumption appears not to hold. For example, the perceived motion of a feature can be influenced by the motion of other features connected to it along a contour (Hildreth 1984, Chen 1985). Ramachandran & Anstis (1985) created a display in which a local pattern of dots whose motion was two-way ambiguous was repreated in a large array. Each local subpattern could in principle be perceived as moving in either of two directions, but observers always perceived the array of patterns as moving in the same direction. The correspondence established within one subpattern of the display could influence the correspondence of dots in neighboring subpatterns.

To summarize, much is known about the matching elements used in long-range correspondence, and the rules or constraints used to match elements. Many recent perceptual studies were motivated by computational models of the correspondence process. Still lacking, however, are computational models that adequately account for all of the long-range motion phenomena observed in perceptual studies. Recent physiological studies that explored the response of MT neurons to apparent movement stimuli (Newsome et al 1982, 1986, Mikama et al 1986) suggest that area MT might provide some of the neural substrate for the interpretation of long-range motion.

The Detection of Motion Discontinuities

If two adjacent surfaces undergo different motions, a discontinuity generally occurs in the optical flow or velocity field along their boundary. The explicit detection of motion discontinuities allows the detection and localization of object boundaries in the scene. Other cues to the presence of boundaries often exist as well, such as sharp changes in stereo disparity or texture, but perceptual studies suggest the possibility of detecting object boundaries on the basis of motion information alone (Anstis 1970, Regan & Spekreijse 1970, Julesz 1971) and using the relative motions in the vicinity of these boundaries to infer the relative locations of surfaces in depth (Kaplan 1969, Nakayama & Loomis 1974, Mutch & Thompson 1985).

Detection of motion discontinuities as early as possible is advantageous for two reasons. First, the fast detection of a sudden relative movement in the environment can serve as an early warning system, alerting the observer to a possible prey or predator, or to the sudden movement of an object toward the viewer. It is essential not only to detect the presence of movement but also to identify the outline of the object. A second reason for detecting motion discontinuities early is that they facilitate the subsequent measurement of 2-D motion in the image. The computation of a velocity field requires the integration of local measurements of the perpendicular components of motion. Motion measurements should only be combined within single surfaces, as the combination of measurements across object boundaries will generally yield errors in the velocity field. If detected early, the motion discontinuities can define regions of the image within which the local motion measurements should be combined.

With regard to computational schemes, one issue that arises is the question of what stage in the analysis of the image should discontinuities first be detected. Three alternatives present themselves. First, motion discontinuities could be localized prior to the computation of the full velocity field, just after the initial measurements of the perpendicular components of motion in the image (for example, Schunck & Horn 1981, Hildreth 1984). Schunck & Horn used simple heuristics to avoid combining motion measurements likely to occur on surfaces undergoing different motions. Hildreth presented a scheme to detect sudden changes in the perpendicular components of motion, which uses techniques that were previously used for edge detection (Marr & Hildreth 1980). H. Bülthoff and T. Poggio (1986, personal communication) use the binary output of simple correlation-like detectors, signaling motion to the left or the right, to localize discontinuities in dense random-dot patterns. Surprisingly, such a simple measure gives a fairly accurate assessment of discontinuities, at least for random-dot stimuli.

A second possible stage at which boundaries can be detected is after the velocity field has been computed explicitly everywhere. For example, Nakayama & Loomis (1974) proposed a local center-surround operator to detect boundaries in optical flow fields. Similar ideas are incorporated in models suggested by Clocksin (1980) and Thompson et al (1982, 1985, Mutch & Thompson 1985), which use a Laplacian operator applied to components of the optical flow field. In other schemes explored, for example, by Potter (1977) and Fennema & Thompson (1979, Thompson 1980), region-growing techniques are used to group together elements of similar velocities.

Finally, the velocity field and its discontinuities could be computed simultaneously. In a scheme suggested by Wohn (1984) and Waxman (1986), the motion segmentation problem is approached by detecting "boundaries of analyticity" at which an approximation of the local image flow by second-order polynomials breaks down. The boundaries are located within the process that models the local motion field. Koch et al (1986a) have proposed that binary line processes, first introduced in the solution of vision problems by Geman & Geman (1984), can successfully demarcate motion boundaries. At locations at which this line process is set, an unobservable line or edge is postulated to interrupt the otherwise smooth velocity field, segmenting the image into its natural components. The appropriate algorithm can be formulated as an energy minimization problem that maps naturally into simple analog networks (Koch et al 1986a).

A detailed neural circuitry for the detection of motion discontinuities by the housefly was proposed by Reichardt et al (1983, Reichardt & Poggio 1979). Large field binocular "pool" cells summate the output of a retinotopic array of small field elementary movement detectors (EMD) over a large part of the visual field of the two compound eyes. The EMD signal movement in one of two directions: progressive, i.e. movement from front to back, and regressive, i.e. movement from back to front. The pool cells inhibit in turn, via a silent or shunting inhibition (see the section on circuitry and biophysics), the signals provided by the EMD, irrespective of their preferred direction. After inhibition of each channel, all signals from the EMD feed into a large field output cell. This circuit shows two important properties: It detects relative motion of a moving figure superimposed on a stationary background of the same texture as the figure, and its output, the optomotor response, is independent of the size of moving figure. Motion discontinuities are signaled by significant activity in the output cells. The model agrees well with behavioral data from the fly. Moreover, elements of the proposed citcuitry can be identified with anatomically and physiologically characterized cells in the visual system of the fly (Egelhaaf 1985).

Physiological studies have revealed center-surround mechanisms that are organized antagonistically for direction of motion in many vertebrate species (for example, Sterling & Wickelgren 1969, Collett 1972, Bridgeman 1972, Frost 1978, Frost et al 1981, Frost & Nakayama 1983). Motion-sensitive cells with this organization have been found in area MT of the owl monkey (Miezin et al 1982, Allman et al 1985) and in striate cortex of the cat (Orban et al 1986). The existence of center-surround relative motion detection mechanisms across such a range of species suggests that a similar strategy may be utilized in the underlying computations. Richards &

Lieberman (1982) show in psychophysical studies that some viewers are "blind" to shearing motions, and suggest that the neural substrate for detecting such discontinuous motions may be independent from mechanisms detecting other motion boundaries.

Psychophysical studies of motion discontinuities have mainly used dynamic random dot patterns, in which only motion cues signal the presence of boundaries. Braddick's (1974, 1980) studies revealed a limit on the spatial and temporal displacements required to perceive coherent motion in dense random-dot patterns, and showed that a boundary between coherent and incoherent fields of motion can be detected. Experiments by Baker & Braddick (1982a) and van Doorn & Koenderink (1982, 1983) suggest that the detection of discontinuities is not based on a computation that explicitly measures only relative movement; rather, an absolute measurement of motion takes place first, followed by a process that compares nearby motions to locate discontinuities. Baker & Braddick (1982b) showed that the ability to discriminate the orientation of a patch that moves against an uncorrelated background varies little with dot density and increases with the patch size (see also Chang & Julesz 1983). In general, the size of a patch of moving dots that can be discriminated against a differentially moving background increases with larger displacements of the dots between frames (Hildreth 1984). This phenomenon may reflect the limitations of multiple spatial frequency channels involved in the early detection of motion. Other perceptual studies have shown that spatial frequency plays a role in determining the maximum displacements that allow the perception of coherent motion in random dot patterns (Chang & Julesz 1983, Nakayama & Silverman 1984b).

It is important to draw a distinction between the ability to detect differences in motion and the ability to localize a boundary between surfaces undergoing different motions. For example, if two adjacent fields are undergoing motion in the same direction, a 5% difference in speed is sufficient to detect relative movement (McKee 1981, Nakayama 1981). To localize a boundary, however, requires much larger differences in speed, between 50–60% (van Doorn & Koenderink 1982, 1983, Hildreth 1984). If two adjacent surfaces undergo motions with similar speeds but different directions, then an angular change in direction of at least 20° is required to localize the position of the boundary (Hildreth 1984).

Experimental studies have provided much insight into the nature of the mechanisms that underlies the detection of motion discontinuities in biological systems. Many fundamental questions still remain, however. Perhaps the most basic open question concerns at what stage in the analysis of motion the discontinuities are first detected. It is not known, for example, what representation of motion forms the input to the center-surround mechanisms observed in area MT by Allman et al (1985). These mechanisms may operate directly on the perpendicular components of motion, or they may operate on the real 2-D directions of image motion. Psychophysical studies have not yet addressed this issue directly. Furthermore, while physiological studies reveal that some sort of center-surround mechanisms are involved in the detection of relative movement, little is known about what these mechanisms really compute and how they compute this information. Further computational studies are needed to examine possible algorithms for detecting motion boundaries that may utilize these center-surround mechanisms.

THE RECOVERY OF THREE-DIMENSIONAL STRUCTURE FROM MOTION

The Computational Problem and Related Perceptual Studies

When an object moves in space, the motions of individual points on the object differ in a way that conveys information about its 3-D structure, as illustrated in Figure 5a. The directions of motion in this case are all horizontal, but the speed of movement varies in a way that depends on the structure of the object. Using wire-frame objects such as that shown in Figure 5a, Wallach & O'Connell (1953) showed that the human visual system can derive the correct 3-D structure of moving objects from their changing 2-D projection alone. Other perceptual studies also demonstrated this remarkable ability (for example, Green 1961, Braunstein 1962, 1976, Johansson 1973, 1975, Rogers & Graham 1979, Ullman 1979, Cutting 1982, Cutting & Proffitt 1982). Relative motion in the image is also created by movement of the observer relative to the environment, and can be used to infer observer motion from the changing image (Gibson 1950, Lee & Aronson 1974, Johansson 1971, Lee 1980).

Theoretically, the two problems of (*a*) recovering the 3-D structure and movement of objects in the environment and (*b*) recovering the 3-D motion of the observer from the changing image are closely related. The main difficulty faced by both is that infinitely many combinations of 3-D structure and motion could give rise to any particular 2-D image. To resolve this inherent ambiguity, some additional constraint must be imposed to rule out most 3-D interpretations, leaving one that is most plausible from a physical standpoint. Computational studies have used the *rigidity assumption* to derive a unique 3-D structure and motion; they assume that if the changing 2-D image can be interpreted as the projection of a rigid 3-D object in motion, then such an interpretation should be chosen (for example, Ullman 1979, 1983, Clocksin 1980, Prazdny 1980, 1983, Longuet-

Higgins 1981, Longuet-Higgins & Prazdny 1981, Tsai & Huang 1981, Hoffman & Flinchbaugh 1982, Bobick 1983, Mitiche 1984, 1986, Mitiche et al 1985, Waxman & Ullman 1985, Grzywacz & Yuille 1986). When the rigidity assumption is used in this way, the recovery of structure from motion requires the computation of the rigid 3-D object that would project onto a given 2-D image. The rigidity assumption was suggested by perceptual studies that described a tendency for the human visual system to choose a rigid interpretation of moving elements (Wallach & O'Connell 1953, Gibson & Gibson 1957, Green 1961, Jansson & Johansson 1973, Johansson 1975, 1977).

Computational studies have shown the rigidity assumption can yield a unique 3-D structure from the changing 2-D image. Furthermore, this unique 3-D interpretation can be derived by integrating image information over a limited extent in space and in time. For example, suppose that a rigid object in motion is projected onto an image plane by using orthographic projection. Three distinct views of four points on the moving object are sufficient to compute a unique rigid 3-D structure for the points (Ullman 1979). In general, if only two views of the moving points are considered or fewer points are observed, multiple rigid 3-D structures are consistent with the changing 2-D projection. If a perspective projection of objects onto the image is used instead, then two distinct views of seven or eight points in motion are usually sufficient to compute a unique 3-D structure for the points (Longuet-Higgins 1981, Tsai & Huang 1981). If the instantaneous velocity of movement in the image is known at discrete points, then under perspective projection, the position and velocity at five points may be sufficient to derive a unique structure (Prazdny 1980, Roach & Aggarwal 1980). Longuet-Higgins & Prazdny (1981) originally showed that if the continuous velocity field is known everywhere within a region of the image, then the velocity field together with its first and second spatial derivatives at a point is consistent with at most three possible surface orientations at that point. Waxman, Kamgar-Parsi & Subbarao (see Waxman 1986) have recently shown that a unique solution can usually be determined in this case. Finally, for the case of orthographic projection, 3-D structure can be recovered uniquely if both the velocity and acceleration fields are known within a region (Hoffman 1982). Additional theoretical results have been obtained for classes of restricted motion, such as planar surfaces in motion (Hay 1966, Koenderink & van Doorn 1976, Buxton et al 1984, Longuet-Higgins 1984, Murray & Buxton 1984, Kanatani 1985, Waxman & Ullman 1985, Negahdaripour & Horn 1985, Subbarao & Waxman 1985), pure translatory motion of the observer (Clocksin 1980, Lawton 1983, Jerian & Jain 1984), planar or fixed axis rotation (Hoffman & Flinchbaugh 1982, Webb & Aggarwal 1981, Bobick 1983,

Bennett & Hoffman 1984, Sugie & Inagaki 1984), translation perpendicular to the rotation axis (Longuet-Higgins 1983), and motion of quadratic surfaces (Waxman & Ullman 1985, Waxman & Wohn 1985). A review of the theoretical results regarding the recovery of structure from motion can be found in Ullman (1983).

The theoretical results summarized above are important for the study of the recovery of structure from motion in biological vision systems, for at least two reasons. First, they show that by using the rigidity assumption, unique structure can be recovered from motion information alone; no further physical assumptions are needed to obtain a unique solution. Second, these results show that it is possible to recover 3-D structure by integrating image information over a small extent in space and time. This second observation could bear on the neural mechanisms that compute structure from motion; in principle, they need only integrate motion information over a limited area of the visual field and a limited extent in time.

The above computational studies of the recovery of structure from motion also provide algorithms for deriving the structure of moving objects. Typically, measurements of the positions or velocities of image features give rise to a set of mathematical equations whose solution represents the desired 3-D structure. The algorithms generally derive this structure from motion information extracted over a limited area of the image and a limited extent in time. Testing of these algorithms reveals that although this strategy is possible in theory, it is not reliable in practice. A small amount of error in the image measurements can lead to very different (and often incorrect) 3-D structures. This behavior is due in part to the observation that over a small extent in space and time, very different objects can induce almost identical patterns of motion in the image (Ullman 1983, 1984).

This sensitivity to error inherent in algorithms that integrate motion information only over a small extent in space and time suggests that a robust scheme for deriving structure should use image information that is more extended in space, time, or both. This conclusion is supported in recent computational studies (Bruss & Horn 1983, Lawton 1983, Ullman 1984, Adiv 1985, Negahdaripour & Horn 1985, Waxman & Wohn 1985, Wohn & Waxman 1985). Lawton (1983) showed that recovery of the translatory motion of an observer could be coupled with the solution to the motion correspondence problem over an extended region of the image, to yield a robust solution. Adiv (1985) presented an algorithm for recovering the motion parameters for several moving objects, which assumes that object surfaces are piece-wise planar. The extraction of the motion parameters uses a least-squares approach that minimizes the deviation between the measured flow field (at a large number of points) and that predicted from the estimated motion and structure (Bruss & Horn 1983). Negahdaripour & Horn (1985) also addressed the recovery of the motion of an observer relative to a stationary planar surface, and showed that a robust recovery of the observer motion and the orientation of the plane is possible when dense measurements of the spatial and temporal derivatives of image brightness are integrated over a large region of the changing image. Thus, consideration of motion information that is more extended in space can lead to a stable recovery of structure. The study by Ullman (1984), elaborated below, demonstrated that a robust recovery of structure is also possible when motion information is integrated over an extended period of time. The extension in time can be achieved, for example, by considering a large number of discrete frames or by observing continuous motion over a significant temporal extent.

With regard to the human visual system, the dependence of perceived structure on the spatial and temporal extent of the viewed motion has not yet been studied systematically, but the following informal observations have been made. Regarding spatial extent, two or three points undergoing relative motion are sufficient to elicit a perception of 3-D structure (Borjesson & von Hofsten 1973, Johansson 1975), although theoretically the recovery of structure is less constrained for two points in motion, and perceptually the sensation of structure is weaker. An increase in the number of moving elements in view appears to have little effect on the quality of perceived structure (for example, Petersik 1980). Regarding the temporal extent of viewed motion, Johansson (1975) showed that a brief observation of patterns of moving lights generated by human figures moving in the dark (commonly referred to as biological motion displays) can lead to a perception of the 3-D motion and structure of the figures. Other perceptual studies indicate that the human visual system requires an extended time period to reach an accurate perception of 3-D structure (Wallach & O'Connell 1953, White & Mueser 1960, Green 1961, Doner et al 1984, Inada et al 1986). It is not known, however, whether this implies an algorithm with an extended "convergence" time, i.e. many iterations, or whether eye movements are necessary for the recovery of 3-D structure. A brief observation of a moving pattern sometimes yields an impression of structure that is "flatter" than the true structure of the moving object. Thus, the human visual system is capable of deriving some sense of structure from motion information that is integrated over a small extent in space and time. An accurate perception of structure may, however, require a more extended viewing period.

Most methods compute a 3-D structure from motion only when the changing image can be interpreted as the projection of a rigid object in motion. They otherwise yield no interpretation of structure or yield a solution that is incorrect or unstable. Algorithms that are exceptions to this can interpret only restricted classes of nonrigid motions (Bennett & Hoffman 1984, Hoffman & Flinchbaugh 1982, Koenderink & van Doorn 1986, Grzywacz & Yuille 1986). The human visual system, however, can derive some sense of structure for a wide range of nonrigid motions, including stretching, bending, and more complex types of deformation (Johansson 1964, Jansson & Johansson 1973, Todd 1982, 1984). Furthermore, displays a rigid objects in motion sometimes give rise to the perception of somewhat distorting objects (Wallach et al 1956, White & Mueser 1960, Green 1961, Braunstein 1962, Sperling et al 1983, Braunstein & Andersen 1984, Hildreth 1984, Adelson 1985). These observations suggest that while the human visual system tends to choose rigid interpretations of a changing image, it probably does not use the rigidity assumption in the strict way that previous computational studies have suggested.

Ullman (1984) proposed a more flexible method for deriving structure from motion that interprets both rigid and nonrigid motion. Referred to as the *incremental rigidity scheme*, this algorithm uses the rigidity assumption in a way different from previous studies. It maintains an internal model of the structure of a moving object that consists of the estimated 3-D coordinates of points on the object. The model is continually updated as new positions of image features are considered. Initially, the object is assumed to be flat, if no other cues to 3-D structure are present. Otherwise, its initial structure may be determined by other cues available, from stereopsis, shading, texture, or perspective. As each new view of the moving object appears, the algorithm computes a new set of 3-D coordinates for points on the object that maximizes the rigidity in the transformation from the current model to the new positions. This is achieved by minimizing the change in the 3-D distances between points in the model. Thus the algorithm interprets the changing 2-D image as the projection of a moving 3-D object that changes as little as possible from one moment to the next. Through a process of repeatedly considering new views of objects in motion and updating the current model of their structure, the algorithm builds up and maintains a 3-D model of the objects. If objects deform over time, the 3-D model computed by the algorithm also changes over time. Other models have been proposed that impose rigidity by requiring that the 3-D distances between points in space change very little from one moment to the next (for example, Mitiche 1984, 1986, Mitiche et al 1985), although these models do not build up a 3-D model incrementally as in Ullman's proposed scheme.

The method proposed by Ullman (1984) was motivated partly by the limitations of previous computer algorithms and partly by knowledge of the human visual system. The method has overcome limitations of previous computational studies in two ways. First, it provides a reliable recovery of structure in the presence of error in the image measurements, by integrating image information over an extended time period. Second, it allows the interpretation of nonrigid motions. These are essential qualities for any method that is proposed as a viable model for the recovery of structure from motion by the human visual system. This method also has other attributes that are consistent with human perceptual behavior: (*a*) it sometimes yields a nonrigid interpretation of rigid structures in motion, (*b*) a brief viewing time results in a structure that is "flatter" than the true structure of the object, (*c*) it allows a 3-D interpretation of scenes containing as few as two points in motion (Borjesson & von Hofsten 1973, Johansson 1975), and (*d*) it provides a natural means for integrating multiple sources of 3-D information.

A recent computational study by Grzywacz & Hildreth (1985) has extended Ullman's incremental rigidity scheme, presenting a formulation of the algorithm that makes direct use of instantaneous velocity information over an extended time and showing how the algorithm can be modified to use perspective projection of the scene onto the image. With regard to the use of velocities, previous studies had suggested that the recovery of 3-D structure from velocity information at a single moment is inherently unstable (Prazdny 1980, Ullman 1983). Through computer simulations and a theoretical analysis, Grzywacz & Hildreth showed that the integration of velocity information over an extended time does not overcome this problem of instability. The velocity-based formulation of the incremental rigidity scheme does not yield a robust computation of structure over an extended time; rather, the solution oscillates between good and poor estimates of the 3-D structure of a moving object. More generally, if discrete views of moving elements are used instead, the incremental rigidity scheme performs best when the spatial changes between views are large. For example, if an object is rotating, the algorithm computes a better 3-D structure for the object if larger angular rotations between discrete frames are considered.

With regard to the human visual system, discrete movie-like "snapshots" are unlikely to form a direct input to the recovery of 3-D structure from motion. Second, if a short-range motion measurement system exists and provides essentially instantaneous measurements of movement in the changing image, these measurements should be used in some way to interpret the 3-D structure of the scene. These short-range measurements may, however, form the input to a longer-range tracking operation that integrates image motion information over a more extended time for the accurate recovery of 3-D structure. In any case, the short-range measurements can also be used to identify motion discontinuities, which are likely to indicate the locations of object boundaries in the scene. Knowledge of object boundaries can improve the overall recovery of structure from motion.

This discussion of the structure-from-motion problem illustrates a number of important points that often arise in the computational study of other problems in the early stages of vision. First, a single solution to the problem cannot be obtained from information in the image alone; some additional constraint is required. Second, theoretical studies can be used to show that a general physical assumption such as rigidity is sufficient to solve the structure-from-motion problem uniquely. Third, an assumption such as rigidity can be incorporated in many ways into an algorithm to recover structure. The development of a reliable algorithm requires an iterative process of computer implementation, testing, and refinement. Finally, perceptual studies can suggest and test particular assumptions and reveal aspects of the algorithm used by the human visual system for solving a given problem. A typical characteristic of computational studies is that the initial methods proposed for solving a problem only loosely consider the detailed observations of biological systems. These first studies uncover useful aspects of the problems, however. Later studies then combine this knowledge of the problem with observations of biological systems to derive models that more closely reflect the computations carried out in biological systems.

Physiological Studies of the Recovery of Structure from Motion

Physiological studies have uncovered neurons in higher cortical areas that are sensitive to properties of the motion field that may be relevant to the recovery of the 3-D structure and motion of surfaces in the environment, or to the recovery of the motion of the observer relative to the scene. Many studies have revealed neurons sensitive to uniform expansion or contraction of the visual field, a property that is correlated either with translation of the observer forward or backward, or equivalently, motion of an object toward or away from the observer. Such neurons have been found, for example, in the posterior parietal cortex of the monkey (Motter & Mountcastle 1981, Andersen 1986). Other neurons have been found that are sensitive to global rotations in the visual field (Andersen 1986, Sakata et al 1985). All of these neurons have large receptive fields, so they probably lack the spatial sensitivity required to derive the detailed shape of an object surface from relative motion. In the human visual system, the accurate recovery of object shape from motion may be an ability that is restricted to the central region of the eye; the ability to interpret 2-D structure-from-motion displays seems to degrade rapidly as one moves away from the fovea (S. Ullman, personal communication). Siegel & Andersen (1986) showed that motion processing in area MT is critical to the recovery of structure from motion.

The neurons sensitive to relative movement that were discussed in the context of motion discontinuities may also contribute to the recovery of 3-D structure. Certainly the detection and localization of object boundaries is essential to the construction of a 3-D representation of surfaces in the scene. Mechanisms such as the "convexity" detector suggested by Nakayama & Loomis (1974) may also derive information about the relative depths of surfaces on either side of a motion boundary. The computational study by Mutch & Thompson (1985) also addressed this issue.

Regan & Beverley (1979, 1983) have hypothesized the existence of "changing-size" detectors (analogous to detectors of uniform expansion or contraction in the visual field) based on psychophysical evidence from adaptation studies. They also suggested that the changing-size detectors may be distinct from neural mechanisms signaling motion in depth (Beverley & Regan 1979). Neurons exist in area 18 of the cat visual cortex (for example, Cynader & Regan 1978, 1982) and area V1 of the primate visual cortex (Poggio & Talbot 1981) that appear to be selective for direction of movement in depth. These studies on cells responsive to movement in depth used binocularly viewed moving bars, however, so they may address the interaction between binocular stereopsis and motion measurement for the recovery of movement in space, rather than the recovery of structure from motion alone.

Finally, Grzywacz & Yuille (1986) have proposed a neuronal network implementation for the motion correspondence problem. They derive a nonconvex energy expression for both the matching and the recovery of 3-D structure, which they minimize by using analog networks, similar to Koch et al (1986).

CONCLUDING REMARKS

We have tried to integrate studies from computation, psychophysics, physiology, and biophysics into a computational framework. The interaction among these different approaches promises to be fruitful in furthering our understanding of motion analysis in biological vision systems, because the various perspectives each provide valuable and different insight into how vision systems analyze motion information.

Perceptual studies, for example, help to define the problems in motion analysis that are solved and reveal the quantitative ability with which the human visual system can solve these problems. We have seen that many problems in motion analysis do not have a unique solution, and additional constraint must be imposed to solve them. Often, different choices for the assumptions can be embodied in the underlying computations, which critical perceptual experiments can attempt to distinguish. Many algorithms can solve a given problem, and different algorithms might fail in different ways. Again, critical perceptual experiments can be designed to determine whether the human visual system fails in the same way. Perceptual studies often provide initial hints about the strategies used in the underlying computations.

Studies from physiology and biophysics can reveal what parts of the visual system are involved in a particular computation, and what the elementary operations are that neurons use in processing motion information. Properties of the underlying hardware also constrain the nature of the algorithms and representations that are used in motion computations. Detailed computer models of neuronal networks subserving motion measurement have helped to focus further experimental questions regarding physiological and biophysical behavior. Finally, physiological methods can help eliminate ambiguities in perceptual studies. Since the primate visual system may have evolved a variety of different algorithms to cope with a particular problem, a psychophysical paradigm may be unable to distinguish between these different algorithms, while single-cell recordings may do so.

Computational studies help to focus questions for perceptual studies about the assumptions, representations, and algorithms used by the human visual system to analyze motion. Implementations of proposed algorithms have provided powerful predictive tools for making hypotheses about what the behavior of the system ought to be if it is performing motion computations in particular ways. In the case of physiological studies, by elucidating the problems that need to be solved in motion analysis, computational studies can aid the initial exploration of the function of neurons in motion-sensitive areas in the visual pathway. By elucidating possible methods by which computations can be performed, computational studies can help to refine our understanding of how neurons function and by what mechanisms.

ACKNOWLEDGMENTS

We thank Tomaso Poggio and Shimon Ullman for useful comments on a draft of this manuscript. This article describes research done within the Artificial Intelligence Laboratory and the Center for Biological Information Processing (Whitaker College) at the Massachusetts Institute of Technology. Support for the A.I. Laboratory's artificial intelligence research is provided in part by the Advanced Research Projects Agency of the Department of Defense under Office of Naval Research contract N00014-80-C-0505. Support for this research is also provided by a grant from the Office of Naval Research, Engineering Psychology Division.

Literature Cited

Adelson, E. H. 1984. Binocular disparity and the computation of two-dimensional motion. *J. Opt. Soc. Am. A* 1: 1266

Adelson, E. H. 1985. Rigid objects that appear highly non-rigid. *Invest. Ophthalmol. Vision Sci. Suppl.* 26: 56

Adelson, E. H., Bergen, J. R. 1985. Spatiotemporal energy models for the perception of motion. *J. Opt. Soc. Am. A* 2: 284–99

Adelson, E. H., Movshon, J. A. 1982. Phenomenal coherence of moving visual patterns. *Nature* 300: 523–25

Adiv, G. 1985. Determining three-dimensional motion and structure from optical flow generated by several moving objects. *IEEE Trans. Pattern Anal. Machine Intell.* PAMI-7: 384–401

Allman, J., Miezin, F., McGuinness, E. 1985. Direction- and velocity-specific responses from beyond the classical receptive field in the middle temporal area (MT). *Perception* 14: 105–26

Amthor, F. R., Oyster, C. W., Takahashi, E. S. 1984. Morphology of on-off direction-selective ganglion cells in the rabbit retina. *Brain Res.* 298: 187–90

Anandan, P., Weiss, R. 1985. Introducing a smoothness constraint in a matching approach for the computation of optical flow fields. *Proc. IEEE Workshop on Computer Vision: Representation and Control, Bellaire, Mich., October*, pp. 186–94

Andersen, R. A. 1986. The anatomy and physiology of the inferior parietal lobule. In *Development of Spatial Relations*, ed. U. Belugi. Chicago: Univ. Chicago Press. In press

Anstis, S. M. 1970. Phi movement as a subtraction process. *Vision Res.* 10: 1411–30

Anstis, S. M. 1980. The perception of apparent motion. *Philos. Trans. R. Soc. London Ser. B* 290: 153–68

Anstis, S. M., Mather, G. 1985. Effects of luminance and contrast on direction of ambiguous apparent motion. *Perception* 14: 167–79

Anstis, S. M., Rogers, B. J. 1975. Illusory reversal of visual depth and movement during changes of contrast. *Vision Res.* 15: 957–61

Ariel, M., Adolph, A. R. 1985. Neurotransmitter inputs to directionally sensitive turtle retinal ganglion cells. *J. Neurophysiol.* 54: 1123–43

Ariel, M., Daw, N. W. 1982. Pharmacological analysis of directionally sensitive rabbit retinal ganglion cells. *J. Physiol.* 324: 161–85

Baker, C. L., Braddick, O. J. 1982a. Does segregation of differently moving areas depend on relative or absolute displacement. *Vision Res.* 7: 851–56

Baker, C. L., Braddick, O. J. 1982b. The basis of area and dot number effects in random dot motion perception. *Vision Res.* 10: 1253–60

Ballard, D. H. 1986. Cortical connections and parallel processing: Structure and function. *Behav. Brain Sci.* 9: 67–120

Barlow, H. B., Hill, R. M., Levick, R. E. 1964. Retinal ganglion cells responding selectively to direction and speed of image motion in the retina. *J. Physiol.* 173: 377–407

Barlow, H. B., Levick, R. W. 1965. The mechanism of directional selectivity in the rabbit's retina. *J. Physiol.* 173: 477–504

Barron, J. 1984. A survey of approaches for determining optic flow, environmental layout and egomotion. *Univ. Toronto Tech. Rep. Res. Biol. Comp. Vision* RBCV-TR-84-5

Bennett, B. M., Hoffman, D. D. 1984. The computation of structure from fixed axis motions: Nonrigid structures. *Biol. Cybern.* 51: 293–300

Beverley, K. I., Regan, D. 1979. Separable aftereffects of changing-size and motion-in-depth: Different neural mechanisms? *Vision Res.* 19: 727–32

Bobick, A. 1983. A hybrid approach to structure-from-motion. *Proc. ACM Interdisc. Workshop on Motion: Representation and Perception, Toronto, Canada*, pp. 91–109

Borjesson, E., von Hofsten, C. 1973. Visual perception of motion in depth: Application of a vector model to three-dot motion

patterns. *Percept. Psychophys.* 13: 169–79
Boycott, B. B., Wässle, H. 1974. The morphological types of ganglion cells of the domestic cat's retina. *J. Physiol.* 240: 397–419
Braddick, O. J. 1974. A short-range process in apparent motion. *Vision Res.* 14: 519–27
Braddick, O. J. 1980. Low-level and high-level processes in apparent motion. *Philos. Trans. R. Soc. London Ser. B* 290: 137–51
Braunstein, M. L. 1962. Depth perception in rotation dot patterns: Effects of numerosity and perspective. *J. Exp. Psychol.* 6: 41–420
Braunstein, M. L. 1976. *Depth Perception Through Motion.* New York: Academic
Braunstein, M. L., Andersen, G. J. 1984. A counterexample to the rigidity assumption in the visual perception of structure from motion. *Perception* 13: 213–17
Bridgeman, B. 1972. Visual receptive fields sensitive to absolute and relative motion during tracking. *Science* 178: 1106–8
Bruss, A., Horn, B. K. P. 1983. Passive navigation. *Comput. Vision Graph. Image Proc.* 21: 3–20
Bülthoff, H., Bülthoff, I. 1986. GABA-antagonist inverts movement and object detection in flies. *Brain Res.* In press
Burr, D. C., Ross, J. 1982. Contrast sensitivity at high velocities. *Vision Res.* 22: 479–84
Burt, P., Sperling, G. 1981. Time, distance, and feature trade-offs in visual apparent motion. *Psych. Rev.* 88: 171–95
Buxton, B. F., Buxton, H., Murray, D. W., Williams, N. S. 1984. 3D solutions to the aperture problem. In *Advances in Artificial Intelligence*, ed. T. O'Shea. Amsterdam: Elsevier
Caldwell, J. H., Daw, N. W., Wyatt, H. J. 1978. Effects of picrotoxin and strychnine on rabbit retinal ganglion cells: Lateral interactions for cells with more complex receptive fields. *J. Physiol.* 276: 277–98
Chang, J. J., Julesz, B. 1983. Displacement limits, directional anisotropy and direction versus form discrimination in random-dot cinematograms. *Vision Res.* 23: 639–46
Chen, L. 1985. Topological structure in the perception of apparent motion. *Perception* 14: 137–208
Clatworthy, J. L., Frisby, J. P. 1973. Real and apparent movement: Evidence for unitary mechanism. *Perception* 2: 161–64
Cleland, B. G., Levick, W. R. 1974. Properties of rarely encountered types of ganglion cells in the cat's retina and an overall classification. *J. Physiol.* 240: 457–92
Clementi, E. 1985. Ab initio computational chemistry. *J. Phys. Chem.* 89: 4426–36
Clocksin, W. F. 1980. Perception of surface slant and edge labels from optical flow: A computational approach. *Perception* 9: 253–69
Collett, T. 1972. Visual neurons in the anterior optic tract of the privet hawk moth. *J. Comp. Physiol.* 78: 396–433
Cutting, J. E. 1982. Blowing in the wind: Perceiving structure in trees and bushes. *Cognition* 12: 25–44
Cutting, J. E., Proffitt, D. R. 1982. The minimum principle and the perception of absolute, common, and relative motions. *Cognit. Psychol.* 14: 211–246
Cynader, M., Regan, D. 1978. Neurons in the cat parastriate cortex sensitive to the direction of motion in three-dimensional space. *J. Physiol.* 274: 549–69
Cynader, M., Regan, D. 1982. Neurons in cat visual cortex tuned to the direction of motion in depth: Effect of positional disparity. *Vision Res.* 22: 967–82
Davis, L., Wu, Z., Sun, H. 1983. Contour-based motion estimation. *Comput. Vision Graphics Image Proc.* 23: 313–26
DeVoe, R. D., Guy, R. G., Criswell, M. H. 1985. Directionally selective cells of the inner nuclear layer in the turtle retina. *Invest. Ophthalmol. Vis. Sci. Suppl.* 26: 311
Diener, H. C., Wist, E. R., Dichgans, J., Brandt, Th. 1976. The spatial frequency effect on perceived velocity. *Vision Res.* 16: 169–76
Doner, J., Lappin, J. S., Perfetto, G. 1984. Detection of three-dimensional structure in moving optical patterns. *J. Exp. Psychol. Human Percept. Perform.* 10: 1–11
Dow, B. M. 1974. Functional classes of cells and their laminar distribution in monkey visual cortex. *J. Neurophysiol.* 37: 927–46
Dowling, J. E. 1979. Information processing by local circuits: The vertebrate retina as a model system. In *The Neurosciences: Fourth Study Program*, ed. F. O. Schmitt, F. G. Worden, pp. 163–81. Cambridge: MIT Press
Eckert, H. E. 1973. Optomotorische Untersuchungen am visuellen System der Stubenfliege. *Musca Domest. L., Kybernet.* 14: 1–23
Egelhaaf, M. 1985. On the neuronal basis of figure-ground discrimination by relative motion in the visual systems of the fly: III. Possible input circuitries and behavioural significance of the FD-cells. *Biol. Cybernet.* 52: 267–80
Famiglietti, E. V. 1983. On and off pathways through amacrine cells in mammalian retina: The synaptic connections of "star-

burst" amacrine cells. *Vision Res.* 23: 1265–79
Famiglietti, E. V., Kolb, H. 1976. Structural basis for on- and off-center responses in retinal ganglion cells. *Science* 194: 193–95
Fennema, C. L., Thompson, W. B. 1979. Velocity determination in scenes containing several moving objects. *Comput. Graph. Image Proc.* 9: 301–15
Fermi, G., Reichardt, W. 1963. Optomotorische Reaktionen der Fliege. *Musca Domest. L. Kybernet.* 2: 15–28
Frisby, J. P. 1972. The effect of stimulus orientation on the phi phenomenon. *Vision Res.* 12: 1145–66
Frost, B. J. 1978. Moving background patterns after directionally specific responses of pigeon tectal neurons. *Brain Res.* 151: 599–603
Frost, B. J., Nakayama, K. 1983. Single visual neurons code opposing motion independent of direction. *Science* 220: 744–45
Frost, B. J., Scilley, P. L., Wong, S. C. P. 1981. Moving background patterns reveal double opponency of directionally specific pigeon tectal neurons. *Exp. Brain Res.* 43: 173–85
Geman, S., Geman, D. 1984. Stochastic relaxation, Gibbs distribution, and the Bayesian Restoration of Images. *IEEE Trans. Pattern Anal. Machine Intell.* 6: 721–41
Gibson, J. J. 1950. *The Perception of the Visual World.* Boston: Houghton Miffin
Gibson, J. J., Gibson, E. J. 1957. Continuous perceptive transformations and the perception of rigid motion. *J. Exp. Psychol.* 54: 129–38
Götz, K. G. 1965. Die optischen Übertragungseigenschaften der Komplexaugen von *Drosophila. Kybernetik* 2: 215–21
Götz, K. G. 1968. Flight control in *Drosophila* by visual perception of motion. *Kybernetik* 4: 199–208
Götz, K. G. 1972. Principles of optomotor reactions in insects. *Bibl. Ophthal.* 82: 251–59
Green, B. F. 1961. Figure coherence in the kinetic depth effect. *J. Exp. Psychol.* 62: 272–82
Green, M., von Grünau, M. 1983. Real and apparent motion: One mechanism or two? *Proc. ACM Interdisc. Workshop on Motion: Representation and Perception, Toronto, Canada*, pp. 17–22
Gregory, R. L. 1985. Movement nulling: For heterochromatic photometry and isolating channels for 'real' and 'apparent' motion. *Perception* 14: 193–96
Grzywacz, N. M., Hildreth, E. C. 1985. The incremental rigidity scheme for recovering structure from motion: Positions vs. velocity based formulations. *MIT Artif. Intell. Memo 845*
Grzywacz, N. M., Yuille, A. 1986. Motion correspondence and analog networks. *MIT Artif. Intell. Memo 888*
Hassenstein, B., Reichardt, W. E. 1956. Functional structure of a mechanism of perception of optical movement. In *Proc. 1st Intl. Congr. Cybernet. Namar*, pp. 797–801
Hausen, K. 1982a. Movement sensitive interneurons in the optomotor system of the fly. I. The horizontal cells: Structure and signals. *Biol. Cybernet.* 45: 143–56
Hausen, K. 1982b. Movement sensitive interneurons in the optomotor system of the fly. II. The horizontal cells: Receptive field organization and response characteristics. *Biol. Cybernet.* 46: 67–79
Hay, C. J. 1966. Optical motions and space perception—an extension of Gibson's analysis. *Psychol. Rev.* 73: 550–65
Hildreth, E. C. 1984. *The Measurement of Visual Motion.* Cambridge: MIT Press
Hildreth, E. C., Hollerbach, J. M. 1985. The computational approach to vision and motor control. *MIT Artif. Intell. Memo 846.* (Also in 1987. *Handb. Physiol.* 5 (Sect. 1). In press
Hoffman, D. D. 1982. Inferring local surface orientation from motion fields. *J. Opt. Soc. Am.* 72: 888–92
Hoffman, D. D., Flinchbaugh, B. E. 1982. The interpretation of biological motion. *Biol. Cybernet.* 42: 195–204
Holden, A. L. 1977. Responses of directional ganglion cells in the pigeon retina. *J. Physiol.* 270: 253–69
Horn, B. K. P., Schunck, B. G. 1981. Determining optical flow. *Artif. Intell.* 17: 185–203
Hubel, D. H., Wiesel, T. N. 1959. Receptive fields of single neurons in the cat's striate cortex. *J. Physiol.* 148: 574–91
Hubel, D. H., Wiesel, T. N. 1962. Receptive fields, binocular interaction and functional architecture in the cat's visual cortex. *J. Physiol.* 160: 106–54
Inada, V. K., Hildreth, E. C., Grzywacz, N. M., Adelson, E. H. 1986. The perceptual buildup of three-dimensional structure from motion. *Invest. Ophthal. Visual Sci. Suppl.* 27: 142
Ito, M. 1984. *The Cerebellum and Neural Control.* New York: Raven
Jansson, G., Johansson, G. 1973. Visual perception of bending motion. *Perception* 2: 321–26
Jensen, R. J., DeVoe, R. D. 1983. Comparisons of directionally selective with other ganglion cells of the turtle retina: Intracellular recording and staining. *J. Comp. Neurol.* 217: 271–87
Jerian, C., Jain, R. 1984. Determining

motion parameters for schemes with translation and rotation. *IEEE Trans. Pattern Anal. Machine Intell.* PAMI-6: 523–30

Johansson, G. 1964. Perception of motion and changing form. *Scand. J. Psychol.* 5: 181–208

Johansson, G. 1971. Studies on visual perception of locomotion. *Perception* 6: 365–76

Johansson, G. 1973. Visual perception of biological motion and a model for its analysis. *Percept. Psychophys.* 14: 201–11

Johansson, G. 1975. Visual motion perception. *Sci. Am.* 232: 76–88

Johansson, G. 1977. Spatial constancy and motion in visual perception. In *Stability and Constancy in Visual Perception*, ed. W. Epstein. New York: Wiley

Jones, E. G., Powell, T. P. S. 1969. Morphological variations in the dendritic spines of the neocortex. *J. Cell. Sci.* 5: 509–29

Julesz, B. 1971. *Foundations of Cyclopean Perception.* Chicago: Univ. Chicago Press

Kanatani, K. 1985. Structure from motion without correspondence: General principle. *Proc. Image Understanding Workshop, Miami, Fla.*, pp. 107–16

Kaplan, G. A. 1969. Kinetic disruption of optical texture: The perception of depth at an edge. *Percept. Psychophys.* 6: 193–98

Kirschfeld, K. 1972. The visual system of Musca: Studies on optics, structure, and function. In *Information Processing in the Visual System of Arthropods*, ed. R. Welmer, pp. 61–74. Berlin: Springer

Koch, C. 1986. What's in the term connectionist? *Behav. Brain Sci.* 9: 100–1

Koch, C., Marroquin, J., Yuille, A. 1986a. Analog "neuronal" networks in early vision. *Proc. Natl. Acad. Sci. USA* 83: 4263–67

Koch, C., Poggio, T. 1983. A theoretical analysis of electrical properties of spines. *Proc. R. Soc. London Ser. B* 218: 455–77

Koch, C., Poggio, T. 1985. The synaptic veto mechanism: Does it underlie direction and orientation selectivity in the visual cortex? In *Models of the Visual Cortex*, ed. D. Rose, V. Dobson, pp. 408–19. Sussex: Wiley

Koch, C., Poggio, T. 1986. Biophysics of computational systems: Neurons, synapses and membranes. In *New Insights into Synaptic Function*, ed. G. M. Edelman, W. E. Gall, W. M. Cowan. New York: Neurosci. Res. Found./Wiley. In press

Koch, C., Poggio, T., Torre, V. 1982. Retinal ganglion cells: A functional interpretation of dendritic morphology. *Philos. Trans. R. Soc. B* 298: 227–64

Koch, C., Poggio, T., Torre, V. 1983. Nonlinear interaction in a dendritic tree: Localization, timing and role in information processing. *Proc. Natl. Acad. Sci. USA* 80: 2799–2802

Koch, C., Torre, V., Poggio, T. 1986b. Computations in the vertebrate retina: Motion discrimination, gain enhancement and differentiation. *Trends Neurosci.* 9: 204–11

Koenderink, J. J., van Doorn, A. J. 1976. Local structure of movement parallax of the plane. *J. Opt. Soc. Am.* 66: 717–23

Koenderink, J. J., van Doorn, A. J. 1986. Depth and shape from differential perspective in the presence of bending deformations. *J. Opt. Soc. Am. A* 3: 242–49

Kolers, P. A. 1972. *Aspects of Motion Perception.* New York: Pergamon

Lappin, J. S., Bell, H. H. 1976. The detection of coherence in moving random dot patterns. *Vision Res.* 16: 161–68

Lawton, D. T. 1983. Processing translational motion sequences. *Comput. Vision Graph. Image Proc.* 22: 116–44

Lee, D. N. 1980. The optic flow field: The foundation of vision. *Philos. Trans. R. Soc. London B* 290: 169–79

Lee, D. N., Aronson, E. 1974. Visual proprioceptive control of standing in human infants. *Percept. Psychophys.* 15: 529–32

Levinson, E., Sekuler, R. 1976. Adaptation alters perceived direction of motion. *Vision Res.* 16: 779–81

Limb, J. O., Murphy, J. A. 1975. Estimating the velocity of moving images in television signals. *Comput. Graph. Image Proc.* 4: 311–27

Longuet-Higgins, H. C. 1981. A computer algorithm for reconstructing a scene from two projections. *Nature* 293: 133–35

Longuet-Higgins, H. C. 1983. The role of the vertical dimension in stereoscopic vision. *Perception* 11: 377–86

Longuet-Higgins, H. C. 1984. The visual ambiguity of a moving plane. *Proc. R. Soc. London Ser. B* 223: 165–75

Longuet-Higgins, H. C., Prazdny, K. 1981. The interpretation of moving retinal images. *Proc. R. Soc. London Ser. B* 208: 385–97

Marchiafava, P. L. 1979. The responses of retinal ganglion cells to stationary and moving visual stimuli. *Vis. Res.* 19: 1203–11

Marr, D. 1969. A theory of cerebellar cortex. *J. Physiol.* 202: 437–70

Marr, D. 1982. *Vision.* San Francisco: Freeman

Marr, D., Hildreth, E. C. 1980. Theory of edge detection. *Proc. R. Soc. London Ser. B* 207: 187–217

Marr, D., Poggio, T. 1977. From understanding computation to understanding neural circuitry. *Neurosci. Res. Prog. Bull.* 15: 470–88

Marr, D., Ullman, S. 1981. Directional selectivity and its use in early visual processing. *Proc. R. Soc. London Ser. B* 211: 151–80

Masland, R. H., Mills, W., Cassidy, C. 1984. The functions of acetylcholine in the rabbit retina. *Proc. R. Soc. London Ser. B* 223: 121–39

Massey, S. C., Neal, M. J. 1979. The light evoked release of ACh from the rabbit retina in vivo and its inhibition by GABA. *J. Neurochem.* 32: 1327–29

Mather, G., Cavanagh, P., Anstis, S. M. 1985. A moving display which opposes short-range and long-range signals. *Perception* 14: 163–66

Maturana, H., Frenk, S. 1963. Directional movement and horizontal edge detectors in the pigeon retina. *Science* 142: 977–79

Maturana, H. R., Lettvin, J. Y., McCulloch, W. S., Pitts, W. H. 1960. Anatomy and physiology of vision in the frog (*Rana pipiens*). *J. Gen. Physiol.* 43: (Suppl. 2): 129–71

Maunsell, J. H. R., Van Essen, D. C. 1983. Functional properties of neurons in middle temporal visual area of the macaque monkey. I. Selectivity for stimulus direction, speed and orientation. *J. Neurophysiol.* 49: 1127–47

Michael, C. R. 1966. Receptive fields of directionally selective units in the optic nerve of the ground squirrel. *Science* 152: 1092–95

McCulloch, W. S., Pitts, W. 1943. A logical calculus of ideas immanent in neural nets. *Bull. Math. Biophys.* 5: 115–37

McKee, S. 1981. A local mechanism for differential velocity detection. *Vision Res.* 21: 491–500

Miezin, F., McGuinness, E., Allman, J. 1982. Antagonistic direction specific mechanisms in area MT in the owl monkey. *Neurosci. Abstr.* 8: 681

Mikami, A., Newsome, W. T., Wurtz, R. H. 1986. Motion selectivity in macaque visual cortex. II. Spatio-temporal range of directional interactions in MT and V1. *J. Neurophys.* 55: 1308–27

Miller, R. F., Bloomfield, S. A. 1983. Electroanatomy of an unique amacrine cell in the rabbit retina. *Proc. Natl. Acad. Sci. USA* 80: 3069–73

Minsky, M., Papert, S. 1969. *Perceptrons: An Introduction to Computational Geometry*. Cambridge: MIT Press

Mitiche, A. 1984. Computation of optical flow and rigid motion. *Proc. Workshop on Computer Vision: Representation and Control, Annapolis, Md.*, pp. 63–71

Mitiche, A. 1986. On kineopsis and computation of structure and motion. *IEEE Trans. Pattern Anal. Machine Intell.* PAMI-8: 109–12

Mitiche, A., Seida, S., Aggarwal, J. K. 1985. Determining position and displacement in space from images. *Proc. IEEE Conf. on Computer Vision and Pattern Recognition, San Francisco, June*, pp. 504–9

Morgan, M. J. 1985. Computational vision. In *Textbook Series in Psychology*. London: Brit. Psychol. Soc.

Motter, B. C., Mountcastle, V. B. 1981. The functional properties of the light-sensitive neurons in the posterior parietal cortex studied in waking monkeys: Foveal spacing and opponent vector organization. *J. Neurosci.* 1: 3–26

Moulden, B., Begg, H. 1986. Some tests of the Marr-Ullman model of movement detection. *Perception* 15: 139–55

Movshon, J. A., Adelson, E. H., Gizzi, M. S., Newsome, W. T. 1985. The analysis of moving visual patterns. In *Pattern Recognition Mechanisms*, ed. C. Chagas, R. Gattas, C. G. Gross. Rome: Vatican Press

Murray, D. W., Buxton, B. F. 1984. Reconstructing the optic flow field from edge motion: An examination of two different approaches. *Proc. 1st Conf. AI Applications, Denver*

Mutch, K., Smith, I. M., Yonas, A. 1983. The effect of two-dimensional and three-dimensional distance on apparent motion. *Perception* 12: 305–12

Mutch, K. M., Thompson, W. B. 1985. Analysis of accretion and deletion at boundaries in dynamic scenes. *IEEE Trans. Patt. Anal. Mach. Intell.* PAMI-7: 133–38

Nagel, H.-H. 1984. Recent advances in image sequence analysis. *Proc. Premier Colloque Image—Traitement, Synthese, Technologie et Applications, Biarritz, France, May*, pp. 545–58

Nagel, H.-H., Enkelmann, W. 1984. Towards the estimation of displacement vector fields by "oriented smoothness" constraints. *Proc. 7th Int. Conf. Pattern Recognition, Montreal, Canada, July*, pp. 6–8

Nagel, H.-H., Enkelmann, W. 1986. An investigation of smoothness constraints for the estimation of displacement vector fields for image sequences. *IEEE Trans. Patt. Anal. Machine Intell.* PAMI-8: 565–93

Nakayama, K. 1981. Differential motion hyperacuity under conditions of common image motion. *Vision Res.* 21: 1475–82

Nakayama, K. 1985. Biological motion

processing: A review. *Vision Res.* 25: 625–60

Nakayama, K., Loomis, J. M. 1974. Optical velocity patterns, velocity-sensitive neurons, and space perception: A hypothesis. *Perception* 3: 63–80

Nakayama, K., Silverman, G. H. 1983. Perception of moving sinusoidal lines. *J. Opt. Soc. Am. A* 72

Nakayama, K., Silverman, G. H. 1984a. Propagation of velocity information along moving contours. *J. Opt. Soc. Am. A* 1: 1266

Nakayama, K., Silverman, G. H. 1984b. Temporal and spatial characteristics of the upper displacement limit for motion in random dots. *Vision Res.* 24: 293–99

Navon, D. 1976. Irrelevance of figural identity for resolving ambiguities in apparent motion. *J. Exp. Psychol., Human Percep. Perform.* 2: 130–38

Negahdaripour, S., Horn, B. K. P. 1985. Direct passive navigation. *MIT Artif. Intell. Memo 821*

Newsome, W. T., Mikami, A., Wurtz, R. H. 1982. Direction selective responses to sequentially flashed stimuli in extrastriate area MT in the awake macaque monkey. *Neurosci. Abstr.* 8: 812

Newsome, W. T., Mikami, A., Wurtz, R. H. 1986. Motion selectivity in macaque visual cortex: III. Psychophysics and physiology of apparent motion. *J. Neurophys.* 55: 1340–58

Newsome, W. T., Wurtz, R. H., Dursteler, M. R., Mikami, A. 1985. Deficits in visual motion processing following ibotenic acid lesions of the middle temporal visual area of the macaque monkey. *J. Neurosci.* 5: 825–40

O'Donnell, P., Koch, C., Poggio, T. 1985. Demonstrating the nonlinear interaction between excitation and inhibition in dendritic trees using computer-generated color graphics: A film. *Neurosci. Abstr.* 11: 142.1

Orban, G. A., Gulyas, B., Vogels, R. 1986. Influence of a moving textured background on direction-selectivity of cat striate neurons. *J. Neurophysiol.* Submitted

Orban, G. A., Kennedy, H., Maes, H. 1981. Responses to movement of neurons in areas 17 and 18 of the cat: Direction selectivity. *J. Neurophysiol.* 45: 1059–73

Oyster, C. W., Takahashi, E., Collewijn, H. 1972. Direction-selective retinal ganglion cells and control of optokinetic nystagmus in the rabbit. *Vision Res.* 13: 183–93

Pantle, A. J., Picciano, L. 1976. A multistable display: Evidence for two separate motion systems in human vision. *Science* 193: 500–2

Petersik, J. T. 1980. The effect of spatial and temporal factors on the perception of stroboscopic rotation stimulations. *Perception* 9: 271–83

Petersik, J. T., Pantle, A. 1979. Factors controlling the competing sensations produced by a bistable stroboscopic motion display. *Vision Res.* 19: 143–54

Poggio, G. F., Talbot, W. H. 1981. Mechanisms of static and dynamic stereopsis in foveal cortex of the rhesus monkey. *J. Physiol.* 315: 469–92

Poggio, T. 1982. Visual algorithms. In *Physical and Biological Processing of Images*, ed. O. J. Braddick, A. C. Sleigh, pp. 128–53. Berlin: Springer-Verlag

Poggio, T. 1984. Vision by man and machine. *Sci. Am.* 250: 106–15

Poggio, T., Koch, C. 1985. Ill-posed problems in early vision: from computational theory to analog networks. *Proc. R. Soc. London Ser. B* 226: 303–23

Poggio, T., Reichardt, W. E. 1973. Considerations on models of movement detection. *Kybernetiks* 13: 223–27

Poggio, T., Reichardt, W. 1976. Visual control of orientation behaviour in the fly: Part II: Towards the underlying neural interactions. *Q. Rev. Biophys.* 9: 377–438

Poggio, T., Torre, V., Koch, C. 1985. Computational vision and regularization theory. *Nature* 317: 314–19

Potter, J. L. 1977. Scene segmentation using motion information. *Comput. Graph. Image Proc.* 6: 558–81

Prazdny, K. 1980. Egomotion and relative depth map from optical flow. *Biol. Cybernet.* 36: 87–102

Prazdny, K. 1983. On the information in optical flows. *Comput. Vision Graph. Image Proc.* 22: 239–59

Ramachandran, V. S. 1985. Apparent motion of subjective surfaces. *Perception* 14: 127–34

Ramachandran, V. S., Anstis, S. M. 1983. Extrapolation of motion path in human visual perception. *Vision Res.* 23: 83–85

Ramachandran, V. S., Anstis, S. M. 1985. Perceptual organization in multistable apparent motion. *Perception* 14: 135–43

Ramachandran, V. S., Cronin-Golomb, A., Myers, J. J. 1986. Perception of apparent motion by commissurotomy patients. *Nature* 320: 358–59

Ramachandran, V. S., Gregory, R. L. 1978. *Nature* 275: 55–56

Ramachandran, V. S., Rao, V. M., Vidyasagar, T. R. 1973. Apparent motion with subjective contours. *Vision Res.* 13: 1399–1401

Regan, D., Beverley, K. I. 1979. Visually guided locomotion: Psychophysical evidence for neural mechanisms sensitive to flow patterns. *Science* 205: 311–13

Regan, D., Beverley, K. I. 1983. Visual fields for frontal plane motion and changing size. *Vision Res.* 23: 673–76

Regan, D., Sperkreijse, H. 1970. Electrophysiological correlate of binocular depth perception in man. *Nature* 225: 92–94

Reichardt, W. 1969. Movement perception in insects. In *Processing of Optical Data by Organisms and Machines*, ed. W. Reichardt, pp. 465–93. London/New York: Academic

Reichardt, W., Guo, A.-K. 1986. Elementary pattern discrimination (behavioral experiments with the fly *Musca domestica*). *Biol. Cybernet.* 53: 285–306

Reichardt, W., Poggio, T. 1979. Figure-ground discrimination by relative movement in the visual system of the fly. *Biol. Cybernet.* 35: 81–100

Reichardt, W., Poggio, T., Hausen, K. 1983. Figure-ground discrimination by relative movement in the visual system of the fly. Part II: Towards the neural circuitry. *Biol. Cybernet.* 46: 1–30

Richards, W., Lieberman, H. R. 1982. Velocity blindness during shearing motion. *Vision Res.* 22: 97–100

Riley, M. D. 1981. The representation of image texture. *MIT Artif. Intell. Tech. Rep. AI-TR-649*

Roach, J. W., Aggarwal, J. K. 1980. Determining the movement of objects from a sequence of images. *IEEE Proc. Pattern Anal. Machine Intell.* PAMI-2: 554–62

Rodieck, R. W. 1979. Visual pathways. *Ann. Rev. Neurosci.* 2: 193–226

Rogers, B. J., Graham, M. 1979. Motion parallax as an independent cue for depth perception. *Perception* 8: 125–34

Saito, H.-A., Yukie, M., Tanaka, K., Hikosaka, K. Fukuda, Y., Iwai, E. 1986. Interaction of direction signals of image motion in the superior temporal sulcus of the macaque monkey. *J. Neurosci.* 6: 145–57

Sakata, H., Shibutani, H., Kawano, K., Harrington, T. L. 1985. Neural mechanisms of space vision in the parietal association cortex of the monkey. *Vis. Res.* 25: 453–63

Schiller, P. H. 1982. Central connections of the retinal ON and OFF pathways. *Nature* 297: 580–83

Schiller, P. H., Finlay, B. L., Volman, S. F. 1976. Quantitative studies of single-cell properties in monkey striate cortex. I. Spatiotemporal organization of receptive fields. *J. Neurophysiol.* 49: 1288–1319

Schunck, B. G. 1984. The motion constraint equation for optical flow. *Intl. Conf. Pattern Recognition, Montreal, Canada*

Schunck, B. G., Horn, B. K. P. 1981. Constraints on optical flow computation. *Proc. IEEE Conf. Pattern Recognition Image Proc., Aug.*, pp. 205–10

Scott, G. L. 1986. *Local and global interpretation of moving images*. PhD thesis, Univ. Sussex

Segev, I., Fleshman, J. W., Miller, J. P., Bunow, B. 1985. Modeling the electrical behavior of anatomically complex neurons using a network analysis program: Passive membrane. *Biol. Cybernet.* 53: 27–40

Segev, I., Parnas, I. 1983. Synaptic integration mechanisms. *Biophys. J.* 41: 41–50

Shepard, R. N., Judd, S. A. 1976. Perceptual illusion of rotation of three-dimensional objects. *Science* 191: 952–54

Sherk, H., Horton, J. C. 1984. Receptive-field properties in the cat's area 17 in the absence of on-center geniculate input. *J. Neurosci.* 4: 374–80

Siegel, R. M., Anderson, R. A. 1986. Motion perceptual deficits following ibotenic acid lesions of the middle temporal area (MT) in the behaving Rhesus monkey. *Neurosci. Abstr.* 12: 324.8

Sillito, A. M. 1977. Inhibitory processes underlying the directional specificity of simple, complex and hypercomplex cells in the cat's visual cortex. *J. Physiol.* 271: 699–720

Sillito, A. M., Kemp, J. A., Milson, J. A., Berardi, N. 1980. A re-evaluation of the mechanism underlying simple cell orientation selectivity. *Brain Res.* 194: 517–20

Simpson, J. I. 1984. The accessory optic system. *Ann. Rev. Neurosci.* 7: 13–41

Sloper, J. J., Powell, T. P. S. 1979. An experimental electron microscopic study of afferent connections to the primate motor and somatic sensory cortices. *Philos. Trans. R. Soc. London B* 285: 199–226

Sperling, G., Pavel, M., Cohen, Y., Landy, M. S., Schwartz, B. J. 1983. Image processing in perception and cognition. In *Physical and Biological Processing of Images*, ed. O. J. Braddick, A. C. Sleigh. Berlin: Springer-Verlag

Sterling, P., Wickelgren, B. G. 1969. Visual receptive fields in the superior colliculus of the cat. *J. Neurophysiol.* 32: 1–15

Stone, J., Fabian, M. 1966. Specialized receptive fields of the cat's retina. *Science* 152: 1277–79

Strauss, G., von Seelen, W. 1986. Contribution of suprasylvian cortex to pattern recognition in the cat. Manuscript in preparation

Subbarao, M., Waxman, A. M. 1985. On the uniqueness of image flow solutions for planar surfaces in motion. *Proc. Workshop on Computer Vision. Representation and Control, Bellaire, Mich., Oct.*, pp. 129–40

Sugie, N., Inagaki, H. 1984. A computational aspect of kinetic depth effect. *Biol. Cybernet.* 50: 431–36

Tarr, M. J., Pinker, S. P. 1985. Nearest neighbors in apparent motion: Two or three dimensions? *Proc. Ann. Meet. Psychonom. Soc., Boston, Nov.*, p. 19

Tauchi, M., Masland, R. H. 1984. The shape and arrangement of the cholinergic neurons in the rabbit retina. *Proc. R. Soc. London Ser. B* 223: 101–19

Ternus, J. 1926. Experimentelle Untersuchung über phänomenale Identität. *Psychol. Forsch.* 7: 81–136. (Trans. in *A Source Book of Gestalt Psychology*, ed. W. D. Ellis. New York: Humanities Press. 1967)

Thompson, W. B. 1980. Combining motion and contrast for segmentation. *IEEE Trans. Pattern Anal. Machine Intell.* PAMI-2: 543–49

Thompson, W. B., Barnard, S. T. 1981. Lower-level estimation and interpretation of visual motion. *IEEE Comput.* 14: 20–28

Thompson, W. B., Mutch, K. M., Berzins, V. 1982. Edge detection in optical flow fields. *Proc. 2nd Natl. Conf. Artif. Intell., Aug 26–29*

Thompson, W. B., Mutch, K. M., Berzins, V. 1985. Dynamic occlusion analysis in optical flow fields. *IEEE Trans. Pattern Anal. Machine Intell.* PAMI-7: 374–83

Todd, J. T. 1982. Visual information about rigid and nonrigid motion: A geometric analysis. *J. Exp. Psychol.* 8: 238–52

Todd, J. T. 1984. The perception of three-dimensional structure from rigid and nonrigid motion. *Percept. Psychophys.* 36: 97–103

Torre, V., Poggio, T. 1978. A synaptic mechanism possibly underlying directional selectivity to motion. *Proc. R. Soc. London Ser. B* 202: 409–16

Traub, R. D., Knowles, W. D., Miles, R., Wong, R. K. S. 1984. Synchronized afterdischarges in the hippocampus: Simulation studies of the cellular mechanism. *Neuroscience* 12: 1191–1200

Tsai, R. Y., Huang, T. S. 1981. Uniqueness and estimation of three-dimensional motion parameters of rigid objects with curved surfaces. *Univ. Illinois Urbana-Champaign, Coord. Sci. Lab. Rep. R-921*

Ullman, S. 1979. *The Interpretation of Visual Motion.* Cambridge: MIT Press

Ullman, S. 1981a. Analysis of visual motion by biological and computer systems. *IEEE Comput.* 14: 57–69

Ullman, S. 1981b. The effect of similarity between line segments on the correspondence strength in apparent motion. *Perception* 9: 617–26

Ullman, S. 1983. Computational studies in the interpretation of structure and motion: Summary and extension. In *Human and Machine Vision*, ed. J. Beck, B. Hope, A. Rosenfeld. New York: Academic

Ullman, S. 1984. Maximizing rigidity: The incremental recovery of 3-D structure from rigid and rubbery motion. *Perception* 13: 255–74

Ullman, S. 1985. The optical flow of planar surfaces. *MIT Artif. Intell. Memo 870*

Ullman, S. 1986. Artificial intelligence and the brain: Computational studies of the visual system. *Ann. Rev. Neurosci.* 9: 1–26

van Doorn, A. J., Koenderink, J. J. 1982. Visibility of movement gradients. *Biol. Cybernet.* 44: 167–75

van Doorn, A. J., Koenderink, J. J. 1983. Detectability of velocity gradients in moving random-dot patterns. *Vision Res.* 23: 799–804

Van Essen, D. C., Maunsell, J. H. R. 1983. Hierarchical organization and functional streams in the visual cortex. *Trends Neurosci.* 6: 370–75

Van Santen, J. P. H., Sperling, G. 1984. A temporal covariance model of motion perception. *J. Opt. Soc. Am. A* 1: 451–73

Van Santen, J. P. H., Sperling, G. 1985. Elaborated Reichardt detectors. *J. Opt. Soc. Am. A* 2: 300–20

Varju, D., Reichardt, W. 1967. Übertragungseigenschaften im Auswertesystem für das Bewegungssehen. *Z. Naturforsch.* 22b: 1343–51

Vaughn, J. E., Famiglietti, E. V., Barber, R. P., Saito, K., Roberts, E., Ribak, C. E. 1981. GABAergic amacrine cells in rat retina: Immunocytochemical identification and synaptic connectivity. *J. Comp. Neurol.* 197: 113–27

Wallach, H. 1976. On perceived identity: 1. The direction of motion of straight lines. In *On Perception*, ed. H. Wallach. New York: Quadrangle

Wallach, H., O'Connell, D. N. 1953. The kinetic depth effect. *J. Exp. Psych.* 45: 205–17

Wallach, H., Weisz, A., Adams, P. A. 1956. Circles and derived figures in rotation. *Am. J. Psych.* 69: 48–59

Warren, S., Hamalainen, H. A., Gardner, E. P. 1986. Coding of the spatial period of gratings rolled across the receptive fields of somatosensory cortical neurons in awake monkeys. *J. Neurophysiol.* In press

Watanabe, S.-I., Murakami, M. 1984. Synaptic mechanisms of directional selectivity in ganglion cells of frog retina as revealed by intracellular recordings. *Jpn. J. Physiol.* 34: 497–511

Watson, A. B., Ahumada, A. J. 1985. Model

of human visual-motion sensing. *J. Opt. Soc. Am. A* 2: 322–41

Waxman, A. M. 1986. Image flow theory: A framework for 3-D inference from time-varying imagery. In *Advances in Computer Vision*, ed. C. Brown. New Jersey: Erlbaum. In press

Waxman, A. M., Wohn, K. 1985. Contour evolution, neighborhood deformation and global image flow: Planar surfaces in motion. *Int. J. Robotics Res.* 4: 95–108

Waxman, A. M., Ullman, S. 1985. Surface structure and three-dimensional motion from image flow kinematics. *J. Robotics Res.* 4: 72–94

Webb, J. A., Aggarwal, J. K. 1981. Visually interpreting the motions of objects in space. *Computer* 14: 40–49

Werblin, F. S. 1970. Response of retinal cells to moving spots: intracellular recording in *Necturus maculosis*. *J. Neurophysiol.* 33: 342–50

White, B. W., Mueser, G. E. 1960. Accuracy in reconstructing the arrangement of elements generating kinetic depth displays. *J. Exp. Psychol.* 60: 1–11

Wohn, K. 1984. *A contour-based approach to image flow*. PhD thesis, Univ. Md., Dept. Comput. Sci.

Wohn, K., Waxman, A. M. 1985. Contour evolution, neighborhood deformation and local image flow: Curved surfaces in motion. *Univ. Md. Cent. Automation Res. Tech. Rep. 134*, July

Ann. Rev. Neurosci. 1987. 10: 535–594

MOLECULAR GENETIC INSIGHTS INTO NEUROLOGIC DISEASES

Xandra O. Breakefield and Franca Cambi

Department of Molecular Neurogenetics, E. K. Shriver Center, Waltham, Massachusetts 02254, and Neuroscience Program (Neurology), Harvard Medical School, Boston, Massachusetts 02115

INTRODUCTION

New techniques in molecular biology can be used to identify gene defects causing neurologic disease in humans. These involve finding the position of defective genes in the genome and elucidating structural differences in them that underlie altered function. Two of the new approaches that have promoted active research in this area are (*a*) carrying out linkage analysis in human pedigrees by using normal variations in DNA sequence to mark alleles at gene loci (Botstein et al 1980), and (*b*) looking for alterations in gene function caused by substantial rearrangements in DNA sequence. The advantage of carrying out analysis with DNA is that inherited gene lesions are present in all cells of the body, so that nervous tissue does not need to be obtained and blood samples can provide sufficient experimental material. Further, the same recombinant technology can be applied to diseases involving very different neurochemical anomalies. For other reviews in this field the reader is referred to Rosenberg 1984, Rosenberg et al 1985, Gusella et al 1984, White et al 1985, and Roses et al 1983.

Linkage analysis is based on the observation that the closer alleles for two different genes are to each other within a chromosome the more likely they are to segregate together during meiosis, or in other words, the less likely it is that a recombinational event will occur between them. By using cloned DNA probes homologous to unique sequence DNA (present only once or a few times per haploid genome), it is possible to identify "gene" loci throughout the genome. Further, alleles for these loci can be dis-

0147–006X/87/0301–0535$02.00

tinguished on the basis of restriction fragment length polymorphisms. These are variations in the size of genomic DNA fragments that hybridize to labeled probes following digestion of DNA with restriction endonucleases, resolution of fragments by electrophoresis in agarose, and transfer to nitrocellulose (genomic blots). The term *polymorphism* is a relative one in genetics meaning that the variation occurs at least once in the same gene from 50 people (100 alleles). The usefulness of this analysis in human disease increases with the availability of sets of probes that are spaced at appropriate intervals along chromosomes, ideally at 10–20 centiMorgan (cM) intervals (equivalent to 10–20% recombination between loci), so that an unknown gene will fall within 5–10 cM of a marker probe (Botstein et al 1980). Linked sets of probes also allow co-inheritance of a disease locus to be assessed simultaneously with several marker loci by using multipoint analysis, which expands the chromosomal region examined (Lathrop et al 1984). These marker probes are most informative when they recognize highly polymorphic regions in the genome. These regions are characterized by the presence of multiple polymorphic restriction endonuclease sites within them (e.g. the beta-globin gene; Orkin & Kazazian 1984) or through short sequences repeated in tandem a variable number of times among alleles (Jeffreys et al 1985).

Elucidation of rearrangements in DNA sequence associated with altered function has taken several paths. Marker DNA probes have been used to demonstrate the absence (deletion) of specific DNA sequences by hybridizing them to genomic blots or directly to metaphase chromosomes. In a similar manner one can demonstrate the transposition of DNA sequences to other chromosomes through breakage and reunion of chromosomes (translocation) or by replication as extrachromosomal elements followed by their reintegration into random sites in the genome (as in the formation of double minute chromosomes and homogeneously stained regions; Stark & Wahl 1984). The latter process is also involved in "gene" amplification, or an increase in the copy number of specific DNA sequences. These amplified sequences can be identified by a faster rate of rehybridization as compared to less abundant sequences (Kohl et al 1983, Trent et al 1986).

The neurologic diseases discussed here were chosen because they illustrate different and successful ways molecular genetics can be used to understand them. They are discussed, roughly in order, from diseases in which the defective gene is unknown to ones in which it has been identified and sequenced. Huntington disease and Duchenne muscular dystrophy are diseases in which the genes are unknown, and linkage mapping, as well as the use of deletions and translocations, has been used to find their chromosomal location. (These diseases are discussed only briefly in the introduction, as they will be included in future articles in the *Annual Review* series.) Familial dysautonomia serves as an example of an inherited disease for which protein candidates exist and cloned DNA probes for them have been used to exclude the structural genes encoding them as the primary genetic lesions. Studies of inherited retinoblastoma demonstrate how different somatic cell events, including mitotic recombination, can lead to loss of normal alleles for a gene and serve as the initiating event in tumorigenesis. In sporadic neural tumors, such as neuroblastoma, genetic studies show that the malignancy of a tumor can be enhanced by amplification of specific onc gene sequences. In inherited amyloidoses, such as familial amyloid polyneuropathy, biochemical analysis of the accumulated protein has revealed discrete amino acid substitutions that correspond to altered restriction sites in the DNA encoding them and can be used for direct DNA diagnosis. Genetic analysis has also revealed that in an animal model of acquired amyloidosis, scrapie, the amyloid protein that accumulates in the brain is encoded in a normal cellular message found in many tissues and thus may represent a reaction to, rather than a cause of, the disease state. A number of components of peripheral and central myelin have been cloned and used in understanding the coordinated regulation of their gene expression as well as differences in the structure of myelin proteins. The cloning of proteolipid protein has provided a candidate gene for an inherited human disease, Pelizaeus-Merzbacher, affecting CNS myelin, as well as mouse models of this disease. Moreover, the molecular defect in the myelin basic protein gene in a mouse mutant, *shiverer*, has been elucidated. And lastly, for Lesch-Nyhan syndrome the responsible gene has been cloned and a number of DNA rearrangements associated with the disease state have been elucidated.

Molecular genetic studies of Huntington disease (HD) have been at the forefront in demonstrating the power of linkage analysis using DNA polymorphisms. The advantages of this disease for linkage studies are that it follows a clear-cut autosomal dominant pattern of inheritance, with essentially complete gene penetrance (i.e. everyone who inherits the defective gene will have the disease if they live long enough). In addition, there are essentially no phenocopies of the disease state, and very large affected pedigrees are available. Gusella and co-workers (1983, 1984) demonstrated that a random genomic probe, G8, hybridized to DNA sequences that are linked to the Huntington disease locus at a distance of about 4 cM. Alleles at the G8 locus can be distinguished by several restriction endonuclease site polymorphisms (Figure 1). Within a family, a particular allele co-inherits with the HD gene about 96% of the time; this is not to say that the same G8 allele is always linked to the Huntington disease gene (Figure 2). The G8 locus was mapped to the short arm (p) of chromosome 4 by somatic cell genetic techniques (Naylor et al 1984). The chromosomal

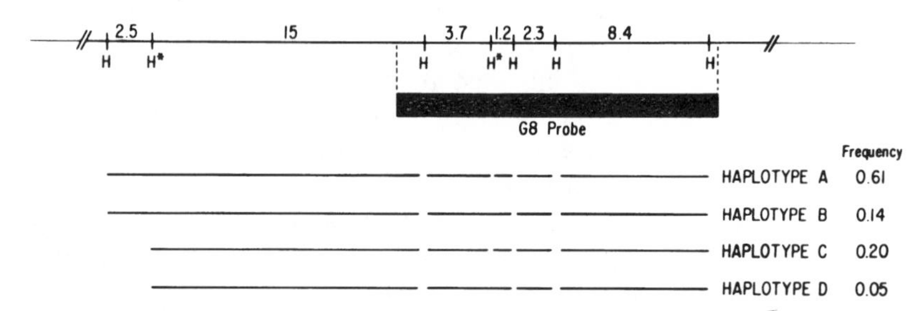

Figure 1 Haplotypes for the G8 locus. The G8 probe corresponds to 17.6 kilobases (kb) of human DNA on chromosome 4 that is linked to the Huntington's disease locus. On Southern blots of genomic DNA cut with Hind III, this probe hybridizes to the fragments of the size shown in the *upper map*. The 2.3 and 8.4 kb Hind III fragments are invariant. The other fragments generated depend on the presence or absence of the two polymorphic Hind III sites marked by *asterisks*. The population frequencies for each of the four allelic haplotypes A, B, C, and D are shown. (Reproduced with permission from Gusella et al 1983.)

location was further defined by analyzing DNA from individuals with the Wolf-Hirschhorn syndrome, in which deletions in the terminal region of chromosome 4p in the heterozygous condition lead to mental retardation and dysmorphology (Figure 3; Gusella et al 1985). Alleles at the G8 locus were found to be lost in all these patients, including one missing just the terminal band of this chromosome, thus indicating that the G8 locus is in this terminal band. Linkage analysis using this probe has clarified two genetic issues in Huntington disease. First, when two affected individuals have produced affected offspring, no clinical difference appears between individuals with one or two copies of the Huntington disease gene (Wexler et al 1985). Second, in at least three families from America and Europe and one from Venezuela, the gene causing Huntington disease appears to be the same, as it is linked to G8, although whether the same mutation is involved is not yet clear (J. F. Gusella, personal communication).

Similar studies have been carried out for Duchenne muscular dystrophy. Here it was known by the mode of inheritance that the defective gene was on the X chromosome (which represents about 8% of the coding capacity of the human genome). By carrying out linkage analysis with probes for the X chromosome, the gene has been localized to band 21 of the p arm (Murray et al 1982). Further, it has been shown that another form of muscular dystrophy, Becker, long thought to be a distinct clinical entity, is caused by a defective gene in the same location (possibly a different mutation in the same gene; Brown et al 1985). The location of the Duchenne gene has been demarcated further to a 2000–5000 kb sequence by

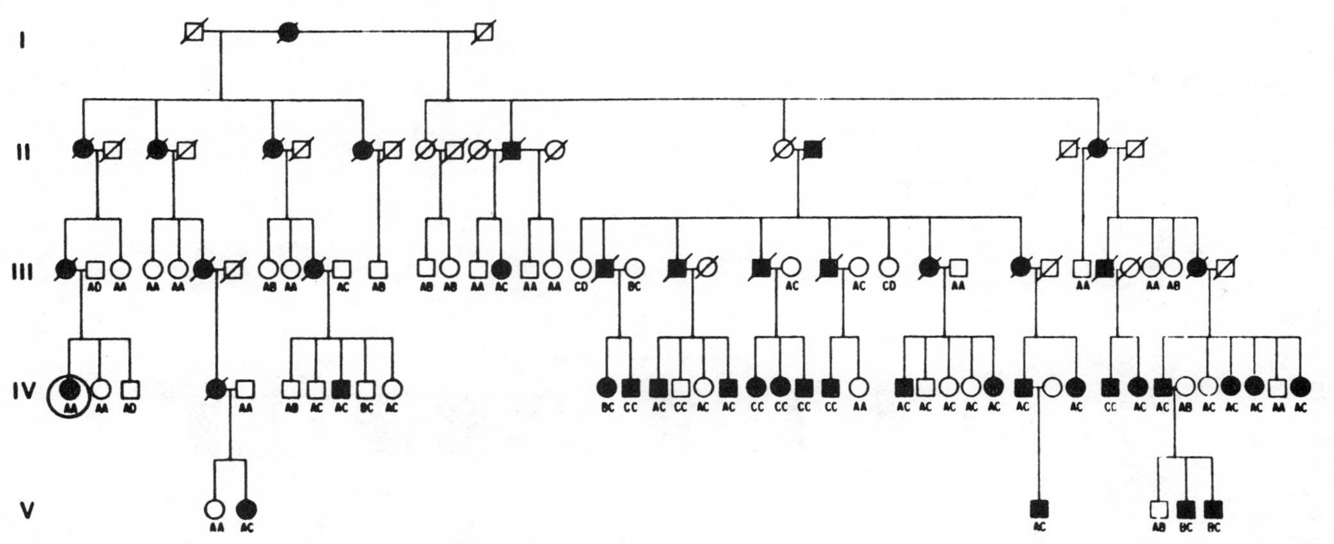

Figure 2 Portion of the Venezuela pedigree showing the inheritance of Huntington's disease and alleles at the G8 locus. The HD gene co-segregates with the C haplotype at the G8 locus, although the C haplotype also occurs in association with a normal allele at the HD locus. A single recombinational event was detected (*circle*). *Solid symbols* represent affected individuals. *Slashed symbols* indicate deceased individuals. (Reproduced with permission from Gusella et al 1984.)

determining which probes for this region of the X chromosome are deleted in some Duchenne muscular dystrophy patients (Monaco et al 1985; Figure 4). Another approach to characterizing this locus has been to analyze DNA from rare females who express the disease state due to a break in one of their X chromosomes within the normal counterpart of the Duchenne gene and in an autosome, followed by a reciprocal translocation between these chromosomal fragments and subsequent non-

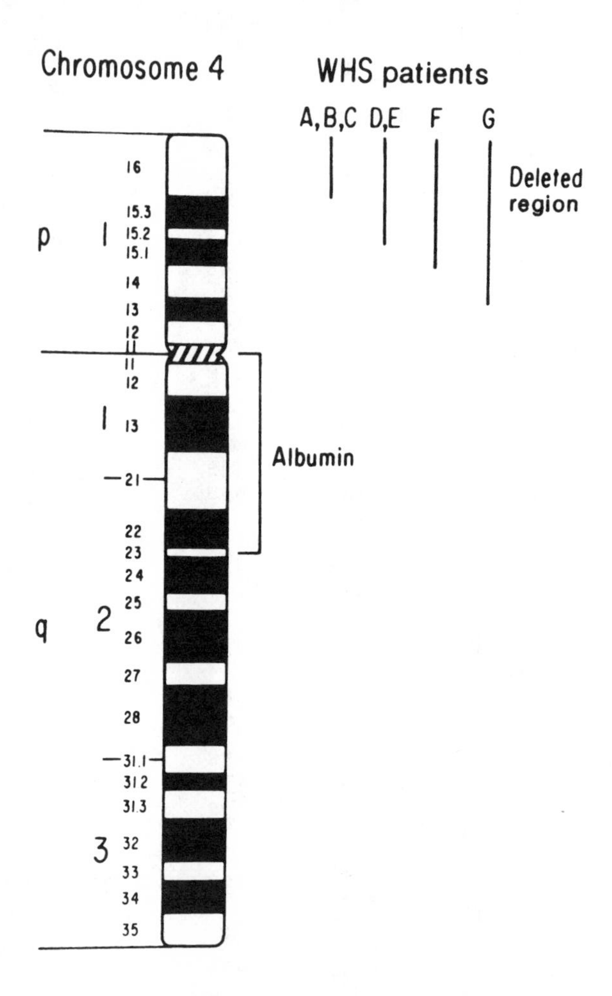

Figure 3 Extent of chromosome 4 deletion in Wolf-Hirschhorn syndrome (WHS) patients. Regions deleted in seven patients (A–G) are shown on *right*. The standard trypsin-Giemsa banding pattern of chromosome 4 is shown on the *left*. (Reproduced with permission from Gusella et al 1985).

random inactivation of the remaining normal X chromosome. Thus the break itself causes the lesion in one copy of the Duchenne gene, and the normal copy is not expressed. Presumably if the translocated X chromosome sequences were inactivated, adjacent autosomal material would also be inactivated and lead to cell death. Using DNA from one female in whom Duchenne gene sequences were translocated to a region of chromosome 21 containing repeated ribosomal genes, Worton and co-workers (Ray et al 1985) were able to use cloned ribosomal genes to identify and clone the translocation breakpoint that presumably is in the Duchenne gene (Figure 5).

The studies described here demonstrate the flexibility and wide applications of molecular genetic techniques to human disease. Further, they illustrate how an understanding of mutations and genetic events can change our perspective of disease processes and provide new means for studying them. Still, our knowledge of the molecular pathology in these neurologic diseases at the cellular and tissue levels is frequently poor, and a clearer understanding will require a close collaboration among molecular geneticists, neuroscientists, and neurologists.

NERVE GROWTH FACTORS: FAMILIAL DYSAUTONOMIA

Overview

Familial dysautonomia (the Riley-Day syndrome) is a rare, inherited neurologic disease associated with severe depletion of neurons in sympathetic and sensory ganglia. This autosomal recessive condition appears to result from loss of activity of a neurotrophic substance, such as nerve growth factor (NGF), that is needed for development and survival of these neuronal populations. A possible autoimmune model for this disease has been created by "depriving developing rodent fetuses of NGF via transplacentally transferred maternal antibodies" (Johnson et al 1986) (Figure 6). Analysis of the structure of NGF and its receptor in this disease has been hindered, as a source of these proteins from normal and affected individuals has yet to be identified. The availability of cloned DNA probes for the genes encoding these proteins, however, combined with linkage analysis has provided a means to assess their role in this disease. These studies serve as a model for genetic investigations into other inherited neurologic diseases for which candidate proteins have been identified and cloned probes exist.

Neurologic Symptoms

All classic cases of familial dysautonomia have been reported in the Ashkenazic Jewish population. Criteria for diagnosis used in the neonatal period include lack of an axon flare response to histamine, absence of fungiform papillae on the tongue, contracture of the pupil in response to methacholine, diminished deep tendon reflexes, and a lack of overflow tears (Axelrod & Pearson 1984). Other early features include poor feeding and failure to thrive. The many other symptoms in this disease include vomiting crises, cardiovascular instability, insensitivity to pain and temperature, skin blotching, excessive sweating, abnormal responses to levels of oxygen and carbon dioxide in the blood (resulting in apnea and drowning), scoliosis, and progressive loss of renal function. Symptoms tend to worsen with age, especially perception of pain and temperature (Pearson et al 1978), and cardiac and renal functions. In general, verbal and motor milestones are delayed, but intelligence and personality characteristics are within the normal range, with some tendency toward emotional instability and periodic lapses in judgement (Clayson et al 1980, Welton et al 1979). The mean survival age of these patients is 20–30 years; death usually results from aspiration pneumonia secondary to vomiting crises, cardiac arrest, or renal failure. Since few affected individuals have survived past 40 years of age, it has not been possible to evaluate the process of aging in them.

Familial dysautonomia falls into a group of diseases called *congenital sensory neuropathies* that manifest diminished pain sensitivity with or without autonomic dysfunction (Axelrod & Pearson 1984). Five additional disease states can be distinguished on the basis of the presence or absence of visceral pain, sensation of touch, overflow tears, and mental retardation, as well as the extent of sensory loss. Common features include lack of the histamine flare response, decreased deep tendon and corneal reflexes, and no fungiform papillae. This set of diseases may result from different defects affecting the action of the same neuronal growth factor (or developmental determinant) or lesions in different developmental factors that have overlapping neuronal specificities, including action on sensory neurons.

Figure 4 Region of DNA on the X chromosome containing the Duchenne muscular dystrophy locus as determined by deletion mutations. (*a*) Schematic map of 41 kb of genomic DNA from the Xp21 region, showing the regions that represent unique sequences as *open blocks*, numerically labeled. *Horizontal arrows* show five overlapping bacteriophage clones isolated from human DNA recombinant libraries.

(*b*) Southern blot analysis of Pst I-digested genomic DNA from seven males with Duchenne muscular dystrophy. The probes used are three of the unique sequences shown above pERT87-18, 8 and 1, (indicated on *left*), as well as another unique sequence clone pERT55-2 from a nondeleted region of Xp21. In five individuals (lanes 2–6) there is a deletion of the DNA that covers the 38 kb in the region hybridizing to the pERT87 probes. In these same individuals sequences corresponding to pERT55 are present. (Reproduced with permission from Monaco et al 1985.)

Neuropathology

A number of marked changes in the peripheral nervous system have been noted in familial dysautonomia; the central nervous system has not yet been examined extensively. So far essentially all changes have been documented in the youngest patients examined, supporting the concept of a defect in neuronal development. Dorsal root ganglia show a marked

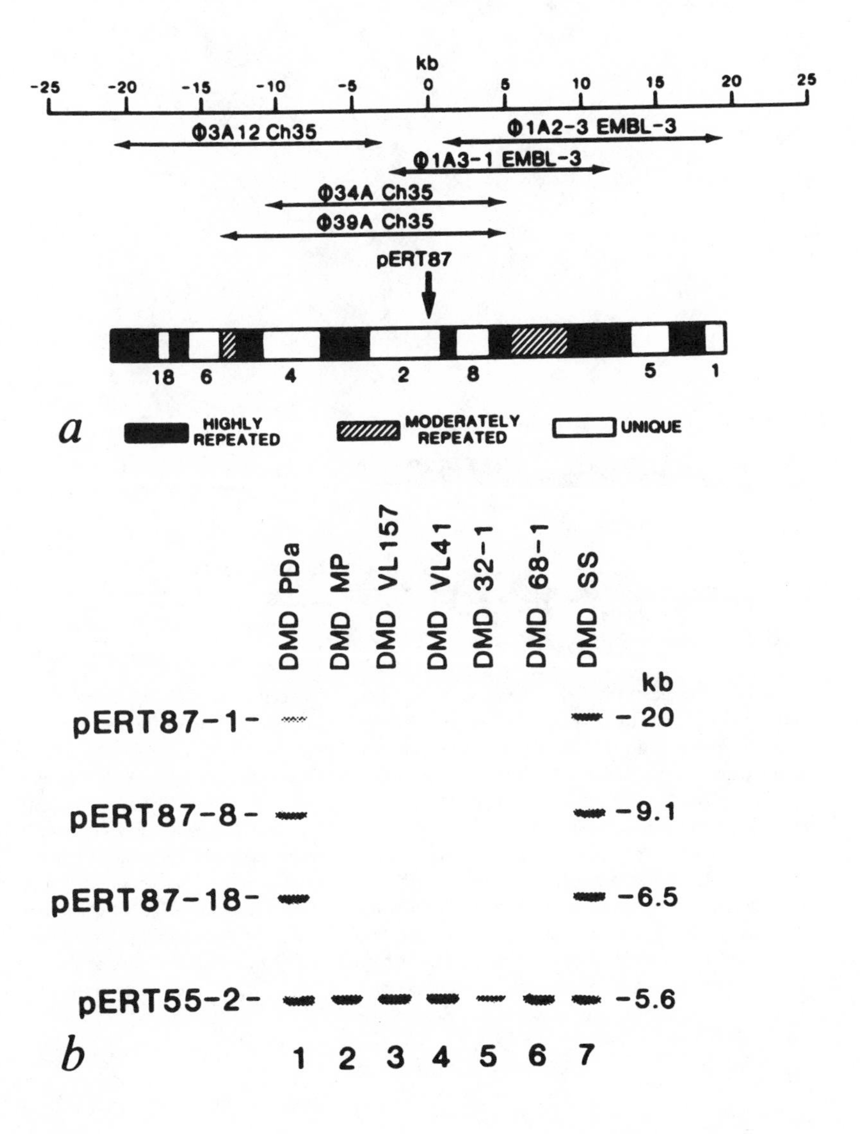

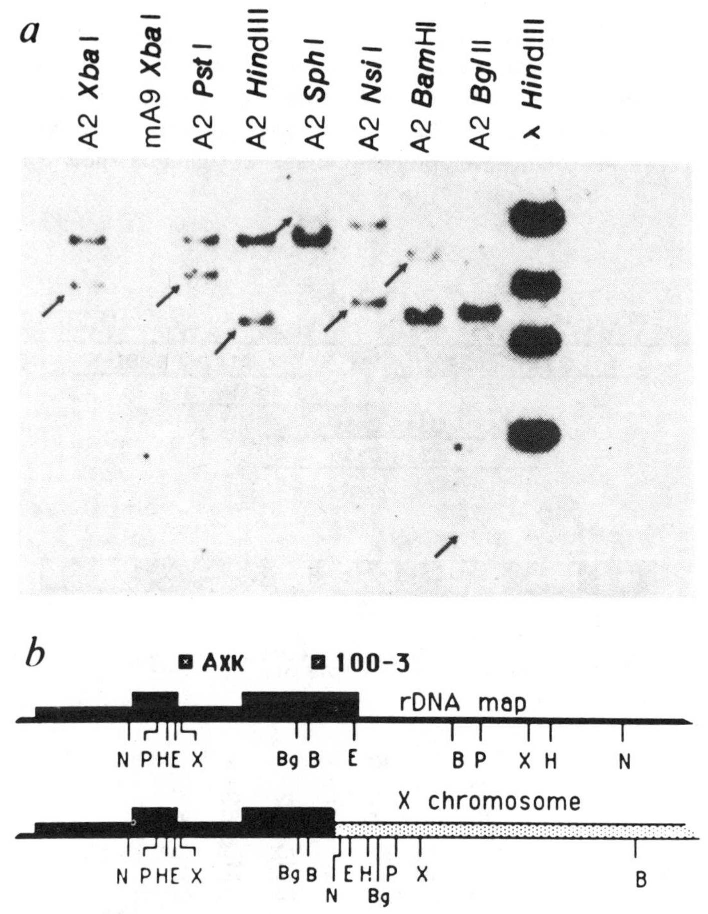

Figure 5 Identification of the chromosomal junction between X and 21 in a female with Duchenne muscular dystrophy who carries a translocation of a portion of the short arm of the X chromosome to the short arm of chromosome 21. (*a*) Southern blot of DNA from a human-mouse somatic cell hybrid (A2) containing the derivative chromosome X; 21 cut with different restriction enzymes, and from the parental mouse DNA (mA9) cut with XbaI. Human-specific ribosomal-DNA (rDNA) sequences, AXK and 100-3, which hybridize to sequences on chromosome 21, were used as probes. The *arrows* indicate new unique human-specific bands derived from the translocation break point. The *other bands* represent fragments derived from the normal ribosomal DNA repeat units. Lambda DNA cut with Hind III is used as molecular weight markers. (*b*) Restriction maps of the normal rDNA repeat unit on chromosome 21 (*top*) and the translocation junction (*bottom*). Ribosomal sequences from 21, *solid box*, X chromosomal sequences, *stippled box*. (Reproduced with permission from Ray et al 1985.)

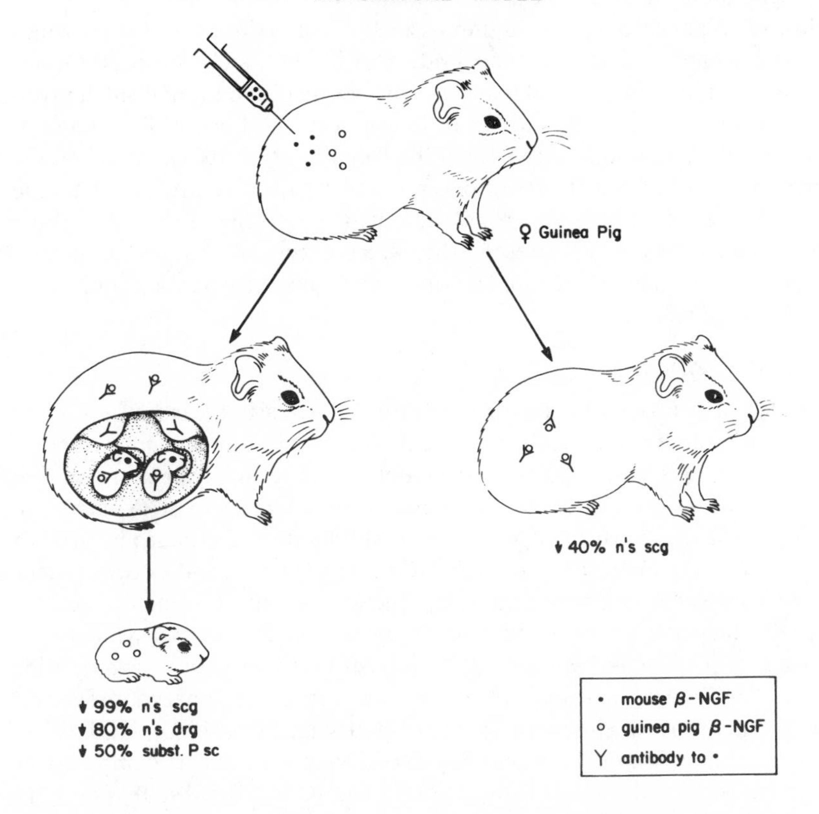

Figure 6 Autoimmune model for familial dysautonomia. Female guines pigs were immunized with purified mouse beta-NGF. This results (*right side*) in a 40% depletion of neurons (n's) in the superior cervical ganglion (scg) after six months. Passage of the anti-NGF antibodies through the placenta to the offspring interfered with the normal development of sensory and autonomic ganglia (*left side*). Pups showed severe neuronal depletion in scg and dorsal root ganglia (drg). In addition, levels of substance P (subst. P) were reduced in the spinal cord (sc). (Adapted from Gorin & Johnson 1980 with permission.)

reduction in size, and neurons are markedly depleted (10% of controls), with a slow continuing loss with age (Pearson et al 1978). Sympathetic ganglia are also smaller and have about a 90% loss of neurons as compared to age-matched controls (Pearson & Pytel 1978b). Of the two parasympathetic ganglia examined, the ciliary and sphenopalatine, the former has only a 20% diminution in neurons whereas the latter is depleted by greater than 95% (Pearson & Pytel 1978a). Other neuropathological changes observed in these patients are reduced numbers of sympathetic preganglionic neurons in the intermediolateral columns of the spinal cord; loss of spinal cord dorsal column tissue, including dorsal root entry zones and Lissauer's tracts; poor innervation of blood vessels; decreased numbers of lingual submucosal neurons; and severe diminution of small myelinated and unmyelinated axons in the sensory sural nerve (Pearson et al 1979, 1982, Pearson & Pytel 1978b, Aguayo et al 1971). Most of the symptoms observed in dysautonomia can be directly attributed to the absence of peripheral neurons that mediate normal functions. Some neuronal losses may represent the direct effects of the genetic defect, whereas others may be caused secondarily by loss of synaptic connections to affected neurons.

Neurochemistry

A shared neurotrophic factor is considered a feasible candidate for the common cause of death of the different types of neurons seen in this disease (Purves & Lichtman 1985). Unfortunately, although many lines of evidence indicate that a large number of neurotrophic factors exist (Thoenen & Edgar 1985), our knowledge of them is still limited. The neuronal growth factor about which most is known, NGF, is important for the development and survival of sensory and sympathetic neurons (Greene & Shooter 1980, Harper & Thoenen 1980) and may exert effects on parasympathetic neurons (Collins & Dawson 1983). Recent evidence indicates that NGF also modulates neuronal properties in the central nervous system, especially those of cholinergic neurons (Korsching et al 1985). NGF exists as a 7S complex in the mouse submaxillary gland, and is composed of alpha, beta, and gamma subunits (Greene & Shooter 1980). The beta subunit contains all the neurotrophic activity of this complex. Whether or not other tissues that produce beta-NGF also synthesize these other subunits is unknown (Pantazis et al 1977).

Since a defect in NGF action could explain most of the neuronal deficits in familial dysautonomia, it is important to assess the role NGF and its receptor in this disease state. Although a complete loss of NGF action would seem to be incompatible with life, it is possible that defective molecules retain some residual activity. Evaluation of the integrity of NGF and the NGF receptor in dysautonomia has been complicated by the lack of a reliable source of these proteins from normal human tissues. The presence of NGF-like molecules has been described in human serum (Siggers et al 1976), placenta (Goldstein et al 1978), and cultured skin fibroblasts (Schwartz & Breakefield 1980). However, the assays used, especially the single-site radioimmunoassay, can be unreliable (Suda et al 1978) and the presence of an authentic human NGF protein has yet to be confirmed. It is difficult then to evaluate reported differences in NGF-like molecules in serum (Siggers et al 1976) and fibroblasts (Schwartz & Breakefield 1980) from patients as compared to controls. In fact, the presence of NGF in serum at all is disputed (Skaper & Varon 1982). Although the human NGF receptor has been well-characterized in melanoma cells (Grob et al 1985), some lines of which have very high levels of expression of this protein, no source of the receptor from normal human tissues has been identified.

Changes in other biochemical properties in dysautonomia patients appear to represent the results rather than the cause of neuronal loss. These changes include a decrease in substance P in the substantia gelatinosa of the spinal cord, which receives input from dorsal root ganglia neurons, with no change of this neuropeptide in other areas of the CNS (Pearson et al 1982). This finding is consistent with the use of substance P in transmission of pain by sensory neurons. Reduced levels of dopamine beta-hydroxylase and norepinephrine in the sera of these patients can be accounted for by the paucity of sympathetic innervation to blood vessels (Zeigler et al 1976). No loss of tyrosine hydroxylase immunoreactivity has been observed in surviving neurons of the sympathetic ganglia or in other areas of the CNS, thus indicating that this enzyme is not directly affected by the lesion (Pearson et al 1979).

Neuroscience

Nerve growth factor exerts a large number of effects on responsive neurons, including survival at critical stages in development and throughout life; extension and regeneration of neurites; regulation of a number of metabolic properties, such as tyrosine hydroxylase activity, RNA transcription, electrical excitability, and expression of cell surface glycoproteins; and chemotaxis of neurite outgrowth (Purves & Lichtman 1985, Greene 1984). These processes are mediated by the interaction of NGF with membrane receptors and can be dependent on or independent of new gene transcription. NGF receptors exist in both high and low affinity forms and their relationship to each other is not clear (Sutter et al 1984). It is generally believed that the high affinity receptors mediate most cellular responses to NGF (Bernd & Greene 1984). Following interaction with its receptor, NGF is taken up by receptor-mediated endocytosis. The NGF itself appears to be degraded and the receptor recycled during subsequent fusion with lysosomal vesicles.

A role for defective NGF action in dysautonomia has been supported to some extent by the large number of similarities in neuronal pathology between rodents exposed to NGF antibodies and patients. One of the most informative model for this comparison was developed by Gorin & Johnson (1980; Figure 6). Female rats, rabbits, or guinea pigs are immunized with purified mouse NGF; this in many cases leads to production of high titer antibodies that crossreact with and compromise the function of endogenous NGF. These antibodies can disrupt NGF action, depending on the extent to which they infiltrate peripheral areas where NGF is released and block NGF binding to its receptor (presumably they do not enter the CNS). Immunized adult animals show a 30–40% reduction in neuronal number in sympathetic ganglia over a six month period, thus indicating that NGF continues to be needed by some neurons throughout life (Gorin & Johnson 1980). The normal action of NGF can also be assessed by examining neuronal development in fetuses carried by immunized mothers. Here the ability of these antibodies to bind to endogenous NGF also depends on the period of time in gestation that they can cross the placenta, but here access to the CNS may be possible. In guinea pigs such offspring showed a dramatic reduction in neuronal numbers (>90%) in sympathetic and dorsal root ganglia, but not in nodose or sphenopalatine ganglia (Gorin & Johnson 1979, Pearson et al 1983). Behavioral analysis of such offspring in rats showed severe deficits in response to stress (as measured by stomach ulceration and corticosteroid levels), subnormal vocalization to footshocks (presumably an indication of loss of sensitivity to pain), poor control of body temperature, and decreased tactile discrimination; while a number of motor skills as well as taste appeared to be normal (Bell et al 1982). These rats also demonstrated depletion of substance P in the spinal cord, and cataracts, failure to thrive, and aspiration pneumonia (Bell et al 1982, Pearson et al 1982). The most notable difference between these NGF-compromised rats and dysautonomia patients was the dramatic loss of neurons in nodose and sphenopalatine ganglia seen in the latter, but not the former. These differences may be explained in several ways; by differences in neuronal development in rodents and humans; the inability of the NGF antibodies to cross the rodent placenta very early in development and/or to access certain regions of the embryos; and the possible role of another neuronal growth factor in the human disease. A defect in NGF action in dysautonomia could result from altered synthesis and expression of NGF or its receptor, or in problems in relaying cellular messages mediated by their interaction.

Genetics

Molecular genetic techniques have provided insight into both the structure of NGF and its receptor, as well as the possible role of genes encoding these proteins in familial dysautonomia. Sequencing of a cDNA for beta-NGF from the male mouse submaxillary gland revealed that the 13.5 kD polypeptide was contained in the carboxy terminal portion of a 27 or 34 kD precursor molecule and could be liberated from it by proteolytic cleavage at dibasic amino acids (Ullrich et al 1983, Scott et al 1983). The precursor molecule appears to have a signal sequence and thus to be processed through the endothelial reticulum and Golgi. No other known neuroactive peptides are contained in the precursor sequences. The mouse cDNA for beta-NGF was used to identify the human gene equivalent, which has greater than 90% homology in its coding sequences and consists of at least two exons, with the entire beta-NGF sequence being at the 3′ end of one exon (Ullrich et al 1983, Figure 7). The human gene is present in a single copy per haploid genome on chromosome 1p (Francke et al 1983).

cDNAs for the mouse alpha- and gamma-NGF subunits reveal that they are highly homologous members of the serine protease family, with the alpha sequences having undergone a discrete change in the active site region that renders the protein enzymatically inactive (Isackson et al 1984, Ullrich et al 1984a). Genes for these subunits are part of a family of over 30 members in the mouse (Mason et al 1983) and an unknown number in humans.

Recently, cloning and the partial characterization of the human gene for the NGF receptor has been achieved by Chao and co-workers (1986)

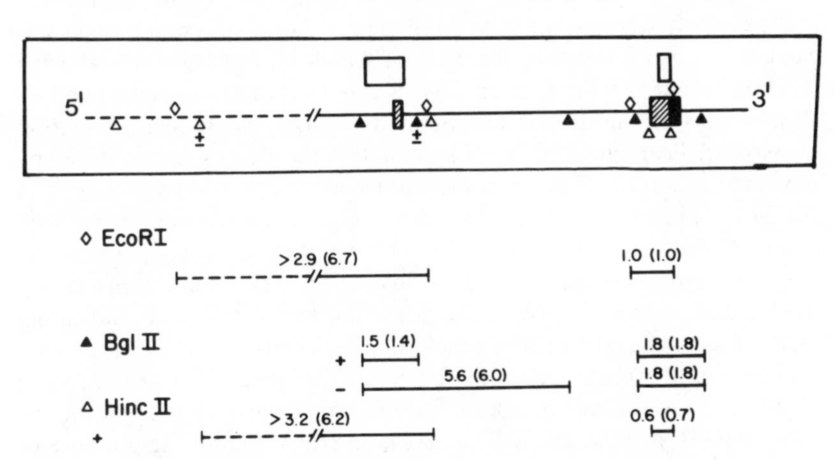

Figure 7 Diagram of the human beta-NGF gene showing polymorphic restriction sites. *Solid box*, coding region for active beta-NGF; *hatched boxes*, coding regions of precursor sequences. The *solid line* indicates noncoding regions sequenced; the *dashed line*, region not sequenced. The positions of restriction endonuclease sites and predicted (or observed) sizes (in kb) of fragments of genomic DNA cut with EcoRI, Bgl II, and Hinc II and hybridized to the cloned probes (*open boxes*) are indicated. An additional polymorphic Taq I site (not shown in the Figure) is present about 6 kb 5′ to the polymorphic Bgl II site (Darby et al 1985). (Reproduced with permission from Breakefield et al 1984.)

by transfer of genomic DNA from human melanoma cells into rat fibroblasts and screening for expression of the receptor on the cell surface. Their studies have revealed that the 70–80 kD receptor polypeptide is encoded in a 3.8 kb mRNA; presumably both high and low affinity receptors are encoded in the same gene. The gene encoding this receptor maps to the long arm (q) of human chromosome 17 (Huebner et al 1986).

The availability of these cloned probes for NGF-related genes has allowed their role in dysautonomia to be evaluated. Although to isolate these genes from patients and compare their sequences with those from controls is theoretically possible, it would be a laborious procedure. Further, since differences exist even among normal alleles in gene sequences, especially in noncoding regions, interpretation of the significance of differences found between genes from patients and controls might be difficult. An easier way to establish the role of a particular gene in a disease process is to determine whether an allele for it co-inherits with the disease state or carrier status in affected families. This method is an abbreviated form of linkage analysis looking for identity or essentially no recombination between the marker sequence and disease gene. The advantages of this approach are that it simultaneously assesses the integrity of the entire gene in coding, noncoding, and flanking sequences, and it can be informative in small pedigrees with multiple affected members.

Linkage analysis for the beta NGF gene in familial dysautonomia was carried out using five families, four of which had at least two affected children (Figure 8). Since the disease is rare (1/20,000 live births) and all classic cases of the disease appear in the Askenazic Jewish population (Axelrod & Pearson 1984), it is likely that all these cases are produced by the same mutation. Thus linkage information can be summed among families. Three restriction endonuclease site polymorphisms have been described for the NGF gene (Figure 7). One of these proved informative in the dysautonomia families under study. Affected children within a family were found to carry different alleles for the beta-NGF gene, indicating independent assortment of this gene and the one causing the disease (Figure 8). These data excluded not only the beta-NGF gene but also a sequence of about 3 cM on either side of it. Similar analysis of the role of the gamma-NGF gene in this disease has been frustrated by the presence of more than one member of this gene family, including pseudogenes, in the human genome (A. Bowcock and L. L. Cavalli-Sforza, personal communication). Preliminary analysis of the NGF receptor gene also appears to exclude a role for it in this disease (X. O. Breakefield, L. Ozelius, M. Bothwell, J. F. Gusella, A. Ross, F. B. Axelrod, R. Kramer, K. Kidd, and M. V. Chao, unpublished data).

Genetic linkage analysis then has proven useful in excluding a role for

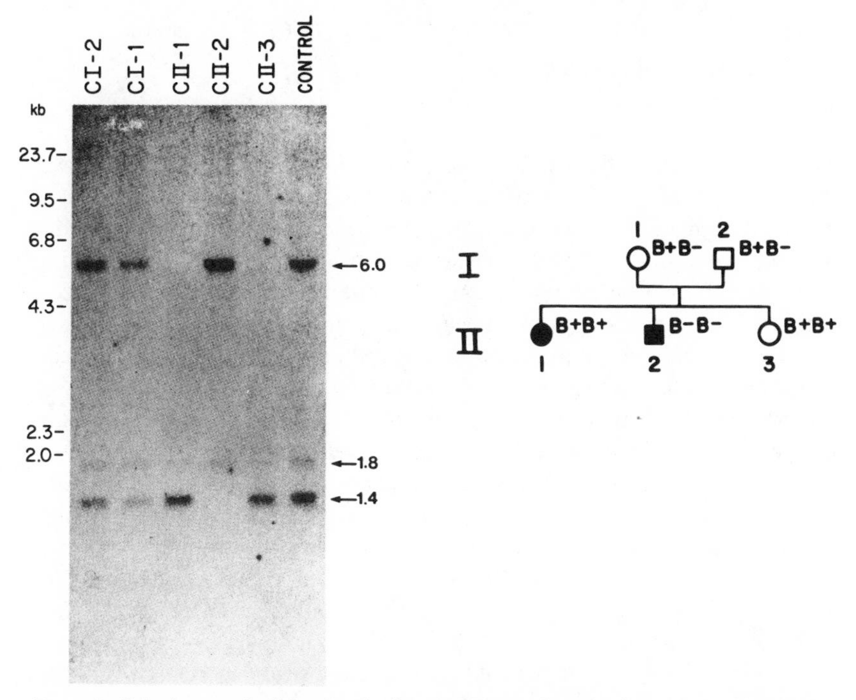

Figure 8 Inheritance of alleles for the beta-NGF gene in a family with familial dysautonomia. *Right*: Pedigree of a dysautonomia family. Affected individuals are shown as *solid symbols*. The alleles at the beta-NGF locus are designated as B+, B−, according to the presence or absence of the variable Bgl II site, as deduced from the fragment lengths observed in genomic blots. *Left*: Genomic blot of Bgl II digested DNA from members of the family shown on the right hybridized to NGF probes (Figure 7). The 1.8 kb band is invariant. The 6.0 and 1.4 kb bands are visualized when the variable Bgl II site is absent or present, respectively. Note, two affected children have inherited different alleles at the beta-NGF locus. (Reproduced with permission from Breakefield et al 1984.)

structural genes for NGF and the NGF receptor in dysautonomia when it has not been possible to do this by biochemical means. These results do not exclude a defect in NGF action, however, as other genes involved in synthesis, expression, or responses to NGF may be defective. Linkage analysis could also be used to search the genome for the defective gene, although this is difficult to do in a recessive disease in which only a small number of families with multiple affected individuals are available. As a rough estimate, a minimum of ten families with two or more affected individuals in each would be needed to carry out linkage over 10 cM intervals in the genome.

NEURAL TUMORS: RETINOBLASTOMA AND NEUROBLASTOMA

Overview

Tumors of neural cells, as well as other cell types, originate and metastasize through changes in the structure and expression of DNA sequences involved in growth regulation. Molecular genetic studies have revealed a number of changes in DNA associated with neural tumors, including rearrangements, loss, amplification, and mutation of genomic sequences. A limited set of genes appears to be involved in these events, including onc genes, the altered counterpart of normal cellular genes (proto-onc genes), which can confer a transformed phenotype onto normal cells. The loss of growth regulation by cells can be accompanied by a series of changes in the genome. In general, benign tumors have fewer alterations in DNA structure than malignant ones. A few genes and chromosomal regions appear to have an important role in the formation of tumors in the nervous system, although none appears to be uniquely involved in neural tumors. For additional reviews see Breakefield & Stern 1986, Schwab 1986, Weinberg 1985.

Here we focus on three genetic mechanisms involved in neural tumor formation. These include a hereditary predisposition to certain types of tumors combined with somatic cell events, leading to homozygosity at the defective gene locus in retinoblastoma and central neurofibromatosis; gene amplification and chromosomal rearrangements involving onc gene sequences in neuroblastoma and glioblastoma; and discrete mutational changes in the *neu* onc gene in experimentally induced neuroglioblastomas in rats.

Neurologic Symptoms and Neuropathology

Tumors of neurons *per se* are largely a phenomenon of childhood, and this probably reflects the increased susceptibility of dividing cells to changes in DNA sequence. Two of these tumors, retinoblastoma and neuroblastoma, can be highly metastatic. Tumors of other cell types in the nervous system, such as astrocytes, Schwann cells, oligodendrocytes, and ependymal cells, can form throughout life, presumably reflecting the fact that these cells retain the ability to divide. These tumors usually remain confined to the brain.

Retinoblastomas form, possibly from a retinal/glial precursor cell, during a critical period in the normal differentiation of the eye. These tumors are seen first as a whitish mass in the vitreous humor and must be removed before they spread down the optic nerve to other parts of the body. The cells from the tumor appear to have antigenic features of both retinal cells and glia (Kyritsis et al 1984). In some cases they assume a rosette formation, which may reflect a type of cellular differentiation (Hittner et al 1979). Most cases of retinoblastoma occur sporadically and are unilateral, but about 15% of the cases are hereditary and usually bilateral (Murphree & Benedict 1984). Interestingly, individuals with the hereditary form of this tumor also have a higher predisposition to have other forms of cancer later in life, most frequently osteosarcomas (Hansen et al 1985).

Another class of hereditary tumors that can be considered together with retinoblastomas, as they share common features, is acoustic neuromas in the central form of neurofibromatosis (Kanter et al 1980, Martuza & Ojemann 1982, Huson & Thrush 1985). These are essentially pure Schwann cell tumors of the vestibular nerve that form in the internal auditory foramen, at the junction between the peripheral and central nervous systems. Again sporadic cases are most common and tend to be unilateral, whereas hereditary cases are rare and frequently bilateral. These tumors are usually benign but may be associated with other malignant tumors of the nervous system, particularly meningiomas.

Neuroblastomas usually derive from neuroblast precursors of sympathetic neurons. The tumors are highly malignant but in a small number of cases, especially when they form before 1 yr of age, may spontaneously regress. Within a tumor, cells may manifest different levels of differentiation, suggesting that the genetic constitution varies from cell to cell (Schwab et al 1984a). Although hereditary cases of neuroblastoma have been described (Roberts & Lee 1975), they are rare. These tumors may share pathogenetic mechanisms with glial tumors, such as glioblastoma multiforme, which occur sporadically and usually in adults. These glioblastomas are also highly malignant and demonstrate a mixed population with respect to glial cell differentiation (Libermann et al 1985).

An animal model of neural tumor formation that has been extensively studied by molecular genetic techniques is mutagen-induced neuro/glioblastomas in fetal rats. When pregnant rats are given injections of *N*-nitrosourea, a mutagen that causes single base pair substitutions in DNA, at 15 days gestation, their offspring have a high frequency of brain tumors (Schechter et al 1984). The tumor cells have features of both neurons and glia, and may derive from a stem cell precursor of both cell types. Why tumors form preferentially in the brain remains unclear, since the mutagen is present throughout the embryo. The selectivity of this agent may reflect a decreased ability of brain cells, as compared to other cells of the body, to inactivate it metabolically, in combination with a very high rate of proliferation of neural cells at this time in development (Rajewsky 1983).

Biochemistry

A number of proto-onc genes and *onc* genes have been implicated in tumors of the nervous system. Since expression of these genes has been studied both at the protein and mRNA levels, they are discussed thus in this section. So far in hereditary neural tumors, such as retinoblastoma, the chromosomal region containing the defective genes (see below) does not contain any known proto-onc genes. However, some retinoblastoma tumors do show increased expression of mRNA for, sometimes accompanied by amplification of, N-*myc* sequences (Lee et al 1984). These sequences were discovered by their homology to a viral onc gene, v-*myc*, and a proto-onc gene, c-*myc* (Schwab et al 1983). Whether N-*myc* acts as an "onc" gene through alterations in its normal sequence or by increased levels of expression of a normal gene product remains unclear (Schwab 1986). The protein encoded in N-*myc* has an apparent molecular weight of 65 kD and, like the protein encoded in c-*myc*, is a nuclear protein that binds to DNA (Ramsay et al 1986). The N-*myc* mRNA is normally found at low levels in many tissues in the developing mouse and at somewhat higher levels in embryonic brain and newborn intestine (Zimmerman et al 1986). C-*myc* sequences have also been shown to be amplified in a number of different types of tumors, including a human glioblastoma (Trent et al 1986).

In neuroblastoma tumors, an increase in N-*myc* expression, usually in combination with gene amplification (see below), has been associated with progression toward malignancy (Schwab 1986). Studies of more than 100 tumors revealed low or no amplification in tumors that were still confined to their site of origin or had not crossed the midline, whereas amplification was a consistent feature of tumors that had metastasized. A rough correlation is found between the number of gene copies and the level of N-*myc* expression (Schwab et al 1984a). Within tumors, highest expression of N-*myc* mRNA also appears to occur in undifferentiated, dividing cells (Schwab et al 1984a). As a corollary to this, retinoic acid inhibits cell division and decreases expression of N-*myc* in cultured neuroblastoma cells (Thiele et al 1985). Of all the "onc" genes characterized so far, N-*myc* seems to be the most neural specific and may have a pivotal role in neuronal differentiation.

Some onc genes or their proto-onc gene counterparts code for growth factors, their receptors, or related proteins. Several of these genes have been implicated in tumors of the nervous system. The proto-onc gene, c-*erb*B, encodes the epidermal growth factor (EGF) receptor (Ullrich et al 1984b). A number of different types of human brain tumors contain high levels of an immunoreactive EGF receptor protein with tyrosine kinase activity (Libermann et al 1984). This activity is especially high in astrocytomas and glioblastoma multiforme, with lower levels in some neuroblastomas and meningiomas. High levels of activity correlate with increased levels of mRNA for and amplification of the *erb*B gene, as well as with the presence of a homologous mRNA of smaller size that may represent the product of an altered *erb*B gene (Libermann et al 1985). Overexpression of a platelet-derived growth factor (PDGF)-like protein and a glial growth factor (GGF)-like protein have also been reported in human glioblastoma cell lines (Pantazis et al 1985) and acoustic neuromas (Brockes et al 1986). These findings suggest the possibility that alterations in the expression or structure of growth factors or their receptors may promote and/or sustain proliferation of certain neural cell types.

Other studies point to structurally altered onc gene proteins in neural tumors. The *ras* gene family codes for GTP-binding proteins that may link extracellular stimuli to second messenger systems, similar to the G proteins that link hormone receptors to adenylate cyclase (Hurley et al 1984, Newbold 1984). *Ras* genes have been found to be activated in a number of different tumor types, including neuroblastomas. Activation can be affected by single base pair substitutions in these gene products (Taparowsky et al 1983), some of which inhibit the GTPase activity that normally serves to regulate cellular responses to hormones (Weinberg 1985). Another onc gene that appears to undergo structural alterations associated with tumor formation is *neu*, which is the transforming agent in mutagen-induced neuro/glioblastomas in rats (Hung et al 1986). This gene has homology to but is distinct from *erb*B, and is thought to encode a receptor for an as yet unidentified growth factor (Schechter et al 1984). The *neu* gene is able to transform mouse 3T3 cells, which then display a new 185 kD cell surface antigen that cross-reacts with antisera against the EGF receptor (Schechter et al 1984). The proto-onc gene counterpart of this onc gene encodes a 138 kD protein with all the features of a cell surface receptor and is expressed by a number of normal tissues (Coussens et al 1985).

Genetics

Analysis of changes in the genomic composition of neural tumors has revealed a number of genetic mechanisms involved in tumor formation and progression. Here we focus on genes that predispose individuals to neural tumors and the somatic cell events that unmask them, and on amplification of DNA sequences in cells as a means of increasing the expression of "onc" genes.

Hereditary forms of neural tumors include retinoblastoma, peripheral and central forms of neurofibromastosis (involving Schwann cells and possibly other cell types), and Von Hippel Lindau syndrome (involving endothelial cells). Retinoblastoma may serve as a model for the genetic mechanisms involved in these syndromes. Hereditary cases of this disease are transmitted in an autosomal dominant mode with incomplete gene penetrance (10% of individuals with the defective gene escape the disease state). The defective gene itself is recessive, that is no tumors form when it is paired with a normal allele at the same locus, and the hereditary defect may be loss of this gene (Benedict et al 1983; for review see Phillips & Gallie 1984). This locus, RB1, was located to chromosome 13q band 14 by tight linkage to the esterase D locus and by deletions of DNA in this region associated with retinoblastoma. Tumors form when retinoblasts with one defective copy of the RB1 gene undergo a somatic "mutational" event or "second hit" (Knudson 1978) whereby the remaining normal copy of this gene is inactivated or lost (Cavenee et al 1983). The genetic mechanisms responsible for "loss" of the normal allele were elucidated by using DNA probes homologous to single copy DNA sequences spaced along the q arm of chromosome 13 (Cavenee et al 1983, Dryja et al 1984). "Alleles" in homologous sequences on chromosome 13q were distinguished by restriction fragment length polymorphisms and compared in lymphocytes and tumor tissue (or cell lines derived from tumors) from a patient, as well as in lymphocytes from his/her parents.

Events that can knock out the normal RB1 gene are illustrated in Figure 9. They include non-disjunction at mitosis of the replicated normal chromosome 13, such that one daughter cell becomes triploid for chromosome 13 with 2 copies of the normal gene and one of the mutant gene (the fate of this cell is unknown), and the other is haploid for chromosome 13 with only a copy of the mutant gene (this cell can form a retinoblastoma). This latter cell, in turn, can undergo another non-disjunction event of chromosome 13 such that one daughter contains no chromosome 13 (cell fate unknown, presumably it dies) and the other contains two copies of the mutant gene (this cell can also give rise to a retinoblastoma). Another way the normal gene can be lost is by recombination between homologous, replicated chromosome 13's at mitosis, such that one daughter cell ends up with two normal copies of the RB1 gene and the other with two mutant copies. Alternately, but less frequently, the normal gene can be lost by a mutational event that affects it directly or through an unbalanced translocation between part of the long arm of 13 and another chromosome. These events also occur in the formation of osteosarcomas in these same individuals, if they survive the retinoblastoma (Hansen et al 1985). Studies in retinoblastoma suggest that "mistakes" in mitosis leading to homozygosity or hemizygosity of DNA sequences are not uncommon during cell proliferation. Further, there appears to be a discrete region on chromosome 13q containing a gene(s) that when missing

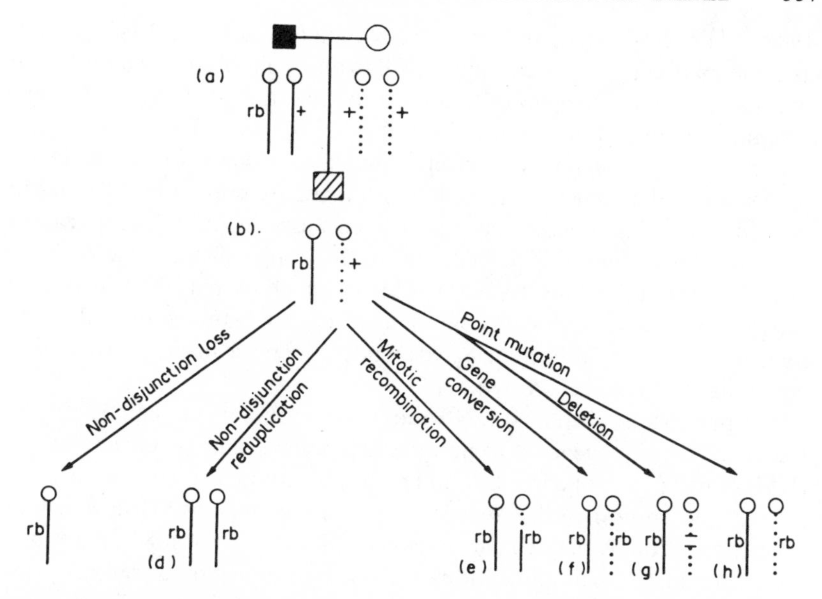

Figure 9 Chromosomal mechanisms that can unmask recessive mutations at the retinoblastoma locus. The recessive defect is designated as rb and the normal allele at this locus as +. (*a*) Family with affected father (*solid box*) and son at risk (*hatched box*). (*b*) Different somatic cell events that can lead to retinoblastoma formation in a retinoblast cell by elimination of the dominant wild-type allele at this locus. (Reproduced with permission from Cavenee et al 1983.)

or abnormal can lead to tumor formation of a few cell types. A cDNA encoded in this chromosomal region and apparently representing the retinoblastoma gene has recently been identified (T. Dryja and R. Weinberg, personal communication). Interestingly, in retinoblasts there appears to be a relatively small time window in development when disruption of this gene(s) can cause tumors; this is approximately when retinal precursor cells stop dividing and begin to differentiate. Recent studies on acoustic neuromas (both spontaneous and hereditary), as well as meningiomas, suggest that homozygosity of sequences on chromosome 22 may be associated with tumorigenesis and/or progressive growth of the tumor cells (Seitzinger et al 1986).

Another genetic mechanism that appears to have an important role in neural tumor formation is amplification of onc genes. This includes amplification of N-*myc*, which has been described in about half of advanced stage neuroblastomas (Schwab 1986), a few retinoblastomas (Lee et al

1984, Squire et al 1985), and one astrocytoma (Garson et al 1985), as well as some small cell lung carcinomas; *erb*B in many glioblastoma multiforme tumors as well as other brain tumors (Libermann et al 1985); and c-*myc* in a glioblastoma (Trent et al 1986).

Gene amplification refers to selective increases in copies of specific DNA sequences so that genes contained within them are present many times in diploid cells. This process may result from mistakes in DNA replication such that DNA synthesis precedes more than once off the same replicon in a cell cycle (Varshavsky 1981). This extra DNA may be lost during subsequent mitoses unless it confers a selective advantage on cells. One such advantage could be a dosage dependent increase in the expression of genes that promote cell growth.

The presence of amplified DNA sequences in tumor cells can usually be visualized by cytogenetic techniques as double minute chromosomes (DMs, small pieces of chromosomal material without centromeres present in multiple copies) or homogeneously staining regions (HSRs) within otherwise normally banded chromosomes (Stark & Wahl 1984). Although the events in DNA amplifications are not well understood, several principles have been noted: Variable lengths of DNA sequence can be involved from 100 to 1000 kb; rearrangements within the amplified sequence may occur; amplification can be progressive with increases ranging from 5–700-fold; a chromosomal translocation is frequently associated with HSR formation; and the same amplified sequences may interchange between DMs and HSRs (Stark & Wahl 1984).

N-*myc* is located normally on chromosome 2 p23–24 and can appear in HSRs on several different chromosomes (Schwab et al 1984b, Shiloh et al 1985; Figure 10). The amplification unit containing N-*myc* in eight independent neuroblastomas was about 290 to 430 kb, a value that may reflect the size of the replicon within which this gene is contained (Schwab 1986). Although amplification of *erb*B in glial tumors is less well characterized, it appears to follow a scheme similar to that of N-*myc* (Libermann et al 1985).

Gene amplification appears to be a random process that leads to increased expression of specific genes that, in turn, can augment the malignant phenotype of tumor cells. Thus amplification of N-*myc* may promote growth of neuronal cells and *erb*B of glia cells. This genetic event is not thought to be an initial event in tumorigenesis, but rather a subsequent and continuing one associated with tumor progression (Schwab 1986). On the other hand, loss of some gene loci, such as the RB1 gene, can serve as the initial step in tumorigenesis but may not directly facilitate cell growth. In both cases, tumor cells have a different genetic constitution as compared to normal cells in the same individual.

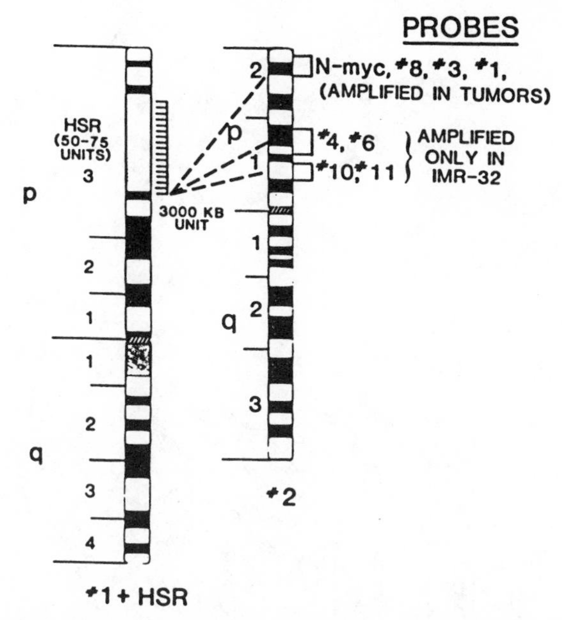

Figure 10 Amplification and translocation of N-myc in neuroblastoma cells. A schematic drawing showing the three different domains, as defined by hybridization to different probes (*brackets* on right of the figure), in chromosome 2p that are transposed and amplified on chromosome 1p in a neuroblastoma cell line, IMR-32. Only some of these regions numbers 4, 6, 10, and 11, have been found to be amplified only in IMR-32 cells, whereas others, N-myc, numbers 8, 3, and 1, are amplified in many primary neuroblastoma tumors. HSR = homogenously staining region. (Reproduced with permission from Shiloh et al 1985.)

AMYLOID: FAMILIAL AMYLOID POLYNEUROPATHY, SLOW VIRAL INFECTIONS, AND ALZHEIMER DISEASE

Overview

Some diseases of the nervous system result from disruption of cellular function by the accumulation of insoluble proteinaceous material in the form of amyloid. Amyloid can form from any protein or polypeptide fragment that can form a substantial amount of beta pleated sheet configuration such that amino acid sequences can stack in antiparallel arrays (Figure 11) (Glenner 1980). Theoretically, amyloid forms when such a protein reaches a critical concentration or has a "seed" structure upon which to align. This, in turn, can result from a number of causes: an increase in the synthesis of this protein, a decrease in its degradation, changes in its susceptibility to proteolytic enzymes, or alterations in its primary structure that serve to increase the extent of beta pleated sheet

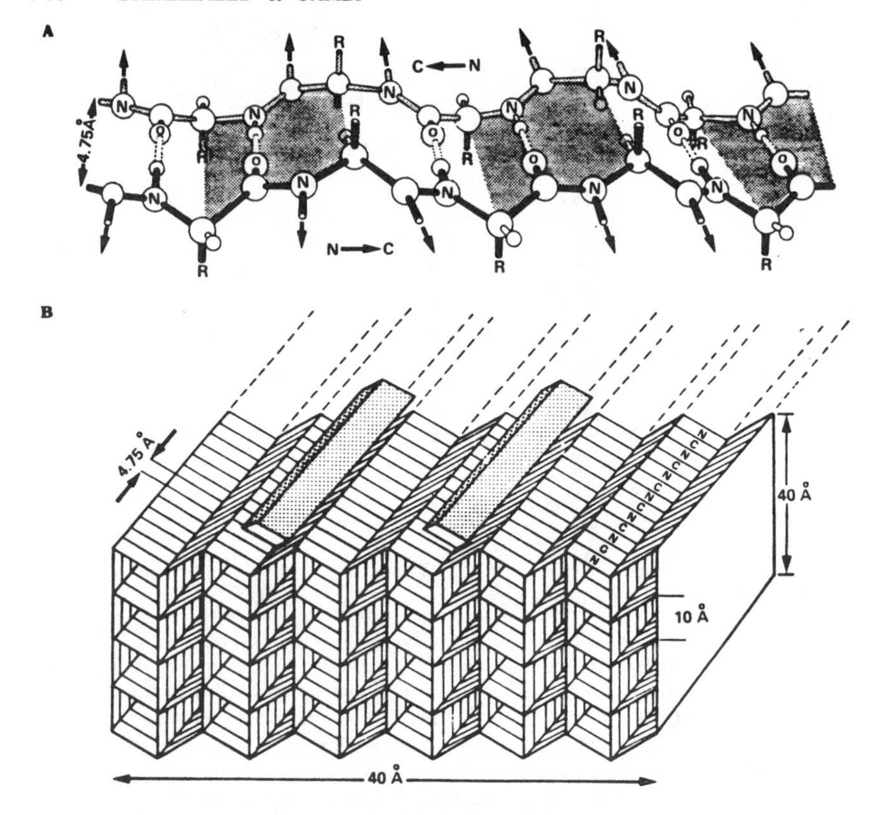

Figure 11 Polypeptides in the beta-pleated sheet configuration in amyloid. (*a*) Formation of beta pleated sheets through hydrogen bonding between two antiparallel chains of amino acids. (*b*) Formation of amyloid by stacking of beta pleated protein sheets. Reproduced with permission from Sack et al (1981).

configuration. Amyloid accumulation can be systemic or tissue-limited, and can occur from a number of causes: acute and chronic infections and other inflammatory conditions, viral infections, tumors, and single gene lesions. In the nervous system, accumulation of amyloid has been found in familial amyloid polyneuropathies, "slow viral" infections, and aging. Here we concentrate on genetic studies of a few of these diseases: Portuguese familial amyloid polyneuropathy, Creutzfeldt-Jakob disease (and the possible animal model, scrapie), and Alzheimer disease.

Neurologic Symptoms

The clinical manifestations of amyloid deposits in nervous tissue correlate with the loss of neuronal function due to pressure atrophy and death. The location of amyloid deposits in the nervous system may reflect sites where the protein is synthesized, degraded, or associated with other molecules or structures. The rate of amyloid accumulation varies depending on its protein composition and source. Presumably, symptoms appear only when the amyloid deposit is large enough to compromise neuronal function, and the loss of some neurons may lead in chain-like fashion to the demise of other neurons.

Several types of hereditary amyloid polyneuropathy have been described that affect the peripheral nervous system (Sack et al 1981). They are distinguished by the regions in which amyloid accumulates, the specific symptoms, and the ethnic group in which they occur. In the severe, Portuguese form there is progressive lower limb neuropathy with onset in the second to fourth decade, followed by a rapid decline in sensory and autonomic functions, leading to death. In a Jewish variant of this form, there is also progressive blindness. Although the primary symptoms are neurologic, amyloid accumulates in organs and compromises functions throughout the body. By comparison, in the milder Swiss/German and Finnish forms, symptoms begin in fifth decade and include sensorimotor impairments in the hands and face. Here progression is slower, and again non-neuronal tissues are also affected.

Amyloid also accumulates in the CNS in a variety of neurodegenerative diseases. Both disorders discussed here, Alzheimer disease and Creutzfeld-Jakob disease, are characterized clinically by dementia, but each one has very distinctive features. Patients affected by Alzheimer disease undergo a premature and steadily progressive deterioration of intellectual functions in a characteristic temporal order. Memory loss is the first and most striking symptom, followed by loss of other cognitive functions and later by personality changes. Motor disabilities, including myoclonus and rigidity, occur only in the very late stages of the disease. Creutzfeld-Jakob disease is characterized by a rapidly progressive dementia associated with motor abnormalities, mainly myoclonous, ataxia, paralysis and rigidity.

Neuropathology

Amyloid can occur in two forms: as fibrils, typically paired and twisted with a 10 nm diameter, or as plaques, which are masses of tightly packed fibrils (Glenner 1980). The morphology of these structures can vary depending on the nature of the protein comprising them and the environmental conditions under which they form. Amyloid deposits are typically extracellular and can be identified by their green birefringence under polarized light following staining with Congo red.

In familial amyloid polyneuropathy of the Portuguese type, amyloid deposits are found along peripheral nerve trunks in dorsal roots of the spinal cord and in autonomic ganglia (Andrade 1952). The deposits are frequently associated with connective tissue elements, particularly the basal lamina of Schwann cells adjacent to unmyelinated axons. Deposits are also found along blood vessel walls and in the kidneys, gastrointestinal tract, heart, and thyroid where they can comprise up to 5% of the total organ weight. It seems clear in this disease that the amyloid itself is the pathogenic agent.

Amyloid formation also occurs in slow viral infections or subacute spongiform viral encephalopathies, such as scrapie in sheep and Creutzfeld-Jakob and kuru in humans. These diseases are characterized by the dense accumulation of proteinaceous rods or filaments in the brain, diffuse degeneration of the cerebral cortex, and glial proliferation. Amyloid plaques similar to those in Alzheimer disease have also been described in some patients with kuru and Creutzfeld-Jakob syndrome (Perry & Perry 1985, Gajdusek 1985).

Several neuropathologic features serve to define Alzheimer disease (Price et al 1986). These include intraneuronal neurofibrillary tangles composed primarily of 10 nm paired helical filaments; neuritic (senile) plaques consisting of enlarged synaptic terminals, abnormal dendritic processes, and variable numbers of astroglia and microglia surrounding an amyloid plaque; and amyloid plaques adjacent to cerebral blood vessels. These features are seen primarily in the basal forebrain, amygdala, hippocampus and neocortex and are similar to those found in brains of elderly individuals. The paired helical filaments resemble morphologically those found in scrapie "infection" (Merz et al 1984).

Neurochemistry and Neuroscience

The proteins that constitute the amyloid in these neurologic diseases have been described to varying extents. In the case of the Portuguese form of familial amyloid polyneuropathy, the deposits are composed of 4–5, 8–9, and 14 kD fragments derived from a circulating protein referred to as *prealbumin* based on its electrophoretic mobility (Costa et al 1978). This protein is composed of four identical 14 kD subunits that have about 50% of their structure in the beta pleated sheet configuration. This protein is synthesized predominantly by liver, but probably also by other cell types. It appears to function as a carrier for retinol-binding protein, and hence vitamin A, as well as thyroxin (Costa et al 1978). Genetic variants of prealbumin from affected individuals in different ethnic groups have been sequenced. The Portuguese, Swedish, and Japanese variants have a single amino acid substitution of methionine for valine at position 30, as compared to the normal protein (Skinner & Cohen 1981, Sasaki et al 1984), whereas the Jewish variant has a single change of glycine for threonine at position 49 (Pras et al 1983). It is not clear how these changes in amino acid sequence lead to a greater tendency toward amyloid formation by the protein. The nature of the protein altered in other ethnic forms of familial amyloid polyneuropathy is not known.

Although relatively little is known about the amyloid protein in Creutzfeld-Jakob and kuru, it has antigenic cross-reactivity with the amyloid protein (or prion) in scrapie (De Armond et al 1985, Botton et al 1985). In scrapie, amyloid rods and filaments are composed of a glycoprotein of apparent molecular weight 27–30 kD, which is presumably derived from a larger precursor protein. This protein appears to be a normal constituent of brain and other tissues (see below). Amyloid formation in scrapie may result from alterations in translation, post-translational processing, intracellular localization, or degradation of this protein related to the infectious process. In these slow viral infections it is not clear whether the amyloid itself is harmful to cells.

In Alzheimer disease a 4 kD protein has been purified from neurofibrillary tangles, neuritic plaques, and cerebrovascular amyloid deposits. Whether this is the primary component of all these structures and whether it is derived from another larger protein by proteolytic cleavage during isolation remains to be established, however. Recent studies indicate that the N-terminal sequences of proteins isolated from these three structures (Figure 12) are identical to each other and to a similar protein from neuritic

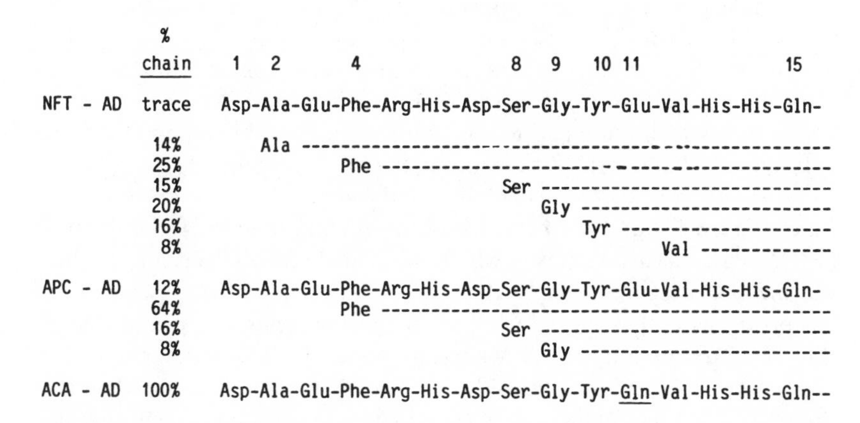

Figure 12 N-terminal amino acid sequences of the 4 kD proteins isolated from NFT (neurofibrillary tangles), APC (amyloid-plaque cores), and ACA (cerebral amyloid) of Alzheimer's disease (AD) patients. The *percentage* refers to the relative amount of protein ending at a particular amino acid. The *underlined* amino acid refers to the only discrepancy in amino acid sequence between proteins isolated from the three sources. (Reproduced with permission from Masters et al 1985a.)

and cerebrovascular amyloid of adult Down patients with dementia (Masters et al 1985b, Wong et al 1985). This sequence is not similar to any protein sequences now known, including those for neurofilaments, microtubule-associated protein, and prions. Further, although some disagreement exists about whether the neurofibrillary tangles in Alzheimer disease are composed of the same protein as the neuritic plaques and cerebrovascular deposits, recent evidence indicates that these neurofibrils do have a beta pleated sheet configuration and hence could form amyloid (Kirschner et al 1986). Other issues that remain to be resolved are whether the source of the amyloid protein is vascular (Wong et al 1985) or neuronal (Masters et al 1985a), and to what extent cytoskeletal elements, e.g. neurofilaments and microtubule-associated proteins, contribute to the neurofibrillary tangles (Perry et al 1985). Since neurofibrillary tangles lead to swelling, and disruption of neurons and amyloid plaques accumulate at nerve endings, as well as other areas, it is tempting to infer that these two structures share a common molecular mechanism.

Genetics

In diseases where the amyloid protein has been identified, molecular genetic techniques can be used in several ways. In the case of Portuguese familial amyloid polyneuropathy, a cDNA for prealbumin has been cloned from a human liver library by using synthetic oligonucleotides corresponding to known amino acid sequences (Sasaki et al 1984). Prealbumin is encoded in a single gene on human chromosome 18q (Wallace et al 1985). The amino acid substitution in this disease predicts a single base pair change of G to A, which, in turn, generates two new restriction endonuclease sites for Nsi I and Bal I (Sasaki et al 1984). By digesting DNA from patients and controls with these restriction endonucleases and examining the size of fragments that hybridize to prealbumin cDNA on genomic blots, Sasaki and co-workers (1984) have demonstrated fragments unique to the disease state (Figure 13) and used this assay in prenatal diagnosis (Sasaki et al 1985). In a case such as this in which the gene directly affects a restriction enzyme site, diagnosis can be done on isolated individuals. In addition, this probe, as well as those for other amyloid proteins, e.g. prions (Oesch et al 1985), serum amyloid P (Mantzouranis et al 1985), serum amyloid A (Sipe et al 1985) and the amyloid protein in Alzheimer disease, when it is cloned, can be used for linkage analysis in hereditary amyloid neuropathies to help identify the responsible proteins.

Susceptibility to "slow viral" infections (Gajdusek 1985) may be modulated by genetic factors, as illustrated by the differential sensitivity of different mouse strains to scrapie infection (Bruce & Dickinson 1985). Molecular biological techniques have made a major contribution to our

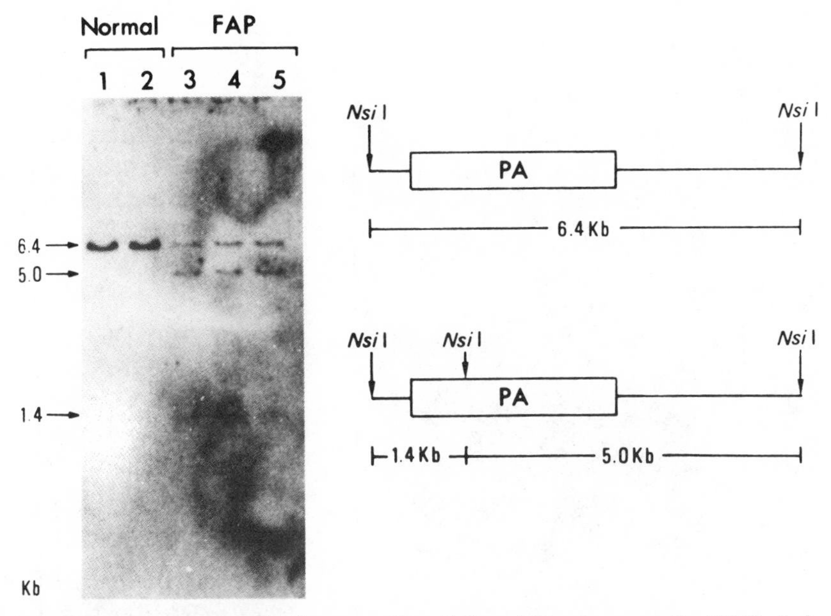

Figure 13 Molecular genetic diagnosis of familial amyloid polyneuropathy (FAP). *Left*: Southern blot of Nsi I digested DNA from two normal individuals and three FAP patients hybridized to a cDNA for prealbumin (PA). A new band of 5.0 kb is detected in the FAP patients because of a single bp substitution that creates a new Nsi I site. *Right*: Schematic drawing of the chromosomal prealbumin gene indicating the positions of the Nsi I sites in normal (*top*) and mutated (*bottom*) genes. (Reproduced with permission from Sasaki et al 1984.)

understanding of scrapie infection by demonstrating that the prion polypeptide is part of a larger protein (65 kD) encoded in an mRNA normally present in brain, as well as other tissues (Oesch et al 1985; Figure 14). The amount of this mRNA remains essentially constant during the period of prion amplification in hamster brain, and no homologous nucleotide sequences are present in isolated, "infectious" preparations of prion particles. It is possible that the prion is autocatalytic and forms a more stable version of itself (Oesch et al 1985) or that the infectious process is mediated by a virus, with accumulation of prion amyloid fibrils representing a response to infection (Braig & Diringer 1985).

The etiology of Alzheimer disease is unclear and many pathogenic mechanisms have been proposed, including aluminum or metal toxicity, viruses, and amyloid formation. In fact, the same set of neuropathologic changes may be caused by a number of different agents that kill neurons (Gajdusek 1985). A subset of Alzheimer disease cases, about 10%, appear to be inherited as an autosomal dominant gene lesion (Heston et al 1966,

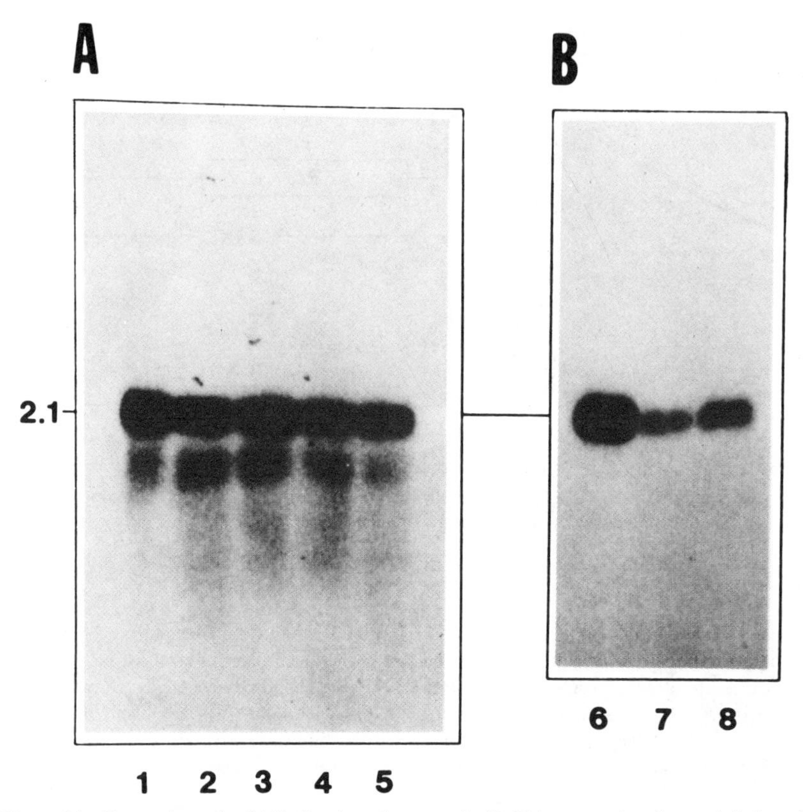

Figure 14 Expression of mRNA for the prion protein (PrP) in normal and scrapie infected hamsters. (*a*) Northern blot analysis of brain RNA in normal lane (1) and scrapie-infected hamsters lanes (2–5), 24 hrs (2), 20 (3), 40 (4), and 60 (5) days after inoculation of PrP. The PrP message (2.1 kb) is present in normal brain and its expression does not change after infection with the scrapie agent. (*b*) Northern blot analysis of RNA from brain lane (6), heart (7), and lung (8) showing the presence of the PrP message in all these tissues. (Reproduced with permission from Oesch et al 1985.)

Goudsmit et al 1981, Heyman et al 1983, Cook et al 1979). In these families it will be possible to carry out genetic linkage analysis using cloned probes for proteins known to form amyloid, as well as for those associated with the cytoskeleton. The main limitations of this type of analysis in Alzheimer disease are the late onset of the disease and the need for neuropathologic examination for definite diagnosis, which restricts the number of informative individuals. A possible clue to the chromosomal location of an Alzheimer gene lies in the fact that individuals with Down syndrome typically develop Alzheimer neuropathology in their 40s (Wong et al 1985). It is possible that the gene encoding the amyloid in these diseases maps to chromosome 21q, so that in Down syndrome an extra dose of it accelerates its accumulation above the normal rate. A number of DNA probes covering most of 21q are available (Watkins et al 1985).

DEFECTS IN MYELINATION: PELIZAEUS-MERZBACHER DISEASE

Overview

The focus of this section is the application of recombinant DNA technology to the understanding of myelin formation and its pathology. DNA sequences encoding the major proteins of myelin from the central nervous system (CNS) and peripheral nervous system (PNS) have been cloned (see below). These cDNAs have been used to elucidate gene defects in some dysmyelinating mouse mutants and provide a very promising tool for the study of human inherited disorders of myelin.

The review focuses on one of these human diseases, Pelizaeus-Merzbacher disease (PMD), since a candidate gene is now available. PMD is characterized clinically by a slow progression of neurologic symptoms that start early in life and histologically by patches of hypomyelination in the CNS. This disease follows an X-linked recessive mode of inheritance (Figure 15). The mouse mutants, *jimpy* and *myelin synthesis deficient* (*msd*) are considered animal models for PMD. The mutants are caused by allelic mutations and also have an X-linked recessive mode of inheritance. The recent cloning of DNA sequences encoding proteolipid protein (PLP), the major protein of CNS myelin, has opened a new level of analysis for PMD and these mouse mutants (Milner et al 1985, Naismith et al 1985, Willard & Riordan 1985). The PLP gene maps to the X-chromosome in humans and mice, and mutations at the PLP locus may be the genetic defect in both PMD and the mouse mutants.

Neurologic Symptoms

PMD is a very rare disorder that affects almost exclusively males. The heterozygous (carrier) females are rarely affected. The clinical picture is quite distinctive and permits a tentative diagnosis. Typically, the initial symptoms are nystagmoid (rapid and irregular) eye movement followed by delayed motor and somatic development. Mentation is variably affected. The time of onset and duration of symptoms vary, but two main patterns can be defined. In some families (Seitelberger 1970), onset is during the early neonatal period, and affected individuals never develop to the point of being able to stand or talk (Seitelberger's variant). In other families, the onset is not until the first few months of life or later (Merzbacher 1910; classical PMD). The progression of this variant is

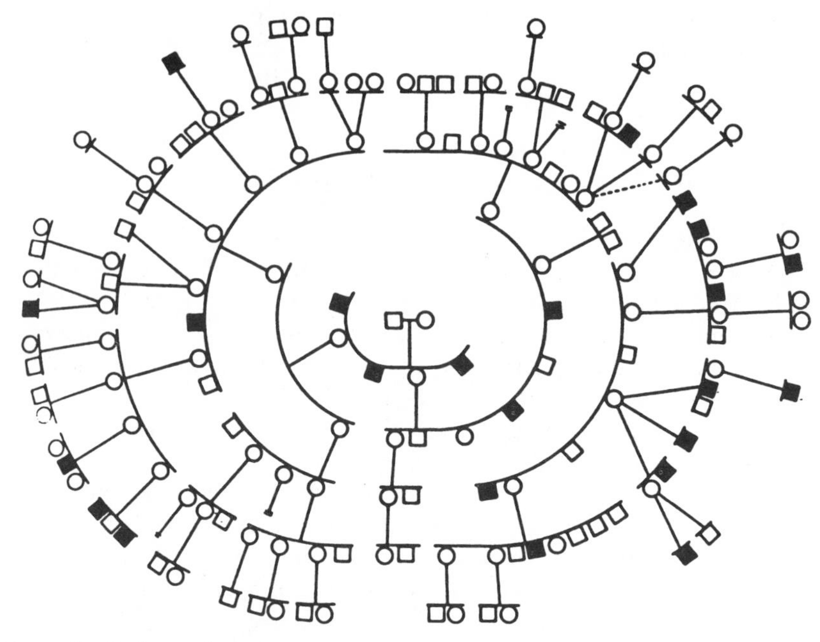

Figure 15 Pedigree of a large family with Pelizaeus-Merzbacher disease showing the X-linked mode of inheritance. There are 23 affected males (*solid symbols*) in six successive generations. (Reproduced with permission from Watanabe et al 1973.)

relatively slow; the affected individuals are sometimes able to sit, talk, and feed themselves (Watanabe et al 1973); and intellectual functions can be quite well preserved. In either type of PMD, patients rarely survive past 5–7 years, but some have lived until the second or third decade, with variable degrees of mental retardation, dysarthria (faulty enunciation of words), and spastic ataxia (Watanabe et al 1973).

Neuropathology and Neurochemistry

Pathological features of classical PMD and the Seitelberger variant suggest a developmental failure to myelinate. In humans, myelination occurs almost exclusively in the postnatal period and proceeds in a caudocranial sequence, occurring first in the PNS, then in the spinal cord, and finally in the brain. The brain is almost completely myelinated by the end of the second year of life, although in some cortical areas myelination continues until the second decade. Gross examination of the brains of PMD patients shows cerebral and cerebellar atrophy. At the microscopic level the myelin lesions are localized mainly in these areas, but the brainstem is also involved. The pathological findings vary from patient to patient and can include either a total absence of myelin sheaths in the CNS (in the Seitelberger's variant), or focal preservation of "myelin islands" (in classical PMD). In both types of the disease, all the neural cell types, except the oligodendrocytes, are normal and there is no invasion of macrophages or microglia. The oligodendrocytes are diminished in number and have membranous whorls with dense inclusions that contain myelin components (Watanabe et al 1969, 1973). The impairment of myelination in this disease can either involve all the oligodendrocytes (Seitelberger's variant) or be limited to only certain groups of these cells (classical PMD).

The biochemical data in PMD suggest a primary defect in PLP expression. PLP has been shown to be totally absent in the brains of two patients by immunoblot analysis, whereas trace amounts of myelin basic protein (MBP) and myelin-associated glycoprotein were detectable (Koeppen 1986). The lipid components of myelin and the enzymes involved in lipid metabolism in these brains range from normal to greatly reduce levels (Watanabe et al 1969, 1973, Bourre et al 1978).

Neuroscience

The basic defect in PMD is probably an impairment of the developmentally regulated program of myelination. This is a complex process that involves the temporally and spatially coordinated expression of several genes coding for structural proteins of myelin and for enzymes involved in lipid metabolism. Myelin is a multilamellar membrane that wraps around the axons and represents a specialized extension of oligodendrocytes in the CNS and Schwann cells in the PNS. The embryonic origin, as well as the biologic and metabolic properties of these two cell types are different, as is the protein composition of the myelin that they produce. CNS myelin contains as major proteins MBP, an extrinsic protein located on the cytoplasmic side of the membrane (Martenson et al 1970), and PLP, an intrinsic membrane protein that spans the lipid bilayer (Folch-Pi & Lees 1951, Lees & Brostoff 1984). In the PNS the protein components of myelin are a major intrinsic protein, PO, a glycoprotein accounting for about 50% of the protein content, and basic proteins equivalent to MBP (Greenfield et al 1973).

The dysmyelinating mouse mutants provide useful tools to study the control and regulation of myelin formation. The autosomal recessive mutant, *quaking*, and the two allelic X-linked mutants, *jimpy* and *msd*, are considered animal models for PMD. We discuss only the *quaking* and *jimpy* mutants, as *jimpy* and *msd* share the same phenotype. In *quaking* mice, myelination proceeds normally for a short period and then stops. The white matter in the CNS is affected in a cranial-caudal gradient and patches of relatively normal myelin remain. PNS myelin is also somewhat abnormal (Appeldoorn et al 1975, Costantino-Ceccarini et al 1980). The clinical symptoms are evident ten days postnatally in the *quaking* mouse, with trembling of the trunk and the limbs during motion, followed by tonic seizures. The animals survive to adulthood, but males are sterile. The *jimpy* mutants show a total lack of CNS myelin, with no effect in the PNS. The symptoms in *jimpy* are similar to those of *quaking* but more severe, and the animals die by 30–35 days of age, usually during a seizure.

In both *quaking* and *jimpy* mutants, cerebrosides, sulfatides, and the enzymes involved in lipid metabolism are reduced, but to a greater extent in *jimpy*. Biochemical analyses of the composition of the myelin proteins have shown a reduction of both PLP (Nussbaum & Mandel 1973) and MBP (Delassalle et al 1981). In *quaking*, the synthesis of MBP, as assayed by "in vitro polysome runoffs" (in vitro translation of polysomes isolated from brain), occurs at a normal rate. Nevertheless the steady state level of MBP in these brains as demonstrated by immunoblot analysis is only 5–20% of normal values. In addition a delay occurs in the developmental expression of mRNA for MBP, as demonstrated by Northern blot and RNA dot analysis (Roth et al 1985, Campagnoni 1985). During the normal process of myelination, mRNA encoding MBP starts to accumulate at four days postnatally and peaks at 18 days (Zeller et al 1984). In *quaking*, the MBP-mRNA levels start to accumulate at 4 days, but are only 40% of the control values at day 12 and reach control levels only at 21 days. Further, the synthesis of PLP in *quaking*, as assayed by in vitro translation, is reduced to about 15% of the control values, and very little of this protein is incorporated into the myelin membrane (Sorg et al 1986). Data relative to the quantitation of mRNA encoding PLP in *quaking* and *jimpy* mice are not currently available.

In the *jimpy* mouse the most striking findings are the reduction of PLP synthesis, assayed in polysome runoffs to about 3% of the control values, and the total absence of immunoreactive PLP protein, as assayed by immunoblot analysis (Sorg et al 1986). In addition, in *jimpy* mice the steady-state levels of MBP are about 2–10% of normal, and the synthesis of MBP in polysome runoff experiments is reduced to about 25% of control values (Campagnoni et al 1984, Sorg et al 1986). Quantitation of the MBP-mRNA also shows a reduction to about 25% of control levels and an alteration in its developmental expression, as the mRNA peaks at 15 days postnatally at 40% of control values (Roth et al 1985, Campagnoni 1985).

Recently, Willard and co-workers (Willard & Riordan 1985) have mapped the PLP locus by somatic cell hybrid techniques to the region on the mouse X chromosome containing the *jimpy* locus (see below). This suggests the possibility that the *jimpy/msd* mutations may lie within the murine PLP locus. Such defects would be consistent with the fact that in *jimpy* only CNS myelin is affected. In the *quaking* mutants, however, both Schwann cells and oligodendrocytes fail to myelinate properly (Aguayo et al 1980), suggesting that mutations in this autosomal gene may cause a generalized defect in the regulation of myelination. The *quaking* locus maps to chromosome 17 in the mouse near the t locus (Hammerberg & Klein 1975); thus far none of the myelin proteins map to this region.

Genetics

cDNA clones for mouse and rat MBP are available in several laboratories (Roach et al 1983, Zeller et al 1984) and have permitted the study of the genomic organization of the MBP gene and its expression during myelination in both species (Takahashi et al 1985, De Ferra et al 1985). Apparently a single gene per haploid genome maps to chromosome 18 in mouse (Roach et al 1985) and humans (Saxe et al 1986). This gene spans 32 kb of DNA and contains seven exons of variable length (Figure 16). In rodents there are four MBPs of molecular weight 21.5, 18.5, 17, and 14 kD and four distinct mRNAs coding for these proteins. These mRNAs originate through alternate splicing modes of a unique primary transcript (Takahashi et al 1985, De Ferra et al 1985). Exons 2 and 6 are of particular interest, since they contain the sequences found in the mRNAs coding for the large forms of MBP and are spliced out to obtain the lower molecular

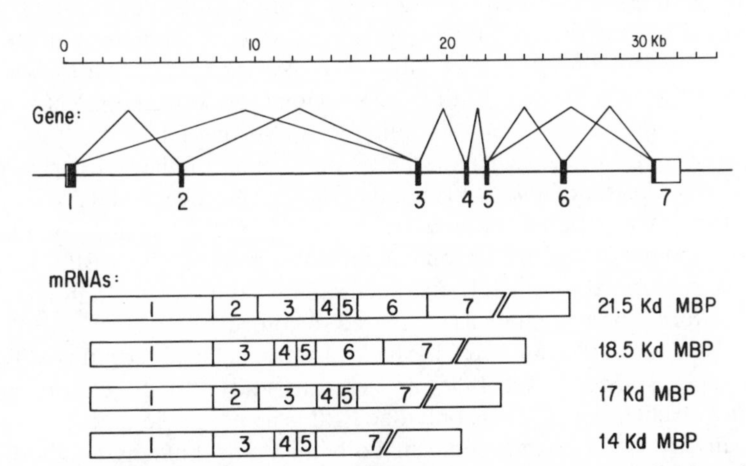

Figure 16 Mouse myelin basic protein (MBP) gene. *Top*: Genomic organization of the mouse MBP gene. *Boxes* indicate exons; *lines*, introns. *Open boxes* show the untranslated portions of the first and last exons; *solid boxes*, the translated portions. *Bottom*: MBP mRNAs generated through alternate splicing modes of exons 2 and 6 which code for the four protein species of MBP (shown to the *right*). (Adapted from Takahashi et al 1985 and De Ferra et al 1985.)

weight forms. Specifically, exon 2 encodes the 26 amino acids present at the N-terminus of the 21.5 and 17 kD MBPs, and exon 6 encodes the 41 amino acids present at the C-terminus of the 21.5 and 18.5 kD proteins (Figure 16). On Northern blots the cDNA clones hybridize to a broad mRNA band ranging from 2.1 to 2.3 kb that corresponds to these four mRNA species. The message sizes are much larger than the 600 bp needed to code for even the largest MBP. In fact, these mRNAs contain very long 3′ untranslated regions, 1.4 kb in length. This seems to be a common feature of brain mRNAs, but its significance is unknown (Milner & Sutcliffe 1983, Milner et al 1985).

The developmental expression of MBP-mRNA has been studied quantitatively by using dot blot analysis. The levels of these mRNAs parallel that of the protein they encode, with the first appearance of MBP at 5-7 days after birth, a peak at 18 days, and then maintenance of steady state levels through adulthood. The kinetics of accumulation and the molar ratios among the four MBPs do vary during development (Barbarese et al 1978). An interesting, yet unproven, possibility is that these variations might derive from a developmentally regulated splicing mechanism of the primary MBP transcript (Breitbart et al 1985).

The availability of the full length MBP gene has allowed Roach and coworkers (1983, 1985) to define the molecular defect in a myelin deficient mouse mutant, *shiverer*. This mutant is characterized by the total lack of MBP, with levels less than 0.1% and 0.4% of control values in the CNS and the PNS, respectively. By using classical genetics (Sidman et al 1985), the *shiverer* mutation has been mapped to mouse chromosome 18, where the MBP gene is located. With a cDNA clone for MBP, no mRNA was detectable in the brain of this mutant by Northern blot analysis. By hybridizing probes specific for the 5′ or 3′ ends of the MBP gene to restriction digests of *shiverer* genomic DNA on Southern blots, a major deletion of the 3′ end of the gene was elucidated (Roach et al 1985). Only exon 1 and part of the first intron remain in this mutant. The MBP-mRNA that is transcribed is probably degraded rapidly, as its concentration is 16-fold lower in these mutants as compared to controls.

cDNAs for bovine and rat PLP have also been cloned recently (Milner et al 1985, Naismith et al 1985) and used to study PLP gene expression during myelination. There is only one PLP gene per haploid genome and it maps to the X chromosome in both mice and humans (Willard & Riordan 1985). Its genomic organization has not yet been defined. Two major mRNA species in brain of 3.2 and 1.6 kb are detected on Northern blots. These messages have the same coding capacity (277 amino acids) and are similar at their 5′ ends, but differ at their 3′ end because of different sites of polyadenylation of the same primary RNA transcript. There is also some heterogeneity at the 5′ end, in that three different genomic regions within a 75 bp stretch are used as transcription initiation sites with comparable efficiency. Both PLP-mRNAs appear at 10 days postnatally, increase coordinately until 25–30 days after birth, and then decline to adult levels, which are three- to fivefold less than peak levels (Milner et al 1985) (Figure 17, *left*). These data are in agreement with protein data analyses showing the appearance of PLP at 10 days postnatally, with a peak at about 25–30 days after birth (Sorg et al 1986). It appears that for both MBP and PLP, developmental expression of the proteins is regulated largely at the level of transcription.

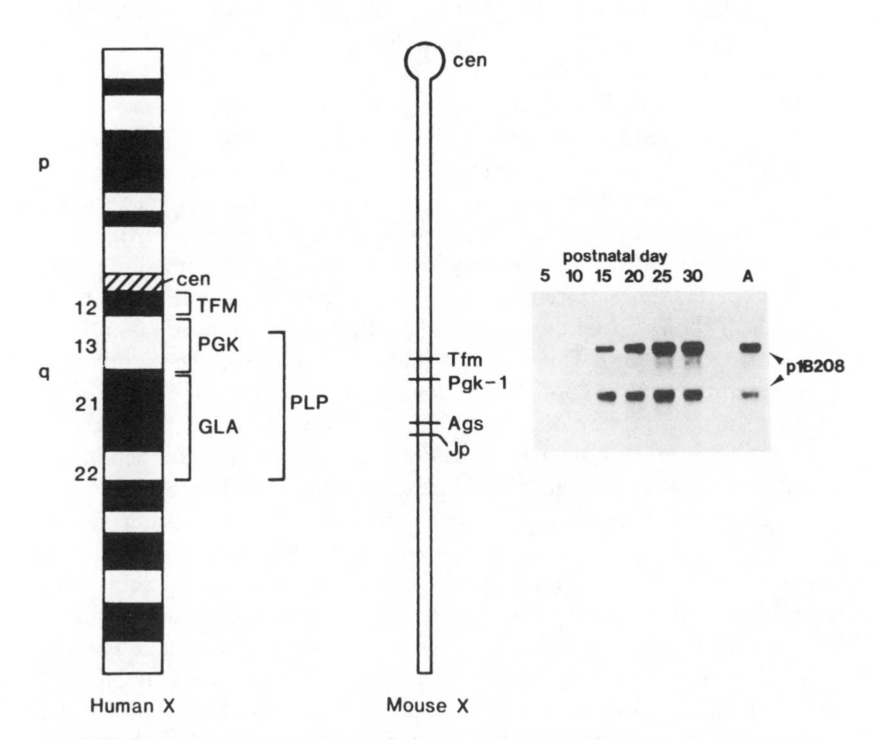

Figure 17 Location of the proteolipid protein (PLP) gene on the human and mouse X chromosomes and developmental expression of mRNA for PLP. *Left*: Schematic representations of human and mouse X chromosomes: cen, centromere; TFM and tfm, testicular feminization gene loci; PGK and Pgk-1, phosphoglycerate kinase loci; GLA and Ags, alpha-galactosidase loci. *Right*: Northern blot analysis of the rat brain RNA showing the developmental expression of two molecular species of PLP-mRNA during the postnatal period and in the adult (A). [Reproduced with permission from Milner et al 1985 (*right*) and Willard & Riordan 1985 (*left*).]

Interestingly, the PLP and *jimpy* genes map to a region of the X chromosome, Xq13–Xq22, that represents an evolutionarily conserved linkage group and lines within 2 cM of the alpha-galactosidase locus (Figure 17, *right*; Willard & Riordan 1985). Studies of gene structure analogous to those carried out in *shiverer*, as well as DNA sequencing, should elucidate whether a mutation at the PLP locus is the genetic defect in *jimpy*. Although the position of the PMD locus on the X chromosome is still unknown, this disease may also represent a genetic defect at the PLP locus. The availability of the cDNA for PLP will allow linkage studies to be carried out using the inclusion/exclusion mapping approach (Breakefield et al 1984). If the PLP gene is indeed the disease locus, essentially no recombination should occur between marked alleles for this gene and the PMD locus in affected families. In addition, autopsy or biopsy samples from brains of affected patients could be analysed by in situ hybridization to establish whether the PLP gene is expressed.

PO is the major protein of PNS myelin, and a cDNA clone encoding the rat protein has become available recently (Lemke & Axel 1985). Southern blot analyses of genomic DNA indicate that there is a single gene per haploid genome. Northern blot analysis shows a single mRNA species of 1.9–2.0 kb. PO-mRNA in the rat sciatic nerve is present at birth, peaks at 14 days, and falls to steady state levels in the adult. Expression of the mRNA for HMG-CoA reductase, the rate-limiting enzyme in de novo synthesis of cholesterol, follows a parallel time course to that of PO-mRNA (Lemke & Axel 1985). This suggests that these and other genes involved in myelination may be coordinately regulated.

The primary structure of PO protein, until now unknown, was derived from the nucleotide sequence of the cDNA. It shows a signal peptide of 28 amino acids and three domains: an extracellular one (1–125) containing stretches of hydrophobic amino acids and an asparagine residue for *N*-glycosylation, a transmembrane domain (125–150), and a cytoplasmic domain (151–219) containing mainly polar and basic amino acids. The fairly large cytoplasmic domain may be important in promoting the apposition of the cytoplasmic phases of the membrane and may take the role of MBP in the *shiverer* mutant where peripheral myelin is functionally and morphologically normal. The availability of the cDNA for PO should allow elucidation of the role of this gene in inherited disorders that affect only peripheral myelin. Such is the case in another dysmyelinating mouse mutant, *trembler*, where the defect in myelination is confined to the Schwann cells (Aguayo et al 1980). The *trembler* locus is on chromosome 11. It will be important to determine the location of the PO gene in order to evaluate its possible role in this mutant.

PURINE METABOLISM: LESCH-NYHAN SYNDROME

Overview

The Lesch-Nyhan syndrome is an X-linked disease that affects several aspects of neural function, including development and the control of movement as well as intelligence. The most unique and devastating feature of this syndrome is compulsive self-mutilatory behavior (SMB), usually not associated with loss of pain sensation. This behavior is presumably caused by altered neurotransmission in the brain, although the nature of these alterations is still unclear. Neurochemical analysis of the brain of Lesch-Nyhan patients and animal models of SMB suggest two possible mechanisms, which are not mutually exclusive, involving alterations in dopamine and/or purinergic transmission. The lack of a clear understanding of the neurological aspects of this disease stands in sharp contrast to the well understood biochemical and molecular lesions. Individuals with the Lesch-Nyhan disease manifest an almost complete loss of activity of hypoxanthine phosphoribosyltransferase (HPRT), an enzyme important in the purine salvage pathway (Figure 18). HPRT is present in all cells of the body, albeit at about four-fold higher levels in the brain. (For further reviews of this syndrome see Baumeister & Frye 1985, Wilson et al 1983, Stout & Caskey 1985).

Neurologic Symptoms

Almost all affected individuals are male and appear normal at birth. Motor development is notably delayed by five months of age, with choreic and writhing movements appearing in the first year (Baumeister & Frye 1985, Watts et al 1982). Muscular hypo- and hypertonia, spasticity, dysarthria, dystonia, and SMB usually occur during the first few years of life. Most patients are mentally retarded. Self-mutilatory behavior is most often seen as biting of the lips and fingers, although some patients also bang their heads and may strike out or use abusive language (Christie et al 1982).

Neurologic symptoms are present when HPRT activity in a patient is less than 1% of normal, although even then symptoms may vary among individuals. Individuals with 1–50% of normal activity usually do not show any neurologic symptoms, although they frequently suffer from gout due to hyperuricemia and the resulting accumulation of uric acid crystals. Several individuals with less than 1% HPRT activity have been reported with normal intelligence and/or mild neurologic involvement. This may reflect the type of mutation in the HPRT gene. In some (Bakay et al 1979) but not all (Fattal 1984) of these atypical patients, the enzyme has higher

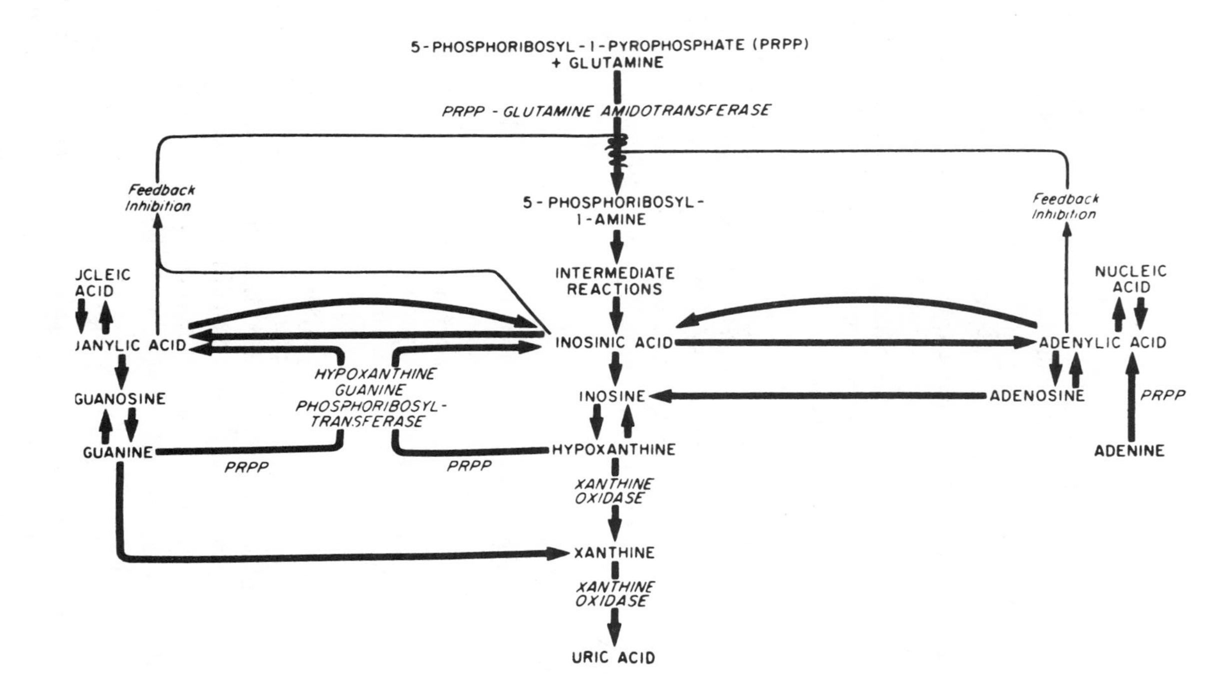

Figure 18 Pathway of purine metabolism. In the case of HPRT deficiency there can be an accumulation of inosine, hypoxanthine, xanthine, and uric acid because of both lack of conversion of hypoxanthine to inosinic and quanylic acids and, in turn, decreased feed-back inhibition on the rate-limiting enzyme (PRPP-glutamine amidotransferase) in purine synthesis. (Reproduced with permission from Baker & Baker 1983.)

activity when measured in intact cultured fibroblasts than in homogenates of erythrocytes, thus suggesting that it may have higher activity in vivo than in vitro.

Neuropathology

Although brains from Lesch-Nyhan patients have been examined at the histopathologic and electron microscopic levels, no neuroanatomical changes have been noted (Watts et al 1982). Studies of neurotransmitter metabolism in Lesch-Nyhan brains and neuropharmacologic responses in an animal model for SMB (see below) suggest, however, that more extensive, focused examination of the basal ganglia may reveal differences in dopaminergic innervation. This area of the brain is also implicated by the abnormal movements seen in this disorder.

Biochemistry

Hypoxanthine phosphoribosyltransferase converts hypoxanthine and guanine to inosinic acid and guanylic acid, respectively, using the cofactor phosphoribosylpyrophosphate (PRPP) (Figure 18). In the absence of HPRT activity there is an increase in the cellular and extracellular concentrations of hypoxanthine, xanthine, and uric acid in most tissues. This accumulation is also accompanied by a slight increase in PRPP concentration. Thus loss of HPRT unbalances regulated synthesis of purines, as well as causing the possibly toxic accumulation of purine degradatory products.

The purine concentration in a given tissue depends on the availability of extracellular purines and their precursors, as well as the relative use of *de novo* and salvage pathways of synthesis. As noted by Baumeister & Frye (1985), the brain appears to be relatively deficient in amidophosphoribosyl transferase activity, which catalyzes the initial and rate-limiting step in *de novo* purine synthesis, and relatively high in HPRT activity, the salvage pathway, as compared to other tissues. This relationship is especially pronounced in the basal ganglia, which has the highest level of HPRT activity in the body. In addition, the brain is relatively low in xanthine oxidase activity, so xanthine and uric acid do not accumulate to an appreciable degree. One would anticipate then that HPRT deficiency might result in an increase in hypoxanthine concentrations in the brain, and, in fact, four-fold higher levels have been reported in CSF from patients as compared to controls, as well as in a decrease in purine concentrations, although this deficit has not been demonstrated convincingly. Either of these metabolic imbalances could contribute to neurologic dysfunction.

Neuroscience and Neurochemistry

Several hypotheses have been proposed to explain why the brain is particularly susceptible to HPRT deficiency. One hypothesis highlights a critical role for dopaminergic neurons, and the other the role of purines as neurotransmitters. Based on neurochemical findings in Lesch-Nyhan brains and a rat model for SMB, the hypothesis has been forwarded that dopaminergic neurons in the substantia nigra are especially sensitive to HPRT deficiency. Hornykiewicz and co-workers (Lloyd et al 1981, Kish et al 1985) have analyzed a number of neurochemical parameters in the brains of Lesch-Nyhan patients. Markedly lower values than normal were observed in different areas of the basal ganglia of these patients for various aspects of catecholamine metabolism, including concentrations of dopamine and homovanillic acid, as well as activities of dopa decarboxylase and tyrosine hydroxylase. Levels of the catecholamine degradative enzymes, monoamine oxidase and catechol-0-methyltransferase, were not abnormal, nor were levels of norepinephrine and dopamine beta-hydroxylase. Collectively, these data suggest a deficit in dopaminergic innervations to the basal ganglia. Since the dopamine concentration was not also low in the substantia nigra of these patients, the area containing the dopaminergic cell bodies that innervate the basal ganglia, it is possible that the terminal arborization of these dopamine neurons is deficient.

The reason that alterations in purine metabolism should affect dopaminergic innervation is not clear. Baumeister & Frye (1985) have proposed that this effect reflects a combination of the high degree of dependence of the basal ganglia on the purine salvage pathway and the rapid rate of growth of this region, and hence the need for purines for DNA and RNA synthesis, during a particular stage of postnatal development. Developmental defects in the basal ganglia could affect synaptic connections of dopamine neurons located in the substantia nigra. It is also possible that altered purine metabolism disrupts the function of neurons that use purines as second messengers or neurotransmitters; this, in turn, may directly or secondarily affect dopaminergic neurons.

Deficits in dopaminergic transmission in the Lesch-Nyhan syndrome is also supported by the work of Breese and co-workers (1985a,b). Neonatal rats exposed to 6-hydroxydopamine undergo a loss of catecholamine neurons, including dopaminergic neurons located in the substantia nigra. If these rats are later given L-DOPA they respond with a number of stereotyped behaviors, including SMB. Extensive pharmacologic studies indicate that this behavior is mediated primarily by supersensitivity of D1 dopamine receptors in the basal ganglia as a result of denervation supersensitivity. Interestingly, when 6-hydroxydopamine is administered to adult rats this behavioral response to L-DOPA does not ensue. In adult rats, continuous administration of amphetamine decreased dopaminergic innervation of the caudate (Ellison et al 1978) and produced SMB, which could be blocked with haloperidol, a D2 dopamine receptor antagonist

(Mueller et al 1982). These studies support the idea that the "stage" for SMB can be set by decreased dopaminergic innervation of the caudate, leading to receptor supersensitivity. "Overstimulation" of these receptors by dopamine may then trigger SMB. An interesting aspect of these studies is the difference between responses in neonates and adults. Possibly if the neurotoxic agent is present during brain development, it can affect the formation of neuronal circuitry.

Other hypotheses focus on possible interactions of purines and their metabolites with receptors for various neurotransmitters or modulators. Hypoxanthine, which accumulates in the brains of Lesch-Nyhan patients, may interfere with binding of the as yet unidentified natural ligand for the benzodiazepine receptor (Kopin 1981). However, the binding affinity of hypoxanthine to this receptor is quite low (Asano & Spector 1979). Evaluation of benzodiazepine receptors in patients is complicated because they are given benzodiazepines as tranquilizers. Studies by Hornykiewicz and co-workers (Kish et al 1985) using washed brain tissue from patients found no change in the density of these receptors in the cerebral cortex, but the receptors did show a slightly enhanced affinity for benzodiazepine and a marked reduction in GABA-stimulated benzodiazepine binding. Purines and their metabolites may also act through binding to purine receptors such as for adenosine. Methylxanthines themselves have been shown to produce SMB, e.g. caffeine at high doses in rats (Lloyd et al 1981) and theophylline in combination with clonidine at low doses in neonatal mice (Katsuragi et al 1984).

Genetics

There is a single active gene in the human haploid genome encoding HPRT, which is located near the end of the long arm of the X chromosome (for review see Wilson et al 1983). Females are rarely affected with the Lesch-Nyhan syndrome as they would have to have inherited two defective HPRT alleles or be unfortunate enough to have the X chromosome containing the normal allele be inactivated in most of their cells. Carrier females can be detected by the presence of about 50% normal levels of HPRT activity in lymphocytes, and either normal or absent HPRT activity in the clonal progeny of hair follicles (due to random X-inactivation). Interestingly, there appears to be a selective disadvantage to survival of some cell types with HPRT deficiency, for example HPRT activity measured in red blood cells of carrier females tends to be normal.

Affected males do not reproduce, and more than one third of the affected males represent new mutations occurring in the germ line of the mother or maternal grandparents (Francke et al 1976). Thus although the same gene locus is defective in all patients, many different mutational events account for these defects. Biochemical analyses of HPRT in cases of the Lesch-Nyhan syndrome and gout, where the mutant protein is present, have identified four different amino acid substitutions that can be accounted for by single base pair substitutions in HPRT gene (Wilson et al 1983). These mutant enzymes show decreases in intracellular concentrations of immunoreactive protein, in reaction velocities, and/or in binding affinity for hypoxanthine or 5-phosphoribosyl-1-pyrophosphate. The fact that the enzyme can be altered in a number of different ways may contribute to clinical variability in neurologic symptoms.

The availability of cloned DNA sequences for HPRT has allowed a more detailed analysis of mutations in the HPRT locus, some of which result in no detectable enzyme protein being produced (Yang et al 1984). More is known now about alterations in DNA sequence that cause this disease than any other inherited neurologic condition. The active HPRT gene is about 44 kb in length, contains 9 exons (Figure 19), and codes for a 1.6 kb poly A+ mRNA. Southern blot analysis of genomic DNA from lymphocytes of 28 unrelated Lesch-Nyhan patients reveals that five of them (18%) contained rearrangements in DNA that resulted in substantial alterations or deletions of sequences in the HPRT gene (Figure 19). None of the deletion mutations produced detectable HPRT-mRNA by Northern blot analysis using mRNA from patient's lymphocytes, and neither did at least one of the mutations with no substantial rearrangement. As further analysis of these HPRT mutations proceeds, we can anticipate that in addition to rearrangements, defective expression of the HPRT gene will be associated with discrete changes in the flanking exon and intron regions of the gene, as has been documented for the beta-globin gene (Orkin & Kazazian 1984). These studies demonstrate the many different types of mutations that can be associated with loss of gene function.

Changes in DNA sequences associated with HPRT deficiency can also be used to track the mutant gene in families. One of the DNA rearrangements in the HPRT gene represents an apparent partial duplication of genetic material and can be identified by the appearance of a unique hybridizing fragment on Southern blots (Figure 20; Yang et al 1984). Analysis of DNA from members of the affected family revealed that the mutation took place in the germ line of the maternal grandmother or grandfather. Another base pair substitution in the HPRT gene causing gout alters a specific restriction endonuclease site and thus causes a change in the size of genomic fragments hybridizing to the HPRT probe; this difference can also be used to identify carrier females in affected families (Wilson et al 1983). Further, a common restriction endonuclease polymorphism near the HPRT gene can be used to follow the inheritance of HPRT alleles in a number of families (Nussbaum et al 1983).

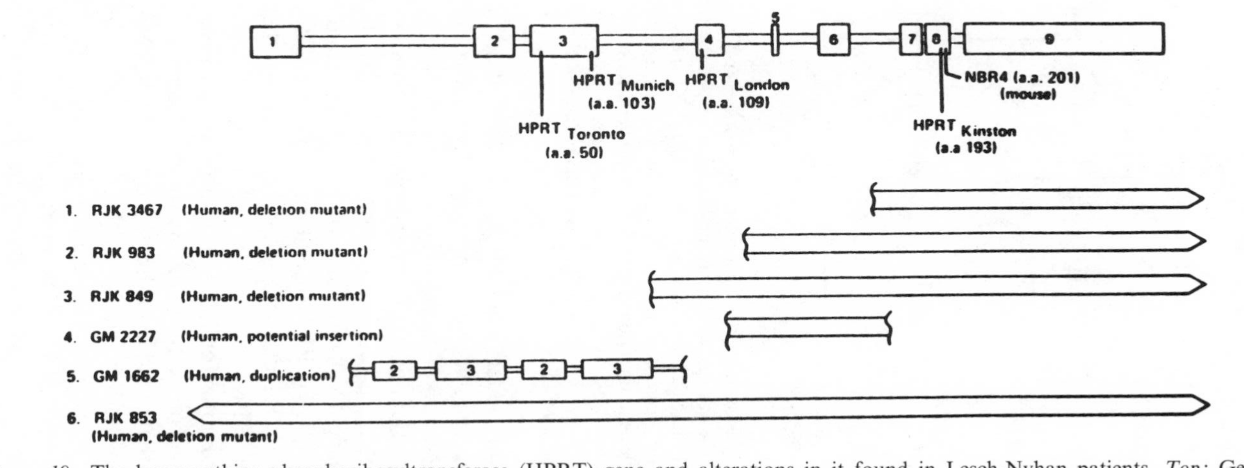

Figure 19 The hypoxanthine phosphoribosyltransferase (HPRT) gene and alterations in it found in Lesch-Nyhan patients. *Top*: Genomic organization of the human HPRT gene and positions within the exons of point mutations shown to cause HPRT deficiency in humans and mice. *Bottom*: Deletions and rearrangements of the HPRT locus in a number of other unrelated Lesch-Nyhan patients. (Reproduced with permission from Stout & Caskey 1985.)

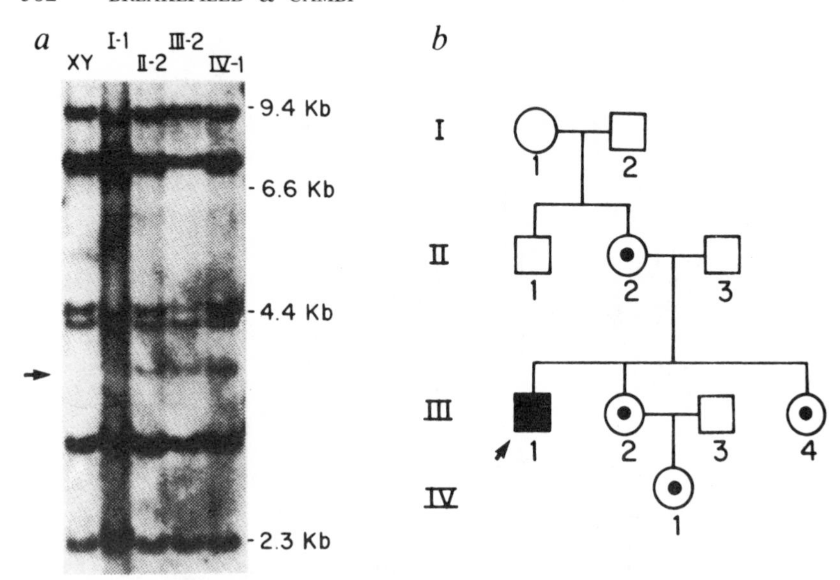

Figure 20 Molecular analysis of the HPRT gene alteration in a Lesch-Nyhan family. The affected member (III) inherited the gene from his mother, who represents a new mutation. (*a*) Southern blot of Bgl II digested DNA from a normal male (XY) and several members of the family shown on the *right*. The *arrow* indicates the appearance of a new 4.1 kb fragment that represents the gene alteration in the proband. (*b*) Pedigree of the family where the new mutation has occurred. *Filled symbols*, affected male; *open symbols*, unaffected males and noncarrier females; *dotted symbols*, carrier females. (Reproduced with permission from Yang et al 1984.)

There are no animal models available for the HPRT deficiency state, but several genetic strategies have been proposed for generating them. One strategy involves generation of mosaic mice containing both normal cells and mutant cells, the latter derived from embryonic carcinoma cells selected for HPRT deficiency in culture (Dewey et al 1977). If such mice contained HPRT-deficient cells in the germ line, they could be used to generate carrier female or HPRT-deficient male mice that, in turn, could be used to foster an HPRT-deficient mouse colony. Another approach would be to insert an HPRT gene or cDNA, in the reverse orientation with respect to its promoter for RNA transcription, into the genome of fertilized mouse eggs via microinjection. This inserted gene would produce an "antisense" RNA that could hybridize to the normal mRNA for HPRT and block its processing and/or translation (Izant & Weintraub 1984, Stout & Caskey 1986). The resulting transgenic mice would be functionally HPRT deficient, although the DNA change causing this deficiency would act as a dominant allele and not necessarily be on the X chromosome. However, a decrease in HPRT activity of $>98\%$ would be necessary to produce a biochemical situation comparable to the Lesch-Nyhan syndrome, which can be difficult to achieve with antisense RNA.

Another interesting aspect of the HPRT deficiency state is the possibility of the recovery of normal activity through introduction of a functional HPRT gene into the genome (gene therapy). Transfer of active HPRT cDNA into the genome of deficient cells in culture has been achieved through DNA uptake (Brennand et al 1983) and retroviral mediated transfection (Miller et al 1983). Expression of exogenous HPRT sequences has also been achieved in mice through retroviral mediated integration of its cDNA into isolated bone marrow cells, followed by the introduction of these cells back into mice, previously irradiated to kill off their own marrow cells (Miller et al 1983, Willis et al 1983, Williams et al 1984). Continued expression of foreign HPRT activity indicated that hemapoietic stem cells had been transformed and were able to give rise to progeny cells expressing this new phenotype. In other studies, foreign HPRT cDNA has been injected into fertilized mouse eggs, resulting in mice that express this activity in most or all tissues of their body, with highest expression in the brain (Stout & Caskey 1985).

The issues of how and whether to confer HPRT activity onto brain cells of Lesch-Nyhan patients remain to be resolved. It might be possible to introduce the gene via a replication-defective retroviral vector into discrete areas of the brain, such as the basal ganglia, but it would be difficult to regulate the level of expression. Further, it is not clear whether delivery of HPRT activity to this area of the brain would alleviate neurologic symptoms and whether too much HPRT activity might be as damaging as too little. Transgenic mice that express abnormally high levels of HPRT activity in the nervous system are phenotypically normal at the gross developmental and motor levels (C. T. Caskey, personal communication). It does appear that metabolic cooperation can occur between cells with and without HPRT activity within a limited tissue domain, so that not all cells in an area would have to be transfected (Gruber et al 1984). Clearly, an animal model of HPRT deficiency is needed to resolve these issues. The Lesch-Nyhan syndrome illustrates the many ways molecular genetic techniques can be used to explore and intervene in an inherited neurologic disease once the gene defect is known.

CONCLUSION

The preceding sections describe how molecular genetic techniques have been used to find the chromosomal location of genes responsible for neurologic diseases and to identify structural changes in these genes. This information is useful clinically for carrier detection, prenatal diagnosis, and disease classification, as well as in evaluation of tumor types and their state of malignancy. As yet, with the exception of onc genes in neural tumors, these techniques have not yet elucidated the molecular mechanisms involved in human disease processes further than was known from biochemical techniques, although their potential to do so is clear. In the case of diseases in experimental animals, scrapie infection, and the mouse mutant *shiverer*, insights into the molecular lesions have already been forthcoming. Recombinant DNA technology combined with other disciplines in the neurosciences has the power to determine the molecular defects in many neurologic diseases. As so clearly demonstrated by the Lesch-Nyhan syndrome, however, even a complete understanding of the defective gene and its products may not necessarily explain its devastating effects on the nervous system. Molecular genetic techniques then must serve as just one of the many tools used by neuroscientists in trying to unravel the complex, interrelated actions in the nervous system.

The excitement of molecular genetic techniques comes from the speed with which they can provide new information, which is sometimes inaccessible by other approaches, and their potential for developing new strategies. Dramatic progress can be anticipated in four areas: (*a*) finding and marking defective genes in the human genome; (*b*) identifying and characterizing defective genes; (*c*) creating animal models for human diseases; and (*d*) developing methods for gene transfer. In the first area of linkage analysis, DNA probes for highly polymorphic sequences spaced at appropriate intervals along the entire human genome should become available over the next few years. This will allow linked markers to be established for many defective genes.

Still, the best way to proceed from positional information to identification of the defective gene remains unclear. For both Huntington disease and Duchenne muscular dystrophy, a sequence of DNA containing the gene has already been defined. Methods being used to find the gene within this larger sequence include "walking" or "leaping" from marker sequences closer to the gene by using overlapping genomic clones, searching for cDNAs that are encoded in it, and trying to clone open-reading frames within this sequence (Gusella et al 1984). In the later case, the open-reading frames could be converted into amino acid sequences by inserting them in an expression vector or synthesizing corresponding polypeptides. The polypeptides could in turn be used to generate antibodies for use in immunocytochemical localization of such proteins in the nervous system (Sutcliffe et al 1983). In addition, these DNA sequences could be used for in situ hybridization and compared to known sequences for other proteins. The number of known sequences is increasing rapidly as more and more cDNAs for neural proteins and brain-specific mRNAs are being cloned.

Candidate genes, once they have been cloned, can also be used directly to screen genomic DNA from patients for alterations in gene structure. Studies on the Lesch-Nyhan syndrome show that 10–20% of gene lesions may be caused by relatively large deletions or rearrangements, which will result in altered sizes of hybridizing fragments on genomic blots (Yang et al 1984). Methods have also been developed whereby a single base pair mismatch between defective and normal genes can be visualized by altered mobility of hybrid complexes under agarose gel electrophoresis (Novack et al 1986, Myers et al 1985). Since even among normal alleles there is a substantial amount of sequence variation at a given gene locus in non-coding regions, this approach will be most useful in comparing coding sequences.

Two new techniques are available for the creation of animal models for inherited human diseases. These techniques depend on first knowing and cloning a probe for a candidate gene. Cloned probes are available, for example, for the HPRT gene in the Lesch-Nyhan syndrome (Yang et al 1984), prealbumin in familial amyloid polyneuropathy (Sasaki et al 1984), and alpha and beta subunits of hexosaminidase in Tay Sachs and Sandhoff's disease (Myerowitz et al 1985, O'Dowd et al 1985). In addition, a large number of cloned probes are available for other neural proteins that can be used to assess how defects in the corresponding genes would affect the nervous system. Genetic manipulations can be carried out on very early embryos so as to create experimental animals in which most cells, including the germ line, contain the genetic alteration. One approach would be to introduce a gene or cDNA sequence into the genome in the reverse orientation with respect to a strong promoter so that a large excess of antisense RNA is generated that hybridizes to and blocks expression of the endogenous mRNA. Another approach would be to carry out site-directed mutagenesis (Izant & Weintraub 1984). Recent studies indicate that although the process occurs at a relatively low frequency, exogenous sequences transferred into cells can recognize and recombine with homologous sequences in the mammalian genome. This process then may be used to insert defects (or correct defects) in specific genes (Smithies et al 1985). A somewhat more laborious approach to creating animal models of diseases is to introduce retroviral sequences at random into the genome, where they create mutations at the site of the insertion, and then to produce offspring homozygous at these loci and screen them for behavioral abnormalities (Jaenisch et al 1983).

Methods are also being developed to transfer genes into tissues or embryos to correct genetic defects. One of the most promising of these is to use replication-defective retroviral particles that contain a normal cDNA under the control of a strong and/or tissue-specific promoter (Mann et al

1983). The advantage of this approach is that gene transfer is efficient and can be carried out in vivo. Special problems will be faced, however, in trying to carry out gene transfer into the nervous system because of its relative inaccessibility and the lack of division of neurons. These gene transfer methods may also prove useful in assessing the role of certain genes in specific regions of the nervous system during development homeostasis and aging. The rate of increase of new genetic approaches useful in the study of the nervous system has been so rapid in recent years that it can only be assumed that a host of new ones will be added in the near future.

ACKNOWLEDGMENTS

We thank Anne Marie Cullin for skilled and patient preparation of this manuscript, and Marjorie Lees, George Breese, Timothy Stout, George Glenner, Brad Evans, and David Corey for their helpful advice. X. O. B. is supported by NIH grants NS21921 (Senator Jacob Javits Neuroscience Investigator Award) and GM34536.

Literature Cited

Aguayo, A., Navi, C. P. V., Bray, G. M. 1971. Peripheral abnormalities in the Riley-Day syndrome. *Arch. Neurol.* 24: 106–16

Aguayo, A., Bray, G., Perkins, S., Duncan, I. 1980. Experimental strategies for the study of disorders of myelination in nerves of mouse mutants. In *Neurological Mutations Affecting Myelination, INSERM Symp.* 14: 87–98

Andrade, C. 1952. A peculiar form of peripheral neuropathy: Familial atypical generalized amyloidosis with special involvement of peripheral nerves. *Brain* 75: 408–27

Appeldoorn, B. J., Chandross, R. J., Bear, R. S., Hogan, E. L. 1975. An X-ray diffraction study of central and peripheral myelination in *Jimpy* and *Quaking* mice. *Brain Res.* 85: 517–21

Asano, K., Spector, S. 1979. Identification of inosine and hypoxanthine as endogenous ligands for the brain benzodiazepine-binding sites. *Proc. Natl. Acad. Sci. USA* 76: 977–81

Axelrod, F. B., Pearson, J. 1984. Congenital sensory neuropathies. *Am. J. Dis. Child.* 138: 947–54

Bakay, B., Nissinen, E., Sweetman, L., Francke, U., Nyhan, W. L. 1979. Utilization of purines by HGPRT variant in an intelligent, nonmutilative patient with features of the Lesch-Nyhan syndrome. *Pediat. Res.* 13: 1365–70

Baker, A. B., Baker, L. H. 1983. *Clinical Neurology*, Vol. 4, Chap. 56, p. 150. Philadelphia: Harper & Row

Barbarese, E., Carson, J. H., Braun, P. E. 1978. Accumulation of the four myelin basic proteins in mouse brain during development. *J. Neurochem.* 31: 779–82

Baumeister, A. A., Frye, G. D. 1985. The biochemical basis of the behavioral disorder in the Lesch-Nyhan syndrome. *Neurol. Biobehav. Rev.* 9: 169–78

Bell, J., Gruenthal, M., Finger, S., Lundberg, P., Johnson, E. 1982. Behavioral effects of early deprivation of nerve growth factor: Some similarities to familial dysautonomia. *Brain Res.* 234: 409–21

Benedict, W. F., Murphree, A. L., Banerjee, A., Spina, C. A., Sparkes, M. C., et al. 1983. Patient with 13 chromosome deletion: Evidence that the retinoblastoma gene is a recessive cancer gene. *Science* 219: 973–75

Bernd, P., Greene, L. A. 1984. Association of ^{125}I-nerve growth factor with PC12. *J. Biol. Chem.* 259: 15509–16

Botstein, D., White, R. L., Skolnick, M., Davis, R. W. 1980. Construction of a genetic linkage map in man using restriction fragment length polymorphisms. *Am. J. Hum. Genet.* 32: 314–31

Botton, D. C., Meyer, R. K., Prusiner, S. B. 1985. Scrapie PrP27-30 is a sialoglycoprotein. *J. Virol.* 53: 596–606

Bourre, J. M., Bornhofen, J. H., Araoz, C. A., Daudu, O., Baumann, N. A. 1978. Pelizaeus-Merzbacher disease: Brain lipid and fatty acid composition. *J. Neurochem.* 30: 719–27

Braig, H. R., Diringer, H. 1985. Scrapie: Concept of a virus-induced amyloidosis of the brain. *EMBO J.* 4: 2309–12

Breakefield, X. O., Orloff, G., Castiglione, C., Coussens, L., Axelrod, F. B., et al. 1984. The gene for beta-nerve growth factor is not defective in familial dysautonomia. *Proc. Natl. Acad. Sci. USA* 81: 4213–16

Breakefield, X. O., Stern, D. F. 1986. Oncogenes in neural tumors. *Trends Neurosci.* 9: 150–54

Breese, G. R., Baumeister, A., Napier, T. C., Frye, G. D., Mueller, R. A. 1985a. Evidence that D-1 dopamine receptors contribute to the supersensitive behavioral responses induced by L-dihydroxyphenylalanine in rats treated neonatally with 6-hydroxydopamine. *J. Pharm. Exp. Ther.* 235: 287–95

Breese, G. R., Napier, T. C., Mueller, R. A. 1985b. Dopamine agonist-induced locomotor activity in rats treated with 6-hydroxydopamine at different ages: Functional supersensitivity of D-1 dopamine receptors in neonatally lesioned rats. *J. Pharm. Exp. Ther.* 234: 447–55

Breitbart, R. E., Nguyen, H. T., Medford, R. M., Destree, A. T., Mamdavi, V., et al. 1985. Intricate combinatorial patterns of exon splicing generate multiple regulated troponin T isoforms from a single gene. *Cell* 41: 67–82

Brennand, J., Konecki, D. S., Caskey, C. T. 1983. Expression of human and Chinese hamster hypoxanthine-guanine phosphoribosyltransferase cDNA recombinants in cultured Lesch-Nyhan and Chinese hamster fibroblasts. *J. Biol. Chem.* 258: 9593–96

Brockes, J. P., Breakefield, X. O., Martuza, R. L. 1986. Glial growth factor-like activity in Schwann cell tumors. *Ann. Neurol.* 20: 317–22

Brown, C. S., Thomas, N. S., Sarfarazi, M., Davies, K. E., Kunkel, L., et al. 1985. Genetic linkage relationships of seven DNA probes with Duchenne and Becker muscular dystrophy. *Hum. Genet.* 71: 62–74

Bruce, M. E., Dickinson, A. G. 1985. Genetic control of amyloid plaque production and incubation period in scrapie-infected mice. *J. Neuropathol. Exp. Neurol.* 44: 285–94

Campagnoni, A. T. 1985. Molecular biology of myelination: Gene expression of myelin basic protein. In *Gene Expression in Brain*, ed. C. Zomzely-Neurath, W. A. Walker, pp. 206–33. New York: Wiley

Campagnoni, A. T., Campagnoni, C. W., Bourre, J. M., Jacque, C., Baumann, N. 1984. Cell-free synthesis of myelin basic proteins in normal and dysmyelinating mutant mice. *J. Neurochem.* 42: 733–39

Cavenee, W. K., Dryja, T. P., Phillips, R. A., Benedict, W. F., Godbout, R., et al. 1983. Expression of recessive alleles by chromosomal mechanisms in retinoblastoma. *Nature* 305: 779–84

Chao, M. V., Bothwell, M. A., Ross, A. H., Koprowski, H., Lavahan, A. A., et al. 1986. Gene transfer and molecular cloning of the human NGF receptor. *Science* 232: 518–21

Christie, R., Bay, C., Kaufman, I. A., Bakay, B., Borden, M., et al. 1982. Lesch-Nyhan disease: Clinical experience with nineteen patients. *Devel. Med. Child Neurol.* 24: 293–306

Clayson, D., Welton, W., Axelrod, F. B. 1980. Personality development and familial dysautonomia. *Pediatrics* 65: 269–74

Collins, F., Dawson, A. 1983. An effect of nerve growth factor on parasympathetic neurite outgrowth. *Proc. Natl. Acad. Sci. USA* 80: 2091–94

Cook, R. H., Ward, B. E., Austin, J. H. 1979. Studies in aging of the brain: IV. Familial Alzheimer disease: Relation to transmissible dementia, aneuploidy, and microtubular defects. *Neurology* 29: 1402–12

Costa, P. P., Figueira, A. S., Bravo, F. R. 1978. Amyloid fibril protein related to prealbumin in familial amyloid polyneuropathy. *Proc. Natl. Acad. Sci. USA* 75: 4499–4503

Costantino-Ceccarini, E., Matthieu, J.-M., Reigner, J. 1980. Biosynthesis of glycolipids in the peripheral nervous system of the mutant mice, quaking and jimpy. In *Neurological Mutations Affecting Myelination. INSERM Symp.* 14: 237–40

Coussens, L., Yang-Feng, T. L., Liao, Y.-C., Chen, E., Gray, A., et al. 1985. Tyrosine kinase receptor with extensive homology to EGF receptor shares chromosomal location with the *neu* oncogene. *Science* 230: 1132–39

Darby, J. K., Feder, J., Selby, M., Riccardi, V., Ferrell, R., et al. 1985. A discordant sibship analysis between beta-NGF and neurofibromatosis. *Am. J. Hum. Genet.* 37. 52–59

De Armond, S. J., McKinley, M. P., Barry, R. A., Braunfeld, M. B., McColloch, J. R., et al. 1985. Identification of prion amyloid filaments in scrapie-infected brain. *Cell* 41: 221–35

De Ferra, F., Engh, H., Hudson, L., Kamholz, J., Puckett, C., et al. 1985. Alternative splicing accounts for the four forms of myelin basic protein. *Cell* 43: 721–27

Delassalle, A., Zalc, B., Lachapelle, F., Raoul, M., Collier, P., et al. 1981. Regional distribution of myelin basic protein in the central nervous system of quaking, jimpy, and normal mice during development and aging. *J. Neurosci. Res.* 6: 303–13

Dewey, M. J., Martin, D. W. Jr., Martin, G. R., Mintz, B. 1977. Mosaic mice with teratocarcinoma-derived mutant cells deficient in hypoxanthine phosphoribosyltransferase. *Proc. Natl. Acad. Sci. USA* 74: 5564–68

Dryja, T. P., Cavenee, W., White, R., Rapaport, J. M., Petersen, R., et al. 1984. Homozygosity of chromosome 13 in retinoblastoma. *New Engl. J. Med.* 310: 550–53

Ellison, G., Eison, M. S., Huberman, H. S., Daniel, F. 1978. Long-term changes in dopaminergic innervation of caudate nucleus after continuous amphetamine administration. *Science* 201: 276–78

Fattal, J., Spirer, Z., Zoref-Shani, E., Sperling, O. 1984. Lesch-Nyhan syndrome: Biochemical characterization of a case with attenuated behavioral manifestation. *Enzyme* 31: 55–60

Folch-Pi, Lees, M. B. 1951. Proteolipids, a new type of tissue lipoproteins. Their isolation from brain. *J. Biol. Chem.* 191: 807–17

Francke, U., Felstein, J., Gartler, S. M., Migeon, B. R., Dancis, J., et al. 1976. The occurrence of new mutants in the X-linked recessive Lesch-Nyhan disease. *Am. J. Hum. Genet.* 28: 123–37

Francke, U., de Martinville, B., Coussens, L., Ullrich, A. 1983. The human gene for the beta subunit of nerve growth factor is located on the proximal short arm of chromosome 1. *Science* 222: 1248–51

Gajdusek, D. C. 1985. Hypothesis: Interference with axonal transport of neurofilament as a common pathogenic mechanism in certain diseases of the central nervous system. *New England J. Med.* 312: 714–19

Garson, J. A., McIntyre, P. G., Kemshead, J. T. 1985. N-*myc* amplification in a malignant astrocytoma. *Lancet* 28: 718–19

Glenner, G. G. 1980. Amyloid deposits and amyloidosis. The beta fibrilloses. *New Engl. J. Med.* 302: 1283–92

Goldstein, L. D., Reynolds, C. P., Perez-Polo, J. R. 1978. Isolation of human nerve growth factor from placental tissue. *Neurochem. Res.* 3: 175–80

Gorin, P. D., Johnson, E. M. 1979. Experimental autoimmune model of nerve growth factor deprivation: Effects on developing peripheral sympathetic and sensory neurons. *Proc. Natl. Acad. Sci. USA* 76: 5382–86

Gorin, P. D., Johnson, Jr. E. M. 1980. Effects of long-term nerve growth factor deprivation on the nervous system of the adult rat: An experimental autoimmune approach. *Brain Res.* 198: 27–42

Goudsmit, J., White, B. J., Weitkamp, L. R., Krats, B. J. B., Morrow, C. H., et al. 1981. Familial Alzheimer's disease in two kindreds of the same geographic and ethnic origin. A clinical and genetic study. *J. Neurol. Sci.* 49: 79–89

Greene, L. A. 1984. The importance of both early and delayed responses in the biological actions of nerve growth factor. *Trends Neurosci.* 7: 91–94

Greene, L. A., Shooter, E. M. 1980. The nerve growth factor: Biochemistry, synthesis and mechanisms of action. *Ann. Rev. Neurosci.* 3: 353–402

Greenfield, S., Brostoff, S., Eylar, E. H., Morell, P. 1973. Protein composition of myelin of the peripheral nervous system. *J. Neurochem.* 20: 1207–16

Grob, P. M., Ross, A. H., Koprowski, H., Bothwell, M. 1985. Characterization of the human melanoma nerve growth factor receptor. *J. Biol. Chem.* 260: 8044–49

Gruber, H. E., Vuchinich, M., Marlow, T. A., Plent, M. M., Willis, R. C., et al. 1984. Clinical and biochemical correlates of a new HPRT mutation. *Adv. Exp. Med. Biol.* 165: 19–22

Gusella, J. F., Wexler, N. S., Conneally, P. M., Naylor, S. L., Anderson, M. A. 1983. A polymorphic DNA marker genetically linked to Huntington's disease. *Nature* 306: 234–38

Gusella, J. F., Tanzi, R. E., Anderson, M. A., Hobbs, W., Gibbons, K., et al. 1984. DNA markers for nervous system diseases. *Science* 225: 1320–26

Gusella, J. F., Tanzi, R. E., Bader, P. I., Phelan, M. C., Stevenson, R., et al. 1985. Deletion of Huntington's disease-linked G8 (D4S10) locus in Wolf-Hirschhorn syndrome. *Nature* 318: 75–78

Hammerberg, C., Klein, J. 1975. Linkage relationships of markers on chromosome 17 of the house mouse. *Genet. Res.* 26: 203–13

Hansen, M. F., Koufos, A., Gallie, B. L., Phillips, R. A., Fodstad, O., et al. 1985. Osteosarcoma and retinoblastoma: A shared chromosome mechanism revealing predisposition. *Proc. Natl. Acad. Sci. USA* 82: 1–5

Harper, G. P., Thoenen, H. 1980. The nerve

growth factor: Biological significance, measurement and distribution. *J. Neurochem.* 34: 5–16

Heston, L. L., Lowther, D. L. W., Leventhal, C. M. 1966. Alzheimer's disease: A family study. *Arch. Neurol.* 15: 225–33

Heyman, A., Wilkinson, W. E., Hurwitz, B. J., Schmechel, D., Sigmon, A. H., et al. 1983. Alzheimer's disease: Genetic aspects and associated clinical disorders. *Ann. Neurol.* 14: 507–15

Hittner, H. M., Riccardi, V., Kretzer, F. L., Levy, C. H., Moura, R. A. 1979. Two-step mutation theory for retinoblastoma: Ultrastructural support. *Docum. Ophthal.* 48: 345–62

Huebner, K., Isobe, M., Chao, M., Bothwell, M., Ross, A. H., et al. 1986. The Nerve growth factor receptor gene is at human chromosome region 17q12/17q22, distal to the chromosome 17 breakpoint in acute leukemias. *Proc. Natl. Acad. Sci. USA* 83: 1403–7

Hung, M. C., Schechter, A. L., Chevray, P. Y., Stern, D. F., Weinberg, R. A. 1986. Molecular cloning of the neu gene: Absence of gross structural alteration in oncogenic alleles. *Proc. Natl. Acad. Sci. USA* 83: 261–64

Hurley, J. B., Simon, M. I., Teplow, D. B., Robishaw, J. D., Gilman, A. G. 1984. Homologies between signal transducing G proteins and ras gene products. *Science* 226: 860–62

Huson, S. M., Thrush, D. C. 1985. Central neurofibromatosis. *Quant. J. Med.* 55: 213–24

Izant, J. G., Weintraub, H. 1984. Inhibition of thymidine kinase gene expression by anti-sense RNA: A molecular approach to genetic analysis. *Cell* 36: 1007–15

Isackson, P. J., Ullrich, A., Bradshaw, R. A. 1984. Mouse 7S nerve growth factor: Complete sequence of a cDNA coding for the alpha-subunit precursor and its relationship to serine proteases. *Biochemistry* 23: 5997–6002

Jaenisch, R., Harbers, K., Schnieke, A., Lohler, J., Chumakov, I., et al. 1983. Germline integration of moloney murine leukemia virus at the mov 13 locus leads to recessive lethal mutation and early embryonic death. *Cell* 32: 209–16

Jeffreys, A. J., Wilson, V., Lay, T. 1985. Hypervariable "minisatellite" regions in human DNA. *Nature* 314: 67–73

Johnson, E. M. Jr., Rich, K. M., Yip, H. K. 1986. The role of NGF in sensory neurons *in vivo*. *Trends Neurosci.* 9: 33–37

Kanter, W. R., Eldridge, R., Fabricant, R., Allen, J. C., Koerber, T. 1980. Central neurofibromatosis with bilateral acoustic neuroma: Genetic, clinical and biochemical distinctions from peripheral neurifibromatosis. *Neurology* 30: 851–59

Katsuragi, T., Ushijima, I., Furukawa, T. 1984. The clonidine-induced self-injurious behavior of mice involves purinergic mechanisms. *Pharmacol. Biochem. Behav.* 20: 943–46

Kirschner, D. A., Abraham, C., Selkoe, D. J. 1986. X-ray diffraction from intraneuronal paired helical filaments and extraneuronal amyloid fibers in Alzheimer disease indicates cross-beta conformation. *Proc. Natl. Acad. Sci. USA* 83: 503–7

Kish, S. J., Fox, I. H., Kapur, B. M., Lloyd, K., Hornykiewicz, O. 1985. Brain benzodiazepine receptor binding and purine concentration in Lesch-Nyhan syndrome. *Brain Res.* 336: 117–23

Knudson, A. G. Jr. 1978. Retinoblastoma: A prototypic hereditary neoplasia. *Semin. Oncol.* 5: 57–60

Koeppen, A. H. 1986. The genetic defect in Pelizaeus-Merzbacher disease. *7th Int. Congr. Hum. Genet.* In press. (Abstr.)

Kohl, N. E., Kanda, N., Schreck, R. R., Bruns, G., Latt, S., et al. 1983. Transposition and amplification of oncogene-related sequences in human neuroblastomas. *Cell* 35: 359–67

Kopin, I. J. 1981. Neurotransmitters and the Lesch-Nyhan syndrome. *New Engl. J. Med.* 305: 1148–49

Korsching, S., Auburger, G., Heumann, R., Scott, J., Thoenen, H. 1985. Levels of nerve growth factor and its mRNA in the central nervous system of the rat correlate with cholinergic innervation. *EMBO J.* 4: 1389–93

Kyritsis, A. P., Tsokos, M., Triche, T. J., Chader, G. J. 1984. Retinoblastoma-origin from a primitive neuroectodermal cell? *Nature* 307: 471–73

Lathrop, G. M., Lalouel, J. M., Julier, C., Ott, J. 1984. Strategies for multilocus linkage analysis in humans. *Proc. Natl. Acad. Sci. USA* 81: 3443–46

Lee, W.-H., Murphree, L., Benedict, W. F. 1984. Expression and amplification of the N-myc gene in primary retinoblastoma. *Nature* 309: 458–60

Lees, M. B., Brostoff, S. W. 1984. Proteins of myelin. In *Myelin*, ed. P. Morell, pp. 197–224. New York: Plenum

Lemke, G., Axel, R. 1985. Isolation and sequence of a cDNA encoding the major structural protein of peripheral myelin. *Cell* 42: 501–8

Libermann, T. A., Nusbaum, H. R., Razon, N., Kris, R., Lax, I., et al. 1985. Amplification, enhanced expression and possible rearrangement of EGF receptor gene in primary human brain tumours of glial origin. *Nature* 313: 144–47

Libermann, T. A., Razon, N., Bartal, A. D., Yarden, Y., Schlessinger, J., et al. 1984. Expression of epidermal growth factor receptors in human brain tumors. *Cancer Res.* 44: 753–60

Lloyd, K. G., Hornykiewicz, O., Davidson, L., Shannak, K., Farley, I., et al. 1981. Biochemical evidence of dysfunction of brain neurotransmitters in the Lesch-Nyhan syndrome. *New Engl. J. Med.* 305: 1106–11

Mann, R., Mulligan, R. C., Baltimore, D. 1983. Construction of a retrovirus packaging mutant and its use to produce helper-free defective retrovirus. *Cell* 33: 153–59

Mantzouranis, E. C., Dowton, S. B., Whitehead, A. S., Edge, M. D., Bruns, G. A., et al. 1985. Human serum amyloid P component. cDNA isolation, complete sequence of pre-serum amyloid P component, and localization of the gene to chromosome 1. *J. Biol. Chem.* 260: 7752–56

Martenson, R. E., Deibler, G. E., Kies, M. W. 1970. Myelin basic proteins of the rat central nervous system. *Biochem. Biophys. Acta.* 200: 353–62

Martuza, R. L., Ojemann, R. G. 1982. Bilateral acoustic neuromas: Clinical aspects, pathogenesis, and treatment. *Neurosurgery* 10: 1–12

Mason, A. J., Evans, B. A., Cox, D. R., Shine, J., Richards, R. I. 1983. Structure of mouse kallikrein gene family suggests a role in specific processing of biologically active peptides. *Nature* 303: 300–7

Masters, C. L., Multhaup, G., Simms, G., Pottgiesser, J., Martins, R. N., et al. 1985a. Neuronal origin of a cerebral amyloid: Neurofibrillary tangles of Alzheimer's disease contain the same protein as the amyloid of plaque cores and blood vessels. *EMBO J.* 4: 2757–63

Masters, C. L., Simms, G., Weinman, N. A., Multhaup, G., McDonald, B. L., et al. 1985b. Amyloid plaque core protein in Alzheimer disease and Down syndrome. *Proc. Natl. Acad. Sci. USA* 82: 4245–49

Merz, P. A., Rohwer, R. G., Kascsak, R., Wisniewski, H. M., Somerville, R. A., et al. 1984. Infection-specific particle from the unconventional slow virus diseases. *Science* 225: 437–40

Merzbacher, L. 1910. Eine eigenartige familiare Erkrankeengsform (Aplasia axialis extracorticalis congenita). *Z. Ges. Neurol. Psychiat.* 3: 1–138

Miller, A. D., Jolly, D. J., Friedmann, T., Verma, I. M. 1983. A transmissible retrovirus expressing human hypoxanthine posphoribosyltransferase (HPRT): Gene transfer into cells obtained from humans deficient in HPRT. *Proc. Natl. Acad. Sci. USA* 80: 4709–13

Milner, R. J., Sutcliffe, J. G. 1983. Gene expression in rat brain. *Nucl. Acids Res.* 11: 5497–5520

Milner, R. J., Lai, C., Nave, K.-A., Lenoir, D., Ogata, J., et al. 1985. Nucleotide sequences of two mRNAs for rat brain myelin proteolipid protein. *Cell* 42: 931–39

Monaco, A. P., Bertelson, C. J., Middlesworth, W., Colletti, C. A., Aldridge, J., et al. 1985. Detection of deletions spanning the Duchenne muscular dystrophy locus using a tightly linked DNA segment. *Nature* 316: 842–45

Mueller, K., Saboda, S., Palmour, R., Nyhan, W. L. 1982. Self-injurious behavior produced in rats by daily caffeine and continuous amphetamine. *Pharmacol. Biochem. Behav.* 17: 613–17

Murphree, A. L., Benedict, W. 1984. Retinoblastoma: Clues to human oncogenesis. *Science* 233: 1028–33

Murray, J. M., Davies, K. E., Harper, P. S., Meredith, L., Mueller, C. R., et al. 1982. Linkage relationship of a cloned DNA sequence on the short arm of the X chromosome to Duchenne muscular dystrophy. *Nature* 300: 69–71

Myerowitz, R., Piekarz, R., Neufeld, E. F., Shows, T. B., Suzuki, K. 1985. Human beta hexosaminidase alpha chain: Coding sequence and homology with the beta chain. *Proc. Natl. Acad. Sci. USA* 82: 7830–34

Myers, R. M., Lumelsky, N., Lerman, L. S., Maniatis, T. 1985. Detection of single base substitutions in total genomic DNA. *Nature* 313: 495–98

Naismith, A. L., Hoffman-Chudzik, E., Tsui, L. C., Riordan, J. R. 1985. Study of the expression of myelin proteolipid protein (lipophilin) using a cloned complementary DNA. *Nucleic Acids Res.* 13: 7413–25

Naylor, S. L., Gusella, J., Sakaguchi, A. Y. 1984. Mapping DNA polymorphisms to human chromosomes. *Cytogen. Cell Gen.* 37: 553–54

Newbold, R. 1984. Cancer: Mutant *ras* proteins and cell transformation. *Nature* 310: 628–29

Novack, D. F., Casna, N. J., Fischer, S. G., Ford, J. P. 1986. Detection of single base-pair mismatches in DNA by chemical modification followed by electrophoresis in 15% polyacrylamide gel. *Proc. Natl. Acad. Sci. USA* 83: 586–90

Nussbaum, J. L., Mandel, P. 1973. Brain proteolipids in neurological mutant mice. *Brain Res.* 61: 295–310

Nussbaum, R., Brennand, J., Chinault, C., Fuscoe, J., Konecki, D., et al. 1983.

Molecular analysis of the hypoxanthine phosphoribosyltransferase locus. In *Recombinant DNA Application to Human Disease*, ed. C. T. Caskey, R. L. White, pp. 81–89. Cold Spring Harbor, NY: Cold Spring Harbor Press

O'Dowd, B. F., Quan, F., Willard, H. F., Lamhonwah, A.-M., Korneluk, R. G., et al. 1985. Isolation of cDNA clones coding for the beta subunit of human beta hexosaminidase. *Proc. Natl. Acad. Sci. USA* 82: 1184–88

Oesch, B., Westaway, D., Walchli, M., McKinley, M. P., Kent, S. B. H., et al. 1985. A cellular gene encodes scrapie PrP 27–30 protein. *Cell* 40: 735–46

Orkin, S. H., Kazazian, H. H. 1984. The mutation and polymorphism of the human beta-globin gene and its surrounding DNA. *Ann. Rev. Genet.* 18: 131–71

Pantazis, N. J., Blanchard, M. H., Arnason, B. G. W., Young, M. 1977. Molecular properties of the nerve growth factor secreted by L cells. *Proc. Natl. Acad. Sci. USA* 74: 1492–96

Pantazis, P., Pelicci, P. G., Dalla-Favera, R., Antoniades, H. N. 1985. Synthesis and secretion of proteins resembling platelet-derived growth factor by human glioblastoma and fibrosarcoma cells in culture. *Proc. Natl. Acad. Sci. USA* 82: 2404–8

Pearson, J., Pytel, B. 1978a. Quantitative studies of ciliary and sphenopalatine ganglia in familial dysautonomia. *J. Neurol. Sci.* 39: 123–30

Pearson, J., Pytel, B. 1978b. Quantitative studies of sympathetic ganglia and spinal cord intermedio-lateral gray columns in familial dysautonomia. *J. Neurol. Sci.* 39: 47–59

Pearson, J., Brandeis, L., Goldstein, M. 1979. Tyrosine hydroxylase immunoreactivity in familial dysautonomia. *Science* 206: 71–72

Pearson, J., Brandeis, L., Cuello, A. C. 1982. Depletion of substance P-containing axons in substantia gelatinosa of patients with diminished pain sensitivity. *Nature* 295: 61–63

Pearson, J., Pytel, B. A., Grover-Johnson, N., Axelrod, F., Dancis, J. 1978. Quantitative studies of dorsal root ganglia and neuropathologic observations on spinal cords in familial dysautonomia. *J. Neurol. Sci.* 35: 77–92

Pearson, J., Johnson, E. M., Brandeis, L. 1983. Effects of antibodies to nerve growth factor on intrauterine development of derivatives of cranial neural crest and placode in the guinea pig. *Dev. Biol.* 96: 32–36

Perry, E., Perry, R. H. 1985. New insights into the nature of senile (Alzheimer-type) plaques. *Trends Neurosci.* 8: 301–5

Perry, G., Rizzuto, N., Autilio-Gambetti, L., Gambetti, P. 1985. Paired helical filaments from Alzheimer disease patients contain cytoskeletal components. *Proc. Natl. Acad. Sci. USA* 82: 3916–20

Phillips, R. A., Gallie, B. L. 1984. Retinoblastoma: Importance of recessive mutations in tumorigenesis. *J. Cell. Physiol. Suppl.* 3: 79–85

Pras, M., Prelli, F., Franklin, E. C., Frangione, B. 1983. Primary structure of an amyloid prealbumin variant in familial polyneuropathy of Jewish origin. *Proc. Natl. Acad. Sci. USA* 80: 539–42

Price, D. L., Whitehouse, P. J., Struble, R. G. 1986. Cellular pathology in Alzheimer's and Parkinson's diseases. *Trends Neurosci.* 9: 29–33

Purves, D., Lichtmann, J. W. 1985. *Principles of Neural Development*. Sunderland, Mass.: Sinauer Assoc.

Ramsay, G., Stanton, L., Schwab, M., Bishop, J. M. 1986. N-*myc* encodes a nuclear protein that binds to DNA. *Molec. Cell. Biol.* In press

Rajewski, M. F. 1983. Structural modifications and repair of DNA in neurooncogenesis by *N*-ethyl-*N*-nitrosurea. *Recent Results Cancer Res.* 84: 63–76

Ray, P. N., Belfall, B., Duff, C., Logan, C., Kean, V., et al. 1985. Cloning of the breakpoint of an X; 21 translocation associated with Duchenne muscular dystrophy. *Nature* 318: 672–75

Roach, A., Boylan, K., Horvath, S., Prusiner, S. B., Hood, L. E. 1983. Characterization of cloned cDNA representing rat myelin basic protein: Absence of expression in brain of shiverer mutant mice. *Cell* 34: 799–806

Roach, A., Takahashi, N., Pravtcheva, D., Ruddle, F., Hood, L. 1985. Chromosomal mapping of mouse myelin basic protein gene and structure and transcription of the partially deleted gene in shiverer mutant mice. *Cell* 42: 149–55

Roberts, F. F., Lee, K. R. 1975. Familial neuroblastoma presenting as multiple tumors. *Radiology* 116: 133–36

Rosenberg, R. N. 1984. Molecular genetics, recombinant DNA techniques, and genetic neurological disease. *Ann. Neurol.* 15: 511–20

Rosenberg, M. B., Hansen, C., Breakefield, X. O. 1985. Molecular genetic approaches to neurologic and psychiatric diseases. *Prog. Neurobiol.* 24: 95–140

Roses, A. D., Pericak-Vance, M. A., Yamaoka, L. H., Stubblefield, E., Stajich, J., et al. 1983. Recombinant DNA strategies in genetic neurological diseases. *Muscle Nerve* 6: 339–55

Roth, H. J., Hunkeler, M. J., Campagnoni, A. T. 1985. Expression of myelin basic protein genes in several dysmelinating mouse mutants during early postnatal brain development. *J. Neurochem.* 45: 572–80

Sack, G. H. Jr., Dumars, K. W., Gummerson, K. S., Law, A., McKusick, V. A. 1981. Three forms of dominant amyloid neuropathy. *Johns Hopkins Med. J.* 149: 239–47

Sasaki, H., Sasaki, Y., Matsuo, H., Goto, I., Kuroiwa, Y., et al. 1984. Diagnosis of familial amyloidotic polyneuropathy by recombinant DNA techniques. *Biochem. Biophys. Res. Commun.* 125: 636–42

Sasaki, H., Sasaki, Y., Takagi, Y., Sahashi, K., Takahashi, A., et al. 1985. Presymptomatic diagnosis of heterozygosity for familial amyloidotic polyneuropathy by recombinant DNA techniques. *Lancet* 1: 100

Saxe, D. F., Takahashi, N., Hood, L., Simon, I. 1986. Localization of the human myelin basic protein gene (MBP) to region 18q22 Ter by *in situ* hybridization. *Cytogenet. Cell Genet.* In press

Schechter, A. L., Stern, D. F., Vaidyanathan, L., Decker, S. J., Drebin, J. A. 1984. The neu oncogene: An erb-B-related gene encoding a 185,000-Mr tumour antigen. *Nature* 312: 513–16

Schwab, M. 1986. Amplification of proto-oncogenes may contribute to tumor progression. In *Oncogenes and Growth Control*, ed. T. Graf, P. Kahn. New York: Springer-Verlag. In press

Schwab, M., Alitalo, K., Klempnauer, K.-H., Varmus, H. E., Bishop, J. M., et al. 1983. Amplified DNA with limited homology to *myc* cellular oncogene is shared by human neuroblastoma cell lines and a neuroblastoma tumour. *Nature* 305: 245–48

Schwab, M., Ellison, J., Bush, M., Rosenau, W., Varmus, H. E., et al. 1984a. Enhanced expression of the human gene N-*myc* consequent to amplification of DNA may contribute to malignant progression of neuroblastoma. *Proc. Natl. Acad. Sci. USA* 81: 4940–44

Schwab, M., Varmus, H. E., Bishop, J. M., Grzeschile, K. H., Naylor, S., et al. 1984b. Chromosome localization in normal human cells and neuroblastoma of a gene related to c-*myc*. *Nature* 308: 288–91

Schwartz, J. P., Breakefield, X. O. 1980. Altered nerve growth factor in fibroblasts from patients with familial dysautonomia. *Proc. Natl. Acad. Sci. USA* 77: 1154–58

Scott, J., Selby, M., Urdea, M., Quiroga, M., Bell, G. I., et al. 1983. Isolation and nucleotide sequence of a cDNA encoding the precursor of mouse nerve growth factor. *Nature* 302: 538–40

Seitelberger, F. 1970. Leukodystrophies and poliodystrophies. *Handb. Clin. Neurol.* 10: 151–202

Seitzinger, B. R., Martuza, R. L., Gusella, J. F. 1986. Loss of genes on chromosome 22 in tumorigenesis of human acoustic neuroma. *Nature* 322: 644–47

Shiloh, Y., Shipley, J., Brodeur, G. M., Bruns, G., Korf, B., et al. 1985. Differential amplification, assembly, and relocation of multiple DNA sequences in human neuroblastomas and neuroblastoma cell lines. *Proc. Natl. Acad. Sci. USA* 82: 3761–65

Sidman, R. L., Conover, C. S., Carson, J. H. 1985. Shiverer gene maps near the distal end of chromosome 18 in the house mouse. *Cytogenet. Cell Genet.* 39: 241–45

Siggers, D. C., Rogers, J. G., Boyer, S. H., Margolet, G., Dorkin, H., et al. 1976. Increased nerve growth factor beta-chain cross reacting material in familial dysautonomia. *New Engl. J. Med.* 295: 629–34

Sipe, J. D., Colten, H. R., Goldberger, G., Edge, M. D., Tack, B. F., et al. 1985. Human serum amyloid A (SAA): Biosynthesis and postsynthetic processing of pre-SAA and structural variants defined by complementary DNA. *Biochemistry* 24: 2931–36

Skaper, S. D., Varon, S. 1982. Three independent biological assays for nerve growth factor: No measurable activity in human sera. *Exp. Neurol.* 76: 655–65

Skinner, M., Cohen, A. S. 1981. The prealbumin nature of the amyloid protein in familial amyloid polyneuropathy (FAP)-Swedish variety. *Biochem. Biophys. Res. Commun.* 99: 1326–32

Smithies, O., Gregg, R. G., Boggs, S. S., Koraleski, M. A., Kucherlapati, R. S. 1985. Insertion of DNA sequences into human chromosomal beta-globin locus by homologous recombination. *Nature* 317: 230–34

Sorg, B. J. A., Agrawal, D., Agrawal, H. C., Campagnoni, A. T. 1986. Expression of myelin proteolipid protein and basic protein in normal and dysmyelinating mutant mice. *J. Neurochem.* 46: 379–87

Stark, G. R., Wahl, G. M. 1984. Gene amplification. *Ann. Rev. Biochem.* 53: 447–91

Stout, J. T., Caskey, C. T. 1985. HPRT: Gene structure, expression, and mutation. *Ann. Rev. Genet.* 19: 127–48

Stout, J. T., Caskey, C. T. 1986. Anti-sense RNA inhibition of endogenous genes. *Methods Enzymol.* In press

Suda, K., Barde, Y. A., Thoenen, H. 1978. Nerve growth factor in mouse and rat serum: Correlation between bioassay and radioimmunoassay determinations. *Proc. Natl. Acad. Sci. USA* 75: 4042–46

Sutcliffe, J. G., Milner, R. J., Shinnick, T. M., Bloom, F. E. 1983. Identifying the protein products of brain-specific genes with antibodies to chemically synthesized peptides. *Cell* 33: 671–82

Sutter, A., Hosang, M., Vale, R. D., Shooter, E. M. 1984. The interaction of nerve growth factor with its specific receptor. In *Cellular and Molecular Biology of Neuronal Development*, ed. I. B. Black, pp. 201–14. New York: Plenum

Takahashi, N., Roach, A., Teplow, D. B., Prusiner, S. B., Hood, L. 1985. Cloning and characterization of the myelin basic protein gene from mouse: One gene can encode both 14 kd and 18.5 kd MBPs by alternate use of exons. *Cell* 42: 139–48

Taparowsky, E., Shimizu, K., Goldfarb, M., Wigler, M. 1983. Structure and activation of the human N-*ras* gene. *Cell* 34: 581–86

Thiele, C. J., Reynolds, C. P., Israel, M. A. 1985. Decreased expression of N-*myc* precedes retinoic acid-induced morphological differentiation of human neuroblastoma. *Nature* 313: 404–6

Thoenen, H., Edgar, D. 1985. Neurotrophic factors. *Science* 229: 238–42

Trent, J., Meltzer, P., Rosenblum, M., Harsh, G., Kinsler, K., et al. 1986. Evidence for rearrangement, amplification, and expression of c-myc in a human glioblastoma. *Proc. Natl. Acad. Sci. USA* 83: 47073

Ullrich, A., Gray, A., Wood, W. I., Hayflick, J., Seeburg, P. H. 1984a. Isolation of a cDNA clone coding for the gamma-subunit of mouse nerve growth factor using a high-stringency selection procedure. *DNA* 3: 387–92

Ullrich, A., Coussens, L., Hayflick, J. S., Dull, T. J., Gray, A., et al. 1984b. Human epidermal growth factor receptor cDNA sequence and aberrant expression of the amplified gene in A431 epidermoid carcinoma cells. *Nature* 309: 418–25

Ullrich, A., Gray, A., Berman, C., Dull, T. J. 1983. The nucleotide sequence of human nerve growth factor beta subunit gene is highly homologous to that of mouse. *Nature* 303: 821–25

Varshavsky, A. 1981. On the possibility of metabolic control of replicon misfiring: Relationship to emergence of malignant phenotypes in mammalian cell lineages. *Proc. Natl. Acad. Sci. USA* 78: 3673–77

Wallace, M. R., Naylor, S. L., Kluve-Beckerman, B., Long, G., McDonald, L., et al. 1985. Localization of the human prealbumin gene to chromosome 18. *Biochem. Biophys. Res. Commun.* 129: 753–58

Watanabe, I., McCaman, R., Dyken, P., Zeman, W. 1969. Absence of cerebral myelin sheaths in a case of presumed Pelizaeus-Merzbacher disease: Electron microscopic and biochemical studies. *J. Neuropath. Exp. Neurol.* 28: 243–56

Watanabe, I., Patel, V., Goebel, H. H., Siakotos, A. N., Zeman, W., et al. 1973. Early lesion of Pelizaeus-Merzbacher disease: Electron microscopic and biochemical study. *J. Neuropath. Exp. Neurol.* 32: 313–33

Watkins, P. C., Tanzi, R. E., Gibbons, K. P., Tricoli, J. V., Landis, G., et al. 1985. Isolation of polymorphic DNA segments from human chromosome 21. *Nucleic Acid Res.* 13: 6075–88

Watts, R. W. E., Spellacy, E., Gibbs, D. A., Allsop, J., McKeran, R. O., et al. 1982. Clinical, post-mortem, biochemical and therapeutic observations on the Lesch-Nyhan syndrome with particular reference to the neurological manifestations. *Q. J. Med.* 201: 43–78

Weinberg, R. A. 1985. The action of oncogenes in the cytoplasm and nucleus. *Science* 230: 770–76

Welton, W., Clayson, D., Axelrod, F. B., Levine, D. B. 1979. Intellectual development and familial dysautonomia. *Pediatrics* 63: 708–12

Wexler, N. S., Young, A., Tanzi, R., Starosta, S., Gomez, F., et al. 1985. Huntington's disease heterozygotes detected. *Am. J. Hum. Genet.* 37: a82

White, R., Leppert, M., Bishop, D. T., Barker, D., Berkowitz, J., et al. 1985. Construction of linkage maps with DNA markers for human chromosomes. *Nature* 313: 101–5

Willard, H. F., Riordan, J. R. 1985. Assignment of the gene for myelin proteolipid protein to the X chromosome: Implications for X-linked myelin disorders. *Science* 230: 940–42

Williams, D. A., Lemischka, I. R., Nathan, D. J., Mulligan, R. C. 1984. Introduction of new genetic material into pluripotent haematopoietic stem cells of the mouse. *Nature* 310: 476–80

Willis, R. C., Jolly, D. J., Miller, A. D., Plent, M. M., Esty, A. C., et al. 1983. Partial phenotypic correction of human Lesch-Nyhan (hypoxanthine-guanine phosphoribosyltransferase-deficient) lymphoblasts with a transmissible retroviral vector. *J. Biol. Chem.* 259: 7842–47

Wilson, J. M., Young, A. B., Kelley, W. N. 1983. Hypoxanthine-guanine phosphoribosyltransferase deficiency. *New Engl. J. Med.* 309: 900–10

Wong, C. W., Quaranta, V., Glenner, G. G.

1985. Neuritic plaques and cerebrovascular amyloid in Alzheimer disease are antigenically related. *Proc. Natl. Acad. Sci. USA* 82: 8729–32

Yang, T. P., Patel, P. I., Chinault, A. C., Stout, J. T., Jackson, J. G., et al. 1984. Molecular evidence for new mutation at the hprt locus in Lesch-Nyhan patients. *Nature* 310: 412–13

Zeigler, M. G., Lake, R. C., Kopin, I. J. 1976. Deficient sympathetic nervous response in familial dysautonomia. *New Engl. J. Med.* 294: 630–33

Zeller, N. K., Hunkeler, M. J., Campagnoni, A. T., Sprague, J., Lazzarini, R. A. 1984. Characterization of mouse myelin basic protein messenger RNAs with a myelin basic protein cDNA clone. *Proc. Natl. Acad. Sci. USA* 81: 18–22

Zimmerman, K. A., Yancopoulos, G. D., Collum, R. G., Smith, R. K., Kohl, N. E., et al. 1986. Differential expression of myc family genes during murine development. *Nature* 319: 780–83

Ann. Rev. Neurosci. 1987. 10: 595–632

GUSTATORY NEURAL PROCESSING IN THE HINDBRAIN

Joseph B. Travers,[1] *Susan P. Travers,*[1] *and Ralph Norgren*

Department of Behavioral Science, College of Medicine, The Pennsylvania State University, Hershey, Pennsylvania 17033

INTRODUCTION

Sensory neurophysiology must account for sensation and perception. Given the richness and complexity of our auditory and visual worlds, this remains a daunting charge even after decades of sophisticated visual and auditory neurophysiology. Taste, on the other hand, should be easy. It contains little of the spatial and temporal dimensions found in vision, audition, or somesthesis, and humans report only a few taste categories or qualities. In English, the most common are sweet, salty, sour, and bitter. Given this apparently simpler task, how does gustatory neurophysiology acquit itself? Gustatory neurophysiological data, primarily from rodents, match human quality judgments moderately well, but the degree of this success or failure depends upon the theoretical bias of the beholder, as well as on the scope of the gustatory responses included. Taste *per se* may have only a few sensory categories, but other attributes contribute complexity and richness to the sensory experience that leaves present neurophysiological data far behind.

The absence of obvious stimulus dimensions continues to impede neurophysiological analysis of gustatory sensibility. The molecules that elicit similar verbal reports of taste from humans and neural activity in gustatory afferent nerves often bear little chemical relation to one another. Sugars,

[1] Present address: Department of Oral Biology, College of Dentistry, The Ohio State University, Columbus, Ohio 43210-1241.

0147–006X/87/0301–0595$02.00

of course, form a chemical family, and many mono- and disaccharides taste sweet to humans. Not all sugars, however, taste particularly sweet (deoxyribose; Moskowitz 1974) and many other, unrelated molecules, such as glycine and alanine, do. The lack of a rational physico-chemical stimulus dimension dictates another approach to deciphering the gustatory system; that is, deriving organizing principles for the stimulus domain either from neural responses or from behavior (Erickson et al 1965, McBurney & Gent 1979). Even the nature of taste sensations, however, is subject to debate. Many argue for a limited number, usually four (McBurney & Gent 1979). Others, however, find the data supporting the four-taste position unconvincing and suggest the possibility of a continuum of sensations (Erickson 1985). Thus, the choice of stimulating chemicals and their concentrations remains a theoretical issue, rather than a simple starting point for analysis.

Moderate concentrations of sucrose, NaCl, HCl, and quinine HCl (QHCl) produce qualitatively different sensations in humans. In some species, particularly laboratory rats, behavioral evidence indicates that these same stimuli also are highly discriminable (Nowlis et al 1980) and support preference and aversion responses similar to those of humans (Pfaffmann 1960). Despite the limitations inherent in using only four chemicals to span an unknown domain, these prototypical stimuli have been used so often in studies at all levels of the peripheral and central gustatory system that, perforce, they have come to be considered standard. As standards, they facilitate comparisons among physiology, behavior, species, and levels of the neuraxis. A contemporary account of the mammalian gustatory system cannot ignore them.

The generality of the human perceptual categories can be supported with a biological or evolutionary rationale summarized as follows. The primary biological function of the gustatory system is to assess some chemical characteristics of food and fluid. Molecules that elicit unambiguous category responses from humans are found commonly in the natural foods of many species and often provide important nutritional information. Alkaloids are both poisonous and bitter. Sodium ions are both required for life and salty. Sugars are both calorically dense and sweet. Acids pose a modest problem for this approach. Although hydrogen ions are both common in foods and taste sour to humans, their nutritional value is minimal. The ratio of sugar to acid often defines ripeness in fruit, but it seems unlikely that this relationship carries a potential for selective pressure equivalent to the ability to detect potentially dangerous alkaloids.

The functional basis for gustatory perceptual categories also emphasizes the special relationship between taste and the regulation of water, energy, and electrolyte balance in general, and feeding and drinking behavior in particular. Within this domain, taste effects range from the control of salivation and neurally mediated insulin release to long-term body weight regulation. Nevertheless, theories of gustatory coding normally attempt to account for human perceptual responses, i.e. the four standard categories. Other behavioral and autonomic responses elicited or modified by taste stimuli seldom are correlated with gustatory neural activity. Unlike perceptual categories, many of these other responses to gustatory stimuli appear to be organized within the hindbrain (Grill & Norgren 1978b, Norgren & Grill 1982). Thus, the analysis of gustatory neural activity from the hindbrain may have been correlated with responses that are inappropriate to that level of the neuraxis.

The most complete gustatory neurophysiological data base derives from peripheral taste axons and single neurons in the first two central relays, the nucleus of the solitary tract and the parabrachial nuclei, both of which are in the hindbrain. This review therefore concentrates on recent gustatory neurophysiological data from the periphery and the hindbrain. After summarizing the data, we assess how some analyses of that data account for taste qualities. In addition, we provide a synopsis of those few neurophysiological studies germaine to other taste related responses, such as ingestion and rejection, salivation, and swallowing. Other responses, such as the cephalic phase of insulin release, have a documented gustatory component, but the neural mechanisms of the relationship await more analysis (Grill et al 1984, Norgren 1984). Other recent reviews of gustatory function cover different topics (Grill & Berridge 1985, Norgren 1985, Yamamoto 1985) or have a different perspective (Norgren 1984, Scott & Chang 1984).

PERIPHERAL ORGANIZATION AND SENSITIVITY

Taste receptor cells cluster to form morphologically distinct structures, the taste buds. Taste buds occur in spatially discrete subpopulations that are widely distributed in the oral cavity. The distribution and innervation of taste receptor subpopulations in the rat provide the most complete description available and are summarized in Table 1. This distribution is similar in other mammalian species although details may vary (Bradley 1971, Klein & Schroeder 1979, Lalonde & Eglitis 1961, Miller & Smith 1984).

These distinct concentrations of taste receptors and their attendant anatomical specializations coupled with differences in chemical sensitivity bespeak different functional roles. Different functions have been proposed (Atema 1971, Nowlis 1973) but, to date, denervation experiments fail to provide strong support for any of them (Jacquin 1983, Pfaffmann 1952;

Table 1 Gustatory receptor subpopulations in the rat: Innervation and relative distribution[a]

Receptor subpopulation	Number	Percentage[b]	Innervation[c]
Anterior tongue			
Fungiform papillae	185	13.7	CT
Posterior tongue			
Foliate papillae	460	34.2	IX
Circumvallate papilla	350	26.0	IX
Anterior (hard) palate			
Nasoincisor ducts	67	5.0	GSP
Posterior (soft) palate			
Geshmacksstrieffen	66	4.9	GSP
Posterior palatal field	88	6.5	GSP
Buccal wall	46	3.4	CT
Sublingal organ	34	2.5	CT
Epiglottis	50	3.7	SLN

[a] Derived from a summary by Miller (1977) and description by Iida et al (1983) and Ooishi-Iida (1979).

[b] Refers to the percentage of the total taste receptor population.

[c] Abbreviations as follows: Chorda tympani branch of the facial nerve (CT), glossopharyngeal nerve (IX), greater superficial petrosal branch of the facial nerve (GSP), superior laryngeal branch of the vagus nerve (SLN).

J. Travers, H. Grill, R. Norgren, submitted). Part of the difficulty in assigning functional roles to the various gustatory receptor subpopulations arises from the paucity of electrophysiological data for any but those on the anterior tongue.

Receptor Cells

Taste receptor cells respond with slow depolarizing or hyperpolarizing potentials when chemical stimuli come in contact with the microvilli, finger-like projections of the receptor cell extending into the fluid on the oral cavity surface (Kimura & Beidler 1961, Sato & Beidler 1982, 1983, Tonosaki & Funakoshi 1984). Action potentials do not occur in the receptor cells themselves; instead they are first detected in the peripheral taste nerves.

Intracellular recordings from mammals have been obtained from receptor cells in only one of the subpopulations listed in Table 1, those of the anterior tongue, permitting only an initial estimate of the chemical specificity at this level of the gustatory pathway. In the rat, only 10% to 17% of the cells respond exclusively to one of the four standard taste compounds: sucrose, NaCl, HCl or QHCl (Ozeki & Sato 1972, Sato & Beidler 1982). The remaining cells respond to two, three, or all four chemicals. In addition, the probability that any two stimuli will depolarize (or hyperpolarize) a receptor cell is given by the product of the probability that either stimulus will influence the cell. In other words, sensitivity to the four standard stimuli is distributed randomly among receptor cells (Ozeki & Sato 1972, Sato & Beidler 1983).

The summation of depolarizing and hyperpolarizing receptor potentials for generating peripheral nerve action potentials is suggested by comparing the gustatory sensitivity of the chorda tympani nerve with calculations consisting of the sum of depolarizations, or depolarizations plus hyperpolarizations (Sato & Beidler 1983). When hyperpolarizing responses to a given gustatory stimulus are summed together with depolarizing potentials, the calculated response better approximates the effectiveness of that stimulus for the chorda tympani nerve than when only depolarizing potentials are considered. In the rat, this effect is most pronounced for QHCl and sucrose. Based on the ability of sucrose and QHCl to depolarize receptor cells, these stimuli should be nearly as effective as NaCl and HCl for stimulating the chorda tympani nerve. Sucrose and QHCl, however, elicit larger hyperpolarizing potentials in receptor cells than do NaCl or HCl. Sucrose and QHCl are relatively ineffective stimuli for the chorda tympani of the rat, suggesting that depolarizing responses elicited by these chemicals are counteracted significantly by the presence of these hyperpolarizing potentials. In mouse anterior tongue receptor cells, sucrose elicited only depolarizing potentials (Tonosaki & Funakoshi 1984), and the mouse chorda tympani nerve responds well to this chemical (Ninomiya et al 1984).

Anatomical and physiological evidence suggests that most taste receptor cells in the fungiform papillae are innervated by more than one chorda tympani fiber and that a single fiber innervates more than one taste bud (Beidler 1969, Boudreau et al 1985, Miller 1971, 1974, Pfaffmann 1970). The physiological significance of these innervation patterns is only beginning to be understood. For example, chorda tympani responses in the rat can be either enhanced or depressed by stimulating different regions of a fiber's receptive field (Miller 1971).

Chemosensitivities of Peripheral Fibers

Because of the technical difficulty of recording intracellularly from gustatory receptor cells, extracellular recordings from the gustatory afferent fibers that innervate the receptor cells have produced a more complete characterization of peripheral chemosensitivity. What emerges from these studies is that each of the taste nerves or nerve branches has a different pattern of sensitivity to the four standard gustatory stimuli. For instance, in most species, the chorda tympani responds well to salts and acids but the response to QHCl and sugars is more species specific. Within each of the gustatory nerves, single fibers display a variety of sensitivity patterns, some highly specific to a single stimulus (or class of stimuli), others broadly responsive to stimuli with distinct tastes. Species differences in these sensitivity patterns are also evident. An understanding of how central synapses alter these sensitivity profiles depends on a complete characterization in the periphery.

CHORDA TYMPANI NERVE In studies of single chorda tympani fibers, patterns of multiple sensitivity to different tasting compounds are common, but striking examples of stimulus specificity also occur. Stimulus specificity can be quantified by application of a metric, derived from information theory, designed to quantify the breadth of responsiveness to four standard taste stimuli (*H*; Smith & Travers 1979). An *H* value of 0.0 is derived from a cell that responds exclusively to one (of four) stimuli; a value of 1.0 from a cell that responds identically to all four stimuli. Table 2 lists the breadth of responsiveness (*H*) for four groups of neurons, classified according to which of the four standard stimuli elicited the greatest response, i.e. "best-stimulus" classes (Frank 1973). These data from neural populations at different levels in a variety of species provide one uniform basis of comparison.

Single units highly specific to sodium or lithium cations have been described in several species (Boudreau et al 1982, 1983, Frank et al 1983, Ninomiya et al 1984). In mice, depending on the strain, between 14–27% of the single chorda tympani fibers responded relatively specifically to NaCl, with the associated *H* value ranging from 0.182–0.235 (Ninomiya et al 1984). In the rat, nearly 50% of individual chorda tympani (CT) fibers responded best to NaCl but with minor responses to sucrose, HCl, or QHCl (Frank et al 1983, Ogawa et al 1968). Accordingly, Table 2 indicates that the average breadth of responsiveness for NaCl-best fibers in the rat CT (0.402) is slightly higher than for NaCl-best fibers in the mouse. Most single chorda tympani fibers responsive to chemicals containing the sodium cation in the hamster, cat, and macaque monkey, however, are less specific and also respond to acids (Boudreau & Alev 1973, Frank 1973, Ogawa et al 1968, Pfaffmann 1941, Sato et al 1975). In macaque and hamster, average breadth of responsiveness values of 0.576–0.625 for NaCl-best units indicate cells responding well to two (out of four) stimuli.

Chorda tympani fibers responsive specifically to sucrose stimulation also have been reported in several species (Frank 1973, Ninomiya et al 1984, Ogawa et al 1968, Pfaffmann et al 1976, Sato et al 1975). In the mouse, approximately 30% of the sampled units responded best, and highly specifically, to sucrose, with an associated *H* of 0.116–0.336, again depend-

Table 2 Breadth of responsiveness (H)[a] to 4 standard taste stimuli for best stimulus classes of neurons in the periphery and hindbrain

Level	Species	(Strain)	Source[c]	Best stimulus types[b] S	N	H	Q	Average
	mouse[d]	(BALB)	CT	0.336	0.182	0.666	0.480	0.405
		(C3H)	CT	0.275	0.180	0.596	0.284	0.327
		(C57BL)	CT	0.116	0.235	0.577	0.367	0.284
		average		0.242	0.199	0.611	0.377	0.339
	hamster[e]		CT	0.411	0.640	0.751		0.608
PNS	hamster[f]		CT	0.363	0.610	0.820		0.523
		average		0.387	0.625	0.786		0.566
	rat[g]		CT	0.611	0.403	0.820		0.561
	rat[h]		CT	0.547	0.401	0.634		0.521
		average		0.579	0.402	0.727		0.541
	monkey[i]		CT	0.448	0.576	0.760	0.525	0.565
	rat[j]		IX	0.281	0.306	0.418	0.211	0.298
		Level average		0.387	0.422	0.660	0.223	0.462
	hamster[k]		NST	0.643	0.730	0.743	0.236	0.698
	hamster[l]		NST	0.609	0.572	0.775	0.342	0.610
CNS		average		0.626	0.651	0.759	0.289	0.654
	rat[m]		NST	0.624	0.670	0.801		0.671
	monkey[n]		NST	0.796	0.821	0.832	0.843	0.821
		NST average		0.682	0.714	0.797	0.566	0.717
	hamster[o]		PBN	0.743	0.603	0.651		0.652
	rabbit[p]		PBN	0.550	0.570			0.656[q]

[a] The breadth of responsiveness (H) is calculated for each neuron and averaged within the designated class. $H = K\Sigma p_i \log p_i$ where K is a scaling factor, and p_i are the proportions for each response to the total response to all four stimuli (Smith & Travers 1979). Derived calculations are from published figures.

[b] Neurons from each study were divided into classes according to which of the four standard taste stimuli elicited the greatest response per unit time. Sucrose (S), NaCl (N), HCl (H), QHCl (Q).

[c] Abbreviations are as follows: Chorda tympani branch of the facial nerve (CT), glossopharyngeal nerve (IX), greater superficial petrosal branch of the facial nerve (GSP), superior laryngeal branch of the vagus nerve (SLN), nucleus of the solitary tract (NST), parabrachial nuclei (PBN).

[d] Ninomiya et al (1984).

[e] Derived from Frank (1973).

[f] Derived from Ogawa et al (1968).

[g] Derived from Ogawa et al (1968).

[h] Derived from Frank et al (1983).

[i] Derived from Sato et al (1975).

[j] Derived from Nowlis & Frank (1981).

[k] Travers & Smith (1979).

[l] Sweazey (1983).

[m] Dervied from Travers et al (1986).

[n] Derived from Scott et al (1986).

[o] Van Buskirk & Smith (1981).

[p] Di Lorenzo & Schwartzbaum (1982).

[q] A large number of broadly tuned neurons were not divided into best-stimulus classes.

ing on the strain (Ninomiya et al 1984). The hamster chorda tympani nerve also revealed a good sucrose sensitivity, with 25–43% of the sampled units responding better to sucrose than to the other three standard taste stimuli (Frank 1973, Ogawa et al 1968). The breadth of sensitivity for sucrose-best units in the hamster is slightly higher than those reported for the mouse (Table 2), consistent with the observation that, on average, NaCl stimulated hamster sucrose-best cells 25% as well as sucrose (Frank 1973).

In several species, QHCl and other bitter-tasting chemicals are relatively ineffective stimuli for the chorda tympani nerve. Only one or two QHCl-best units were reported in either the hamster or rat (Frank 1973, Frank et al 1983, Ogawa et al 1968). In these two species, the limited chorda tympani quinine sensitivity typically occurs in units more sensitive to either HCl or NaCl (Frank 1973, Ogawa et al 1968). In both the macaque monkey (Sato et al 1975) and mouse (Ninomiya et al 1984), significant numbers of chorda tympani fibers responded better to QHCl than to the other three standard stimuli (16% and 25%, respectively). These units are relatively narrow (Table 2), with only minor sensitivities to acids and salts. In all species, QHCl and sucrose sensitivities are rarely found in the same unit.

Hydrochloric acid is a highly effective stimulus in the chorda tympani of most species, but rarely activates individual units specifically. More often, a sensitivity to HCl is associated with a sensitivity to other electrolytes, including both salts and acids. Table 2 indicates that in four species, HCl-best units are the most broadly tuned chorda tympani fibers. In the rat, HCl-best units show some stimulus specificity by responding highly preferentially to L-malic acid (Boudreau et al 1983) but also clearly respond well to NaCl and QHCl (Frank et al 1983, Ogawa et al 1968). Only in the cat have relatively specific acid cells been reported (Boudreau & Alev 1973, Pfaffmann 1941).

GLOSSOPHARYNGEAL NERVE Although it innervates approximately 60% of all taste buds (Table 1), the response properties of the glossopharyngeal nerve have not been studied as thoroughly as those of the chorda tympani, perhaps because of the mechanical difficulty of stimulating posterior tongue receptors. Across several species, the glossopharyngeal nerve displays a good sensitivity to QHCl, relative to the chorda tympani (Oakley 1967, Pfaffmann et al 1967, Pfaffmann 1941, Yamada 1967, Oakley et al 1979). In the rat, this enhanced sensitivity generalizes to some other substances (e.g. nicotine) that are bitter to man (Iwasaki & Sato 1981).

Investigations of single glossopharyngeal nerve fibers in the rat corroborate the quinine sensitivity apparent from whole nerve investigations (Frank 1975, Frank & Pfaffmann 1969, Nowlis & Frank 1981). Of the four standard taste stimuli, QHCl was most effective in stimulating over

30% of the fibers in the glossopharyngeal nerve, with many responding at high rates (200–400 impulses/5 s). Conversely, only 15% of the glossopharyngeal units responded maximally to NaCl, in contrast to 50% of chorda tympani nerve fibers. Table 2 indicates that the average breadth of responsiveness for the QHCl-best class of fibers was quite narrow (0.211).

GREATER SUPERFICIAL PETROSAL Only scant information on the chemosensitivity of the greater superficial petrosal nerve (GSP) is available. Integrated responses from the whole nerve evoked by stimulating palatal receptors have been recorded in both the rat and hamster (Beidler & Nejad 1985, Nejad 1986). The order of effectiveness of the four standard stimuli for the rat GSP was sucrose > citric acid > NaCl > QHCl. In the hamster, citric acid switched with sucrose. In both these species, the relative effectiveness of the four standard stimuli for the greater superficial petrosal differed from that for the chorda tympani. This difference was much more striking for the rat, however, since the optimal stimulus for GSP stimulation, sucrose, is quite ineffective in activating the chorda tympani in this species.

SUPERIOR LARYNGEAL NERVE The chemosensitive fibers traveling in the superior laryngeal nerve innervate taste buds located on the laryngeal surface of the epiglottis. They are thus in a unique position relative to taste receptors in the oral cavity in that they are not likely to come in contact with foodstuff during the course of normal ingestion. The function of epiglottal chemoreceptors is presumably one of protecting the airway, rather than monitoring the chemical composition of food (Bradley et al 1980). Although the response properties of superior laryngeal fibers overlap those of other gustatory nerves, they have several distinct characteristics.

For example, although some fibers specifically responsive to chemical stimulation have been reported (Harding et al 1978), many chemosensitive fibers also respond to *low-threshold* mechanical stimulation (Bradley et al 1983, Harding et al 1978, Stedman et al 1980, Storey & Johnson 1975), a characteristic seldom observed in other chemosensitive nerves. A second distinctive property of superior laryngeal fibers, evident across a number of species, is a frequent sensitivity to water stimulation (Bradley et al 1983, Boushey et al 1974, Harding et al 1978, Shingai 1977, 1980, Shingai & Beidler 1985, Stedman et al 1980, Storey & Johnson 1975). Although responses to water have been reported for other taste nerves, these responses have been discounted as being a result of adaptation to physiological saline (0.15 M), which was used as a rinse in these studies (Bartoshuk 1978). Physiological saline is also usually employed both as a solvent and rinse in studies of superior laryngeal sensitivity, making it possible that the mechanism underlying the water response is the same for all these nerves. However, NaCl is a poor stimulus for superior laryngeal fibers, relative to the effectiveness of this chemical for other taste nerves in most species (Bradley et al 1983, Boushey et al 1974, Harding et al 1978, Shingai 1977, 1980, Shingai & Beidler 1985, Stedman et al 1980, Storey & Johnson 1975). This suggests that the superior laryngeal nerve displays a distinctive pattern of chemosensitivity to salt and water. Parametric studies of the chemosensitivity of individual superior laryngeal fibers to a battery of chemicals have only been carried out in two species, cat (Stedman et al 1980) and sheep (Bradley et al 1983). For both species, KCl was the most effective stimulus for epiglottal stimulation.

Organization of Peripheral Chemosensitivities

The relative effectiveness of chemical stimuli differentiate among each of the gustatory nerves and nerve branches. Likewise, distinctive patterns of chemosensitivity differentiate among single fibers comprising the chorda tympani, glossopharyngeal, and superior laryngeal nerves. A peripheral fiber can usually be categorized by determining which of the four standard taste compounds optimally stimulates the cell. If more than one stimulus is effective for an afferent fiber, criteria in addition to the best-stimulus classification may be necessary for resolving the neuron type. For example, NaCl is a potent stimulus for most fibers of the rat chorda tympani nerve. When units are rank ordered according to their NaCl response magnitude, the sensitivity to NaCl appears continuously distributed across the population of cells, with no obvious break points for delineating types (Frank et al 1983). Nevertheless, within this population two types of electrolyte-sensitive cells are evident. Some cells that respond well to NaCl are quite unresponsive to acid stimuli (particularly malic acid). In addition, many of these sodium-specific units actually have a higher threshold and a steeper concentration-response function for NaCl than do those that are also sensitive to acids (Boudreau et al 1983, Frank et al 1983). Thus, in the rat CT there is one type of fiber relatively specific for the sodium ion and a second that responds well to both acid and sodium.

A sensitivity to thermal stimuli further differentiates among some gustatory afferent fibers. For example, data from species with a robust chorda tympani sucrose response provide evidence for sucrose-best fiber types and, in several strains of mice, these sucrose-best fibers are quite narrowly tuned. In the hamster and macaque monkey, however, sucrose-best fibers are also sensitive to other classes of gustatory stimuli. Warm water, however, selectively stimulates sucrose-best units, further differentiating them from fibers optimally sensitive to electrolytes, which are more likely to respond to lingual cooling (Ogawa et al 1968, Sato et al 1975).

Not all peripheral gustatory units can be characterized adequately by using the four standard stimuli. Clearly, single fibers in the superior laryngeal nerve have maximal sensitivities to other chemical stimuli, and preliminary data suggest that chemoreceptors associated with the soft palate of the rat are not characterized easily by the use of just the four standard stimuli (Travers et al 1986).

CENTRAL REPRESENTATION

Nucleus of the Solitary Tract

Gustatory afferent fibers from the facial, glossopharyngeal, and vagus nerves terminate in a rostral to caudal sequence within the ipsilateral nucleus of the solitary tract (NST) (reviewed in Norgren 1984). Complete anatomical segregation of the nerves is not maintained, however, suggesting either that some second-order neurons receive afferent input from more than one nerve or that second-order cells innervated by different nerves are intermingled. Trigeminal afferent axons also overlap gustatory terminal fields, providing a basis for convergence of gustatory and somatosensory information (Beckstead & Norgren 1979, Contreras et al 1982, Hamilton & Norgren 1984, Torvik 1956). Further, both gustatory and somatosensory afferent fibers from the posterior tongue travel in the glossopharyngeal nerve, providing an additional possible substrate for convergence between taste and somatosensory neural activity in the NST.

RECEPTOR FIELD TOPOGRAPHY In the anterior one third of the rat NST, gustatory neurons have receptor fields consistent with the projections of the gustatory primary afferent nerves (Travers et al 1986). Neurons responding to stimulation of receptor subpopulations in the anterior oral cavity (anterior tongue or nasoincisor ducts) were more likely to be located in the anterior half of the distribution; those that responded to stimulation of the posterior oral cavity (soft palate or foliate papillae) were encountered further caudally. A similar topographical organization was evident in the hamster NST; neurons responding specifically to anterior tongue stimulation were anterior to those that responded solely to stimulation of the remaining oral cavity (Sweazey 1983).

CHEMOSENSITIVITY The chemosensitivity of NST neurons corresponds closely to that of the peripheral fibers innervating the region of the oral cavity that is stimulated. Thus, in the rat, anterior tongue stimulation reveals a population of NST neurons predominantly sensitive to salts and acids but much less so to sugars or bitter-tasting stimuli (Doetsch &

Erickson 1970, Ganchrow & Erickson 1970, Travers et al 1986, Woolston & Erickson 1979). When the nasoincisor ducts are stimulated, however, sucrose elicits large responses from neurons in the same area of NST (Travers et al 1986). Figure 1 compares responses evoked by sucrose and NaCl for a population of NST neurons whose receptor field included both the anterior tongue and nasoincisor ducts. The sensitivities evinced by these two receptor subpopulations are quite distinct. An enhanced sensitivity to sucrose in the rat NST has also been seen in some studies using indiscriminate "whole-mouth" stimulation, which probably included stimulation of the nasoincisor ducts (Ogawa et al 1984b, Chang & Scott 1984).

Similarly, the sensitivity of hamster NST neurons reflects this species' peripheral sensitivity. Consequently, anterior tongue stimulation with sucrose, NaCl, and HCl evokes robust responses that are relatively equivalent to each other (Sweazey 1983, Travers & Smith 1979). The greater sensitivity to sucrose in the hamster is also reflected in the high proportion of sucrose-best neurons in this species. As in the rat, the average response to QHCl in the hamster NST was quite low. Single units in the NST of the alert, behaving cynomologous monkey (in which all taste receptors were presumably stimulated) responded well to glucose, NaCl, and QHCl, but weakly to HCl. The poor sensitivity to HCl was probably due to the reported difficulty in isolating neurons in the posterior NST, where cells most responsive to this stimulus were located (Scott et al 1986).

CENTRAL PROCESSING Although the overall gustatory sensitivities of second-order NST cells closely match the sensitivities of the terminating afferent fibers, there is also evidence for synaptic processing. Second-order NST cells are more broadly responsive to gustatory stimuli than primary afferent fibers, as indicated by higher scores for breadth of responsiveness to the four standard taste stimuli in the rat, hamster, and monkey (Table 2) and by higher across-neuron correlations between all pairs of stimuli for the rat and hamster (Doetsch & Erickson 1970, Travers & Smith 1979). Convergence of afferent fibers onto the same second-order neuron, from the same nerve or from different nerves (in cases of whole-mouth stimulation), could provide a mechanism for the apparent increases in breadth. Indeed, several studies have demonstrated such convergence while stimulating different subpopulations of taste receptors or different gustatory nerves (Hayama et al 1985, Ishiko & Akagi 1972, Ogawa & Kaisaku 1982, Ogawa & Hayama 1984, Travers et al 1986, Sweazey 1983).

In the hamster NST, 67% of the neurons were activated by stimulating both the anterior tongue and the remaining oral cavity (Sweazey 1983). Varying degrees of convergence were also apparent in the rat NST. Ogawa and his colleagues used whole-mouth gustatory stimulation to identify the

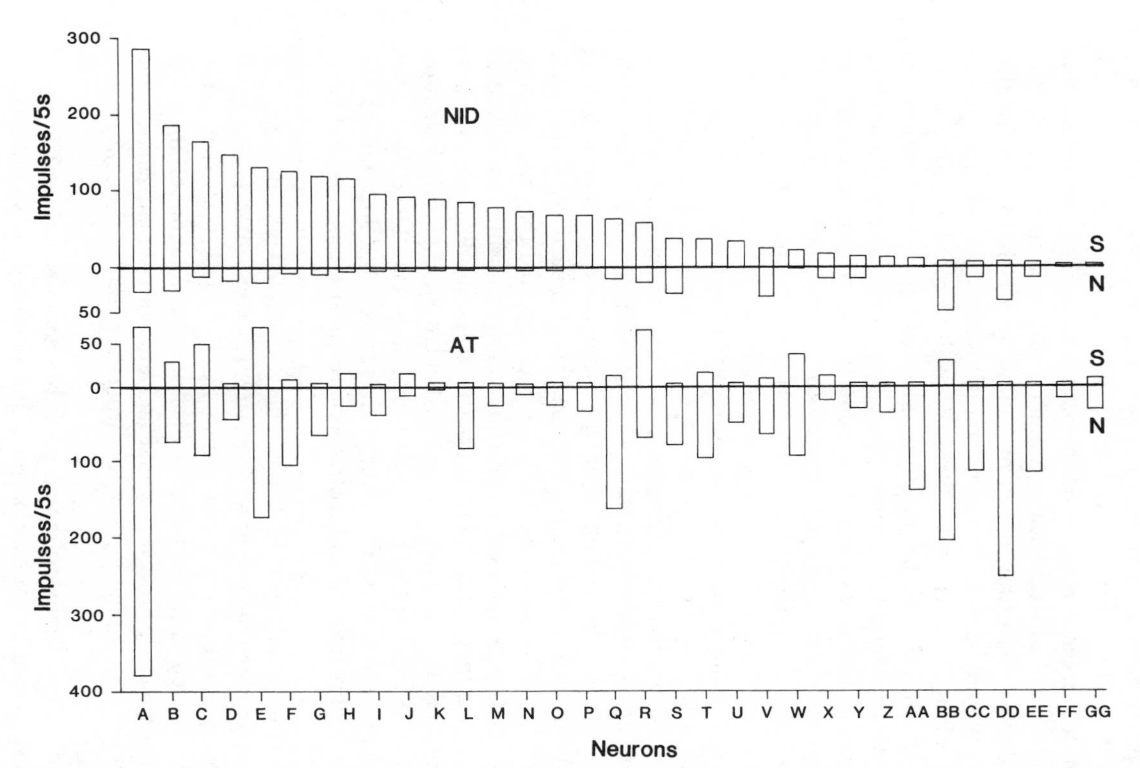

Figure 1 Responses of 33 neurons in the solitary nucleus of the rat to stimulation of the nasoincisor ducts (NID) and anterior tongue (AT) with sucrose (S) and NaCl (N). The neurons are ranked in descending order of their response to sucrose applied to the nasoincisor ducts; all four responses in a single column are excitatory responses from one neuron. All 33 neurons depicted responded to sapid stimulation of both the anterior tongue and nasoincisor ducts. For a few of the cells, some other chemicals besides sucrose or salt were the effective stimuli.

most effective stimulus for a neuron, prior to using that best stimulus to identify the neuron's receptive field. With this procedure, relatively little convergence was found (6%) in the anesthetized rat (Ogawa & Hayama 1984), although more convergence was apparent (35%) in an unanesthetized, decerebrate preparation (Hayama et al 1985). A greater incidence of convergence is reported, however, when punctate stimulation with a mixture of different-tasting stimuli is used to identify the receptor subpopulations activating a neuron, allowing for the possibility that different subpopulations of taste receptors are activated by qualitatively different stimuli (Travers et al 1986). With this procedure, 51% of the neurons in the NST of the intact, anesthetized rat received convergent input, with the majority of convergence occurring between the anterior tongue and nasoincisor ducts. Neurons in the solitary nucleus that responded robustly to sucrose applied to the nasoincisor ducts often responded well to NaCl on the anterior tongue (Figure 1). Convergence between afferent axons with divergent sensitivities is consistent with the increase in breadth of responsiveness observed in second-order cells.

As discussed above, gustatory-responsive peripheral fibers often respond to somatosensory stimulation, and anatomical data suggest the possibility of gustatory/somatosensory convergence in NST neurons. In the rat, 88% of a population of neurons responsive to anterior tongue taste stimulation also responded to cooling of the same tissue (Makous et al 1963). Further, Ogawa and his colleagues reported that responses to gustatory and intraoral mechanical stimulation frequently converged in single neurons in the rat NST (Ogawa & Hayama 1984, Ogawa et al 1984b). Quantitative measures of the threshold for mechanical stimulation were not reported, but there was some suggestion that the cells were not activated by light touch. The level of detail available does not permit comparison of peripheral and central somatosensory responses, but neurophysiological investigations using electrical stimulation of the chorda tympani, glossopharyngeal, and lingual nerves revealed NST neurons that responded to more than one nerve (Ogawa & Kaisaku 1982). This suggests that convergence between somatosensory and gustatory afferent axons occurs in second-order taste cells. Electrical stimulation of the glossopharyngeal nerve, however, would activate both somatosensory and gustatory fibers in the nerve, thus obscuring the nature of the convergence in this case.

Several studies demonstrate that gustatory neurons in the NST can be modulated by additional exteroceptive and interoceptive stimuli. Taste-evoked responses in the NST can be influenced by gastric distension (Glenn & Erickson 1976), blood glucose levels (Giza & Scott 1983), and odorants (Van Buskirk & Erickson 1977). Solitary nucleus neurons also receive central projections. Anatomical data show that the cortex, hypothalamus, and amygdala all send efferents to the NST (van der Kooy et al 1984). In some NST cells, electrical stimulation in the lateral hypothalamus facilitates or inhibits responses driven by either sapid stimulation of the anterior tongue (Matsuo et al 1984) or electrical stimulation of the chorda tympani nerve (Bereiter et al 1980). The significance of the central modulation of NST taste responses is not known, but both the gustatory system and limbic areas have powerful influences on ingestion (Norgren 1983).

EFFERENT ORGANIZATION Anatomical data demonstrate that neurons in the rostral NST of the rat and hamster contribute to both ascending and descending pathways (Norgren 1978, Travers 1979). In these species, ascending axons terminate in the ipsilateral pontine parabrachial nuclei, with the descending pathway primarily to the caudal, nongustatory NST and subjacent reticular formation. Only 21% of the neurons in the NST that responded to electrical stimulation of the lingual, chorda tympani, or glossopharyngeal nerves could be antidromically activated from the parabrachial nuclei (Ogawa & Kaisaku 1982). Similarly, only a minority (34%) of NST neurons driven by whole-mouth gustatory stimulation could be antidromically invaded from the pons. These cells had higher response rates and tended to be more narrowly tuned to the four standard taste stimuli than did the nonprojecting cells (Ogawa et al 1984b).

Parabrachial Nuclei

Although ascending axons arising from the rostral NST synapse upon cells in the parabrachial nuclei (PBN) of the pons in both rodents and lagomorphs (Norgren & Leonard 1973, Travers 1979, Block & Schwartzbaum 1983), in primates, only fibers from the caudal, visceral afferent NST synapse in the pons. Those from the rostral, gustatory NST bypass the PBN to synapse in the ventroposteromedial nucleus of the thalamus (Beckstead et al 1980). Thus, the location of the second-order relay nucleus for taste varies between species.

RECEPTOR FIELD TOPOGRAPHY There is some evidence that taste-responsive neurons in PBN are topographically organized along a dorso-ventral axis, in contrast to the rostro-caudal organization in NST. In the rat, parabrachial responses to stimulation of the anterior tongue were maximal at sites ventral to the maximal responses evoked by stimulating the remainder of the oral cavity (Norgren & Pfaffmann 1975). In species other than the rat, the organization is unknown.

CHEMOSENSITIVITY The chemosensitivity of PBN neurons is similar to that of lower order neurons in the taste pathway. In the rat, stimulating only the anterior tongue evokes much larger responses to salts and acids than to sucrose and QHCl (Perrotto & Scott 1976, Scott & Perrotto 1980); stimulating additional regions of the oral cavity elicits sucrose responses more equivalent to those produced by the electrolytes (Ogawa et al 1984a). In the hamster, anterior tongue stimulation with sugars, as well as salts and acids, evokes robust responses in PBN neurons (Van Buskirk & Smith 1981, Travers & Smith 1984), again reflecting the chemosensitivity of the chorda tympani nerve in this species (Frank 1973, Ogawa et al 1968). Regardless of the stimulation method, as in the NST, robust PBN responses to QHCl are conspicuously absent in acute preparations in both species. In studies that flooded the whole mouth or posterior oral cavity, it is unclear whether the small QHCl responses resulted from an inadequate sampling of neurons or from ineffective stimulation of posterior tongue receptors (Chang & Scott 1984, Maes & Erickson 1984, Ogawa et al 1984b). Given the location of these taste buds deep in the trenches and folds associated with the circumvallate and foliate papillae, the latter possibility seems likely.

Parabrachial taste responses generally increase with increases in stimulus concentration, similar to those in the periphery and NST (Di Lorenzo & Schwartzbaum 1982, Frank 1973, Ganchrow & Erickson 1970, Ogawa et al 1968, Pfaffmann 1941, Scott & Perrotto 1980, Stedman et al 1980, Travers & Smith 1979, Van Buskirk & Smith 1981, Yamada 1966, 1967). The frequent responses to intraoral mechanical and thermal stimulation observed in PBN taste neurons (Ogawa et al 1982, Norgren & Pfaffmann 1975, Travers & Smith 1984) also occurs at lower levels of the gustatory pathway. Unfortunately, the data available do not permit quantitative comparison of the degree of somatosensory responsiveness at various levels.

CENTRAL PROCESSING On average, both the spontaneous rate and evoked responses of hamster PBN neurons are greater than those observed in the NST; those in the NST are higher than observed in the periphery. Altogether, a sixfold increase in spontaneous rate and 1.8-fold increase in the average evoked response rate occurs between the periphery and PBN in this species (Van Buskirk & Smith 1981). Because systematic information on neither flow rate nor spontaneous rate is available for this species, the reported amplification followed by deamplification in the evoked discharge rate of rat gustatory neurons at successive neural levels is difficult to interpret (Ganchrow & Erickson 1970, Perrotto & Scott 1976). In both species, response complexity increases from NST to PBN, as exemplified by a greater incidence of taste-contingent water responses and/or inhibition (Travers & Smith 1979, Van Buskirk & Smith 1981, Perrotto & Scott 1976).

In the rat, across-neuron correlations in the PBN show an overall increase (Perrotto & Scott 1976) suggesting that individual pontine cells respond more equally to a battery of stimuli than do their counterparts in the medulla. In the hamster PBN, the mean breadth of responsiveness across all neuron classes does not increase but sucrose-best pontine cells are significantly less specific (Table 2). Inspection of Table 2 suggests a trend toward more broadly responsive neurons at three successive levels of the gustatory system : peripheral nerve, NST, and PBN.

In the rat, stimulation of different taste receptor populations demonstrates that some single PBN neurons receive gustatory information from multiple regions of the oral cavity (Norgren & Pfaffmann 1975, Ogawa et al 1982). Estimates of the proportion of convergent neurons range from 50–61%, similar to most estimates of convergence in the NST. Because of discrepancies in the results obtained from various NST studies, as well as differences in the stimulation techniques used at the two levels, the question of whether convergence increases at the pontine level remains unanswered. Hermann and her colleagues (Hermann et al 1983, Hermann & Rogers 1985) have suggested that vagal and gustatory afferent input are separate in the NST but converge in PBN. Approximately 64% of a sample of PBN neurons that responded to sapid stimulation of the anterior tongue also responded to electrical stimulation of the vagus nerve (Hermann & Rogers 1985), whereas no NST neurons exhibited this dual sensitivity (Hermann et al 1983). Another investigation, however, has reported that electrical stimulation of the CT and vagus nerves do co-activate single NST neurons (Bereiter et al 1981). The reason for this discrepancy is unclear.

The PBN also receives input from rostral structures, including the hypothalamus (Takeuchi & Hopkins 1984, Berk & Finkelstein 1982), amygdala (Takeuchi et al 1982), bed nucleus of the stria terminalis (Holstege et al 1985), and insular cortex (Shipley & Sanders 1982, Norgren & Grill 1976). The physiological influence of these areas upon taste-responsive neurons has not yet been investigated.

EFFERENT ORGANIZATION Neurons in the PBN project to the ventroposteromedial nucleus of the thalamus, the hypothalamus, the central nucleus of the amygdala, and to the bed nucleus of the stria terminalis (Norgren & Leonard 1973, Norgren 1976). Antidromic activation of PBN taste neurons has documented that some of these axonal projections carry gustatory afferent information (Norgren 1974, 1976, Ogawa et al 1984a). Anatomical evidence also supports weaker parabrachial projections to substantia nigra, the preoptic area, and insular cortex, but physiological confirmation of gustatory function is lacking (Shipley & Sanders 1982,

Saper & Loewy 1980). Similarly, neurons in or near the PBN project locally into the reticular formation and sparsely back to the NST, but their gustatory character remains to be established (Fulwiler & Saper 1984, Norgren 1976, Saper & Loewy 1980, Travers & Norgren 1983).

Organization of Hindbrain Chemosensitivity

Describing the chemosensitivity of brainstem gustatory neurons engenders little disagreement, but the interpretation of these data remains controversial. In particular, the patterns of responsiveness in individual cells may represent discrete types or points along a continuum. In the hindbrain, gustatory neurons of the rat and hamster have been categorized according to which of the four standard gustatory stimuli they respond most vigorously, i.e. their "best-stimulus" (Ogawa et al 1984b, Travers & Smith 1979, Travers & Smith 1984, Travers et al 1986, Van Buskirk & Smith 1981) or by their response patterns to a wider array of stimuli (Scott & Chang 1984, Smith et al 1983a). Nevertheless, other studies in these and other species have failed to report neuron types (Gill & Erickson 1985, Woolston & Erickson 1979, Scott et al 1986). Differences in methodological and statistical procedures as well as stimulus concentration may account for some of the apparent discrepancies.

The degree of similarity between the responses of two neurons to the same stimulus battery can be quantified by calculating the correlation coefficient (usually Pearson's r) between their responses (Figure 2). The similarities among a sample of neurons can be appreciated by calculating correlations between all possible pairs of cells and displaying these similarities by using multidimensional scaling techniques as shown in Figure 3a (Doetsch & Erickson 1970, Woolston & Erickson 1979, Smith et al 1983a). Neurons that respond similarly across a stimulus array will correlate highly and be located near one another in the "neuron space," allowing a visual assessment of whether or not neurons segregate into groups. The degree of grouping can be assessed more quantitatively with hierarchical cluster analysis (Figure 3b). Both multidimensional scaling and cluster analysis, however, merely guide the investigator in deciding whether neuron types exist; they do not provide a probabilistic statement concerning the existence of neuron clusters (Bieber & Smith 1986).

Using responses evoked by 18 compounds, and the multivariate statistical techniques described above, Smith et al (1983a) decided that gustatory neurons clustered into three groups in both the NST and PBN. These groups corresponded well with the classification of the neurons into sucrose-, NaCl-, or HCl-best groups, as determined by their responses to the four standard taste stimuli. The four standard stimuli thus provide

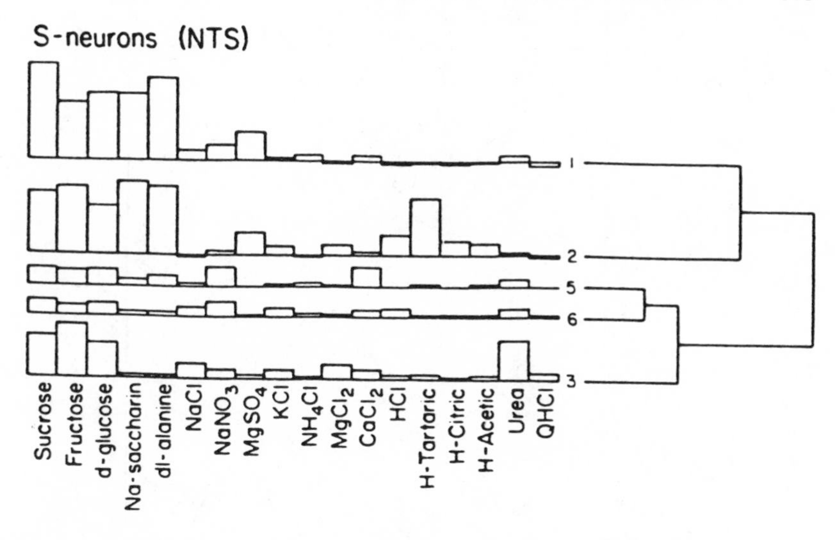

Figure 2 Histograms comparing the responses of five neurons in the hamster solitary nucleus (NTS) to 18 chemicals applied to the anterior tongue. Responses from a single neuron appear on the same *line*, responses to a particular stimulus in the same *column*. The height of the *bars* represents response magnitude (the response of neuron 1 to sucrose was 155 impulses/5 s). These cells each responded well to sucrose (S-neurons) and had similar (i.e. highly correlated) response profiles across the 18 stimuli. For example, the correlation (Pearson's r) between neurons 1 and 2 was +0.81. *Brackets* at right are a cluster analysis of these data (see Figure 3b). The relationship of these neurons to others in this sample is depicted in Figure 3. Reproduced with permission from Smith et al (1983a).

a reasonably accurate prediction of the pattern of responsiveness to a wider array of compounds. In another study also using responses of hamster PBN cells, multidimensional scaling techniques failed to provide clear evidence for neuron types (Gill & Erickson 1985). The latter study used higher stimulus concentrations (from 0.5 to 1.0 log step higher), and other studies have shown that PBN neurons become significantly more broadly tuned with increases in concentration (Van Buskirk & Smith 1981). This suggests that the ability to distinguish neuron types may depend upon stimulus concentration.

Based on factor and cluster analysis, four types of chemosensitive neurons were evident in the rabbit PBN (Di Lorenzo & Schwartzbaum 1982). Only two of these types, however, were associated with best-stimulus designations (NaCl- and sucrose-best). The two other types were more nonspecific, responding nearly equally to NaCl and HCl or to sucrose, NaCl, and HCl.

In the rat NST, multidimensional scaling and cluster analysis give a less

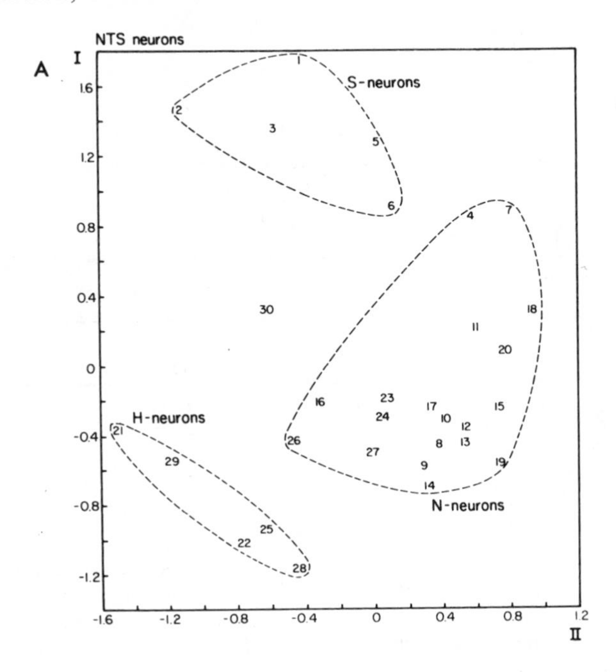
NTS neurons
A
I
II
S-neurons
H-neurons
N-neurons
1.6
1.2
0.8
0.4
0
-0.4
-0.8
-1.2
-1.6
-1.2
-0.8
-0.4
0
0.4
0.8
1.2
1
2
3
5
6
4
7
30
18
11
20
16
23
17
15
24
10
12
21
26
8
13
27
29
9
19
14
25
22
28

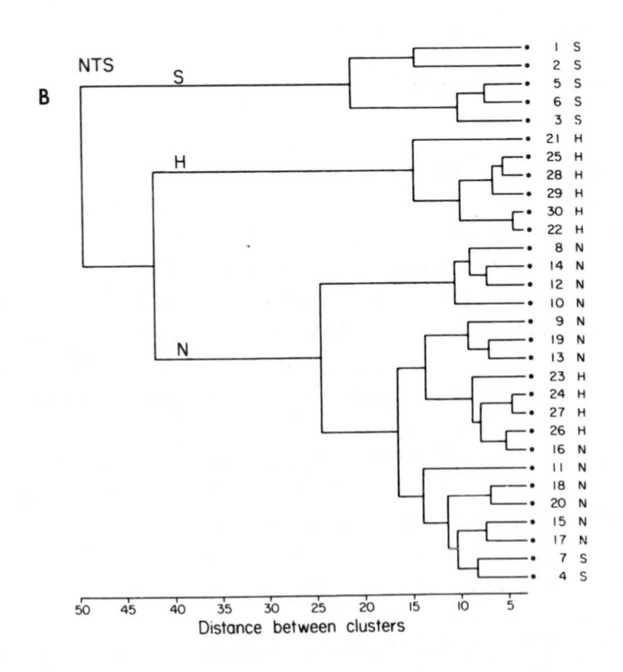
B
NTS
S
H
N
1 S
2 S
5 S
6 S
3 S
21 H
25 H
28 H
29 H
30 H
22 H
8 N
14 N
12 N
10 N
9 N
19 N
13 N
23 H
24 H
27 H
26 H
16 N
11 N
18 N
20 N
15 N
17 N
7 S
4 S
50 45 40 35 30 25 20 15 10 5
Distance between clusters

clear indication for neuron types (Doetsch & Erickson 1970, Woolston & Erickson 1979). Failure to find types in these studies, however, may result partially from the predominant electrolyte sensitivity of rat NST neurons, when only the anterior tongue is stimulated, combined with the use of stimuli at a single concentration. Some electrolyte-sensitive fibers in the chorda tympani respond well to both NaCl and HCl and can be most unambiguously categorized when additional concentrations are used and when other response measures, e.g. threshold, are considered (Frank et al 1983). Further, approximately 50% of rat NST cells that are driven by anterior tongue stimulation also respond well to nasoincisor duct stimulation (usually with sucrose) (Travers et al 1986). The exclusive use of anterior tongue stimulation would miss this important attribute of NST cells and, thus, types may not be as apparent. Indeed, Chang & Scott (1984) found clearer evidence for neuron types in this species, perhaps as a result of employing "whole-mouth" taste stimulation.

The least compelling evidence for gustatory neuron types in the brainstem comes from the monkey NST (Scott et al 1986). These neurons, recorded in an awake behaving preparation, are extremely broadly responsive (Table 2), with little evidence for types based on either multidimensional scaling or cluster analysis. The neurons in this study were isolated from a restricted area within NST and, on average, had very low spontaneous and evoked activity rates. These characteristics may reflect a peculiar sampling bias imposed by the difficulty of maintaining electrophysiological isolation. Thus, conclusions about gustatory neuron types in the primate NST must remain tentative.

The typing of neurons based only on their pattern of responsiveness to a battery of stimuli at a single concentration appears vulnerable to methodological procedures, including site of stimulus delivery and stimulus concentration. In addition, many multivariate techniques used to resolve neuron types begin with a definition of similarity based on the Pearson product-moment correlation. With this metric, unresponsive cells may generate spurious correlations, making the neuron space less inter-

←

Figure 3 *A*: Two-dimensional "neuron space" for 30 neurons in the hamster solitary nucleus (NTS). The distance between neurons indicates the similarity in response profiles, quantitified using correlation coefficients. The S-neurons are those shown in Figure 2. The *dotted line* indicates the clusters determined by hierarchical cluster analysis. *B*: Hierarchical cluster analysis of same neurons shown above. The length of the horizontal lines linking neurons represents the degree of difference between the response profiles of the bracketed cells at successive stages of the clustering process. Letter symbol to *right* of neuron number designates best stimulus classification: sucrose (S), HCl(H), NaCl (N). Reproduced with permission from Smith et al (1983a).

pretable (Gill & Erickson 1985). Finally, basing a definition of types solely on responses to stimuli presented at a single concentration forfeits the use of other potential differentiating characteristics, such as spontaneous rate, latency, threshold, concentration-response slope, response time course, receptive field, thermosensitivity, and anatomical connectivity (Boudreau et al 1985, Frank et al 1983, Ogawa et al 1984b, Travers & Smith 1984, Travers et al 1986). The categorization of gustatory neurons remains an ongoing enterprise.

The controversy surrounding gustatory neuron types should not obscure the inherent orderliness of neuron sensitivities. When responses are scored only for being above background levels, the distribution of sensitivities to the four standard stimuli are usually random in both peripheral and brainstem neurons (Frank 1973, Frank & Pfaffmann 1969, Ogawa et al 1968, Sato et al 1975, Travers & Smith 1979). Occasional deviations from randomness are observed, however, and these are systematic. The most frequent deviations are for sucrose sensitivity to be lacking in neurons that respond to NaCl, HCl, or QHCl and for the electrolytes to be jointly stimulatory more frequently than predicted from chance.

When response magnitude is considered, orderly associations between responses to stimuli are more pronounced. Further, these associations are evident both with the use of statistical procedures that type neurons and those that do not. For instance, when neurons are classified into best-stimulus categories, not only do the average response profiles for each class peak at the best-stimulus (by definition), but the average response to adjacent stimuli systematically decreases when the stimuli are ordered sucrose, NaCl, HCl, and QHCl along the abscissa (Figure 4) (Frank 1973,

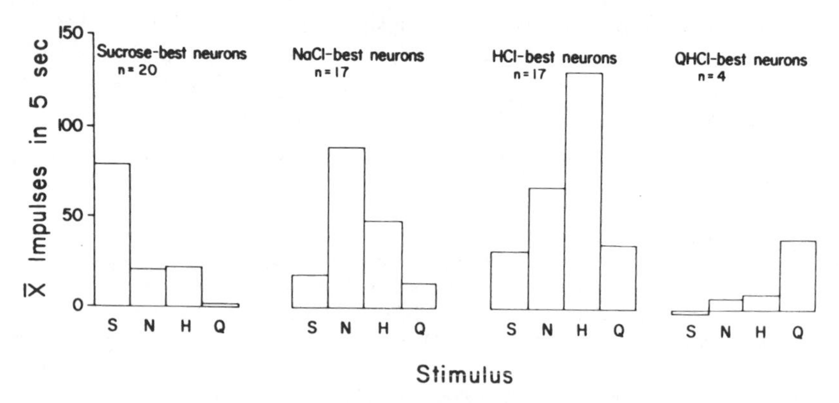

Figure 4 Average response profiles of neurons in the hamster NST classified according to which stimulus elicits the greatest response. Sucrose (S), NaCl (N), HCl (H), QHCl (Q). Reproduced with permission from Travers & Smith (1979).

Nowlis & Frank 1981, Travers & Smith 1979, Travers & Smith 1984, Travers et al 1986, Van Buskirk & Smith 1981, Ogawa et al 1984b). This tendency for associations between particular stimuli is also evident with statistical procedures not dependent on dividing neurons into categories. Calculating the correlation between the responses evoked by two chemicals across individual neurons in a sample (across-neuron correlations) provides a measure of the degree of similarity in the relative rates of activity that the two compounds evoke across that sample. When across-neuron correlations between the four standard stimuli are calculated, NaCl and HCl are often much more highly correlated than sucrose and QHCl (Chang & Scott 1984, Frank 1973, 1975, Smith et al 1983b, Travers & Smith 1979, Woolston & Erickson 1979), implying that sucrose and QHCl rarely activate the same neuron but that responses to NaCl and HCl occur in the same cell more frequently. These same relationships are evident in the average responses of the best-stimulus classes. For example, in the hamster NST, QHCl was the poorest stimulus for sucrose-best cells and vice versa, whereas NaCl elicited significant activity in both NaCl- and HCl-best neurons (Figure 4). The two stimuli least likely to co-activate individual neurons elicit strikingly different behavioral responses; sucrose elicits ingestion and quinine elicits rejection (Grill & Norgren 1978a, Schwartz & Grill 1984, Travers & Norgren 1986).

Hindbrain Taste Responses and Discrimination

Psychophysical studies in both humans and animals indicate that the four standard taste stimuli are very distinct from one another (Bartoshuk 1978, Nowlis et al 1980, Smith et al 1979). Most neurophysiological studies, however, have demonstrated that individual gustatory elements typically respond to more than one of these qualitatively distinct stimuli. This finding originally prompted Pfaffmann (1941) to propose that taste quality was encoded by the pattern of activity across a population of neurons, rather than by activity in any one set of cells, specifically responsive to that quality. This scheme for taste quality coding, dubbed the "across-neuron pattern theory," was elaborated and quantified by Erickson (1963, 1968, 1974). Across-neuron patterns function well in predicting behavior; stimuli that taste similar evoke similar patterns of activity across the population of responding neurons, as quantified by their across-neuron correlations (Erickson 1963, Scott 1981, Smith et al 1979).

For many years, the across-neuron pattern theory was not seriously challenged. As discussed above, however, more recent studies (Frank 1973, Nowlis & Frank 1981, Smith et al 1983a) argued for the existence of discrete classes of neurons, each maximally sensitive to one of the four standard stimuli. The identification of these types, particularly in the periphery where they are most narrowly tuned, led to the hypothesis that however elicited, activity in a given class of neurons signals the taste quality evoked by that class's optimal stimulus (Pfaffmann 1974, Pfaffmann et al 1976, Nowlis et al 1980, Nowlis & Frank 1981). For example, activity in sucrose-best neurons signals "sweetness." This "labeled-line" theory assumes that activity elicited in the appropriate best-stimulus class is signal, in contrast to submaximal activity evoked by the same stimulus in the other classes, which is "noise." Smith and his colleagues (1979) directly compared the efficacy of the labeled-line and across-neuron pattern theories by using the responses of neurons in the hamster hindbrain for predicting behavioral responses to an array of 18 compounds: both quality-coding schemes functioned well.

The success of both the labeled-line and across-neuron pattern theories in explaining behavior reflects an assumption common to both. Both theories posit that quality coding is determined by which neurons in the population are responding. The relationship between these two hypotheses was explored systematically by determining the role the various best-stimulus classes play in establishing across-neuron patterns for particular stimuli (Smith et al 1983b). The across-neuron pattern for a stimulus was found to be dominated by the activity of neurons in the appropriate best-stimulus class, e.g. the neurons with the highest response rates in the across-neuron pattern for sucrose were those classified as sucrose-best. Excluding different fiber groups from the across-neuron patterns also suggested that a particular best-stimulus group was primarily responsible for establishing the across-neuron pattern for that quality, but further showed that a single best-stimulus class could not establish distinct patterns for dissimilar stimuli. These analyses clarify the relationship between the two theories but do not, as pointed out by the authors, distinguish between them. For example, sucrose-best neurons were demonstrated to be the crucial group in establishing similar patterns for the sweet-tasting stimuli, a finding predicted from labeled-line theory but not at odds with a patterning notion. It was also shown that distinguishing between sweet-tasting stimuli and nonsweet stimuli required the presence of neurons outside the sucrose-best class. Thus, these analyses demonstrate the importance of information from multiple neurons, a notion emphasized by patterning theory but not incompatible with labeled-line theory, given the lack of response specificity in individual gustatory cells.

The controversy concerning neuron types has implications for quality coding, but even the resolution of this issue may not put the quality coding debate to rest (Erickson 1985). Clearly, the existence of types is a requirement for a labeled-line mechanism. However, the converse is not true; the existence of neuron types does not dictate coding by labeled lines.

Rather, taste quality could be coded in a manner similar to the code for color in visual receptors, with the code for taste quality determined by the pattern of activity across broadly tuned types (Smith et al 1983b).

The fundamental challenge to the labeled-line theory, of course, rests upon broad tuning in gustatory neurons: Even stimuli that elicit relatively pure taste quality reports evoke some activity in inappropriate channels. As can be seen in Figure 4, the greatest response to a particular stimulus typically occurs within the appropriate category of cells, but responses also occur in inappropriate classes of neurons (see also Nowlis et al 1980). Proponents of labeled-line coding would posit that the sensation evoked by the greater activity in the appropriate channel simply dominates that evoked by the lesser activity in inappropriate ones (Nowlis et al 1980, Pfaffmann et al 1976). The issue of sideband sensitivity is not as problematic in the periphery, where neurons are more narrowly tuned. Sideband sensitivity becomes more prominent, and therefore problematic, in central neurons (Table 2). It is certainly possible that peripheral fibers code by labeled lines, whereas central neurons code by using patterns. At any rate, taste quality discrimination is only one function that brainstem gustatory neurons must perform. As discussed below, other functions (e.g. acceptance and rejection) usually require some kind of discrimination, but often fewer or different stimulus categories must be resolved. Indeed, as discussed above, antidromic stimulation studies suggest that less than half of the gustatory-responsive neurons in NST appear to project rostrally, an observation consistent with the notion that substantial numbers of hindbrain gustatory neurons are involved in functions other than perceptual discrimination.

HINDBRAIN TASTE RESPONSES AND BEHAVIOR

As discussed in the introduction, gustatory stimuli elicit a number of behavioral and physiological responses related to feeding. These responses range from the reflexive and stereotypic to more complex motivational and associative behaviors. Following chronic, precollicular decerebration, a number of these responses remain intact, including salivation, licking, swallowing, and active rejection (Berntson & Micco 1976, Grill & Norgren 1978b). The effective stimuli are virtually identical in both normal and decerebrate preparations. These observations support the inference that the neural substrate mediating the gustatory control of these responses is largely complete in the hindbrain. Anatomical and physiological studies indicate that the pathways mediating these responses are polysynaptic, and that, therefore, gustatory responsive cells in neither the solitary nor parabrachial nuclei directly control the efferent response (Halpern 1985,

Matsuo et al 1982, Travers & Norgren 1983). Where possible, comparison of afferent activity with efferent responses suggests intervening processing of gustatory information.

Salivation

Salivary flow is modified both by the site and quality of a gustatory stimulus. Salivary flow was augmented preferentially on the side involved with either mastication (Anderson et al 1985) or punctate gustatory stimulation (Eshel & Korczyn 1978). In addition, sublingual and submandibular flow was influenced differentially by anterior tongue stimulation, parotid flow by posterior tongue stimulation (Eshel & Korczyn 1978). Nevertheless, following gustatory stimulation, all of the salivary glands participate to some degree. With concentrations of four standard taste stimuli adjusted to be equally intensive to human subjects, tartaric acid was the most effective stimulus for eliciting salivary flow, followed by QHCl, NaCl, and lastly sucrose (Funakoshi & Kawamura 1967). In addition, increasing the stimulus concentration increased the salivary flow. Aversive gustatory or somatosensory stimuli are generally more potent elicitors of salivary flow than are nonaversive stimuli (Kawamura & Yamamoto 1977). In contrast, sweet stimuli preferentially induced amylase secretion (Newbrun 1962) from the parotid compared with either the sublingual or submaxillary glands (Schneyer 1956). A specific gustatory influence on amylase secretion was further supported by a study in the rabbit in which cutting the glossopharyngeal nerve eliminated amylase secretion elicited by sugars, but not by mastication of hard food pellets (Gjorstrup 1980a,b).

Central salivary neurons and preganglionic efferent fibers respond to sapid stimulation of the anterior tongue (Kawamura & Yamamoto 1978, Matsuo et al 1982, Yamamoto & Kawamura 1977a,b). In the decerebrate rabbit, taste-elicited activity in chorda tympani fibers exhibited a linear relationship to the response rate of preganglionic efferents to the submandibular gland (Kawamura & Yamamoto 1978, Yamamoto & Kawamura 1977a,b). Nevertheless, salivatory efferent responses to sapid stimuli differ in several respects from chorda tympani fiber responses to the same chemicals. Compared with taste afferent activity, salivatory efferent responses have no initial fast transient phase, the magnitude of their sustained discharge is attenuated by a factor of five, and they are generally less specific. This reduced specificity is reflected in the higher correlations between pairs of stimuli calculated from the responses of salivatory efferents compared with the same stimulus correlations generated from taste afferent activity.

An analysis of across-fiber correlations generated by responses of preganglionic salivatory axons to pairs of gustatory stimili showed that sucrose was highly correlated with NaCl and that HCl was highly correlated with QHCl, but that the sucrose-NaCl pair and the QHCl-HCl pair were negatively correlated (Kawamura & Yamamoto 1978). These patterns of across-fiber correlations contrasted with those from the chorda tympani nerve in which QHCl and HCl were the only correlated stimulus pair. The investigators conclude that synaptic processing between afferent and efferent fibers results in a regrouping of gustatory sensitivities along a hedonic (acceptance/rejection) dimension.

Lingual and Masticatory Efferents

The use of intraoral cannulae and videotaping made possible the detailed behavioral analysis of ingestion and rejection (Grill & Norgren 1978a). These techniques permitted the documentation of gustatory control of ingestion in chronically decerebrate preparations, as well as new perspectives on concepts such as palatability (Grill & Berridge 1985). Both single unit and multiunit studies have begun to characterize the lingual and masticatory motor responses defining these behaviors (Schwartzbaum 1983a, Travers & Norgren 1986, Yamamoto et al 1982a,b). In chronic EMG studies using fluid stimuli (Travers & Norgren 1986), ingestive sequences were characterized by rhythmic alternations (licking) between lingual protruder (genioglossus) and retractor muscles (styloglossus), with intermittent pharyngeal contractions indicating swallowing (Figure 5a). Stimulation with NaCl or sucrose could not be differentiated based upon either the EMG signature of a single lick cycle or the duration of the response. Nevertheless, the concentration of gustatory stimuli differentially influenced the duration of a licking bout. Higher concentrations of sucrose prolonged both a voluntary licking bout (Davis 1973) and licking induced by the intraoral delivery of a constant fluid volume (Travers & Norgren 1986). In lightly anesthetized rats, sucrose and NaCl activated single lingual efferent units with a bursting pattern and higher concentrations prolonged the burst sequence, even after stimulus removal (Yamamoto et al 1982a). Quinine hydrochloride and HCl were significantly less effective stimuli in producing a bursting pattern. In similar preparations, gustatory stimuli, particularly sucrose, occasionally elicit bursting from peripheral axons (Mistretta 1972, Nagai & Ueda 1981, Ogawa et al 1974, Sato et al 1975), but this pattern rarely has been reported in sucrose-best neurons in the hindbrain.

Chronic recording in the rabbit parabrachial nuclei revealed gustatory units whose activity correlated in various ways with oro-lingual movement (Schwartzbaum 1983b). One type of "movement-taste sensitive" unit accelerated during oro-lingual movement and responded significantly more to sucrose and NaCl than to HCl or QHCl. Other units differentially respon-

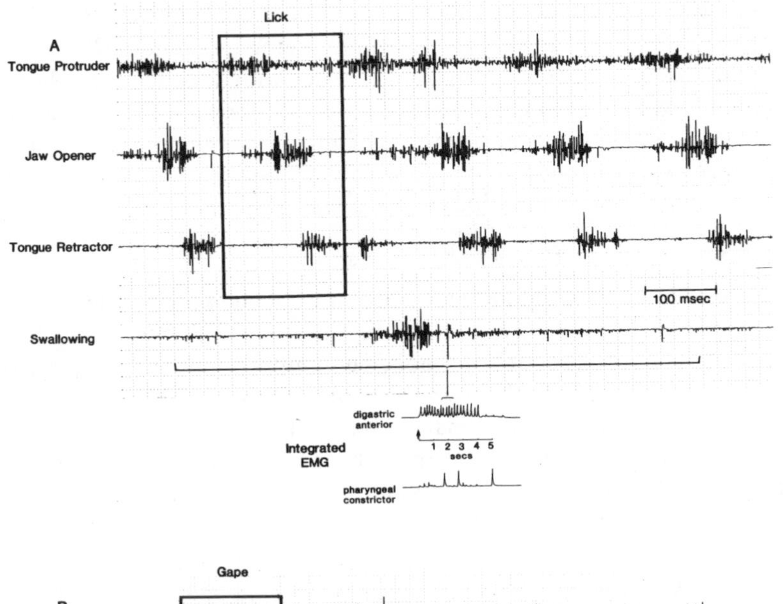

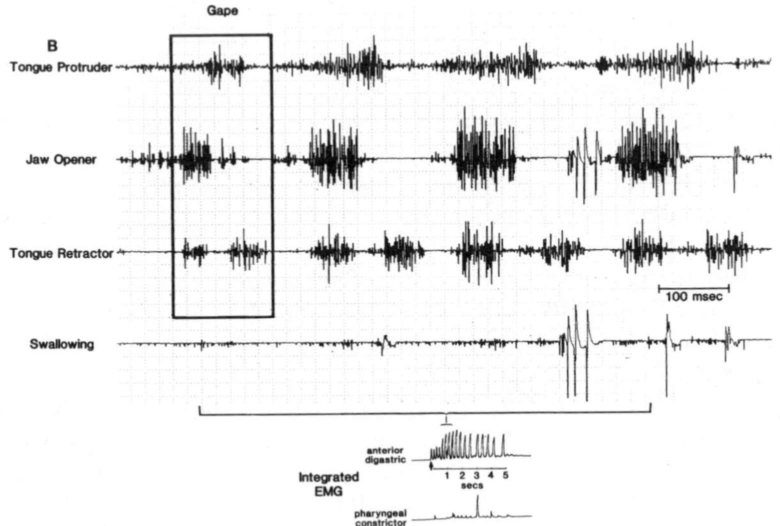

Figure 5 EMG activity of four muscles in the awake behaving rat in response to intraoral delivery of 50 μl of 0.3 M sucrose (*A*) or 0.03 M QHCl (*B*). Reproduced with permission from Travers & Norgren (1986).

sive to taste stimuli were phase locked during orolingual movements, but showed brief response suppression to the fluid stimuli. Not all gustatory units showed hedonic features. For example, some responded well to sucrose, NaCl, and QHCl. Nevertheless, the existence of neurons that apparently respond on hedonic dimensions, as well as with movement, suggest a role for the parabrachial nuclei in generating or maintaining orolingual responses to tastants. Some neurons in the lateral part of the medial PBN and the adjacent Kolliker-Fuse nucleus project to all of the oro-motor nuclei in the rat (Borke et al 1983, Hinrichsen et al 1983, Travers & Norgren 1983, Vornov & Sutin 1983).

Rejection responses elicited by sapid stimuli require much the same musculature as ingestion responses. Both the duration and phase relationships of the muscle contractions, however, differed for the two behaviors (Travers & Norgren 1986, Yamamoto et al 1982b). Thus, the alternation between tongue protrusion and retraction that characterized ingestion gave way during rejection of QHCl to a large amplitude jaw opening (gape response), in which the tongue initially retracts (Figure 5b). This retraction was followed by a long duration tongue protrusion and second tongue retraction with the mouth closed. This entire sequence repeated several times at a rate of 4–5/sec, during which swallowing was reduced but not eliminated. Most species show a high sensitivity to QHCl from posterior tongue receptors. In the rat, bilateral removal of the glossopharyngeal nerves innervating the posterior tongue attenuates the number of gapes in response to QHCl by nearly 50% (J. Travers, H. Grill, and R. Norgren, submitted). The oro-motor rejection response (gapes) often is followed by a sequence of fixed action patterns that include chin rubbing, head shaking, paw wiping, and forelimb flailing (Grill & Norgren 1978a). This sequence figures prominently in the behavioral measurement of rejection, but has yet to be subjected to EMG or neurophysiological analysis.

Swallowing

The neural substrates for swallowing interdigitate with those of the gustatory system. Neurons in the solitary nucleus respond to stimuli that elicit swallowing and appear to be involved in its motor control as well (Jean 1972, Jean et al 1975). Neurons in the pons also respond during swallowing, but at this level the evidence implies a strictly sensory function rather than involvement in sequencing the motor activity (Car & Amri 1982, Amri & Car 1982, Jean et al 1975). Neurophysiological studies usually use electrical stimulation of the superior laryngeal nerve to elicit swallowing, and the convergence of gustatory input on swallowing interneurons has yet to be demonstrated. Gustatory afferent fibers in the seventh and ninth nerves synapse further rostral in the NST than cells identified with swallowing activity (Hamilton & Norgren 1984, Jean 1972). Nevertheless, several studies have demonstrated the effects of gustatory stimuli on efferents controlling the pharyngeal phase of swallowing. In behaving rats, increasing concentrations of a fixed volume of either sucrose or NaCl produced increasing numbers of swallows (Travers & Norgren 1986). Increasing stimulus concentration was also associated with longer licking bouts, however, thus obscuring any direct (central) involvement of gustatory stimuli with swallowing. Moreover, at five different concentrations of these stimuli, the latency to the first swallow was relatively constant, suggesting that mechanoreceptors sensitive to volume triggered the first swallow. Subsequent swallows may have been indirectly influenced by the gustatory stimulus, because it governs the amount of orolingual activity (licking), which, in turn, serves as an adequate stimulus for generating swallows. Increased orolingual activity or gustatory stimulation itself also may have increased salivation (Kawamura & Yamamoto 1978, Murakami et al 1983), which could then serve to facilitate swallowing (Mansson & Sandberg 1975). Other studies in the anesthetized rabbit have suggested a facilitory effect on the latency to swallow by gustatory stimuli compared to water (Shingai & Shimada 1976).

Stimulation with QHCl produced significant inhibition in the latency to the first swallow in both the chronic rat and anesthetized rabbit (Shingai & Shimada 1976, Travers & Norgren 1986). Again, it is unclear whether this inhibition is a direct (central) effect or reflects ongoing orolingual activity. Nevertheless, a direct influence of taste on swallowing has been demonstrated in the rat, because section of both glossopharyngeal nerves decreased the degree of inhibition to the first swallow following QHCl stimulation (J. Travers, H. Grill, and R. Norgren, submitted). In the rabbit, cutting the glossopharyngeal nerve as well as the superior laryngeal and pharyngeal branches of the vagus eliminated swallowing induced by chemical but not by mechanical stimulation (Shingai & Shimada 1976).

Complex Responses

Although taste stimuli influence a variety of behavioral and autonomic responses related to ingestion, those responses vary depending upon the organism's metabolic condition and past experience. Humans identify glucose solutions correctly under most circumstances, but their hedonic ratings of the solution can reverse, depending upon how much of it they have ingested (Cabanac 1971, Thompson & Campbell 1977). Sodium appetite and conditioned taste aversions are other examples of similar response reversals. The neural underpinnings of these response alterations are poorly understood, but one obvious possibility is that some neural or humoral consequence of the metabolic or mnemonic condition directly alters the gustatory neural code. Some metabolic feedback requires afferent axons of the vagus (Smith et al 1985), and anatomical evidence exists for convergence between the gustatory and vagal afferent systems within the brain (Hermann et al 1983, Norgren 1983, 1985). As cited above, prior reports of gustatory and visceral afferent influence converging on single neurons in and near the nucleus of the solitary tract have been contradicted recently (Bereiter et al 1981, Hermann et al 1983). In the pons, however, both electrical stimulation of the vagus and hypertonic saline infused into the hepatic portal vein activated some neurons that responded to sapid stimuli (Hermann & Rogers 1985, Rogers et al 1979). The influence of this visceral afferent activity on the code for gustatory quality or intensity remains to be assessed.

Although the evidence for convergence of gustatory and vagal afferent activity in the NST is equivocal, manipulations that alter behavioral responses to sapid stimuli can influence gustatory neural activity in the solitary nucleus. Acute increases in blood glucose levels differentially reduced multiunit taste responses in rostral NST (Giza & Scott 1983). Responsiveness to oral glucose was reduced most. Reductions to NaCl and HCl were equivalent to one another, but somewhat less than the decrease to oral glucose. Responses to quinine were unaffected, but the magnitude of this activity was low to begin with. In a somewhat more complicated, across-subjects design, Chang & Scott (1984) demonstrated that in rats previously poisoned with LiCl after ingesting saccharin (0.0025 M Na sac), sucrose-best NST neurons had augmented, but longer latency, responses to sweet tasting chemicals. Both these results imply that past or present physiological factors influence the gustatory neural code, but in neither case is the mechanism of that influence specifiable.

The most well-documented induced change in the gustatory neural code occurs in peripheral neurons in response to dietary sodium deficiency or adrenalectomy. Both conditions elicit a robust salt appetite and reduce the peak response to the sodium ion of those rat chorda tympani neurons for which sodium is the best stimulus (Contreras 1977, Contreras & Frank 1979, Kosten & Contreras 1985). The mechanism for this change in sensitivity also is unknown, but some evidence implies a reduction in the number of sodium ion receptor sites or a change in their average binding constant (Contreras et al 1984). It would be instructive to know whether salt appetite induced with systematically or centrally applied hormones is accompanied by similar changes in peripheral gustatory sensitivity to sodium salts.

CONCLUSION

Several analytic approaches to data from the periphery and hindbrain can reconstruct plausible representations of taste quality and intensity coding. Less attention has been devoted to correlating gustatory neural activity with autonomic or motor responses related to ingestion. Nevertheless, a few observations are possible. For instance, the gustatory system and the sensory control of salivation are both predominantly ipsilateral. Further, the neural code for concentration remains relatively constant from the peripheral taste nerves through the efferent control of salivary volume, licking bout duration, and the number of gapes. In general, both afferent and efferent components of these responses increase with increasing stimulus concentration. Finally, the overlap of gustatory neural responses with those to intraoral somatosensory stimulation parallels the dual control of many ingestive behaviors by both these modalities.

Gustatory stimuli elicit at least four distinct quality responses from humans and similar discriminations can be made by other animals. Other gustatory functions represented in the hindbrain, however, may not require as fine a differentiation of quality information. Even when the interactions are complex, ingestion, rejection, salivation, and swallowing appear to respond to gustatory stimuli on not more than two dimensions (Grill & Berridge 1985). In awake behaving rabbits, responses of some pontine gustatory neurons were better correlated with the animal's normal preference or aversion to a sapid stimulus than with the taste quality associated with that chemical (Schwartzbaum 1983b). Taken with evidence that, rather than projecting to the forebrain, a majority of medullary gustatory neurons terminate locally (Ogawa et al 1984b), this suggests that the code for gustatory quality may have different dimensions for different subpopulations of gustatory neurons. Proper assessment of this possibility requires analyzing gustatory afferent activity with respect to a more complete range of the responses influenced by taste.

ACKNOWLEDGMENT

We wish to thank Melanie Newman for help with the preparation of this review. This work was supported by NS 20397 and NS 20477.

Literature Cited

Amri, M., Car, A. 1982. Etude des neurones deglutiteurs pontiques chez la brebis. II. Effets de la stimulation des afferences peripheriques et du cortex fronto-orbitaire. *Exp. Brain Res.* 48: 355–61

Anderson, D. J., Hector, M. P., Linden, R. W. A. 1985. The possible relation between mastication and parotid secretion in the rabbit. *J. Physiol.* 364: 19–29

Atema, J. 1971. Structures and functions of

the sense of taste in the catfish (*Ictalurus natalis*). *Brain Behav. Evol.* 4: 273–94

Bartoshuk, L. M. 1978. Gustatory system. *Handb. Behav. Neurobiol.* 1: 503–67

Beckstead, R. M., Norgren, R. 1979. An autoradiographic examination of the central distribution of the trigeminal, facial, glossopharyngeal, and vagal nerves in the monkey. *J. Comp. Neurol.* 184: 455–72

Beckstead, R. M., Morse, J. R., Norgren, R. 1980. The nucleus of the solitary tract in the monkey: Projections to the thalamus and brain stem nuclei. *J. Comp. Neurol.* 190: 259–82

Beidler, L. M. 1969. Innervation of rat fungiform papilla. In *Olfaction and Taste III*, ed. C. Pfaffmann, pp. 352–69. New York: Rockefeller

Beidler, L. M., Nejad, M. S. 1985. The comparative gustatory response profile of the greater superficial petrosal and chorda tympani nerves of the hamster and the rat to some chemical stimuli. *Abstr. 7th Ann. Meet. Assoc. Chemorecept. Sci. No. 16* (Abstr.)

Bereiter, D. A., Berthoud, H.-R., Jeanreaud, B. 1980. Hypothalamic input to brain stem neurons responsive to oropharyngeal stimulation. *Exp. Brain Res.* 39: 33–39

Bereiter, D. A., Berthoud, H.-R., Jeanreaud, B. 1981. Chorda tympani and vagus nerve convergence onto caudal brain stem neurons in the rat. *Brain Res. Bull.* 7: 261–66

Berk, M. L., Finkelstein, J. A. 1982. Efferent connections of the lateral hypothalamic area of the rat: An autoradiographic investigation. *Brain Res. Bull.* 8: 511–26

Berntson, G. G., Micco, D. J. 1976. Organization of brainstem behavioral systems. *Brain Res. Bull.* 1: 471–83

Bieber, S. L., Smith, D. V. 1986. Multivariate analysis of sensory data: A comparison of methods. *Chem. Senses* 11: 19–47

Block, C. H., Schwartzbaum, J. S. 1983. Ascending efferent projections of the gustatory parabrachial nuclei in the rabbit. *Brain Res.* 259: 1–9

Borke, R. C., Nau, M. E., Ringler, R. L. Jr. 1983. Brain stem afferents of hypoglossal neurons in the rat. *Brain Res.* 269: 47–55

Boudreau, J. C., Alev, N. 1973. Classification of chemoresponsive tongue units of the cat geniculate ganglion. *Brain Res.* 54: 157–75

Boudreau, J. C., Hoang, N. K., Oravec, J., Do, L. T. 1983. Rat neurophysiological taste responses to salt solutions. *Chem. Senses* 8: 131–50

Boudreau, J. C., Oravec, J. J., Hoang, N. K. 1982. Taste systems of goat geniculate ganglion. *J. Neurophysiol.* 48: 1226–42

Boudreau, J. C., Sivakumar, L., Do, L. T., White, T. D., Oravec, J., Hoang, N. K. 1985. Neurophysiology of geniculate ganglion (facial nerve) taste systems: Species comparisons. *Chem. Senses* 10: 89–127

Boushey, H. A., Richardson, P. S., Widdicombe, J. G., Wise, J. C. M. 1974. The response of laryngeal afferent fibres to mechanical and chemical stimuli. *J. Physiol. London* 240: 153–75

Bradley, R. M. 1971. Tongue Topography. *Handb. Sensory Physiol.* 4(1): 1–30

Bradley, R. M., Cheal, M. L., Kim, Y. H. 1980. Quantitative analysis of developing epiglottal taste buds in sheep. *J. Anat.* 130: 25–32

Bradley, R. M., Stedman, H. M., Mistretta, C. M. 1983. Superior laryngeal nerve response patterns to chemical stimulation of sheep epiglottis. *Brain Res.* 276: 81–93

Cabanac, M. 1971. Physiological role of pleasure. *Science* 173: 1103–7

Car, A., Amri, M. 1982. Etudes des neurones deglutiteurs pontiques chez la brebis. I. Activite et localisation. *Exp. Brain Res.* 48: 345–54

Chang, F.-C. T., Scott, T. R. 1984. Conditioned taste aversions modify neural responses in the rat nucleus tractus solitarius. *J. Neurosci.* 4: 1850–62

Contreras, R. J. 1977. Changes in gustatory nerve discharges with sodium deficiency: A single unit analysis. *Brain Res.* 121: 373–78

Contreras, R. J., Beckstead, R. M., Norgren, R. 1982. The central projections of the trigeminal, facial, glossopharyngeal and vagus nerves: An autoradiographic study in the rat. *J. Autonom. Nerv. Syst.* 6: 303–22

Contreras, R. J., Frank, M. E. 1979. Sodium deprivation alters neural responses to gustatory stimuli. *J. Gen. Physiol.* 73: 569–94

Contreras, R. J., Kosten, T., Frank, M. E. 1984. Activity in salt taste fibers: Peripheral mechanism for mediating changes in salt intake. *Chem. Senses* 8: 275–88

Davis, J. D. 1973. The effectiveness of some sugars in stimulating licking behavior in the rat. *Physiol. Behav.* 11: 39–45

Di Lorenzo, P. M., Schwartzbaum, J. S. 1982. Coding of gustatory information in the pontine parabrachial nuclei of the rabbit: Magnitude of neural response. *Brain Res.* 251: 229–44

Doetsch, G. S., Erickson, R. P. 1970. Synaptic processing of taste-quality information in the nucleus tractus solitarius of the rat. *J. Neurophys.* 33: 490–507

Erickson, R. P. 1963. Sensory neural patterns and gustation. In *Olfaction and Taste*, ed. Y. Zotterman, pp. 205–13. New York: Pergamon

Erickson, R. P. 1968. Stimulus coding in topographic and non-topographic afferent modalities: On the significance of the activity of individual sensory neurons. *Psychol. Rev.* 75: 447–65

Erickson, R. P. 1974. Parallel "population" neural coding in feature extraction. In *The Neurosciences Third Study Program*, ed. F. O. Schmitt, F. G. Worden, Ch. 15, pp. 155–69. Cambridge: MIT Press

Erickson, R. P. 1985. Definitions: A matter of taste. In *Taste, Olfaction, and the Central Nervous System*, ed. D. W. Pfaff, Ch. 6, pp. 129–50. New York: Rockefeller Univ. Press

Erickson, R. P., Doetsch, G. S., Marshall, D. A. 1965. The gustatory neural response function. *J. Gen. Physiol.* 49: 247–63

Eshel, Y., Korczyn, A. D. 1978. Central connections of the salivation reflex in man. *Exp. Neurol.* 62: 497–99

Frank, M. 1973. An analysis of hamster afferent taste nerve response functions. *J. Gen. Physiol.* 61: 588–618

Frank, M. 1975. Response patterns of rat glossopharyngeal taste neurons. In *Olfaction and Taste V*, ed. D. A. Denton, J. P. Coghlan, pp. 59–64. New York: Academic

Frank, M., Contreras, R. J., Hettinger, T. P. 1983. Nerve fibers sensitive to ionic taste stimuli in chorda tympani of the rat. *J. Neurophysiol.* 50: 941–60

Frank, M. E., Pfaffmann, C. 1969. Taste nerve fibers: A random distribution of sensitivities to four tastes. *Science* 164: 1183–85

Fulwiler, C. E., Saper, C. B. 1984. Subnuclear organization of the efferent connections of the parabrachial nucleus in the rat. *Brain Res. Rev.* 7: 229–59

Funakoshi, M., Kawamura, Y. 1967. Relations between taste qualities and parotid gland secretion rate. In *Olfaction and Taste II*, ed. T. Hayashi, pp. 281–87. London: Pergamon

Ganchrow, J. R., Erickson, R. P. 1970. Neural correlates of gustatory intensity and quality. *J. Neurophysiol.* 33: 768–83

Gill, J. M., Erickson, R. P. 1985. Neural mass differences in gustation. *Chem. Senses* 10: 531–48

Giza, B. K., Scott, T. R. 1983. Blood glucose selectively affects taste-evoked activity in rat nucleus tractus solitarius. *Physiol. Behav.* 31: 643–50

Gjorstrup, P. 1980a. Parotid secretion of fluid and amylase in rabbits during feeding. *J. Physiol.* 309: 101–16

Gjorstrup, P. 1980b. Taste and chewing as stimuli for the secretion of amylase from the parotid gland of the rabbit. *Acta Physiol. Scand.* 110: 295–301

Glenn, J. F., Erickson, R. P. 1976. Gastric modulation of gustatory afferent activity. *Physiol. Behav.* 16: 561–68

Grill, H. J., Berridge, K. C. 1985. Taste reactivity as a measure of the neural control of palatability. In *Progress in Psychobiology and Physiological Psychology*, ed. J. M. Sprague, A. N. Epstein, 11: 1–61. New York: Academic

Grill, H. J., Berridge, K. C., Ganster, D. J. 1984. Oral glucose is the prime elicitor of preabsorptive insulin secretion. *Am. J. Physiol.* 246: R88–R95

Grill, H. J., Norgren, R. 1978a. The taste reactivity test. I. Mimetic responses to gustatory stimuli in neurologically normal rats. *Brain Res.* 143: 263–79

Grill, H. J., Norgren, R. 1978b. The taste reactivity test. II. Mimetic responses to gustatory stimuli in chronic thalamic and chronic decerebrate rats. *Brain Res.* 143: 281–97

Halpern, B. P. 1985. Time as a factor in gustation: Temporal patterns of stimulation and response. See Erickson 1985, Ch. 8, pp. 181–209

Hamilton, R. B., Norgren, R. 1984. Central projections of gustatory nerves in the rat. *J. Comp. Neurol.* 222: 560–77

Harding, R., Johnson, P., McClelland, M. E. 1978. Liquid-sensitive laryngeal receptors in the developing sheep, cat and monkey. *J. Physiol.* 277: 409–22

Hayama, T., Ito, S., Ogawa, H. 1985. Responses of solitary tract nucleus neurons to taste and mechanical stimulations of the oral cavity in decerebrate rats. *Exp. Brain Res.* 60: 235–42

Hermann, G. E., Kohlerman, N. J., Rogers, R. C. 1983. Hepatic-vagal and gustatory afferent interactions in the brainstem of the rat. *J. Autonom. Nerv. Syst.* 9: 477–95

Hermann, G. E., Rogers, R. C. 1985. Convergence of vagal and gustatory afferent input within the parabrachial nucleus of the rat. *J. Autonom. Nerv. Syst.* 13: 1–17

Hinrichsen, C. F. L., Watson, C. D. 1983. Brain stem projections to the facial nucleus of the rat. *Brain Behav. Evol.* 22: 153–63

Holstege, G., Meiners, L., Tan, K. 1985. Projections of the bed nucleus of the stria terminalis to the mesencephalon, pons, and medulla oblongata in the cat. *Exp. Brain Res.* 58: 379–91

Iida, M., Yoshioka, I., Muto, H. 1983. Taste bud papillae on the retromolar mucosa of the rat, mouse and golden hamster. *Acta Anat.* 117: 374–81

Ishiko, N., Akagi, T. 1972. Topographical organization of gustatory nervous system.

In *Olfaction and Taste IV*, ed. D. Schneider, pp. 343–49. Stuttgart: Wissenschaftliche

Iwasaki, K., Sato, M. 1981. Neural responses and aversion to bitter stimuli in rats. *Chem. Senses* 6: 119–28

Jacquin, M. F. 1983. Gustation and ingestive behavior in the rat. *Behav. Neurosci.* 97: 98–109

Jean, A. 1972. Localisation et activite des neurones deglutiteurs bulbaires. *J. Physiol. Paris* 64: 227–68

Jean, A., Car, A., Roman, C. 1975. Comparison of activity in pontine versus medullary neurones during swallowing. *Exp. Brain Res.* 22: 211–20

Kawamura, Y., Yamamoto, T. 1977. Salivary secretion to noxious stimulation of the trigeminal area. In *Pain in the Trigeminal Region*, ed. D. J. Anderson, B. Matthews, pp. 395–404. Holland: Elsevier Biomed. Press

Kawamura, Y., Yamamoto, T. 1978. Studies on neural mechanisms of the gustatory-salivary reflex in rabbits. *J. Physiol.* 285: 35–47

Kimura, K., Beidler, L. M. 1961. Microelectrode study of taste receptors of rat and hamster. *J. Cell. Comp. Physiol.* 58: 131–39

Klein, P. B., Schroeder, H. E. 1979. Epithelial differentiation and taste buds in the soft palate of the monkey, *Macaca irus*. *Cell Tissue Res.* 196: 181–88

Kosten, T., Contreras, R. J. 1985. Adrenalectomy reduces peripheral neural responses to gustatory stimuli in the rat. *Behav. Neurosci.* 99: 734–41

Lalonde, E. R., Eglitis, J. A. 1961. Number and distribution of taste buds on the epiglottis, pharynx, larynx, soft palate and uvula in a human newborn. *Anatom. Rec.* 140: 91–95

Maes, F. W., Erickson, R. P. 1984. Gustatory intensity discrimination in rat NTS: A tool for the evaluation of neural coding theories. *J. Comp. Physiol. A* 155: 271–82

Makous, W., Nord, S., Oakley, B., Pfaffmann, C. 1963. The gustatory relay in the medulla. In *Olfaction and Taste*, ed. Y. Zotterman, pp. 381–93. New York: Pergamon

Mansson, I., Sandberg, N. 1975. Salivary stimulus and swallowing reflex in man. *Acta Otolaryngol.* 79: 445–50

Matsuo, R., Shimizu, N., Kusano, K. 1984. Lateral hypothalamic modulation of oral sensory afferent activity in nucleus tractus solitarius neurons of rats. *J. Neurosci.* 4: 1201–7

Matsuo, R., Yamamoto, T., Kawamura, Y. 1982. Responses of salivatory neurons in the medulla oblongata of rabbit. *Jpn. J. Physiol.* 32: 309–13

McBurney, D. H., Gent, J. F. 1979. On the nature of taste qualities. *Psychol. Bull.* 86: 151–67

Miller, I. J. Jr. 1971. Peripheral interactions among single papilla inputs to gustatory nerve fibers. *J. Gen. Physiol.* 57: 1–25

Miller, I. J. Jr. 1974. Branched chorda tympani neurons and interactions among taste receptors. *J. Comp. Neurol.* 158: 155–66

Miller, I. J. Jr. 1977. Gustatory receptors of the palate. In *Food Intake and Chemical Senses*, ed. Y. Katsuki, M. Sato, S. F. Takagi, Y. Oomura, pp. 173–85. Tokyo: Univ. Tokyo Press

Miller, I. J. Jr., Smith, D. V. 1984. Quantitative taste bud distribution in the hamster. *Physiol. Behav.* 32: 275–85

Mistretta, C. M. 1972. A quantative analysis of rat chorda tympani fiber discharge patterns. In *Olfaction and Taste IV*, ed. D. Schneider, pp. 294–300. Stuttgart: Wissenschaftliche

Moskowitz, H. R. 1974. The psychology of sweetness. In *Sugars in Nutrition*, ed. H. L. Sipple, K. W. McNutt, Ch. 4, pp. 37–64. New York: Academic

Murakami, T., Ishizuka, K., Yoshihara, M., Uchiyama, M. 1983. Reflex responses of single salivatory neurons to stimulation of trigeminal sensory branches in the cat. *Brain Res.* 280: 233–37

Nagai, T., Ueda, K. 1981. Stochastic properties of gustatory impulse discharges in rat chorda tympani fibers. *J. Neurophysiol.* 45: 574–92

Nejad, M. S. 1986. The neural activities of the greater superficial petrosal nerve of the rat in response to chemical stimulation of the palate. *Chem. Senses* 11: 283–94

Newbrun, E. 1962. Observations on the amylase content and flow rate of human saliva following gustatory stimulation. *J. Dent. Res.* 41: 459–65

Ninomiya, Y., Mizukoshi, T., Higashi, T., Katsukawa, H., Funakoshi, M. 1984. Gustatory neural responses in three different strains of mice. *Brain Res.* 302: 305–14

Norgren, R. 1974. Gustatory afferents to ventral forebrain. *Brain Res.* 81: 285–95

Norgren, R. 1976. Taste pathways to hypothalamus and amygdala. *J. Comp. Neurol.* 166: 17–30

Norgren, R. 1978. Projections from the nucleus of the solitary tract in the rat. *Neuroscience* 3: 207–18

Norgren, R. 1983. Afferent interactions of cranial nerves involved in ingestion. *J. Autonom. Nerv. Syst.* 9: 67–77

Norgren, R. 1984. Central neural mechanisms of taste. In *Handbook of Physiology—The Nervous System*, Vol. 3, ed.

I. Darian-Smith, Ch. 24, pp. 1087–1128. Bethesda: Am. Physiol. Soc.

Norgren, R. 1985. Taste and the autonomic nervous system. *Chem. Senses* 10: 143–61

Norgren, R., Grill, H. J. 1976. Efferent distribution from the cortical gustatory area in rats. *2nd Ann. Meet. Soc. Neurosci.* 176: 124 (Abstr.)

Norgren, R., Grill, H. J. 1982. Brain-stem control of ingestive behavior. In *The Physiological Mechanisms of Motivation*, ed. D. W. Pfaff, pp. 131. New York: Rockefeller Univ. Press

Norgren, R., Leonard, C. M. 1973. Ascending central gustatory pathways. *J. Comp. Neurol.* 150: 217–38

Norgren, R., Pfaffmann, C. 1975. The pontine taste area in the rat. *Brain Res.* 91: 99–117

Nowlis, G. H. 1973. Taste-elicited tongue movements in human newborn infants: An approach to palatability. In *4th Symp. Oral Sensation Percept.*, ed. J. F. Bosma, pp. 292–303. Washington DC: Superintendent of Documents, US GPO

Nowlis, G. H., Frank, M. E. 1981. Quality coding in gustatory systems of rats and hamsters. In *Perception of Behavioral Chemicals*, ed. D. M. Norris, pp. 59–80. Amsterdam: Elsevier/North Holland

Nowlis, G. H., Frank, M. E., Pfaffmann, C. 1980. Specificity of acquired aversions to taste qualities in hamsters and rats. *J. Comp. Physiol. Psychol.* 94: 932–42

Oakley, B. 1967. Altered taste responses from cross-regenerated taste nerves in the rat. In *Olfaction and Taste II*, ed. T. Hayashi, pp. 535–47. London: Pergamon

Oakley, B., Jones, L. B., Kaliszewski, J. M. 1979. Taste responses of the gerbil IXth nerve. *Chem. Senses Flavour* 4: 79–87

Ogawa, H., Hayama, T. 1984. Receptive fields of solitario-parabrachial neurons responsive to natural stimulation of the oral cavity in rats. *Exp. Brain Res.* 54: 359–66

Ogawa, H., Hayama, T., Ito, S. 1982. Convergence of input from tongue and palate to the parabrachial nucleus neurons of rats. *Neurosci. Lett.* 28: 9–14

Ogawa, H., Hayama, T., Ito, S. 1984a. Location and taste responses of parabrachio-thalamic relay neurons in rats. *Exp. Neurol.* 83: 507–17

Ogawa, H., Imoto, T., Hayama, T. 1984b. Responsiveness of solitario-parabrachial relay neurons to taste and mechanical stimulation applied to the oral cavity in rats. *Exp. Brain Res.* 54: 349–58

Ogawa, H., Kaisaku, J. 1982. Physiological characteristics of the solitario-parabrachial relay neurons with tongue afferent inputs in rats. *Exp. Brain Res.* 48: 362–68

Ogawa, H., Sato, M., Yamashita, S. 1968. Multiple sensitivity of chorda tympani fibres of the rat and hamster to gustatory and thermal stimuli. *J. Physiol.* 199: 223–40

Ogawa, H., Yamashita, S., Sato, M. 1974. Variation in gustatory nerve fiber discharge pattern with change in stimulus concentration and quality. *J. Neurophysiol.* 37: 443–57

Ooishi-Iida, M. 1979. Taste areas of the oral cavity base of the rat and mouse, with special reference to the taste-bud papillae on the caruncula sublingualis. *Okajimas Filia Anat. Jpn.* 55: 329–40

Ozeki, M., Sato, M. 1972. Responses of gustatory cells in the tongue of rat to stimuli representing four taste qualities. *Comp. Biochem. Physiol.* 41A: 391–407

Perrotto, R. S., Scott, T. R. 1976. Gustatory neural coding in the pons. *Brain Res.* 110: 283–300

Pfaffmann, C. 1941. Gustatory afferent impulses. *J. Cell. and Comp. Physiol.* 17: 243–58

Pfaffmann, C. 1952. Taste preference and aversion following lingual denervation. *J. Comp. Physiol. Psychol.* 45: 393–400

Pfaffmann, C. 1960. The pleasures of sensation. *Psychol. Rev.* 67: 253–68

Pfaffmann, C. 1970. Physiological and behavioral processes of the sense of taste. In *Ciba Found. Symp. Taste Smell Vertebrates*, ed. G. E. W. Wolstenholme, J. Knight, pp. 31–55. London: Churchill

Pfaffmann, C. 1974. Specificity of the sweet receptors of the squirrel monkey. *Chem. Senses Flavour* 1: 61-67

Pfaffmann, C., Fisher, G. L., Frank, M. 1967. The sensory and behavioral factors in taste preference. In *Olfaction and Taste II*, ed. T. Hayashi, pp. 361–81. London: Pergamon

Pfaffmann, C., Frank, M., Bartoshuk, L. M., Snell, T. C. 1976. Coding gustatory information in the squirrel monkey chorda tympani. *Progr. Psychobiol. Physiol. Psychol.* 6: 1–27

Rogers, R. C., Novin, D., Butcher, L. L. 1979. Electrophysiological and neuroanatomical studies of hepatic portal osmo- and sodium receptive afferent projections within the brain.*J. Autonom. Nerv. Syst.* 1: 183–202

Saper, C. B., Loewy, A. D. 1980. Efferent connections of the parabrachial nucleus in the rat. *Brain Res.* 197: 291–317

Sato, T., Beidler, L. M. 1982. The response characteristics of rat taste cells to four basic taste stimuli. *Comp. Biochem. Physiol.* 73A: 1–10

Sato, T., Beidler, L. M. 1983. Dependence of

gustatory neural response on depolarizing and hyperpolarizing receptor potentials of taste cells in the rat. *Comp. Biochem. Physiol.* 75A: 131–37

Sato, M., Ogawa, H., Yamashita, S. 1975. Response properties of macaque monkey chorda tympani fibers. *J. Gen. Physiol.* 66: 781–810

Schneyer, L. H. 1956. Amylase content of separate salivary gland secretions of man. *J. Appl. Physiol.* 9: 453–55

Schwartz, G. J., Grill, H. J. 1984. Relationships between taste reactivity and intake in the neurologically intact rat. *Chem. Senses* 9: 249–72

Schwartzbaum, J. S. 1983a. Operant licking in rabbits for intraoral injection of basic types of tastants. *Physiol. Behav.* 31: 445–51

Scott, T. R. 1981. Brain stem and forebrain involvement in the gustatory neural code. In *Brain Mechanisms of Sensation*, ed. Y. Katsuki, R. Norgren, M. Sato, Ch. 12, pp. 177–96. New York: Wiley

Scott, T. R., Chang, F.-C. T. 1984. The state of gustatory neural coding. *Chem. Senses* 8: 297–314

Scott, T. R., Perrotto, R. S. 1980. Intensity coding in pontine taste area: Gustatory information is processed similarly throughout rat's brain stem. *J. Neurophysiol.* 44: 739–50

Scott, T. R., Yaxley, S., Sienkiewicz, Z. J., Rolls, E. T. 1986. Gustatory responses in the nucleus tractus solitarius of the alert cynomolgus monkey. *J. Neurophysiol.* 55: 182–200

Shingai, T. 1977. Ionic mechanism of water receptors in the laryngeal mucosa of the rabbit. *Jpn. J. Physiol.* 27: 27–42

Shingai, T. 1980. Water fibers in the superior laryngeal nerve of the rat. *Jpn. J. Physiol.* 30: 305–7

Shingai, T., Beidler, L. M. 1985. Response characteristics of three taste nerves in mice. *Brain Res.* 335: 245–49

Shingai, T., Shimada, K. 1976. Reflex swallowing elicited by water and chemical substances applied in the oral cavity, pharynx, and larynx of the rabbit. *Jpn. J. Physiol.* 26: 455–69

Shipley, M. T., Sanders, M. S. 1982. Special senses are really special: Evidence for a reciprocal, bilateral pathway between insular cortex and nucleus parabrachialis. *Brain Res. Bull.* 8: 493–501

Smith, G. P., Jerome, C., Norgren, R. 1985. Afferent axons in abdominal vagus mediate satiety effect of cholecystokinin in rats. *Am. J. Physiol.* 249: R638–41

Smith, D. V., Travers, J. B. 1979. A metric for the breadth of tuning of gustatory neurons. *Chem. Senses Flavour* 4: 215–29

Smith, D. V., Travers, J. B., Van Buskirk, R. L. 1979. *Brain Res. Bull.* 4: 359–72

Smith, D. V., Van Buskirk, R. L., Travers, J. B., Bieber, S. L. 1983a. Gustatory neuron types in hamster brain stem. *J. Neurophysiol.* 50: 522–40

Smith, D. V., Van Buskirk, R. L., Travers, J. B., Bieber, S. L. 1983b. Coding of taste stimuli by hamster brain stem neurons. *J. Neurophysiol.* 50: 541–58

Stedman, H. M., Bradley, R. M., Mistretta, C. M., Bradley, B. E. 1980. Chemosensitive responses from the cat epiglottis. *Chem. Senses* 5: 233–45

Storey, A. T., Johnson, P. 1975. Laryngeal water receptors initiating apnea in the lamb. *Exp. Neurol.* 47: 42–55

Sweazey, R. D. 1983. *Separate receptor populations drive activity in hamster medullary taste neurons*. Unpubl. dissertation, Univ. Wyoming, Laramie

Takeuchi, Y., Hopkins, D. A. 1984. Light and electron microscopic demonstration of hypothalamic projections to the parabrachial nuclei in the cat. *Neurosci. Lett.* 46: 53–58

Takeuchi, Y., McClean, J. H., Hopkins, D. A. 1982. Reciprocal connections between the amygdala and parabrachial nuclei: Ultrastructural demonstration by degeneration and axonal transport of horseradish peroxidase in the cat. *Brain Res.* 239: 583–88

Thompson, D. A., Campbell, R. G. 1977. Hunger in humans induced by 2-deoxy-D-glucose: Glucoprivic control of taste preference and food intake. *Science* 198: 1065–68

Tonosaki, K., Funakoshi, M. 1984. Intracellular taste cell responses of mouse. *Comp. Biochem. Physiol.* 78A: 651–56

Torvik, A. 1956. Afferent connections to the sensory trigeminal nuclei, the nucleus of the solitary tract and adjacent structures. *J. Comp. Neurol.* 106: 51–141

Travers, J. B. 1979. *Projections from the anterior nucleus tractus solitarius of the hamster demonstrated with anterograde and retrograde tracing techniques*. Unpubl. dissertation, Univ. Wyoming, Laramie

Travers, J. B., Norgren, R. 1983. Afferent projections to the oral motor nuclei in the rat. *J. Comp. Neurol.* 220: 280–98

Travers, J. B., Norgren, R. 1986. An electromyographic analysis of the ingestion and rejection of sapid stimuli in the rat. *Behav. Neurosci.* 100: 544–55

Travers, J. B., Smith, D. V. 1979. Gustatory sensitivities in neurons of the hamster nucleus tractus solitarius. *Sens. Processes* 3: 1–26

Travers, S. P., Pfaffmann, C., Norgren, R. 1986. Convergence of lingual and palatal gustatory neural activity in the nucleus of the solitary tract. *Brain Res.* 365 : 305–20

Travers, S. P., Smith, D. V. 1984. Responsiveness of neurons in the hamster parabrachial nuclei to taste mixtures. *J. Gen. Physiol.* 84 : 221–50

Van Buskirk, R. L., Erickson, R. P. 1977. Odorant responses in taste neurons of the rat NTS. *Brain Res.* 135 : 287–303

Van Buskirk, R. L., Smith, D. V. 1981. Taste sensitivity of hamster parabrachial pontine neurons. *J. Neurophysiol.* 45 : 144–71

van der Kooy, D., Koda, L. Y., McGinty, J. F., Gerfen, C. R., Bloom, F. E. 1984. The organization of projections from the cortex, amygdala, and hypothalamus to the nucleus of the solitary tract in rat. *J. Comp. Neurol.* 224 : 1–24

Vornov, J. J., Sutin, J. 1983. Brainstem projections to the normal and noradrenergically hyperinnervated trigeminal motor nucleus. *J. Comp. Neurol.* 214 : 198–208

Woolston, D. C., Erickson, R. P. 1979. Concept of neuron types in gustation in the rat. *J. Neurophysiol.* 42 : 1390–1409

Yamada, K. 1966. Gustatory and thermal responses in the glossopharyngeal nerve of the rat. *Jpn. J. Physiol.* 16 : 599–611

Yamada, K. 1967. Gustatory and thermal responses in the glossopharyngeal nerve of the rabbit and cat. *Jpn. J. Physiol.* 17 : 94–110

Yamamoto, T. 1985. Taste responses of cortical neurons. *Prog. Neurobiol.* 23 : 273–315

Yamamoto, T., Fujiwara, T., Matsuo, R., Kawamura, Y. 1982a. Hypoglossal motor nerve activity elicited by taste and thermal stimuli applied to the tongue in rats. *Brain Res.* 238 : 89–104

Yamamoto, T., Kawamura, Y. 1977a. Gustatory-salivary reflex in the rabbit . In *Food Intake and Chemical Senses*, ed. Y. Katsuki, M. Sato, S. Takagi, Y. Oomura, pp. 211–221. Tokyo : Tokyo Univ. Press

Yamamoto, T., Kawamura, Y. 1977b. Responses of the submandibular secretory nerve to taste stimuli. *Brain Res.* 130 : 152–55

Yamamoto, T., Matsuo, R., Fujiwara, T., Kawamura, Y. 1982b. EMG activities of masticatory muscles during licking in rats. *Physiol. Behav.* 29 : 905–13

Reference added in proof:

Schwartzbaum, J. S. 1983b. Electrophysiology of taste-mediated functions in parabrachial nuclei of behaving rabbit. *Brain Res. Bull.* 11 : 61–89

Ann. Rev. Neurosci. 1987. 10: 633–93

CALCIUM ACTION IN SYNAPTIC TRANSMITTER RELEASE

George J. Augustine

Section of Neurobiology, Department of Biological Sciences, University of Southern California, Los Angeles, California 90089-0371

Milton P. Charlton

Department of Physiology, University of Toronto, Toronto, Ontario M5S 1A8

Stephen J Smith

Howard Hughes Medical Institute, Section of Molecular Neurobiology, Yale Medical School, New Haven, Connecticut 06510

INTRODUCTION

In this review we are concerned with the mechanisms by which the electrical potential across the membrane of presynaptic nerve terminals regulates the release of neurotransmitter substances into the synaptic cleft. Ca ions play a central role in this process: Release occurs when a voltage-dependent Ca channel in the presynaptic membrane opens, thus permitting an influx of Ca ions and diffusion of Ca in cytoplasm, followed by binding of Ca at some cytoplasmic site that triggers the exocytotic release of quanta of neurotransmitter. These concepts, introduced mainly by the work of Katz and associates (Katz 1969), comprise the "Ca hypothesis" of transmitter release, a hypothesis now widely accepted.

Although the general role of Ca as a presynaptic messenger is well supported, many important specifics of its action remain to be resolved. We review progress with these details in three parts, corresponding to

0147–006X/87/0301–0633$02.00

three successive steps in presynaptic excitation–release coupling. First, we describe progress in characterizing the synaptic voltage-dependent Ca channel. Second, we discuss studies of the cytoplasmic Ca diffusion step. Finally, we review work on identification of the cytoplasmic Ca binding site and the molecules that translate Ca binding into exocytosis.

We touch on many results from nonsynaptic cells and membranes. Because technical limitations have impeded direct study of presynaptic nerve terminals, much of our current view of Ca action in presynaptic terminal function is based on generalizations from nonsynaptic systems more amenable to experimental analysis. For instance, our understanding of presynaptic ion channels, including the Ca channel, is mainly an extrapolation from ion channels studied on neuronal axons, cell bodies, and dendrites, as well as from heart muscle and other nonneuronal cells. Likewise, many of our ideas about Ca-secretion coupling are extrapolations from findings in secretory cells, beginning with the work of Douglas (1968) on Ca action in catecholamine secretion by chromaffin cells of the adrenal medulla.

Because most of our information now comes from nonsynaptic membranes and cell types, one of the main challenges for future research on synaptic mechanisms is the development of methods for direct study of presynaptic terminals. Although the supposition that Ca channels in cell bodies are homologous to those in synaptic terminals seems eminently reasonable, in no case has it been proved. Likewise, even though homologies between nerve terminals and secretory cells are apparent, the homologies are imperfect, and only direct studies of nerve terminals will resolve the true extent of these similarities. In this review, we wish to combine an appreciation of the hints and lessons from a broad range of cell biology with a critical focus on the need to confirm these hints through direct studies of the presynaptic terminal.

The literature relevant to Ca action in synaptic transmitter release is immense, so arbitrary boundaries have been necessary to circumscribe the material for a single review. We have chosen to focus on particular themes within each area addressed. In treating the presynaptic Ca channel, we focus on the prospects for characterizing individual presynaptic Ca channels. In our treatment of the cytoplasmic Ca diffusion step, we focus on recent attempts to investigate mathematically the implications of diffusive coupling between quantized Ca influx events (i.e. single channel openings) and quantized release events (i.e. vesicle exocytosis). In our treatment of the mechanism by which cytoplasmic Ca triggers release events, we focus on attempts to identify the specific molecules that bind Ca and mediate the subsequent effector actions. This rather specialized approach is warranted by the wealth of more general summaries of the role of calcium ions in transmitter release (Katz 1969, Martin 1977, Rahamimoff et al 1978a, 1980, Kelly et al 1979, Reichardt & Kelly 1983, Silinsky 1985).

PRESYNAPTIC Ca CHANNEL

Ca channels function at presynaptic terminals to transduce changes in membrane potential into an influx of Ca ions down their electrochemical gradient. In this section we discuss what is known about these voltage-dependent ion channels. A much larger literature is available on Ca channels in general and even on Ca channels in neurons, but here we focus on the Ca channels of presynaptic terminals in particular. For general reviews of Ca channels, see Hagiwara & Byerly (1981), Kostyuk (1981), and Tsien (1983, 1986).

Understanding of Ca channel function has progressed rapidly in the last few years, due largely to the advent of single-channel recording methods. Unfortunately, single channel recordings of a positively identified presynaptic Ca channel have not yet been reported, so one can only extrapolate from other Ca channels that have been studied at the single-channel level. Ca currents have been measured macroscopically at a giant presynaptic terminal of the squid (Llinas 1982), however, and Lemos et al (1986) have published a preliminary report of whole-cell patch recording from synaptosomes prepared from crab secretory terminals.

The paucity of studies of presynaptic Ca channels results from the difficulty of studying most terminals electrophysiologically. Presynaptic terminals are usually very small (1–2 μm in diameter) and completely surrounded by other tissues, such as glia and postsynaptic cells. Strong generalizations about presynaptic Ca channels on the basis of the few direct measurements currently available are obviously dangerous, now particularly so because of recent demonstrations of Ca channel diversity.

Many different types of Ca channels appear to exist (Hagiwara & Byerly 1981, Reuter 1985, Miller 1985a, Tsien 1986). As many as three types of Ca channels have been demonstrated in single neurons from chick dorsal root ganglia (Nowycky et al 1985). These channel types, called L, N, and T, can be distinguished by their activation and inactivation kinetics, voltage-dependence, single-channel conductance, and sensitivity to pharmacological antagonists (Tsien 1986, Fox et al 1986). Many more neuronal Ca channel variants may await classification.

Two avenues to knowledge of presynaptic Ca channel properties are available: direct experimental studies of presynaptic terminal membranes, and generalization from results on extrasynaptic Ca channels of one kind or another. Neither avenue is completely satisfactory. The first approach is limited both in the types of synapses that have been studied and the types of information available. The second approach is limited by the hazards posed by the diversity of Ca channel types. We consider these two approaches separately below.

Studies of Ca Channels in Presynaptic Membranes

Detailed electrical measurements of Ca current at an actual presynaptic terminal have so far been reported only for the giant synapse of the squid (Llinas et al 1981a, Charlton et al 1982, Augustine & Eckert 1984a, Augustine et al 1985a). Here accurate recording of macroscopic Ca current is possible simultaneously with measurement of the postsynaptic responses caused by neurotransmitter release. Extracellular currents can be measured from motor nerve terminals (Gundersen et al 1982, Brigant & Mallart 1982, Mallart 1984), but properties of any one ionic current are difficult to infer from such recordings (Konishi 1985). Measurement of synaptic Ca current derived from a synaptosome source may also be possible (Umbach et al 1984, Nelson et al 1984), but caution is necessary because of the heterogeneity of most synaptosomal preparations and because of the possible presence of Ca channels from extrasynaptic membrane. Synaptosomal preparations have also been used in tracer studies of Ca channels (Nachshen & Blaustein 1979, 1980, 1982).

VOLTAGE-DEPENDENT GATING For studies of voltage-dependent channel gating, no other method approaches the resolution and precision of electrophysiological measurement. As an electrophysiological analysis of presynaptic Ca current has so far been reported only for the squid giant synapse, this analysis is summarized here as a point of reference. Other synapses, however, may have Ca channels that behave quite differently.

Current-voltage relationship In common with most other Ca currents studied (Hagiwara & Byerly 1981), Ca currents at squid presynaptic terminals have a bell-shaped current-voltage relationship (Llinas et al 1981a, Augustine & Eckert 1984a, Augustine et al 1985a). The peak Ca currents are observed at around −10 to 0 mV. Measurable Ca current flows with depolarizations from the resting potential (near −70 mV) as small as 20 mV, and a gating voltage dependence as steep as 5 mV/e-fold has been shown. Analysis of Ca "tail" currents indicates that these presynaptic Ca channels are maximally activated at potentials above +20 mV, with half-maximal activation occurring at −15 mV.

Activation kinetics Squid presynaptic Ca currents activate with a sigmoidal time course that is slow compared to Na current activation (Llinas et al 1981a). This sigmoidal activation time course indicates that Ca channels have more than one closed state. The rate of activation is voltage-dependent, with half-activation times on the order of a few milliseconds or less (Augustine et al 1985a). Following the end of a depolarizing pulse, the Ca channels deactivate rapidly. The resultant "tail" currents, which are at the temporal limit of present measurement methods, decay over several hundred microseconds (Llinas et al 1982; Augustine et al 1985a).

Inactivation Inactivation of presynaptic Ca channels during prolonged depolarization appears to be slow and incomplete. At the squid synapse, inactivation is prominent only during depolarizations lasting several seconds (Augustine & Eckert 1982). This inactivation appears to be due to Ca-dependent inactivation of the presynaptic Ca channels (Eckert & Chad 1984), because the rate of inactivation depends upon the amount of Ca entry and the identity of the ion that is permeating through the channel (Augustine & Eckert 1984b). Ca flux measurements on synaptosomes also reveal very slow inactivation of presynaptic Ca influx (Nachshen & Blaustein 1980, 1982), but it is not clear whether this is due to Ca-dependent or voltage-dependent inactivation (Nachshen 1985a).

PERMEATION AND SELECTIVITY Presynaptic Ca channels so far appear similar to many other Ca channels in ion permeation mechanism. At the squid synapse, current amplitudes saturate as a function of $[Ca]_0$ with a dissociation constant (K_d) of approximately 60 mM (Augustine & Charlton 1986). Ca fluxes into synaptosomes are also saturable, but have a lower K_d (Nachshen & Blaustein 1982). These results are consistent with models that require that Ca bind at one (Hagiwara & Takahashi 1967) or more (Kostyuk et al 1983, Hess & Tsien 1984, Almers & McCleskey 1984) sites within the channel in order to permeate. These channels are also permeable to Sr and Ba, with Ba currents usually somewhat larger than currents carried by the other two divalents (Augustine & Eckert 1984a). The larger size of presynaptic Ba currents is not attributable to differences in membrane surface charges (Augustine & Eckert 1984a). Nachshen & Blaustein (1982) used tracer flux measurements to examine the relative selectivity of synaptosomal Ca channels and found a permeability sequence of Ca > Sr > Ba.

PHARMACOLOGICAL CHANNEL BLOCKERS A number of inorganic cations reduce presynaptic Ca influx, as assessed by direct measurement of presynaptic Ca fluxes or by blockade of transmitter release. Most divalent or trivalent cations that block Ca channels (reviewed in Hagiwara & Byerly 1981) have been found to block evoked transmitter release (reviewed in Silinsky 1985). The relative potency of these ions in blocking release is approximately La > Cd > Pb > Ni > Co = Mn > Mg. A similar sequence is seen for inhibition of Ca fluxes into depolarized synaptosomes (Nachshen 1984). Although systematic study is lacking, Cd, Co, and Mn have been shown to block presynaptic Ca currents at the squid synapse (Llinas et al 1981a, Charlton et al 1982, Augustine & Eckert 1984a, Augustine et al 1985a, Augustine & Charlton 1986). Cd has also been shown to block Ca currents in neurosecretory terminals (Lemos et al 1986).

In recent years an enormous literature has emerged on organic compounds that appear to bind to Ca channels with very high affinity. Verapamil and related compounds, as well as dihydropyridine antagonists and agonists, have very potent actions on Ca channel currents in a number of tissues (Reuter 1983, Fox et al 1986). Verapamil apparently has only a very weak ability to block Ca entry into motor nerve terminals (Van der Kloot & Kita 1975, Gotgilf & Magazanik 1977, Publicover & Duncan 1979) and synaptosomes (Nachshen & Blaustein 1979). Predepolarization has been reported to enhance the efficacy of verapamil on synaptosomes (Nachshen 1985a). Whether or not the dihydropyridines are capable of interacting with presynaptic Ca channels has been a subject of controversy. Several groups have demonstrated high-affinity binding of dihydropyridines to nervous tissue (Gould et al 1983) and have proposed that these dihydropyridine receptors are Ca channels. However, it has been difficult to demonstrate that these compounds block Ca influx. For example, several studies have been unable to demonstrate dihydropyridine block of Ca influx into synaptosomes (reviewed in Miller & Freedman 1984) or block of Ca currents in squid presynaptic terminals (G. Augustine, unpublished). This discrepancy suggests that the dihydropyridine binding sites may not be Ca channels (Miller & Freedman 1984, Schwartz et al 1985). However, more recent studies have found that dihydropyridines affect presynaptic Ca fluxes and/or transmitter release (Ogura & Takahashi 1984 Albus et al 1985, Turner & Goldin 1985, Middlemiss & Spedding 1985). Part of the reason for this diversity of responses to the dihydropyridines may be a diversity of presynaptic Ca channels.

Several neurotoxins also have been reported to interact with presynaptic Ca channels. The first toxin identified was leptinotarsin, a 57 kD protein extracted from beetle hemolymph (Crosland et al 1984). Leptinotarsin causes increases in transmitter release that are abolished or reduced if Ca ions are removed from the external medium (McClure et al 1980, Crosland et al 1984, Stimers 1982). A number of results suggest that this toxin directly opens presynaptic Ca channels, but this possibility has not yet been tested directly. Leptinotarsin is active on mammalian brain synaptosomes (McClure et al 1980, Crosland et al 1984), mammalian neuromuscular junctions (McClure et al 1980, Stimers 1982), and elasmobranch electric organ synaptosomes (Yeager et al 1987), but it is ineffective at frog neuromuscular synapses (Stimers 1982) and squid optic lobe synaptosomes (Koenig 1985).

A second toxin reported to block presynaptic Ca channels is ω-CgTX, a 27 amino acid peptide from *Conus geographus* snails (Olivera et al 1985). This toxin blocks transmitter release from frog neuromuscular junctions, Ca-dependent action potentials recorded from cultured chick dorsal root ganglion neurons (Kerr & Yoshikami 1984), and transmitter release and Ca influx into electric organ synaptosomes (Yeager et al 1987). Because the toxin blocks Ca channel currents recorded from chick dorsal root ganglion neurons (Feldman & Yoshikami 1985), the other actions mentioned above are probably also due to Ca channel blockade. A third potential Ca channel-specific toxin is maitotoxin, a 145 kD nonpeptide molecule that enhances Ca influx and norepinepherine release from PC12 cells (Takahashi et al 1982). Little is known about the mechanism of action of this toxin.

Barbiturates reduce presynaptic Ca spikes (Morgan & Bryant 1977) and Ca influx into synaptosomes (Blaustein & Ector 1975). Whether this reduction is a direct action upon presynaptic Ca channels or is indirectly mediated is not known.

MODULATION BY NEUROTRANSMITTERS AND HORMONES Because transmitter release is exquisitely sensitive to the amount of Ca that enters the presynaptic terminal (Augustine et al 1985b), any modulation of presynaptic Ca channel current will have large effects on synaptic transmission. A wealth of literature demonstrates the presynaptic actions of neurotransmitters upon transmitter release (e.g. Kuba 1970, O'Shea & Evans 1979, Glusman & Kravitz 1982). Many of these actions are probably due to direct or indirect effects upon presynaptic Ca channels, although it remains unproven. For example, Kretz et al (1986) have shown that presynaptic inhibition at an *Aplysia* synapse is correlated with a decrease in the Ca current recorded in the cell body of the presynaptic neuron. Perhaps the best case can be made for the neuropeptide, enkephalin, which decreases transmitter release at many synapses (Macdonald & Nelson 1978, Konishi et al 1979, Bixby & Spitzer 1983a), decreases somatic Ca-dependent action potentials (Mudge et al 1979, Werz & Macdonald 1982, Bixby & Spitzer 1983b), and also decreases Ca influx into electric organ synaptosomes (Michaelson et al 1984). Although the evidence supporting actions of transmitters or hormones upon presynaptic Ca channels is incomplete, such actions are probably an important regulatory mechanism in the nervous system (Klein et al 1980). Neurotransmitters or hormones also modulate extrasynaptic Ca currents (Dunlap & Fischbach 1981, Forscher & Oxford 1985, Siegelbaum & Tsien 1983, Tsien 1986, Fox et al 1986).

Generalizations from Studies of Extrasynaptic Ca Channels

Although the studies described above have yielded a considerable amount of information regarding the properties of presynaptic Ca channels, key information about single channel current characteristics and many other properties remains to be understood. In this section we review some of the characteristics of Ca channel currents that have been considered only in nonsynaptic preparations, and we try to relate these results to presynaptic Ca channels.

COMPARISON OF CHANNEL TYPES Table 1 compares certain properties of the best-known presynaptic and extrasynaptic Ca channels. Such comparison is acknowledged as a risky endeavor, because no single channel measurements have been reported for an identified presynaptic Ca channel, and different presynaptic terminals may well have different, or even multiple, Ca channel types. Nonetheless, we consider Ca channels responsible for transmitter release at frog neuromuscular synapses, synaptosomes prepared from mammalian brain, and squid "giant" synapse. Extrasynaptic channels represented are the three Ca channel types, summarized by Nowycky et al (1985), and the Ca channels of bovine adrenal medullary chromaffin cells, secretory cells that are similar in many ways to sympathetic neurons. Properties are shown in Table 1 when they are known or can be inferred without undue risk. Based on the criteria shown in Table 1, frog and squid Ca channels clearly do not correspond to any of the channel types studied in extrasynaptic membranes. Mammalian synaptosome Ca channels may be similar to the L channel and the chromaffin cell channel; if so, they should be blocked by ω-CgTX. Therefore, at least certain presynaptic Ca channels probably do not fit into existing classification schemes. This means that it is not possible to generalize presynaptic Ca channel characteristics completely from available studies on nonsynaptic systems.

Ca CHANNELS IN CHROMAFFIN CELL MEMBRANES The chromaffin cell Ca channel merits special consideration here because, though extrasynaptic, this channel likely can be assigned a role analogous to that played by synaptic Ca channels, namely rapid triggering of exocytotic secretion. All available evidence points to one class of Ca channels in chromaffin cells, so any Ca channel found would presumably be of the type that regulates secretion. As shown in Table 1, the gating properties of chromaffin cell Ca channels are reasonably similar to those of the squid synapse, so homologies in the channels are possible. Chromaffin cell Ca channels have been shown to have a unitary current of 0.9 pA (at -5 mV) with isotonic $BaCl_2$

Table 1 Comparisons of selected neuronal Ca channel types

	Channel type						
	Extrasynaptic				Presynaptic		
Property	T	N	L	Chromaffin	Frog	Mammalian synaptosome	Squid
Inactivates with small depolarizations	yes	yes	no	no	no	??	no
Inactivation kinetics	fast	fast	slow	slow	slow	slow	slow
Sensitivity to inorganic blockers	low	high	high	high	high	high	high
Sensitivity to dihydropyridines	low	low	high	high	low	high(?)	low
Sensitivity to ω-CgTX	low	high	high	?	high	?	?
Activation range (mV)	>-70	>-10	>-10	>-40	>-70(?)	?	>-50

outside, which might correspond to 0.03 pA at a physiological extracellular Ca concentration of 1 mM (Fenwick et al 1982). With depolarization to near 0 mV, Ca channels appear to remain open for average times on the order of 1 msec, so that roughly 100 Ca ions enter a single open channel exposed to physiological saline. With return to a negative resting potential, Ca channels appear to close in a small fraction of a millisecond. Detailed analysis of channel opening statistics reveals the presence of multiple closed states (Hoshi & Smith 1987). These data may represent the best guess that can currently be made about the properties of single presynaptic Ca channels. One major difference between the gating kinetics of chromaffin cell and squid synapse Ca channels has been reported, however. Ca channels at the squid giant synapse do not appear to facilitate in their responses to multiple pulse voltage-clamp stimulation (Charlton et al 1982), whereas some facilitation of chromaffin cell Ca current has been described (Hoshi et al 1984). The possibility of facilitation of Ca channels is of special interest because use-dependent facilitation has been found to be a characteristic of transmitter release at many synapses (see Ca Trigger Molecules, below).

PRESYNAPTIC Ca DIFFUSION

To trigger release of neurotransmitter, Ca ions must diffuse in cytoplasm from an open Ca channel to a binding site that regulates exocytosis. In this section we discuss this process of intracellular signal propagation. The Ca diffusion step is of interest because cytoplasm contains a high concentration of Ca binding and transporting sites that may shape the Ca signal during propagation to any regulatory binding site. Furthermore, diffusion theory indicates that large spatial and temporal gradients of Ca occur during evoked release. The specific spatial and temporal forms of the triggering intracellular Ca ($[Ca]_i$) concentration transient are therefore likely to influence the amount and timing of transmitter release. In addition, there have been suggestions that the $[Ca]_i$ diffusion step may play a central role in in use-dependent changes in synaptic function (see below). Ca diffusion may also serve to coordinate a multiplicity of spatially distributed functions associated with release, e.g. vesicle transport and recycling, or energy and transmitter metabolism. Finally, knowledge of Ca diffusion will be required to define the magnitude of the $[Ca]_i$ changes that trigger release, a definition necessary to establish criteria for identification of the Ca receptor involved in release.

Direct measurements of intracellular Ca concentration have so far provided only limited information about the presynaptic Ca concentration transient. Calcium entry into synaptic cytoplasm has been ascertained (Llinas & Nicholson 1975, Miledi & Parker 1981, Charlton et al 1982, Stockbridge & Ross 1984), and the expected quantitative relation with the voltage-dependent Ca current has been confirmed (Augustine et al 1985a). Unfortunately, techniques with both the temporal and spatial resolution required for measurements of transients occurring during a millisecond-long evoked release event have been lacking.

A detailed view of Ca movement in presynaptic cytoplasm is now obtainable only by theoretical extrapolation from more readily measured parameters. A well-developed mathematical theory for diffusion can be applied to Ca movements in cytoplasm. Simple applications of this theory demonstrate the need to consider diffusion kinetics in presynaptic function. More elaborate applications have been used to simulate or reconstruct Ca transients, given specific assumptions about the composition of the presynaptic terminal. This section reviews the applications of diffusion theory, with special attention to the experimental basis for the many assumptions that have been necessary as the enterprise has grown more ambitious.

Characteristics of the Diffusion Process

Diffusion can be described mathematically by Fick's first law:

$$J = -D\frac{\partial C}{\partial x}, \tag{1}$$

which states that the flux J of a substance along an axis is proportional to the gradient of concentration C for that substance. The constant of proportionality D is called the *diffusion coefficient*. It is typically expressed in units of cm^2/sec. Values of D for small ions in water are on the order of 10^{-6} to 10^{-5} cm^2/sec. Any well-defined diffusion problem can be expressed as a partial differential equation based on Eq. 1 and solved at least approximately. For a comprehensive development of macroscopic diffusion theory, see the monograph of Crank (1975). Berg (1983) offers an interesting account of diffusion theory from a statistical or microscopic point of view.

The importance of diffusion kinetics to synaptic function can be grasped by considering one of the simplest cases of diffusion: that of a substance released instantaneously at a point source into an infinite surrounding volume. This case represents a characteristic diffusion response for a given substance in a given medium, and the mathematical solution takes a simple form (see Crank 1975, Eq. 3.5; Berg 1983, Eq. 2.8). The solution for diffusion of M moles released at the point $r = 0$ and the time $t = 0$ is:

$$C = M/8(\pi Dt)^{3/2} \exp(-r^2/4Dt), \tag{2}$$

where r is the radial coordinate for the resultant spherically-symmetrical diffusion. Note that r appears only in the exponential term in Eq. 2: This term alone expresses the radial distribution of the diffusing substance at any given time after the instant the substance was released. Thus, the concentration will fall to $1/e$ of its current value at the origin at a distance $(4Dt)^{1/2}$ from the origin. A distance proportional to the square root of time is characteristic of diffusive spreading and is sometimes called the *square-root law of diffusion*. Note that for values of D typical for small molecules in water, the $1/e$ concentration distance is approximately 1 μ at 1 msec. This provides a memorable yardstick for rough estimation of diffusion distances, as long as the square root law is also remembered: thus, 1 μ at 1 msec, 10 μ at 100 msec, 100 μ at 10 sec, and so on. Concentration profiles for real sources that are extended in time or space can be calculated by integration of Eq. 2. Many specific examples are discussed in the references on general diffusion theory noted above.

Some specific numbers relevant to presynaptic Ca diffusion are a minimum synaptic delay of 200 μsec at the squid giant synapse (Llinas et al 1981a) and the diffusion coefficient D for Ca ion in water, 6×10^{-6} cm^2/sec. The 200 μsec places an upper limit on the time available for diffusion of Ca from the Ca channel to an effective trigger-binding site. The numbers imply that by the beginning of transmitter release, Ca ions spread into a volume with a radius of about 2/3 μ from any one Ca channel. Since a typical synaptic terminal is 1 μ or more in size, this number indicates that spatial gradients within the terminal may be substantial at the beginning of transmitter release, depending on the spatial distribution of Ca channels. The simple calculation is imprecise: The terminal bouton would not behave as an infinite volume and, since the Ca channel remains open for a finite time, Eq. 2 should be integrated temporally. Nonetheless, a formulation more realistic in these respects does not change the conclusion very much. This simple case still probably represents only an upper limit to actual diffusion distances for synaptic trigger Ca, however. The studies discussed below show that as theoretical simulations take more realistic account of cytoplasmic diffusion conditions, expected diffusion distances can shrink considerably. Finally, for the case of sustained influx at a point source, very steep gradients near the source may persist for longer periods of time, depending on the volume of the synaptic terminal and the characteristics of any Ca sinks present (pumps, organelles, and other removal agents).

Composition and Architecture of Presynaptic Terminals

SYNAPTIC ACTIVE ZONES Accurate theoretical reconstruction of the presynaptic Ca concentration transient requires information about the cytoplasmic molecules and structures that transport, bind, or exclude Ca ions. These include Ca channels, Ca transporters, Ca binding proteins, and both internal and surface membranes. The localization of each of these components must also be defined with respect to the sites of transmitter release within presynaptic active zones [see Peters et al (1970) and Heuser & Reese (1977) for reviews of the functional morphology of presynaptic terminals]. Figure 1 shows two different views of active zones at the squid giant synapse and the frog neuromuscular junction.

Ca CHANNELS Ca channels are located in the plasma membrane of squid nerve terminals at an estimated average density of $10/\mu m^2$ (Pumplin et al

→

Figure 1 Morphology of active zones at two presynaptic terminals. *Top*, thin section through the squid giant synapse. Presynaptic terminal contains numerous synaptic vesicles (V) that are in the vicinity of electron-dense material closely associated with the presynaptic plasma membrane. Arrows indicate the narrow synaptic cleft which separates pre- and postsynaptic elements. Micrograph from an unpublished study by J. Buchanan. *Bottom*, freeze-fracture micrograph of the plasma membrane of a presynaptic terminal of a frog neuromuscular junction. Active zone (az) can be seen as two double rows of large intramembranous particles, which have been proposed to be calcium channels. Dimples near these rows of particles (see *inset*) are exocytotic events caused by synaptic vesicles fusing with the presynaptic membrane. A fortuitous cross-fracture through an active zone and into the presynaptic cytoplasm (at *arrow*) reveals synaptic vesicles and other organizational features similar to those shown in the upper micrograph. From Ko (1984).

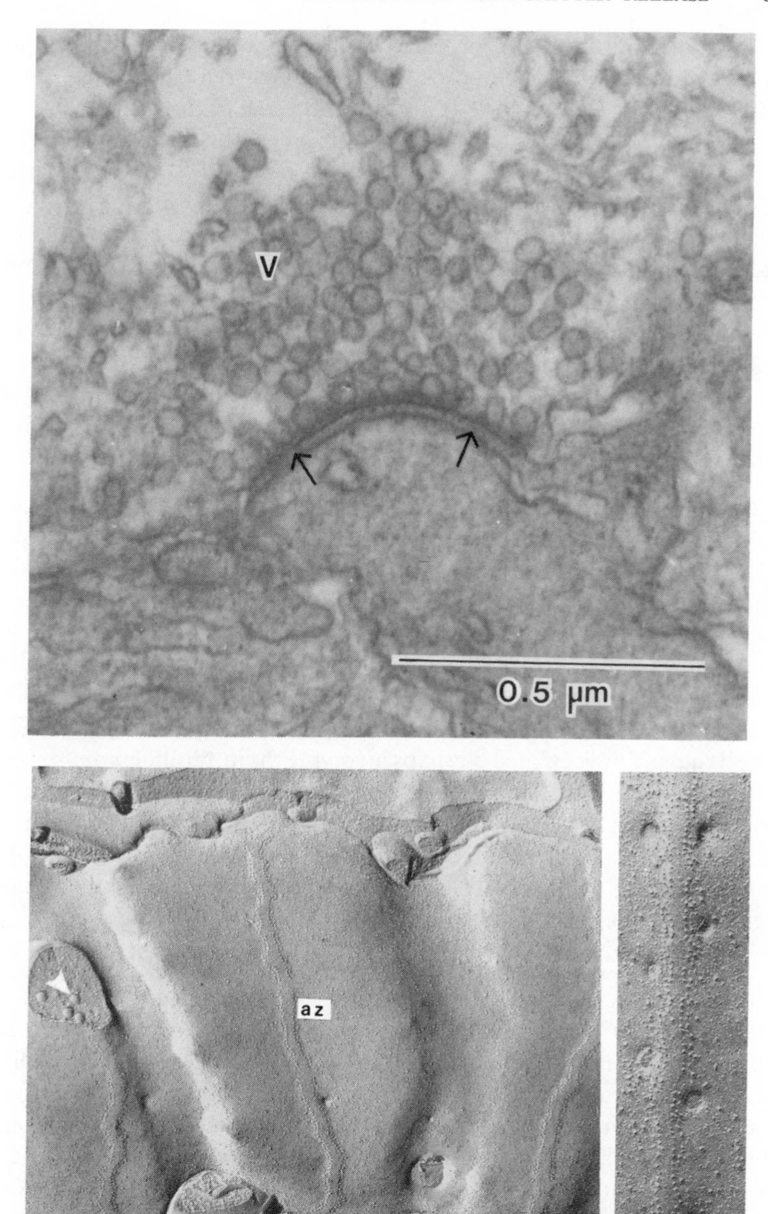
V
0.5 μm
az

1981, Simon & Llinas 1985). Since Ca channels are concentrated in synaptic (as opposed to axonal) membrane (Katz & Miledi 1969, Miledi & Parker 1981, Stockbridge & Ross 1984), further clustering of these channels may occur within presynaptic terminals. The largest particles clustered in presynaptic membranes near sites of release (see Figure 1, *bottom*) have been proposed to be Ca channels (Heuser et al 1974, Pumplin et al 1981). A structural probe that labels presynaptic Ca channels would be very useful in determining the precise distribution of Ca channels within the presynaptic membrane. The toxins described in Presynaptic Ca Channel, above, are candidates for such a probe. If Ca channels are clustered within active zones, then channel opening will result in increases in local $[Ca]_i$ that are larger and faster than if the same channels were evenly distributed over the terminal membrane.

Ca TRANSPORTERS Active Ca removal by metabolically driven transport molecules accounts for the low resting Ca concentration, on the order of 10^{-7} M characteristic of cytoplasm (DiPolo & Beauge 1983). This active transport may occur at the plasma membrane or at the membranes of intracellular organelles. Although the specific mechanisms responsible for regulation of presynaptic $[Ca]_i$ are not known, several likely candidates have been discussed by Blaustein et al (1980), Rahamimoff et al (1980), and Nachshen (1985b). Ca fluxes across internal membranes are possibly quite dynamically regulated, perhaps involving the intracellular messenger, inositol trisphosphate (Burgess et al 1984, Berridge & Irvine 1984). Structures reminiscent of sarcoplasmic reticulum have been described in neuronal cytoplasm (Henkart 1980), and certain physiological signs of a dynamically regulated intracellular Ca sequestration compartment have been reported (Kuba 1980, Smith et al 1983).

CYTOPLASMIC Ca BINDING SITES Neuronal cytoplasm contains a concentration of Ca binding sites sufficient to buffer strongly against changes in free Ca concentration. Since information on cytoplasmic Ca binding sites in synaptic terminals is fragmentary, such properties must be derived from studies on the cytoplasm of axons, somata, and paraneuronal secretory cells. Most cytoplasmic Ca binding sites probably also have specific regulatory functions as well, but their Ca buffering action alone must exert substantial effects on transmitter release. Quantitative estimates of cytoplasmic Ca buffering have been made in several different ways: (*a*) from equilibrium dialysis measurements on isolated cytoplasm (Baker & Schlaepfer 1978); (*b*) from estimates of the cytoplasmic concentrations of known Ca-binding molecules (e.g. calmodulin, see Cheung 1980); (*c*) from measured effects of adding known quantities of an exogenous buffer to cytoplasm (Neher 1986), and (*d*) from direct, in situ measurements using Ca indicators and known Ca loads (Brinley 1978, Smith & Zucker 1980, Connor & Ahmed 1984). These methods have produced rather diverse estimates of cytoplasmic Ca buffering strength but also complete agreement that the vast majority (90–99.99%) of cytoplasmic Ca ions are in complexed rather than free form. The localization of Ca binding sites in relation to the active zone structure is also probably very important in determining the impact of the Ca buffering action on transmitter release, but little information on this point is currently available.

TRIGGER Ca BINDING SITE Localization of the particular Ca binding site that triggers transmitter secretion is clearly of central interest, but again little is known about this point. The trigger site may be a soluble cytoplasmic molecule, a molecule anchored to the presynaptic cytoskeleton, a molecule intrinsic to surface membrane, or a molecule intrinsic to vesicle membrane (see Ca Trigger Molecules, below). The corresponding differences in localization may lead to very different expectations about the arrival of Ca by diffusion from a surface membrane channel.

Ca Diffusion in Cytoplasm

Ca ions diffuse much slower in cytoplasm than they do in water (Hodgkin & Keynes 1957). This is probably due to binding and sequestration of Ca ions at relatively immobile cytoplasmic sites. The Ca buffering action of reversible cytoplasmic Ca binding will tend not only to reduce the amplitude of transient changes in Ca concentration but also the apparent value of the diffusion coefficient D (see Crank 1975, p. 326). To a first approximation, the effective diffusion coefficient will be reduced by a factor proportional to the ratio of Ca free to Ca bound. Most modeling of intracellular Ca transients has represented large amounts of cytoplasmic Ca buffering at fixed or diffusible sites. As a round number, 10^{-7} cm^2/sec has been used as the reduced value of D, reflecting a ratio of free : bound Ca of 1 : 60 in reduction of D from a nominal aqueous value, 6×10^{-6} cm^2/sec. Better measurements of presynaptic Ca buffering are needed to satisfy the requirements posed by the theoretical reconstruction of the brief $[Ca]_i$ changes responsible for transmitter release evoked by presynaptic action potentials, including sets of rate constants for formation and breakdown of Ca complexes. Other factors that may influence Ca diffusion include pumping or leaking of Ca across surface or internal membranes, and tortuosity or excluded volume (Nicholson & Phillips 1981) associated with vesicles or other internal membrane systems in the cytoplasm. Although putting quantitative limits on any of these factors is still difficult, diffusion with the reduced D of 10^{-7} cm^2/sec and diffusion with the aqueous D of 6×10^{-6} cm^2/sec may serve as lower and upper limits to the diffusion of Ca in cytoplasm.

Reconstruction of Cytoplasmic Ca Transients

Models based on diffusion theory have been developed for diffusion conditions chosen to approximate more or less closely presynaptic cytoplasm. Uncertainties about presynaptic architecture and composition presently preclude a theoretical treatment of any great precision or certainty, but studies to date at least serve to direct our curiosity. Published models fall into three categories, depending on the structure of the Ca source represented. Ca has been modeled as diffusing into cytoplasm from (*a*) a homogeneous surface membrane source (Zucker & Stockbridge 1983, Stockbridge & Moore 1984), (*b*) a localized active site complex (Llinas et al 1981b), or (*c*) individual Ca channels (Chad & Eckert 1984, Simon & Llinas 1985, Fogelson & Zucker 1985, Zucker & Fogelson 1986). With the availability of electrophysiological data on single Ca channels, attention has focused on diffusion from single channels.

Chad & Eckert (1984) have described a general model for diffusion of Ca from channels to regulatory sites and have introduced the term "domain" to refer to the volume of cytoplasm over which the $[Ca]_i$ transient produced by a single Ca channel exerts some regulatory action. Simon & Llinas (1985) have developed their own model for diffusion from Ca channels specifically to investigate Ca action in triggering transmitter release. Fogelson & Zucker (1985) and Zucker & Fogelson (1986) have carried out a similar analysis, focusing on both brief and longer-lasting transmission events. Because all three mathematical models are based on diffusion of Ca ions from single Ca channels, their results follow the characteristics of diffusion from a point source that was described above. Each model includes strong cytoplasmic Ca buffering, so that diffusion of Ca within the presynaptic cytoplasm is greatly restricted compared to simple aqueous diffusion. Under these conditions, each study has found that individual Ca channel domains may sometimes be small in comparison to estimated distances between active Ca channels. The resultant lack of overlap between individual $[Ca]_i$ domains means that Ca channels could act independently to trigger transmitter release or other phenomena that are regulated by $[Ca]_i$.

Presently available measurements of presynaptic Ca current are derived from macroscopic voltage-clamp experiments and thus reflect both the number of Ca channels open and the current through each open channel. Within an isolated domain of a single open channel, however, the expected elevation of Ca concentration would reflect only the single channel current, not the number of channels open. With a small depolarization from the resting potential, a small number of channels may open with a large Ca flux per channel. With larger depolarizations, a given macroscopic Ca current would be associated with a smaller flux of Ca per open channel but a larger number of open channels. The domain modeling studies have pointed out that if there is a nonlinear dependence of transmitter release on cytoplasmic Ca concentration, the same macroscopic Ca currents could lead to very different activation of release, depending on the current through single active channels. Transmitter release could thus appear to depend directly on membrane potential in addition to macroscopically measured Ca current.

Specific predictions from the modeling of Ca channel domains depend very much on the absolute distance of Ca diffusion relative to the spacing between single channels. If Ca diffuses far enough before binding to the release-triggering site, then the Ca concentration at that site may reflect summation of Ca from many channels rather than a single domain. In this case, Ca concentration would more closely reflect total macroscopic Ca current than the current in a single open channel. Unfortunately, the values of diffusion parameters that determine this crucial distinction between Ca action in isolated single domains versus overlapping multiple domains appear to lie within the range of uncertainty. The diffusion values chosen by Simon & Llinas (1985) probably reflect an extreme of buffer-restricted diffusion. Their model predicts a very minimal extent of functional overlap between adjacent channel domains. This in turn leads to extreme apparent voltage dependence of the release process when release is assumed to be a nonlinear function of intracellular Ca concentration. Zucker & Fogelson (1986) assumed a somewhat more modest restriction from aqueous diffusion. With their parameters, domains are large enough to overlap considerably, and only minor apparent voltage dependence arose.

The available experimental measurements of synaptic currents indicate that most of the voltage dependence of transmitter release can be accounted for by the voltage dependence of macroscopically measured Ca current (Augustine et al 1985b). The strong additional dependence of transmitter release on membrane potential predicted by some of the channel domain calculations is not apparent. This may imply, as suggested by Simon & Llinas (1985), that the dependence of transmitter release on cytoplasmic Ca concentration is effectively a linear one. Linear Ca trigger sites would integrate the flux per channel over the total number of active channels in the same fashion as the macroscopic Ca current measurement, so no apparent direct voltage dependence of release would be expected. However, considerable evidence indicates that Ca triggering may not be even approximately linear (see below), but rather may follow power function kinetics of precisely the type that did lead to voltage dependence of release in domain diffusion calculations. Ca may not act predominantly in isolated domains, and the restriction of Ca diffusion to isolated domains possibly has been overestimated. Again, if Ca channel domains overlap at individual Ca trigger sites, summation of Ca from multiple channels would cause the Ca concentration to resemble macroscopic Ca current measurements rather that the flux per open channel. This would account for the lack of apparent direct voltage dependence of release, even though the release process itself still might have a highly nonlinear dependence on Ca. Additional experimental data are needed to place more realistic constraints upon models of presynaptic Ca diffusion, including more information about the localization and functional properties of target molecules as well as about presynaptic Ca buffering and removal mechanisms. Neher (1986) has recently described one ingenious approach to obtaining such information for a related problem in potassium channel activation by intracellular Ca ions. In addition, new methods for direct measurements of intracellular Ca transients based on digital video microscopy with fluorescent Ca indicators (Williams et al 1985) offer improved spatial resolution of $[Ca]_i$ transients. These approaches might be extended to the question of transmitter release, to clarify the role of presynaptic Ca diffusion in this process.

Ca TRIGGER MOLECULES

Although an intracellular $[Ca]_i$ transient is generally accepted as the trigger for transmitter release, the molecular mechanisms underlying this triggering action of Ca are unknown. Intracellular Ca ions presumably bind to a receptor molecule within the presynaptic terminal, and the binding of Ca to this receptor somehow promotes the fusion of synaptic vesicles with the presynaptic membrane at the active zone (Heuser et al 1979). In this section we discuss a variety of candidates for the intracellular Ca receptor and a number of molecular reactions that Ca-receptor association might initiate. We maintain this dichotomy between Ca receptor molecules and Ca-dependent effector molecules only as a convenient organizing principle. Although such a dichotomy may be common and useful conceptually, in some models of release the postulated receptor and effector are the same molecule.

The identity of the intracellular receptor, and the consequences of its binding Ca ions, have remained enigmatic even though a Ca ion requirement for transmitter release has been recognized for over 30 years (del Castillo & Katz 1956, Katz 1969, 1971). This lingering enigma is due to the technical difficulty of obtaining access to the cytoplasmic constituents of the presynaptic terminal, where Ca is acting, while maintaining and measuring transmitter release. Traditionally, access to intracellular molecular constituents has required subcellular fractionation methods, which disrupt the release process. Conversely, transmitter release has been measured from more or less intact presynaptic terminals, where it is difficult to gain access to the cytoplasmic domain of the presynaptic terminal. Recent advances in cell permeabilization methods (Baker & Knight 1984) and use of electrophysiological methods that permit introduction of molecular probes into the cytoplasm while assaying release [such as microinjection (Miledi 1973), liposomes (Rahamimoff et al 1978b), and whole-cell patch clamp (Lindau & Fernandez 1986)] promise to overcome some of the obstacles that have hindered elucidation of the molecular mechanisms responsible for Ca-dependent transmitter release.

Transmitter release represents the culmination of a series of events that are specifically intended to prepare the presynaptic terminal for secretion. These events, which occur in advance of the stimulus, include formation of synaptic vesicles (as well as other components of the active zone), packaging of transmitter into vesicles, and the movement and docking of vesicles at the sites of fusion with the presynaptic plasma membrane. When looking at transmitter release, one is actually examining the net sum of all of these processes. Experimental treatments that affect release may not always be working specifically on the Ca binding or exocytosis steps that are the topic of this section. Furthermore, following these events, retrieval of vesicular membrane occurs via endocytosis (Ceccarelli & Hurlbut 1980a, Heuser & Reese 1981, Miller & Heuser 1984). This also may be Ca-dependent (Ceccarelli & Hurlbut 1980b, Meldolesi & Ceccarelli 1981) and thus also may be affected by treatments that manipulate $[Ca]_i$ in attempts to cause transmitter release.

Despite these technical difficulties, several approaches have been used to understand the triggering of release. Three of these strategies have proven particularly valuable and are described below.

Strategies for Identification of Relevant Molecules

IDENTIFICATION OF Ca-BINDING MOLECULES This approach takes advantage of the Ca requirement of release to ask which molecules within the presynaptic terminal bind Ca ions, and has identified a number of molecules within presynaptic terminals and other secretory cells that either bind Ca or undergo Ca-dependent changes. Many of these molecules are described below. The drawback of identifying molecules in this way is that, because Ca is involved in so many cellular functions (Kretsinger 1981,

Rasmussen & Barrett 1984), the molecular entities found may not be unique to release.

IDENTIFICATION OF PROBES THAT BLOCK OR STIMULATE RELEASE Another approach is to identify molecules that inhibit or promote the release of neurotransmitters and to use these molecules as probes to extract information about the molecules responsible for release. Two categories of molecules have been used. First, probes known to interact with particular molecules are applied to synapses to ask whether the probes alter release. Although this approach ultimately is likely to be the most useful of the two, it suffers from the fundamental difficulty listed above, namely the ability to monitor release while introducing the probe molecule into the presynaptic cytoplasm. One way to circumvent this difficulty is to choose probe molecules that are sufficiently hydrophobic to permeate through the membranes of intact cells. Unfortunately, such hydrophobic compounds often are capable of a wide spectrum of nonspecific effects. More specific probes, such as antibodies, stimulatory or inhibitory proteins, and substrates, seldom are capable of unaided entry into cellular interiors. New techniques for gaining access to the cytoplasm of intact presynaptic terminals should make this approach increasingly valuable in the future. Thus, when formulating molecular hypotheses to account for release, it will be important to identify specific molecular probes that may be used to test these hypotheses.

A second category of molecules, which include neurotoxins, have been identified based on their ability to alter transmitter release. The strategy is to then identify their target molecule, which one hopes is directly involved in mediating release. This approach has been quite popular, yielding an extensive list of molecules that increase or decrease transmitter release at a number of different synapses (Ceccarelli & Clementi 1979, Kelly et al 1979, Howard & Gundersen 1980). Although the identification of such probes has progressed enviably, tracking down the molecular target of any of them has proven rather difficult. Molecules likely to be of special interest include botulinum toxin (Gundersen 1980, Simpson 1981, Knight et al 1985), tetanus toxin (Mellanby & Green 1981, Bevan & Wendon 1984), and α-latrotoxin (Hurlbut & Ceccarelli 1979, Meldolesi et al 1984). Part of the difficulty with this strategy is that many potential targets exist within the presynaptic terminal for such compounds. For example, any molecule that interferes with presynaptic metabolism is likely to affect release, by blocking energy-dependent $[Ca]_i$ regulation and thus increasing $[Ca]_i$ (Glagoleva et al 1970, Alnaes & Rahamimoff 1975, Adams et al 1985). Difficulties in measuring presynaptic $[Ca]_i$, summarized above, make such mechanisms difficult to eliminate.

IDENTIFICATION VIA FUNCTIONAL CRITERIA Another useful strategy in thinking about release has been to characterize the functional attributes of release in intact synapses and to use these attributes as criteria for identifying the properties of the underlying molecular mechanisms. Several important functional criteria have been established.

Ca dependence Beyond the basic observations that revealed the necessity of Ca ions for release, physiological experiments also have demonstrated additional features of the Ca-dependence of release. One is that release has a high-order dependence upon $[Ca]_o$ (Dodge & Rahamimoff 1967, Martin 1977). This is not due to a high-order dependence of presynaptic Ca entry upon $[Ca]_o$ (Augustine & Charlton 1986), but reflects a steep dependence of release upon the Ca ions that enter during depolarization (Smith et al 1985, Augustine et al 1985b). Dodge & Rahamimoff (1967) suggested that this steep dependence upon $[Ca]_o$ reflects a "cooperative" action of Ca, meaning that multiple Ca ions must act in concert in order to trigger release. The "cooperative" hypothesis has attracted a great deal of attention, because it has been interpreted as meaning that the intracellular receptor binds multiple Ca ions in order to trigger release.

A problem with this interpretation is that several steps occur between entry of Ca ions and transmitter release; thus "cooperativity" need not be an attribute of the receptor. For example, one mathematical model of synaptic transmission can explain such behavior without invoking multiple Ca-binding events (Simon & Llinas 1985), although other models cannot (Zucker & Stockbridge 1983, Stockbridge & Moore 1984, Zucker & Fogelson 1986). Elevation of resting $[Ca]_i$, via liposomes or by microinjection of Ca ions, provides some support for a cooperative hypothesis. Such increases in $[Ca]_i$ increases evoked release (Rahamimoff et al 1978b, Charlton et al 1982), an increase that is predicted from the cooperative hypothesis because some Ca binding sites would be occupied by Ca in the resting condition and thus enhance release caused by a constant, stimulus-induced $[Ca]_i$ transient.

A more direct approach would be to determine the dependence of release upon $[Ca]_i$. Although the spatial resolution required to measure $[Ca]_i$ at local sites of transmitter release has not yet been obtained, $[Ca]_i$ has been manipulated in permeabilized non-neural cells to examine the relationship between $[Ca]_i$ and secretion. A quantitative study of the Ca-dependence of secretion from permeabilized chromaffin cells yielded a Hill coefficient of 2.2 (Knight & Baker 1982), suggesting that a high-order binding of Ca is required for secretion.

Studies on permeabilized non-neural cells also have permitted a direct determination of affinity of the secretory apparatus for $[Ca]_i$. Most studies have yielded K_d values in the low micromolar range (Bennett et al 1981, Steinhardt & Alderton 1982, Knight & Baker 1982, Whitaker & Baker 1983, Dunn & Holz 1983, Curran & Brodwick 1984), but Wilson & Kirshner (1983) have also found a second, larger component of release in chromaffin cells that requires millimolar Ca ion concentrations. Buffer injection into intact sea urchin eggs also has yielded k_d estimates for secretion in the micromolar range (Hamaguchi & Hiramoto 1981, Zucker et al 1978). Similar studies in permeabilized nerve terminals would be worthwhile, as well as injecting Ca buffers to determine both the affinity and stoichiometry of release for $[Ca]_i$.

Kinetics Another clue comes from studies of the temporal relationship between Ca influx and postsynaptic response. The "synaptic delay" between presynaptic action potentials and postsynaptic response is typically 1 msec or more at chemical synapses (Katz 1969). At least part of this delay is due to the time required for Ca channel activation (Llinas et al 1981a, 1982). Examination of the timing of release is possible, without Ca channel gating delays, by looking at release produced by Ca tail currents. In such conditions the delay between presynaptic Ca influx and postsynaptic response is as short as a few hundred microseconds (Llinas et al 1981b), although most release takes place roughly 2 msec after a Ca tail current (Augustine et al 1985b). This release appears to be much faster than secretion from non-neural cells (Fernandez et al 1984), so the assumption that both neural and non-neural secretion occur via the same mechanism may be an over-generalization. However, perhaps secretion from other cells is slower simply because their $[Ca]_i$ transients are slower or smaller.

The timing of release places rather stringent temporal constraints on molecular models, particularly since the synaptic delay includes diffusion of $[Ca]_i$ and the transmitter, the Ca-dependent reactions responsible for exocytosis, and postsynaptic channel gating properties. Thus, the rapidity of release may conflict with hypothetical molecular mechanisms that require multi-enzyme cascades, although perhaps not single enzyme schemes. In general, the relevant turnover numbers for most candidate enzymes under presynaptic physiological conditions remain unclear. Some enzymes have turnover rates as high as 10^6 sec^{-1} (see Hille 1984, p. 202), so single enzyme mechanisms, at least, should not be dismissed off-hand, particularly when considering events at a unimolecular level. If one quantal release event requires only a single enzyme-catalyzed reaction, then perhaps 200 μsec is a long time.

Divalent selectivity Evoked transmitter release depends upon the divalent cation present in the extracellular medium. Release evoked by single presynaptic action potentials generally is supported by divalents in the sequence Ca > Sr > Ba (Dodge et al 1969, Silinsky 1977). These differences are not due to differences in ability of these ions to enter the presynaptic terminal, but probably reflect the selectivity of the intracellular receptor for divalent cations (Augustine & Eckert 1984a).

Curiously, slowly loading presynaptic terminals with these ions produces the opposite finding. Tetanic stimulation of frog neuromuscular terminals causes increases in mepp frequency that are greatest for Ba and least for Ca (Silinsky 1978, Zengel & Magleby 1981). Similar results are obtained when liposomes are used to load divalents into the presynaptic interior (Mellow et al 1982, Crosland et al 1983). This paradox is probably caused by differences in the ability of presynaptic terminals to handle divalent loads. Intracellular removal of Ba is very ineffective compared to Ca (Ahmed & Connor 1979, Tillotson & Gorman 1984), so that intracellular Ba may accumulate to very high concentrations during prolonged loading and thereby elicit some release. Consistent with this hypothesis, slow delivery of divalents to permeabilized mast cells causes secretion that has the same Ca > Sr > Ba sequence seen for evoked transmitter release (M. Brodwick, personal communication).

Mg and many other Ca channel blockers are unable to support evoked release, a finding not surprising, since they are unable to enter the presynaptic cytoplasm via Ca channels (Hagiwara & Byerly 1981, Augustine & Eckert 1984a). Mn is capable of weakly entering synaptosomes and has a limited ability to cause transmitter release (Drapeau & Nachshen 1984). Kharasch et al (1981) have used liposomes containing Mg and Ca to deliver these ions to frog neuromuscular terminals. They found that inclusion of Mg produced a slight increase in release over that produced by liposomes containing only Ca. The physiological relevance of this stimulation of release by Mg loading is not clear, since intracellular Mg concentrations are thought to be on the order of 1 mM or more (Alvarez-Leefmans et al 1984). High concentrations of Mg seem to inhibit secretion from permeabilized chromaffin cells (Knight & Baker 1982, Wilson & Kirshner 1983). Further use of permeabilized cells should make it possible to sidestep the limitations imposed by Ca channel permeability and compare a more extensive series of divalents.

Temperature dependence Transmitter release is extremely sensitive to temperature changes, with lower temperatures reducing evoked transmitter release (Weight & Erulkar 1976, Barrett et al 1978). This sensitivity suggests that one or more steps in the release process possess a very high energy of activation. At least some of the temperature sensitivity is due to the temperature-dependence of Ca channel gating (Charlton & Atwood

1979). However, the steps following Ca entry are also likely to be very temperature-sensitive, because of the involvement of cytoplasmic Ca binding proteins and sequestration mechanisms. The release step conferring temperature sensitivity will be difficult to resolve in intact presynaptic terminals, because lower temperatures may reduce $[Ca]_i$ buffering (Duncan & Statham 1977) and thus increase the $[Ca]_i$ transient. Therefore, examining the temperature-dependence of release produced by a constant, known $[Ca]_i$ transient would be informative. Wilson & Kirshner (1983) have initiated such a study by showing that the rate of secretion from permeabilized chromaffin cells is greatly reduced by decreased temperatures.

ATP dependence Release from permeabilized non-neural cells seems to have an absolute requirement for Mg-ATP (Baker & Whitaker 1978, Knight & Baker 1982, Dunn & Holz 1983, Wilson & Kirshner 1983, Curran & Brodwick 1984). Whether or not transmitter release has a similar requirement for ATP is not known. Exposure of intact terminals to metabolic inhibitors initially causes release to increase (Glagoleva et al 1970, Alnaes & Rahamimoff 1975) but eventually release declines (Glagoleva et al 1970, Adams et al 1985). These complex effects of metabolic inhibitors are presumably due to the influence of ATP levels on $[Ca]_i$ regulation. Adams et al (1985) have provided a nice demonstration of this by showing that evoked release could be restored to inhibitor-treated presynaptic terminals by microinjecting the Ca buffer, EGTA.

Anion requirements in permeabilized cells Secretion from permeabilized chromaffin cells depends upon which intracellular anion is present, exocytosis being supported in the sequence glutamate > acetate > Cl- > Br- > SCN- (Knight & Baker 1982). Similar observations have been made in permeabilized sea urchin eggs, and have been proposed to be caused by anion-dependent loss of a 100 kDa MW protein (Sasaki 1984). No analogous studies have been performed on synapses.

Use dependence The amount of neurotransmitter released from a presynaptic terminal is plastic and depends upon the amount of previous activity that the synapse has experienced. These forms of plasticity can range in time course from milliseconds to years. The two forms of plasticity studied most extensively are synaptic facilitation and depression. At the squid giant synapse, neither of these relatively short-lasting forms of plasticity are due to use-dependent changes in the amount of presynaptic Ca influx or average intracellular $[Ca]_i$ transient (Charlton et al 1982). Facilitation has been proposed to be caused by persistent action of $[Ca]_i$

that outlasts a presynaptic stimulus and enhances the response to a subsequent stimulus (Katz & Miledi 1968, Rahamimoff 1968, Charlton et al 1982). This persistent action could simply be a prolonged increase in $[Ca]_i$ or it could be something more elaborate, such as activation of a Ca-dependent enzyme that enhances the response of the terminal to a subsequent stimulus. Synaptic depression has been proposed to result from a local depletion of transmitter quanta available for release (Kusano & Landau 1975, Llinas et al 1981b, Charlton et al 1982). Any molecular model of Ca-dependent transmitter release ultimately must account for these and other use-dependent phenomena.

Voltage independence Dudel et al (1983) recently proposed that release is dependent upon the presynaptic membrane potential rather than (or in addition to) $[Ca]_i$. This suggestion largely is based on studies performed at the crayfish neuromuscular junction (Dudel et al 1983). Since measurement of the presynaptic membrane potential or $[Ca]_i$ in these experiments was not possible, Dudel and co-workers' conclusions are indirect at best. Zucker & Lando (1986) proposed that most of these results may be explained by poor spatial control of the presynaptic membrane potential. More direct experiments at the squid giant synapse suggest that release has little (Llinas et al 1981b) or no (Augustine et al 1985b) voltage dependence, beyond the expected effects upon Ca channel gating.

Sensitivity to protein-modifying reagents Secretion in permeabilized non-neural cells is sensitive to protein-modifying reagents. The sulfhydryl reagent, *N*-ethylmaleimide, blocks secretion in permeabilized chromaffin cells (Knight & Baker 1982, Wilson & Kirshner 1983, Frye & Holz 1985) and sea urchin eggs (Jackson et al 1985), suggesting that sulfhydryl groups may have an important function in a protein involved in release.

In contrast, *N*-ethylmaleimide and disulfide reagents increase release from intact neuromuscular synapses (Carlen et al 1976, Carmody 1978), synaptosomes (Baba et al 1979), tissue slices (Wade et al 1981), and PC12 cells (Pozzan et al 1984). The results on permeabilized cells mentioned in the previous paragraph make it unlikely that these effects are due to a specific attack on a pivotal sulfhydryl group involved in release, unless secretion from non-neural cells has a different molecular basis. More likely the effects of disulfide reagents on intact nerve terminals are due to increases in $[Ca]_i$ and/or general cytotoxic actions (Pozzan et al 1984).

Proteolytic enzymes, such as trypsin and pronase, also affect secretion in permeabilized sea urchin eggs (Jackson et al 1985). Collectively, these results indicate that proteins are likely to be involved in some step of the secretory process.

Possible Ca Receptors

A number of Ca receptor candidates have been proposed to initiate transmitter release. Below we attempt to list several of the more popular candidates. Most of these candidates have been proposed on the basis of their presence in presynaptic terminals and their ability to bind Ca. Additional evidence for or against their involvement in release also is described below.

PHOSPHOLIPIDS Many membrane phospholipids and certain sugar moieties associated with membrane glycolipids or glycoproteins have net negative charges. Possibly Ca can bind to or screen these negative charges to initiate secretion (Blioch et al 1968). Ca binding to phospholipids has been amply documented in the literature (Papahadjoupolos 1978).

The primary evidence against this hypothesis is that divalent binding to surface charges seems to have neither the selectivity nor high affinity expected of the receptor (Hauser et al 1976). Although some studies have proposed that Ca has some ability to bind to membrane surface charges, an ability not shared by other divalents (McLaughlin et al 1971, Newton et al 1978), such effects are most prominent at millimolar Ca levels, which are probably too high to be involved in release. Ca and other divalents can initiate membrane fusion in a number of model systems at millimolar concentrations, but, as described above, at least some forms of secretion occur at micromolar Ca concentrations. Zimmerberg et al (1980) showed that incorporation of an unidentified Ca-binding protein (not one of those described below) was able to confer micromolar affinity and the appropriate selectivity on fusion of phospholipid vesicles to planar bilayers. This argues that the receptor may be a protein, rather than a membrane lipid. Such an interpretation also would be consistent with observations that secretion from permeabilized cells is altered by protein-modifying reagents (see above). In general, models of secretion based exclusively on membrane charge neutralization have fallen out of favor (Llinas & Heuser 1977, Kelly et al 1979, Reichardt & Kelly 1983, Wilschut & Hoekstra 1984).

Baux et al (1979) have addressed the role of a specific molecule in release that could provide negative surface charges, namely sialic acid. They injected ruthenium red and neuraminidase into the cell bodies of *Aplysia* neurons and found that transmitter release was blocked by both of these agents. They proposed that blockade was caused by interference with intracellular sugar residues that play a role in triggering release, specifically by providing negative charges to which Ca ions bind.

These results, although interesting, cannot be regarded as definitive evidence for the involvement of such negatively-charged moieties in release. Perhaps the most severe problem is that sugar moieties have been found only on the extracellular surface of biological membranes (Hakomori 1981). Ruthenium red is reported to affect a number of ion channels, including Ca channels (Alnaes & Rahamimoff 1975, Stimers & Byerly 1982, Hermann & Gorman 1982). The compound also inhibits mitochondrial Ca uptake (Moore 1971), and this presumably influences $[Ca]_i$. Such effects may explain the ability of ruthenium red to block release, although Baux et al (1979) were unable to detect any effect of the injections on somatic action potentials. Intracellular consequences of neuraminidase are unknown. Further work will be required to understand these results.

CALMODULIN Calmodulin is a protein of 16.7 kDa molecular weight (MW) that is found in virtually every eukaryotic cell (Cheung 1980, Means & Dedman 1980, Klee et al 1980, Means et al 1982). The molecule has been sequenced (Kretsinger 1980, Lagace et al 1983, Putkey et al 1983, Marshak et al 1984) and found to consist of 148 amino acids arranged in a single polypeptide chain with four Ca-binding domains. Crystallographic studies on the three-dimensional structure of calmodulin reveal that it is a dumbbell-shaped molecule, with two paired Ca-binding domains connected by a long central alpha-helical region (Babu et al 1985). Binding of Ca to the Ca-binding sites may expose the central helix, which may be the conformational change that exposes the hydrophobic domain responsible for activation of target proteins (Dedman et al 1977, LaPorte et al 1980, Tanaka & Hikada 1980) or interacting with phenothiazine drugs (Weiss & Levin 1978). Ca apparently must be bound at three or more binding sites in order for this conformational change to occur (Huang et al 1981). The gene coding for calmodulin has been cloned (Lagace et al 1983, Putkey et al 1983, Simmen et al 1985), thus paving the way for detailed fine-structural studies on this important intracellular Ca receptor.

By virtue of its widespread role in a number of cellular functions, calmodulin has received much attention as the receptor that initiates transmitter release. Several lines of evidence support a role for calmodulin in release. Biochemical studies have shown that calmodulin is present in terminals and binds to synaptic vesicles with high affinity (Moskowitz et al 1983, Hooper & Kelly 1984a,b). Immunocytochemical studies have demonstrated calmodulin at synapses, but immunoreactivity was much more highly concentrated in postsynaptic structures (Wood et al 1980). Calmodulin has a selectivity for divalent cations that closely parallels that of release (Teo & Wang 1973, Lin et al 1974), and its ability to bind multiple Ca ions could explain the apparent high-order sensitivity of release to Ca. Addition of calmodulin to isolated synaptic vesicles causes the vesicles to release their transmitter (DeLorenzo et al 1979). Trifluoperazine (TFP), a phenothiazine that prevents calmodulin from initiating cellular responses (Weiss & Levin 1978), blocks release of transmitter from synaptosomes (Schweitzer & Kelly 1982) and a wide spectrum of non-neural secretory cells (Knight & Baker 1982, Kenigsberg et al 1982, Douglas & Nemeth 1982, Sand et al 1983, Clapham & Neher 1984).

Although evidence supporting calmodulin as the intracellular Ca receptor responsible for release is far more abundant than for any other candidate, several problems leave the issue open. For example, the presence of calmodulin in presynaptic terminals is not itself a strong argument because calmodulin is present in all cells, and probably in most regions of these cells. Further, its role in a wide spectrum of cellular functions means that it could have presynaptic functions not directly related to release. The relevance of its ability to trigger release from isolated vesicles is unclear, since release from intact synapses requires vesicle-membrane fusion rather than vesicle-vesicle interactions. In fact, "compound" exocytosis, caused by vesicle-vesicle fusion, occurs in certain non-neural secretory cells but never has been reported at synapses.

The ability of TFP to block release merits special attention, particularly since TFP has been reported not to block release at the frog neuromuscular synapse (Cheng et al 1981) and the squid giant synapse (M. P. Charlton, unpublished) and high concentrations of TFP increase basal secretion from permeabilized chromaffin cells (Knight & Baker 1982). The hydrophobic nature of the TFP molecule means that it has a wide variety of actions upon other molecules with hydrophobic domains. Such actions could be responsible for many of the reported effects of TFP on release. For example, TFP blocks Ca influx into a number of cells (Sand et al 1983, Wada et al 1983, Clapham & Neher 1984). The ability of TFP to block depolarization-evoked release from electric organ synaptosomes may be due to such an action, because TFP does not block release triggered by the Ca ionophore, A-23187 (R. Yeager and G. Miljanich, personal communication). However, in some cases inhibition of secretion or phosphorylation appears not to be attributable to block of Ca influx (Kenigsberg et al 1982, Clapham & Neher 1984, Robinson et al 1984). TFP also reduces voltage-gated Na currents in perfused squid giant axons (G. S. Oxford, personal communication), alters acetylcholine-gated channels (Cheng et al 1981, Clapham & Neher 1984), inhibits the Na/K pump (Luthra 1982), and has general membrane stabilization effects (Seeman & Weinstein 1966). TFP also blocks other Ca-binding molecules, such as protein kinase C (Wise et al 1982), synexin (Pollard et al 1981), and the mitochondrial Ca transporter (M. Hirata et al 1982). Thus, caution is advised when interpreting studies that use this compound as a specific inhibitor of calmodulin.

A more convincing approach would be to use a wide spectrum of inhibitors to examine the role of calmodulin in release. Many other membrane-permeant compounds have also been reported to inhibit calmodulin (Hidaka et al 1981). The specificity of many of these compounds may be superior to TFP but warrants closer scrutiny. For example, one of these compounds, W-7, inhibits Ca channels (Hennessey & Kung 1984) and blocks kinase C (Schatzman et al 1983). One class of molecular probes with very high specificity are antibodies directed against calmodulin (Means et al 1982). Steinhardt & Alderton (1982) have used such antibodies to inhibit Ca-dependent exocytosis from permeabilized sea urchin eggs, and Kenigsberg & Trifaro (1985) have shown that microinjecting these antibodies into intact chromaffin cells blocks secretion. Proteolytic fragments of calmodulin may also be useful as specific inhibitors of calmodulin (Newton et al 1984).

Even if highly specific inhibitors of calmodulin are used and shown to inhibit release in intact synapses, the interpretation of such experiments may be difficult. Because of the multiple roles of calmodulin in cell function, interfering with calmodulin could inhibit these other functions and perhaps underlie their effect (or lack of effect) on release. For example, calmodulin stimulates transmembrane Ca pumping (Larsen & Vincenzi 1979, Carafoli 1981). Inhibiting this pump will increase $[Ca]_i$, which of itself will eventually inhibit evoked transmitter release (Kusano 1970, Charlton et al 1982, Adams et al 1985).

Such considerations point out the general difficulty in approaching the intracellular Ca receptor, namely gaining access to intracellular sites while maintaining essential cellular functions. Use of permeabilized cells may make such perturbation experiments more interpretable by putting more parameters under experimental control (Baker & Knight 1984). It is encouraging that secretion from certain permeabilized cell preparations is blocked by agents that inhibit calmodulin (Steinhardt & Alderton 1982, Knight & Baker 1982; but see Wilson & Kirshner 1983).

Finally, if calmodulin is the receptor, it could directly mediate exocytosis, perhaps as a consequence of exposing its hydrophobic central helix (see the section below on Ca-dependent effector actions). Calmodulin was not, however, capable of causing fusion between artificial membranes in the experiments of Zimmerberg et al (1980). Alternatively, calmodulin could work by activating calmodulin-binding proteins. One class of such proteins, Ca/calmodulin-dependent protein kinases, has been identified in presynaptic terminals and is discussed below. Other calmodulin-binding proteins have been identified in neural (Andreasen et al 1983) and non-neural (Hikita et al 1984, Bader et al 1985) tissues, but their roles in transmitter release are unknown.

TROPONIN C AND OTHER CALMODULIN-RELATED PROTEINS Calmodulin is only one member of a rather large family of Ca-binding proteins with similar molecular weights and acidic pKs. The Ca-binding sites of these molecules appear to be highly conserved (Kretsinger 1980). Other members of this family include troponin C, parvalbumin, intestinal Ca-binding protein, and S-100. Of these, troponin C has received the most attention (Kretsinger 1980, Tsalkova & Privalov 1985) because of its role in regulating muscle contraction (Ebashi & Endo 1968, Weber & Murray 1973).

Very little evidence supports the involvement of any of these proteins in release. Because of their similar Ca-binding sites, all of them are likely to have the same divalent selectivity as calmodulin, which, as mentioned above, is the same as that of transmitter release. Similarity of Ca-binding sites may make it difficult to distinguish among these possibilities. Troponin C (Puszkin & Kochwa 1974), S-100 (Moore 1965, Calissano et al 1969, Donato 1983), and parvalbumin (Baron et al 1975, Celio 1986) have been identified in neurons, but further work will be necessary to evaluate the role of these molecules in release.

Ca-DEPENDENT POTASSIUM CHANNEL Stanley & Ehrenstein (1985) have proposed that a Ca-activated potassium channel may be the receptor to which Ca binds in order to initiate release. Their model proposes that elevating intracellular $[Ca]_i$ causes Ca-activated K channels within the vesicular membrane to open, creating an influx of K, down a presumed electrochemical gradient, into the vesicle. This influx is accompanied by a parallel influx of anions, which prevents rapid dissipation of the K electrochemical gradient. The net influx of these ions is then accompanied by an influx of water; the resulting osmotic swelling causes vesicular fusion and, thus, release.

This hypothesis has certain attractive features. The Ca-activated K channel has the same cation selectivity as transmitter release (Gorman & Hermann 1979), it is activated by Ca in the micromolar range (Barrett et al 1982), and it requires several Ca ions to bind in order to cause channel opening (Barrett et al 1982). Vesicle fractions from neurosecretory nerve terminals also have been reported to contain Ca-dependent channels (Stanley et al 1986), although the fractions also contain other membranes that could be a source of these channels (Russell 1981).

There are several arguments against this hypothesis as a general mechanism for secretion. First, no evidence shows that vesicles are capable of maintaining an inwardly-directed electrochemical gradient for K. For example, the K concentration within chromaffin granules is similar to that of the cytoplasm (Ornberg & Leapman 1986). Since chromaffin granules have an inside-positive potential of +50 mV (Holz 1979, Johnson & Scarpa 1979), opening K channels in the granule membrane might cause an *efflux* of K ions from the granule. Another problem is that chromaffin cells can secrete when intracellular K has been replaced by Na (Clapham & Neher 1984) or other monovalent cations (Knight & Baker 1982) incapable of flowing through K channels (Latorre & Miller 1983). Thus, in these cells secretion can occur in the absence of an inwardly directed electrochemical gradient for K. Further, dissipation of the vesicular membrane potential, which would open Ca-dependent K channels (Barrett et al 1982, Latorre & Miller 1983), has little effect on secretion from chromaffin cells (Knight & Baker 1982, Holz et al 1983). This hypothesis therefore seems incapable of explaining secretion from chromaffin cells. Other osmotic models of secretion may be tenable and are discussed in the section on Ca-dependent effector actions.

Few direct tests of this hypothesis have been made so far. Transmitter release occurs in squid giant terminals that have been microinjected with the K channel blocker, TEA (Kusano et al 1967, Katz & Miledi 1967), but intracellular TEA has relatively little effect on the Ca-activated K channels in this preparation (Augustine & Eckert 1982). Other quanternary ammonium compounds which are more potent at blocking Ca-activated K channels do block evoked transmitter release, but this appears to be due to blockade of Ca channels rather than an effect upon Ca-dependent secretion (Augustine et al 1986a). Further evaluation of this hypothesis awaits more critical experimental tests.

VESICLE-ASSOCIATED Ca-BINDING PROTEINS Another approach is to identify Ca-binding proteins associated with synaptic vesicles. A candidate for such a molecule, named calelectrin, has been identified in *Torpedo* electric organ synaptosomes (Walker 1982). Calelectrin has a MW of 34 kDa and, in the presence of 10 μM or higher Ca, binds to cholinergic synaptic vesicles and various other membranes, including liposomes prepared from chromaffin granules (Walker 1982, Südhof et al 1982). Ca and Ba are able to bind to calelectrin and cause it to aggregate, whereas Sr cannot (Südhof et al 1982). This selectivity for divalent cations differs from that of transmitter release, making calelectrin an unlikely candidate for the receptor mediating release.

Creutz et al (1983) have extended this approach to chromaffin granules and have discovered 22 soluble proteins of MW 15–66 kDa that bind to chromaffin granules in the presence of Ca. These proteins, termed chromobindins, vary widely in their affinity for Ca. Some require 40 μM or more Ca to cause binding; many others require between 40 and 0.1 μM Ca. Three chromobindin proteins, of 70, 36, and 32.5 kDa MW, cross-react with anti-calelectrin antibodies (Geisow et al 1984). Because the 36 and 32.5 kDa proteins and calelectrin contain similar 17-amino-acid-long sequences, they have been proposed to be members of a unique family of

Ca-binding proteins (Geisow et al 1986). Other chromobindins include calmodulin, synexin, protein kinase C, and caldesmon. Synexin is described next, and protein kinase C and caldesmon are discussed in the following section.

Collectively, these chromobindin proteins provide many potential targets for Ca during exocytosis, but their roles in secretion are largely unknown. With the exception of calmodulin and kinase C, it is not even clear that they are present in synapses. Further work will be required to determine their function in transmitter release.

SYNEXIN Synexin is a protein of 47 kDa MW that binds to chromaffin granules in a Ca-dependent manner (Creutz et al 1983). Synexin has attracted attention because it causes a Ca-dependent aggregation of chromaffin granules, with a K_d of approximately 200 μM and a Hill coefficient of 2.3 (Pollard et al 1981).

The evidence supporting a role for synexin in secretion from chromaffin cells is that agents that block synexin-induced granule aggregation, such as TFP and promethazine, inhibit synexin-induced secretion from intact chromaffin cells. The evidence against its involvement in chromaffin cell secretion is that its Ca affinity seems low, given that secretion in permeabilized chromaffin cells appears to have a much lower K_d for Ca (Knight & Baker 1982, Dunn & Holz 1983, Wilson & Kirshner 1983). Pollard et al (1981) have proposed that synexin-induced granule aggregation may be useful as a model for compound exocytosis, although it is not known whether the Ca-dependence of compound exocytosis is different from that seen for net release from the permeabilized cell experiments.

Synexin has a unique divalent cation selectivity, with Sr and Ba incapable of supporting granule aggregation (Pollard et al 1981). Because transmitter release is supported, to some extent, by Sr and Ba, synexin is probably not the receptor mediating neurotransmitter release.

Possible Ca-Dependent Effector Actions

In this section we consider some of the hypotheses proposed to explain the triggering of release that follows binding of Ca to its receptor.

TRANSLOCATION One possible mode of action of Ca in triggering release would be for Ca to initiate the translocation of synaptic vesicles toward potential sites of release. This hypothesis was one of the earliest proposed to account for release (Berl et al 1973) and has been lent some credence by observations that cytoskeletal elements are capable of undergoing a wide spectrum of Ca-dependent changes. However, direct examination of secretion from chromaffin (Edwards et al 1984) and mast (Curran et al 1984) cells has failed to detect any gross movement of secretory vesicles during exocytosis. No such experiments have been performed at synaptic terminals, in part because most synaptic vesicles are too small to be resolved with this method. Even if such Ca-dependent translocation events are not the actual trigger for release, they may play other roles in preparing active zones for release. Possible cytoskeletal mediators of synaptic vesicle translocation are listed below.

Actin Actin has been most closely scrutinized as a possible source of vesicle movement. Actin is a 43 kDa molecule that can undergo rapid polymerization/depolymerization reactions to form filamentous assemblies (microfilaments) that are 8 nm in diameter and up to several μm in length (Korn 1982). Polymerization/depolymerization reactions, as well as cross-linking or bundling of assembled filaments, appear to be regulated by a number of accessory proteins (Weeds 1982). Many of these accessory proteins are Ca-regulated.

Actin was an integral player in one of the first specific molecular hypotheses proposed for transmitter release. Berl et al (1973) demonstrated actin and myosin in brain and proposed that an actomyosin-mediated contractile event triggered release. In this case, by analogy with skeletal muscle, troponin C presumably would be the actual Ca receptor that initiates the reaction. Subsequent work has shown that actin is associated with secretory vesicles (Burridge & Phillips 1975) and that chromaffin granules bind and cross-link actin (Fowler & Pollard 1982).

This hypothesis has not found much favor in the intervening years. One of the objections to this specific model is that the kinetics of myosin-mediated actin filament sliding are too slow. Muscle contraction, which is known to occur via this mechanism, requires several milliseconds after action potential initiation in order to reach peak tension. However, part of this delay is due to the time required for the intracellular Ca signal to occur (Baylor et al 1983). Further, the time required for a single actin-myosin crossbridge cycle to occur is much less than several milliseconds (Adelstein & Eisenberg 1980). Parsegian (1977) has calculated that an actomyosin-based translocation could only move vesicles a few Ångstroms during the brief delay between Ca influx and transmitter release. Another objection is that microfilaments have not been found within active zones. Instead, a smaller filament (3–4 nm diameter) has been reported (Landis & Reese 1984). Finally, cytochalasin B, a membrane-permeant alkaloid that prevents actin polymerization (Korn 1982) and thus results in eventual depolymerization of actin filaments (Carter 1967), does not eliminate transmitter release at the frog neuromuscular junction (Katz 1972). Cytochalasin and other agents that alter actin-based functions also have little effect on secretion from permeabilized chromaffin cells (Knight & Baker 1982) or sea urchin eggs (Whitaker & Baker 1983).

While these results argue against a role for actin filaments in mediating release, the results are not inconsistent with more subtle roles for actin, such as maintainence of active zone organization, maintainence of terminal shape, and perhaps movement or docking of vesicles or other organelles. Like calmodulin, actin plays a number of roles in most cells and it is unlikely that nerve terminals are an exception (Lasek & Shelanski 1981, Bray & Gilbert 1981, Porter 1984).

Recent studies have implicated several actin-associated proteins in secretion from chromaffin cells. Burgoyne et al (1986) have identified a 70 kDa protein that binds to chromaffin granules in a Ca-dependent manner as caldesmon, a calmodulin-regulated, actin-binding protein. The binding of caldesmon to actin is inhibited by high concentrations of Ca (Bretscher & Lynch 1985). If this occurs in chromaffin cells during secretion, it could perhaps cause a decrease in the viscosity of submembranous cytoplasm, thus in turn providing chromaffin granules with access to the presynaptic plasma membrane. Bader et al (1986) have identified three chromaffin cell proteins that cause Ca-dependent fragmentation of actin filaments. They have proposed that these proteins could cause a breakdown of a microfilament network, again decreasing cytoplasmic viscosity and, perhaps, triggering secretion. Another cytoskeletal element, fodrin, is redistributed within chromaffin cells during exocytosis (Perrin & Aunis 1985). Fodrin redistribution could also be involved in secretion from these cells, although a specific mechanism has not been found. It is perhaps not surprising that cytoskeletal elements undergo changes during secretion, given that $[Ca]_i$ is known to increase during secretion and that actin and its associated regulatory proteins are known to undergo a wide range of Ca-dependent reactions. The relationship of these events to transmitter release and other forms of secretion remains unknown.

Tubulin Tubulin is the name of two 50 kDa proteins that polymerize to form hollow, 25 nm diameter tubes called *microtubules*. Microtubules are involved in a number of intracellular translocation activities, including axoplasmic transport (Thoenen & Kreutzberg 1981, Schnapp et al 1985). Because tubulin is associated with synaptic vesicles (Zisapel et al 1980, Gray 1983) and is phosphorylated in the presence of Ca [by the Ca/calmodulin-dependent protein kinase described below (DeLorenzo 1982, Vallano et al 1985)], DeLorenzo (1982) has proposed that Ca-dependent phosphorylation of tubulin may trigger transmitter release.

Currently, little additional evidence supports or refutes this hypothesis. Experiments exploring the general role of Ca/calmodulin-dependent phosphorylation in release are described below. Intact, polymerized microtubules are unlikely to be required for transmitter release, because release can occur at frog neuromuscular synapses treated with microtubule-depolymerizing drugs, such as colchicine and vinblastine (Katz 1972, Pecot-Dechavassine 1976). Similar noneffects of such agents have been reported for permeabilized cells (Knight & Baker 1982, Whitaker & Baker 1983). However, these agents do reduce secretion from certain intact cells (see examples in Soifer 1975), perhaps by inhibiting the delivery of secretory vesicles to release sites or by a more general disruption of cytoskeletal organization (Hoffstein et al 1977, Hoffstein & Weissmann 1978).

Vesicle-associated proteins If neurotransmitter release involves a vesicular exocytosis event, then the molecule responsible for this event should, at some point in the process, interact with vesicles. Thus, identification of vesicle-associated proteins may provide a clue to molecules responsible for vesicle fusion, translocation, and/or docking. Wagner & Kelly (1979) have prepared highly purified synaptic vesicles and have described their associated proteins. Most of these proteins, with the exception of those discussed below, have not been considered from the standpoint of possible mediators of release.

1. Synapsin: At the moment, the most interesting protein associated with synaptic vesicles is synapsin I (previously called protein I). Synapsin I is composed of two polypeptides of 80 and 86 kDa MW. Synapsin I has not yet been sequenced but this should be forthcoming because cDNA clones complementary to synapsin I have been identified (Kilimann & DeGennaro 1985). Synapsin I has a globular head and a collagen-like tail region (Ueda & Greengard 1977). This molecule is associated with small, clear-cored (but not dense-cored) synaptic vesicles in a number of tissues (Navone et al 1984). Two interesting locations where synapsin appears to be absent are *Torpedo* electric organ (Palfrey et al 1983) and chromaffin cells (De Camilli et al 1979).

Synapsin I was first identified because it acts as a substrate for the cAMP-dependent and two Ca/calmodulin-dependent protein kinases (Nairn et al 1985). These kinases phosphorylate synapsin I at two sites (Huttner et al 1981). Phosphorylation of one of these sites by the type II Ca/calmodulin-dependent kinase causes synapsin I to dissociate from synaptic vesicles (Huttner et al 1983).

Navone et al (1984) have proposed that synapsin I may provide a link between synaptic vesicles and cytoskeletal elements. Consistent with this suggestion, synapsin I has been reported to bind to microtubules (Baines & Bennett 1986) and spectrin, another cytoskeletal protein (Baines & Bennett 1985). If synapsin I links vesicles to the cytoskeleton, then phosphorylation-dependent dissociation of synapsin I from the vesicles could allow Ca to mediate or modulate synaptic transmission by regulating vesicle availability for exocytosis. Direct support for this proposal comes from the experiments of Llinas et al (1986), who showed that microinjection of the type II-Ca/calmodulin-dependent kinase into squid giant nerve terminals enhanced evoked release, whereas injection of dephosphorylated synapsin I reduced release. Although this mechanism has received some support, it cannot be regarded as universal because it clearly does not apply to secretory tissues that do not contain synapsin and/or synapses that use transmitters secreted from dense-cored vesicles. However, it should (and undoubtedly will) receive further attention.

2. Other vesicle-associated proteins: Immunochemical methods have recently been used to study several vesicle-associated glycoproteins (Matthew et al 1981, Carlson & Kelly 1983, Buckley & Kelly 1985, Wiedenmann & Franke 1985). All of these proteins have been identified via antibodies generated against purified synaptic vesicle preparations. These antibodies have been used to localize the glycoproteins and have shown that all four are present in a wide range of neural and non-neural secretory cells. Such distributions suggest that the glycoproteins may play some role in vesicular function, although the functions are still unclear. An interesting first step toward elucidating the functions of these glycoproteins would be to microinject the antibodies into presynaptic cytoplasm to determine whether any of them interfered with transmitter release.

PROTEIN KINASE ACTIVATION In recent years we have seen an explosion in the identification and characterization of neuronal protein kinases (Nestler & Greengard 1985, Nairn et al 1985). A number of protein kinases have been indentified, including a cyclic AMP-dependent kinase, a cyclic GMP-dependent kinase, several Ca/calmodulin-dependent kinases, a Ca- and phospholipid-dependent kinase termed *kinase C*, and other kinases not activated by any of the above messengers.

An important goal of this research is to understand the functional significance of the plethora of phosphorylation activities (Greengard 1978, Kennedy 1983, Nairn et al 1985, Levitan 1985). One such physiological function might be to trigger or regulate transmitter release. If so, this function could account for the ATP requirement for secretion in permeabilized non-neural cells. Although protein kinases are thought to have such slow turnover rates (perhaps 2–20 sec^{-1}; Erondu & Kennedy 1985) that they may not be fast enough to mediate release, several studies demonstrate a correlation between protein phosphorylation (or dephosphorylation) and secretion in neural tissue (Kruger et al 1977, Form & Greengard 1978, Mistler & Greengard 1980, DeLorenzo 1982) and non-neural secretory cells (Sieghart et al 1978, Nishikawa et al 1980, Amy & Kirschner 1981, Gilligan & Satir 1982, Roberts & Butcher 1983, Pocotte et al 1985, Quissell et al 1985, Zieseniss & Plattner 1985). However, more direct evidence of a role for protein phosphorylation/dephosphorylation in transmitter release has been sparse (Miller 1985b). The following kinases have generated some interest relative to secretion.

Kinase C Although one of the most recently identified, the Ca- and phospholipid-dependent protein kinase C rapidly is becoming one of the most exhaustively studied enzymes in the nervous system (Nishizuka 1984). Kinase C has a molecular weight of approximately 51 kDa and is operationally distinguished from Ca/calmodulin-dependent kinases by its requirement for phosphatidylserine or related phospholipids and diacylglycerol and its lack of activation by calmodulin. Calmodulin has, in fact, been reported to inhibit kinase C (Albert et al 1984). The activity of purified kinase C molecules can be enhanced by augmentation of the ambient Ca level, but significant activity may be observed at 10^{-7} M concentrations of Ca (Kishimoto et al 1980). Thus this kinase may be capable of phosphorylating its substrates at the low levels of Ca ions expected in resting cells.

Kinase C activity normally is dependent upon intracellular levels of the second messenger, diacylglycerol (Nishizuka 1984). This action of diacylglycerol is mimicked by a number of membrane-permeant, tumor-promoting phorbol esters (Castagna et al 1982). These esters have been particularly helpful as molecular probes of cellular kinase C actions. It also would be useful to have reliable inhibitors of this kinase. Kuo et al (1985) have described a polyclonal antibody that is somewhat effective in blocking kinase C activity, but more potent (and, perhaps, membrane permeant) inhibitors would be valuable. Kuo et al (1983) also have described a few inhibitors that show some selectivity for kinase C over other protein kinases.

There are many suggestions that kinase C plays some role in secretion. A rapidly growing number of studies have shown that treatment of intact, non-neural cells with phorbol esters causes these cells to secrete more of their product than usual (reviewed in Nishizuka 1984). Extension of these studies to permeabilized non-neural cells has resulted in the revelation that kinase C activation enhances the Ca-sensitivity of the secretory process (Knight & Baker 1983, Knight & Scrutton 1984, Pocotte et al 1985). In some cases this shift in Ca-sensitivity may allow secretion to occur without a detectable increase in average $[Ca]_i$ (Rink et al 1983, Pozzan et al 1983, Di Virgilio et al 1984).

As pointed out by Baker (1984), the results described above do not make clear whether kinase C is a mediator or modulator of secretion. Because kinase C is Ca-sensitive, it conceivably could be the receptor/effector that is activated by Ca to trigger release. However, it also could be a parallel pathway that normally is not activated by the $[Ca]_i$ transient, or it could be a regulatory pathway that modulates the Ca sensitivity of the molecules responsible for Ca-triggered release.

The case for an involvement of kinase C in neurotransmitter release from presynaptic terminals is less compelling. Meldolesi et al (1984) have found that α-latrotoxin, when applied in media containing no Ca ions, can increase release from PC12 cells and synaptosomes without increasing $[Ca]_i$. That the toxin also increased phosphoinositide turnover (Vicentini & Meldolesi 1984), and phorbol esters increase Ca ionophore-induced transmitter release from these cells (Pozzan et al 1984), suggests that the toxin-induced release was due to kinase C activation and subsequent enhancement of the Ca-sensitivity of release, analogous to the situation for non-neural cells described above.

To examine the possible role of kinase C in transmitter release further, a number of studies have asked whether phorbol esters increase transmission at intact synapses. At the guinea pig ileum (Tanaka et al 1984), frog neuromuscular junction (Haimann 1985 and R. Rahamimoff, personal communication), and squid giant synapse (Osses et al 1986), phorbol esters increase transmitter release evoked by presynaptic action potentials. Augustine et al (1986b) have examined the mechanism of action of the phorbol esters at the squid synapse and see no increase in transmission when the presynaptic terminal is voltage clamped. They suggest that the ability of kinase C activation to augment release may be due to a broadening of the presynaptic action potential, perhaps because of a reduction in a presynaptic K current. Phorbol esters also do not enhance depolarization-induced secretion from intact chromaffin cells, even though they do enhance Ca ionophore-induced secretion from these cells (Morita et al 1985). The negative results indicate that, under these conditions, kinase C does not affect the $[Ca]_i$ sensitivity of transmitter release. This may mean that kinase C is not a mediator of transmitter release, because phorbol esters should activate the enzyme and enhance release, even in terminals under voltage clamp. An alternative explanation is that kinase C does mediate release, but the presynaptic $[Ca]_i$ transient already is producing maximal activation of the kinase within individual $[Ca]_i$ domains. Increases in release observed above would then reflect an increased number of domains. Further work will be required to elucidate the role of kinase C in transmitter release.

Calmodulin-dependent protein kinases Many studies have addressed the role of Ca/calmodulin-dependent protein kinases in synaptic organization, in particular concentrating on the type II Ca/calmodulin-dependent kinase.

This enzyme is composed of two subunits of 50–60 kDa MW, which assemble to produce a holoenzyme of 300–700 kDa MW. The properties of this kinase, and other Ca/calmodulin-dependent protein kinases, have been reviewed most recently by Kennedy et al (1986). One particularly interesting property of the type II kinase is that it is capable of autophosphorylating itself, with the phosphorylated form capable of autonomous, Ca-independent kinase activity (Saitoh & Schwartz 1985, Miller & Kennedy 1986). This property may permit the kinase to be a Ca-triggered, long-lasting internal switch (Kennedy et al 1986).

Several lines of evidence suggest a role for this kinase in transmitter release. It is abundant in synaptic structures (Nairn et al 1985) and its substrates, such as synapsin I, are phosphorylated under conditions that promote release (for example, Kruger et al 1977). Most convincing are the microinjection experiments of Llinas et al (1985), already described above, which demonstrate that introduction of the type II kinase into squid presynaptic terminals dramatically enhances release.

Although this evidence provides important information about the ability of the type II kinase to enhance release, whether the kinase would be a mediator or modulator of the process is again not clear. The rather slow turnover rates of this kinase may make it more likely to serve a modulatory role. As such, it may be involved in forms synaptic plasticity, such as facilitation, post-tetanic potentiation, or long-term potentiation, which enhance transmitter release. If so, the "switch" capability thought to be conferred by autophosphorylation may permit the kinase to generate such long-lasting phenomena without requiring continual elevation of $[Ca]_i$.

Cyclic AMP-dependent protein kinase The cyclic AMP-dependent protein kinase is the first neuronal kinase to receive close scrutiny. This kinase consists of two pairs of subunits, the catalytic subunits (40 kDa MW) that are responsible for phosphorylation, and the regulatory subunits (49–55 kDa MW) that inhibit catalytic subunit activity in the absence of cyclic AMP (cAMP) but dissociate from the catalytic subunits in the presence of cAMP (Nairn et al 1985).

Many studies have shown that agents that are thought to increase intracellular cAMP levels stimulate release at a number of synapses (Standaert & Dretchen 1979, Kandel & Schwartz 1982). Since all known effects of intracellular cAMP are mediated via the cAMP-dependent kinase, these effects on release probably are as well. Not all synapses are influenced by these cAMP probe molecules; release at the squid synapse, for example, is not enhanced by membrane-permeant cAMP analogs or the adenylate cyclase activator, forskolin (G. Augustine, M. Charlton, A. Gurney, S. Smith, unpublished). This lack of generality suggests that the cAMP-dependent kinase plays a modulatory role, which may vary in different synapses, rather than being a general mediator of release.

The mechanisms underlying the modulatory actions of cAMP at different synapses are largely unknown. Only at a sensory-motor synapse in *Aplysia* has the mechanism been pursued so far. In this preparation release is enhanced by an effect on a unique K channel (Camardo et al 1983) and/or the $[Ca]_i$ transient evoked by presynaptic depolarization (Boyle et al 1984), rather than on the exocytotic apparatus directly. This mechanism contrasts with that of platelets, where high intracellular cAMP concentrations ($IC_{50} > 10\ \mu M$) seem to decrease the Ca sensitivity of secretion (Knight & Scrutton 1984). Whether these are general mechanisms for the effects of cAMP on synapses remains unknown.

Cyclic GMP-dependent protein kinases Although a cyclic GMP-dependent protein kinase has been identified in the nervous system (Nairn et al 1985) and is known to phosphorylate substrates, at least in cerebellar tissue (Aswad & Greengard 1981), it has not been unambiguously implicated in any physiological response in the nervous system. Cyclic GMP may be involved in transduction in vertebrate photoreceptors, but apparently this is not a consequence of cyclic GMP-dependent phosphorylation (Fesenko et al 1985).

At present, no reason exists to propose a role for these kinases in transmitter release. Elevation of intracellular GTP increases secretion from mast cells, but since this increase is thought to be a consequence of activation of GTP-dependent regulatory proteins, rather than of formation of cyclic GMP and subsequent activation of the cyclic GMP-dependent protein kinase, these experiments are discussed in a subsequent section.

PHOSPHOLIPID MODIFICATION Given that secretion involves two structures that are largely composed of phospholipids, enzymatic modification of these lipids could be involved in secretion. Several possible reactions have been proposed, largely based on evidence obtained from non-neural cells.

Phospholipase C activation Phospholipases C are a collection of Ca-sensitive enzymes that break down a number of types of phospholipids. These phospholipases range in MW from 23–290 kDa (Shukla 1982, Irvine 1982, Low et al 1984). Interest in the involvement of phospholipases in secretion virtually co-developed with the discovery of these enzymes, since Hokin & Hokin (1953) showed that phospholipase C-induced turnover of phosphatidylinositol coincided with secretion.

Phospholipase C activation can cleave phosphatidylinositol bisphosphate to produce two products that are intracellular messengers potentially involved in secretion. One product is diacylglycerol, which can stimulate kinase C, as described above. Beyond its role as an activator of kinase C, diacylglycerol is known to be a potent membrane fusogen (Ahkong et al 1973). This ability to fuse membranes could indicate a more direct role for diacylglycerol in exocytosis (Whitaker & Aitchison 1985). The other product of phosphatidylinositol bisphosphate breakdown is inositol trisphosphate (IP_3), which can increase $[Ca]_i$ by releasing it from intracellular storage sites, such as smooth endoplasmic reticulum (Burgess et al 1984, Berridge & Irvine 1984). IP_3-induced increases in $[Ca]_i$ could then activate the Ca receptor that initiates secretion (Whitaker & Irvine 1984, Busa et al 1985, Rubin 1982) or could be used to promote further phospholipase C activation (Whitaker & Aitchison 1985). IP_3-mediated release of intracellular Ca has not yet been demonstrated in any synapse, but has been shown for *Limulus* photoreceptor neurons (Fein et al 1984, Brown et al 1984).

That phospholipase C activation is a general mechanism for Ca-dependent secretion, though possible, seems unlikely. In chromaffin cells, Fisher et al (1981) showed that phosphatidylinositol turnover is not required for secretion, and thus the two are distinct processes. Phospholipase C instead may represent a parallel, indirect pathway for triggering release.

Phospholipase A_2 activation Another phospholipase, phospholipase A_2, has also been proposed as a mediator of secretion. This phospholipase has a molecular weight of 12–75 kDa in different tissues (Van den Bosch 1980) and is Ca-sensitive, perhaps conferred by calmodulin activation (Wong & Cheung 1979).

Phospholipase A_2 has been proposed as a potential trigger of release because its products, fatty acids such as arachidonic acid, promote the fusion of isolated chromaffin granules (Creutz 1981). Consistent with this proposal, a close correlation has been demonstrated between the production and release of arachidonic acid and secretion (Bills et al 1976, Naor & Catt 1981, Frye & Holz 1984, 1985). Although these parallels are interesting, they do not provide unambiguous support for the hypothesis. Part of the uncertainty arises because diacylglycerol, which is produced by phospholipase C, also is degraded to arachidonic acid (Berridge & Irvine 1984). Thus arachidonic acid production is not necessarily diagnostic of phospholipase A_2 activity. In activated platelets, for example, both phospholipases C and A_2 may be producing arachidonic acid (Bell et al 1979), although the experiments of Frye & Holz (1985) on chromaffin cells seemed to be examining arachidonic acid produced primarily by phospholipase A_2.

Further work will be needed to evaluate this hypothesis and apply it to the problem of transmitter release. The only evidence supporting a role for this phospholipase in release is that β-bungarotoxin, a toxin that stimulates neurotransmitter release, also has phospholipase A_2 activity (Kelly et al 1975, Chang et al 1977). Proteins that inhibit phospholipase A_2 activity (F. Hirata et al 1982) may prove useful as molecular probes to determine whether phospholipase A_2 is involved in transmitter release. Among these proteins is lipomodulin (also called lipocortin), a protein of 16 kDa MW that only inhibits the phospholipase when dephosphorylated (Hirata 1981). Independent of its potential utility as a probe, regulation of lipomodulin's inhibitory activity by phosphorylation provides another potential link between protein phosphorylation and secretion.

Lipid methylation Two phospholipid methyltransferases found in the brain and other tissues can alter the properties of membrane phospholipids by methylation (Hirata & Axelrod 1980). Some evidence suggests that methylation is involved in histamine secretion from mast cells (Hirata & Axelrod 1980), perhaps as a consequence of methylation enhancing membrane fluidity (Hirata & Axelrod 1978). However, secretion in permeabilized chromaffin cells does not require S-adenosyl-L-methionine, the preferred methyl donor for both enzymes, and therefore probably does not rely on methylation (Knight & Baker 1982). The involvement of lipid methylation in neurotransmitter release is not known.

PROTEASE ACTIVATION Four classes of proteases have been classified on the basis of the details of their enzymatic mechanisms. These categories are metal-dependent, thiol-dependent, serine-dependent, and acid-dependent (Barrett 1977). Of these proteases, the first two have received some attention as possible mediators of secretion. The evidence supporting such roles are described below.

Metalloendoprotease activation The Ca-dependent fusion of myoblasts apparently requires the activity of a metalloendoprotease (Couch & Strittmatter 1983, 1984). The mechanism by which metalloendoprotease activity enhances fusion is not known. This protease requires Zn or other heavy metals, and thus is inhibited by heavy metal chelators. It also hydrolyzes peptide bonds on the amino side of hydrophobic amino acids. Therefore, synthetic dipeptides that are capable of acting as substrates for this enzyme may be used as probes, for they will compete with cellular substrates that may mediate physiological responses.

Because of the parallels between cell-cell fusion and exocytosis, metalloendoprotease activity may also be required for exocytosis. Several experiments support this notion. The heavy metal chelator, 1,10-phenanthroline, blocks histamine secretion from intact mast cells (Mundy & Strittmatter

1985). Synthetic peptides that are suitable substrates for the protease block mast cell secretion (Mundy & Strittmatter 1985), end-plate potentials at the rat neuromuscular junction (Baxter et al 1983), and depolarization-induced release of glycine from *Xenopus* retina (Frederick et al 1984). Finally, other inhibitors of metalloendoproteases block secretion in these three preparations. The only evidence against this hypothesis is that the synthetic dipeptide substrates do not appear to affect transmission at the squid giant synapse (G. Augustine and M. Brodwick, unpublished). This hypothesis merits further attention.

Activation of Ca-dependent thiol proteases Ca-activated proteases are widely distributed in cells, including neurons (Pant & Gainer 1980, Siman et al 1983, Malik et al 1983, Zimmerman & Schlaepfer 1984, Kamakura et al 1985). This class of protease is thought to act on neurofilaments (Pant & Gainer 1980, Schlaepfer et al 1985), a property that may be the basis for the well-known ability of Ca ions to liquefy axoplasm (Hodgkin & Katz 1949). Chad & Eckert (1986) have suggested that a Ca-activated protease may partially underlie the "washout" of Ca currents that has been reported in internally dialyzed neurons (Byerly & Hagiwara 1982).

Aside from the fact that both release and the protease are Ca-sensitive, there currently is little evidence that a Ca-activated protease is involved in release. The sensitivity of these proteases to Ca usually is rather low, requiring 10–1000 μM free Ca levels; this may make them poor candidates for a Ca-sensitive trigger for release. Knight & Baker (1982) reported that secretion from permeabilized chromaffin cells was not inhibited by an inhibitor of Ca-activated proteases. However, Ca-dependent proteolysis has been observed in platelets during conditions that stimulate secretion (Fox et al 1983).

ALTERATION IN NUCLEOTIDE BINDING Several studies in non-neural cells have demonstrated that secretion rates are influenced by the intracellular concentration of GTP. Perhaps most dramatic is the demonstration, by Fernandez et al (1984), that intracellular dialysis of a nonhydrolyzable GTP analog is capable of triggering secretion from mast cells whose $[Ca]_i$ was controlled by high concentrations of Ca buffers. They conclude that GTP, rather than Ca influx, may trigger secretion from these cells (Lindau & Fernandez 1986).

Perhaps the simplest interpretation of these studies is that GTP is influencing secretion via a GTP-dependent regulatory protein (Cockcroft & Gomperts 1985). Several such proteins have been identified in a number of cell types, where they serve as molecular transducers, often coupling occupancy of an extracellular receptor to activation of an intracellular enzyme (Rodbell 1980, Gilman 1984). Among the several enzymes known to be influenced by GTP-binding proteins are adenylate cyclase, phospholipase C, and cGMP phosphodiesterase. The availability of toxins, such as pertussis toxin and cholera toxin, and antibodies (Mumby et al 1986) that are targeted against different GTP-binding proteins should help test this and other hypotheses regarding the role of GTP is secretion at synapses and non-neural cells. Several potential GTP-binding proteins have been identified in *Torpedo* synaptosomes by Lester et al (1982).

INCREASE IN PROTEIN HYDROPHOBICITY The best-understood example of membrane fusion is the fusion between virus and host membranes (White et al 1983). In this fusion event, H^+ in the host's organelles triggers a conformational change in a 70 kDa glycoprotein called hemagglutinin. This conformational change exposes a hydrophobic domain in one of the two proteins that make up hemagglutinin, and exposure of this hydrophobic domain somehow promotes membrane fusion (Doms et al 1985). Site-directed mutagenesis of the hemmaglutinin gene (Gething et al 1980) can influence the ability of hemagglutinin to induce fusion, and it is being used to elucidate the molecular mechanism of fusion (Gething et al 1986). Although a H^+-induced conformational change in hemagglutinin is unlikely to underlie transmitter release, it is interesting that lowering the pH of presynaptic cytoplasm causes a dramatic increase in spontaneous transmitter release without increasing average $[Ca]_i$ (Nachshen & Drapeau 1986). Regardless of its potential role in transmitter release, the hemagglutinin system should serve as a very useful model in attempting to understand how Ca^{2+} promotes exocytosis at synapses.

OSMOTIC ACTIVATION At a somewhat different level, substantial evidence has supported the involvement of osmotic forces in vesicular fusion. Fusion of vesicles to planar bilayers depends upon a transvesicular osmotic gradient (Finkelstein et al 1986), with the intravesicular osmotic pressure needing to be higher than that outside the vesicle. Similar results have been reported for sea urchin eggs (Zimmerberg & Whitaker 1985, Zimmerberg et al 1985) and permeabilized chromaffin cells (Knight & Baker 1982, Holz 1986). Further, swelling of secretory vesicles has been observed during exocytosis (Curran et al 1984, Zimmerberg et al 1985). An osmotic gradient may thus provide the force required for membrane fusion and/or expulsion of vesicular contents to take place.

Few tests of this hypothesis have been made at synapses. At the crayfish neuromuscular junction, raising the external osmotic pressure reduced evoked release in the manner predicted from this hypothesis (Niles & Smith 1982). However, at other neuromuscular junctions, increasing external osmotic pressure increases transmitter release (Furshpan 1956, Kita & Van der Kloot 1977, Kita et al 1982). This is opposite the result expected from the osmotic hypothesis, but the increased release observed at these synapses in high osmotic pressure medium may be a secondary consequence of elevating $[Ca]_i$ (Shimoni et al 1977). Further work will be required to evaluate the role of osmotic forces in transmitter release. Should this mechanism prove to be involved, the Ca receptor would need to initiate release by increasing the intravesicular osmotic pressure. The Ca-dependent K channel model presently is the only one that includes an explicit provision for this potential requirement.

CONCLUDING REMARKS

We have reviewed progress in three separate areas related to the action of Ca ions in triggering transmitter release at chemical synapses. Each area has shown a great deal of progress since the initial concepts were introduced by Katz and associates.

Understanding of the voltage-dependent Ca current has been propelled by the dramatic progress in the understanding of ion channels that has accompanied development of single channel recording methods, combined with new pharmacological tools and key studies of the unique presynaptic terminal of the squid giant synapse. Intracellular Ca diffusion has been illuminated as the result of intense interest in the biochemistry of intracellular Ca metabolism in general and theoretical methods that have allowed extrapolation to a microscopic realm presently beyond direct observation. Understanding of the molecular basis of transmitter release triggering by cytoplasmic Ca ions has advanced with the identification of a host of possible transduction molecules in presynaptic terminals.

Much remains to be done in each of these three areas, however. More must be learned about the specific properties of synaptic Ca channels and how these properties relate to those of certain extrasynaptic Ca channels about which more is presently known. Single channel recordings from identified synaptic channels and a high-affinity chemical probe of defined specificity are especially needed. Given the present pace at which molecular biological techniques are being applied to well-defined channel molecules, such knowledge may lead rather directly to determination of amino acid sequences of synaptic Ca channels, and thence to detailed study of structure/function relationships. Better understanding of the cytoplasmic Ca concentration transient will require improved biochemical characterization of presynaptic cytoplasm, more refined experimental measurements of cytoplasmic Ca transients, and more sophisticated and realistic theoretical reconstruction of the cytoplasmic Ca signal. Such refinements should permit a glimpse of a new level of physiological and developmental regulation of synaptic function. Finally, and perhaps most important, the specific molecule or molecules that mediate the Ca-sensitivity of transmitter release must be identified. We hope that the relevant molecules are among the bewildering array of possibilities we have reviewed above, but an unexpected contender remains a possibility. In either case, genuine elucidation of the molecular basis of synaptic transmitter release urgently awaits identification of this molecule or set of molecules.

Acknowledgments

We thank L. Byerly, G. Miljanich, and R. Yeager for commenting on this review, J. Buchanan, D. Gray, and D. Nguyen for helping to organize our literature citations, K. Douglas for performing computer literature searches, R. Strickler for telecommunications assistance, J. Buchanan and C.-P. Ko for contributing electron micrographs, and numerous colleagues for sending us reprints or preprints of their papers. Financial support was provided by NIH grant NS 21624 to G. Augustine, an MRC (Canada) grant to M. Charlton, and NIH grant NS 11671 to S. Smith.

Literature Cited

Adams, D. J., Takeda, K., Umbach, J. A. 1985. Inhibitors of calcium buffering depress evoked transmitter release at the squid giant synapse. *J. Physiol.* 369: 145–59

Adelstein, R. S., Eisenberg, E. 1980. Regulation and kinetics of the actin-myosin-ATP interaction. *Ann. Rev. Biochem.* 49: 921–56

Ahkong, Q. F., Fisher, D., Tampion, W., Lucy, J. A. 1973. The fusion of erythrocytes by fatty acids, esters, retinol and α-tocopherol. *Biochem. J.* 136: 147–55

Ahmed, Z., Connor, J. A. 1979. Measurement of calcium influx under voltage clamp in molluscan neurons using the metallochromic dye arsenazo III. *J. Physiol.* 286: 61–82

Albert, K. A., Wu, W. C.-S., Nairn, A. C., Greengard, P. 1984. Inhibition by calmodulin of calcium/phospholipid-dependent protein phosphorylation. *Proc. Natl. Acad. Sci. USA* 81: 3622–25

Albus, U., Habermann, E., Ferry, D. R., Glossmann, H. 1984. Novel 1,4-dihydropyridine (Bay K–8644) facilitates calcium-dependent [^{3}H] noradernaline release from PC–12 cells. *J. Neurochem.* 42: 1186–89

Almers, W., McCleskey, E. W. 1984. Nonselective conductance in calcium channels of frog muscle: Calcium selectivity in a single-file pore. *J. Physiol.* 353: 585–608

Alnaes, E., Rahamimoff, R. 1975. On the role of mitochondria in transmitter release from motor nerve terminals. *J. Physiol.* 248: 285–306

Alvarez-Leefmans, F. J., Gamino, S. M., Rink, T. J. 1984. Intracellular free magnesium in neurones of *Helix aspersa* measured with ion-selective microelectrodes. *J. Physiol.* 354: 303–17

Amy, C. M., Kirshner, N. 1981. Phosphorylation of adrenal medulla cell proteins in conjunction with stimulation of catecholamine secretion. *J. Neurochem.* 36: 847–54

Andreasen, T. J., Luetje, C. W., Heideman, W., Storm, D. R. 1983. Purification of a novel calmodulin binding protein from bovine cerebral cortex membranes. *Biochemistry* 22: 4615–18

Aswad, D. W., Greengard, P. 1981. A specific substrate from rabbit cerebellum for guanosine 3′: 5′-monophosphate-dependent protein kinase. *J. Biol. Chem.* 256: 3487–93

Augustine, G. J., Charlton, M. P. 1986. Calcium dependence of presynaptic calcium current and post-synaptic response at the squid giant synapse. *J. Physiol.* In press

Augustine, G. J., Charlton, M. P., Horn, R. 1986a. Potassium channel blockers reduce transmitter release at the squid giant synapse. *Biol. Bull.* 171: 489–90

Augustine, G. J., Charlton, M. P., Smith, S. J. 1985a. Calcium entry into voltage-clamped presynaptic terminals of squid. *J. Physiol.* 367: 143–62

Augustine, G. J., Charlton, M. P., Smith, S. J. 1985b. Calcium entry and transmitter release at voltage-clamped nerve terminals of squid. *J. Physiol.* 367: 163–81

Augustine, G. J., Eckert, R. 1982. Calcium-dependent potassium current in squid presynaptic nerve terminals. *Biol. Bull.* 163: 397

Augustine, G. J., Eckert, R. 1984a. Divalent cations differentially support transmitter release at the squid synapse. *J. Physiol.* 346: 257–71

Augustine, G. J., Eckert, R. 1984b. Calcium-dependent inactivation of presynaptic calcium channels. *Soc. Neurosci. Abstr.* 10: 194

Augustine, G. J., Osses, L. R., Barry, S. R., Charlton, M. P. 1986b. Presynaptic mechanism of kinase C activators at the squid giant synapse. *Soc. Neurosci. Abstr.* 12: 821

Baba, A., Fisherman, J. S., Cooper, J. R. 1979. Action of sulfhydryl reagents on cholinergic mechanisms in synaptosomes. *Biochem. Pharmacol.* 28: 1879–83

Babu, Y. S., Sack, J. S., Greenhough, T. J., Bugg, C. E., Means, A. R., Cook, W. J. 1985. Three-dimensional structure of calmodulin. *Nature* 315: 37–40

Bader, M. F., Hikita, T., Trifaro, J. M. 1985. Calcium-dependent calmodulin binding to chromaffin granule membranes: Presence of a 65-kilodalton calmodulin-binding protein. *J. Neurochem.* 44: 526–39

Bader, M. F., Trifaró, J.-M., Langley, O. K., Thiersé, D., Aunis, D. 1986. Secretory cell actin-binding proteins: Identification of a gelsolin-like protein in chromaffin cells. *J. Cell Biol.* 102: 636–46

Baines, A. J., Bennett, V. 1985. Synapsin–I is a spectrin-binding protein immunologically related to erythrocyte protein–4.1. *Nature* 315: 410–13

Baines, A. J., Bennett, V. 1986. Synapsin–I is a microtubule-bundling protein. *Nature* 319: 145–47

Baker, P. F. 1984. Multiple controls for secretion? *Nature* 310: 629–30

Baker, P. F., Knight, D. E. 1984. Calcium control of exocytosis in bovine adrenal medullary cells. *Trends Neurosci.* 7: 120–26

Baker, P. F., Schlaepfer, W. W. 1978. Uptake and binding of calcium by axoplasm isolated from giant axons of *Loligo* and *Myxicola*. *J. Physiol.* 276: 103–25

Baker, P. F., Whitaker, M. J. 1978. Influence of ATP and calcium on the cortical reaction in sea urchin eggs. *Nature* 276: 513–15

Baron, G., Demaille, J., Dutruge, E. 1975. The distribution of parvalbumins in muscle and in other tissues. *FEBS Lett.* 56: 156–60

Barrett, A. J. 1977. Introduction to the history and classification of tissue proteinases. In *Proteinases in Mammalian Cells and Tissues*, ed. A. J. Barrett, pp. 1–55. Amsterdam: Elsevier

Barrett, E. F., Barrett, J. N., Botz, D., Chang, D. B., Mahaffey, D. 1978. Temperature-sensitive aspects of evoked and spontaneous transmitter release at the frog neuromuscular junction. *J. Physiol.* 279: 253–73

Barrett, J. N., Magleby, K. L., Pallotta, B. S. 1982. Properties of single calcium-activated potassium channels in cultured rat muscle. *J. Physiol.* 331: 211–30

Baux, G., Simonneau, M., Tauc, L. 1979. Transmitter release: Ruthenium red used to demonstrate a possible role of sialic acid containing substrates. *J. Physiol.* 291: 161–78

Baylor, S. M., Chandler, W. K., Marshall, M. W. 1983. Sarcoplasmic reticulum calcium release in frog skeletal muscle fibres estimated from Arsenazo–III calcium transients. *J. Physiol.* 344: 625–66

Baxter, D. A., Johnston, D., Strittmatter, W. J. 1983. Protease inhibitors implicate metalloendoprotease in synaptic transmission at the mammalian neuromuscular junction. *Proc. Natl. Acad. Sci. USA* 80: 4174–78

Bell, R. L., Kennerly, D. A., Stanford, N., Majerus, P. W. 1979. Diglyceride lipase: A pathway for arachidonate release from human platelets. *Proc. Natl. Acad. Sci. USA* 76: 3238–41

Bennett, J. P., Cockcroft, S., Gomperts, B. D. 1981. Rat mast cells permeabilized with ATP secrete histamine in response to calcium ions buffered in the micromolar range. *J. Physiol.* 317: 335–45

Berg, H. C. 1983. *Random Walks in Biology*. Princeton, NJ: Princeton Univ. Press

Berl, S., Puszkin, S., Nicklas, W. J. 1973. Actomyosin-like protein in brain. *Science* 179: 441–46

Berridge, M. J., Irvine, R. F. 1984. Inositol trisphosphate, a novel second messenger in cellular signal transduction. *Nature* 312: 315–21

Bevan, S., Wendon, L. M. B. 1984. A study of the action of tetanus toxin at rat soleus neuromuscular junctions. *J. Physiol.* 348: 1–17

Bills, T. K., Smith, J. B., Silver, M. J. 1976. Metabolism of [^{14}C] arachidonic acid by human platelets. *Biochim. Biophys. Acta* 424: 303–14

Bixby, J. L., Spitzer, N. C. 1983a. Enkephalin reduces quantal content at the frog neuromuscular junction. *Nature* 301: 431–32

Bixby, J. L., Spitzer, N. C. 1983b. Enkephalin reduces calcium action potentials in Rohon-Beard neurons *in vivo*. *J. Neurosci.* 3: 1014–18

Blaustein, M. P., Ector, A. C. 1975. Barbiturate inhibition of calcium uptake by depolarized nerve terminals *in vitro*. *Mol. Pharmacol.* 11: 369–78

Blaustein, M. P., Ratzlaff, R. W., Schweitzer, E. S. 1980. Control of intracellular calcium in presynaptic nerve terminals. *Fed. Proc.* 39: 2790–95

Blioch, Z. L., Glagoleva, I. M., Liberman, E. A., Nenashev, V. A. 1968. A study of the mechanism of quantal transmitter release at a chemical synapse. *J. Physiol.* 199: 11–35

Boyle, M. B., Klein, M., Smith, S. J., Kandel, E. R. 1984. Serotonin increases intracellular Ca^{2+} transients in voltage-clamped sensory neurons of *Aplysia californica*. *Proc. Natl. Acad. Sci. USA* 81: 7642–46

Bray, D., Gilbert, D. 1981. Cytoskeletal elements in neurons. *Ann. Rev. Neurosci.* 4: 505–23

Bretscher, A., Lynch, W. 1985. Identification and localization of immunoreactive forms of caldesmon in smooth and nonmuscle cells: A comparison with the distributions of tropomyosin and alpha-actin. *J. Cell Biol.* 100: 1656–63

Brigant, J. L., Mallart, A. 1982. Presynaptic currents in mouse motor endings. *J. Physiol.* 333: 619–36

Brinley, F. J. Jr. 1978. Calcium buffering in squid axons. *Ann. Rev. Biophys. Bioeng.* 7: 363–92

Brown, J. E., Rubin, L. J., Ghalayini, A. J., Tarver, A. P., Irvine, R. F., Berridge, M. J., Anderson, R. E. 1984. *myo*-inositol polyphosphate may be a messenger for visual excitation in *Limulus* photoreceptors. *Nature* 311: 160–63

Buckley, K., Kelly, R. B. 1985. Identification of a transmembrane glycoprotein specific for secretory vesicles of neural and endocrine cells. *J. Cell Biol.* 100: 1284–94

Burgess, G. M., Godfrey, P. P., McKinney, J. S., Berridge, M. J., Irvine, R. F., Putney, J. W. Jr. 1984. The second messenger linking receptor activation to internal Ca release in liver. *Nature* 309: 63–66

Burgoyne, R. D., Cheek, T. R., Norman, K. M. 1986. Identification of a secretory granule-binding protein as caldesmon. *Nature* 319: 68–70

Burridge, K., Phillips, J. H. 1975. Association of actin and myosin with secretory granule membranes. *Nature* 254: 526–29

Busa, W. B., Ferguson, J. E., Joseph, S. K., Williamson, J. R., Nuccitelli, R. 1985. Activation of frog (*Xenopus laevis*) eggs by inositol trisphosphate. I. Characterization of Ca^{2+} release from intracellular stores. *J. Cell Biol.* 101: 677–82

Byerly, L., Hagiwara, S. 1982. Calcium currents in internally perfused nerve cell bodies of *Limnaea stagnalis*. *J. Physiol.* 322: 503–28

Calissano, P., Moore, B. W., Friesen, A. 1969. Effect of calcium ion on S-100, a protein of the nervous system. *Biochemistry* 8: 4318–26

Camardo, J. S., Shuster, M. J., Siegelbaum, S. A., Kandel, E. R. 1983. Modulation of a specific potassium channel in sensory neurons of *Aplysia* by serotonin and cAMP-dependent protein phosphorylation. *Cold Spring Harbor Symp. Quant. Biol.* 48: 213–20

Carafoli, E. 1981. Calmodulin in the membrane transport of Ca^{++}. *Cell Calcium* 2: 353–63

Carlen, P. L., Kosower, E. M., Werman, R. 1976. The thiol-oxidating agent diamide increases transmitter release by decreasing calcium requirements for neuromuscular transmission in the frog. *Brain Res.* 117: 257–76

Carlson, S. S., Kelly, R. B. 1983. A highly antigenic proteoglycan-like component of cholinergic synaptic vesicles. *J. Biol. Chem.* 258: 11082–91

Carmody, J. J. 1978. Enhancement of acetylcholine secretion by two sulfhydryl reagents. *Eur. J. Pharmacol.* 47: 457–60

Carter, S. B. 1967. Effects of cytochalasins on mammalian cells. *Nature* 213: 261–64

Castagna, M., Takai, Y., Kaibuchi, K., Sano, K., Kikkawa, U., Nishizuka, Y. 1982. Direct activation of calcium-activated, phospholipid-dependent protein kinase by tumor-promoting phorbol esters. *J. Biol. Chem.* 257: 7847–51

Ceccarelli, B., Clementi, F., eds. 1979. *Advances in Cytopharmacology*, Vol. 3: *Neurotoxins: Tools in Neurobiology*. New York: Raven

Ceccarelli, B., Hurlbut, W. P. 1980a. Vesicle hypothesis of the release of quanta of acetylcholine. *Physiol. Rev.* 60: 396–441

Ceccarelli, B., Hurlbut, W. P. 1980b. Ca^{2+}-dependent recycling of synaptic vesicles at the frog neuromuscular junction. *J. Cell Biol.* 87: 297–303

Celio, M. R. 1986. Parvalbumin in most γ-aminobutyric acid-containing neurons of the rat cerebral cortex. *Science* 231: 995–97

Chad, J. E., Eckert, R. O. 1984. Calcium domains associated with individual channels can account for anomalous voltage relations of Ca-dependent responses. *Biophys. J.* 45 : 993–99

Chad, J. E., Eckert, R. O. 1986. An enzymatic mechanism for calcium current inactivation in dialysed *Helix* neurones. *J. Physiol.* 378 : 31–51

Chang, C. C., Su, M. J., Lee, J. D., Eaker, D. 1977. Effects of Sr^{2+} and Mg^{2+} on the phospholipase A and the presynaptic neuromuscular blocking actions of β-bungarotoxin, crotoxin and taipoxin. *Naunyn-Schmiedeberg's Arch. Pharmacol.* 299 : 155–61

Charlton, M. P., Atwood, H. L. 1979. Synaptic transmission: temperature-sensitivity of calcium entry in presynaptic terminals. *Brain Res.* 170 : 543–46

Charlton, M. P., Smith, S. J., Zucker, R. S. 1982. Role of presynaptic calcium ions and channels in synaptic facilitation and depression at the squid giant synapse. *J. Physiol.* 323 : 173–93

Cheng, K.-C., Lambert, J. J., Henderson, E. G., Smilowitz, H., Epstein, P. M. 1981. Postsynaptic inhibition of neuromuscular transmission by trifluoperazine. *J. Pharmacol. Exp. Ther.* 217 : 44–50

Cheung, W. Y. 1980. Calmodulin plays a pivotal role in cellular regulation. *Science* 207 : 19–27

Clapham, D. E., Neher, E. 1984. Trifluoperazine reduces inward ionic currents and secretion by separate mechanisms in bovine chromaffin cells. *J. Physiol.* 353 : 541–64

Cockcroft, S., Gomperts, B. D. 1985. Role of guanine nucleotide binding protein in the activation of polyphosphoinositide phosphodiesterase. *Nature* 314 : 534–36

Connor, J. A., Ahmed, Z. 1984. Diffusion of ions and indicator dyes in neural cytoplasm. *Cell. Mol. Neurobiol.* 4 : 53–66

Couch, C. B., Strittmatter, W. J. 1983. Rat myoblast fusion requires metalloendoprotease activity. *Cell* 32 : 257–65

Couch, C. B., Strittmatter, W. J. 1984. Specific blockers of myoblast fusion inhibit a soluble and not the membrane-associated metalloendoprotease in myoblasts. *J. Biol. Chem.* 259 : 5396–99

Crank, J. 1975. *The Mathematics of Diffusion*. Oxford : Clarendon. 2nd ed.

Creutz, C. E. 1981. *cis*-Unsaturated fatty acids induce the fusion of chromaffin granules aggregated by synexin. *J. Cell Biol.* 91 : 247–56

Creutz, C. E., Dowling, L. G., Sando, J. J., Villar-Palasi, C., Whipple, J. H., Zaks, W. J. 1983. Characterization of the chromobindins. Soluble proteins that bind to the chromaffin granule membrane in the presence of Ca^{2+}. *J. Biol. Chem.* 258 : 14664–74

Crosland, R. D., Hsiao, T. H., McClure, W. O. 1984. Purification and characterization of β-Leptinotarsin-h, an activator of presynaptic calcium channels. *Biochemistry* 23 : 734–41

Crosland, R. D., Martin, J. V., McClure, W. O. 1983. Effect of liposomes containing various divalent cations on the release of acetylcholine from synaptosomes. *J. Neurochem.* 40 : 681–87

Curran, M. J., Brodwick, M. S. 1984. Exocytosis in permeabilized mast cells. *Biophys. J.* 45 : 170a

Curran, M. J., Brodwick, M. S., Edwards, C. 1984. Direct visualization of exocytosis in mast cells. *Biophys. J.* 45 : 170a

De Camilli, P., Ueda, T., Bloom, F. E., Battenberg, E., Greengard, P. 1979. Widespread distribution of protein I in the central and peripheral nervous systems. *Proc. Natl. Acad. Sci. USA* 76 : 5977–81

Dedman, J. R., Potter, J. D., Jackson, R. L., Johnson, J. D., Means, A. R. 1977. Physicochemical properties of rat testis Ca^{2+}-dependent regulator protein of cyclic nucleotide phosphodiesterase: relationship of Ca^{2+}-binding, conformational changes and phosphodiesterase activity. *J. Biol. Chem.* 252 : 8415–22

del Castillo, J., Katz, B. 1956. Biophysical aspects of neuro-muscular transmission. *Prog. Biophys. Biophys. Chem.* 6 : 121–70

DeLorenzo, R. J. 1982. Calmodulin in neurotransmitter release and synaptic function. *Fed. Proc.* 41 : 2265–72

DeLorenzo, R. J., Freedman, S. D., Yohe, W. B., Maurer, S. C. 1979. Stimulation of Ca^{2+}-dependent neurotransmitter release and presynaptic nerve terminal protein phosphorylation by calmodulin and a calmodulin-like protein isolated from synaptic vesicles. *Proc. Natl. Acad. Sci. USA* 76 : 1838–42

DiPolo, R., Beaugé, L. 1983. The calcium pump and sodium-calcium exchange in squid axons. *Ann. Rev. Physiol.* 45 : 313–24

Di Virgilio, F., Lew, D. P., Pozzan, T. 1984. Protein kinase C activation of physiological processes in human neutrophils at vanishingly small cytosolic Ca^{2+} levels. *Nature* 310 : 691–93

Dodge, F. A. Jr., Rahamimoff, R. 1967. Cooperative action of calcium ions in transmitter release at the neuromuscular junction. *J. Physiol.* 193 : 419–32

Dodge, F. A. Jr., Miledi, R., Rahamimoff, R. 1969. Strontium and quantal release of transmitter at the neuromuscular junction. *J. Physiol.* 200 : 267–83

Doms, R. W., Helenius, A., White, J. 1985. Membrane fusion activity of the influenza virus hemagglutinin. *J. Biol. Chem.* 260 : 2973–81

Donato, R. 1983. Biochemical and physicochemical properties of the solubilized S-100 protein binding activity of synaptosomal particulate fractions. *Cell. Mol. Neurobiol.* 3: 239–54

Douglas, W. W. 1968. Stimulus-secretion coupling: the concept and clues from chromaffin and other cells. *Br. J. Pharmacol.* 34: 451–74

Douglas, W. W., Nemeth, E. F. 1982. On the calcium receptor activating exocytosis: Inhibitory effects of calmodulin-interacting drugs on rat mast cells. *J. Physiol.* 323: 229–44

Drapeau, P., Nachshen, D. A. 1984. Manganese fluxes and manganese-dependent neurotransmitter release in presynaptic nerve endings isolated from rat brain. *J. Physiol.* 348: 493–510

Dudel, J., Parnas, I., Parnas, H. 1983. Neurotransmitter release and its facilitation in crayfish muscle. VI. Release determined by both, intracellular calcium concentration and depolarization of the nerve terminal. *Pflügers Arch.* 399: 1–10

Duncan, C. J., Statham, H. E. 1977. Interacting effects of temperature and extracellular calcium on the spontaneous release of transmitter at the frog neuromuscular junction. *J. Physiol.* 268: 319–33

Dunlap, K., Fischbach, G. D. 1981. Neurotransmitters decrease the calcium conductance activated by depolarization of embryonic chick sensory neurones. *J. Physiol.* 317: 519–35

Dunn, L. A., Holz, R. W. 1983. Catecholamine secretion from digitonin-treated adrenal medullary chromaffin cells. *J. Biol. Chem.* 258: 4989–93

Ebashi, S., Endo, M. 1968. Calcium ion and muscle contraction. *Prog. Biophys. Mol. Biol.* 18: 123–83

Eckert, R., Chad, J. E. 1984. Inactivation of Ca channels. *Prog. Biophys. Mol. Biol.* 44: 215–67

Edwards, C., Englert, D., Lotshaw, D., Ye, H. Z. 1984. Light microscopic observations on the release of vesicles by isolated chromaffin cells. *Cell Motility* 4: 297–303

Erondu, N. E., Kennedy, M. B. 1985. Regional distribution of type II Ca^{2+}/calmodulin-dependent protein kinase in rat brain. *J. Neurosci.* 5: 3270–77

Fein, A., Payne, R., Corson, D. W., Berridge, M. J., Irvine, R. F. 1984. Photoreceptor excitation and adaptation by inositol 1,4,5-triphosphate. *Nature* 311: 157–60

Feldman, D. H., Yoshikami, D. 1985. A peptide toxin from *Conus geographus* blocks voltage-gated calcium channels. *Soc. Neurosci. Abstr.* 11: 517

Fenwick, E. M., Marty, A., Neher, E. 1982. Sodium and calcium channels in bovine chromaffin cells. *J. Physiol.* 331: 599–635

Fernandez, J. M., Neher, E., Gomperts, B. D. 1984. Capacitance measurements reveal stepwise fusion events in degranulating mast cells. *Nature* 312: 453–55

Fesenko, E. E., Kolesnikov, S. S., Lyubarsky, A. L. 1985. Induction by cyclic GMP of cationic conductance in plasma membrane of retinal rod outer segment. *Nature* 313: 310–13

Fisher, S. K., Holz, R. W., Agranoff, B. W. 1981. Muscarinic receptors in chromaffin cell cultures mediate enhanced phospholipid labeling but not catecholamine secretion. *J. Neurochem.* 37: 491–97

Finkelstein, A., Zimmerberg, J., Cohen, F. S. 1986. Osmotic swelling of vesicles: Its role in the fusion of vesicles with planar phospholipid bilayer membranes and its possible role in exocytosis. *Ann. Rev. Physiol.* 48: 163–74

Fogelson, A. L., Zucker, R. S. 1985. Presynaptic calcium diffusion from various arrays of single channels. Implications for transmitter release and synaptic facilitation. *Biophys. J.* 48: 1003–17

Forn, J., Greengard, P. 1978. Depolarizing agents and cyclic nucleotides regulate the phosphorylation of specific neuronal proteins in rat cerebral cortex slices. *Proc. Natl. Acad. Sci. USA* 75: 5195–99

Forscher, P., Oxford, G. S. 1985. Modulation of calcium channels by norepinephrine in internally dialyzed avian sensory neurons. *J. Gen. Physiol.* 85: 743–63

Fowler, V. M., Pollard, H. B. 1982. Chromaffin granule membrane-F-actin interactions are calcium sensitive. *Nature* 295: 336–39

Fox, A. P., Hess, P., Lansman, J. B., Nilius, B., Nowycky, M. C., Tsien, R. W. 1986. Shifts between modes of calcium channel gating as a basis for pharmacological modulation of calcium influx in cardiac, neuronal, and smooth-muscle-derived cells. In *New Insights Into Cell Membrane Transport Processes*, ed. G. Poste, S. T. Crooke, pp. 99–124. New York: Plenum

Fox, J. E. B., Reynolds, C. C., Phillips, D. R. 1983. Calcium-dependent proteolysis occurs during platelet aggregation. *J. Biol. Chem.* 258: 9973–81

Frederick, J. M., Hollyfield, J. G., Strittmatter, W. J. 1984. Inhibitors of metalloendoprotease activity prevent K^+-stimulated neurotransmitter release from the retina of *Xenopus laevis*. *J. Neurosci.* 4: 3112–19

Frye, R. A., Holz, R. W. 1984. The relationship between arachidonic acid release and catecholamine secretion from cultured bovine adrenal chromaffin cells. *J. Neurochem.* 43: 146–50

Frye, R. A., Holz, R. W. 1985. Arachidonic acid release and catecholamine secretion from digitonin-treated chromaffin cells: Effects of micromolar calcium, phorbol ester, and protein aklylating agents. *J. Neurochem.* 44: 265–73

Furshpan, E. J. 1956. The effects of osmotic pressure changes on the spontaneous activity at motor nerve endings. *J. Physiol.* 134: 689–97

Geisow, M. J., Childs, J., Dash, B., Harris, A., Panayotou, G., Südhof, T., Walker, J. H. 1984. Cellular distribution of three mammalian Ca^{2+}-binding proteins related to *Torpedo* calelectrin. *EMBO J.* 3: 2969–74

Geisow, M. J., Fritsche, U., Hexham, J. M., Dash, B., Johnson, T. 1986. A consensus amino-acid sequence repeat in *Torpedo* and mammalian Ca^{2+}-dependent membrane-binding proteins. *Nature* 320: 636–38

Gething, M.-J., Bye, J., Skehel, J., Waterfield, M. 1980. Cloning and DNA sequence of double-stranded copies of haemagglutinin genes from H2 and H3 strains elucidates antigenic shift and drift in human influenza virus. *Nature* 287: 301–6

Gething, M.-J., Doms, R. W., York, D., White, J. 1986. Studies on the mechanism of membrane fusion: Site-specific mutagenesis of the hemagglutinin of influenza virus. *J. Cell Biol.* 102: 11–23

Gilligan, D. M., Satir, B. H. 1982. Protein phosphorylation/dephosophorylation and stimulus-secretion coupling in wild type and mutant *Paramecium. J. Biol. Chem.* 257: 13903–6

Gilman, A. G. 1984. G proteins and dual control of adenylate cyclase. *Cell* 36: 577–79

Glagoleva, I. M., Liberman, Y. A., Khashayev, Z. K. M. 1970. Effect of uncoupling agents of oxidative phosphorylation on the release of acetylcholine from nerve endings. *Biophysics* 15: 74–82

Glusman, S., Kravitz, E. A. 1982. The action of serotonin on excitatory nerve terminals in lobster nerve-muscle preparations. *J. Physiol.* 325: 223–41

Gorman, A. L. F., Hermann, A. 1979. Internal effects of divalent cations on potassium permeability in molluscan neurones. *J. Physiol.* 296: 393–410

Gotgilf, I. M., Magazanik, L. G. 1977. Action of calcium channels blocking agents (verapamil, D-600 and manganese ions) on transmitter release from motor nerve endings of frog muscle. *Neurophysiology* 9: 415–21

Gould, R. J., Murphy, K. M. M., Snyder, S. H. 1983. Studies on voltage-operated calcium channels using radioligands. *Cold Spring Harbor Symp. Quant. Biol.* 48: 355–62

Gray, E. G. 1983. Neurotransmitter release mechanisms and microtubules. *Proc. R. Soc. London Ser. B* 218: 253–58

Greengard, P. 1978. Phosphorylated proteins as physiological effectors. *Science* 199: 146–52

Gundersen, C. B. 1980. The effects of botulinum toxin on the synthesis, storage and release of acetylcholine. *Prog. Neurobiol.* 14: 99–119

Gundersen, C. B., Katz, B., Miledi, R. 1982. The antagonism between botulinum toxin and calcium in motor-nerve terminals. *Proc. R. Soc. London Ser. B* 216: 369–76

Hagiwara, S., Byerly, L. 1981. Calcium channel. *Ann. Rev. Neurosci.* 4: 69–125

Hagiwara, S., Takahashi, K. 1967. Surface density of calcium ions and calcium spikes in the barnacle muscle fiber membrane. *J. Gen. Physiol.* 50: 583–601

Haimann, C. 1985. Effects of phorbol ester on quantal acetylcholine release at frog neuromuscular junction. *Soc. Neurosci. Abstr.* 11: 846

Hakomori, S.-I. 1981. Glycosphingolipids in cellular interaction, differentiation, and oncogenesis. *Ann. Rev. Biochem.* 50: 733–64

Hamaguchi, Y., Hiramoto, Y. 1981. Activation of sea urchin eggs by microinjection of calcium buffers. *Exp. Cell Res.* 134: 171–79

Hauser, H., Darke, A., Phillips, M. C. 1976. Ion binding to phospholipids. Interaction of calcium with phosphatidylserine. *Eur. J. Biochem.* 62: 335–44

Henkart, M. 1980. Identification and function of intracellular calcium stores in axons and cell bodies of neurons. *Fed. Proc.* 39: 2783–89

Hennessey, T. M., Kung, C. 1984. An anticalmodulin drug, W-7, inhibits the voltage-dependent calcium current in *Paramecium caudatum. J. Exp. Biol.* 110: 169–81

Hermann, A., Gorman, A. L. F. 1982. Ruthenium red blocks Ca^{2+} inward current and Ca^{2+} activated outward K^+ current of molluscan neurons. *Biophys. J.* 37: 183a

Hess, P., Tsien, R. W. 1984. Mechanism of ion permeation through calcium channels. *Nature* 309: 453–56

Heuser, J. E., Reese, T. S. 1977. Structure of the synapse. In *Handbook of Physiology, Section 1: The Nervous System*, ed. E. R. Kandel, 1: 261–94. Bethesda, MD: Am. Physiol. Soc.

Heuser, J. E., Reese, T. S. 1981. Structural changes after transmitter release at the frog neuromuscular junction. *J. Cell Biol.* 88: 564–80

Heuser, J. E., Reese, T. S., Dennis, M. J., Jan, Y., Jan, L., Evans, L. 1979. Synaptic vesicle exocytosis captured by quick freezing and correlated with quantal transmitter release. *J. Cell Biol.* 81: 275–300

Heuser, J. E., Reese, T. S., Landis, D. M. D. 1974. Functional changes in frog neuromuscular junctions studied with freeze-fracture. *J. Neurocytol.* 3: 109–31

Hidaka, H., Asano, M., Tanaka, T. 1981. Activity-structure relationship of calmodulin antagonists: Naphthalenesulfonamide derivitives. *Mol. Pharmacol.* 20: 571–78

Hikita, T., Bader, M. F., Trifaró, J. M. 1984. Adrenal chromaffin cell calmodulin: Its subcellular distribution and binding to chromaffin granule proteins. *J. Neurochem.* 43: 1087–97

Hille, B. 1984. *Ionic Channels of Excitable Membranes.* Sinauer: Sunderland, MA

Hirasawa, K., Nishizuka, Y. 1985. Phosphatidylinositol turnover in receptor mechanism and signal transduction. *Ann. Rev. Pharmacol. Toxicol.* 25: 147–70

Hirata, F. 1981. The regulation of lipomodulin, a phospholipase inhibitory protein, in rabbit neurophils by phosphorylation. *J. Biol. Chem.* 256: 7730–33

Hirata, F., Axelrod, J. 1978. Enzymatic methylation of phosphatidylethanolamine increases erythrocyte membrane fluidity. *Nature* 275: 219–20

Hirata, F., Axelrod, J. 1980. Phospholipid methylation and biological signal transmission. *Science* 209: 1082–90

Hirata, F., Notsu, Y., Iwata, M., Parente, L., DiRosa, M., Flower, R. J. 1982. Identification of several species of phospholipase inhibitory protein(s) by radioimmunoassay for lipomodulin. *Biochem. Biophys. Res. Commun.* 109: 223–30

Hirata, M., Suematsu, E., Koga, T. 1982. Calmodulin antagonists inhibit Ca^{2+} uptake of mitochondria of guinea pig peritoneal macrophages. *Biochem. Biophys. Res. Commun.* 105: 1176–81

Hodgkin, A. L., Katz, B. 1949. The effect of calcium on the axoplasm of giant nerve fibres. *J. Exp. Biol.* 26: 292–94

Hodgkin, A. L., Keynes, R. D. 1957. Movements of labelled calcium in squid giant axons. *J. Physiol.* 138: 253–81

Hoffstein, S., Goldstein, I. M., Weissmann, G. 1977. Role of microtubule assembly in lysosomal enzyme secretion from human polymorphonuclear leucocytes. A reevaluation. *J. Cell Biol.* 73: 242–56

Hoffstein, S., Weismann, G. 1978. Microfilaments and microtubules in calcium ionophore-induced secretion of lysosomal enzymes from human polymorphonuclear leukocytes. *J. Cell Biol.* 78: 769–81

Hokin, M. R., Hokin, L. E. 1953. Enzyme secretion and the incorporation of P^{32} into phospholipids of pancreas slices. *J. Biol. Chem.* 203: 967–77

Holz, R. W. 1979. Measurement of membrane potential of chromaffin granules by the accumulation of triphenylmethylphosphonium cation. *J. Biol. Chem.* 254: 6703–9

Holz, R. W. 1986. The role of osmotic forces in exocytosis from adrenal chromaffin cells. *Ann. Rev. Physiol.* 48: 175–89

Holz, R. W., Senter, R. A., Sharp, R. R. 1983. Evidence that the H^+ electrochemical gradient across membranes of chromaffin granules is not involved in exocytosis. *J. Biol. Chem.* 258: 7506–13

Hooper, J. E., Kelly, R. B. 1984a. Calcium-dependent calmodulin binding to cholinergic synaptic vesicles. *J. Biol. Chem.* 259: 141–47

Hooper, J. E., Kelly, R. B. 1984b. Calmodulin is tightly associated with synaptic vesicles independent of calcium. *J. Biol. Chem.* 259: 148–53

Hoshi, T., Rothlein, J., Smith, S. J. 1984. Facilitation of Ca^{2+}-channel currents in bovine adrenal chromaffin cells. *Proc. Natl. Acad. Sci. USA* 81: 5871–75

Hoshi, T., Smith, S. J. 1987. Large depolarization induces long openings of voltage-dependent calcium channels in adrenal chromaffin cells. *J. Neurosci.* In press

Howard, B. D., Gundersen, C. B. Jr. 1980. Effects and mechanisms of polypeptide neurotoxins that act presynaptically. *Ann. Rev. Pharmacol. Toxicol.* 20: 307–36

Huang, C. Y., Chau, V., Chock, P. B., Wang, J. H., Sharma, R. K. 1981. Mechanism of activation of cyclic nucleotide phosphodiesterase: Requirement of the binding of four Ca^{2+} to calmodulin for activation. *Proc. Natl. Acad. Sci. USA* 78: 871–74

Hurlbut, W. P., Ceccarelli, B. 1979. Use of black widow spider venom to study the release of neurotransmitters. See Ceccarelli & Clementi 1979, pp. 87–115

Huttner, W. B., DeGennaro, L. J., Greengard, P. 1981. Differential phosphorylation of multiple sites of purified protein I by cyclic AMP-dependent and calcium-dependent protein kinases. *J. Biol. Chem.* 256: 1482–88

Huttner, W. B., Schiebler, W., Greengard, P., DeCamilli, P. 1983. Synapsin–I (Protein–I), a nerve terminal-specific phosphoprotein. III. Its association with synaptic vesicles studied in a highly purified synaptic vesicle preparation. *J. Cell Biol.* 96: 1374–88

Irvine, R. F. 1982. The enzymology of stimulated inositol lipid turnover. *Cell Calcium* 3: 295–309

Jackson, R. C., Ward, K. K., Haggerty, J. G. 1985. Mild proteolytic digestion restores exocytotic activity to N-Ethylmaleimide-inactivated cell surface complex from sea urchin eggs. *J. Cell Biol.* 101: 6–11

Johnson, R. G., Scarpa, A. 1979. Protonmotive force and catecholamine transport in isolated chromaffin granules. *J. Biol. Chem.* 254: 3750–60

Kamakura, K., Ishiura, S., Suzuki, K., Sugita, H., Toyokura, Y. 1985. Calcium-activated neutral protease in the peripheral nerve, which requires μM order Ca^{2+}, and its effect on the neurofilament triplet. *J. Neurosci. Res.* 13: 391–403

Kandel, E. R., Schwartz, J. H. 1982. Molecular biology of learning: Modulation of transmitter release. *Science* 218: 433–43

Katz, B. 1969. *The Release of Neural Transmitter Substances*. Liverpool: Liverpool University Press

Katz, B. 1971. Quantal mechanism of neurotransmitter release. *Science* 173: 123–26

Katz, B., Miledi, R. 1967. A study of synaptic transmission in the absence of nerve impulses. *J. Physiol.* 192: 407–36

Katz, B., Miledi, R. 1968. The role of calcium in neuromuscular facilitation. *J. Physiol.* 195: 481–92

Katz, B., Miledi, R. 1969. Tetrodotoxin-resistant electrical activity in presynaptic terminals. *J. Physiol.* 203: 459–87

Katz, N. L. 1972. The effects on frog neuromuscular transmission of agents which act upon microtubules and microfilaments. *Eur. J. Pharmacol.* 19: 88–93

Kelly, R. B., Deutsch, J. W., Carlson, S. S., Wagner, J. A. 1979. Biochemistry of neurotransmitter release. *Ann. Rev. Neurosci.* 2: 399–446

Kelly, R. B., Oberg, S. G., Strong, P. N., Wagner, G. M. 1975. β-bungarotoxin, a phospholipase that stimulates transmitter release. *Cold Spring Harbor Symp. Quant. Biol.* 40: 117–25

Kenigsberg, R. L., Côté, A., Trifaró, J. M. 1982. Trifluoperazine, a calmodulin inhibitor, blocks secretion in cultured chromaffin cells at a step distal from calcium entry. *Neuroscience* 7: 2277–81

Kenigsberg, R. L., Trifaró, J. M. 1985. Microinjection of calmodulin antibodies into chromaffin cells blocks catecholamine release in response to stimulation. *Neuroscience* 14: 335–47

Kennedy, M. B. 1983. Experimental approaches to understanding the role of protein phosphorylation in the regulation of neuronal function. *Ann. Rev. Neurosci.* 6: 493–525

Kennedy, M. B., Bennett, M. K., Erondu, N. E., Miller, S. G. 1986. Calcium/calmodulin-dependent protein kinases. In *Calcium and Cell Function*, Vol. 7, ed. W. Y. Cheung. New York: Academic. In press

Kerr, L. M., Yoshikami, D. 1984. A venom peptide with a novel presynaptic blocking action. *Nature* 308: 282–84

Kharasch, E. D., Mellow, A. M., Silinsky, E. M. 1981. Intracellular magnesium does not antagonize calcium-dependent acetylcholine secretion. *J. Physiol.* 314: 255–63

Kilimann, M. W., DeGennaro, L. J. 1985. Molecular cloning of cDNAs for the nerve-cell specific phosphoprotein, synapsin I. *EMBO J.* 4: 1997–2002

Kishimoto, A., Takai, Y., Mori, T., Kikkawa, U., Nishizuka, Y. 1980. Activation of calcium and phospholipid-dependent protein kinase by diacylglycerol, its possible relation to phosphatidylinositol turnover. *J. Biol. Chem.* 255: 2273–76

Kita, H., Narita, K., Van der Kloot, W. 1982. The relation between tonicity and impulse-evoked transmitter release in the frog. *J. Physiol.* 325: 213–22

Kita, H., Van der Kloot, W. 1977. Time course and magnitude of effects of changes in tonicity on acetylcholine release at frog neuromuscular junction. *J. Neurophysiol.* 40: 212–24

Klee, C. B., Crouch, T. H., Richman, P. G. 1980. Calmodulin. *Ann. Rev. Biochem.* 49: 489–515

Klein, M., Shapiro, E., Kandel, E. R. 1980. Synaptic plasticity and the modulation of the Ca^{2+} current. *J. Exp. Biol.* 89: 117–57

Knight, D. E., Baker, P. F. 1982. Calcium-dependence of catecholamine release from bovine adrenal medullary cells after exposure to intense electric fields. *J. Membr. Biol.* 68: 107–40

Knight, D. E., Baker, P. F. 1983. The phorbol ester TPA increases the affinity of exocytosis for calcium in 'leaky' adrenal medullary cells. *FEBS Lett.* 160: 98–100

Knight, D. E., Scrutton, M. C. 1984. Cyclic nucleotides control a system which regulates Ca^{2+} sensitivity of platelet secretion. *Nature* 309: 66–68

Knight, D. E., Tonge, D. A., Baker, P. F. 1985. Inhibition of exocytosis in bovine adrenal medullary cells by botulinum toxin type D. *Nature* 317: 719–21

Ko, C.-P. 1984. Regeneration of the active zone at the frog neuromuscular junction. *J. Cell Biol.* 98: 1685–95

Koenig, M. L. 1985. *Studies of two naturally occurring compounds which effect release of acetylcholine from synaptosomes*. PhD thesis. Univ. South. Calif., Los Angeles

Konishi, S., Tsunoo, A., Otsuka, M. 1979. Enkephalins presynaptically inhibit cholinergic transmission in sympathetic ganglia. *Nature* 282: 515–16

Konishi, T. 1985. Electrical excitability of motor nerve terminals in the mouse. *J. Physiol.* 366: 411–21

Korn, E. D. 1982. Actin polymerization and its regulation by proteins from nonmuscle cells. *Physiol. Rev.* 62: 672–737

Kostyuk, P. G. 1981. Calcium channels in the neuronal membrane. *Biochim. Biophys. Acta* 650: 128–50

Kostyuk, P. G., Miranov, S. L., Shuba, Y. M. 1983. Two ion-selecting filters in the calcium channel of the somatic membrane of mollusc neurons. *J. Membr. Biol.* 76: 83–93

Kretsinger, R. H. 1980. Structure and evolution of calcium-modulated proteins. *CRC Crit. Rev. Biochem.* 8: 119–74

Kretsinger, R. H. 1981. Mechanisms of selective signalling by calcium. *Neurosci. Res. Program Bull.* 19: 213–328

Kretz, R., Shapiro, E., Kandel, E. R. 1986. Presynaptic inhibition produced by an identified presynaptic inhibitory neuron. I. Physiological mechanisms. *J. Neurophysiol.* 55: 113–30

Krueger, B. K., Forn, J., Greengard, P. 1977. Depolarization-induced phosphorylation of specific proteins, mediated by calcium ion influx, in rat brain synaptosomes. *J. Biol. Chem.* 252: 2764–73

Kuba, K. 1970. The effect of catecholamines on the neuromuscular junction in the rat diaphragm. *J. Physiol.* 211: 551–70

Kuba, K. 1980. Release of calcium ions linked to the activation of potassium conductance in a caffein-treated sympathetic neurone. *J. Physiol.* 298: 251–69

Kuo, J. F., Raynor, R. L., Mazzei, G. J., Schatzman, R. C., Turner, R. S., Kem, W. R. 1983. Cobra polypeptide cytotoxin I and marine worm polypeptide cytotoxin A-IV are potent and selective inhibitors of phospholipid-sensitive Ca^{2+}-dependent protein kinase. *FEBS Lett.* 153: 183–86

Kusano, K. 1970. Influence of ionic environment on the relationship between pre- and postsynaptic potentials. *J. Neurobiol.* 1: 435–57

Kusano, K., Livengood, D. R., Werman, R. 1967. Tetraethylammonium ions: effect of presynaptic injection on synaptic transmission. *Science* 155: 1257–59

Lagace, L., Chandra, T., Woo, S. L. C., Means, A. R. 1983. Identification of multiple species of calmodulin messenger RNA using a full length complementary DNA. *J. Biol. Chem.* 258: 1684–88

Landis, D. M. D., Reese, T. S. 1984. Cytoplasmic filaments at the presynaptic active zone in rapidly frozen cerebellar cortex. *Soc. Neurosci. Abstr.* 10: 545

LaPorte, D. C., Wierman, B. M., Storm, D. R. 1980. Calcium-induced exposure of a hydrophobic surface on calmodulin. *Biochemistry* 19: 3814–19

Larsen, F. L., Vincenzi, F. F. 1979. Calcium transport across the plasma membrane: Stimulation by calmodulin. *Science* 204: 306–8

Lasek, R. J., Shelanski, M. L. 1981. Cytoskeletons and the architecture of nervous system. *Neurosci. Res. Program Bull.* 19: 1–153

Latorre, R., Miller, C. 1983. Conduction and selectivity in potassium channels. *J. Membr. Biol.* 71: 11–30

Lemos, J. R., Nordmann, J. J., Cooke, I. M., Stuenkel, E. L. 1986. Single channels and ionic currents in peptidergic nerve terminals. *Nature* 319: 410–12

Lester, H. A., Steer, M. L., Michaelson, D. M. 1982. ADP-ribosylation of membrane proteins in cholinergic nerve terminals. *J. Neurochem.* 38: 1080–86

Levitan, I. B. 1985. Phosphorylation of ion channels. *J. Membr. Biol.* 87: 177–90

Lin, Y. M., Liu, Y. P., Cheung, W. Y. 1974. Cyclic 3′:5′-nucleotide phosphodiesterase. Purification, characterization, and active form of the protein activator from bovine brain. *J. Biol. Chem.* 249: 4943–54

Lindau, M., Fernandez, J. M. 1986. IgE-mediated degranulation of mast cells does not require opening of ion channels. *Nature* 319: 150–53

Llinas, R. R. 1982. Calcium in synaptic transmission. *Sci. Am.* 247: 56–65

Llinas, R. R., Heuser, J. E. 1977. Depolarization-release coupling systems in neurons. *Neurosci. Res. Program Bull.* 15: 557–687

Llinás, R., McGuinness, T. L., Leonard, C. S., Sugimori, M., Greengard, P. 1985. Intraterminal injection of synapsin I or calcium-calmodulin-dependent protein kinase II alters neurotransmitter release at the squid giant synapse. *Proc. Natl. Acad. Sci. USA* 82: 3035–39

Llinás, R., Nicholson, C. 1975. Calcium role in depolarization-secretion coupling: An aequorin study in squid giant synapse. *Proc. Natl. Acad. Sci. USA* 72: 187–90

Llinás, R., Steinberg, I. Z., Walton, K. 1981a. Presynaptic calcium currents in squid giant synapse. *Biophys. J.* 33: 289–322

Llinás, R., Steinberg, I. Z., Walton, K. 1981b. Relationship between presynaptic calcium current and postsynaptic potential in squid giant synapse. *Biophys. J.* 33: 323–52

Llinás, R., Sugimori, M., Simon, S. M. 1982. Transmission by presynaptic spike-like depolarization in the squid synapse. *Proc. Natl. Acad. Sci. USA* 79: 2415–19

Low, M. G., Carroll, R. C., Weglicki, W. B. 1984. Multiple forms of phosphoinositide-specific phospholipase C of different relative molecular masses in animal tissues. *Biochem. J.* 221: 813–20

Luthra, M. G. 1982. Trifluoperazine inhibition of calmodulin-sensitive Ca^{2+}-ATPase and calmodulin-insensitive (Na^+-K^+) and Mg^{2+}-ATPase activities of human and rat red blood cells. *Biochem. Biophys. Acta* 692: 271–77

Macdonald, R. L., Nelson, P. G. 1978. Specific opiate-induced depression of transmitter release from dorsal root ganglion cells in culture. *Science* 199: 1449–51

Malik, M. N., Fenko, M. D., Iqbal, K., Wisniewski, H. M. 1983. Purification and characterization of two forms of Ca^{2+}-activated neutral protease from calf brain. *J. Biol. Chem.* 258: 8955–62

Mallart, A. 1984. Presynaptic currents in frog motor endings. *Pflügers Arch.* 400: 8–13

Marshak, D. R., Clarke, M., Roberts, D. M., Watterson, D. M. 1984. Structural and functional properties of calmodulin from the eukaryotic microorganism *Dictyostelium discoideum*. *Biochemistry* 23: 2891–99

Martin, A. R. 1977. Junctional transmission. II. Presynaptic mechanisms. See Heuser & Reese, pp. 329–55

Matthew, W. D., Tsavaler, L., Reichardt, L. F. 1981. Identification of a synaptic vesicle–specific membrane protein with a wide distribution in neural and neurosecretory tissue. *J. Cell Biol.* 91: 257–69

McClure, W. O., Abbott, B. C., Baxter, D. E., Hsiao, T. H., Satin, L. S., Siger, A., Yoshino, J. E. 1980. Leptinotarsin: A presynaptic neurotoxin that stimulates release of acetylcholine. *Proc. Natl. Acad. Sci. USA* 77: 1219–23

McLaughlin, S. G. A., Szabo, G., Eisenman, G. 1971. Divalent ions and the surface potential of charged phospholipid membranes. *J. Gen. Physiol.* 58: 667–87

Means, A. R., Dedman, J. R. 1980. Calmodulin—an intracellular calcium receptor. *Nature* 285: 73–77

Means, A. R., Tash, J. S., Chafouleas, J. G. 1982. Physiological implications of the presence, distribution, and regulation of calmodulin in eukaryotic cells. *Physiol. Rev.* 62: 1–39

Meldolesi, J., Ceccarelli, B. 1981. Exocytosis and membrane recycling. *Philos. Trans. R. Soc. London Ser. B* 296: 55–65

Meldolesi, J., Huttner, W. B., Tsien, R. Y., Pozzan, T. 1984. Free cytoplasmic Ca^{2+} and neurotransmitter release. Studies on PC12 cells and synaptosomes exposed to α-latrotoxin. *Proc. Natl. Acad. Sci. USA* 81: 620–24

Mellanby, J., Green, J. 1981. How does tetanus toxin act? *Neuroscience* 6: 281–300

Mellow, A. M., Perry, B. D., Silinsky, E. M. 1982. Effects of calcium and strontium in the process of acetylcholine release from motor nerve endings. *J. Physiol.* 328: 547–62

Michaelson, D. M., McDowall, G., Sarne, Y. 1984. Opiates inhibit acetylcholine release from *Torpedo* nerve terminals by blocking Ca^{2+} influx. *J. Neurochem.* 43: 614–18

Middlemiss, D. N., Spedding, M. 1985. A functional correlate for the dihydropyridine binding site in rat brain. *Nature* 314: 94–96

Miledi, R. 1973. Transmitter release induced by injection of calcium ions into nerve terminals. *Proc. R. Soc. London Ser. B* 183: 421–25

Miledi, R., Parker, I. 1981. Calcium transients recorded with arsenazo III in the presynaptic terminal of the squid giant synapse. *Proc. R. Soc. London Ser. B* 212: 197–211

Miller, R. J. 1985a. How many types of calcium channels exist in neurones? *Trends Neurosci.* 8: 45–47

Miller, R. J. 1985b. Second messengers, phosphorylation and neurotransmitter release. *Trends Neurosci.* 8: 463–65

Miller, R. J., Freedman, S. B. 1984. Are dihydropyridine binding sites voltage sensitive calcium channels? *Life Sci.* 34: 1205–21

Miller, S. G., Kennedy, M. B. 1986. Regulation of brain type II Ca^{2+}/calmodulin-dependent protein kinase by autophosphorylation: A Ca^{2+}-triggered molecular switch. *Cell* 44: 861–70

Miller, T. M., Heuser, J. E. 1984. Endocytosis of synaptic vesicle membrane at the frog neuromuscular junction. *J. Cell Biol.* 98: 685–98

Moore, B. W. 1965. A soluble protein characteristic of the nervous system. *Biochem. Biophys. Res. Commun.* 19: 739–44

Moore, C. L. 1971. Specific inhibition of mitochondrial Ca^{++} transport by ruthenium red. *Biochem. Biophys. Res. Commun.* 42: 298–305

Morgan, K. G., Bryant, S. H. 1977. Pentobarbital: Presynaptic effect in the squid giant synapse. *Experientia* 33: 487–88

Morita, K., Brocklehurst, K. W., Tomares,

S. M., Pollard, H. B. 1985. The phorbol ester TPA enhances A23187—but not carbachol—and high K^+-induced catecholamine secretion from cultured bovine adrenal chromaffin cell. *Biochem. Biophys. Res. Commun.* 129: 511–16

Moskowitz, N., Schook, W., Beckenstein, K., Puszkin, S. 1983. Preliminary characterization of synaptic vesicle calmodulin interaction. *Brain Res.* 263: 243–50

Mudge, A. W., Leeman, S. E., Fischbach, G. D. 1979. Enkephalin inhibits release of substance P from sensory neurons in tissue culture and decreases action potential duration. *Proc. Natl. Acad. Sci. USA* 76: 526–30

Mumby, S. M., Kahn, R. A., Manning, D. R., Gilman, A. G. 1986. Antisera of designed specificity for subunits of guanine nucleotide-binding regulatory proteins. *Proc. Natl. Acad. Sci. USA* 83: 265–69

Mundy, D. L., Strittmatter, W. J. 1985. Requirement for metalloendoprotease in exocytosis: Evidence in mast cells and adrenal chromaffin cells. *Cell* 40: 645–56

Nachshen, D. A. 1984. Selectivity of the Ca binding site in synaptosome Ca channels. Inhibition of Ca influx by multivalent metal cations. *J. Gen. Physiol.* 83: 941–67

Nachshen, D. A. 1985a. The early time course of potassium-stimulated calcium uptake in presynaptic terminals isolated from rat brain. *J. Physiol.* 361: 251–68

Nachshen, D. A. 1985b. Regulation of cytosolic calcium concentration in presynaptic nerve endings isolated from rat brain. *J. Physiol.* 363: 87–101

Nachshen, D. A., Blaustein, M. P. 1979. The effects of some organic calcium antagonists on calcium influx in presynaptic nerve terminals. *Mol. Pharmacol.* 16: 579–86

Nachshen, D. A., Blaustein, M. P. 1980. Some properties of potassium-stimulated calcium influx in presynaptic nerve endings. *J. Gen. Physiol.* 76: 709–28

Nachshen, D. A., Blaustein, M. P. 1982. Influx of calcium, strontium and barium in presynaptic nerve endings. *J. Gen. Physiol.* 79: 1065–87

Nachshen, D. A., Drapeau, P. 1986. Effects of acidification on Ca^{2+}, Ca fluxes, and dopamine release in synaptosomes. *Biophys. J.* 49: 230a

Nairn, A. C., Hemmings, H. C. Jr., Greengard, P. 1985. Protein kinases in the brain. *Ann. Rev. Biochem.* 54: 931–76

Naor, Z., Catt, K. J. 1981. Mechanism of action of gonadotropin-releasing hormone. Involvement of phospholipid turnover in luteinizing hormone release. *J. Biol. Chem.* 256: 2226–29

Navone, F., Greengard, P., De Camilli, P. 1984. Synapsin I in nerve terminals: Selective association with small synaptic vesicles. *Science* 226: 1209–11

Neher, E. 1986. Concentration profiles of intracellular Ca^{++} in the presence of a diffusible chelator. *Exp. Brain Res. Ser.* 4: In press

Nelson, M. T., French, R. J., Krueger, B. K. 1984. Voltage-dependent calcium channels from brain incorporated into planar lipid bilayers. *Nature* 308: 77–80

Nestler, E. J., Greengard, P. 1980. Dopamine and depolarizing agents regulate the state of phosphorylation of protein I in the mammalian superior cervical sympathetic ganglion. *Proc. Natl. Acad. Sci. USA* 77: 7479–83

Nestler, E. J., Greengard, P. 1985. *Protein Phosphorylation in the Nervous System.* New York: Wiley

Newton, C., Pangborn, W., Nir, S., Papahadjopoulos, D. 1978. Specificity of Ca^{2+} and Mg^{2+} binding to phosphatidylserine vesicles and resultant phase changes of bilayer membrane structure. *Biochim. Biophys. Acta* 506: 281–87

Newton, D. L., Oldewurtel, M. D., Krinks, M. H., Shiloach, J., Klee, C. B. 1984. Agonist and antagonist properties of calmodulin fragments. *J. Biol. Chem.* 259: 4419–26

Nicholson, C., Phillips, J. M. 1981. Ion diffusion modified by tortuosity and volume fraction in the extracellular microenvironment of the rat cerebellum. *J. Physiol.* 321: 225–57

Niles, W. D., Smith, D. O. 1982. Effects of hypertonic solutions on quantal transmitter release at the crayfish neuromuscular junction. *J. Physiol.* 329: 185–202

Nishikawa, M., Tanaka, T., Hidaka, H. 1980. Ca^{2+}-calmodulin-dependent phosphorylation and platelet secretion. *Nature* 287: 863–65

Nishizuka, Y. 1984. Turnover of inositol phospholipids and signal transduction. *Science* 225: 1365–70

Nowycky, M. C., Fox, A. P., Tsien, R. W. 1985. Three types of neuronal calcium channel with different calcium agonist sensitivity. *Nature* 316: 440–43

Ogura, A., Takahashi, M. 1984. Differential effect of a dihydropyridine derivative to Ca^{2+} entry pathways in neuronal preparations. *Brain Res.* 301: 323–30

Olivera, B. M., Gray, W. R., Zeikus, R., McIntosh, J. M., Varga, J., Rivier, J., de Santos, V., Cruz, L. J. 1985. Peptide neurotoxins from fish-hunting cone snails. *Science* 230: 1338–43

Ornberg, R. L., Leapman, R. D. 1986. Elemental composition of bovine adrenal chromaffin granules *in situ* and *in vitro*. *Biophys. J.* 49: 179a

O'Shea, M., Evans, P. D. 1979. Potentiation of neuromuscular transmission by an octopaminergic neurone in the locust. *J. Exp. Biol.* 79: 169–90

Osses, L., Barry, S., Augustine, G., Charlton, M. 1986. Protein kinase C activation enhances transmission at the squid giant synapse. *Biophys. J.* 49: 179a

Palfrey, H. C., Rothlein, J. E., Greengard, P. 1983. Calmodulin-dependent protein kinase and associated substrates in *Torpedo* electric organ. *J. Biol. Chem.* 258: 9496–503

Pant, H. C., Gainer, H. 1980. Properties of a calcium-activated protease in squid axoplasm which selectively degrades neurofilament proteins. *J. Neurobiol.* 11: 1–12

Papahadjopoulos, D. 1978. Calcium induced phase changes and fusion in natural and model membranes. *Cell Surf. Rev.* 5: 765–90

Parsegian, V. A. 1977. Considerations in determining the mode of influence of calcium on vesicle-membrane interaction. In *Society for Neuroscience Symposia*, ed. W. M. Cowan, J. A. Ferrendelli, 2: 161–71. Bethesda, MD: Soc. Neurosci.

Pécot-Dechavassine, M. 1976. Action of vinblastine on the spontaneous release of acetylcholine at the frog neuromuscular junction. *J. Physiol.* 261: 31–48

Perrin, D., Aunis, D. 1985. Reorganization of α-fodrin induced by stimulation in secretory cells. *Nature* 315: 589–92

Peters, A., Palay, S. L., Webster, H. D. 1970. *The Fine Structure of the Nervous System: The Neurons and Their Supporting Cells.* Philadelphia: Saunders

Pocotte, S. L., Frye, R. A., Senter, R. A., TerBush, D. R., Lee, S. A., Holtz, R. W. 1985. Effects of phorbol ester on catecholamine secretion and protein phosphorylation in adrenal medullary cell cultures. *Proc. Natl. Acad. Sci. USA* 82: 930–34

Pollard, H. B., Creutz, C. E., Fowler, V., Scott, J., Pazoles, C. J. 1981. Calcium-dependent regulation of chromaffin granule movement, membrane contact, and fusion during exocytosis. *Cold Spring Harbor Symp. Quant. Biol.* 46: 819–34

Porter, K. R., Chm. 1984. The cytoplasmic matrix and the integration of cellular function. *J. Cell Biol.* 99: 1–248s (Conf. proc.)

Pozzan, T., Gatti, G., Dozio, N., Vincentini, L. M., Meldolesi, J. 1984. Ca^{2+}-dependent and -independent release of neurotransmitters from PC12 cells: A role for protein kinase C activation? *J. Cell Biol.* 99: 628–38

Pozzan, T., Lew, D. P., Wollheim, C. B., Tsien, R. Y. 1983. Is cytosolic ionized calcium regulating neutrophil activation? *Science* 221: 1413–15

Publicover, S. J., Duncan, C. J. 1979. The action of verapamil on the rate of spontaneous release of transmitter at the frog neuromuscular junction. *Eur. J. Pharmacol.* 54: 119–27

Pumplin, D. W., Reese, T. S., Llinás, R. 1981. Are the presynaptic membrane particles the calcium channels? *Proc. Natl. Acad. Sci. USA* 78: 7210–13

Puszkin, S., Kochwa, S. 1974. Regulation of neurotransmitter release by a complex of actin with relaxing protein isolated from rat brain synaptosomes. *J. Biol. Chem.* 249: 7711–14

Putkey, J. A., Ts'ui, K. F., Tanaka, T., Lagacé, L., Stein, J. P., Lai, E. C., Means, A. R. 1983. Chicken calmodulin genes. A species comparison of cDNA sequences and isolation of a genomic clone. *J. Biol. Chem.* 258: 11864–70

Quissell, D. O., Deisher, L. M., Barzen, K. A. 1985. The rate-determining step in cAMP-mediated exocytosis in the rat parotid and submandibular glands appears to involve analogous 26-kDa integral membrane phosphoproteins. *Proc. Natl. Acad. Sci. USA* 82: 3237–41

Rahamimoff, R. 1968. A dual effect of calcium ions on neuromuscular facilitation. *J. Physiol.* 195: 471–80

Rahamimoff, R., Erulkar, S. D., Lev-Tov, A., Meiri, H. 1978a. Intracellular and extracellular calcium ions in transmitter release at the neuromuscular synapse. *Ann. NY Acad. Sci.* 307: 583–98

Rahamimoff, R., Lev-Tov, A., Meiri, H. 1980. Primary and secondary regulation of quantal transmitter release: calcium and sodium. *J. Exp. Biol.* 89: 5–18

Rahamimoff, R., Meiri, H., Erulkar, S. D., Barenholz, Y. 1978b. Changes in transmitter release induced by ion containing liposomes. *Proc. Natl. Acad. Sci. USA* 75: 5214–16

Rasmussen, H., Barrett, P. Q. 1984. Calcium messenger system: an integrated view. *Physiol. Rev.* 64: 938–84

Reichardt, L. F., Kelly, R. B. 1983. A molecular description of nerve terminal function. *Ann. Rev. Biochem.* 52: 871–926

Reuter, H. 1983. Calcium channel modulation by neurotransmitters, enzymes and drugs. *Nature* 301: 569–74

Reuter, H. 1985. A variety of calcium channels. *Nature* 316: 391

Rink, T. J., Sanchez, A., Hallam, T. J. 1983. Diacylglycerol and phorbol ester stimulate secretion without raising cytoplasmic free calcium in human platelets. *Nature* 305: 317–19

Roberts, M. L., Butcher, F. R. 1983. The involvement of protein phosphorylation is stimulus-secretion coupling in the mouse exocrine pancreas. *Biochem. J.* 213: 281–88

Robinson, P. J., Jarvie, P. E., Dunkley, P. R. 1984. Depolarisation-dependent protein Phosphorylation in rat cortical synaptosomes is inhibited by fluphenazine at a step after calcium entry. *J. Neurochem.* 43: 659–67

Rodbell, M. 1980. The role of hormone receptors and GTP-regulatory proteins in membrane transduction. *Nature* 284: 17–22

Rubin, R. P. 1982. Calcium-phospholipid interactions in secretory cells: A new perspective on stimulus-secretion coupling. *Fed. Proc.* 41: 2181–87

Russell, J. T. 1981. The isolation of purified neurosecretory vesicles from bovine neurohypophysis using isoosomolar density gradients. *Analyt. Biochem.* 113: 229–38

Saitoh, T., Schwartz, J. H. 1985. Phosphorylation-dependent subcellular translocation of a Ca^{2+}/calmodulin-dependent protein kinase produces an autonomous enzyme in *Aplysia* neurons. *J. Cell Biol.* 100: 835–42

Sand, O., Sletholt, K., Gautvik, K. M., Haug, E. 1983. Trifluoperazine blocks calcium-dependent action potentials and inhibits hormone release from rat pituitary tumour cells. *Eur. J. Pharmacol.* 86: 177–84

Sasaki, H. 1984. Modulation of calcium sensitivity by a specific cortical protein during sea urchin egg cortical vesicle exocytosis. *Dev. Biol.* 101: 125–35

Schatzman, R. C., Raynor, R. L., Kuo, J. F. 1983. N-(6-aminohexyl)-5-chloro-1-naphthalenesulfonamide (W-7), a calmodulin antagonist, also inhibits phospholipid-sensitive calcium-dependent protein kinase. *Biochim. Biophys. Acta* 755: 144–47

Schlaepfer, W. W., Lee, C., Lee, V. M.-Y., Zimmerman, U.-J. P. 1985. An immunoblot study of neurofilament degradation *in situ* and during calcium-activated proteolysis. *J. Neurochem.* 44: 502–9

Schnapp, B. J., Vale, R. D., Sheetz, M. P., Reese, T. S. 1985. Single microtubules from squid axoplasm support bidirectional movement of organelles. *Cell* 40: 455–62

Schwartz, L. M., McCleskey, E. W., Almers, W. 1985. Dihydropyridine receptors in muscle are voltage-dependent but most are not functional calcium channels. *Nature* 314: 747–51

Schweitzer, E. S., Kelly, R. B. 1982. ATP release from cholinergic synapses. *Soc. Neurosci. Abstr.* 8: 493

Seeman, P., Weinstein, J. 1966. Erythrocyte membrane stabilization by tranquilizers and antihistamines. *Biol. Pharm.* 15: 1737–52

Shimoni, Y., Alnaes, E., Rahamimoff, R. 1977. Is hyperosmotic neurosecretion from motor nerve endings a calcium-dependent process? *Nature* 267: 170–72

Shukla, S. D. 1982. Phosphatidylinositol specific phospholipases C. *Life Sci.* 30: 1323–35

Siegelbaum, S. A., Tsien, R. W. 1983. Modulation of gated ion channels as a mode of transmitter action. *Trends Neurosci.* 6: 307–13

Sieghart, W., Theoharides, T. C., Alper, S. L., Douglas, W. W., Greengard, P. 1978. Calcium-dependent protein phosphorylation during secretion by exocytosis in the mast cell. *Nature* 275: 329–31

Silinsky, E. M. 1977. Can barium support the release of acetylcholine by nerve impulses? *Br. J. Pharmacol.* 59: 215–17

Silinsky, E. M. 1978. On the role of barium in supporting the asynchronous release of acetylcholine quanta by motor nerve impulses. *J. Physiol.* 274: 157–71

Silinsky, E. M. 1985. The biophysical pharmacology of calcium-dependent acetylcholine secretion. *Pharmacol. Rev.* 37: 81–132

Siman, R., Baudry, M., Lynch, G. 1983. Purification from synaptosomal plasma membranes of calpain I, a thiol protease activated by micromolar calcium concentrations. *J. Neurochem.* 41: 950–56

Simmen, R. C. M., Tanaka, T., Ts'ui, K. F., Putkey, J. A., Scott, M. J., Lai, E. C., Means, A. R. 1985. The structural organization of the chicken calmodulin gene. *J. Biol. Chem.* 260: 907–12

Simon, S. M., Llinas, R. R. 1985. Compartmentalization of the submembrane calcium activity during calcium influx and its significance in transmitter release. *Biophys. J.* 48: 485–98

Simpson, L. L. 1981. The origin, structure, and pharmacological activity of botulinum toxin. *Pharmacol. Rev.* 33: 155–88

Smith, S. J., Augustine, G. J., Charlton, M. P. 1985. Transmission at voltage-clamped giant synapse of squid: Evidence for cooperativity of presynaptic calcium action. *Proc. Natl. Acad. Sci. USA* 82: 622–25

Smith, S. J., MacDermott, A. B., Weight, F. F. 1983. Detection of intracellular Ca^{2+} transients in sympathetic neurones using arsenazo III. *Nature* 304: 350–52

Smith, S. J., Zucker, R. S. 1980. Aequorin response facilitation and intracellular calcium accumulation in molluscan neurones. *J. Physiol.* 300: 169–96

Soifer, D., ed. 1975. The biology of cytoplasmic microtubules. *Ann. NY Acad. Sci.* 253: 1–848

Standaert, F. G., Dretchen, K. L. 1979. Cyclic nucleotides and neuromuscular transmission. *Fed. Proc.* 38: 2183–92

Stanley, E. F., Ehrenstein, G. 1985. A model for exocytosis based on the opening of calcium-activated potassium channels in vesicles. *Life Sci.* 37: 1985–95

Stanley, E. F., Ehrenstein, G., Russell, J. 1986. Evidence for calcium-activated potassium channels in vesicles of pituitary cells. *Biophys. J.* 49: 19a

Steinhardt, R. A., Alderton, J. M. 1982. Calmodulin confers calcium sensitivity on secretory exocytosis. *Nature* 295: 154–55

Stimers, J. R. 1982. *Evidence for two physiologically distinct pools of quantally releasable acetylcholine at the rat neuromuscular junction.* PhD thesis. Univ. South. Calif., Los Angeles

Stimers, J. R., Byerly, L. 1982. Slowing of sodium current inactivation by ruthenium red in snail neurons. *J. Gen. Physiol.* 80: 485–97

Stockbridge, N., Moore, J. W. 1984. Dynamics of intracellular calcium and its possible relationship to phasic transmitter release and facilitation at the frog neuromuscular junction. *J. Neurosci.* 4: 803–11

Stockbridge, N., Ross, W. N. 1984. Localized Ca^{2+} and calcium-activated potassium conductances in terminals of a barnacle photoreceptor. *Nature* 309: 266–68

Südhof, T. C., Walker, J. H., Obrocki, J. 1982. Calelectrin self-aggregates and promotes membrane aggregation in the presence of calcium. *EMBO J.* 1: 1167–70

Takahashi, M., Ohizumi, Y., Yasumoto, T. 1982. Maitotoxin, a Ca^{2+} channel activator candidate. *J. Biol. Chem.* 257: 7287–89

Tanaka, C., Taniyama, K., Kusunoki, M. 1984. A phorbol ester and A23187 act synergistically to release acetylcholine from the guinea pig ileum. *FEBS Lett.* 175: 165–69

Tanaka, T., Hidaka, H. 1980. Hydrophobic regions function in calmodulin-enzyme(s) interactions. *J. Biol. Chem.* 255: 11078–80

Teo, T. S., Wang, J. H. 1973. Mechanism of activation of a cyclic adenosine 3′:5′-monophosphate phosphodiesterase from bovine heart by calcium ions. *J. Biol. Chem.* 248: 5950–55

Thoenen, H., Kreutzberg, G. W. 1981. The role of fast transport in the nervous system. *Neurosci. Res. Program Bull.* 20: 1–138

Tillotson, D. L., Gorman, A. L. F. 1984. Localization of neuronal Ca^{2+} buffering near plasma membrane studied with different divalent cations. *Cell. Mol. Neurobiol.* 3: 297–310

Tsalkova, T. N., Privalov, P. L. 1985. Thermodynamic study of domain organization in troponin C and calmodulin. *J. Mol. Biol.* 181: 533–44

Tsien, R. W. 1983. Calcium channels in excitable cell membranes. *Ann. Rev. Physiol.* 45: 341–58

Tsien, R. W. 1986. Modulation of calcium currents in heart cells and neurons. In *Neuromodulation*, ed. I. B. Levitan, L. K. Kaczmarek. Oxford: Oxford Univ. Press

Turner, T. J., Goldin, S. M. 1985. Calcium channels in rat brain synaptosomes: Identification and pharmacological characterization. High affinity blockade by organic Ca^{2+} channel blockers. *J. Neurosci.* 5: 841–49

Ueda, T., Greengard, P. 1977. Adenosine 3′:5′-monophosphate-regulated phosphoprotein system of neuronal membranes. I. Solubilization, purification, and some properties of an endogenous phosphoprotein. *J. Biol. Chem.* 252: 5155–63

Umbach, J. A., Gundersen, C. B., Baker, P. F. 1984. Giant synaptosomes. *Nature* 311: 474–77

Vallano, M. L., Goldenring, J. R., Buckholz, T. M., Larson, R. E., DeLorenzo, R. J. 1985. Separation of endogenous calmodulin- and cAMP-dependent kinases from microtubule preparations. *Proc. Natl. Acad. Sci. USA* 82: 3202–26

Van Den Bosch, H. 1980. Intracellular phospholipase A. *Biochim. Biophys. Acta* 604: 191–246

Van Der Kloot, W., Kita, H. 1975. The effects of the "calcium-antagonist" verapamil on muscle action potentials in the frog and crayfish and on neuromuscular transmission in the crayfish. *Comp. Biochem. Physiol.* 50C: 121–25

Vincentini, L. M., Meldolesi, J. 1984. α-Latrotoxin of black widow spider venom binds to a specific receptor coupled to phosphoinositide breakdown in PC12 cells. *Biochem. Biophys. Res Commun.* 121: 538–44

Wada, A., Yanagihara, N., Izumi, F., Sakurai, S., Kobayashi, H. 1983. Trifluoperazine inhibits $^{45}Ca^{2+}$ uptake and catecholamine secretion and synthesis in adrenal medullary cells. *J. Neurochem.* 40: 481–86

Wade, P. D., Fritz, L. C., Siekevitz, P. 1981. The effect of diamide on transmitter release and on synaptic vesicle population at vertebrate synapses. *Brain Res.* 225: 357–72

Wagner, J. A., Kelly, R. B. 1979. Topological organization of proteins in an intracellular secretory organelle: The synaptic vesicle. *Proc. Natl. Acad. Sci. USA* 76: 4126–30

Walker, J. H. 1982. Isolation from cholinergic synapses of a protein that binds to membranes in a calcium-dependent manner. *J. Neurochem.* 39: 815–23

Weber, A., Murray, J. M. 1973. Molecular control mechanisms in muscle contraction. *Physiol. Rev.* 53: 612–73

Weeds, A. 1982. Actin-binding proteins-regulators of cell architecture and motility. *Nature* 296: 811–16

Weight, F. F., Erulkar, S. D. 1976. Synaptic transmission and effects of temperature at the squid giant synapse. *Nature* 261: 720–22

Weiss, B., Levin, R. M. 1978. Mechanism for selectively inhibiting the activation of cyclic nucleotide phosphodiesterase and adenylate cyclase by antipsychotic agent. *Adv. Cyclic Nucl. Res.* 9: 285–303

Werz, M. A., Macdonald, R. L. 1982. Opioid peptides decrease calcium-dependent action potential duration of mouse dorsal root ganglion neurons in cell culture. *Brain Res.* 239: 315–21

Whitaker, M., Aitchison, M. 1985. Calcium-dependent polyphosphoinositide hydrolysis is associated with exocytosis in vitro. *FEBS Lett.* 182: 119–24

Whitaker, M., Baker, P. F. 1983. Calcium-dependent exocytosis in an *in vitro* secretory granule plasma membrane preparation from sea urchin eggs and the effects of some inhibitors of cytoskeletal function. *Proc. R. Soc. London Ser. B* 218: 397–413

Whitaker, M., Irvine, R. F. 1984. Inositol 1,4,5-trisphosphate microinjection activates sea urchin eggs. *Nature* 312: 636–39

White, J., Kielian, M., Helenius, A. 1983. Membrane fusion proteins of enveloped animal viruses. *Q. Rev. Biophys.* 16: 151–95

Wiedenmann, B., Franke, W. W. 1985. Identification and localization of synaptophysin, an integral membrane glycoprotein of M_r 38,000 characteristic of presynaptic vesicles. *Cell* 41: 1017–28

Williams, D. A., Fogarty, K. E., Tsien, R. Y., Fay, F. S. 1985. Calcium gradients in single smooth muscle cells revealed by the digital imaging microscope using Fura-2. *Nature* 318: 558–61

Wilschut, J., Hoekstra, D. 1984. Membrane fusion: from liposomes to biological membranes. *Trends Biochem. Sci.* 9: 479–83

Wilson, S. P., Kirshner, N. 1983. Calcium-evoked secretion from digitonin-permeabilized adrenal medullary chromaffin cells. *J. Biol. Chem.* 258: 4994–5000

Wise, B. C., Glass, D. B., Jen Chou, C.-H., Raynor, R. L., Katoh, N., Schatzman, R. C., Turner, R. S., Kibler, R. F., Kuo, J. F. 1982. Phospholipid-sensitive Ca^{2+}-dependent protein kinase from heart. II. Substrate specificity and inhibition by various agents. *J. Biol. Chem.* 257: 8489–95

Wong, P. Y.-K., Cheung, W. Y. 1979. Calmodulin stimulates human platelet phospholipase A_2. *Biochem. Biophys. Res. Commun.* 90: 473–80

Wood, J. G., Wallace, R. W., Whitaker, J. N., Cheung, W. Y. 1980. Immunocytochemical localization of calmodulin and a heat-labile calmodulin-binding protein ($CaM\text{-}BP_{80}$) in basal ganglia of mouse brain. *J. Cell Biol.* 84: 66–76

Yeager, R. E., Yoshikami, D., River, J., Cruz, L. J., Miljanich, G. P. 1987. Transmitter release from presynaptic terminals of electric organ: Inhibition by the calcium channel antagonist, omega *Conus* toxin. *J. Neurosci.* 7: In press

Zengel, J. E., Magleby, K. L. 1981. Changes in miniature endplate potential frequency during repetitive nerve stimulation in the presence of Ca^{2+}, Ba^{2+}, and Sr^{2+} at the frog neuromuscular junction. *J. Gen. Physiol.* 77: 503–29

Zieseniss, E., Plattner, H. 1985. Synchronous exocytosis in *Paramecium* cells involves very rapid (≤ 1s), reversible dephosphorylation of a 65-KD phosphoprotein in exocytosis-competent strains. *J. Cell Biol.* 101: 2028–35

Zimmerberg, J., Cohen, F. S., Finkelstein, A. 1980. Micromolar Ca^{2+} stimulates fusion of lipid vesicles with planar bilayers containing a calcium-binding protein. *Science* 210: 906–8

Zimmerberg, J., Sardet, C., Epel, D. 1985. Exocytosis of sea urchin egg cortical vesicles in vitro is retarded by hyperosmotic sucrose: Kinetics of fusion monitored by quantitative light-scattering microscopy. *J. Cell Biol.* 101: 2398–2410

Zimmerberg, J., Whitaker, M. 1985. Irreversible swelling of secretory granules during exocytosis caused by calcium. *Nature* 315: 581–84

Zimmerman, U.-J. P., Schlaepfer, W. W. 1984. Multiple forms of Ca-activated protease from rat brain and muscle. *J. Biol. Chem.* 259: 3210–18

Zisapel, N., Levi, M., Gozes, I. 1980. Tubulin: an integral protein of mammalian synaptic vesicle membranes. *J. Neurochem.* 34: 26–32

Zucker, R. S., Fogelson, A. L. 1986. Relationship between transmitter release and presynaptic calcium influx when calcium enters through discrete channels. *Proc. Natl. Acad. Sci. USA* 83 : 3032–36

Zucker, R. S., Landò, L. 1986. Mechanism of transmitter release : Voltage hypothesis and calcium hypothesis. *Science* 231 : 574–79

Zucker, R. S., Steinhardt, R. A. 1978. Prevention of the cortical reaction in fertilized sea urchin eggs by injection of calcium chelating ligands. *Biochim. Biophys. Acta* 541 : 459–66

Zucker, R. S., Stockbridge, N. 1983. Presynaptic calcium diffusion and the time courses of transmitter release and synaptic facilitation at the squid giant synapse. *J. Neurosci.* 3 : 1263–69

SUBJECT INDEX

CUMULATIVE INDEXES

CONTRIBUTING AUTHORS, VOLUMES 6–10

CHAPTER TITLES, VOLUMES 6–10